Lecture Notes in Civil Engineering

Volume 152

Lecture Notes in Civil Engineering (LNCE) publishes the latest developments in Civil Engineering - quickly, informally and in top quality. Though original research reported in proceedings and post-proceedings represents the core of LNCE, edited volumes of exceptionally high quality and interest may also be considered for publication. Volumes published in LNCE embrace all aspects and subfields of, as well as new challenges in, Civil Engineering. Topics in the series include:

- Construction and Structural Mechanics
- Building Materials
- Concrete, Steel and Timber Structures
- Geotechnical Engineering
- Earthquake Engineering
- Coastal Engineering
- Ocean and Offshore Engineering; Ships and Floating Structures
- Hydraulics, Hydrology and Water Resources Engineering
- Environmental Engineering and Sustainability
- Structural Health and Monitoring
- Surveying and Geographical Information Systems
- Indoor Environments
- Transportation and Traffic
- Risk Analysis
- Safety and Security

To submit a proposal or request further information, please contact the appropriate Springer Editor:

– Pierpaolo Riva at pierpaolo.riva@springer.com (Europe and Americas);
– Swati Meherishi at swati.meherishi@springer.com (Asia - except China, and Australia, New Zealand);
– Wayne Hu at wayne.hu@springer.com (China).

All books in the series now indexed by Scopus and EI Compendex database!

More information about this series at http://www.springer.com/series/15087

C. N. V. Satyanarayana Reddy · Sireesh Saride ·
A. Murali Krishna
Editors

Ground Improvement and Reinforced Soil Structures

Proceedings of Indian Geotechnical Conference 2020 Volume 2

Springer

Editors
C. N. V. Satyanarayana Reddy
Geotechnical Engineering Division
Department of Civil Engineering
Andhra University College of Engineering
Visakhapatnam, Andhra Pradesh, India

Sireesh Saride
Geotechnical Engineering Division
Department of Civil Engineering
Indian Institute of Technology
Hyderabad, Telangana, India

A. Murali Krishna
Department of Civil Engineering
Indian Institute of Technology
Tirupati, Andhra Pradesh, India

ISSN 2366-2557 ISSN 2366-2565 (electronic)
Lecture Notes in Civil Engineering
ISBN 978-981-16-1830-7 ISBN 978-981-16-1831-4 (eBook)
https://doi.org/10.1007/978-981-16-1831-4

This Springer imprint is published by the registered company Springer Nature Singapore Pte Ltd.
The registered company address is: 152 Beach Road, #21-01/04 Gateway East, Singapore 189721, Singapore

Contents

About the Editors

Dr. C. N. V. Satyanarayana Reddy is a Professor of Civil Engineering at College of Engineering, Andhra University, Visakhapatnam, India, and has 28 years of teaching, research, and consultancy experience. His expertise and research interests are in the areas of reinforced soils, ground improvement, landfills, soil retention in excavations, deep foundations, and forensic geotechnical engineering. He obtained B.Tech. in Civil Engineering from Nagarjuna University, M.Tech. in Geotechnical Engineering from Indian Institute of Technology (IIT) Madras, M.E. (Structures) from Andhra University, and Ph.D. from National Institute of Technology (NIT) Warangal, India. He has guided 8 Ph.D. scholars and 92 M.Tech. dissertations. He has more than 110 publications in various national and international journals and seminars/conference proceedings. He has organized more than 30 seminars, workshops, and training programs. He is fellow of Indian Geotechnical Society, The Institution of Engineers (India) and Life member of Indian Roads Congress, Indian Concrete Institute, Indian Society for Technical Education, Indian Society for Rock Mechanics and tunneling Technology.

He handled several research projects funded by DST, UGC and AICTE. He received the Engineer of the Year 2006 Award from Govt. of Andhra Pradesh and The Institution of Engineers (India) A.P. State Centre, Best Academician Award for the year 2014 from Andhra University and Andhra Pradesh Scientist Award 2020 in Civil Engineering Discipline from A.P. State Council of Science and Technology, Govt. of A.P. He served as National Executive Committee member of Indian Geotechnical Society during the period 2013-2020. He served as a member in H-4 Committee on Embankment, Ground Improvement and Drainage Committee of Indian Roads Congress (IRC) for the term 2018-2020. He is serving as member of TC 213 on "Scour and Erosion" of International Society for Soil mechanics and Geotechnical Engineering (ISSMGE).

Dr. Sireesh Saride is currently Professor in the Department of Civil Engineering, Indian Institute of Technology (IIT) Hyderabad. His expertise and research interests are in the area of pavement geotechnics, geosynthetics, and ground improvement. He graduated from JNTU College of Engineering, Kakinada, in 2000 and earned his Master's degree integrated with Ph.D. from Indian Institute of Science (IISc),

Bengaluru, in 2006. Prior to joining IIT Hyderabad, he was Postdoctoral Research Fellow at the University of Texas at Arlington. Dr. Saride has published more than 200 referred journals, conference articles, and technical reports. He is Member of ASCE, ICE, TRB, International Geosynthetic Society, Life Member of IRC, and Fellow of Indian Geotechnical Society.

Prof. A. Murali Krishna is currently a faculty member in the department of Civil Engineering, Indian Institute of Technology (IIT) Tirupati. Dr. Krishna obtained his Ph.D. from Indian Institute of Science Bangalore, M.Tech. degree from IIT Kanpur and B.Tech. degree from Sri Venkateswara University College of Engineering, Tirupati. His research interests include earthquake geotechnics, geosynthetics and ground improvement, site characterization, and numerical and physical modelling of geotechnical structures. He was a recipient of BRNS Young Scientist Research award, BOYSCAST fellowship and HERTAGE fellowship. He was a visiting fellow at Bristol University(UK), Surrey University (UK) and University de Torino (Italy). He supervised 7 Doctoral students and 24 Masters students; co-authored nearly 180 publications of technical papers in international/national Journals and conference/seminar proceedings, including book chapters.

Improvement of Soft Clays Using Stone Columns with and Without Encasement

A. Vittalaiah, E. C. Nirmala Peter, and Rathod Ravinder

1 Introduction

The main advantage of the stone column is its ability to adjust itself to the applied load and redistribute the applied load when stress is concentrated on it. This is due to the increase in deformation associated with bulging when the critical vertical stress level is exceeded. This response is different from load sharing response of a pile in soft clay, wherein the pile can offer the resistance mainly by bearing; hence pile of longer length than stone column is required.

Advantages of encased stone columns over stone columns include: (i) the column is confined in such a way that it does not intrude into the soft soil; (ii) a consistent diameter is maintained by the encased material; and (iii) improved shear capacity to the column is provided by the tensile strength of the encased material and increased confinement of the sand or gravel.

2 Literature Review

Castro et al. [1] studied on the consolidation and deformation around end bearing columns under distributed loads and compared the laboratory results with analytical solution and numerical simulation. Shivashankar et al. [3] studied on the behavior of stone column in layered soil consisting of weak soil in the top layer under a series of plate load tests and presented the behavior of remolded kaolin clay reinforced by stone column. It is found that Young's modulus of kaolin clay increases as the cavity

A. Vittalaiah (✉) · R. Ravinder
CE GRIET, Bachupally, Hyderabad, India

E. C. Nirmala Peter
JNTUH College of Engineering, Hyderabad, India

C. N. V. Satyanarayana Reddy et al. (eds.), *Ground Improvement and Reinforced Soil Structures*, Lecture Notes in Civil Engineering 152,
https://doi.org/10.1007/978-981-16-1831-4_1

expansion ratio and consolidation stress increases and the undrained shear strength is more at lower at consolidation stress. It is also noted that the ratio of undrained Young's modulus to undrained shear stress increases when the consolidation stress decreases [2].

3 Materials Used

3.1 Clay

The clay was collected from a field in Patancheru area of Medak district. The properties of clay obtained from various laboratory tests are presented in Table 1.

3.1.1 Ascertainment of Unconfined Compressive Strength at Different Moisture Contents

Unconfined compressive strength tests were conducted at different moisture contents (30–47%, approximately) to obtain the required consistency (soft) for the clay soil.

Table 1 Properties of soft soil

S. no.	Type of test on clay	Results
Grain size distribution		
1	Gravel (%)	0.5
	Coarse sand (%)	1.5
	Medium sand (%)	6
	Fine sand (%)	9
	Silt and clay (%)	83
Atterberg limits		
2	Liquid limit (%)	67
	Plastic limit (%)	32
Standard proctor compaction test		
3	Maximum dry density (kN/m^3)	14.71
	Optimum moisture content (%)	23.50
Modified proctor compaction test		
4	Maximum dry density (kN/m^3)	17.06
	Optimum moisture content (%)	15.60

Table 2 Variation of UCS with moisture content

S. no.	Moisture content (%)	Bulk density (kN/m^3)	UCS (kN/m^2)
1	30.50	13.802	144.28
2	35.10	12.959	83.39
3	39.62	11.968	25.66
4	45.92	11.212	22.56
5	46.35	11.026	19.52

The UCS values are decreasing with an increase in moisture content. Moisture content around 46.2% is used for the preparation of the clay bed for all the load tests (Table 2).

3.2 Gravel

Aggregate having particle size of 12.5–4.75 mm was used for the preparation of the stone column. The gravel for stone columns is classified as uniformly graded using the grain size distribution curve.

3.3 Sand

Medium to fine sand was used in the preparation of stone columns in the proportion of 70% of gravel and 30% of sand to fill in the gaps between the aggregate particles.

3.3.1 Angle of Internal Friction (ϕ) of Gravel–sand Mix

The angle of internal friction of the stone column material was obtained by conducting the direct shear test at different normal stresses on gravel mixed with sand in the proportion of 70–30%, respectively. Figure 1 depicts the stress–strain curves for gravel–sand mixture.

Figure 2 shows the curve of normal stress versus shear stress (strength envelope) at failure. Angle of internal friction for the mixed proportion of gravel and sand was 52.76°.

3.4 Leno-Netted Bag

The leno-netted bag was used as encased material to the stone columns. The thickness of the leno-netted bag is 0.47 mm (Fig. 3).

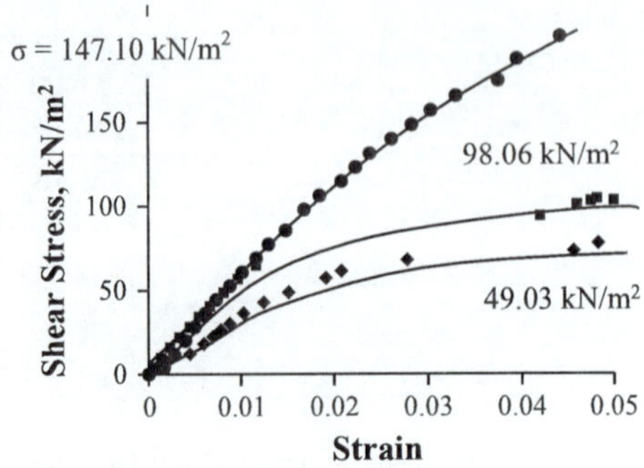

Fig. 1 Stress–strain curves for gravel–sand mix at different normal stresses

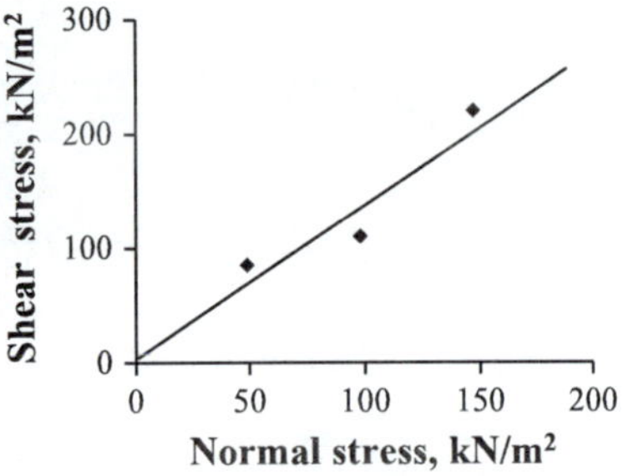

Fig. 2 Normal stress versus shear stress of gravel–sand mix

Fig. 3 Leno-netted bag

4 Test Program

In order to achieve the objectives of the project, load tests were conducted on unreinforced soft clay and soft clay reinforced with stone columns (granular columns). To examine the effect of stiffness on load-carrying capacity, stone columns of diameter 6 cm was used for single-column load tests. The clay bed is prepared by varying density to a thickness of 47 cm.

The properties of the clay and stone column are as follows.

Clay bed: moisture content 46–47%, dry density 10.14–10.25 kN/m^3, thickness of clay bed—approximately 47 cm.

Fig. 4 Cylindrical tank

Stone columns: d = 6.5 cm and 7 cm, L of column = 37 cm, and dry density of column = 18.82 kN/m^3.

Tests were conducted with two sizes of the plates. One exactly equal to the diameter of column and the other 23.5 cm, which is more than three times the diameter of the column covering both the column as well as the soft soil adjacent to the column.

All the tests were conducted in a cylindrical tank (see Fig. 4) of height 50 cm and diameter 40 cm.

4.1 Load Test on Unreinforced Clay Bed

The test procedure for load test on the unreinforced clay bed is as follows.

The clay bed was prepared by compacting in four layers to a total thickness of 47 cm in cylindrical tank using 47% moisture content. A loading plate of diameter 6 cm was placed on the top of the clay bed in the center of plate coinciding with the center of the tank. Two dial gauges were placed on the loading plate to measure the average settlements under load (see Fig. 5). The load was applied gradually on the loading plate using hydraulic jack. The settlements were noted, according to the applied load. The application of the load was ceased when there are negligible settlements. The load was used till the loading plate penetrated completely into the clay bed. The test results were presented in Tables 3, 4 and 5 The stress–settlement curve for this test is shown in Fig. 6.

Fig. 5 Dial gauges on loading plate

Table 3 Load test results for unreinforced clay bed

Stress (kN/m^2)	Average settlement (mm)
0	0
0.82	0.8
1.58	1.2
2.322	2.1
3.896	3.53
5.44	4.96
6.81	6.62
7.18	7.78
9.055	9.95
9.792	11.61

Table 4 Load test results for stone column of diameter 7 cm using a loading plate diameter 6 cm

Stress (kN/m^2)	Average settlement (mm)
0	0
20.169	2.04
40.339	4.95
50.424	7.16
60.509	8.68
70.594	10.85
80.679	11.93
90.764	13.85
100.849	15.65

4.2 Load Tests on Stone Column

Load tests were conducted on a single-stone column of diameter 7 cm using a loading plate of diameter 6 cm and 23.5 cm.

Table 5 Load test results for stone column of diameter 7 cm using a loading platediameter 23.5 cm

Stress (kN/m^2)	Average settlement (mm)
0	0
3.944	0.84
7.889	1.34
11.833	1.83
15.778	2.84
19.722	3.40
23.667	4.52
27.611	5.23
31.556	6.10
35.500	6.91
39.445	8.17
42.075	8.98
44.704	9.88
47.334	11.00
49.964	11.84
52.593	12.90

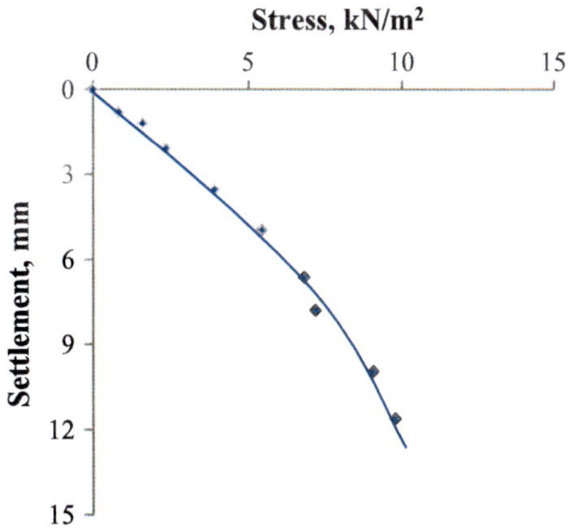

Fig. 6. Stress–settlement curve of unreinforced clay bed using a loading plate diameter 10 cm

(A) *Stone column of diameter 7 cm using a loading plate of diameter 6 cm*

The test procedure for load test on a stone column of diameter 7 cm using a loading plate of diameter 6 cm placed on the top of the stone column is as follows.

Oven-dried clay was mixed thoroughly with a water content of 46.20% taken in the cylindrical tank compacted in four layers to a total thickness of 47 cm. To prepare the stone column, a casing pipe of 6 cm diameter was inserted into the clay bed without disturbing the soil mass as the center of the tank coincides with the center of the pipe. After inserting the casing pipe, the soil sample present in the pipe was removed and replaced with gravel and sand mix at proportions of 70% and 30%, respectively, through the pipe was pulled out gently without disturbing the surrounding soil. The mix proportion was properly compacted with the help of 6 mm diameter rod to maintain the uniform density. After establishment of stone column, the diameter of stone column is changed to a diameter of 7 cm due to removal of pipe. The loading plate placed on the stone column area was shown in Fig. 7b. Two dial gauges arranged on the loading plate to measure the settlements of stone column were shown in Fig. 7c. The load was applied gradually to the plate using hydraulic jack. The settlement of stone column corresponding to applied load was measured. The load was increased after the settlement of stone column has stopped or is negligible. The load was applied up to the loading plate was sunk into the soil. After the test, three samples of testing soil were taken to obtain the moisture content of the soil.

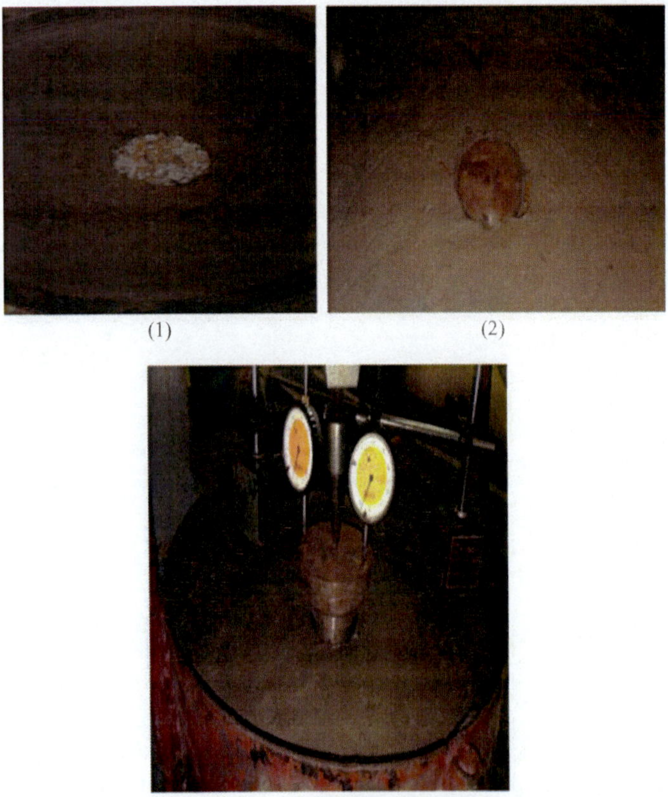

(1) (2)

(3)

Fig. 7 **a** Preparation of a stone column, **b** Loading plate placed on the stone column area, **c** Dial gauges were arranged on loading plate

The test setup for this was shown in Fig. 7. The test results were presented in Tables 3, 4 and Figs. 8 and 9.

(B) *Stone column of diameter 7 cm using a loading plate of diameter 23.5 cm*

The test procedure for load test on a stone column of diameter 7 cm using a loading plate of diameter 23.5 cm placed on the top of the stone column is the same as that of stone column of diameter 7 cm using a loading plate of diameter 6 cm.

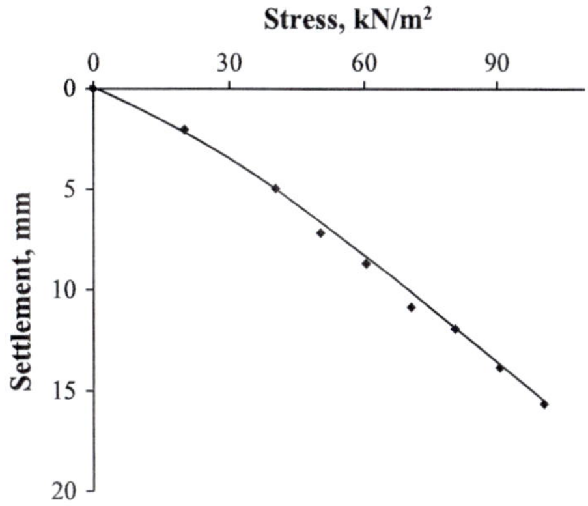

Fig. 8 Stress–settlement curve for stone column of diameter 7 cm using a loading plate diameter 6 cm

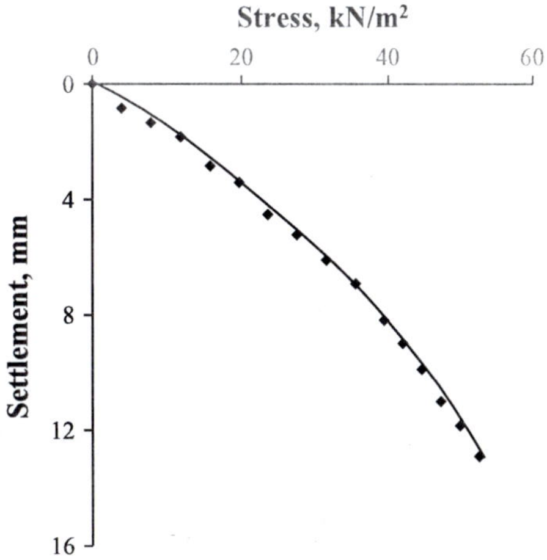

Fig. 9 Stress–settlement curve for stone column of diameter 7 cm using a loading plate diameter 23.5 cm

4.3 Load Tests on Encased Stone Column

The leno-netted bag was used as encased material. The load tests were conducted on single encased stone columns of diameter 7 cm and 6.5 cm using a loading plate of diameter 6 cm and 23.5 cm.

(A) *Encased stone column of diameter 7 cm using a loading plate of diameter 6 cm*

The test procedure for load test on an encased stone column of diameter 7 cm using a loading plate of diameter 6 cm placed on the top of the stone column is as follows.

The clay was taken in the cylindrical tank compacted in four layers to a total thickness of 47 cm, mixed thoroughly with a water content of 46.20%. The stone column was prepared by using a casing pipe of 6 cm diameter through inserting into the clay bed without disturbing the soil mass as the center of the tank coincides with the center of the pipe. After inserting the casing pipe, the soil sample present in the pipe was removed and placed with an empty leno-netted bag. After that the leno-netted bag was filled with gravel and sand mixture at proportions of 70% and 30%, respectively. The casing pipe was pulled out gently without disturbing the surrounding soil. The mix proportion was properly compacted with the help of 10 mm diameter rod to maintain the uniform density. After installation, the diameter of the encased stone column expanded to 7 cm (Fig. 10a). The loading plate was placed only on the encased stone column area.

To measure the settlements of the stone column, two dial gauges were arranged on the loading plate. The load was applied gradually to the plate using hydraulic jack. Corresponding to applied load the settlements of stone column was noted. The load was increased after the settlement of stone column has stopped or is negligible. The load was applied until the loading plate was sunk into the soil. After the test, three samples of testing soil were taken to obtain the moisture content of the soil. The test setup for this was shown in Figs. 10, 11 and Table 6.

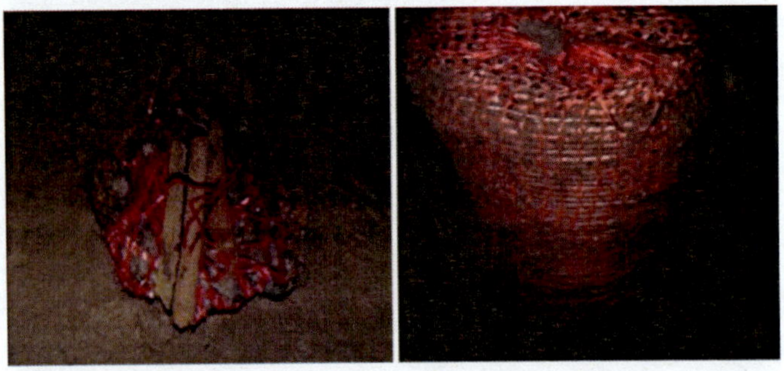

(1) (2)

Fig. 10 **a** Preparation of the encased stone column, **b** Stone column after completion of the test

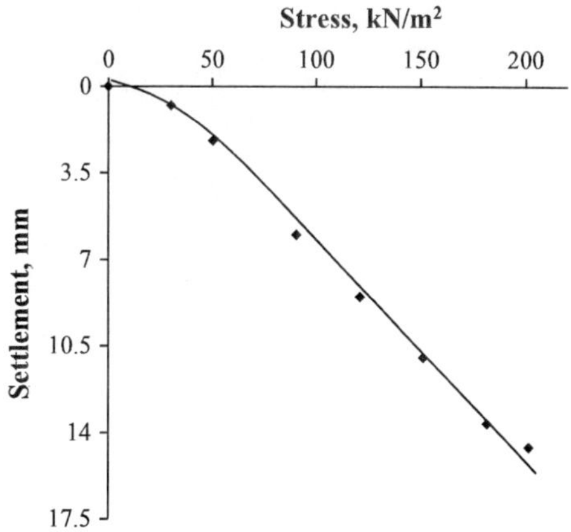

Fig. 11 Stress–settlement curve for stone column of diameter 7 cm using a loading plate diameter 6 cm

Table 6 Load test results for an encased stone column of diameter 7 cm using a loading plate of diameter 6 cm

Stress (kN/m^2)	Average settlement (mm)
0	0
30.25	0.77
50.42	2.20
90.76	5.99
121.01	8.50
151.27	10.97
181.52	13.62
201.69	14.58

(B) *Encased stone column of diameter 6.5 cm using a loading plate of diameter 23.5 cm*

The same test procedure was conducted for encased stone columns of diameter 7 and 6.5 cm using a loading plate of diameter 6 and 23.5 cm (Fig. 12 and Table 7).

5 Analysis and Discussion

This chapter presents the comparison of the load-distortion characteristics of reinforced soft clay by introducing the stone columns. The main focus of this chapter was the comparison of the stiffness and stress settlement behavior of the stone columns with and without encasement installed in soft clay beds. The stiffness and the load-carrying capacity of an encased single-stone column is expected to be increased compared to without encasement.

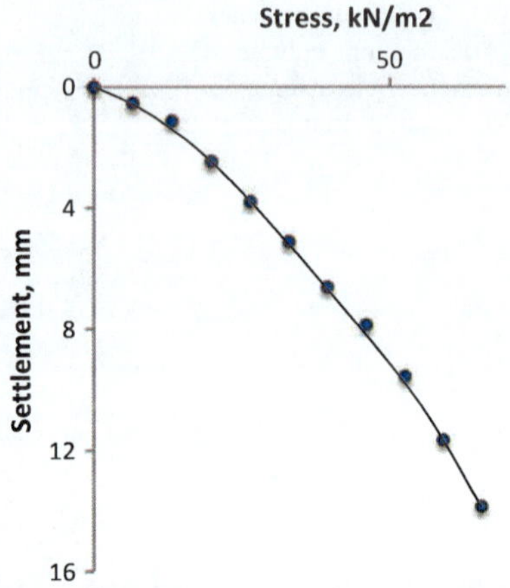

Fig. 12 Stress–settlement curve for stone column of diameter 6.5 cm using a loading plate diameter 23.5 cm

Table 7 Load test results for an encased stone column of diameter 6.5 cm using a loading plate diameter 23.5 cm

Stress (kN/m^2)	Average settlement (mm)
0	0
6.57	0.52
13.14	1.12
19.72	2.47
26.29	3.79
32.87	5.11
39.44	6.61
46.01	7.85
52.59	9.55
59.16	11.65
65.74	13.84

5.1 Modulus of Elasticity of Clay from UCS Test

Stiffness of clay was obtained by conducting UCS tests, at different moisture contents. The results of UCS tests were presented in Table 8. It is observed that with increase in moisture content, the stiffness of clay decreased.

The stress–settlement reactions from load tests on single-stone columns of different diameters are shown in tables and figures. From load test on clay bed at 47% moisture content, the stiffness of clay was E = 1250 kN/m^2.

Table 8 E values for clay at different moisture content

Moisture content (%)	30.50	35.10	39.62	45.92	46.35
E (kN/m^2)	4000	3200	850	800	347

5.2 *Single-Stone Columns of Diameter 6 and 7 cm*

(A) **Using a large diameter of loading plates (23.5 cm)**

The stress–settlement responses were studied for load tests on a stone column of diameter 7 cm and 6.5 cm using a loading plate of 23.5 cm diameter. Table 9 shows the comparison of properties of the stone columns. It was observed that the bulk density of clay and dry density of stone column were almost equal in both test conditions, and the stiffness of the stone column increased with an increase in diameter.

Figure 11 shows the comparison of stress–settlement curves. The settlement of the stone column increases with an increase in stress. For particular stress with increase in diameter the settlements decreases (Fig. 13).

Table 9 Comparison of stone columns of diameter 7 and 6.5 cm using a loading plate diameter 23.5 cm

S. no.	Property	7 cm diameter	6.5 cm diameter
1	Bulk density of clay (kN/m^3)	14.98	14.98
2	Maximum dry density of stone aggregates (kN/m^3)	18.82	17.70
3	Stiffness, E (kN/m^2)	2150.53	3000

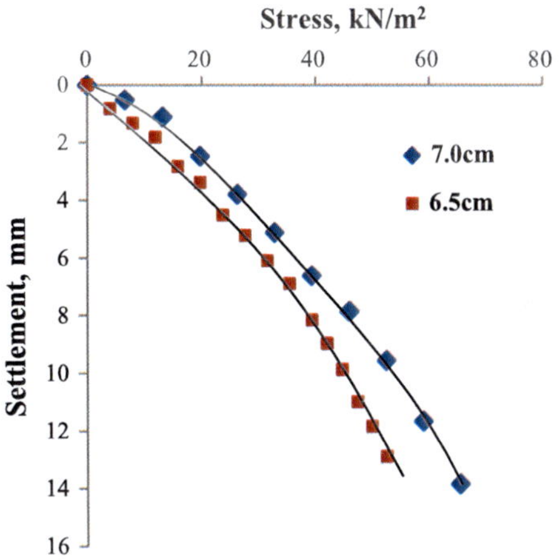

Fig. 13 Stress–settlement responses to the stone column of diameter 7 and 6.5 cm using a loading plate diameter of 23.5 cm

Table 10 Comparison of a stone column with encased stone column of 7 cm diameter

S. no.	Property	Stone column	Encased stone column
1	Bulk density of clay (kN/m^3)	14.83	14.83
2	Maximum dry density of stone aggregates (kN/m^3)	17.70	18.97
3	Modulus of elasticity, E (kN/m^2)	3333.3	7888.89

Fig. 14 Comparison of stress–settlement responses of unreinforced clay bed and stone column of 7 cm diameter

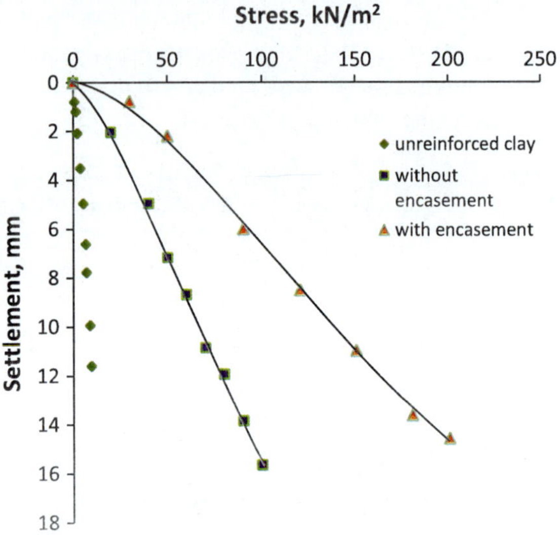

5.3 *Single-Stone Column of 7 cm Diameter Using a Loading Plate of Diameter 6 cm*

The load tests results of a stone column and an encased stone column of diameter 7 cm were presented in Table 9. It was observed that the stiffness of an encased stone column increased (Table 10).

Figure 14 shows the comparison of stress–settlement curves. By applying leno-netted bag as encased material, the settlement of a stone column is decreased.

6 Conclusions

1. A small diameter pile with an encasement can give preferable execution over an uncased pile.
2. The stiffness of an encased stone column increased by 2.37 times when compared with the stiffness of an uncased stone column.

3. If the diameter of loading plate decreases then stiffness of the stone column increases, vice-versa.
4. The stiffness of uncased stone column increased by 2.6 times and the stiffness of encased stone column increased by 6.3 times when compared to the stiffness of unreinforced clay bed.

References

1. Castro, J., et al.: Consolidation and deformation around stone columns: Comparison of theoretical and laboratory results. Comput. Geotech. (2012)
2. Frikha, W., et al.: Observed behaviour of laterally expanded stone column in soft soil. Geotech. Geol. Eng. **31**, 739–752 (2013)
3. Shivashankar, R., et al.: Experimental studies on behaviour of stone columns in layered soils. Geotech. Geol. Eng. **29**, 749–757 (2011)

Geotechnical Properties of Soft Improved Ground from In Situ Time–Settlement Plots

J. Divya, M. Padmavathi, and M. R. Madhav

1 Introduction

Settlement of soft clay consists of immediate, consolidation and creep or secondary settlements. Largest proportion of the settlement of soft clay is attributed to the consolidation process. Because of the very low permeability, the primary consolidation settlement takes a long time to complete. Structures constructed on soft soils face problems of stability and serviceability if measures are not taken to improve them.

Ground improvement techniques are required to overcome these problems. These techniques essentially increase the shear strength and reduce the compressibility of the soil to a desired extent. Prefabricated vertical drain technique is one of the most suitable methods to accelerate the dissipation of excess pore pressures. Application of ground improvement method using prefabricated vertical drains (PVD) coupled with surcharge or preloading can significantly shorten the period of primary settlement. PVDs or sand drains greatly shorten the drainage path.

A method proposed in the paper involves estimation of the coefficient of consolidation for radial flow from in situ time–settlement plots. Schiffman [4] presented the solution for the degree of consolidation, U_r, for radial flow and time-dependent loading for the equal strain case. The degree of consolidation, U_{rt0}, at the end of construction period for no smear (s becomes unity) case is

J. Divya
M.Tech. student, Geotechnical Engineering, JNTU Hyderabad, Hyderabad 500085, Telangana, India

M. Padmavathi (✉)
Associate Professor, Department of Civil Engineering, JNTU Hyderabad, Hyderabad 500085, Telangana, India

M. R. Madhav (✉)
Emeritus Professor, JNTU Hyderabad, Hyderabad 500085, Telangana, India

C. N. V. Satyanarayana Reddy et al. (eds.), *Ground Improvement and Reinforced Soil Structures*, Lecture Notes in Civil Engineering 152,
https://doi.org/10.1007/978-981-16-1831-4_2

$$U_{r(t0)} = 1 - \frac{F(n)}{8T_{R0}}\left[1 - e^{\frac{-8T_{R0}}{F(n)}}\right] \tag{1}$$

where $T_{R0} = \frac{C_{vr} \cdot t_0}{4d_e^2}$ is the time factor at the end of construction

$$F(n, s) = F_1(n, s) + \theta F_2(n, s) \tag{2}$$

$$F_1(n, s) = \frac{n^2}{n^2 - s^2}\ln\left(\frac{n}{s}\right) + \frac{s^2 - 3n^2}{4n^2} \tag{3}$$

$$F_2(n, s) = \frac{n^2}{n^2 - s^2}\ln(s) \tag{4}$$

n is the ratio of diameter of unit cell, d_e, to the diameter of the drain, d_w; $d_e = 1.128$ and 1.05S for drains installed in square and triangular patterns, respectively; S is the spacing between drains. Smear factor, s is the ratio of radius of smear zone to radius of drain and c_{vr} is the coefficient of consolidation for radial flow.

The degree of consolidation, $U_{r(t0)}$ at the end of construction period is plotted with time factor, T_{R0}, for different values of n (the ratio of unit cell, d_e, to that of drain, d_w) (see Fig. 1).

Degree of consolidation, $U_{r(t0)}$ increases (see Fig. 1) with time factor, T_{R0}, corresponding to end of construction but decreases with n, i.e., with larger spacing of the PVDs or sand drains.

Application of the Proposed Method

From a given time–settlement plot, the final/ultimate settlement, S_{ult}, is estimated using Asaoka method [2]. The settlement, S_{t0}, at the end of construction time (t_0) is

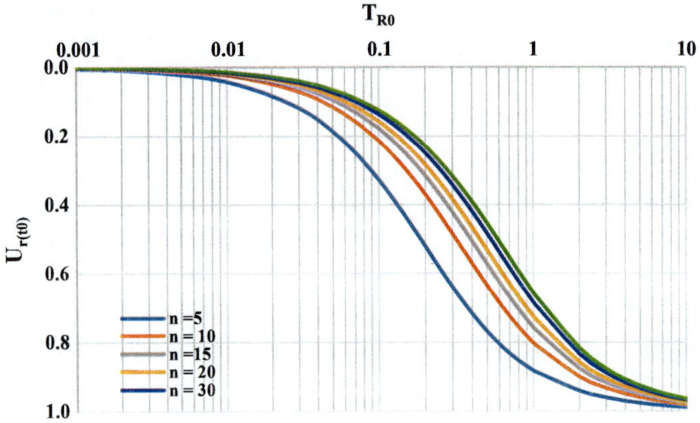

Fig. 1 Degree of consolidation, $U_{r(t0)}$ vs time factor, T_{R0} for different values of 'n'

noted from the in situ time–settlement data. The degree of consolidation, $U_{r(t0)}$, is

$$U_{r(t0)} = \frac{S_{t0}}{S_{\text{ult}}} \tag{5}$$

The time factor, T_{R0}, corresponding to the end of construction, is obtained from Fig. 1 from the estimated $U_{r(t0)}$ and the known value of 'n'. The coefficient of consolidation with radial flow, c_{vr} is then estimated from the known values of t_0, T_{R0} and d_e, as

$$c_{vr} = \frac{T_{R0} \times d_e^2}{t_0} \tag{6}$$

This method is applied to few case studies reported in the literature.

1.1 Mud Flats, South of Iraq [1]

Mud flats, an extension of the deltas of the rivers Tigris and Euphrates, exist at the Arabian Gulf in the South of Iraq. The site is underlain by approximately 10 m of soft, sensitive, normally consolidated clay to silty clay containing seems of silt and/or sand. Unit weight of soil is 1.8 g/cm^3 and plasticity index varies from 20 to 28. Initial void ratio is 0.81 and the compression index is 0.346. Three tanks, A, B and C with diameter of 76.2 m were built.

Sand drains of diameter 0.3 m were installed in triangular pattern with spacing of 2.0, 2.25 and 2.5 m beneath Tanks A, B and C, respectively. Ratios, $n = d_e/d_w$, of diameter of unit cell to diameter of well are 7.0, 7.875 and 8.75, respectively. Observed time versus settlement plots for Tanks A, B and C are shown in Figs. 2, 3 and 4, respectively.

Asaoka plots with 10-day time step are given in Figs. 5, 6 and 7, respectively, for Tanks A, B and C. Ultimate settlements, S_{ult} for the Tanks A, B and C are obtained as 2.7 m, 3.4 m and 2.9 m, respectively, and the slopes of the lines fitted are 0.934, 0.956 and 0.93. Settlements, S_{t0}, and the times, t_0, at the end of construction are 1.14 m and 110 days, 1.18 m and 150 days, and 1.2 m and 100 days. Degree of consolidation at the end of construction, $U_{r(t0)} = \frac{S_{t0}}{S_{ult}}$ are 0.42, 0.35 and 0.41 and the corresponding time factors, T_{R0} are 0.18, 0.16 and 0.2 (from Fig. 1). The coefficients of consolidation, cr for radial flow are c_{vr} = 2.63, 2.17 and 5.05 m^2/year (Eq. 6).

1.2 Embankment on Soft Clay Treated with Sand Drains [3]

A highway embankment was constructed with a height of 6 m and length of 900 m. Strata are very plastic organic clay with a crust approximately 1.0 m thick. The organic compressible soft clay is 10 m thick. Profiles PL1 and PL4 represent different types

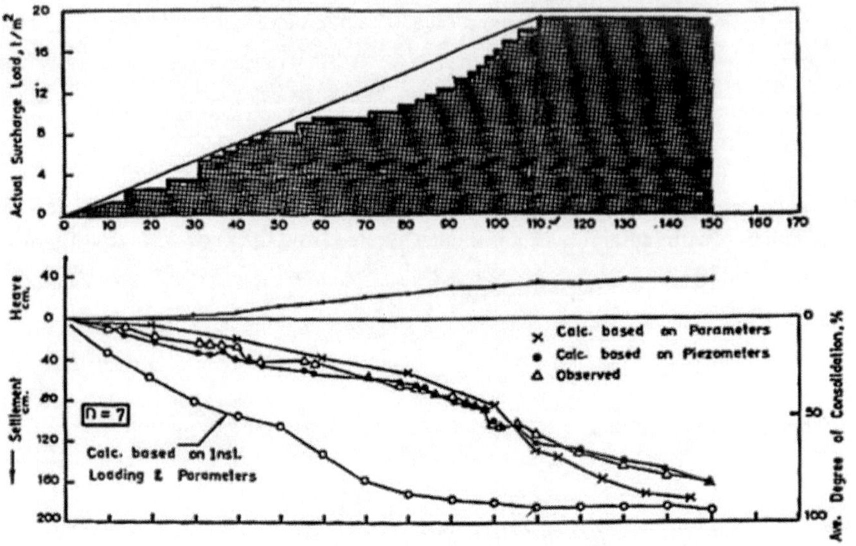

Fig. 2 Observed time versus settlement of Tank A

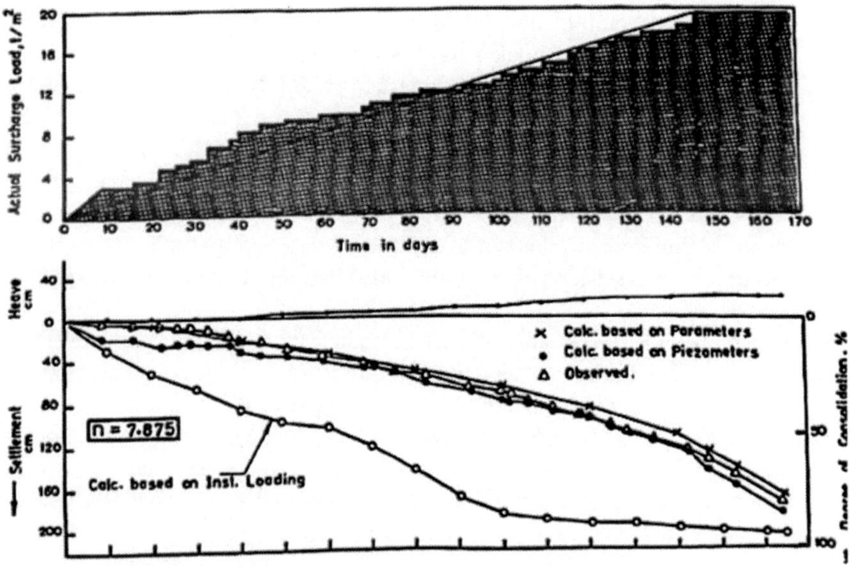

Fig. 3 Observed time versus settlement at Tank B

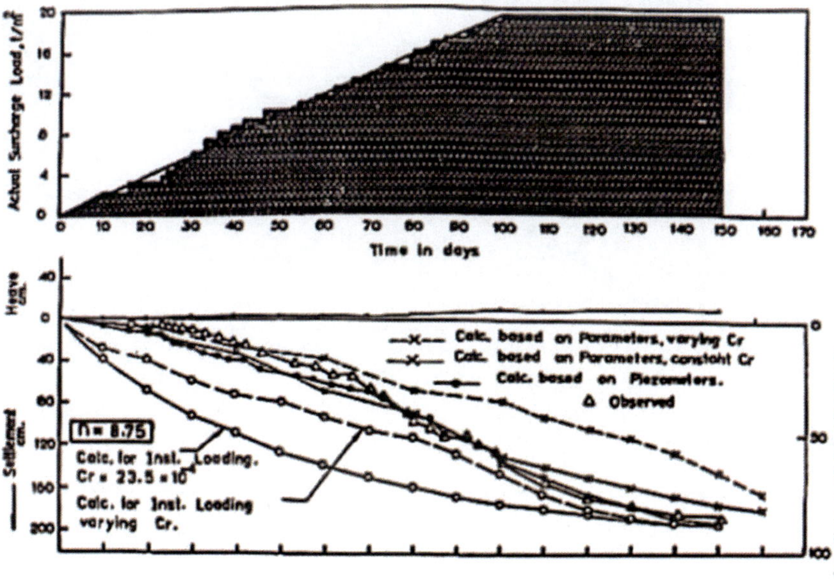

Fig. 4 Observed time versus settlements at Tank C

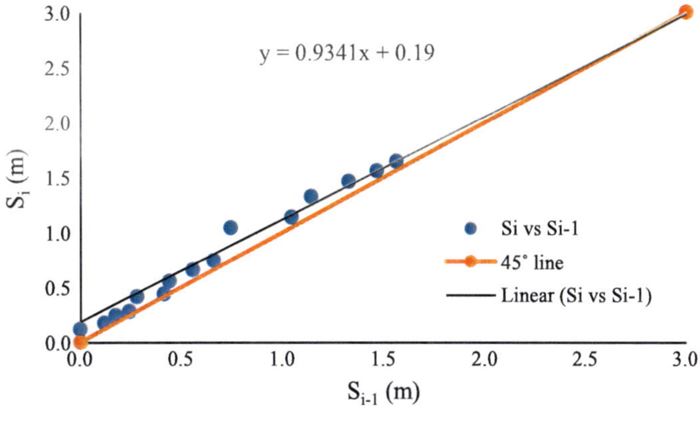

Fig. 5 Asaoka plot for Tank A mud flats

of foundation soils. Sand drains with outer and inner diameters of 0.3 m and 0.1 m, respectively, were installed in a triangular mesh of 2.0 m spacing at both the sites.

Profile PL1

PL1 profile was instrumented with settlement gauges and piezometers. The ratio, n, of diameter of unit cell, d_e, to diameter, d_w, of well is 7.0. Observed time–settlements at points T1, T2 and T3 are shown (see Fig. 8).

Ultimate settlements estimated from Asaoka plots shown in Figs. 9, 10 and 11.

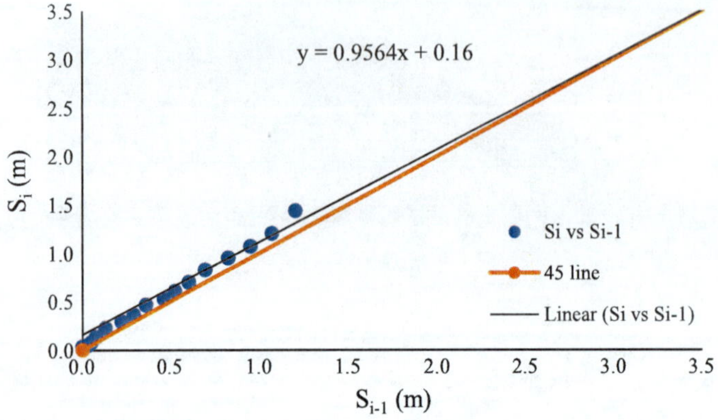

Fig. 6 Asaoka plot for Tank B mud flats

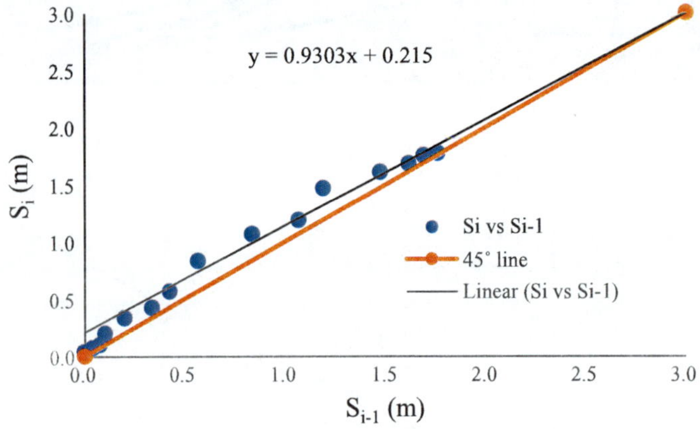

Fig. 7 Asaoka plot at Tank C mud flats

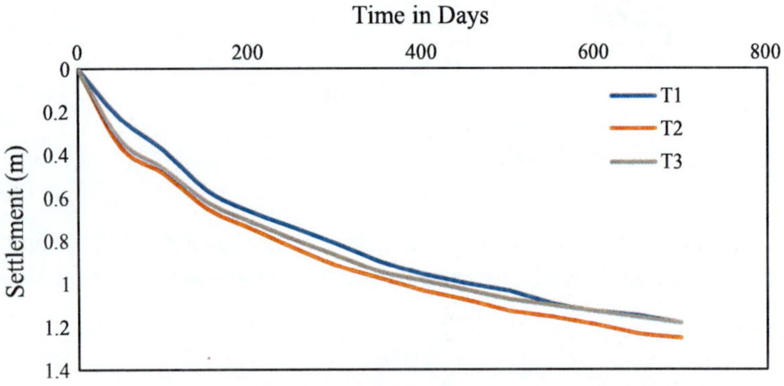

Fig. 8 In situ time–settlement plots at PL1 site [3]

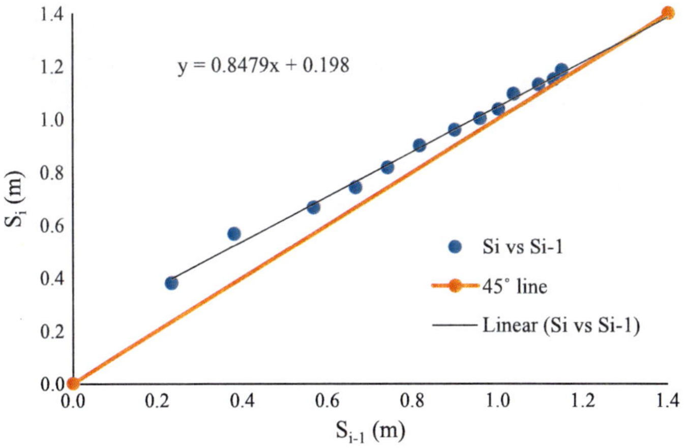

Fig. 9 Asaoka plot at T1, profile PL1

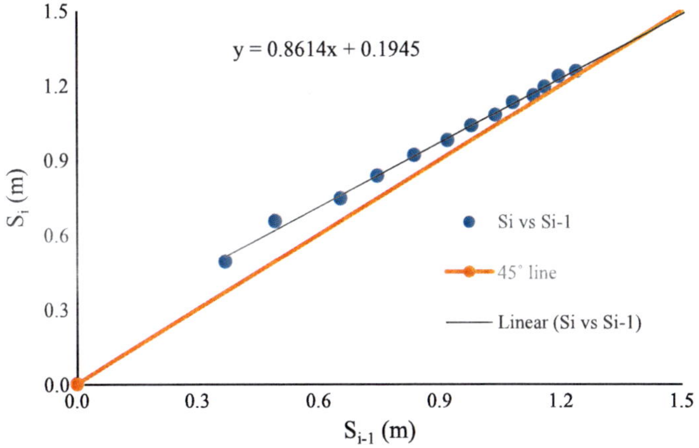

Fig. 10 Asaoka plot at T2, profile PL1

Settlements, S_{t0}, and time t_0, for the end of construction, final settlements, S_{ult}, degrees, $U_{r(t0)}$, of consolidation and time factors, T_{R0}, corresponding to end of construction are determined and listed in Table 1 for the three points for profile PL1.

Profile PL4

At PL4, five surface settlement plates are installed at different points on ground. In situ time–settlement plots at points T1, T2, T3 and T4 for PL4 profile are shown (see Fig. 12).

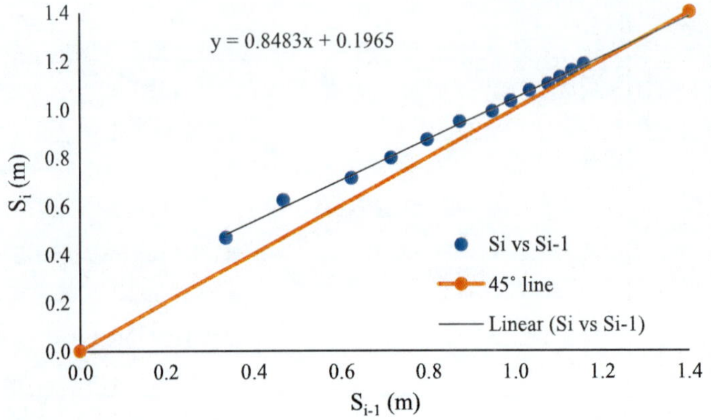

Fig. 11 Asaoka plot at T3, profile PL1

Table 1 Parameters for profile PL1

Profile	S_{t0} (m)	t_0 (days)	S_{ult} (m)	$U_{r(t0)}$	T_{R0}
T1	1.053	500	1.3	0.810	0.80
T2	1.084	500	1.4	0.775	0.68
T3	1.132	500	1.3	0.871	1.25

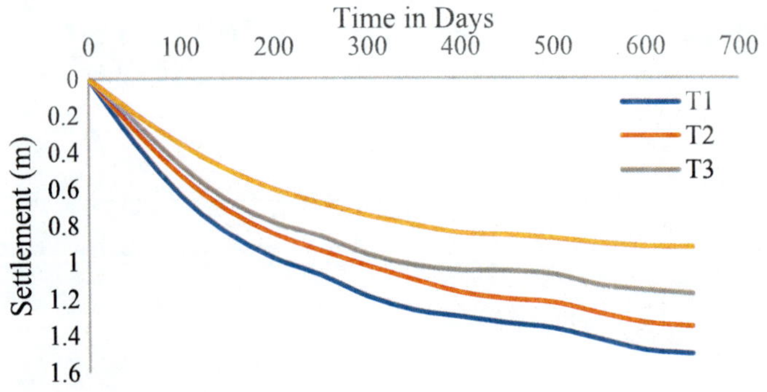

Fig. 12 Observed in situ time–settlement plots at PL4 [3]

Ultimate settlements, S_{ult} estimated using the Asaoka plots for the time–settlement curves for various points T1, T2, T3 and T4 are presented (see Figs. 13, 14, 15 and 16).

Diameter of unit cell $(d_e) = 2.1$ m; diameter of well (d_w) is 0.3 m; ratio, n, of diameter of drain to diameter of well is 7.0; $F(n)$ is 1.598; time interval, Δt, to draw S_i versus S_{i-1} plot, is 50 days. Parameters for profile PL4 are tabulated in Table 2.

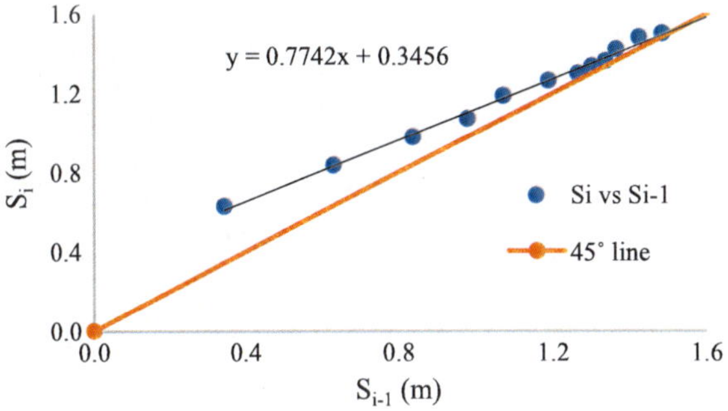

Fig. 13 Asaoka plot at point T1 of profile PL4

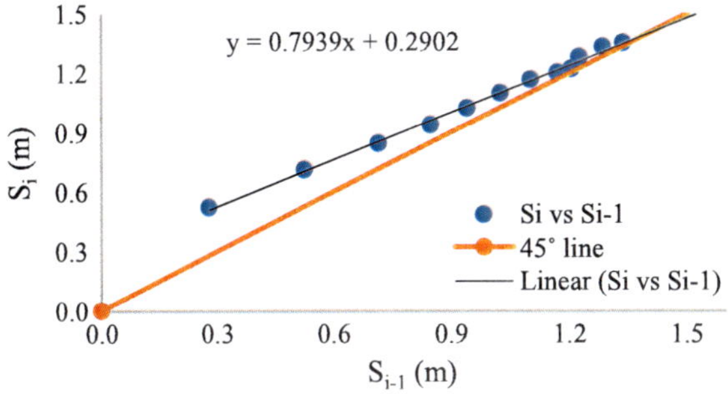

Fig. 14 Asaoka plot at point T2 of profile PL4

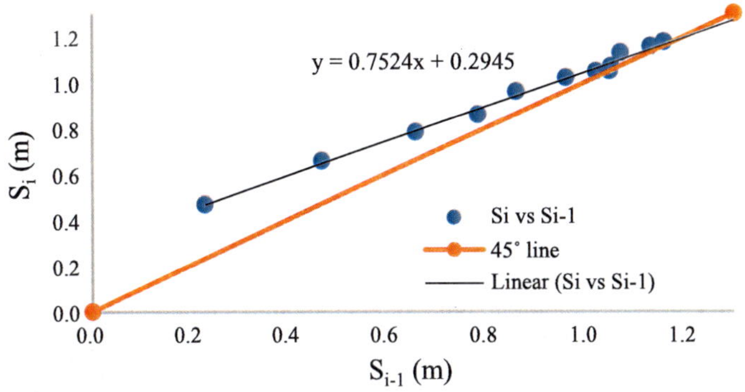

Fig. 15 Asaoka plot at point T3 of profile PL4

Coefficients of consolidation for radial flow, c_{vr} are estimated by substituting the appropriate parameters in Eq. (6), and listed in Table 3 for the two profiles PL1 and PL4.

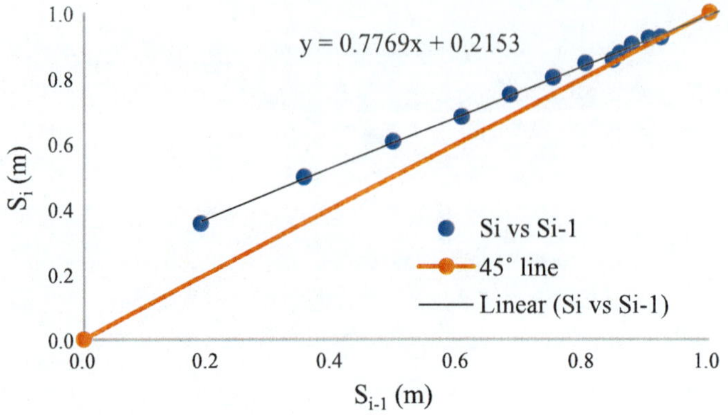

Fig. 16 Asaoka plot at point T4 of profile PL4

Table 2 Parameters of profile PL4

Depth (m)	S_{t0} (m)	t_0 (days)	S_{ult} (m)	$U_{r(t0)}$	T_{R0}
T1	1.364	500	1.50	0.909	1.80
T2	1.226	500	1.36	0.901	1.75
T3	1.071	500	1.16	0.923	2.05
T4	0.868	500	0.96	0.904	1.60

Table 3 c_{vr} for profiles PL1 and PL4

Profile PL1		Profile PL4	
Point	c_{vr} (m^2/year)	Point	c_{vr} (m^2/year)
T1	2.58	T1	5.79
T2	2.19	T2	5.63
T3	4.02	T3	6.60
		T4	5.15

2 Results and Discussion

For the site of mud flats, the range of coefficients of consolidation for radial flow (c_{vr}) from the laboratory test results is from 0.95 to 9.47 m^2/year [1], while the range of in situ coefficients of consolidation for radial flow (c_{vr}) estimated using the proposed degree of consolidation at end of construction is between 2.63 and 5.05 m^2/year, very different from the laboratory determined values.

For the site of treated ground at profiles PL1 and PL4, the range of laboratory coefficients of consolidation is between 4.73 and 37.87 m^2/year [3], while the estimated in situ coefficients of consolidation for radial flow by the proposed new approach, i.e., based on degree of consolidation at the end of construction time, is in the range

2.19–4.02 m^2/year for profile PL1 and 5.15–6.60 m^2/year for profile PL4. Once again the in situ coefficients are very different from the laboratory determined values. The behavior of the ground is very different from that determined in the laboratory based on small size samples.

3 Conclusion

PVDs or sand drains with surcharge helps in accelerating the consolidation process in situ. The proposed method estimates the coefficient of consolidation for radial flow from in situ time–settlement plots at early times. The results estimated from in situ time–settlement plot are different from laboratory determined values.

References

1. Al Alusi, H.R., Al Alusi, A.F.: Prediction of settlements of soft clays caused by sand drains and preloading. In: International Symposium on Soft Clay, Bangkok, Thailand, 195–210 (1977)
2. Asaoka, A.: Observational procedure of settlement prediction. Soils Found. Japanese Soc. Soils Found. Eng. **18**(4), 87–101 (1978)
3. Queyroi, D., Pilot, G., Magman, J.P.: Edification D'un remblai sur foundation d'argile molle traitee par drains de sable. In: International Symposium on Soft Clay, Bangkok, Thailand, pp. 245–259 (1977)
4. Schiffman, R.L.: Consolidation of soil under time dependent loading and varying permeability. In: Proceedings of the Thirty-Seventh Annual Meeting of the Highway Research Board, Washington, DC, 6–10 January 1958, pp. 584–617 (1958)

Settlement of a Square Footing on Dry Sand Bed Reinforced with Stone Columns Under Seismic Conditions: Effect of Frequency

Priti Maheshwari, J. P. Sahoo, and Siddhartha Saxena

1 Introduction

These days use of stone columns is not new to improve the ground, and this technique finds application in various areas such as roads and highways, railways, ports and airports, storage tanks, chemical plants, land reclamations, and residential buildings. Although these are mainly adopted in case of saturated cohesive soils, however, few studies also mention about their effectiveness in densification of loose sands. Raju and Daramalinggam [1] studied different ground improvement techniques and their suitability for different soil conditions. Usefulness of the ground treatment with stone columns was illustrated for improvement in load-carrying capacity and reduction in settlement of foundations for most of the structures. Lot of literature is available pertaining to the foundations on stone column treated ground [2–9]. However, very few studies dealt with the performance of footings on stone columns treated ground under the dynamic loadings. Some of such studies include Adalier et al. [10], Chen et al. [11], Krishna et al. [12], Bouassida and Hazzar [13], Han [14], Ashour [15], and Fahmi et al. [16]. Most of these studies dealt with cyclic loading and used rigid tank for preparation of the physical model of footings on reinforced beds. Rigid tanks have related issues such as boundary effects and side friction between soil and the tank. Some of the experiments have been conducted using laminar box by Maheshwari et al. [17]. However, the current work focuses on the effect of frequency of input motion on the response of a square footing resting on dry sand bed with stone columns inclusion.

P. Maheshwari (✉) · S. Saxena
Indian Institute of Technology Roorkee, Roorkee, India

J. P. Sahoo
Indian Institute of Technology Kanpur, Kanpur, India

C. N. V. Satyanarayana Reddy et al. (eds.), *Ground Improvement and Reinforced Soil Structures*, Lecture Notes in Civil Engineering 152,
https://doi.org/10.1007/978-981-16-1831-4_3

2 Materials and Method

2.1 Experimental Setup

A shaking table facility with only horizontal degree of freedom has been developed in the Geotechnical Engineering Laboratory of IIT Roorkee, in order to simulate the seismic excitation. For varying the frequency of the input motion, an arrangement has been manufactured which could simulate three values of frequency, viz., 1.23, 1.5 and 1.73 Hz. The payload capacity of the developed table is 8 t and it can simulate the acceleration range from 0.05 to 1 g. The facility can have maximum displacement in horizontal direction as ±75 mm.

In order to prepare the model of sand bed, a laminar box is preferred over the rigid tank to overcome the problems of boundary effects and side friction. The box has 23 aluminum layers separated by roller bearings facilitating the relative motion between the layers. The setup has been depicted in Fig. 1.

Fig. 1 Developed facility of shaking table with laminar box

2.2 Instrumentation

Physical models of footing on dry sand bed with stone columns inclusion have been prepared inside the laminar box, and to measure the vertical settlement of the footing, a displacement transducer has been used. An arrangement of slider and roller has been designed to monitor the vertical settlement under horizontal motion and the same has been mounted on the footing. A data acquisition system with 10 channels has been used to record the settlement. This could sample at a sampling speed of 1000 samples/s.

2.3 Materials Used

Poorly graded sand has been used to prepare the sand bed while well-graded stones of sizes between 4.75 and 10 mm have been adopted for making the stone columns. The sand has the following index properties: Specific gravity: 2.57; D_{50}: 0.28 mm; maximum and minimum unit weight: 16.43 and 13.82 kN/m^3, respectively.

2.4 Model Preparation and Testing Procedure

The size of the footing has been taken as 300 mm and all the models have plan dimensions as 1000 mm × 1000 mm. Extent of treatment by stone columns of 50 mm diameter has been kept as 800 mm. After placing the aluminum frames up to 200 mm height, a polythene sheet has been placed to the inner side of the box so that the sand filled inside the box does not flow from the gap between the two frames. The sand has been filled with the desired relative densities. For the construction of stone columns, help of PVC pipes has been taken. Stone columns have been installed in triangular fashion and filled in these pipes with unit weight of 19.62 kN/m^3 in three layers. The pipes have been withdrawn slowly after placement of the stone layer. This procedure has been repeated till the complete bed was constructed. The loading has been applied using dead weights through the square plate. The model has been subjected to the horizontal motion for 30s and vertical settlement-time history has been recorded.

2.5 Law of Similitude

Tests have been conducted at 1 g on the models prepared on reduced scale and the similitude law as proposed by Iai [18] has been adopted with geometrical scale factor of 10.

3 Results and Discussion

The chosen input parameters for studying the effect of frequency of input motion on settlement response of a square footing are as follows: peak ground acceleration (PGA): 0.2 and 0.3 g; frequency: 1.23, 1.5 and 1.73 Hz; spacing to diameter ratio of stone columns: 2, 2.5 and 3; relative density (*RD*) of sand bed: 30 and 50%; applied vertical load: 75 and 90 kg.

The vertical settlements are only because of horizontal motion and initial settlement due to static vertical load has not been included.

Figure 2 depicts the influence of frequency on settlement response of square footing for $d = 50$ mm, PGA $= 0.3$ g, $Q = 75$ kg and $s/d = 2.5$. Both the values of relative densities of 30 and 50% have been considered and tests have been conducted at three values of frequency (1.23, 1.5 and 1.73 Hz). For a relative density of 30%, the final vertical settlement has been found to reduce by 36% as frequency reduces from 1.73 to 1.23 Hz. However, corresponding reduction of about 7% has been observed for 50% relative density of sand bed. At higher relative density of sand bed, frequency does not show significant effect. This may be attributed to the fact that at higher relative density, the sand bed reinforced with stone columns achieves its maximum dry density in 30s of test duration for all the values of frequencies and therefore not much effect of frequency is observed. However, at lower value of relative density, the frequency of input motion significantly affects the settlement response of the footing.

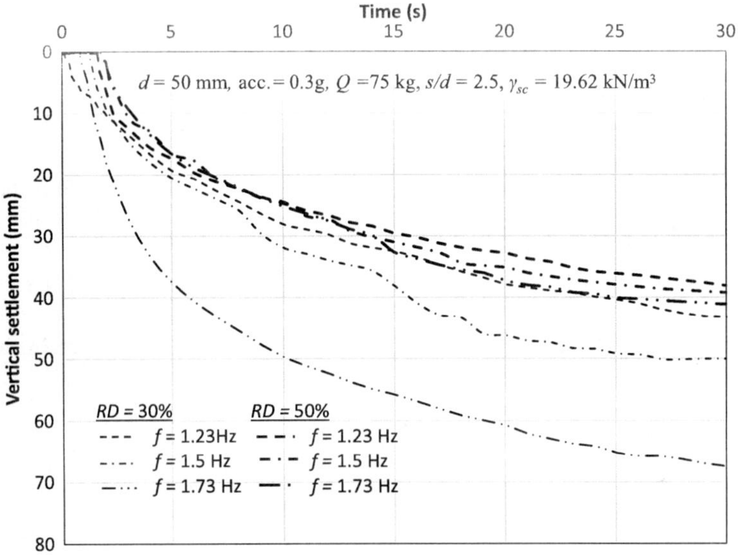

Fig. 2 Influence of frequency for both values of relative density

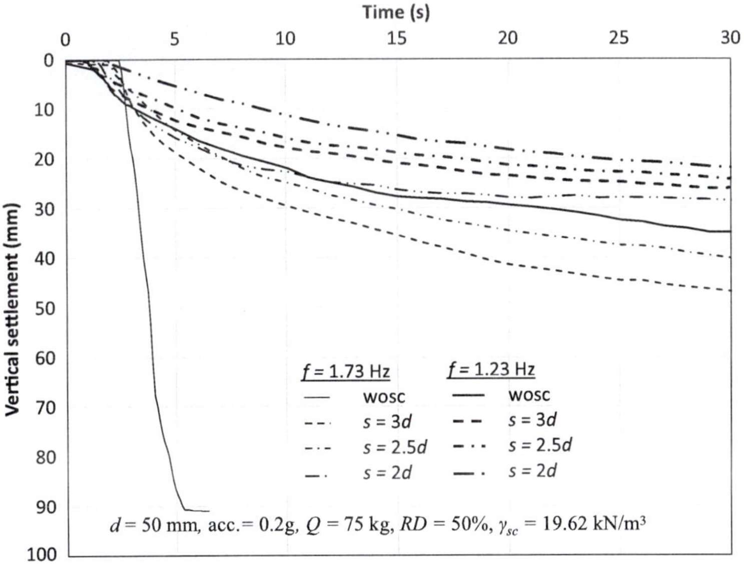

Fig. 3 Influence of frequency: *RD* = 50%; applied load = 75 kg; PGA = 0.2 g

Further, few typical results pertaining to two limiting values of frequency have been reported to show the influence of frequency on response of footing under seismic conditions. Figure 3 shows the effect of frequency on settlement response of footing for applied load level of 75 kg. The PGA and relative density has been considered as 0.2 g and 50%, respectively. It is to be noted that for unreinforced case, i.e., without provision of stone columns, the slope of displacement versus time curve changes drastically along with the magnitude of settlement with reduction in the frequency of input motion. A reduction in frequency of input motion results in reduction of vertical settlement of the footing which has been found to be 62, 24, 41 and 44% for unreinforced case, *s/d* equals to 2, 2.5 and 3, respectively.

The effect for higher value of applied load of 90 kg has been shown in Fig. 4 and similar trend has been observed. Reduction in vertical settlement of the footing corresponding to reduction in frequency from 1.73 to 1.23 Hz has been observed to be 44% for both *s/d* equals to 2, 2.5. The corresponding reduction for unreinforced case and for *s/d* = 3 has been found to be 38% and 55%, respectively.

The corresponding plots exhibiting the influence of frequency on settlement of footing for lower relative density of 30% have been shown in Figs. 5 and 6 for applied loads of 75 and 90 kg, respectively. Similar pattern of results has been obtained. Figure 5 shows a reduction of about 28, 34 and 41% for *s/d* values of 2, 2.5 and 3, respectively, corresponding to 75 kg of applied load level. When the load level is increased to 90 kg, the corresponding reduction has been observed to vary between 51 and 56% for different configurations of stone columns (Fig. 6).

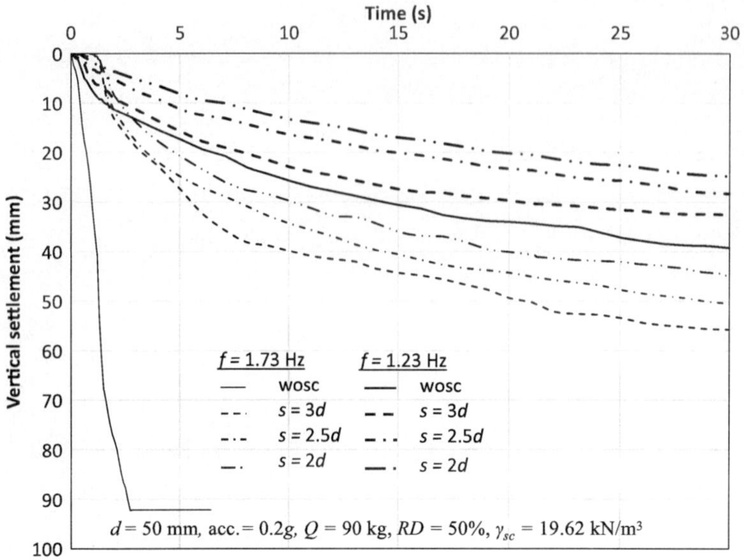

Fig. 4 Influence of frequency: $RD = 50\%$; applied load = 90 kg; PGA = 0.2 g

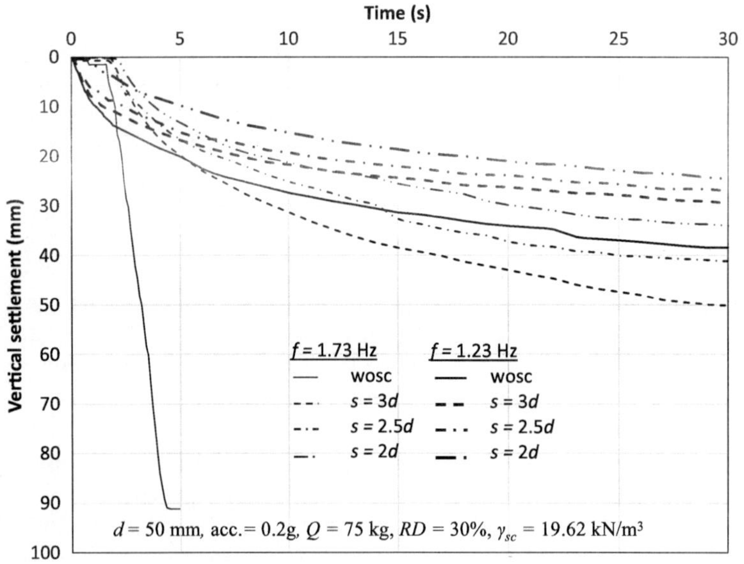

Fig. 5 Influence of frequency: $RD = 30\%$; applied load = 75 kg; PGA = 0.2 g

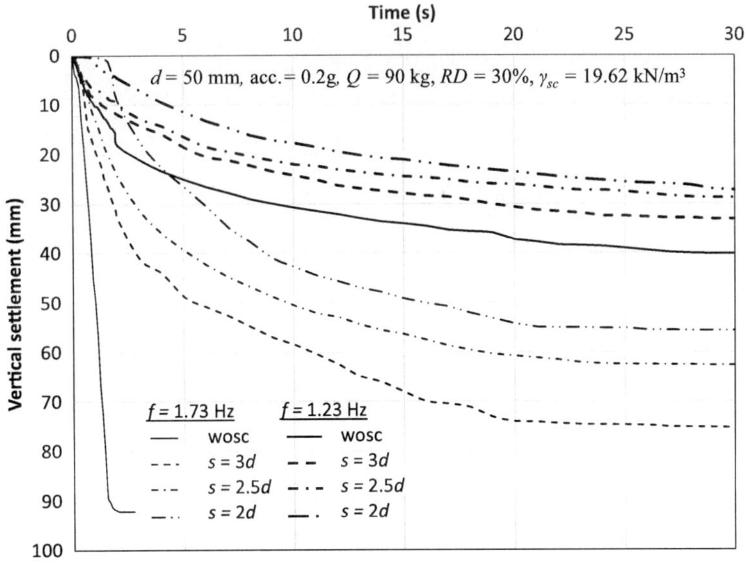

Fig. 6 Influence of frequency: $RD = 30\%$; applied load = 90 kg; PGA = 0.2 g

The effect of frequency of input motion on settlement response of footing has been shown in Figs. 7 and 8 for applied load of 75 kg and 90 kg, respectively. The

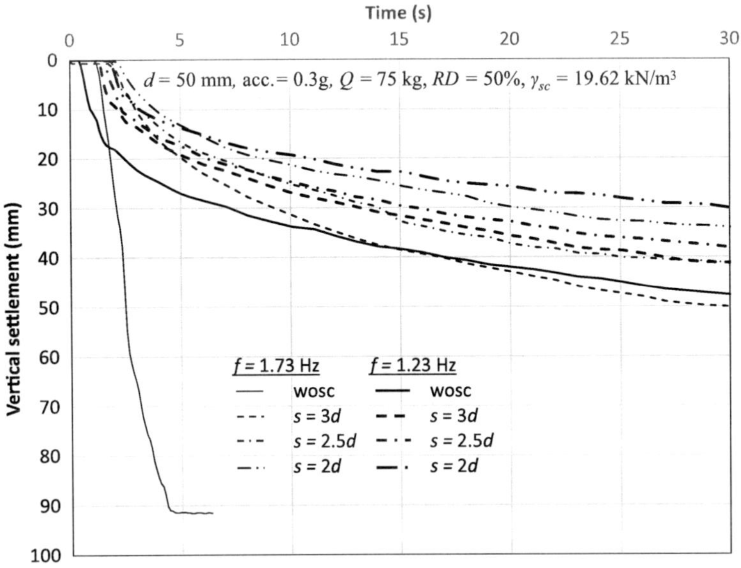

Fig. 7 Influence of frequency: $RD = 50\%$; applied load = 75 kg; PGA = 0.3 g

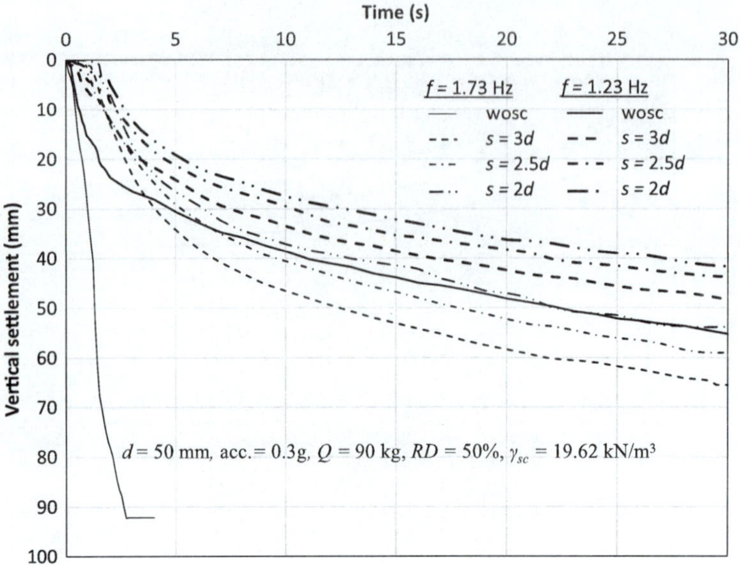

Fig. 8 Influence of frequency: *RD* = 50%; applied load = 90 kg; PGA = 0.3 g

relative density for these tests has been considered as 50% and the tests have been conducted at PGA of 0.3 g. For 75 kg load level, the settlement at 30s has been found to reduce by about 11, 7 and 18% for *s/d* values of 2, 2.5 and 3, respectively (Fig. 7), as the frequency reduces from 1.73 to 1.23 Hz. However, the corresponding reduction for 90 kg load has been found to be about 23–26% (Fig. 8).

Similarly, tests have also been conducted at 30% relative density. Other input parameters have been taken as: PGA = 0.3 g and *d* = 50 mm. For both the load levels of 75 and 90 kg, the effect has been seen and shown in Figs. 9 and 10, respectively. The settlement has been observed to experience a reduction with reduction in frequency of input motion which has been found to be about 30–36% for all the three spacing of stone columns for both levels of applied load.

These figures depict that frequency has significant influence on settlement response of the footing. However, this influence has been found to be more for lower relative density.

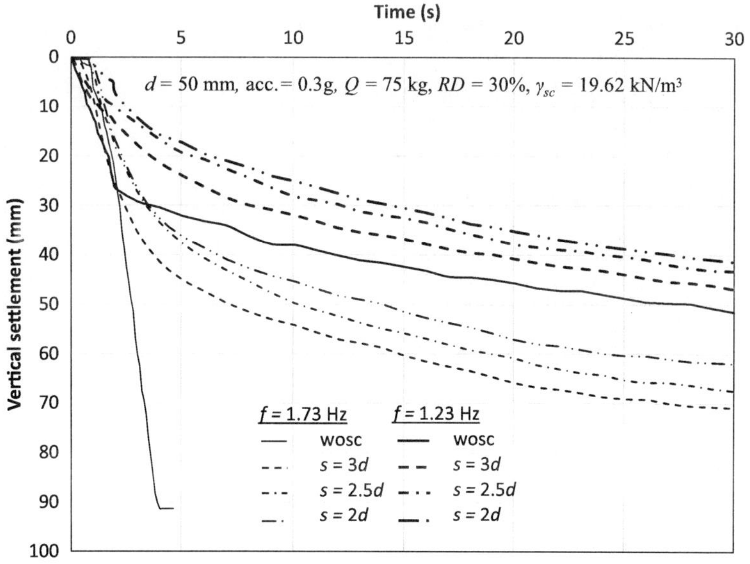

Fig. 9 Influence of frequency: $RD = 30\%$; applied load = 75 kg; PGA = 0.3 g

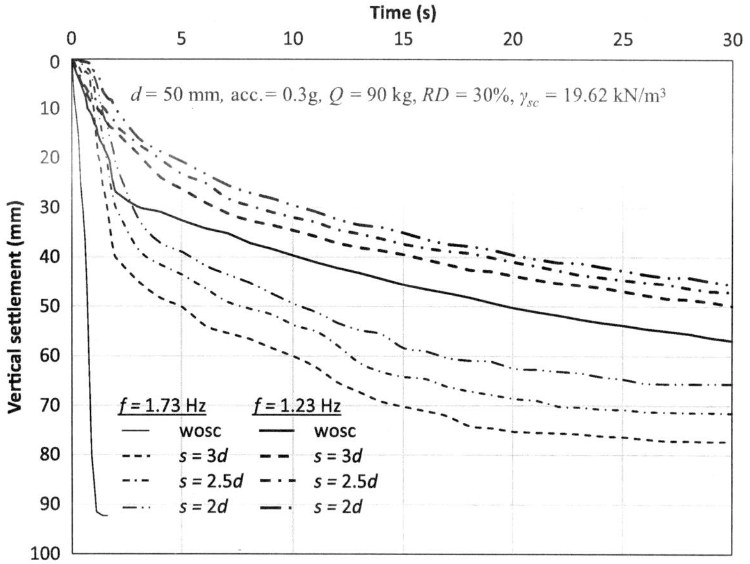

Fig. 10 Influence of frequency: $RD = 30\%$; applied load = 90 kg; PGA = 0.3 g

4 Conclusions

Extensive experimental study has been undertaken to understand the effect frequency of input motion on the settlement response of a square footing placed over stone columns reinforced dry sand.

Frequency of input motion has been found to be significantly influencing the response of footing. It has been observed that for unreinforced case, i.e., without provision of stone columns, the slope of displacement versus time curve changes drastically along with the magnitude of settlement with reduction in the frequency of input motion. In case of few tests on reinforced conditions, the reduction in settlement has been found to be of the order of about 62% as frequency changes from 1.73 to 1.23 Hz.

Acknowledgments The financial support by IIT Roorkee—NBCC Joint R&D Center on "Sustainable Civil Infrastructure" in carrying out this work is highly acknowledged.

References

1. Raju, V.R., Daramalinggam, J.: Ground improvement: principles and applications in Asia. Ground Improv. Proc. Inst. Civil Eng. **165**(GI2), 65–76 (2012)
2. Wood, D.M, Hu, W., Nash, D.F.T: Group effect in stone column foundations: model tests. Géotechnique **50**(6), 689–698 (2000)
3. Ambily, A.P., Gandhi, S.R.: Behavior of stone columns based on experimental and FEM analysis. J. Geotech. Geoenviron. Eng. ASCE **133**(4), 405–415 (2007)
4. Black, J.A., Sivakumar, V., Bell, A.: The settlement performance of stone column foundations. Géotechnique **61**(11), 909–922 (2011)
5. Shivashankar, R., Babu, M.R.D., Nayak, S., Rajathkumar, V.: Experimental studies on behaviour of stone columns in layered soils. Geotech. Geol. Eng. **29**, 749–757 (2011)
6. Fattah, M.Y., Shlash, K.T., Al-Waily, M.J.: Stress concentration ratio of model stone columns in soft clays. Geotech. Test. J. ASTM **34**(1), 50–60 (2011)
7. Fattah, M.Y., Shlash, K.T., Al-Waily, M.J.: Experimental evaluation of stress concentration ratio of model stone columns strengthened by additives. Int. J. Phys. Model. Geotech. **13**(3), 79–98 (2013)
8. Mohanty, P., Samanta, M.: Experimental and numerical studies on response of the stone column in layered soil. Int. J. Geosynthetics Ground Eng. **1**(3), Article 27 (2015)
9. Shivakumar, V., O'Kelly, B.C., Moorhead, C., Madhav, M.R., Mackinnon, P.A.: Effectiveness of granular columns in containing settlement. Proc. Inst. Civil Eng. Geotech. Eng. **167**(GE4), 371–379 (2014)
10. Adalier, K., Elgamal, A., Meneses, J., Baez, J.I.: Stone columns as liquefaction countermeasure in non-plastic silty soils. Soil Dyn. Earthq. Eng. **23**, 571–584 (2003)
11. Chen, C.-H., Ueng, T.-S., Lee, W.-C.: Large scale biaxial shear box tests on shaking table. In: 13th World Conference on Earthquake Engineering, Paper no. 1778, 1–6 August 2004, Vancouver, BC, Canada (2004)
12. Krishna, A.M., Madhav, M.R., Latha, G.M.: Densification effect of granular piles on settlement response of treated ground. Ground Improv. Proc. Inst. Civil Eng. **11**(3), 127–136 (2007)
13. Bouassida, M., Hazzar, L.: Novel tool for optimized design of reinforced soils by columns. Ground Improv. Proc. Inst. Civil Eng. **165**(GI1), 31–40 (2009)

14. Han, J.: Recent research and development of ground column technologies. Ground Improv. Proc. Inst. Civil Eng. **168**(GI4), 246–264 (2015)
15. Ashour, S.: The response of stone columns under cyclic loading. Ph.D. thesis, School of Civil Engineering, College of Engineering and Physical Science, The University of Birmingham (2015)
16. Fahmi, K.S., Fattah, M.Y., Shestakova, A.: Behavior of foundation soil improved by stone column under cyclic load. In: MATEC Web of Conferences, TransSiberia, vol. 239, p. 05015 (2018). https://doi.org/10.1051/matecconf/201823905015
17. Maheshwari, P., Yadav, N.K., Sahoo, J.P.: Seismic performance of footings on stone columns treated dry sand beds. KSCE J. Civil Eng. (2020). https://doi.org/10.1007/s12205-020-2319-9
18. Iai, S.: Similitude for shaking table tests on soil-structure fluid model in 1-g gravitational field. Soils Found. **29**(1), 105–118 (1989)

Experimental Study on Parametric Influences of Stone Column Reinforced Foundation Systems

Arghadeep Biswas, Utpal Mandal, and Agnishek Chakraborty

1 Introduction

Ggranular column in soft soil has been very effective in improving bearing capacity and limiting the settlements of concern geo-structure. It is constructed by filling and compacting the granules in a pre-bored vertical-hollow. The aggregates bear majority of the imposed load and transfer it deeper through interconnections. The column is with high permeability which allows the pore water to dissipate faster and accelerates the consolidation. Studies have revealed various parametric influences, including geometry (diameter, length and column-arrangement), in-fill material (granules) quality, stiffness of surrounding soil (shear strength) and/or encasement type etc., on the performance of granular column. In this regard, articles by Najjar [1] and Ghose et al. [2] would be good reads to have a glimpse of performance of stone column in soft clay.

The natural lateral confinement to granular column is provided by the undrained shear strength (c_u) of surrounding soil; however, its effect is minimal with respect to limiting axial stress [3]. Ambily and Gandhi [3] found that column-aggregates and soft clay squeeze into each other causing disturbance in expected mechanisms. It has been mentioned that the overall ground stiffness depends on spacing between the columns and improves with surcharge loading. Reportedly, the angularity, packing and frictional angle of aggregates influence the behavior of granular columns [4–6]. It is further mentioned that the more the friction angle is, the more will be the column-stiffness, bearing capacity and stability; however, the same has resulted in reduced settlement and lateral bulging. The dependence of column performance on its geometry (i.e., diameter and length) has also been investigated [7–11] and reported

A. Biswas (✉) · A. Chakraborty
Jalpaiguri Government Engineering College, Jalpaiguri 735102, West Bengal, India

U. Mandal
Ramkrishna Mahato Government Engineering College, Purulia 723103, West Bengal, India

C. N. V. Satyanarayana Reddy et al. (eds.), *Ground Improvement and Reinforced Soil Structures*, Lecture Notes in Civil Engineering 152,
https://doi.org/10.1007/978-981-16-1831-4_4

that increase in diameter (while installation) may affect the pre-estimated bearing capacity and drainage function of the granular column. Researchers [12–14] have mentioned a bulge formation for longer columns which limits the overall settlement; however, a punching failure is found for shorter columns which influence bearing capacity [15, 16]. In this regard, there is a mentioned critical length (l_{crit}) of 5–8 times the diameter, beyond which it does not contribute further on bearing capacity.

In this article the performance of model stone column on clay of varying strengths is reported. The investigation program included variation in the strength of surrounded clay (c_u) as 5, 10 and 25 kPa. Two different stone sizes were selected (ranged between 10–4.75 mm and 4.75–1.18 mm) for in-filling the column. The foundation bed-cum-test mold was prepared as per unit cell concept described in Indian Standard [17]. Three column diameters, lengths and footing sizes were selected to simulate the floating and end-bearing column conditions. The initial study was based on physical tests; however, the authors are forced to focus on the numerical simulation (Plaxis3D) due to pandemic condition. As per the observation made till date, it is found that the behavior of stone columns is immensely influenced by the parameters considered.

2 Material and Methodology

In the study locally available clay-soil used for preparing the foundation bed by varying its water content (to vary the shear strength). The Pakur variety stone chips were selected for constructing the granular columns. Basic characterization of soils were performed as per designated Indian Standards and the determined properties are presented in Table 1. The table also includes the parameters required to define materials in the numerical simulation.

2.1 Preparation of Clay

For preparing the clay beds, a relationship is established between the undrained shear strength of clay with water content (Fig. 1), following the procedure described by Biswas [18]. The curve acted as the backbone for selecting clay strengths and

Table 1 Material properties

Material properties	E_{Nu} (kPa)	c_u (kPa)	φ (°)	υ_{Nu}	γ (kN/m^3)
Clay	600 c_u	5, 10, 25	0	0.35	17, 18, 19
Stone chips	50 × 10^3	0	52, 58	0.40	14.55, 14.71

E_{Nu}, υ_{Nu}: Numerical Inputs

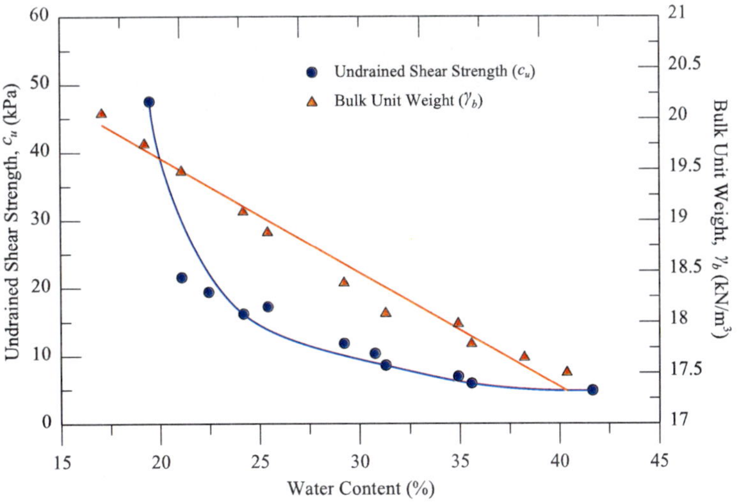

Fig. 1 Calibrated relation between c_u, w (%) and γ of clay

calculating the corresponding water contents (with bulk densities) keeping the clay saturated.

2.2 *Preparation of Clay Bed*

The foundation bed was prepared in a split-mold made of metal sheet of 2 mm thick. The mold was braced with steel clamps (4 mm thick and 20 mm wide flat steel bars) to avoid bending of walls during the bed preparation and while the test progresses. The mold and other accessories are shown schematically in Fig. 2. Thick PVC pipes of different diameters were used to create the column-core. Three nos. of 18 mm thick wooden-plates, having hole at the center (avg. hole diameter = external diameter of corresponding pipe + 2 mm) were fabricated for compacting clay layers (with the pipe inserted) within the mold. Before placing the clay in mold, pipes were made vertical with holders (fabricated and can be fixed strongly with the mold) (Fig. 2). The clay was placed in the mold in layers (about 50 mm thick). After placing clay for a layer, a wooden-plate (with a plastic sheet at the bottom) was placed on the clay surface for compaction. The wooden-plate was supposed to distribute the compactive effort all over the clay surface. A plastic sheet was used to avoid the sticking of clay on the wooden-plate. The layers were compacted with equal number (pre-determined) of blows, with the hammer designated for standard proctor test. After preparing each layer, the clay surface was scratched to get adequate grip with the successive layer. Completion of clay bed preparation allows stone chips to be placed in the PVC pipes in desired density with tamping.

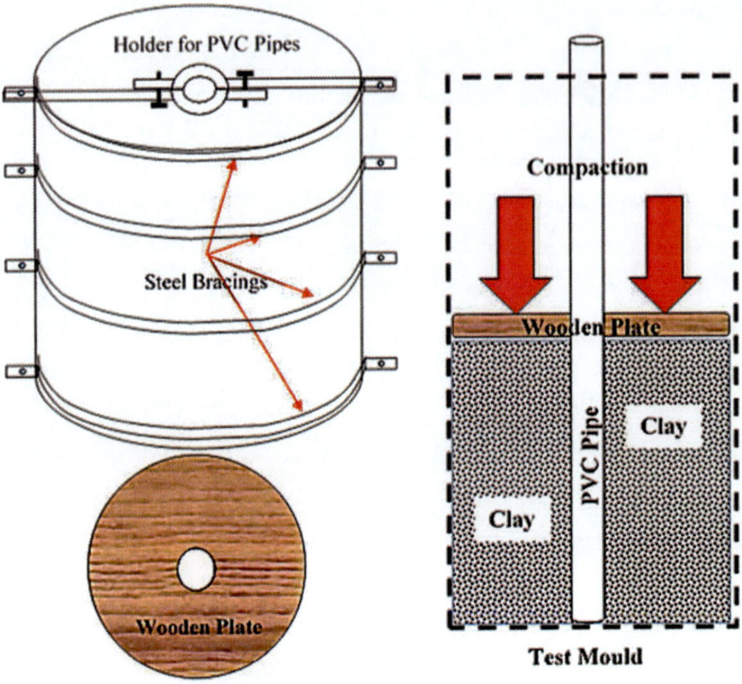

Fig. 2 Schematic of test mold, holder for PVC pipes, wooden-plate for compaction and preparation of test bed

2.3 Preparation of Stone Column

Stone columns were prepared in layers with tamping. The entire length was formed by 50 mm equal lifting. For each layer, the required mass of stone chips was calculated, weighted, poured and compacted until the desired thickness was achieved. The desired effort for tamping was confirmed through several trial tests. After each layer, the PVC pipe was lifted up to the thickness prepared to start preparation of the next layer. Due to softness of surrounding soil, the compaction increased the column volume (length and diameter). Thus, care and trials were made in such a way that the enhancement in column volume should be limited to 5% of the stone weight calculated for each layer. After preparation of stone column footing was placed centrally and load was applied.

2.4 Test Procedure

The schematic test setup is shown in Fig. 3. The load-test was conducted in a universal load frame having 36 nos. strain rates. The prepared test mold, with footing, was

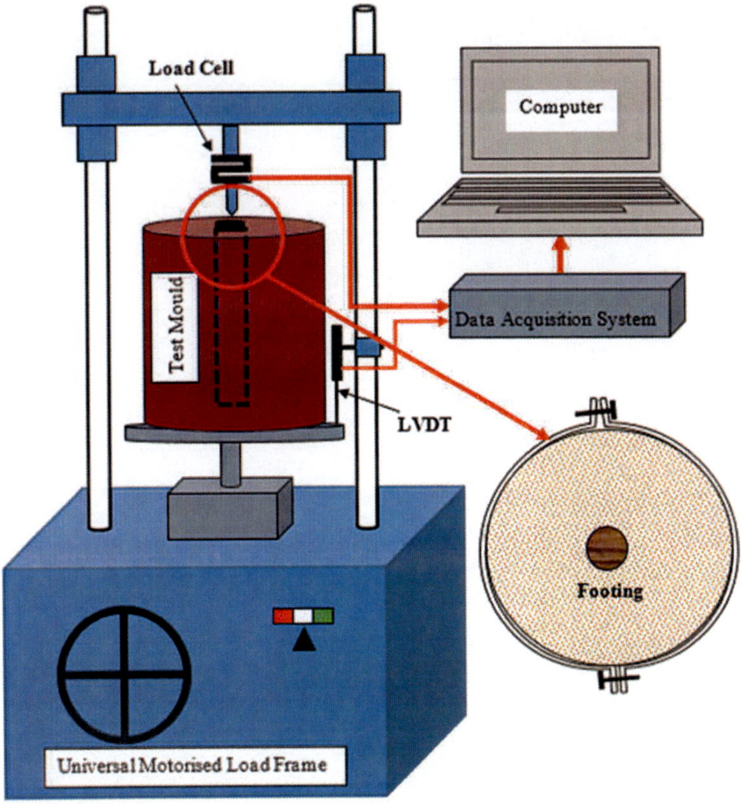

Fig. 3 Schematic of experimental setup and other accessories used

placed under the load frame. The load was applied through a load cell of 10 kN capacity. A LVDT (50 mm run) was used to record footing settlement. The load-settlement was measured and recorded in a data logger. After successful placement of all instruments, test was started with the pre-determined strain rate.

3 Results and Discussion

The tests parameters selected, and aimed for, were footing size (D), diameter of stone column (D_c), length of stone column (L), stone sizes, undrained shear strength of clay (c_u) and the strain rate of load applied. Obstructed by the pandemic situation, the variables were curtailed and only a few tests were possible to perform physically. Behavior of circular footing of different diameters ($D = 30$, 45 and 60 mm) rested on homogeneous clay with $c_u = 5$ kPa were obtained under single strain rate of loading (1.2 mm/min). The load-settlement behavior is presented in Fig. 4. The

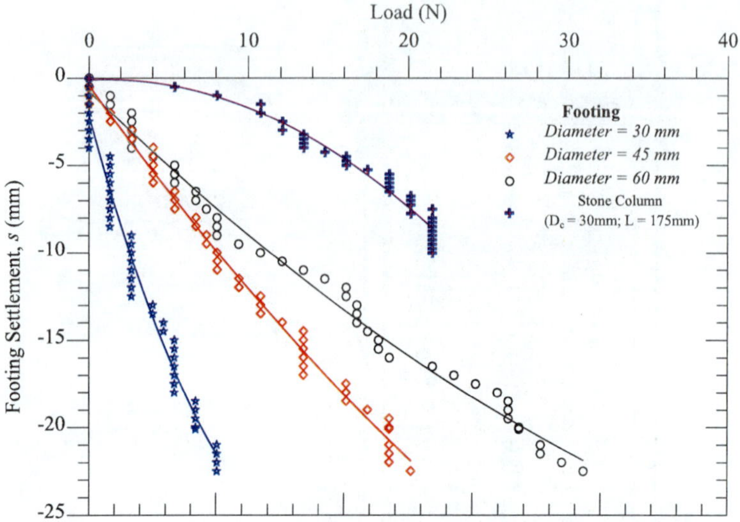

Fig. 4 Responses of physical tests with different footings on homogeneous clay with $c_u = 5$ kPa

responses depicted increase in pressure-settlement behavior with the footing diameter. A test with stone column ($D_c = 30$ mm; $L = 175$ mm) made of "4.75 mm passing and 1.18 mm retained" fraction of stone chips was performed in clay with $c_u = 5$ kPa. Comparing with the corresponding homogeneous response, it can be said that pressure-settlement behavior improved with the installed stone column.

In the numerical simulation, the material properties used were same as determined through laboratory tests. The axisymmetric numerical simulation is performed in Plaxis3D with 60, 90, 120 and 180 mm diameter footing in clay having $c_u = 5$, 10 and 25 kPa (selected from Fig. 1). The laboratory determined properties of two stone chips, designated as 4.75 mm (10 mm passing and 4.75 mm retained) and 1.18 mm (4.75 mm passing and 1.18 mm retained), were used to model the stone column having lengths as 1, 1.5 and 2 m within a $2 \times 2 \times 2$ m^3 soil block. A typical model schematic and corresponding analysis response are shown in Fig. 5.

The simulation results have depicted a considerable increase in load-bearing capacity with increase in undrained shear strength of surrounding clay (Fig. 6). However, it was also found that the short floating column of 1 m length has behaved better than the other two configurations in a clay of same undrained shear strength. This can be attributed to bulging of short column to improve the bearing capacity and failure of slender column before achieving the desired benefits (Fig. 6). A typical comparison of behavior of short (1 m) and slender (2 m) column, in a clay with $c_u = 10$ kPa, is presented in Fig. 7a–d.

A comparison of influence of footing diameter (120 mm and 180 mm) and column length (1, 1.5 and 2 m), keeping the column diameter same (60 mm) with a constant clay consistency of 25 kPa (c_u), is presented in Fig. 8. It is seen that the bearing capacity of the foundation systems improves considerably with increase in footing

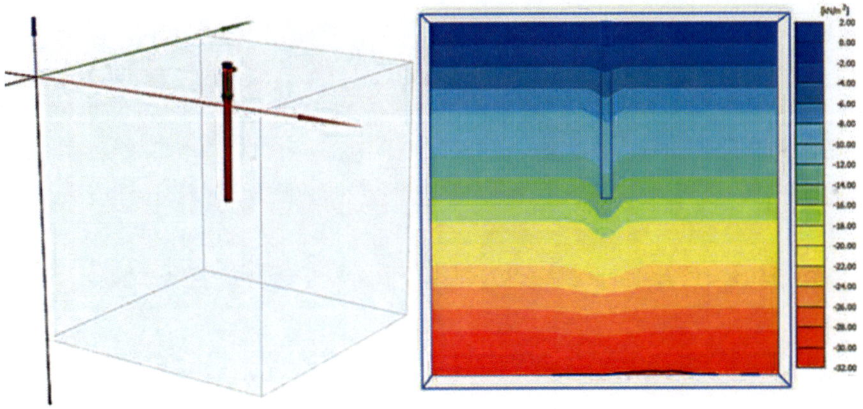

Fig. 5 Typical analysis model and responses of footing having 60 mm diameter rested on stone column of 1 m length with 4.75 mm in-fill material on clay with $c_u = 5$ kPa

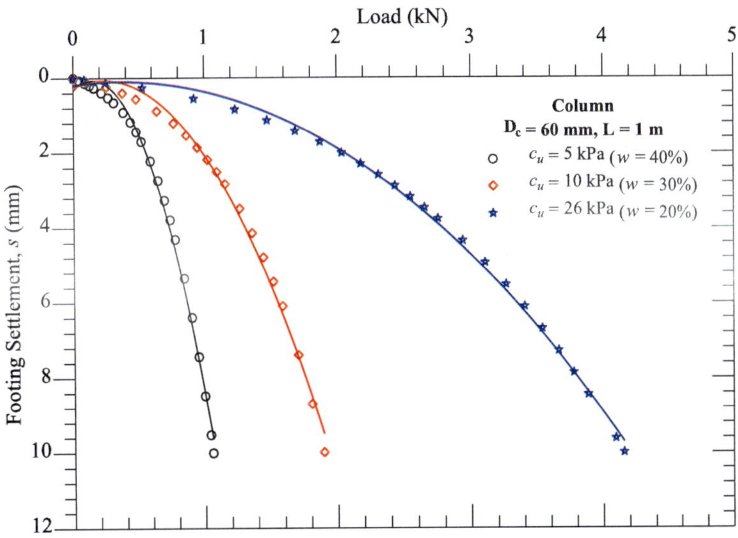

Fig. 6 Responses of circular footing (Dia. = 60 mm) on stone column ($L = 1$ m) on clay bed with different c_u (4.75 mm fraction in-fill material)

diameter and column lengths. It may be attributed to influence of increased footing area supported by the surrounded clay. This in turn enhances the confining surcharge toward the possible bulging of concern stone column. It eventually restrict the column deformation to keep its integrity for load transfer to enhance the bearing capacity of foundation system. However, the influence of column diameter beyond $L = 1.5$ m is found to be negligible.

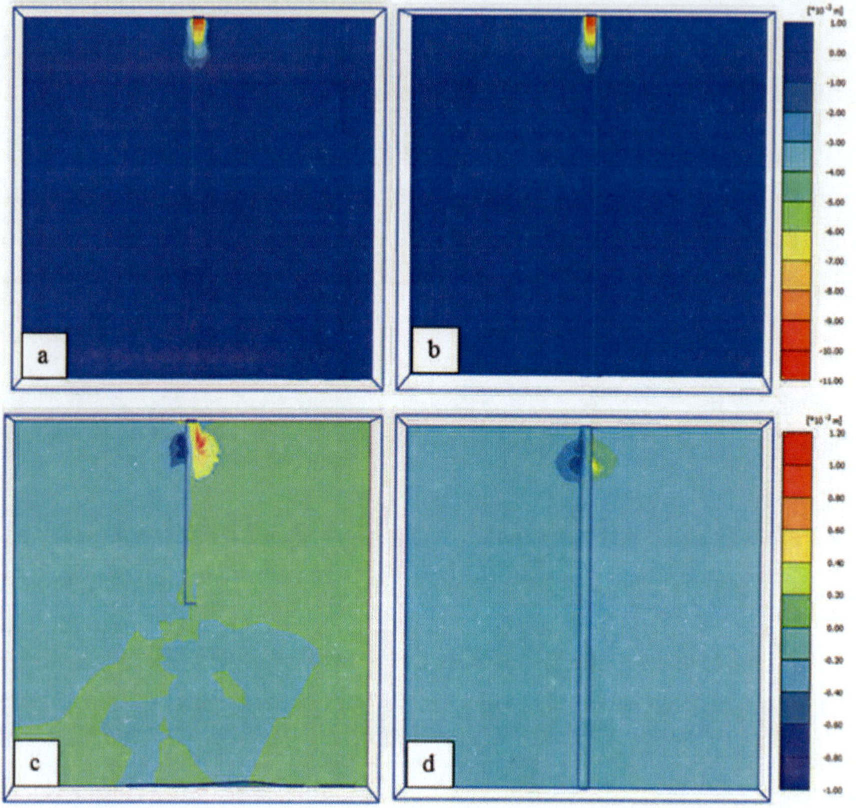

Fig. 7 Response of 60 mm diameter column with 4.75 mm stones in $c_u = 10$ kPa: vertical (**a**) and lateral (**c**) deformation for $L = 1$ m and vertical (**b**) and lateral (**d**) deformation for $L = 2$ m

4 Conclusions

The study investigated responses of footing rested on stone column in clay with varying strengths. Responses indicated considerable effect of parametric variations. The column diameter and length was varied along with the in-fill materials. Improvement in bearing capacity is observed with increase in clay consistency, column length and footing diameter. However, the influence of column diameter beyond 1.5 m was found to be negligible which is attributed to excessive bulging. The increase in loading area through enhanced footing diameter effected in confinement (lateral pressure) of surrounding clay (in addition to direct support provided to the footing). The authors admit that the numerical simulation should have been verified with physical test results; however, the research progress has been badly affected due to the pandemic and the restrictions thereby till date.

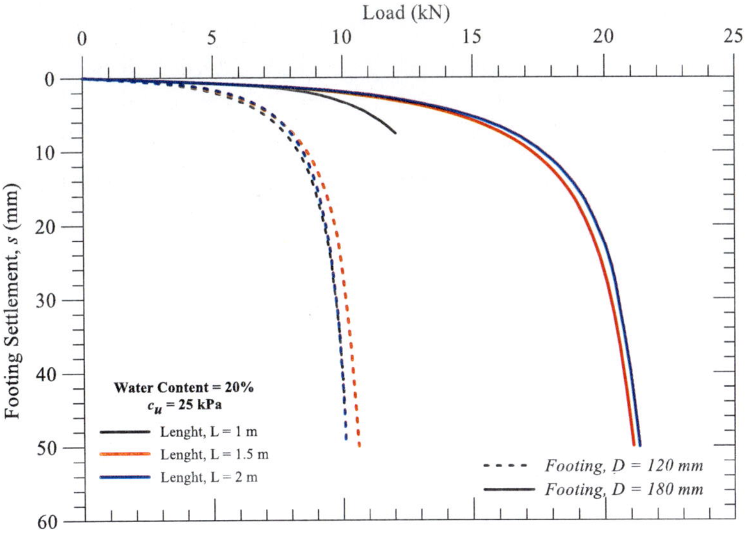

Fig. 8 Comparison of influence of footing diameter and column length with $c_u = 25$ kPa

Acknowledgments The authors sincerely acknowledge the support received from DHESTBT, Govt. of West Bengal, India [Ref. No. 183(Sanc.)/ST/P/S&T/6G-33/2017 dated 16/03/2018].

References

1. Najjar, S.S.: A state-of-the-art review of stone/stand-column reinforced clay systems. Geotech. Geol. Eng. **31**, 355–386 (2013)
2. Ghose, S., Biswas, A., Mandal, U.: Parametric effect on granular columns: a brief review. In: Proceedings of 7th Indian Young Geotechnical Engineers Conference, NIT Silchar, Assam, India, pp. 58–63 (2019)
3. Ambily, A.P., Gandhi, S.R.: Behaviour of stone columns based on experimental and FEM analysis. J. Geotech. Geoenviron. Eng. (ASCE) **133**(4), 405–415 (2007)
4. Bergado, D.T., Lam, F.L.: Full scale load test of granular piles with different densities and different proportions of gravel and sand in the soft Bangkok clay. Soils Found. **27**(1), 86–93 (1987)
5. Malarvizhi, S.N., Ilamparuthi, K.: Comparative study on behaviour of encased stone column and conventional stone column. Soil Found. **47**(5), 873–885 (2007)
6. Keykhosropur, L., Soroush, A., Imam, R.: 3D numerical analyses of geosynthetic encased stone columns. Geotext. Geomembr. **35**, 61–68 (2012)
7. Hughes, J.M.O., Withers, N.J.: Reinforcing of soft cohesive soils with stone columns. Ground Eng. **7**(3), 42–49 (1974)
8. Hughes, J.M.O., Withers, N.J., Greenwood, D.A.: A field trial of the reinforcing effect of a stone column in soil. Geotechnique **25**(1), 31–44 (1975)
9. Murugesan, S., Rajagopal, K.: Model tests on geosynthetic encased stone columns. Geosynthetics International **24**(6), 349–358 (2007)
10. Rao, N.S, Prasad, Y.V.S., Rao, H.V.: Use of stone columns in soft marine clays. In: Proceedings of 45th Canadian Geotechnical Conference, Toronto, Ont, pp. 9/1–9/7 (1992)

11. Wood, M.D., Hu, W., Nash, D.F.T.: Group effects in stone column foundations: model tests. Geotechnique **50**(6), 689–698 (2000)
12. Mckelvey, D., Sivakumar, V., Bell, A., Graham, J.: Modeling of vibrated stone columns in soft clay. Geotech. Eng. **157**(3), 137–149 (2004)
13. Sivakumar, V., McKelvey, D., Graham, J., Hughes, D.: Triaxial tests on model sand columns in clay. Can. Geotech. J. **41**, 299–312 (2004)
14. Black, J.V., Sivakumar, V., Bell, A.: The settlement performance of stone column foundations. Geotechnique **61**(11), 909–922 (2011)
15. Black, J.V., Sivakumar, V., Mckinley, J.D.: Performance of clay samples reinforced with vertical granular columns. Can. Geotech. J. **44**, 89–95 (2007)
16. Ali, K., Shahu, J.T., Sharma, K.G.: Model test on geosynthetics-reinforced stone columns: a comparative study. Geosynthetics Int. **19**(4), 292–305 (2012)
17. IS 15284 (Part 1): Design and construction for ground improvement guidelines: stone columns. Bureau of Indian Standards (2003)
18. Biswas, A.: Comparative performance of different geosynthetics on sandy soil overlying clay subgrades of varying strengths. In: Innovative Infrastructure Solutions. Springer (2019). https://doi.org/10.1007/s41062-019-0204-5

Stabilization of Soft Clay Using Nylon Fiber and Fly Ash

C. V. Nilna and S. Chandrakaran

1 Introduction

Soil has been used as a construction material. Almost all foundations are sited on soil. In places where soil is poor in mechanical property, civil engineers face challenges to improve its properties. The properties of soil vary from site to site. Among soil, clays generally have low shear strength, high compressibility and high volumetric changes. Soft clay is soils with large fractions of fine particles such as silty and clayey soils, which have high moisture content and it is located near or under the water table. The strength of soft clay is very less and it can be molded by light finger pressure. Soft clay is a problematic soil from civil engineering construction point of view. Puppala et al. [2, 4] conducted studies on problematic soils and Correia et al. [9] studied soft clay. To make them more feasible for construction purposes, stabilization has to be done. Soil stabilization is the alteration of any property of soil to improve its engineering performance. Soil stabilization may be physical, chemical, biological or combined methods to meet an engineering purpose. In situations where large area of the soil is covered under construction, the soil stabilization methods become a difficult and costly task. In such situations reliable methods need to be devised. In recent technology more number of traditional additives is used for stabilization purpose. The present study has been undertaken to investigate the strength improvement of soft clay stabilized with fly ash and nylon fiber. Starcher and Liu [1] studied the improvement in cement and cement-fiber-treated soft soil. UCS tests were carried out on cement-improved soils and cement-fiber-improved soils, and the strength gain was found out based on curing time alone and on curing time and vertical curing stress. Unconfined compressive strength was found to be improved. Cement-improved soil behaved as brittle material and introduction of fiber increased the ductility. Kaniraj and Havanagi [3] conducted a study on soil mixed with fly ash stabilized with cement

C. V. Nilna (✉) · S. Chandrakaran
National Institute of Technology, Kozhikode 673601, Kerala, India

C. N. V. Satyanarayana Reddy et al. (eds.), *Ground Improvement and Reinforced Soil Structures*, Lecture Notes in Civil Engineering 152,
https://doi.org/10.1007/978-981-16-1831-4_5

and reinforced with fiber. The soil–fly ash mixture was stabilized using cement and polypropylene fiber. UCS test was carried out in specimen stabilized with cement and also in specimen stabilized with cement and reinforced with fiber. UCS of a fly ash–soil mixture increases due to addition of cement and fibers. The increase in UCS caused by the combined action of cement and fibers is either more than or nearly equal to the sum of the increase caused by them individually based on the period of curing and the material added. Onyejekwe et al. [5] showed that the inclusion of the fibers led to significant improvements in the flexural load carrying capacity. Kafodya and Okonta [8] have shown the result of pre-compression and addition of natural fiber on the mechanical properties of soil stabilized with lime and fly ash. The sisal fibers used were 25 mm length and they were applied at different percentages such as 0.25, 0.5, 0.75 and 1%. Strength of stabilized specimens was determined. Freitag [6] and Attom et al. [7] conducted ucc test on soil with different percentage of fibers. Al Adili et al. [10] considered papyrus fiber as an appropriate soil reinforcement material for strength improvement. Results showed that fiber inclusion improved the strength. Stabilization of soft clay by using fly ash and nylon fiber was studied in this paper.

2 Materials

The soil used in this study was soft clay. The soil was collected from four locations in the paddy field. The paddy field was located near Mavoor–Chathamangalam road. Liquid limit test was carried out on these soils collected from various locations. For each soil the value of liquid limit was different and the soil which got maximum liquid limit of about 97% was selected for the project. The clay was one type of most problematic soil. Table 1 shows the properties of soil. Fly ash was the stabilizer used in this study. Class F fly ash of specific gravity 2.37 was used. Fly ash was collected from Sanjana Traders, Mangalore. The synthetic fiber used in this project was Nylon fiber. Nylon fiber was procured from Go Green Products Company. Nylon fiber of 0.3 mm diameter and 12 mm length was used in this study.

3 Experimental Work

Experimental work included characterization of soil and the investigation of improvement in strength of soil mixed with nylon fiber and fly ash. Standard proctor test, liquid limit test, plastic limit test and UCC test were carried out in the soil–fly ash mix. Added fly ash percentages were 10, 20, 30 and 40% by weight of soil. Test samples for UCC test were prepared by mixing the soil and fly ash at maximum dry density and optimum moisture content obtained from standard proctor test. Both cured and uncured strength were determined. Cured strength was determined for 1, 7 and 28 days. Liquid limit and plastic limit were also determined for cured and uncured samples. Standard proctor test was carried out on soil mixed with various

Table 1 Soil properties

Properties		Values
Natural moisture content, %		129
Specific gravity		2.44
Grain size	Clay, %	38
Distribution	Silt, %	40
	Sand, %	22
IS classification		CH
Liquid limit, %		97
Plastic limit, %		39
Plasticity index, %		58
Shrinkage limit, %		25
Maximum dry density, g/cc		1.26
Optimum moisture content, %		35.52
Unconfined compressive strength, KPa		75

combinations of fly ash and nylon fiber. First the soil was mixed with 0.25% of fiber and four varied percentages of fly ash. Then the soil was mixed with 0.5% of fiber and four varied percentages of fly ash. From standard proctor test, optimum percentage of fly ash and nylon fiber was obtained and the UCC test was carried out for soil mixed with optimum percentage of fly ash and nylon fiber. Both cured and uncured strength of soil–fly ash–fiber mix were determined.

4 Results and Discussion

4.1 Soil Treated with Fly Ash

Compaction Characteristics. Optimum percentage required in the soft clay was determined by conducting light compaction test. Light compaction test was conducted by mixing the soil with 10, 20, 30 and 40% of fly ash by dry weight of soil. From Fig. 1, it is clear that the gradual addition of fly ash in soil increases the dry density up to 30% addition of fly ash and after that it starts reducing. The optimum fly ash percentage was obtained as 30% by dry weight of soil.

Unconfined Compressive Strength of Soil Mixed with Fly Ash. Stress–strain behavior of soil treated with fly ash was determined by conducting unconfined compressive strength test. Test specimens for the unconfined compressive strength test were made with the maximum dry density and optimum moisture content obtained from the respective standard proctor test. For each fly ash percentages such as 10, 20, 30 and 40%, samples were casted and kept for curing. Both uncured

Fig. 1 Variation of maximum dry density with different percentage of fly ash

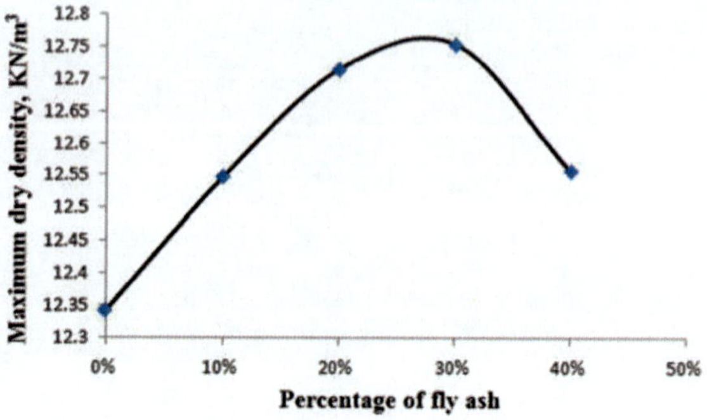

and cured samples were tested to get their unconfined compressive strength. Among the casted samples three were tested after 5 h of casting and remaining samples were kept for curing for 1, 7 and 28 days.

Figure 2 gives the trend of variation of strength of soil mixed with different fly ash percentages. The unconfined compressive strength of soil treated with fly ash was increased with curing period for each percentage of fly ash. The peak strength of specimen increased with increase in fly ash content up to 30% and after that it started reducing. For every percentage the peak strength was increased with curing time.

Plasticity Characteristics. The treatment of soil with admixtures influences the plasticity characteristics of soil. Changes in plasticity characteristics cause changes in strength of the soil. Fly ash was used as the admixture with the soil. Liquid limit and plastic limit tests were carried out as per standard specifications to study the effect of fly ash on plasticity characteristics of soil. The plasticity characteristics of clay got improved on treating with fly ash. Variation of liquid limit of soil treated

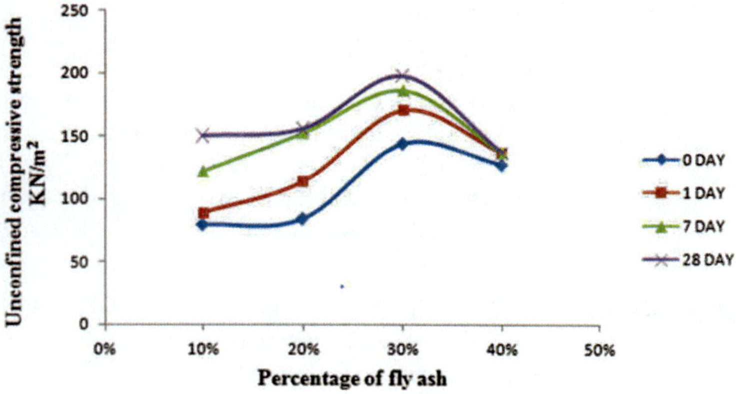

Fig. 2 Unconfined compressive strength of soil mixed with different percentage of fly ash for various curing periods

with fly ash is shown in Fig. 3. The liquid limit was decreased with more addition of fly ash content. Liquid limit was also decreased with the increase in curing period (Fig. 3).

Plastic limit was also decreased with increase in fly ash content (Fig. 4).

Compaction Characteristics. Optimum percentage of fly ash and nylon fiber required to achieve highest strength of soil was determined by conducting light compaction test. The test was conducted by mixing the soil with various combinations of fly ash and nylon fiber. Standard proctor test was carried out in the soil mixed with different combination of fly ash and nylon fiber. Percentages of fly ash were 10, 20, 30 and 40 by dry weight of soil and percentages of fiber were 0.25 and 0.5% by dry weight of soil (Fig. 5).

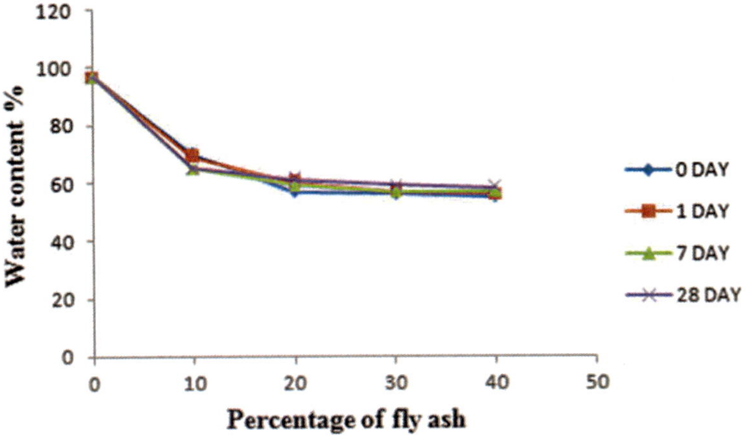

Fig. 3 Liquid limit of soil mixed with different percentage of fly ash for various curing periods

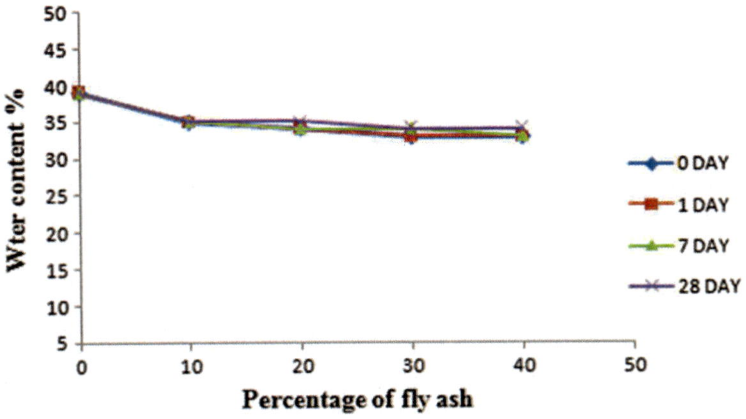

Fig. 4 Plastic limit of soil mixed with different percentage of fly ash for various curing periods

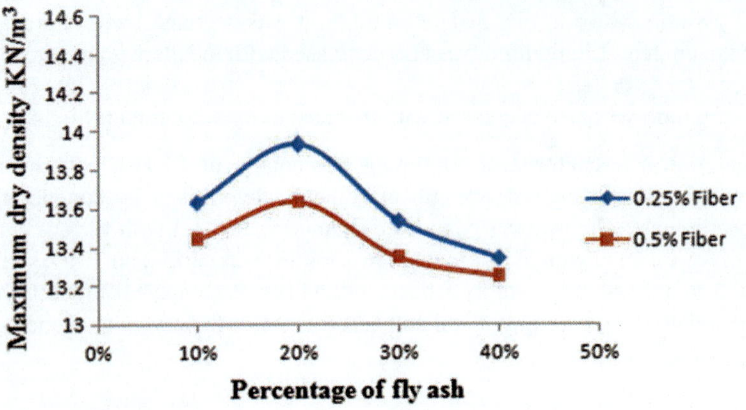

Fig. 5 Maximum dry density of soil treated with various percentage of fly ash and nylon fiber

Compaction test was first done by mixing 0.25% of fiber by weight of soil with four different percentages of fly ash. For 0.25% of fly ash, maximum dry density is increased till 20% addition of fly ash. After that maximum dry density started decreasing. Then 0.5% of fiber by dry weight of soil was mixed with four different percentages of fly ash. The trend of difference in maximum dry density for 0.5% fiber was also the same. While comparing the two percentages of fiber, higher value of maximum dry density was obtained for 0.25% of fiber in the soil. From standard proctor test, optimum fly ash percentage was obtained as 20% by dry weight of soil and optimum percentage of nylon fiber as 0.25%.

Unconfined Compressive Strength of Soil Mixed with Fly Ash and Nylon Fiber. Unconfined compressive strength of soil treated with fly ash and nylon fiber was determined by testing samples in unconfined compression testing machine which were prepared by mixing soil with 20% fly ash, 0.25% nylon fiber and water at a density of 13.93 kN/m^3 and moisture content of 22% which were obtained from light compaction test.

4.2 Comparison of Test Results

1. From standard proctor test, maximum dry density and optimum moisture content of raw soil were obtained as 12.36 kN/m^3 and 35.52%, respectively. For soil mixed with various percentage of fly ash, optimum percentage was obtained as 30%, and maximum dry density and optimum moisture content were changed to 12.75 kN/m^3 and 32.4%, respectively, for soil stabilized with optimum percentage of fly ash. In the case of combination of fly ash and nylon fiber, optimum percentage of fly ash was obtained as 20% and that of nylon fiber was obtained as 0.25%. For soil treated with optimum fly ash and nylon fiber,

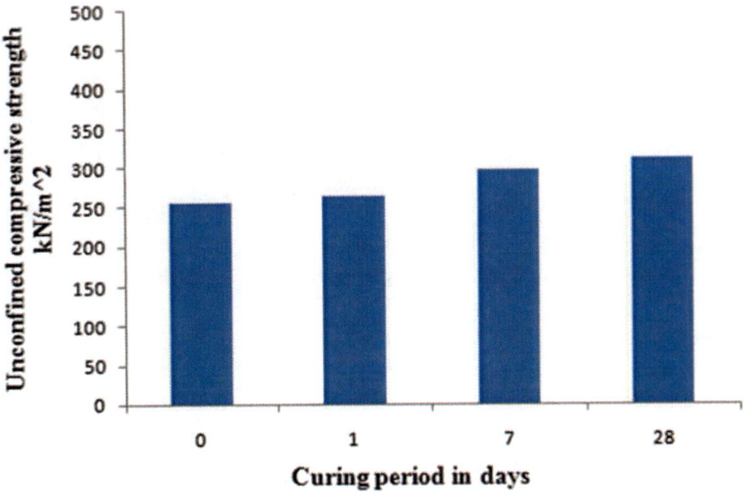

Fig. 6 Variation of unconfined compressive strength of soil mixed with 20% fly ash and 0.25% nylon fiber for various curing periods

maximum dry density and optimum moisture content were changed to 13.93 kN/m^3 and 22%, respectively.

2. For untreated soil unconfined compressive strength was 75 kPa. Strength was increased to 145 kPa for optimum percentage of fly ash without curing. Strength is increased to about 93% for soil mixed with optimum percentage of fly ash without curing. For soil stabilized with optimum percentage of fly ash and nylon fiber, UCS was increased to 256 kPa without curing and the increase was about 77% compared to soil stabilized with fly ash alone.
3. After a curing period of 28 days, UCS was increased to 198 kPa for soil treated with fly ash and strength was increased to 311 kPa for soil treated with both fly ash and nylon fiber. The increase in strength after a curing period of 28 days for soil treated with fly ash alone and soil treated with both fly ash and nylon fiber were 37% and 21%, respectively.

5 Conclusions

The effect of fly ash alone and combined inclusion of nylon fiber and fly ash in soft clay was evaluated in this paper. The major conclusions obtained from laboratory tests and analysis of test results are summarized as follows:

1. The optimum percentage of fly ash was obtained as 30% for soil mixed with fly ash alone, and optimum percentage of fly ash and nylon fiber were 20% and 0.25%, respectively, for soil mixed with both fly ash and nylon fiber. There is

formation of lumps when percentage of nylon fiber is increased after optimum percentage.
2. Addition of fly ash improved the strength of untreated soil. For soil mixed with optimum percentage of fly ash, unconfined compressive strength was increased to about 93% without curing and the strength was increased to about 37% after a curing period of 28 days.
3. Unconfined compressive strength was more in the case of soil mixed with fly ash compared to untreated soil. Strength was maximum for soil put together with optimum fly ash and the strength was increased with increase in period of curing.
4. Plasticity characteristics such as liquid limit and plastic limit were decreased with increase in percentage of fly ash and also with the curing period.
5. For soil stabilized with optimum percentage of fly ash and nylon fiber, UCS was increased to about 77% without curing compared to soil stabilized with fly ash alone and the strength was increased to about 21% after a curing period of 28 days.
6. Strength is more in the case of soil mixed with both fly ash and nylon fiber compared to strength of soil mixed with fly ash alone and untreated soil. Hence combined inclusion gives more strength and inclusion of fiber improved ductility also.

References

1. Starcher, R.D., Liu, C.: Mechanical behavior of cement- and cement-fiber-improved soft soils. Geo-Congr. 2048–2057 (2013)
2. Punthutaecha, K., Puppala, A.J., Vanapalli, S.K., Inyang, H.I.: Volume change behaviors of expansive soils stabilized with recycled ashes and fibers. J. Mater. Civil Eng. **18**(2), 295–306 (2006)
3. Kaniraj, S.R., Havanagi, V.G.: Behavior of cement-stabilized fiber-reinforced fly ash-soil mixtures. J. Geotech. Geoenviron. Eng. **127**(7), 574–584 (2001)
4. Puppala, A.J., Punthutaecha, K., Vanapalli, S.K.: Soil-water characteristic curves of stabilized expansive soils. J. Geotech. Geoenviron. Eng. **132**(6), 736–751 (2006)
5. Onyejekwe, S., Ghataora, G.S.: Effect of fiber inclusions on flexural strength of soils treated with nontraditional additives. J. Mater. Civ. Eng. **26**(8), 1–9 (2014)
6. Freitag, D.R.: Soil randomly reinforced with fibers. J. Geotech. Eng. **112**(8), 823–826 (1986)
7. Attom, M.F, Al-Akhras, N.M., Malkawi, A.H: Effect of fibres on the mechanical properties of clayey soil. Geotech. Eng. **162**(5), 277–282 (2009)
8. Kafodya, I., Okonta, F.: Effects of natural fiber inclusions and pre-compression on the strength properties of lime-fly ash stabilised soil. Constr. Build. Mater. **170**, 737–746 (2018)
9. Correia, A.A.S., Venda Oliveira, P.J., Custodio, D.G.: Effect of polypropylene fibres on the compressive and tensile strength of a soft soil, artificially stabilised with binders. Geotext. Geomembranes **43**, 97–106 (2014)
10. Al Adili, A., Azzam, R., Spagnoli, G., Schrader, J.: Strength of soil reinforced with fiber materials (Papyrus). Soil Mech. Found. Eng. **48**(6), 241–247 (2012)

Experimental Investigation of Expansive Soil Mixed with Shredded Rubber Tyre

Banti Singh, Vishal Kumar Nagar, and Jitendra Kumar Sharma

1 Introduction

The black cotton soil was found in the central and western part of India. Black cotton soil was collected from Agriculture University, Borkhera locality of Kota city in Rajasthan state of India provision. The black cotton soil is also called as Regular soil, which is generally clayey, deep and impermeable in nature. As the black cotton soil expands, it becomes sticky during the rainy season and contracts during summer season, causing deep cracks into the soil. Black cotton soil is formed by lava basaltic rocks. The soil is very dark in colour. They develop cracks during the dry period and swell if moisten, hence they are self-tilling in nature; that's why they are fertile and can hold water for a long time. The whole area cracks up 150 mm wide and formed up to a depth of 3.0–3.5 m. But when the soil is moist it expands, becomes very soft and loses the bearing capacity. Black cotton soil is an inorganic clay of medium to high compressibility and is characterized by high shrinkage and swelling properties. The expansive soil is rich in mineral montmorillonite and few in the elite. Black cotton soil has been a challenge to the geotechnical engineers for a long time in designing the foundation of the building. Soil stabilization improves the engineering properties of the soil and making it more stable. Soil stabilization can be categorized into two main categories, such as mechanical and chemical stabilization. The tyre waste is a hazardous material in the sense of disposing. Nowadays, sustainable use of waste product is necessary. This assignment concentrates on obtaining the optimum amount of crumb rubber tyre for practical work by observations of the effect of crumb rubber tyre on engineering properties of soil. However, the crumb rubber tyre is only

B. Singh (✉) · V. K. Nagar
Rajasthan Technical University, Kota, Rajasthan 324010, India

J. K. Sharma
Department of Civil Engineering, Rajasthan Technical University, Kota, Rajasthan 324010, India
e-mail: jksharma@rtu.ac.in

C. N. V. Satyanarayana Reddy et al. (eds.), *Ground Improvement and Reinforced Soil Structures*, Lecture Notes in Civil Engineering 152,
https://doi.org/10.1007/978-981-16-1831-4_6

used as a partial replacement of adhesive/cementitious materials like cement and lime. It cannot completely take over the cementitious material because crumb rubber tyres have the inherent binding property which is required for longlasting material or durability.

2 Literature Review

Srivastava et al. [1] studied the shredded tyre waste with expensive soil for enhancing the geotechnical properties. In addition to 30–50%, shredded tyre waste decreases the shear strength property. It was concluded that shredded tyre waste helps in reducing its shrinkage and swelling properties. The shear strength was improved in 5% shredded tyre waste. Further, the addition of shredded tyre rubber waste increases the compressibility value. Hambirao et al. [2] investigated the clayey soil stabilization at various percentages (0–15%) of rubber tyre chips. The UCS and CBR values increased with increase in rubber tyre content with an optimum moisture content of 10%. It was remarked that the shredded rubber tyre can be used as a virtuous strengthen material. The investigation shows an increment in CBR values of 2.63–13.79%, which reduces the pavement thickness by 66.66%. Shubha and Balichakra [3] inquired information with proportion of 0, 4, 6, 8, 10 and 15% tyre waste. As the proportion of rubber crumb increases, the maximum dry density tends to starts decreasing. The best enhancement in CBR value was evaluated at 12% of rubber crumb. Kaur and Singh [4] investigated the black cotton soil with various proportion of rubber powder from 0, 5, 10 and 15%. As the rubber crumb was increased by 10% the value of CBR values also increased. They concluded that the optimum rubber tyre was taken to be 10% for the stabilization of black cotton soil. Saini et al. [5] concluded an increase of UCS value up to 8.29% with the increment of 2% tyre powder waste with black cotton soil. Saini et al. [5] concluded that there is slightly an increase of UCS value of up to 8.29% on the addition of 2% shredded tyre waste with black cotton soil. Prasad and Prasada [6] studied the functioning of shredded tyre on the model flexible pavement from the experimental analysis of rock and fly ash material reinforced with different proportion of shredded tyre rubber, and the ideal proportion of waste tyre is equivalent to 5 and 6% of dry unit weight of soil. Sathwik et al. [7] found an increment in CBR values with content up to 8% of shredded rubber. Vinod et al. [8] investigated the optimum proportion for shredded tyre waste as 5% and for sea shell powder it is 12% in expansive soil.

Fig. 1 Black cotton soil

3 Materials Used

3.1 Black Cotton Soil

A large part of Central India was covered with black cotton soil (Fig. 1). The black cotton soil is residual deposit formed from basalt or trap-rocks and possess high plasticity. Soil sample was collected from 2 m below the ground level in Agriculture University of Kota, Rajasthan state. Untreated soil was oven dried for 24 h at 105–110 °C before performing the experiments. The latitude and longitude of the area is 25. 1783° N, 75. 8850° E, respectively. Extremely difficult to work with such soils and is quite suitable for growing cotton.

3.2 Crumb Rubber Tyre (CRT)

Every year greater number of rubber tyres are manufactured and the same number is permanently removed from the vehicle nowadays. These tyres (Fig. 2) are manufactured fundamentally with natural and synthetic rubber and reinforced with the metal element and chemical additives. After the removal of the tread from the tyre, they become useless and the management of this waste becomes crucial. These obsolete tyres are then stockpiled and dumped into landfills and have exposed communities to environmental and health risks. The burning of these tyres is also not possible as they produce toxic gases which make the environment surrounded by hazardous gases. After identification of this problem, a study has been carried out by converting the tyre into powder form and then mixing it into the soil to check the bearing capacity of a rubber tyre. The metal elements were removed while preparing the powder of tyre. The waste tyres were bought from a tyre puncture recovery shop located near Kumher Gate in Bharatpur District of Rajasthan state in the Republic of India. The tyre was then shredded, crushed and then converted to powder form until the

Fig. 2 Crumb tyre rubber

Table 1 Parameters of crumb rubber tyre

S. no.	Parameters	Values found
1	pH	7.26
2	Elect. Conductivity (at 25 °C)	1.113 ms/cm
3	Moisture (at 105 °C)	1.39% w/w
4	Organic Carbon (as OC)	1.42% w/w
5	Sulphur	0.167% w/w
6	Phosphorus	0.23% w/w
7	Potash	1.393% w/w
8	Organic Matter	2.448% w/w

powdery tyre rubber is completely passed from a sieve of 425 μm. After that, the chemical properties of tyre rubber were analysed in the laboratory and are presented in Table 1.

4 Experimental Study

The soil sample collected from Agriculture University, Borkhera was tested as per Indian standard. Various tests were performed to evaluate the index and engineering properties of soil, which are listed in Table 2.

Table 2 List of tests for determining index and engineering properties

S. no.	Experiment performed	Property evaluation
1	Sieve analysis	Soil classification
2	Pycnometer test	Specific gravity
3	Cone penetrometer test	Liquid Limit (LL)
4	Thread test	Plastic Limit (PL)
5	Plasticity index	PI = LL-PL
6	Standard proctor test	(a) OMC (b) MDD
7	Unconfined compressive strength	Cohesion (c')
8	California Bearing Ratio	(a) Unsoaked (b) Soaked

5 Results and Discussion

5.1 *Sieve Analysis*

Sieve analysis test was used for soil classification. It was performed as per IS 2720 (Part-4) [9]. It was observed that particles passing through 75 μm sieve were greater than 50%. So the soil was classified as fine soil. Fine soils are classified as per their Atterberg's limits. The results obtained from sieve analysis are shown in Fig. 3.

5.2 *Specific Gravity*

The specific gravity was governed by pycnometer test apparatus as per IS 2720-3-1 (1980) [10]. It was found that the specific gravity of soil sample is 2.70.

Fig. 3 Grain size distribution curve for virgin soil

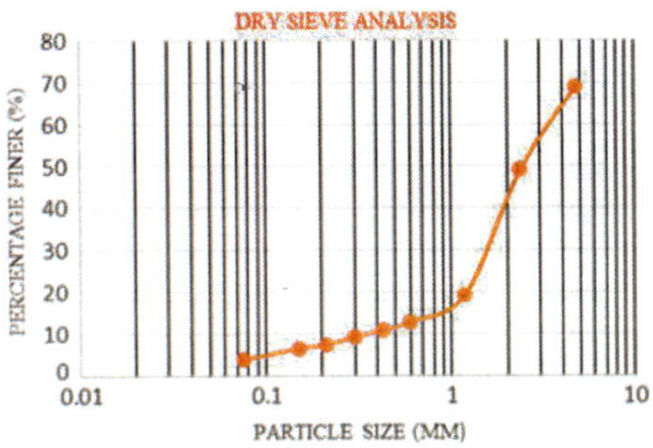

Table 3 Atterberg's limits

Sr. no	Sample	Liquid limit (%)	Plastic limit (%)	Plasticity index (%)
1	BC + 0% CRT	43	29	14
2	BC + 5% CRT	35	23	12
3	BC + 10% CRT	30	21	9
4	BC + 15% CRT	32	22	10
5	BC + 20% CRT	34	23	11

Table 4 Standard proctor test

S. no.	Sample	OMC (in %)	MDD (g/cm^3)
1	BC + 0% CRT	18	1.60
2	BC + 5% CRT	20	1.55
3	BC + 10% CRT	14	1.54
4	BC + 15% CRT	17	1.52
5	BC + 20% CRT	15	1.51

5.3 *Atterberg's Limits*

For determination of liquid limit (LL) cone penetrometer test was performed as per IS 2720-5 [9] and for plastic limit, the test was performed. It was observed that the liquid limit of virgin soil was 43% (Table 3) which is greater than 35% and plasticity index (PI) is 14% so that the soil is classified high plasticity clay (CH) as per IS 1498 [11].

5.4 *Standard Proctor Test*

The standard proctor test was performed for determination of optimum moisture content (OMC) and maximum dry density (MDD) as per IS 2720-7 (1980) [12]. It can be seen from Table 4 that the value of OMC and MDD of soil sample is 18% and 1.60 g/cm^3, respectively. It was observed that by increasing the tyre rubber content MDD is decreasing (Fig. 4), which is because of the light weight of the tyre rubber.

5.5 *Unconfined Compressive Strength Test*

UCS test was conducted as per IS 2720-10 (1991) [13] for determination of shear parameter of soil. The UCS value of untreated soil is 37.51 kN/m^2 (Table 5). The UCS value tends to increase with the addition of tyre rubber (Fig. 5), but after an optimum proportion of 10% tyre rubber it starts decreasing.

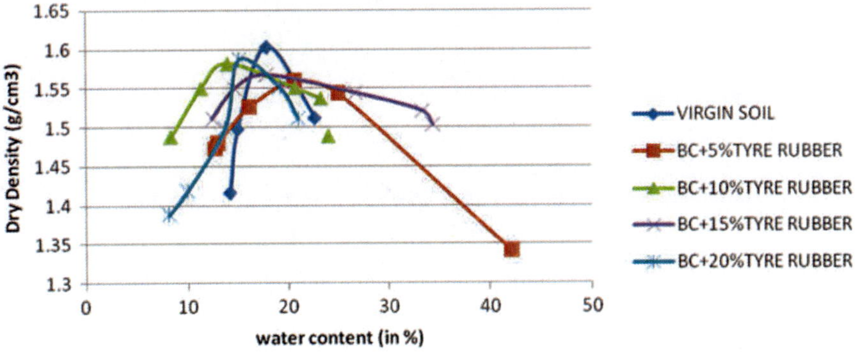

Fig. 4 Standard proctor test

Table 5 Unconfined compressive strength test

S. no.	Sample	UCS value (kN/m^2)	Cohesion (c') (kN/m^2)
1	BC + 0% CRT	37.51	18.75
2	BC + 5% CRT	42.25	21.12
3	BC + 10% CRT	56.83	28.41
4	BC + 15% CRT	46.54	23.27
5	BC + 20% CRT	33.63	16.81

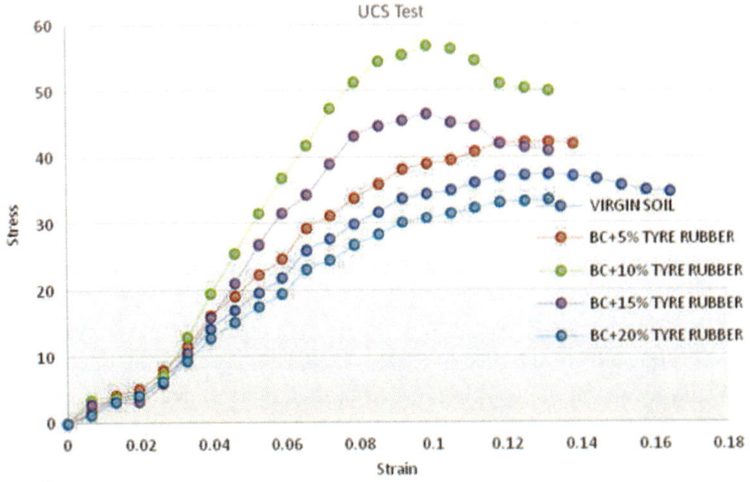

Fig. 5 Unconfined compressive strength test

Table 6 California bearing ratio test

S. no.	Sample	Unsoaked CBR (%)	Soaked CBR (%)
1	BC + 0% CRT	7.74	2.27
2	BC + 5% CRT	7.97	2.58
3	BC + 10% CRT	8.50	3.47
4	BC + 15% CRT	8.23	3.15
5	BC + 20% CRT	7.90	2.48

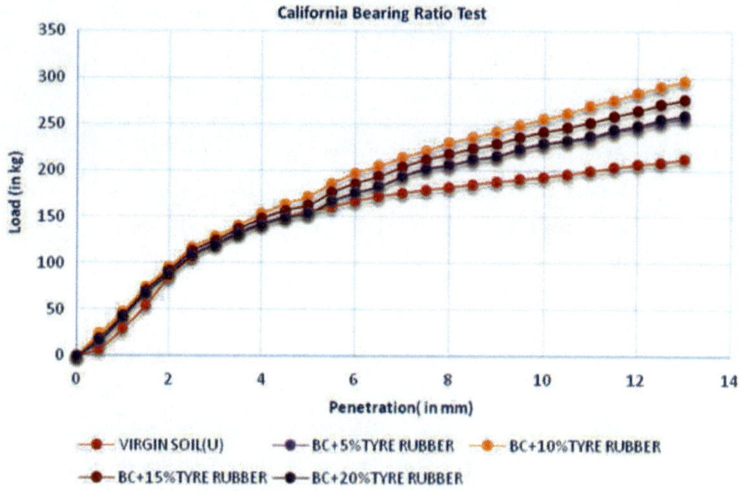

Fig. 6 California bearing ratio (unsoaked) test

5.6 *California Bearing Ratio Test*

CBR test was conducted to evaluate the behaviour of soil in normal (unsoaked) and worst (soaked) condition as per IS 2720-16 [14]. It was observed that the CBR value of virgin soil is 7.74% and 2.27% (Table 6) in unsoaked and soaked condition, respectively. It can be observed that CBR is increasing by addition of tyre rubber up to an optimum content of 10% (Fig. 6). After that, it starts decreasing.

6 Conclusions

1. The optimum percentage of shredded tyre rubber for stabilization of expensive soil was found to be 10% based on the UCS and CBR test results.
2. MDD value shows decrease with increase in shredded rubber content, and OMC values tend to increase.

3. The percentage improvement in soaked and unsoaked CBR was 34% and 7%, respectively. Increase in CBR value of stabilized soil can significantly redoes the overall thickness of pavement.
4. 35% increment in UCS value with 10% shredded tyre was concluded.

References

1. Srivastava, A., Pandey, S., Rana, J.: Use of shredded tyre waste in improving the geotechnical properties of expansive black cotton soil. Geomech. Geoeng. **9**(4), 303–311 (2014)
2. Hambirao, G.S., Rakaraddi, P.G.: Soil stabilization using waste shredded rubber tyre chips. IOSR J. Mech. Civil Eng. **11**(1), 20–27 (2014)
3. Shubha, M., Balichakra, P.K.: Stabilization of Black cotton soil with Tyre Waste. Int. J. Eng. Manag. Res. (IJEMR) **8**(5), 117–119 (2018)
4. Kaur, R., Singh, D.: Tyre rubber powder as well as soil stabilizer. Int. Res. J. Eng. Technol. **6**(6), (2019). ISSN: 2395-0056
5. Saini, L.K., Sehara, R., Kalal, R.: The Geotechnical properties of expansive black cotton soil as well as cohesive soil. J. Basic Appl. Eng. Res. **5**(1), 34–36 (2018)
6. Prasad, D.S.V., Prasada Raju, G.V.R.: Performance of waste tyre rubber on model flexible pavement. ARPN J. Eng. Appl. Sci. **4**(6), 89–92 (2009).
7. Sathwik, A., Arti Sudam, P., Kommu, S.: Behavior of soil by using tyre powder. In: Proceeding of IGC-2016:GEOtrendz, IIT Madras (2016)
8. Vinod, B.R., Shobha, R., Raghavendra, A.B., Rakesh, M., Pallavi, S.: Stabilization on Expansive soil using sea shell powder and Rubber powder. In: IOP Conference Series: Materials Science and Engineering, vol. 814, no. 1, p. 012028. IOP Publishing (2020)
9. IS:2720 (Part V): Methods of tests for soil: determination of liquid limit and plastic limit. Bureau of Indian Standards, New Delhi (1985)
10. IS:2720 (Part I): Methods of test for soil: Preparation of dry soil sample for various tests. Bureau of Indian Standards, New Delhi (1983) (10. IS:2720 (Part III): Methods of tests for soil: determination of specific gravity. Bureau of Indian Standards, New Delhi, 1987)
11. IS:1498: Classification and identification of soils for general engineering purposes. Bureau of Indian Standards, New Delhi (1970)
12. IS:2720 (Part VII): Methods of tests for soil: determination of water content-dry density relation using light compaction. Bureau of Indian Standards, New Delhi (1987)
13. IS:2720 (Part X): Methods of tests for soil: determination of unconfined compressive strength. Bureau of Indian Standards, New Delhi (1973)
14. IS:2720 (Part XVI): Methods of tests for soil: laboratory determination of CBR. Bureau of Indian Standards, New Delhi (1987)

A Laboratory Study on the Stabilized Expansive Soil with Partial Replacement of Fly Ash and Palm Oil Fuel Ash

K. Ramu and R. DayakarBabu

1 Introduction

The expansive soils undergo volume changes with the variation of moisture content in the soil mass. The construction of civil engineering projects in these types of soil deposits is very difficult. Swelling and shrinking of soil with seasonal moisture variation has always created higher problems for light-loaded structures than moderate-loaded structures. Changing its volume along with unequal settlement may damage the foundation systems, structural elements and architectural features, which may result in loss of its function for which the structures are erected. The losses due to extensive damage to the highways running over the expansive sub-grades are estimated to be billions of dollars all over the world [1].

Geotechnical researchers are searching for various options to mitigate the adverse nature of the expansive soil. Volume change characteristics of the expansive soil can be controlled by reducing its plasticity characteristics or by a physical cementing mechanism. Therefore, soil stabilization techniques may be necessary to ensure better stability of soil so that it can successfully sustain the load from the superstructure, especially in the case of soil, which is highly active. Many efforts are being made by the researchers to minimize the effect of the expansive soils on the structures constructed with or within it. Every method has its own limitations. Katti and Katti [2] recommended the use of mechanically stabilized mix (MSM) cushion below the footing with CNS material. This process increases the cost of the foundation. However, studies conducted by Subba Rao [3] on the black cotton soil stabilized with the CNS cushion become ineffective after the first swell–shrink cycle.

K. Ramu (✉)
University College of Engineering, JNTU Kakinada, Kakinada 533003, India
e-mail: kramujntu@hotmail.com

R. DayakarBabu
Department of Civil Engineering, KITS, Divili, India

C. N. V. Satyanarayana Reddy et al. (eds.), *Ground Improvement and Reinforced Soil Structures*, Lecture Notes in Civil Engineering 152,
https://doi.org/10.1007/978-981-16-1831-4_7

This has led to the use of a stabilized fly ash cushion by the researchers, because lime reacts with silica in fly ash to produce cementitious products. Thus, it serves the same purpose as a CNS material. This may also solve the problem of fly ash disposal. The efficacy of lime [4] and cement [5] stabilized fly ash cushions in arresting heave has been established. Cement-stabilized fly ash cushions have yielded satisfactory results in arresting heave [6]. Sreerama Rao et al. [7] observed that the reduction in the heave of the expansive soil stabilized with the lime and fly ash.

The chemical composition of palm oil fuel ash indicates the presence of high amount of silica, which may be useful to stabilize the expansive soils. Muhammad et al. [8] reported that the behavior of the silty soil improved by stabilizing it with the palm oil fuel ash. Ahmad et al. [9] reported that the strength of the peat soil can be improved by the addition of cement and palm oil fuel ash.

Attempts to study about such unpredictable behavior through research on how to bring these problems under control form the backdrop for this project work. Therefore, a number of laboratory experiments were conducted to ascertain host of soil engineering properties of a naturally available expansive soil before and after stabilization. Pre- and post-stabilized results are compared to arrive at a conclusion that can thwart expansive soil problems.

In the present work, an experimental study is conducted to determine the properties of soil, the compaction, unconfined compression test and California bearing ratio (CBR) on expansive soil as well as on the stabilized expansive soil, mixed with different percentages of fly ash and palm oil fuel ash with a view to determine the optimum percentage. Test results show that stabilizing with fly ash and palm oil fuel ash enhance the properties and the strength of the expansive soil as compared to untreated expansive soil.

2 Experimental Program

2.1 Materials Used

Soil. The soil used was a typical black cotton soil collected from 'Mummidivaram' near Amalapuram, in East Godavari District, Andhra Pradesh State, India. The properties of soil are presented in Table 1. All the tests carried out on the soil are as per IS specifications.

Fly ash. Fly ash was collected from the Rajahmundry International Paper Mill, Rajahmundry. The chemical composition of the fly ash is given in Table 2.

Palm oil fuel ash. Palm oil fuel ash used for this study was collected from local oil industries, Samalkot, Mandal, East Godavari district of Andhra Pradesh. The chemical composition of palm oil fuel ash is given in Table 3.

Table 1 Properties of expansive soil

S. no.	Property	Value
1	Differential free swell (%)	120
2	Specific gravity	2.62
3	Grain size analysis Gravel (%)	0
	Sand (%)	8
	Silt (%)	27
	Clay	65
4	Liquid limit (%)	79.3
5	Plastic limit (%)	30.4
6	Plasticity index	48.9
7	Compaction	
	Optimum moisture content (%)	29.1
	Maximum dry density (g/cc)	1.43
8	Unsoaked CBR (%)	3.1
9	Soaked CBR (%)	1.3
10	Unconfined compressive strength (kN/m^2) at MDD and OMC	81

Table 2 Chemical composition of fly ash (Courtesy: RIPM, Rajahmundry)

Name of the chemical	Symbol	Range by percentage of weight
Silica	SiO_2	61–64.29
Alumina	Al_2O_4	21.6–27.04
Ferric oxide	Fe_2O_3	3.09–3.86
Titanium dioxide	TiO_2	1.25–1.69
Manganese oxide	MnO	Up to 0.05
Calcium oxide	CaO	1.02–3.39
Magnesium oxide	MgO	0.5–1.58
Phosphorous	P	0.02–0.14
Sulphur trioxide	SO_3	Up to 0.07
Potassium oxide	K_2O	0.08–1.83
Sodium oxide	Na_2O	0.26–0.48
Loss on ignition		0.20–0.85

2.2 *Sample Preparation*

The soil was initially air dried, pulverized and then was sieved through a 4.75 mm sieve, prior to the testing. The samples were prepared by mixing the required quantity

Table 3 Chemical composition of palm oil fuel ash

Name of the chemical	Symbol	Range by percentage of weight
Silica	SiO_2	55.7–66.91
Alumina	Al_2O_4	0.9–6.44
Ferric oxide	Fe_2O_3	2.0–5.72
Calcium oxide	CaO	5.56–12.5
Magnesium oxide	MgO	3.13–5.10
	P_2O_2	0–3.72
Sulphur trioxide	SO_3	0.33–1.90
Potassium oxide	K_2O	5.2–11.9
Sodium oxide	Na_2O	0.9–1.0
Loss on ignition		

of pulverized soil with the predetermined fly ash in dry condition, followed by the addition of required amount of water to make a consistent mix by thorough mixing.

2.3 Laboratory Experimentation

In the laboratory, the following experiments were conducted on expansive soil with different percentages of fly ash and palm oil fuel ash for various mixes. Differential free swell index (DFSI), liquid limit, plastic limit, compaction, CBR and unconfined compressive tests were conducted with a view to determine the optimum combination of fly ash and palm oil fuel ash in expansive soil.

2.4 Methodology

The experimental work is done in three stages. In the first stage laboratory tests are conducted on the clay soil, fly ash and palm oil fuel ash separately to estimate the individual behaviour of the above materials.

In the second stage, fly ash is blended with expansive soil in different proportions varying from 10 to 25% by weight of dry soil, in increments of 5%. The maximum dry density and optimum moisture content of the corresponding soil and each combination are determined and unconfined compressive strength (UCS) and CBR tests are performed at their respective MDD and OMC. The optimum dosage of fly ash is determined based on the CBR and UCS results.

In the third phase, the optimum percentage of fly ash and palm oil fuel ash with 2.0, 4.0, 6.0 and 8.0% were blended with expansive soil with a view to determine optimum percentage of fly ash and palm oil fuel ash, respectively, to calculate the improvement in the geotechnical properties like compaction, CBR and unconfined compressive strength tests.

3 Results and Discussion

With a view to determine the optimum combination of fly ash and palm oil fuel ash (POFA) as a replacement in expansive soil, free swell index, plasticity, shear strength and CBR tests were conducted on fly ash and POFA expansive soil mix.

3.1 *Effect of Stabilization on Differential Free Swell Index*

From Fig. 1, we can observe the free swell index is decreasing with the increase in fly ash replaced in the expansive soil. The differential free swell index is decreasing from 120 to 108, 95, 86 and 78, respectively, for 5, 10, 15, 20 and 25% fly ash replaced in the expansive soil. Figure 2 shows the variation of free swell index of 20% fly ash, as a replacement in the expansive clay and mixed with different percentages of palm oil fuel ash. The percentage of palm oil fuel ash varied from 0 to 8% with an increment

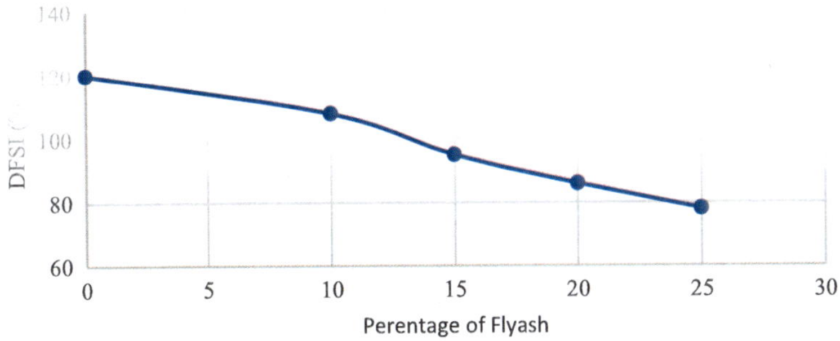

Fig. 1 Variation of differential free swell of stabilized expansive soil with the percentage of replacement of fly ash

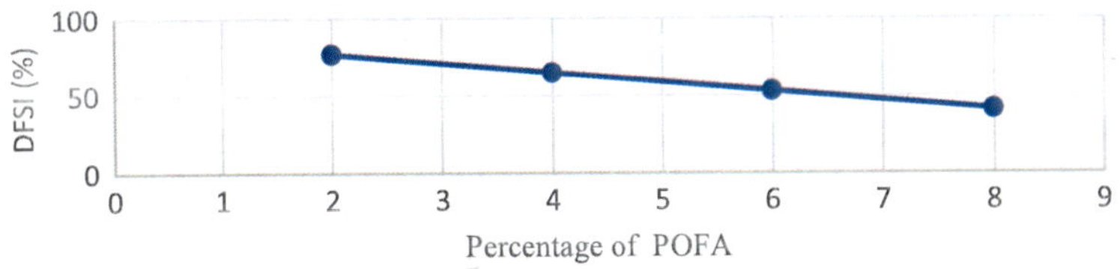

Fig. 2 Variation of differential free swell with the palm oil fuel ash of stabilized expansive soil with the 20% fly ash as replacement

of 2%. The DFS value decreased from 86 to 41 with the addition of palm oil fuel ash by 8%. It can be observed that further reduction in the free swell index with the increase in percentage of palm oil fuel ash in the fly ash expansive soil mix. This phenomenon clearly gives an idea of the improvement of the weak expansive clay when partly replaced with a non-plastic waste, i.e., fly ash and agro waste material, i.e., POFA.

3.2 *Effect of Stabilization on the Atterberg's Limits*

The variation of liquid limit and plastic limit of the stabilized soil mix with the percentage of replacement of fly ash in expansive soil is presented in Fig. 3. With the increase in percentage of replacing fly ash in the stabilized soil mix from 0 to 25%, the liquid limit decreased from 79.3 to 64.2 and the plastic limit increased from 30.4 to 39.7. This phenomenon may have resulted due to some portion of plastic soil replaced with a non-plastic waste material. The effect of palm oil fuel ash as a replacement in expansive soil was presented in Fig. 4. The liquid limit further reduced from 68.5 to 54.1 and the plastic limit increases from 38.2 to 42, with for the stabilized expansive clay with 8% the palm oil fuel ash and 20% fly ash as a replacement in the expansive soil.

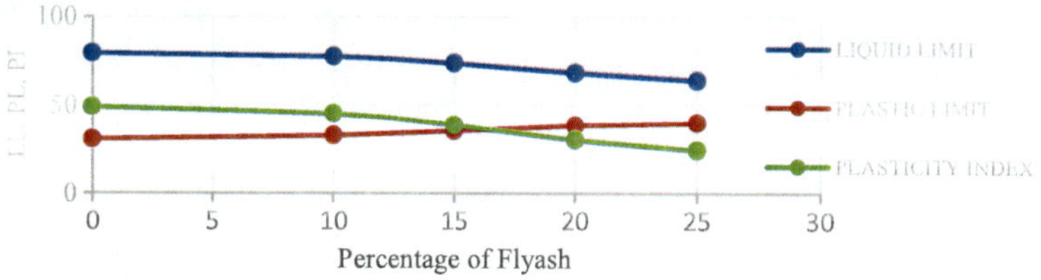

Fig. 3 Variation of liquid limit, plastic limit and plasticity index with the percentage of replacement of fly ash

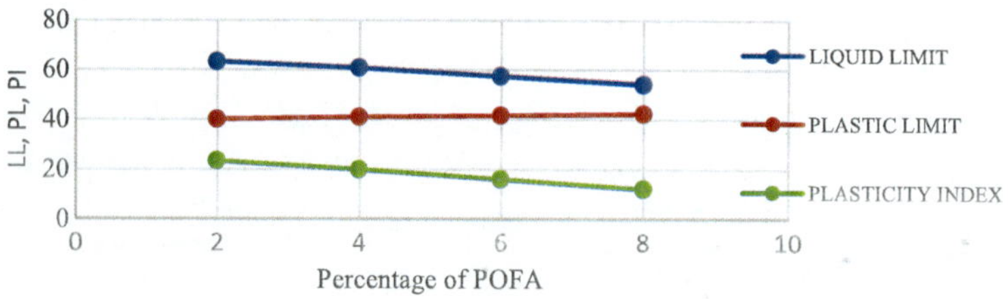

Fig. 4 Variation of liquid limit, plastic limit and plasticity index with the percentage of palm oil fuel ash of stabilized expansive soil with the 20% fly ash as replacement

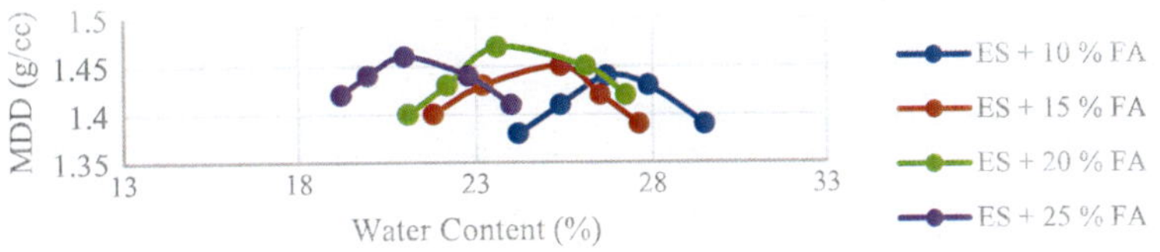

Fig. 5 Variation of dry density and moisture content with the percentage of replacement of fly ash

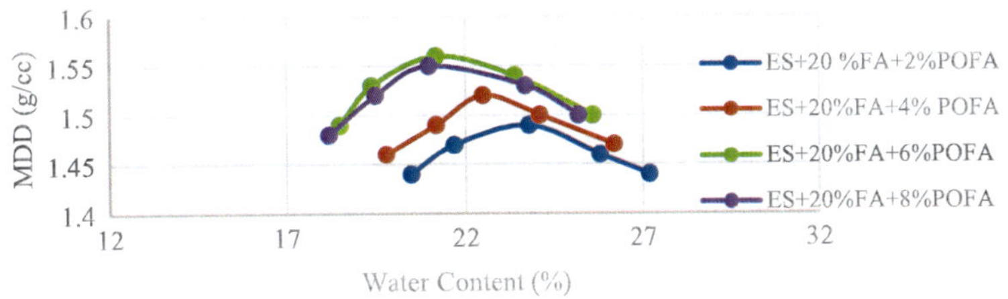

Fig. 6 Variation of dry density and moisture content with the percentage of palm oil fuel ash of stabilized expansive soil with the 20% fly ash as replacement

3.3 *Effect of Stabilization on Compaction Properties*

The variation in maximum dry density and optimum moisture content with the partial replacement in expansive soil with the fly ash are presented in Fig. 5. The maximum dry density is obtained as 1.47 g/cc at 20% partial replacement of fly ash in expansive soil. The corresponding OMC is 25.60% as shown in Fig. 5. The maximum dry density of the stabilized expansive soil further increases from 1.49 to 1.56 and the corresponding optimum moisture content decreases from 23.80 to 21.20 for further replacement with the palm oil fuel ash as shown in Fig. 6 for 20% replacement of the fly ash in the expansive soil.

3.4 *Effect of Stabilization on CBR*

The variation of CBR value with the partial replacement of fly ash in the expansive soil is presented in Fig. 7. It can be observed that the maximum soaked CBR values are obtained for the 20% of fly ash replaced in the expansive soil. The soaked CBR values are increased from 1.30 to 4.20 for a 20% replaced fly ash in the expansive soil. Further replacement of fly ash results in decreasing the CBR value. This trend may be because of more non-plastic material that results in reduction of binding property of the soil. Figure 8 shows the variation of further improvement in CBR values with the further replacement of expansive soil with palm oil fuel ash. The soaked CBR values are further increased from 4.70 to 6.80 for an optimum combination of 20% fly ash and 6% palm oil fuel as replaced in the expansive soil.

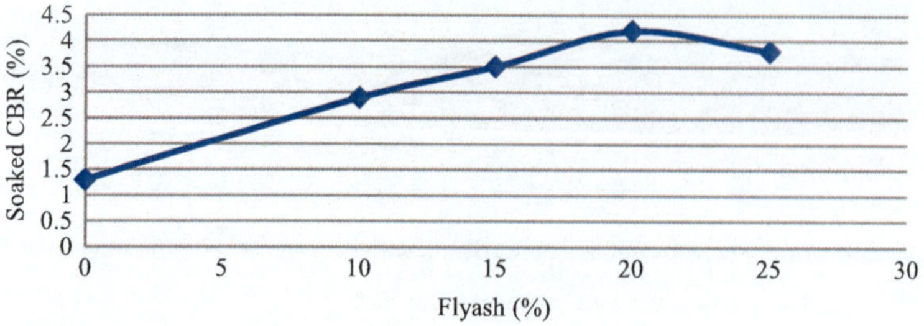

Fig. 7 Variation of soaked CBR with the percentage of replacement of fly ash

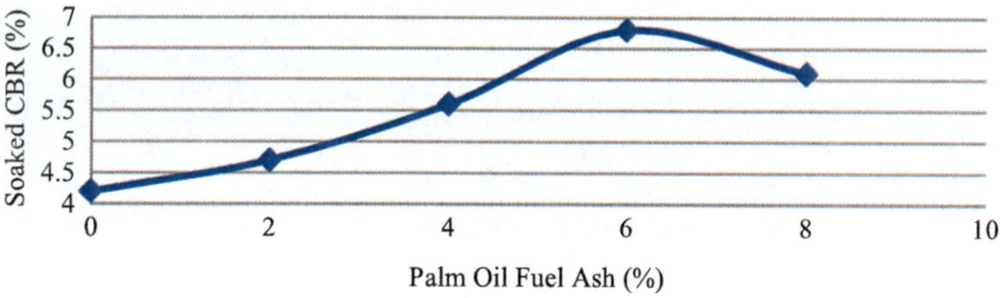

Fig. 8 Variation of soaked CBR with the percentage of palm oil fuel ash of stabilized expansive soil with the 20% fly ash as replacement

3.5 *Effect of Stabilization on Unconfined Compressive Strength*

Figure 9 shows the variation of unconfined strength with the partial replacement of fly ash in the expansive soil. The specimen is prepared at maximum dry density and optimum moisture content. The unconfined compressive strength is increased from 81 to 133 kPa with the partial replacement of fly ash from 0 to 20%. A decrease in unconfined strength is observed in 25% partial replaced fly ash in the expansive soil. The effect on unconfined compressive strength, for the stabilized mix with optimum

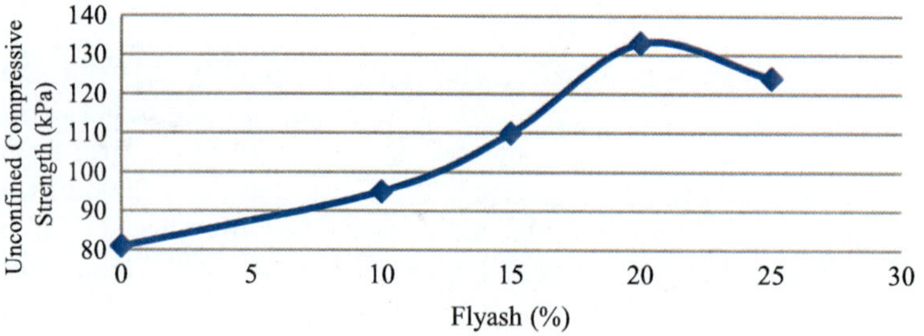

Fig. 9 Variation of unconfined compressive strength with the percentage of replacement of fly ash

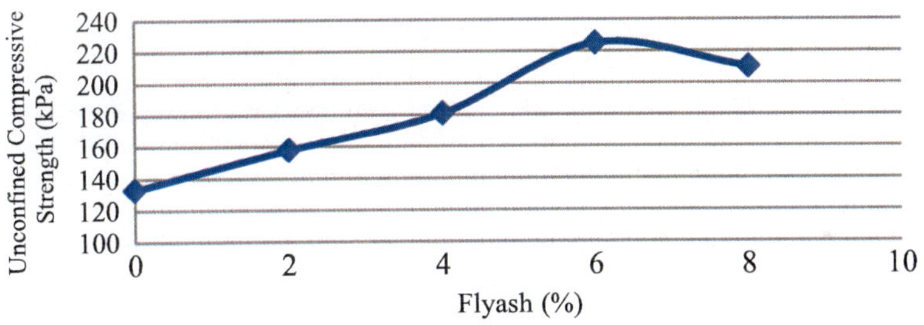

Fig. 10 Variation of unconfined compressive strength with the percentage of palm oil fuel ash of stabilized expansive soil with the 20% fly ash as replacement

fly ash and palm oil fuel ash is shown in Fig. 10. It was observed that on increasing POFA, the unconfined compressive strength is increased from 158 to 225 kPa.

Penetration resistance and the strength of the soil improved may be due to the development of bonding between the plastic and non-plastic soil particles at optimum combination. Hence, it can be summarized from the results of the experimental study to ascertain the objective of improving the weak expansive clay by using fly ash and an agro waste, palm oil fuel ash (POFA), thereby giving a twofold advantage of improving weak expansive clay with a sustainable solution by reusing the waste materials effectively.

4 Conclusions

The following conclusions may be drawn based on the experimental results:

1. The differential free swell is reduced by 28.3% with the 20% fly ash as a replacement in the expansive clay. It can be further reduced by 27.5%, a total of 55.8% reduction with the addition of 6% palm oil fuel ash.
2. The replacement of fly ash by 20% in the expansive soil had reduced the virgin plasticity index of the expansive clay by about 38%, and on further addition of palm oil fuel by 6%, it had further reduced by about 29.2%. This is due to the replacement of plastic soil with a non-plastic waste material.
3. The soaked CBR values have increased by about 223% and 423%, respectively, with 20% fly ash as a replacement and with further addition of palm oil fuel ash by 6%.
4. The unconfined compressive strength is increased by about 64.2%, with 20% of fly ash replaced in expansive soil. The unconfined compressive strength is increased by about 177% with the further addition of 6% palm oil fuel.
5. The present study concluded that problematic expansive clay was improved with a sustainable solution by reusing the waste materials giving a dual advantage.

References

1. Jones, D.E. Holtz, W.G.: Expansive soils-the hidden disaster. Civil Eng. **43**(8) (1973)
2. Katti, A.R., Katti, R.K.: Procedure for design and construction of shallow foundations in expansive clayey soils with CNS and MSM technology. Seminar on partially saturated soils and expansive soils, IGC, Kakinada, pp. 5–14 (1996)
3. Subba Rao, K.S.: Swell-shrink behavior of expansive soils-Geotechnical challenges. In: IGS Annual Lecture, Calcutta (1999)
4. Nanda Kishore, Y.: A study on minimizing swell of expansive clay bed with lime-stabilized fly ash cushion. A thesis submitted to Jawaharlal Nehru Technological University, Kakinada (2003)
5. Prasad, S.: A study on minimizing swell of expansive clay bed with cement-stabilized fly ash cushion. A thesis submitted to Jawaharlal Nehru Technological University, Kakinada (2002)
6. Sreerama Rao, A., Seelam Prasad, M.M., Dayakar Babu, R.: Use of cement-stabilized fly ash cushion in minimizing swell of expansive clays. Geotechnical Engineering for Infrastructural Development, IGC 2003, Roorkee, pp. 455–458 (2003)
7. Sreerama Rao, A., Dayakar Babu, R., Gopal Krishna Reddy, G.: Efficacy of lime stabilized fly ash cushion in arresting heave of expansive clay bed. In: IGC 2004 Warangal (2004)
8. Muhammad, S.A., Muhammad, H.O., Mohd, F.A., Chow, S.H., Damanhuri, J.: Performance of POFA with Lime as Stabilising Agent for Soil Improvement. Esteem, vol. 5, No. 1 (2009)
9. Ahmad, J., Rahman, A.S.A.., Ali, M.R.M., Rahman, K.F.A. Peat soil treatment using POFA. In: IEEE Colloquium on Humanities, Science and Engineering Research, Penang, Dec 5–6 (2011)

Enhancing the Engineering Properties of Black Cotton Soil by Using Magnesium Chloride

Sukanya Sharma, Vijay Kumar, and Ajay Bindlish

1 Introduction

Construction on black cotton soil dependably makes an issue for structural architects due to its swelling behavior for analysis. At the point when the black cotton soil comes in contact with water, then over the top swelling is caused and when water content declines shrinkage happens in the soil. In India, the black cotton soil covers 20–25% land zone and mostly found in areas of Rajasthan, Madhya Pradesh and Andhra Pradesh. The dark cotton soil contains high level of montmorillonite mineral which bestows far-reaching nature to it. Construction of different structures on black cotton soils represents a noteworthy hazard to the structure in view of the expansive level of precariousness in these dark cotton soils. Many researchers are working on different methods to enhance the properties of BC soil. Plastic limit decreases with decrease in amount of magnesium chloride. The effectiveness to cost ratio is better in $MgCl_2$ than other salts like NaCl and $CaCl_2$. One of the methods is transfer of magnesium chloride turning into a stabilizing agent.

2 Literature Review

Abood et al. [1] observed the outcome of distinct chloride compound on engineering property of silty clayey soil. The proportion of soil is taken to be 2%, 4%, 6% and 8%. On addition of NaCl solution maximum UCS was found on 4% concentration, while maximum UCS was found at addition of 8% solution of calcium chloride. On the basis of this study $CaCl_2$ solution found to be more effective than NaCl.

S. Sharma (✉) · V. Kumar · A. Bindlish
Department of Civil Engineering, Rajasthan Technical University, Kota 324010, Rajasthan, India

C. N. V. Satyanarayana Reddy et al. (eds.), *Ground Improvement and Reinforced Soil Structures*, Lecture Notes in Civil Engineering 152,
https://doi.org/10.1007/978-981-16-1831-4_8

Radhakrishnan et al. [2] studied the swelling properties of black cotton soils experimented with chemicals ($MgCl_2$, $AlCl_3$), and fly ash was evaluated as a stabilizer. It was noted that the addition of 1 and 10% of fly ash with $AlCl_3$ was high. The percentage decrement in swell potential is 63, 68% and swell pressure is 69, 73% correspondingly. Venkara Muthyalu et al. [3] evaluated the performance of chemically stabilized expansive soil. It was noted that there was a notable variation in consistency limit. When the chemical added is 1% there was a decrease in liquid limit. Nominal increase in plastic limit and CBR was recorded. Due to the reduction of liquid limit and increment in plastic limit, the net reduction in plasticity index occurs. As the percentage of chemical was increased in untreated soil, a significant increase was observed in the UCS value. Al-Omari et al. [4] estimated the outcome of potassium chloride on cyclic behavior of expansive clays. The liquid limit and plasticity index were decreased by KCl. The OMC decreases as the MDD increases. The best enhancement of free swell and swelling pressure occurs at 5% KCl where they are reduced by 56% and 65%, respectively. Mallika and Ganesh [5] observed the influence of chemical stabilizing medium on strength and swelling properties of soil. The liquid limit of soil decreases and plastic limit slightly increases as the chemical content ($CaCl_2$ and $FeCl_3$) was increased. The percentage reduction in the swelling pressure was 26 and 35% for $CaCl_2$ and $FeCl_3$, respectively, at an optimum value of chemical content. At 1% of $CaCl_2$ the value of UCS was increased by 177 and 203% for $CaCl_2$ and $FeCl_3$, respectively. It was noted that $FeCl_3$ was more effective in improving the properties of BC soil. Ajay Raj et al. [6] examined the stabilization of BC soil by using sodium chloride and fly ash, and came up with the results that the liquid limit decreases when the NaCl and fly ash were added. As the value of OMC decreases, MDD increases, and an improvement in the compaction parameters was seen. As the addition of NaCl (0, 3, 6, 9 and 12%) and fly ash (0, 5, 10, 15 and 20%) increases, the UCS values are also increased. Zumrawi and Eltayeb [7] concluded that there was a significant decrease in plasticity with increasing percent of calcium chloride. At the addition of 5% $CaCl_2$ the shear strength increases. Kolaventi et al. [8] obtained the comparative analysis of black cotton soil using NaCl and $CaCl_2$. The findings indicated that the properties are improved at the addition of 8% $CaCl_2$. Also the author gets better results for calcium chloride as compared to sodium chloride.

Rambabu and Bhavannarayana [9] investigated the experimental study by using expansive soil with banana fiber (BF) along with magnesium chloride. It was observed that with the addition of 1% BF when compared with expansive soil the free swell index and liquid limit of expansive soil have been decreased. The plastic limit, CBR value and static plate load test results improved with the addition of BF along with optimum percentage of $MgCl_2$. Srinivas and Prasada Raju [10] spring up with the findings that swell potential and swell pressure values have been decreased at 1% chemical. The time taken for treated model flexible pavement is nearly two-thirds of the untreated model flexible pavement to attain its maximum heave for ferric chloride.

Fig. 1 Black cotton soil

3 Materials Used

3.1 Black Cotton Soil

It was obtained from the district of Baran, Rajasthan, India. The specimen was extracted from a pit of 1.5 m beneath ground level with the help of Auger. The soil uprooted contains harmful substances and are in various sizes. The properties of the soil have higher proportion of expansive clay which is more problematic type for medium and high constructions (Fig. 1).

3.2 Magnesium Chloride

The magnesium chloride hexahydrate used in this study was collected from K.S. SCIENTIFICS Chemical Shop Kota, Rajasthan. Bischofite with the chemical name of "Magnesium Chloride Hexahydrate" ($MgCl_2 \cdot 6H_2O$) is one of the non-traditional stabilizers which have recently been the center of focus for researchers. Magnesium chloride ($MgCl_2$) was used as a dust control and soil stabilizing agent. The hydrated $MgCl_2$ can be extracted from brine or sea water. As the use of magnesium chloride is becoming more regular especially in pavement industry, its potential to mitigate

Table 1 Composition of magnesium chloride

Property	Value
Molar mass	203.30 g/mol
Appearance	White powder
Minimum assay	99%
Max. limits of impurities	
Sulfate (SO_4)	0.01%
Phosphate (PO_4)	0.004%
Iron (Fe)	0.002%
Calcium (Ca)	0.05%
Density	1.57 g/cm^3
Boiling point	1412 °C

Fig. 2 Magnesium chloride

the swelling potential of expansive soils is getting more attention among researchers. The properties of $MgCl_2$ are shown in Table 1 (Fig. 2).

4 Experimental Program

The $MgCl_2$ was mixed in proportions of 2, 4, 6 and 8% by weight of the dry soil in running water. Atterberg's limit test, sieve analysis, standard procter test, unconfined compression test, swelling pressure and California bearing ratio test are performed to analyze the behavior of expansive soil with $MgCl_2$. The various properties of BC soil are listed in Table 2.

Table 2 Physical properties of BC soil

S. no.	Property	Value
1	Soil classification	CH
2	Liquid limit	63.74%
3	Plastic limit	26.74%
4	Plasticity index	37%
5	Max. Dry density	1.72 kg/cm^3
6	Swelling pressure	1.45 kg/cm^2
7	O.M.C	19.7%
8	Free swell index	64.71%
9	CBR value	1.71%
10	UCS value	1.76 N/cm^2

5 Results and Discussions

5.1 Grain Size Analysis

Particle size analysis was performed as per IS: 2720 (Part-IV)-[11] and this was used to plot the graph of particle size distribution on semi log curve of the sample. The percentage of sample passing each and the percentage mass retained were determined from the data acquired (Fig. 3).

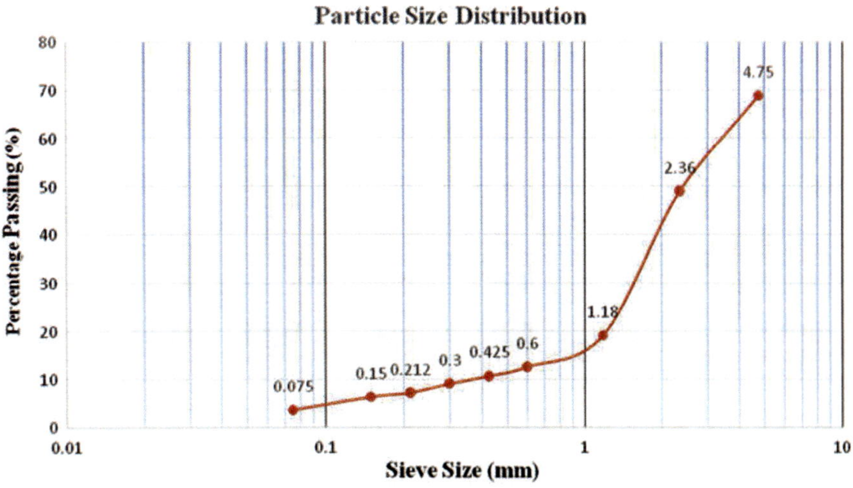

Fig. 3 Particle size distribution curve for BC soil

5.2 *Atterberg's Limit*

The liquid limit, plastic limit and plasticity index were determined as per IS: 2720(Part 5)-[12]. With the addition of $MgCl_2$ the experimental results reveal that the liquid and plastic limits of the soil decrease (Figs. 4, 5 and 6).

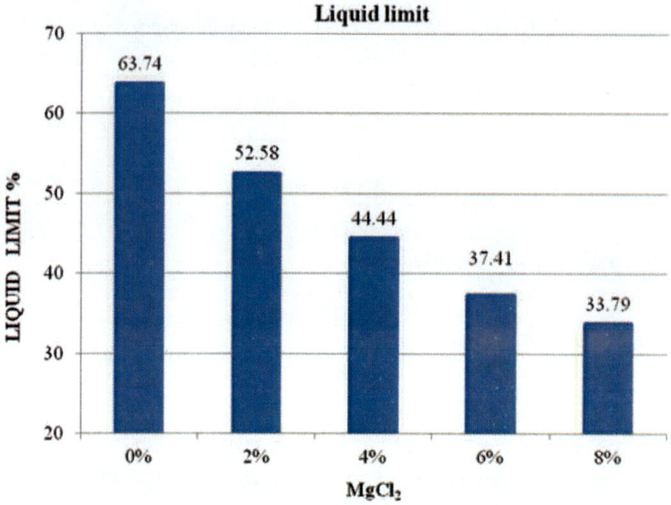

Fig. 4 Liquid limit with $MgCl_2$

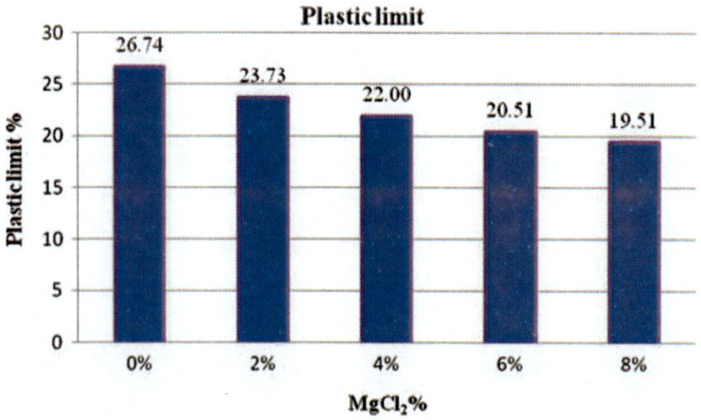

Fig. 5 Plastic limit with $MgCl_2$

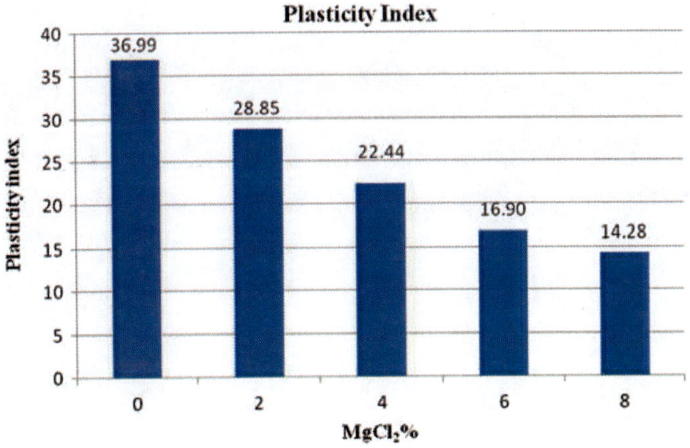

Fig. 6 Plasticity index with $MgCl_2$

5.3 Standard Proctor Compaction Test

The standard proctor test has been analyzed according to IS: 2720(Part 7)-[13] and the curve is plotted between optimum moisture content (OMC) and maximum dry density (MDD). The subsequent result is translated by Fig. 7. It was found that the

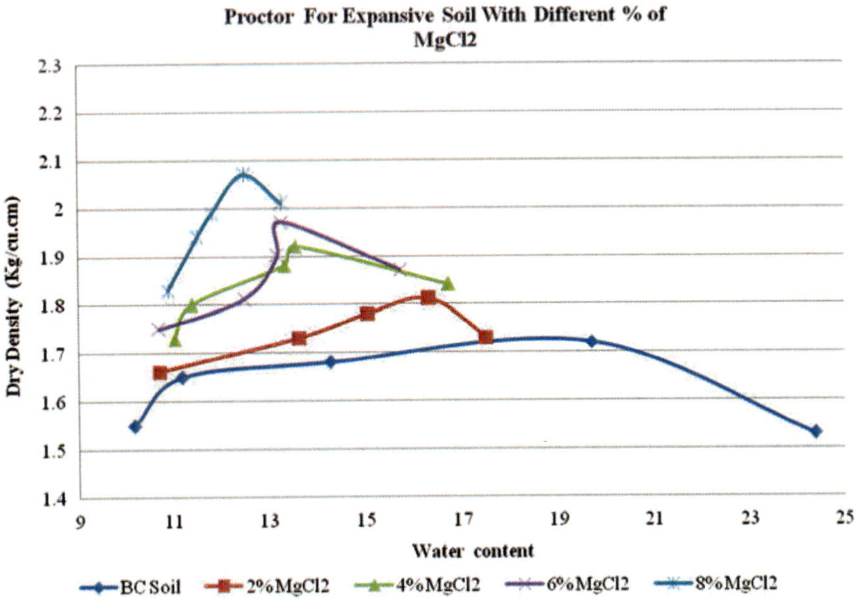

Fig. 7 Compaction test results obtained with various percentage of $MgCl_2$

Fig. 8 UCS testing machine with sample

maximum density of dry soil is 1.72 kg/cm^3. As the soil specimen is mixed with magnesium chloride, the value of OMC decreases from 19.7% to 12.5%. On adding 8% $MgCl_2$ the OMC was decreased by 36.55% and the MDD value was increased from 1.72 g/cc to 2.07 g/cc. In other words, it can also be said that as compared to untreated black cotton soil maximum dry density value was increased by 22.35%.

5.4 *Unconfined Compression Test*

The unconfined compression test (UCS) was executed as per the IS: 2720(Part 10)-[14], and the stress–strain curve is plotted as shown in Fig. 9. It is clear from the figure that the UCS value increased from 1.76 to 3.21% at the addition of 8% $MgCl_2$. On comparing the untreated black cotton soil, the UCS values are increased by 182.39% in treated black cotton soil with 8% $MgCl_2$.

5.5 *California Bearing Ratio Test*

CBR test was performed as per the IS: 2720 (Part 16)-[15]. The value of the soaked CBR was increased from 1.71 to 5.54% of the black cotton soil at 8% addition of $MgCl_2$. The CBR value is increased by 223.98% on adding 8% $MgCl_2$ on comparing with the untreated soil. Load versus penetration curve is plotted and shown in Fig. 11.

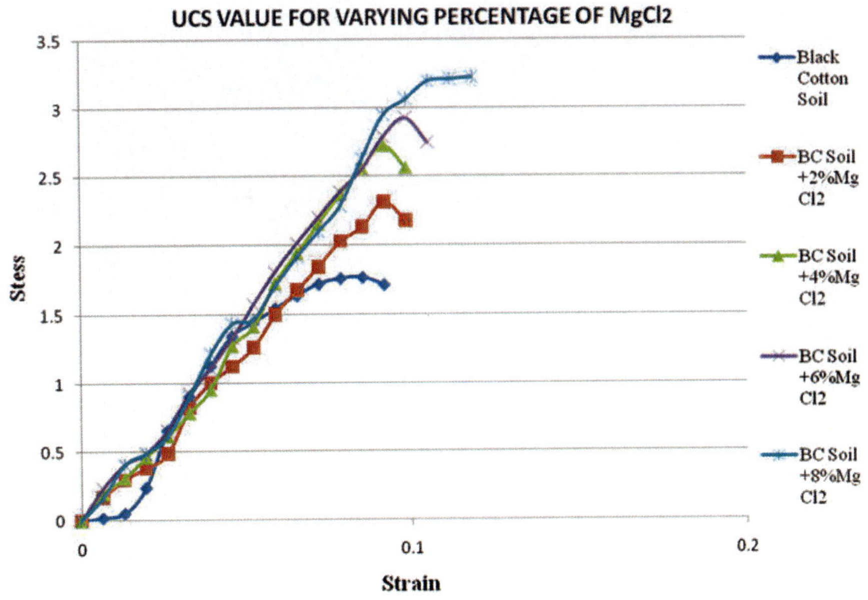

Fig. 9 UCS test results obtained by black cotton soil with magnesium chloride

5.6 *Swelling Pressure Test*

For the observations of swelling pressure of black cotton soil, the Indian standard guidelines were used as per IS: 2720(Part 41)-[16]. It was observed that the swelling pressure test for black cotton soil is 1.45 kg/cm^2. Figure. 12 shows the variation of the swelling pressure with BC soil. The swelling pressure was decreased due to the increase in percentage of $MgCl_2$ in BC soil.

6 Conclusions

The following conclusions can be drawn:

1. With the inclusion of $MgCl_2$ into the soil, the values for liquid limit of the samples were decreasing. It was decreased from 63.74% to 33.79% on adding of 0% to 8% $MgCl_2$ with untreated soil.
2. There is significant reduction in the plastic limit values from 26.74% to 19.51%. The maximum decrease in plastic limit was observed to be 27.04% at 8% $MgCl_2$.
3. The OMC value was decreased by 36.55% from 19.7% to 12.5% on addition of 8% magnesium chloride.
4. The MDD value was raised from 1.72 to 2.07 g/cc and MDD increased by 22.35% on adding 8% $MgCl_2$ when compared with the untreated BC soil.

Fig. 10 California bearing ratio testing machine

5. At 8% $MgCl_2$ the values were increased from 1.71 to 5.54% of the soaked CBR. As compared to untreated black cotton soil, there was an increase of 223.98% in CBR value at 8% $MgCl_2$.
6. At the addition of 8% $MgCl_2$ the UCS value of BC soil was increased from 1.76% to 3.21%.
7. The swelling pressure values are also decreased by 90.95% and 87.67%, respectively, for 8% of $MgCl_2$ treatment.
8. The properties of the BC soil are enhanced at 8% $MgCl_2$.

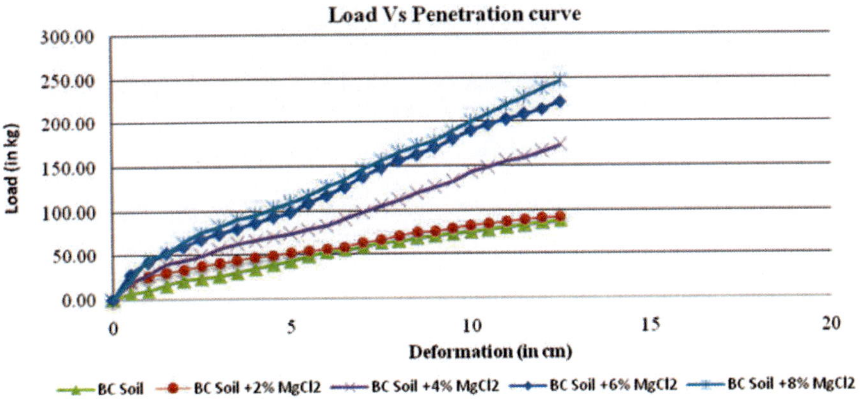

Fig. 11 California bearing ratio results

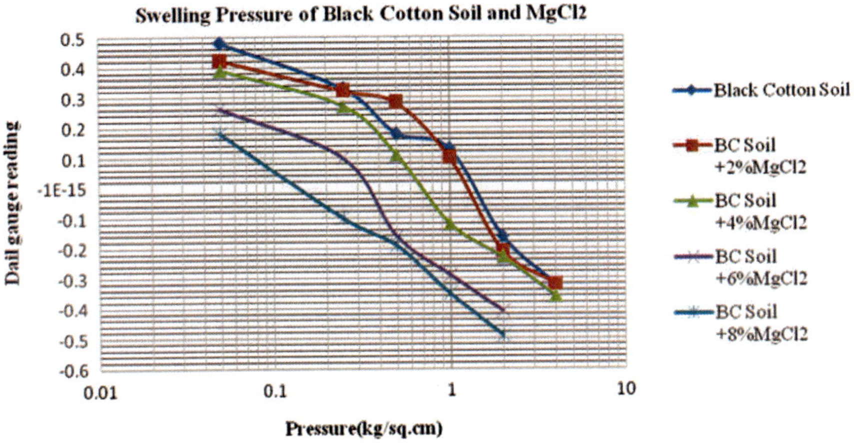

Fig. 12 Swelling pressure results for black cotton soil with magnesium chloride

References

1. Abood, T.T., Kasa, A.B., Chik Z.B.: Stabilisation of silty clay soil using chloride compounds. J. Eng. Sci. Technol. **2**(1), 102–110 (2007)
2. Radhakrishnan, G., Anjan Kumar, M., Raju G.V.R.P.: Swelling properties of expansive soils treated with chemicals and fly ash. Am J Eng Res **3**(4), 245–250 (2014)
3. Venkara Muthyalu, P., Ramu, K., Prasada Raju, G.V.R.: Study on performance of chemically stabilized expansive soil. Int. J. Adv. Eng. Technol. **2**(1), 139–148 (2012)
4. Al-Omari, R., Ibrahim, S., Al-Bayati, I.: Effect of potassium chloride on cyclic behavior of expansive clays. Int. J. Geotech. Eng. **4**(2), 231–239 (2010)
5. Mallika, Ganesh, B.: Study on effect of chemical stabilizing agents on strength and swelling properties of soils. Int. J. Eng. Technol. Sci. Res. **4**(10) (2017)
6. Ajay Raj, M., Ganapathy, C., Vinay, D., Suresh, C.: Stabilization of black cotton soil by using sodium chloride and flyash. Int. J. Pure Appl. Math. **119**(15) (2018). ISSN:1314-3395

7. Zumrawi, M.M.E., Eltayeb, K.A.: Laboratory investigation of expansive soil stabilized with calcium chloride. Int. J. Environ. Chem. Ecol. Geol. Geophys. Eng **10**(2) (2016)
8. Kolaventi, S.S., Venigalla, S.G., Rakesh, D: Stabilization of black cotton soil using salts and their comparative analysis. Int. J. Geotechn. Environ. **4**(2) (2016). ISSN: 2321-9939
9. Rambabu, J., Bhavannarayana, C.H.: An experimental study of expansive soil stabilized with banana fibre and magnesium chloride. Int. J. Innov. Eng. Manag. Res. **8**(11) (2019). ISSN: 2456-5083
10. Srinivas, M., Prasada Raju, G.V.R.: Effect of strong chemicals on the swell properties of expansive clay. In: Proceedings of IGC-2010: GEOtrendz, IIT, Bombay, pp. 613–616 (2010)
11. IS 2720- Part 4: Methods of tests for soils–grain size analysis, Bureau of Indian Standards, New Delhi (1985)
12. IS 2720- Part 5: Method of test for soils–determination of liquid and plastic limit, Bureau of Indian Standards, New Delhi (1985)
13. IS 2720-Part 7: Methods of test for soils–determination of water content-dry density relation using light compaction, Bureau of Indian Standards, New Delhi (1980)
14. IS 2720- Part 10: Method of test for soils–determination of unconfined compressive strength, Bureau of Indian Standards, New Delhi (1991)
15. IS 2720- Part 16: Method of test for soils–laboratory determination of CBR, Bureau of Indian Standards, New Delhi (1987)
16. IS 2720- Part 41: Method of test for soils–measurement of swelling pressure, Bureau of Indian Standards, New Delhi (1977)

Decontamination of Soil by Electro Kinetic Treatment

Muddassir Nadaf, Kaveri D. Jadav, and Vikas Gingine

1 Introduction

In the last so many decades it has become very difficult to extract the contaminants like heavy metals from a contaminated soil and it has been very challenging work for the engineers as well as for the scientists. There exist many processes to clean up the hazardous waste site, however, the technological challenge, cost, and efficiency of these options may vary widely. There are some economical methods like conventional ground burial and land disposal, but these methods do not provide a permanent solution, and in most of the cases they are not necessarily the most effective solution [1]. For removing organic and inorganic contaminants from the solid porous media, the most common ex situ methods employed include soil washing and ligand extraction. As the ex situ methods suffer from several problems, they are not technologically challenged that much. Some of the problems faced in the ex situ methods are excavation of large mass of soil media and transferring it from one place to another and placing it in an external reactor. The above-mentioned processes suffer from several disadvantages. There are several in situ methods, which include vacuum extraction, hydraulic fracturing, thermal desorption, electro kinetic decontamination, immobilization by encapsulation, bio treatment, and placement of barrier systems are already in use for the decontamination of soil. Most of these methods are used for the removal of organics present in the soils. Among these in situ methods electro kinetic decontamination (EKD) is in use for the last few years. The major advantage of electro kinetic decontamination is that it is an in situ process and is suited for the porous soil as well as for the cohesive soil which has low porosity where other processes can be ineffective. From this process accelerated rates contaminant

M. Nadaf (✉) · K. D. Jadav · V. Gingine
Department of Civil Engineering, KLS GIT Belagavi, Belgaum 590008, India

V. Gingine
e-mail: vlgingine@git.edu

C. N. V. Satyanarayana Reddy et al. (eds.), *Ground Improvement and Reinforced Soil Structures*, Lecture Notes in Civil Engineering 152,
https://doi.org/10.1007/978-981-16-1831-4_9

extraction and transport may be achieved. The basic concepts and an overview of the EKD processes and their real-life applications, as of now, in geotechnical and geoenvironmental engineering have been reviewed and presented in this paper.

1.1 Development of Electro Kinetic Process

When a DC current was applied to a mixture of clay-water, Reuss observed the electro kinetic phenomena. Under the electric field he moved the capillary toward the cathode. The flow of water immediately stopped when the electric potential was removed. Electroosmosis is distinguished by Napier, and he found through a membrane, the electric potential difference resulting from streaming potential. In 1879, the first electroosmotic phenomena were analytically treated by Helmholtz. A mathematical basis was provided. Later it was modified to apply to electrophoretic velocity by Pellatt and Smoluchowskic [3]. Casagrande's studies on stabilizing clays by electroosmosis started in the early 1930s. Electrical gradient was introduced 70 years ago, to stabilize the soil mainly by removal of the water. In the early period most of the studies were directed toward soil stabilization from removal of water, and were generally concentrated on the exploration of EK technologies that showed marked interest in the early 1980s for the removal of toxic chemical species in dewatering of fine gravel soils by electroosmosis. In 1960s electro migration was used by Russian researchers in prospecting for metals. The early 1980s showed marked interest in the EK exploration technologies for the removal of chemical toxic species in ionic form in the soil in Europe and the USA [4]. Further researches and field studies are encouraged by the successful applications, resulting in understanding of the in situ remediation of contaminated sands from various processes in EK.

2 Electro Kinetics

The term "electro kinetics" (EK) refers to the introduction of the associate electrical gradient (as critical hydraulic or pressure gradient) within the soil to mobilize or promote the migration of water and/or varied chemical species toward the well-liked conductor. Electro kinetics is a soil redress technology that is comparatively young and has become another procedure for the removal of noxious chemical species in ionic type within the soil in the late 1980s. Electro kinetics could be a method that separates and extracts significant metals, radionuclides, and organic contaminants from saturated or unsaturated soils, sludge, and sediments. A low-intensity electricity is applied across conductor pairs that are implanted within the ground on all sides of the contaminated soil mass. The electrical current causes electron diffusion and particle migration (electro migration) and activity that move the binary compound part contaminants within the submerged from one conductor to the opposite. Contaminants within the binary compound part or contaminants desorbed from the soil surface area are unit transported toward individual electrodes looking at their charge. The contaminants could then be extracted to a recovery system or deposited at the conductor. Surfactants and complexing agents are accustomed to increase the solubility and assist in the movement of the material.

2.1 *Process and Mechanisms*

When an electrical field is formed across a soil volume, it provides a drive that will induce mass movement of particles, almost like the impact of different driving forces, like pressure gradient, concentration gradient, and thermal gradient. Especially, the appliance of an electrical field causes the following main transport phenomena in soils: electroosmosis, electro migration, dielectrolysis of these electro kinetic phenomena square measure extremely influenced by the surface charge densities of the soil particles, and so on by the soil mineralogical composition.

2.2 *Electroosmosis*

Electroosmosis may be a bulk transport of water that flows through the soil as a result of the applied electrical field. The fluid migration happens principally from the anode to the cathode, because of the predominance of an electric charge on the soil particle surfaces. The electroosmotic flow is caused by the very fact that once an electrical field is applied to the soil, the surplus of cations on the brink of soil particles surface (double layer) tend to move toward the cathode. The movement of those ions and the water molecules related to these species (hydration shells) imparts a web strain on the pore fluid encompassing the hydrations shell.

2.3 *Electro Migration*

The second transport mechanism generated by the voltage gradient is electro migration, which is the movement of ions within the pore answer beneath the influence of an electrical field. Positive ions (cations) migrate toward the cathode whereas negative ions (anions) square measure transported toward the anode. Due to electro migration, ions tend to concentrate close to the alternative charged conductor. The particle quality may be a term that describes the speed of migration of a particular ion species beneath a unit field. In soils, the speed of ionic migration will be higher outlined by the effective ionic quality, which conjointly accounts for soil porousness and crookedness, and might considerably affect particle migration.

2.4 Electrophoresis

Electrophoresis consists of the movement of charged particles and colloids below the influence of an electrical field. Once an immediate current (DC) force field is applied across a colloid, charged particles and colloids that are suspended within the pore fluid are electrostatically interested in one among the electrodes and repelled from the opposite. Similarly, to the electro migration method, positively charged particles tend to move towards the cathode and negatively charged particles tend to move towards the anode. Usually, for environmental applications, electrophoresis is a smaller amount vital than electroosmosis and electro migration in terms of mass flux, though in some cases electrophoresis could play a task in removal, for example, if the migrating colloids have the contaminants absorbable on them.

2.5 Factors Affecting Electro Kinetic Technology

Electro migration rates within the submerged rely on grain size, ionic quality, contamination concentration, total ionic concentration, and upon the soil pore water current density and hydrogen ion concentration. The direction and amount of contamination movement is influenced by the contamination concentration (anions versus cations), soil sort and structure, pH, surface chemistry, and current density of the soil pore water. The potency of extraction depends upon many factors, like the sort of species, their solubility within the specific soil, their electrical charge, their concentration relative to alternative species, their location and kind within the soil, and availableness of organic matter within the soil. Electro kinetics is applicable in zones of low hydraulic physical phenomenon, significantly with high clay content. The technology is best once the cation exchange capability (CEC) and the salinity are low, throughout electro kinetic treatment, and electrolysis leads to the formation of H^+ and OH^-. These migrate toward each other by electro kinetic processes. As these two fronts meet, a speedy transition from low to high hydrogen ion concentration happens, making an area of minimum solubility of metals. These sharp discontinuities in hydrogen ion concentration induced inside the soil mass by electro kinetics may end in a deposition front wherever minerals are precipitated in soil pores, by markedly reducing porosity and inhibiting recovery. This could be prevented by flushing the cathode with water or a dilute acid to arrest the migration of the OH^- front into the soil. The mass flux transported throughout the electro kinetic method depends on the transient chemistry that takes place beneath the influence of induced electrical field. Specifically, the sorption–desorption, precipitation–dissolution, and oxidation–reduction behavior of the contaminants throughout the electro kinetic method influence the rectification efficiencies. Activity refers to the partitioning of the contaminants from the answer or pore fluid to the solid section or soil surface. Sorption includes adsorption and ion exchange and it is dependent on (1) the sort of contamination, (2) the sort of soil, and (3) the pore fluid characteristics (desorption natural method natural action) is that the reverse process and is to blame for the discharge of contaminants from the soil surface. Each activity and natural action are suffering from soil hydrogen ion concentration changes caused by the migration of H and Buckeye state ions, which are created by the electrolysis reactions [5]. The hydrogen ion concentration-dependent sorption–natural action behavior is usually determined by playing batch experiments victimization of the soil and contamination of explicit interest. The precipitation and dissolution of the contamination species throughout the electro kinetic method will influence the removal potency of the method [6]. The soil removal method is suffering from the element ions generated at the anode migrating across the contaminated soil and neutralizing the chemical group ions at the cathode. However, in some styles of soils, the migration of the element ions is going to be hindered, thanks to the comparatively high buffering capability of the soil. The presence of the chemical group ions at the cathode can increase the hydrogen ion concentration worth (pH 10–12). During a high hydrogen ion concentration atmosphere, significant metals can precipitate, and the movement of the contaminants are going to be obstructed. The high hydrogen ion concentration and the low significant metals concentration condition at the cathode might also cause the formation of a charged advanced species at the cathode compartment. The movement of those charged advanced species toward the anode and of the significant metals toward the cathode depends upon the relative quality of the element and chemical group ions. For economical contamination removal, it is essential to stop precipitation and to possess the contaminants in dissolved kind throughout the electro kinetic method. Chemical reaction and reduction reactions are vital once coping with antimonial contaminants like Cr. Cr exists most ordinarily in 2 valence states: powerfulness Cr Cr (III) and hexavalent Cr Cr(VI). Cr (III) exists within the cationic hydroxides like Cr $(OH)_2$ and it will migrate toward the cathode throughout electro kinetic rectification. However, Cr (VI) exists within the kind of oxyanions like CrO_4 that migrate toward the anode. The valence state depends on the soil composition, particularly the presence of reducing agents like organic matter and Fe(II) and/or oxidizing agents like Mn(IV); therefore it is vital to grasp the valence state of metals and their possible redox chemistry. Conductor acquisition procedures are necessary to induce favorable chemistry, and as a result, promote bigger rectification potency.

3 Materials Used for the Test

3.1 *Sand*

Sand is a composite material formed by the disintegration of rocks. Sand is a product of several factors such as impact of climate, elevation and slope of terrain, organisms, and the sand's parent materials or original minerals interacting over time. It continuously undergoes numerous physical, chemical, and also biological processes. Sand can be classified into different types based on their particle size and properties.

3.1.1 Properties of Sand

(a) Location: Susodanhatti (Near Desur Belagavi)
(b) Color: Light yellow
(c) Specific gravity: 2.6
(d) Bulk density: 1.6 g/cm^3
(e) Voids ratio: 0.81
(f) Porosity: 0.44

Most sands have a dry bulk density (density of sand taking into account voids when dry) between 1.1 and 1.6 g/cm^3, while the specific gravity of sand should be between 2.6 and 2.85.

3.1.2 Wet Sieve Analysis

Wet sieving is a process used to evaluate particle size distribution of a granular materials. In this method it may be possible that it may undergo sample loss, but the percentage loss is very small, and wet sieve analysis is accurate and efficient compared to dry sieve analysis (Table 1).

3.1.3 Uniformity Coefficient (Cu) and Coefficient of Curvature (Cc) of Sand

The uniformity coefficient (Cu), curvature coefficient (Cc), and the effective size (D10) are the grading characteristics of the sand. These are the geometric properties of a grading curve that describe a particular type of sand.

Table 1 Sieve analysis

S. no.	Particle size (mm)	Mass retained (g)	% Mass retained	Cumulative mass retained	% Finer
1	4.75	1.01	0.239	0.239	99.761
2	2.36	8.81	2.088	2.327	97.673
3	1.18	55.33	13.114	15.441	84.559
4	0.6	80.01	18.963	34.405	65.595
5	0.425	39.05	9.255	43.660	56.340
6	0.3	41.19	9.763	53.422	46.578
7	0.15	124.07	29.406	82.828	17.172
8	0.075	57.25	13.569	96.397	3.603
9	Pan	15.2	3.603	100	0
		421.92			

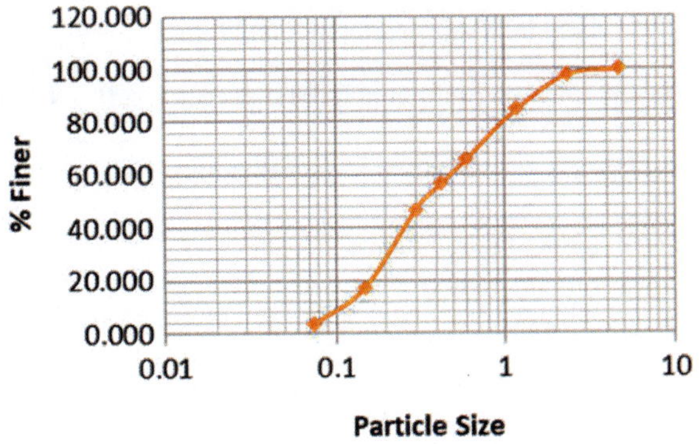

Fig. 1 Particle size distribution curve

3.1.4 Grade Curve Characteristics

Grade size distribution curve is used to identify the type of soil by using different particle sizes: D60, D30, and D10. This graph is plotted on logarithmic scale by taking particle sizes on x-axis and percentage finer on y-axis. This is plotted based on the observations from sieve analysis conducted on the sand sample.

D10 is called as effective particle size. This means that 10% of the particles are finer and 90% of the particles are coarser than D10. This is the size at 10% finer by weight.

Similarly, D60 is the particle size at which 60% of the particles are finer and 40% of the particles are coarser than D60 size. D30 is the size at which 70% of particles are coarser than D30 size and 30% particles are finer by weight. Hence, D10, D30, and D60 are used to determine the measures of grade size distribution curve.

We can know particle size for D10, D30, and D60 from Fig. 1.

D10 = 0.1
D30 = 0.23.
D60 = 0.5

3.1.5 Measures of Gradation

The uniformity coefficient (Cu) and the coefficient of gradation (Cc) are the measures of sand gradation. These coefficients help to classify the sand as well-graded or poorly graded ones.

3.1.6 Uniformity Coefficient (Cu)

The uniformity coefficient (Cu) is defined as the ratio of D60 to D10. A value of Cu greater than 4 to 6 classifies the sand as well-graded. When Cu is less than 4, it is classified as poorly graded or uniformly graded sand.

Cu = D60/D10.
Cu = 0.5/0.1 = 5.

Uniformly graded sand has identical particles with Cu value approximately equal to 1. A uniformity coefficient value of 2 or 3 classifies the sand as poorly graded. Beach sand comes under this category.

Higher value of Cu indicates that the sand mass consists of sand particles with different size ranges.

3.1.7 Coefficient of Curvature (Cc)

Cc = D302/(D10*D60).
Cc = 0.22 / (0.11*0.5) = 1.058.

For the sand to be well graded, the value of Cc must range between 1 and 3.

As we got the values of Cu and Cc in the range 4–5 and 1–3, respectively, we can conclude that the sand is well-graded.

3.2 Leachate

A leachate is a liquid that extracts insoluble and suspended particles while passed through any matter or any other substance of the material. Leachate is a widely used word in the environmental sciences. It has a particular meaning which tells that the liquid has dissolved or entrained environmentally hazard substances that may enter the environment. It is most often used in situations of land-filling of putrescible or industrial waste.

In the tapered environmental situations, leachate is a liquid material that drains from land or amass material and contains significantly raised concentrations of unwanted material derived from the material that it has passed through. Leachate sample is collected from an unsegregated municipal solid waste treatment and disposal facility at No. 19, 40, and 42 of Turamuri village, Belgaum Taluk and District in an area of 66 acres around 10 km away from the city, which is established by city corporation of Belagavi [7]. Figures 2 and 3 show the images of the site.

The steady pH for the leachate that is used lies between 7.6 and 8.9 with total suspended solid (TSS) of 0.22. The pH value is an indicator of the fight of the leachate in aerobic and anaerobic conditions in the waste. Leachate in a mature

Fig. 2 Waste landfill site at Turamuri

Fig. 3 Leachate collection

landfill that exhibits pH value greater than 7, while for leachate that has undergone some stabilization is expected to have neutral pH value [2]. The ammonia content in leachate samples is between 216 mg/L and 288 mg/L. Ammonia is created from the decomposition of waste, mainly proteins. Leachate used in this study is indicated by strong odor and black in color. A similar finding was cited by Jaafar et al. in describing leachate collected from Air Hitam landfill. The measured COD values in leachate ranged between 96 and 112 mg/L (Table 2).

Table 2 Properties of leachate

Parameters	Values
Color	Black
pH	7.6–8.9
TSS	0.22
Ammonia	216–288
COD	2880–3360
Heavy metals	Ni(106), Cr(277.3), Mn(269), Cd(0.16)

4 Methodology

4.1 Electro Kinetic Treatment

- Power supply: A power supply of 30 V has been provided to increase the current flow eventually which is noted down.
- Electro kinetic treatment box

  Dimensions: length: 30 cm, width: 12.7 cm, height: 8.8 cm.

- Graphite electrodes

  Length: 15.24 cm, width: 7.62 cm, thickness: 12 mm (Fig. 4).

Fig. 4 Electro kinetic treatment setup

4.2 *Procedure*

1. Clean the electro kinetic treatment box. Cover the perforated walls of the box with filter paper and a muslin cloth.
2. Pour the slurry of the soil saturated with the leachate to its liquid limit or saturated with leachate, 14% of the soil. Make sure that there are no air voids or soil lumps. If any, they should be eliminated.
3. Pour water on either sides of the box, as shown in the figure.
4. Insert the copper electrodes in the water connected to the DC supply with the help of cables. A constant voltage of 30 V is provided.
5. The water on either sides of the box are changed periodically and the current flow is noted before and after changing the water.
6. The test is carried out for a period of three days with a time interval of 12 h.

Electrolysis reactions (conversion of electrical energy in the form of chemical potential energy) create OH and H_2 at the cathode and O_2 and H^+ at the anode. These reactions create a base front near the cathode and an acid front near the anode that migrates from one to the other. The increase in the mobility of cationic species is aided by acid front, but in some soils, it can retard electroosmosis. The hydroxide front must be controlled to avoid the premature precipitation of some target metal ions. Unfavorable conditions at a site include oil with a high ion exchange capacity, high buffering capacity, high naturally occurring organic content, very low moisture content, and salinity. The presence of subsurface metal structures or utilities can also adversely affect the performance.

5 Results

5.1 *Variation of Current and Voltage*

Figure 5 shows the variation of current versus days. It is observed that initially the current flow is high. With time, the flow of current through the soil layer started to decrease. As can be seen from Fig. 5, the amperage of approximately 100 mA, recorded at the initiation of the experiment, dropped significantly to a level of less than 50 mA, in about 5 days. It shows that the amount of impurities(ions) present in the soil were high at the first day, and as the days pass, the impurities are decreased because of which current flow decreased. Figure 6 shows the variation of voltage drop along the length of the soil. As shown in the graph, voltage drop takes place from anode to cathode.

This variation clearly shows that the migration of ions is taking place and the soil is decontaminating.

Table 3 shows the pH and conductivity variations observed during the test. During the experiment at the end of each day, the pH and conductivity of the solutions present

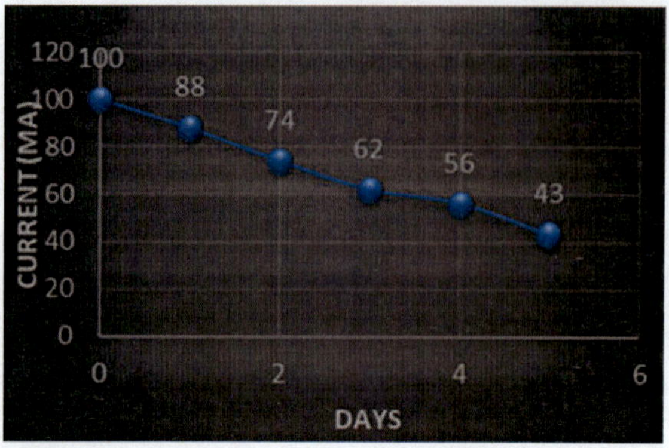

Fig. 5 Variation of current

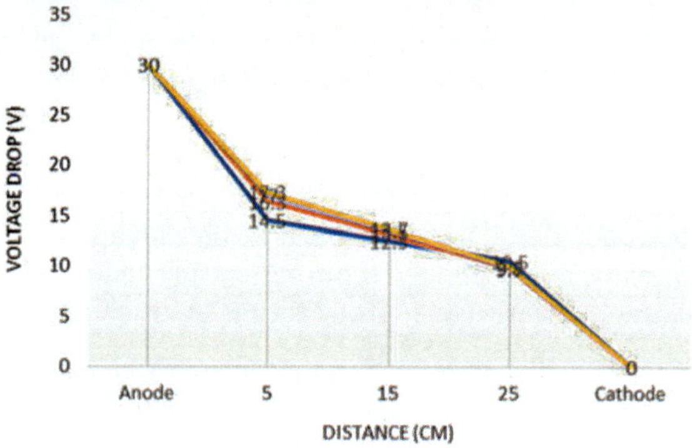

Fig. 6 Variation of voltage drop

Table 3 Variation of pH and conductivity

Parameters		pH	Conductivity
Day 1	Anode	2.45	8.60 ms/cm
	Cathode	12.2	4.65 μs/cm
Day 2	Anode	3.56	4.30 ms/cm
	Cathode	11.84	1.9 μs/cm
Day 3	Anode	4.8	2.86 ms/cm
	Cathode	9.54	0.8 μs/cm

at the anode and cathode are tested and the values are noted down. It is observed that due to electrolysis oxidation occurs at the anode, generating an acid front, while reduction takes place at the cathode producing a base front. These electrolysis reactions cause the pH to decrease at the anode and increase at the cathode. As the anode and cathode compartments solutions are replaced with the distilled water each day, the experiment is continued till the variation in the pH of solution should be very less or nearly equal to the pH of distilled water. Correspondingly, the conductivity values are also noted down and variation is observed. Initially the conductivity values were high and it decreased to low at the end, which shows that the impurities are reducing because the conductivity is decreasing.

6 Conclusions

From the laboratory experiments conducted on the soils, it is evident that this technique may be further developed and used efficiently and effectively as a soil treatment technique. This is especially so when there is an immediate need for treatment and enhancement of physical properties of soil in a locality, within a short time frame, in the maintenance and protection of infrastructure. Similarly, this technique can be used to remediate and clean up already salinized land to prepare for new infrastructure development. Electro kinetic treatment may also support localized revegetation programs, for instance, along roadsides to protect pavements and embankments.

References

1. Pamukcu, S., Wittle, J.K.: Electrokinetic Removal of Selected Heavy Metals from Soil," Environ. Progress, Am.Instit.Chem.Engr.,11(3):241–250 (1992).
2. Lageman, R.: Electro reclamation: applications in The Netherlands. Environ. Sci. Technol. **27**, 2648–2650 (1993)
3. Karim, M.A., Khan, L.I.: Soil Sediment. Contam. **20**(7), 857–875 (2011)
4. Lageman, R., Pool, W., Seffinga, G.: Contaminated soil, electro reclamation: theory and practice. Chem. Ind. **8**, 585–590 (1989)
5. Acar, Y.B., Gale, R.J., Alshawabkeh, A.N., Marks, R.E., Puppala, S., Bricka, M., Parker, R.: Electrokinetic remediation: basics and technology status. J. Hazard Mater. **40**, 117–137 (1995)
6. Acar, Y.B., Gale, R.J., Ugaz, A., Puppala, S., Leonard, C.: Feasibility of removing uranyl, thorium and radium from kaolinite by electro chemical soil processing, Report EK-BR-009-0292, Electrokinetics Inc., Baton Rouge, Louisiana (1992)
7. Inspection Report of Municipal Solid Waste Management site in Belagavi city, CPCB, pp. 1–9 (2015)

Desiccation Cracking Behavior and Strength Characteristics of Areca Fiber-Reinforced Fine Grained Soils

Ravi Diamond and Renjitha Mary Varghese

1 Introduction

Soil stabilization is the adjustment to soils to update their physical properties, like, texture, structure and permeability, and enhance the shear strength of a soil to govern the shrink–swell properties of soil. In this modern era, structures must be frequently based on problematic (weak or expansive) soils. However, stabilizing using chemical additives like lime and cement raises environmental issues, and their usage has been quite limited nowadays. In thermal power plants, the major combustion products are fly ash (FA) and bottom ash (BA). The utilization of bottom ash is limited due to its slow pozzolanic reaction and less reactivity. But utilization of BA has a huge potential in the civil engineering field, which has not been explored widely. Fibers can be used to reinforce the soil. Soil reinforced with fibers goes about as a composite material where fibers improve the tensile strength of the soil. The stabilization of soil utilizing fibers assists in enhancing the engineering properties of the soil [17]. Enhancement in the tensile strength, shear strength, CBR and bearing properties of the soil is observed with the reinforcement of soil with fibers exposed to different sorts of stresses [3, 15]. Stabilization of expansive soils using new natural fiber named areca fiber along with bottom ash has improved the characteristics of soil effectively [11]. And engineering properties of lateritic soil enhanced with areca fiber with cement as the stabilizing agent [7]. The soil in field encounters everyday or occasional rainy and sunny environment changes, and goes through periodical wetting-drying cycles. The subsequent wetting–drying cycles, especially the second cycle, led to a significant rearrangement of soil particles and modification

R. Diamond
National Institute of Technology Calicut, Calicut, India

R. M. Varghese (✉)
National Institute of Technology, Calicut, India
e-mail: renjitha@nitc.ac.in

C. N. V. Satyanarayana Reddy et al. (eds.), *Ground Improvement and Reinforced Soil Structures*, Lecture Notes in Civil Engineering 152,
https://doi.org/10.1007/978-981-16-1831-4_10

of pore system [13, 14]. Mechanical and hydraulic properties gets modified due to the presence of desiccation cracks in soil, thus weakening the performance of soil with respect to various engineering disciplines, particularly geotechnical, geological and environmental engineering. In reality, cracks may build soil compressibility and consolidation rate and lessen the soil strength [9]. A desiccation crack is probably going to show up when the surface tensile stress reaches soil tensile strength or when volume shrinkage is constrained [4]. Be that as it may, progressively quantitative investigations on desiccation cracking behavior of expansive soil are carried out through an image processing technique [1, 2, 5, 6, 8, 12, 16].

This study targets exploring the desiccation cracking behavior at a series of wetting–drying cycles, and strength characteristics of fiber (areca-fiber) and bottom ash (BA) stabilized clayey soil. The geometrical and morphological characteristics of the crack samples were quantitatively analyzed, using an image processing technique.

2 Materials and Methods

2.1 *Soil and Bottom Ash*

The soil utilized in this study is procured from Kolhapur Maharashtra, India. The collected soil has been air-dried and broken into pieces to pass through 4.75 mm. The bottom ash samples used in this study are obtained from the Udupi Power Plant, Mangalore, Karnataka, India. Bottom ash is one of the parts of the non-combustible residue of the combustion process in a furnace. The specific gravity obtained is 2.42 as per IS 2720, Part 3 (BIS 1980; ASTM 2014). The grain size distribution of the soil was found by carrying out both wet sieve and dry sieve analyses as per IS 2720. Figure 1 shows the particle size distribution of clay and bottom ash and indicates that soil is clayey and the bottom ash contains graded silted content. Table 1 shows the physical properties of soil.

2.2 *Areca Fiber*

Areca belongs to the Arecaceae (Palmae) palm family and Aroideae subfamily that grows mostly in East Africa, Asia and Tropical Pacific. In India, Karnataka and Kerala are the largest areca nut-producing states. In this study, the areca fiber is collected from Kannur district, Kerala. Table 2 shows the areca fiber properties. To increase the flexural strength and durability properties of areca fiber it is chemically treated with KOH solution [18]. Initially, the fibers were soaked in water for ten days and after drying then again soaked in KOH solution for ten days. Later they were washed with (CH_3COOH) acetic acid and cleaned with running water if any acid traces remain on fibers and dried for ten days.

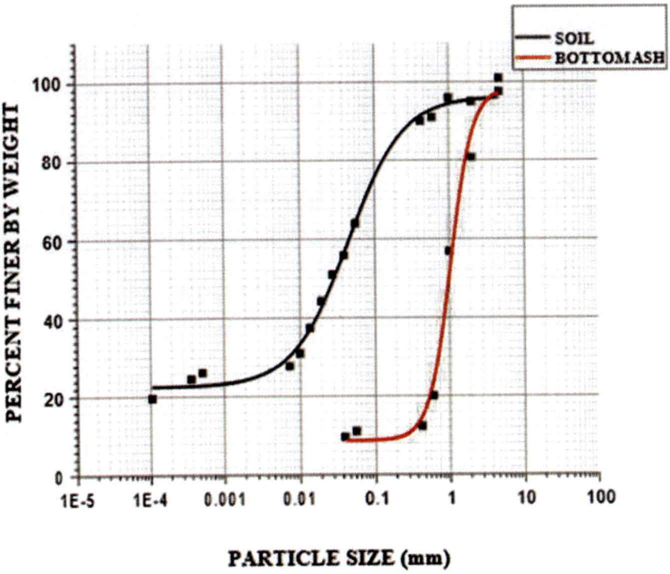

Fig. 1 Grain size distribution curve of soil

Table 1 Soil properties

Parameters	Value
Specific gravity	2.65
Clay (%)	26.8
Silt (%)	43
Sand (%)	30.2
Liquid limit (%)	42.14
Plastic limit (%)	18.2
Shrinkage limit (%)	14.5
Optimum moisture content (%)	28.04
Maximum dry density (g/cc)	1.45
Soil classification (IS)	CI

Table 2 Properties of areca fibers

Parameters	Value
Tensile strength (MPa)	876–948
Elongation of the break (%)	1.47–1.8
Diameter (mm)	0.8–1.8
Density (g/cm^3)	1.34–1.45
Young's modulus (GPa)	42–48

3 Experimental Program

3.1 Compaction Tests

Maximum dry density (MDD) and optimum moisture content (OMC) for different proportions of soil and BA were determined by conducting compaction tests. The compaction apparatus consists of height of 127.3 mm and 100 mm internal diameter with a falling hammer weight of 2.6 kg with 25 blows/layer in three layers for the proctor compaction curve.

3.2 Unconfined Compressive Strength

Three samples were prepared for each test according to the MDD and OMC of the mix obtained from the standard proctor compaction tests. The soil–BA mixture was prepared by adding the required amount of water and mixing it properly in a wet state. The mixing was done by hand, and good care was provided to prepare a uniform mix. The specimens (38 mm inner diameter and 76 mm long) were prepared and tested as per IS 2720, Part 10 (BIS 1991). The prepared mixes are placed inside the mold and compacted. The specimens are then removed from the mold. The UCS tests were conducted on soil–BA mix samples (with varying percentage of BA) to obtain the optimum content of BA.

3.3 Image Analysis

The collected soil is initially air-dried, crushed and then sieved at 2 mm in the laboratory. Saturated slurry specimens are prepared by blending dry soil powder with distilled water. The saturated, slurry soil specimens at different fiber contents of 125 and 250 aspect ratio of fibers and optimum bottom ash content 25% are prepared by blending with distilled water. The water content of the slurry was about equal to the liquid limit. The quantity of slurry mixture was poured onto the 13 cm circular mold. Air bubbles in the mix are removed by placing the glass plate on a vibration machine for 15 min. Specimens were kept in an oven at temperature (45 °C) for drying, and were brought near to the setup where the camera is placed on the platform. Plumb-bob is used to ensure that the camera is vertical to the specimen surface. The camera clicked photographs from a constant height of 30 cm. In this study, the image processing technique is used to characterize the crack pattern quantitatively using ImageJ software.

4 Results and Discussions

4.1 Compaction Tests

Figure 2 shows the variation of optimum moisture content (OMC) and Maximum Dry density (MDD), against the varying percentage of the BA from 0 to 30%. Figure 2a clearly shows that with the increase in BA percentage, there is an increase in the MDD of the mixtures, but after 25% BA, the MDD got decreased. A similar observation is reported by [11]. Addition of BA up to 25% might enhance the gradation of the soil–BA mix and thus improve the MDD, while adding beyond 25% of bottom ash

Fig. 2 Variation of dry density of soil with bottom ash percentage. (**a**) Variation of maximum dry sensity with bottom ash content (%) (**b**) Variation of OMC of soil with bottom ash content (%)

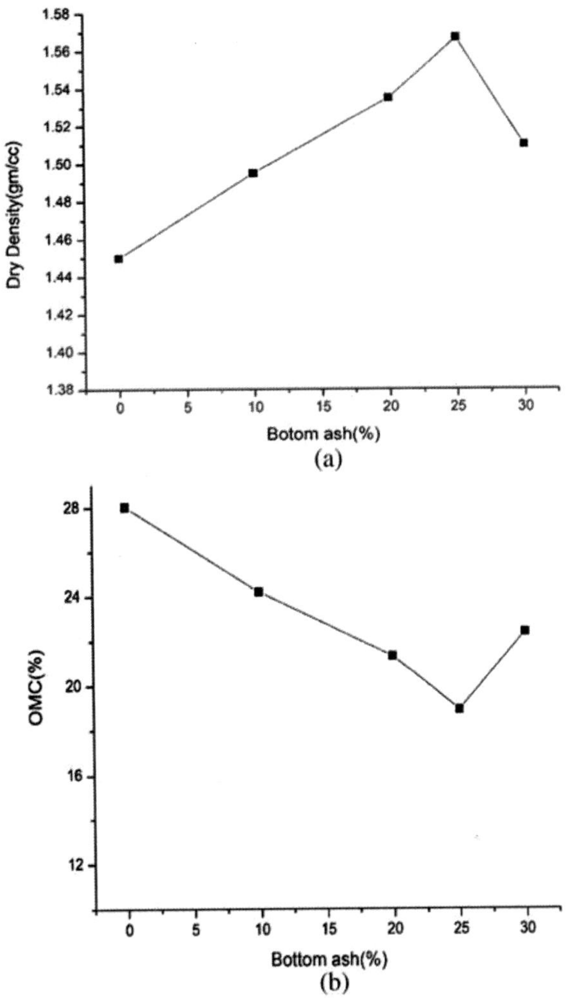

may interrupt the intergranular packing of soil and lead to the reduction in MDD. The substitution of higher content of BA also reduces the MDD as the specific gravity of bottom ash is lower compared to soil.

As the bottom ash content increases, the OMC got decreased till 25% and beyond 25% the OMC increased from Fig. 2b. Similar trends in OMC and MDD were reported by several previous studies [10, 11]. The reduction in optimum water content states that the stabilization of soil can be carried out at lower water content. Besides, the increase in MDD illustrates the eligibility of soil–BA mixtures to be used as subgrade and embankment material.

4.2 Unconfined Compressive Strength Tests

Figure 3 shows the variation of UCS at varying percentages of BA without any curing. The UCS test results show that with the increase in BA content up to 25% the strength of the soil is enhanced, and there is a huge reduction in UCS at 30% of BA. The decrease in UCS after 25% may be due to the reduction in MDD that was noticed in standard proctor compaction test results. The particle packing was reduced with the reduction in density of soil. As a result, the unconfined compressive strength decreased with 30% bottom ash. From the above compaction and UCS test results, 25% BA can be examined as optimum content for further studies.

Figure 4 shows the UCS variation of the soil–BA mix with a varying percentage of fiber for curing periods of 7, 14, 28 days of aspect ratio 150 and 250. The test results indicate that the UCS value improved significantly with the increase in percentages of areca fiber. Particularly, the increase in the UCS till 1% of fiber content is due to the enhancement in the better gradation, cohesion and interlocking or packing of soil and BA, while the UCS decreased with 1.5% of fiber content. At the point

Fig. 3 Variation of UCC value with varying percentages of bottom ash

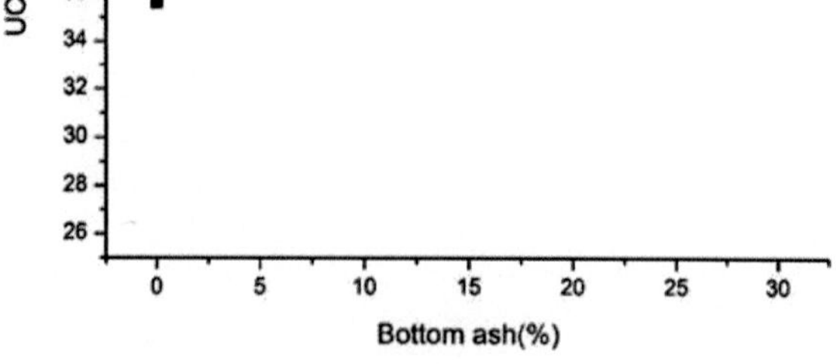

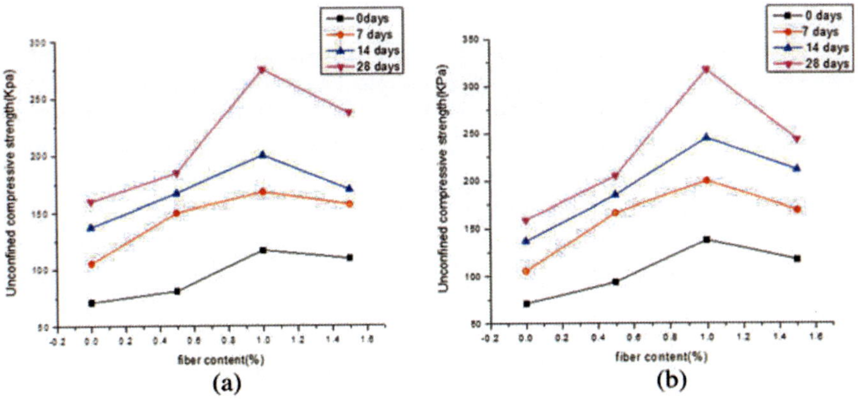

Fig. 4 Variation of shear strength of soil with bottom ash and fiber (%); **a** Aspect ratio 150 and **b** aspect ratio 250

when the fiber content is more than 1%, the appropriation of the fiber in the soil is uneven, compared with the past low content, and some portion of the fibers is locally gathered in soil. The majority of the fibers cannot contact with the clay particles, and accordingly, they cannot completely play the job of reinforcement; these fibers separate the clay particles and the integrity of clay is decimated.

4.3 Image Analysis

The effect of fiber length and drying–wetting (D-W) cycles on fiber-reinforced expansive clay is examined using image analysis technique. The crack parameters like average crack width (wavg), average clod area, crack intensity factor (CIF) and crack density factor (CDF) are measured for every sample. The crack measurements are analyzed in detail using an image analysis program called ImageJ. (\cite{el2017image}.

The shortest possible distance between two arbitrarily chosen points on each boundary of the opposite cells is measured as crack width. The ratio of total area of cracks, Ac, to the total area of specimen, At, is measured as CDF, and the total area of reduced specimen is measured as CIF using image processing technique. The crack patterns obtained for each test were quantitatively analyzed.

Effect of wetting–drying cycles

After the first drying cycle, a regular crack pattern for 150 and 250 aspect ratios is observed. It may very well be seen that the sample surface has been part of a few individual clods by the crack network. The states of crack segments and clods are generally smooth and regular. Clods were generally pentagons and quadrangles. A total detail perception of the crack pattern gives that the greater part of the crack segments is opposite to one another. The crack patterns of fiber-reinforced soil samples at various fiber contents of 125 and 250 aspect ratio and optimum bottom ash content of 25% at each wetting–drying cycle are shown in Figs. 5 and 6.

After the second wetting–drying cycle, prompt breakdown of the clods was observed. The clods got increased marginally. Some new micro cracks showed up on the sample surface. The crack density factor (CDF), crack intensity factor and the number of clods increased, while the average area of clods and width of cracks decreased compared with the first wetting–drying cycle. This occurrence could be imputed due to the elements mentioned below: The additional water exhausted molecule holding capacity and modified soil structure. The disintegration of clods might be due to internally generated forces, such as differential swelling pressure. Entrapment and pressurization of air that arise inside the clods strengthened the destruction of soil structure.

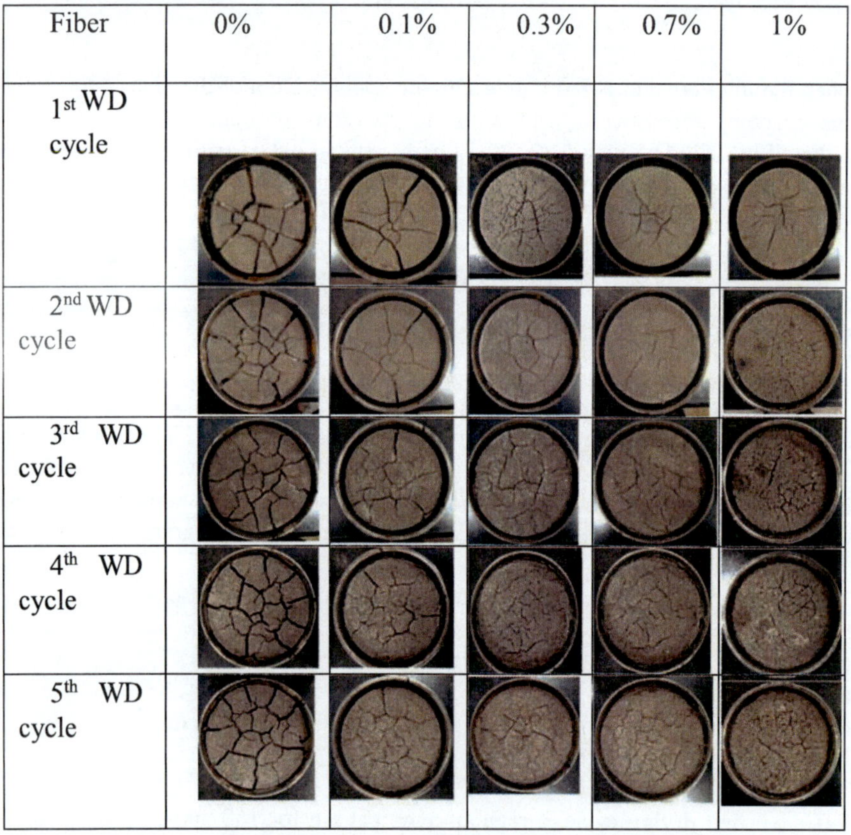

Fig. 5 Crack patterns of aspect ratio 150

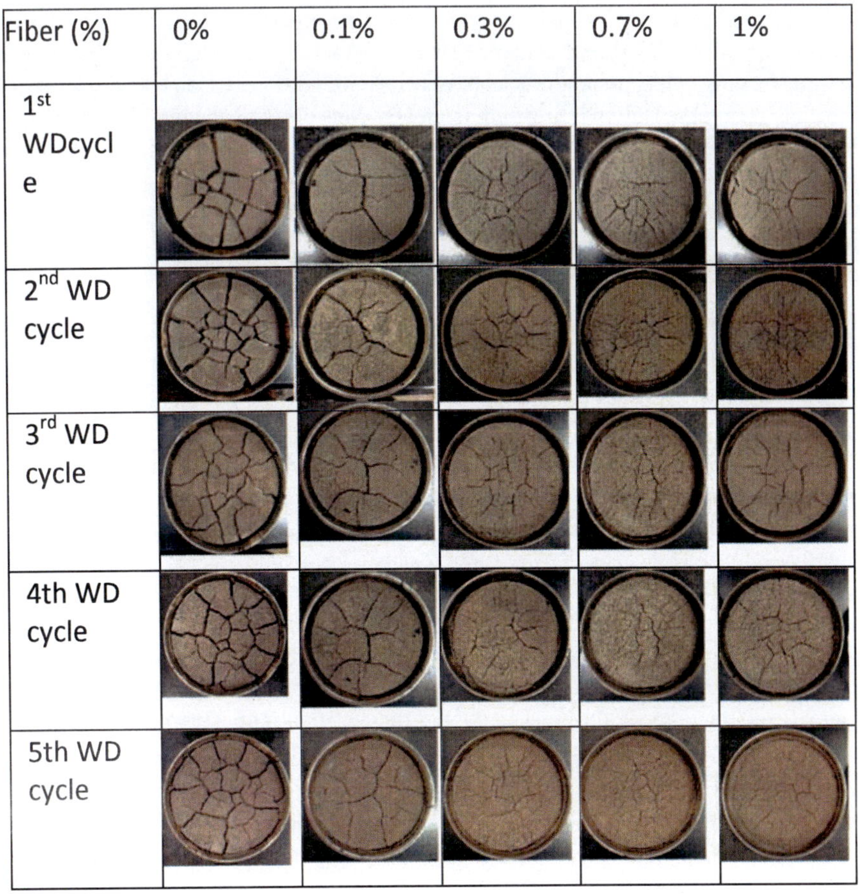

Fig. 6 Crack patterns of aspect ratio 250

At the point when water was placed into the glass cup, it entered inside the clods driven by capillary forces. This fairly severe hydration procedure can prompt entanglement and pressure development of air bubbles. Since capillary initially develops in the smallest pore spaces, the pore air is expelled and slowly assembles in macropores. From one perspective, with a progressive gathering of pore air, the air bubble volume increments and prompts soil destruction. Then again, as hydration continues, the pressure Pa inside the air bubbles increments because of the meniscus imbibition.

Pa (*pressure inside the air bubble*) exceeds the binding strength of neighborhood particles, leads to mini explosions of soil structure and ultimately cause general damage. Sometimes, air bubbles might be locked up and cannot escape outside due to the pressure equilibrium between the air bubbles and soil structure. This phenomenon was confirmed by [13] and the picture is shown in Fig. 6. The entrapped air bubble amplifies the disaggregation of the neighboring soil.

After the third wetting–drying cycle, a similar trend has been observed. First cracks were disappeared entirely, and new microcracks were observed. After a fourth wetting–drying cycle, no new cracks have been noticed. The reason is that stability of the clod during the third wetting path is greatly controlled by the binding strength in between the formed aggregates in the second wetting–drying cycle. There is no significant change in the CDF, CIF, the width of cracks, length of cracks and number of clods average area of clods. In view of the quantification results of the crack patterns, it may very well be presumed that the cracking behavior reached the equilibrium state after the soil experienced four wetting–drying cycles.

Effect of the fiber content

The usual morphological and geometrical features of crack patterns also change with the fiber content. This is imputed due to the "bridging" effect of fibers, which alters the intrinsic development and release of the tensile stress field in a soil specimen and, eventually, gives rise to variation in the mode of crack initiation and propagation and also the distribution characteristics. Quantification of the crack parameters as shown in Figs. 7 and 8 of aspect ratio 150 and 250 demonstrated that the inclusion of fibers greatly decreased the crack density and intensity (Fig. 8a and b), crack width (Fig. 8c) and average clod area (Fig. 8d), while an increase in the number of clods. Similar trends have been observed in 250 aspect ratios shown in Fig. 9. By fiber inclusion, the morphology of the crack patterns was affected greatly. The total number of wide cracks diminished, and the total number of fine cracks expanded altogether with the increase of fiber content; the crack networks changed continuously from a dependable ordinary structure to an irregular structure, and a large number of dead-end cracks or single crack that don't intersect other cracks were recognized. The inclusion of fiber has decreased the crack density factor (CDF), crack intensity factor (CIF) and crack width by increasing the fiber content from 0.1 to 1%.

Effect of aspect ratio

As the aspect ratio increased, the CDF, CIF and average width of cracks have been decreased. This is due to the increase in the surface content of the fiber with the soil. As we increased aspect ratio, the surface contact increases, and the more bonding between the soil and fiber increases. Variation of CDF value, average width and

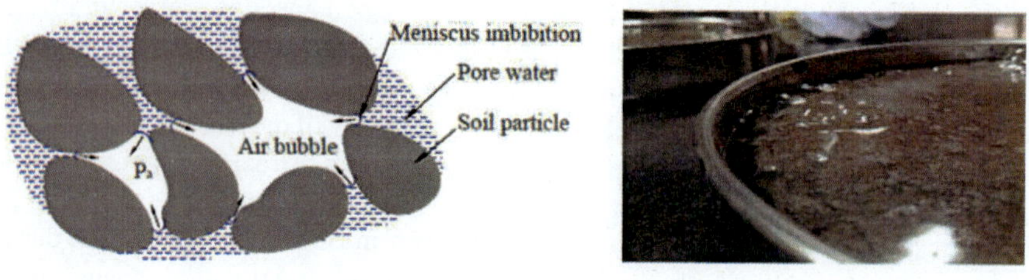

Fig. 7 Meniscus imbibition during wetting

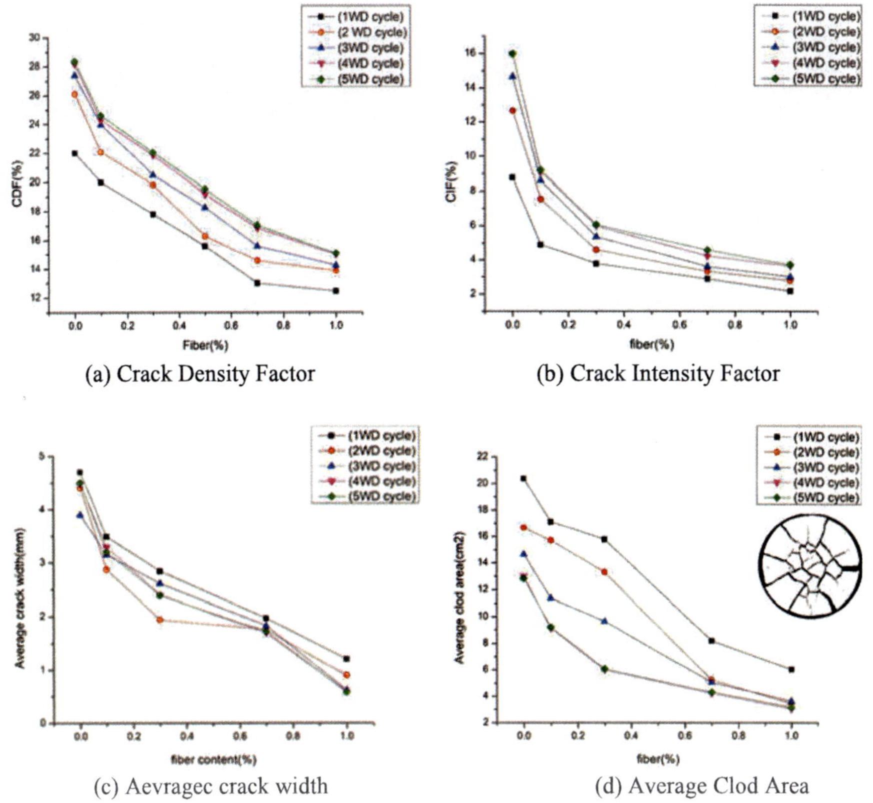

Fig. 8 Variation of crack quantitative parameters with varying percentage of the fiber 150 aspect ratio

average clod area with aspect ratio of first wetting–drying cycle is shown in Fig. 10. Subsequent wetting–drying cycles showed the same trends.

5 Conclusions

1. By the addition of BA to soil till 25% has decreased the OMC, and increased the MDD, but this trend has been reversed with the addition of 30% BA.
2. UCS of the soil–BA mix has been increased up to 25% BA content, and there was a reduction in UCS with 30% BA.
3. The results of UCS tests of the soil–BA mix specimens with varying percentages of areca fibers have shown that the addition of fiber enhanced the strength of the soil–BA mix significantly.

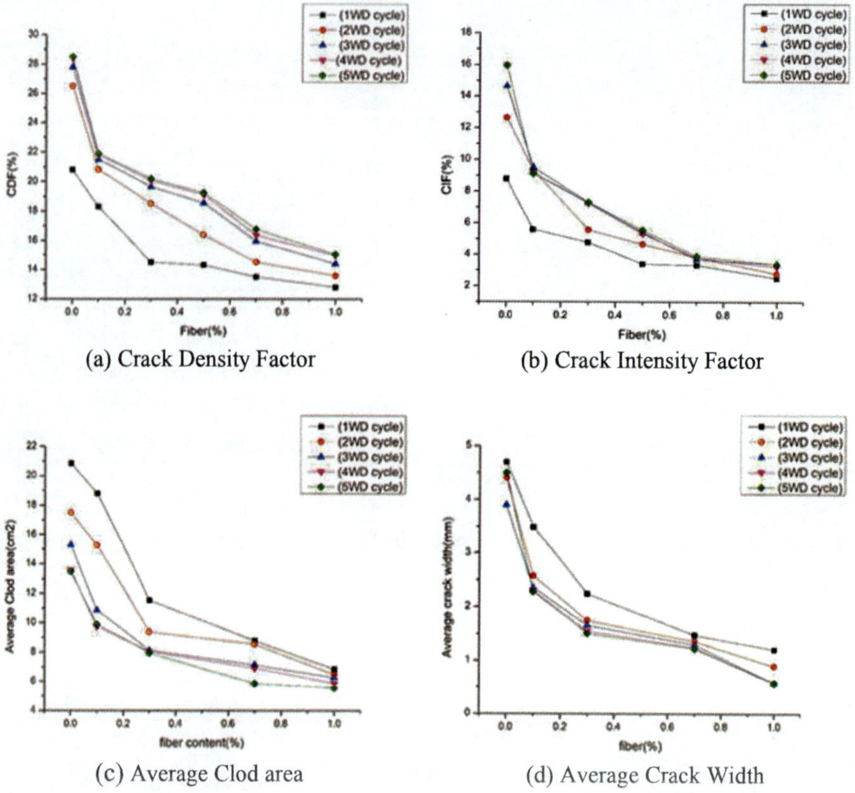

Fig. 9 Variation of crack quantitative parameters with varying percentage of the fiber of 250 aspect ratio

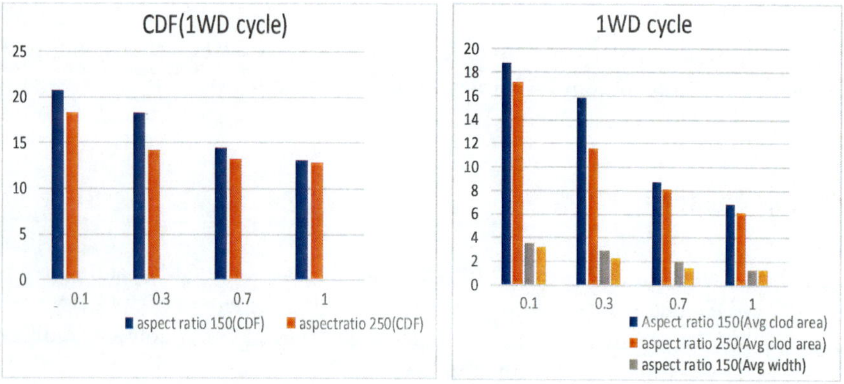

Fig. 10 Variation of crack quantitative parameters with aspect ratio

4. Addition of 1% areca fiber to soil has increased the UCS values for 150 and 250 aspect ratios to 7% and 101%, respectively, for a curing period of 28 days.
5. The inclusion of fiber has improved the crack resistance capacity greatly due to the higher tensile strength of the soil under the range of fiber contents (0–1%) tested in this study, as 1% gives optimum strength, crack reduction in the fiber-reinforced specimens approached 42 and 51.5% reduction of CDF for 150 and 250 aspect ratios. CIF is reduced by 68.8% and 76.8% for 150 and 250 aspect ratio, respectively.
6. The increase in the aspect ratio reduced the CDF by 17.24%, average crack width by 18.33% and average clod area by 11.2% due to increased surface contact and bonding strength between the soil and the fiber.

References

1. Abu-Hejleh, A.N., Znidarčić, D.: Desiccation theory for soft cohesive soils J. Geotech. Eng. **121**(6), 493–502 (1995)
2. Auvray, R., Rosin-Paumier, S., Abdallah, A., Masrouri, F.: Quantification of soft soil cracking during suction cycles by image processing. Eur. J. Environ. Civil Eng. **18**(1), 11–32 (2014)
3. Chaduvula, U., Viswanadham, B., Kodikara, J.: Desiccation cracking behavior of geofiber-reinforced expansive clay. In: Geo-Chicago, pp. 368–377 (2016)
4. Corte, A., Higashi, A.: Experimental research on desiccation cracks in soil. Report no., Cold Regions Research and Engineering Lab Hanover NH (1964)
5. El-Halim, A.A.: Image processing technique to assess the use of sugarcane pith to mitigate clayey soil cracks: laboratory experiment. Soil Tillage Res. **169**, 138–145 (2017)
6. Lakshmikantha, M., Prat, P., Ledesma, A.: Image analysis for the quantification of a developing crack network on a drying soil. Geotech. Test. J. **32**(6), 505–515 (2009)
7. Lekha, B., Goutham, S., Shankar, A.: Evaluation of lateritic soil stabilized with Arecanut coir for low volume pavements. Transp. Geotech. **2**, 20–29 (2015)
8. Miller, C.J., Mi, H., Yesiller, N.: Experimental analysis of desiccation crack propagation in clay liners 1. JAWRA J. Am. Water Resour. Assoc. **34**(3), 677–686 (1998)
9. Morris, P.H., Graham, J., Williams, D.J.: Cracking in drying soils. Can. Geotech. J. **29**(2), 263–277 (1992)
10. Osinubi, K.: Stabilisation of tropical black clay with cement and pulverised coal bottom ash admixture. Advances in Unsaturated Geotechnics, pp. 289–302 (2000)
11. Sudhakaran, S.P., Sharma, A.K., Kolathayar, S.: Soil stabilization using bottom ash and areca fiber: experimental investigations and reliability analysis. J. Mater. Civil Eng. **30**(8), 04018169 (2018)
12. Susanga, C., Kodikara, K.: Investigation of desiccation cracking using automatedd digital photography. In: 10th Australia New Zealand Conference on Geomechanicss, At Brisbane, Australia, vol. 1 (2005)
13. Tang, C.-S., Cui, Y.-J., Shi, B., Tang, A.-M., An, N.: Effect of wetting-drying cycles on soil desiccation cracking behaviour. In: E3S Web of Conferences, vol. 9, EDP Sciences, pp. 12003 (2016)
14. Tang, C.-S., Cui, Y.-J., Shi, B., Tang, A.-M., Liu, C.: Desiccation and cracking behaviour of clay layer from slurry state under wetting–drying cycles. Geoderma **166**(1), 111–118 (2011)
15. Tang, C.-S., Shi, B., Cui, Y.-J., Liu, C., Gu, K.: Desiccation cracking behavior of polypropylene fiber–reinforced clayey soil. Can. Geotech. J. **49**(9), 1088–1101 (2012)

16. Umana, U.E., Davie, C.T., Eminue, O.O.: Investigation of the effect of multiple wetting and drying cycles on the shrinkage and cracking of engineered clay soil. Int. J. Eng. Res. Technol **5**, 24–35 (2016)
17. Velde, B.: Structure of surface cracks in soil and muds. Geoderma **93**(1–2), 101–124 (1999)
18. Venkateshappa, S.C., Bennehalli, B., Kenchappa, M.G., Ranganagowda, R.P.G.: Flexural behaviour of areca fibers composites. BioResources **5**(3), 1846–1858 (2010)

Numerical Analysis and Experimental Study on Reinforcement Using PVC Rod in the Vicinity of Pressure Bulb in Sand

K. Agila, Arvee Sujil Johnson, and Keerthi Kiron

1 Introduction

1.1 General

Improvement of ground is an imperative method before geotechnical engineers everywhere throughout the world. Shortage of land alongside exceptionally frail sub-soil conditions of the current ground constrained the geotechnical engineers to work round the clock to discover appropriate solutions. Idea of support of soil including the usage of suitable material has been utilized since before years. Ground improvement procedures are highly favored methods for load bearing applications as low and moderate intense in soft as well as loose soils. Soil reinforcement is one of the methods in which tensile components can be placed in the soil to improve the stability and consequently control deformation of the soil structure and foundation also; hence thereby reducing the settlement of soil. To be more effective, the reinforcement must pass or meet the potential failure surfaces in the soil mass. Strains in soil mass produce strains in the reinforcements, which in turn generate tensile loads in the reinforcements. These tensile loads act to limit the soil movement and in this manner grant extra shear strength. This outcomes in the composite soil/reinforcement system having altogether more noteworthy shear strength than the soil mass alone. Srivastava et al. (2013) [1], studied on load settlement response of footing placed over buried flexible pipe through a model plate load test and concluded that the ultimate bearing capacity of the buried flexible pipe - soil system is moved forward when the buried flexible pipe exists in loose-medium dense sand. J.N. Jha and S.K. Shukla (2015) [2], studied on bearing capacity and settlement characteristics of sand subgrades with vertical reinforcement supporting a square footing and concluded that substantial improvement of the bearing capacity is a factor of length of reinforcement, spacing

K. Agila · A. S. Johnson (✉) · K. Kiron
Department of Civil Engineering, College of Engineering Trivandrum, Trivandrum, Kerala, India

C. N. V. Satyanarayana Reddy et al. (eds.), *Ground Improvement and Reinforced Soil Structures*, Lecture Notes in Civil Engineering 152,
https://doi.org/10.1007/978-981-16-1831-4_11

of reinforcement, extent of reinforcement and thickness of reinforcement. Craig H. Benson and Milind V. Khire (1994) [3] has studied on reinforcing sand with strips of reclaimed high density polyethylene. They concluded that strips of reclaimed high density polyethylene enhances resistance to deformation and improves strength.

This study intends to investigate the chance of utilizing PVC rods and use it in sandy soils to improve the overall stability and the mechanical properties, like compressibility, shear strength, and density. In general, the increase of strength is proportional to the amount of reinforcing material. Use of PVC rods as reinforcement is a smart thought since these are astounding corrosion resistant and climate safe material, and these rods have high compressive strength. PVC rod is likewise a decent electrical and thermal insulator. These are effectively accessible in bars and tubing. The soil properties can be improved using PVC rods as reinforcement. Inclusion of reinforcement in the vicinity of pressure bulb enriches the shear strength of soil. Also it aids in the use of poorer quality soils as structural component. The stability can be improved by a combination of inclined and vertical insertion of PVC rod. PVC rods can be used to underpin the foundations resting on sandy soil. The surrounding soil achieves strength by the densification process where the soil gets densified into a tighter configuration, thus resulting in increased density. Thus the shear strength of the soil increases and results in the enhancement of engineering properties of the soil. The use of combination of vertical and inclined insertion of the PVC rod gives additional strength when compared to the vertical insertion [4].

2 Experimental Program

In this study PVC rods are inserted as a combination of vertical and inclined reinforcement in two arrays in the vicinity of the pressure bulb. Here, the variables used for the study are the length of reinforcement, diameter of reinforcement, and extent of reinforcement. The plate load test implemented in the laboratory is simulated using 3D software. PVC rods with varying length are compared, and an optimum length is found out. The length was varied to 1.5B, 2B, 2.5B, where B is the base of the footing The load versus settlement graph is evaluated based on plate load test as well as numerical analysis.

2.1 Materials Used

The soil sample was collected from Menamkulam beach, Trivandrum. Sea sand is selected for this study. Various tests are conducted on the sand according to the IS specification. Dry sieve analysis is conducted and the soil is classified as uniformly graded sand as per IS 2720. Table 1 shows the properties of the soil.

PVC rods were purchased from Yogdeep Enterprises Pvt Ltd, Mumbai. The rods are intruded by checking the angle of orientation as constant as 450. PVC rods were

Table 1 Properties of the soil

Soil property	Values
Specific gravity	2.665
Max. dry density (kN/m^3)	17.48
Min. dry density (kN/m^3)	14.69
D_{10}	0.21
D_{30}	0.28
Uniformity coefficient, Cu	1.57
Cc	1.131

Table 2 Properties of PVC rod

PVC rods property	Values	
Density	1580 kg	
Modulus of elasticity	3174 MPa	
Temperature resistance	70 °C	
Yield strength	45 MPa	

placed symmetrically at eight points in inclined manner and vertically at three points (Table 2).

2.2 *Test Setup*

The plate load tests were conducted on sand filled in a circular tank of diameter 0.6 m and 0.4 m height. Cast iron tank was used so that its wall do not deform during tests. A rough mild steel plate of 12 cm diameter and 2 cm thickness was used as footing. The sand was filled in the tank at required density using the raining technique at a height of fall of 30 cm. The points of intrusion of rods were equidistance from the center point of the tank. The sand was filled in the tank in 35% of relative density. Manually operated hydraulic jack is used for loading circular footing. The hydraulic jack was mounted on a self-reacting frame. Proving ring which is pre-calibrated was used to measure the load applied on the footing. The proving ring was placed between the hydraulic jack and the footing with a ball-bearing arrangement. Loading was applied using hydraulic jack of 10 T capacity, and the applied load was measured using a proving ring of 30 kN capacity. Two dial gauges of 0.01 mm least count were placed diametrically opposite to each other for measuring the settlement of plate. The test load was increased gradually till the plate settles at constant rate. The ultimate load-bearing capacity of the soil was calculated by dividing the value of total load on the plate by area of steel plate (Figs. 1, 2, and 3).

Fig. 1 Photograph of the test setup

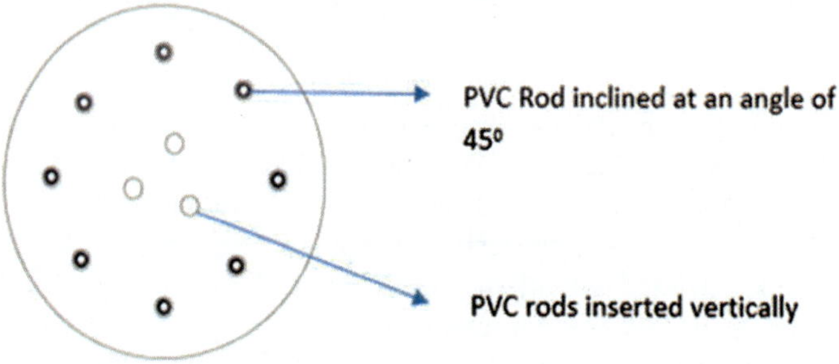

Fig. 2 Arrangement of PVC rods

2.3 Modeling and Analysis

Modeling and analysis is carried out using the ABAQUS 6.14.2-3D Software to study the behavior of sand when combination of vertical and inclined reinforcement with PVC rods are used and also of sand without any reinforcement. The experimental results were also cross checked through finite-element analysis using ABAQUS software. The size of the numerical model is kept equivalent to the trial arrangement, where boundaries were kept five times away from the center of the footing. A versatile completely plastic Mohr–Coulomb constitutive model was considered for sand, and a linear elastic model was picked for the PVC rods. The PVC rods are inserted into the sand using the interaction module. A 3D deformable part is created for the soil specimen, material properties are assigned and section is defined as solid, homogeneous. A uniform pressure load of 300 kPa is applied. The boundary conditions are given such that only vertical movements are allowed (Fig. 4 and Table 3).

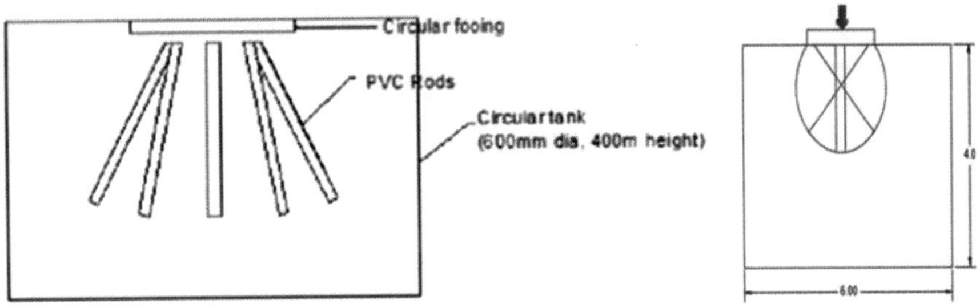

Fig. 3 Schematic diagram of orientation of reinforcing rods into the sand

Fig. 4 PVC rods arranged in two arrays of 1.5B length and 6 mm diameter

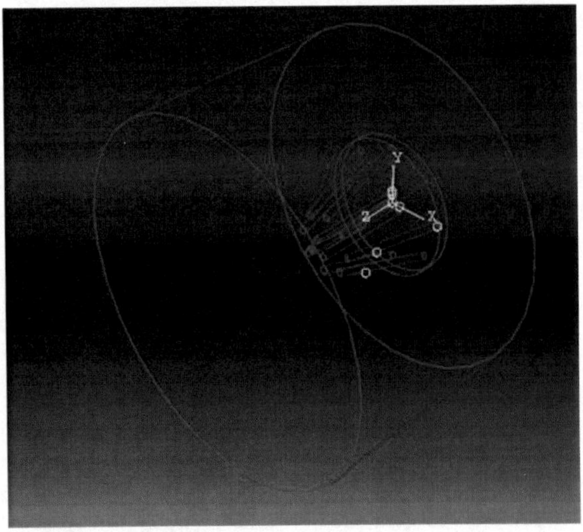

Table 3 Variables of study used

Type of test	Variables
Type of the test tank	Circular
Size of the test tank	0.6 m diameter × 0.4 m
Type of footing	Circular
Size of footing	0.12 m diameter × 0.02 m
Length of reinforcement	1.5B, 2B, 2.5B, 2.75B
Diameter of reinforcement	6, 8 mm

3 Results and Discussion

From the series of plate load test results, load–settlement curves are plotted. The improvement in performance due to the provision of PVC rods is significantly noticeable. The load–settlement response of reinforced and unreinforced sand bed is illustrated in the following figures. Here the reinforcement was applied in the vicinity of the pressure bulb. When the pressure bulb criteria was concerned, it can be seen that we can effectively reduce the cost of consumption of the PVC rods. Since PVC rod being the denser material nearly equal to the sand, we can completely utilize the soil for load transferring.

The main objective of this study is to determine the efficiency of the combined vertical and inclined reinforcement of the PVC rods. The results are presented with the reference of the unreinforced soil. For this purpose the pressure–settlement graphs are plotted. The ultimate bearing capacity was determined by double tangent method since there was no definite failure point. The bearing capacity improvement was represented by a factor called bearing capacity ratio (BCR) defined as bearing capacity of the reinforced soil to the unreinforced soil.

The length of insertion is varied as 1.5B, 2B, 2.5B where B is the base of the footing. The optimum length was found out. Comparing the curves of 6 mm diameter (dia.) and 8 mm dia. reinforcement, curve corresponding to 8 mm dia. bar has lesser slope than 6 mm dia. bar. It indicates that as the diameter of the reinforcing rod increases, the bearing capacity increases and the settlement reduces and enhances the overall stability.

3.1 Effect of Length and Diameter on Pressure–Settlement Response Using Plate Load Test

Length of the PVC rod is a governing factor to influence the behavior of reinforced sand bed. In the cost-effective point of view it helps to reduce the wastage of the reinforcement and thereby reduce the cost (Figs. 5 and 6).

From the figure it can be observed that the slope of the curve decreases with the addition of combination of inclined and vertical reinforcement in the vicinity of the

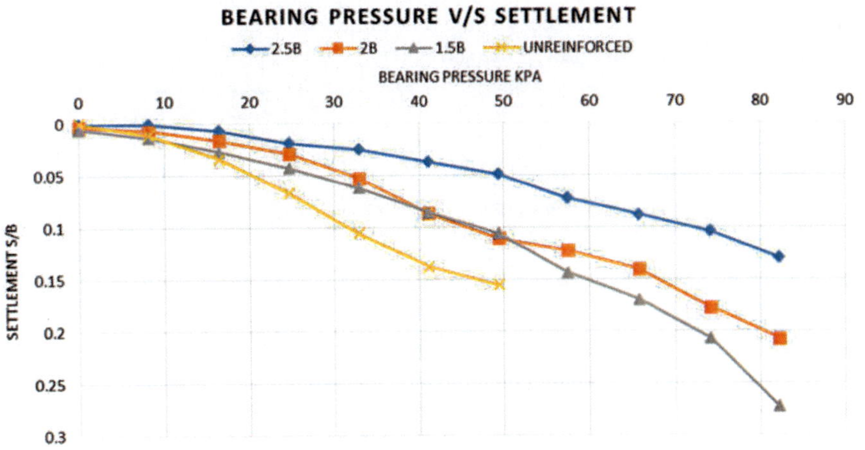

Fig. 5 Pressure–settlement response of footing on reinforced sand bed with varying length having diameter 6 mm

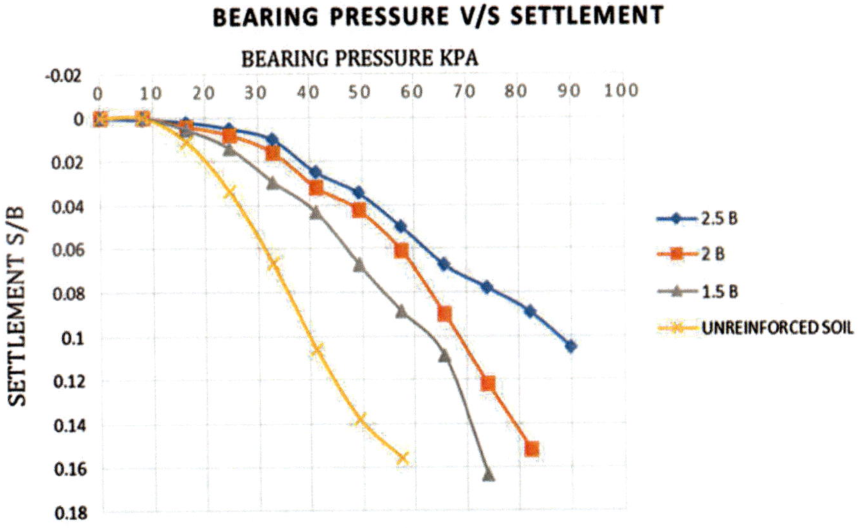

Fig. 6 Pressure–settlement response of footing on reinforced sand bed with varying length having diameter 8 mm

pressure bulb. It indicates that strength has improved and settlement has got reduced with the addition of reinforcement. This may be because of the fact that as a result of the introduction of a stiffer material into the soil the stiffness of the PVC rod–soil system increased, i.e., PVC rod into the loose–medium dense sand.

Diameter of the rod was selected as 6 and 8 mm. From the pressure–settlement graph, it becomes apparent that for the same spacing ratio and test condition reinforcement having thicker diameter gives lower magnitude of displacement. Increase in diameter increases the stiffness of the reinforcing element and reduces the lateral movement which in turn reduces the settlement.

3.2 *Effect of Length and Diameter on Pressure–Settlement Response Using Numerical Analysis*

On comparing the displacement contour of the PVC rod reinforced soil with varying length it can be seen that the embedded part of the reinforcement in the stable underlying soil increases as the length of the reinforcement increases, thus imparting resistance to the lateral movement. This improved resistance enhances the confinement of the soil, resulting in the decrease in the vertical movement.

On further studies it can be inferred that when a length greater than 2.5B was used there was no considerable improvement. Hence in order to reduce the cost of consumption, the optimum length of the PVC rod was taken as 2.5B, where B is the base of the footing.

Stress contours of reinforced and unreinforced soil are given below. It can be seen that when the reinforcement is added to the soil, the stress values get decreased (Figs. 7, 8, 9, 10, 11, 12, 13 and 14).

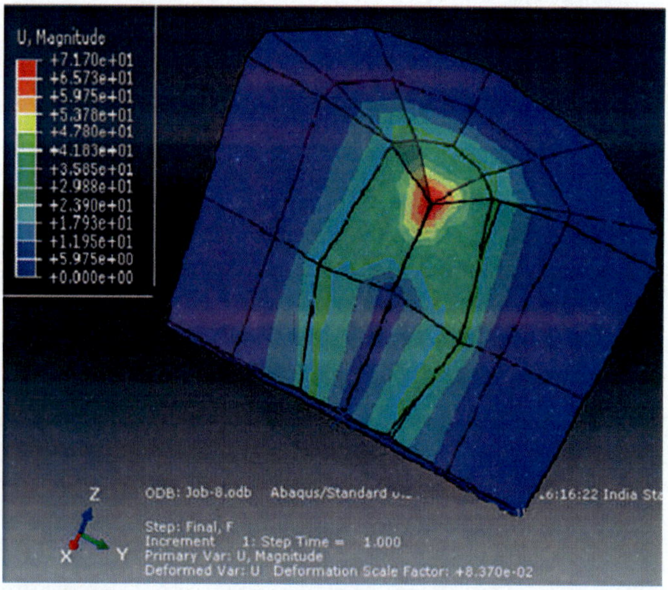

Fig. 7 Displacement contour of the Unreinforced soil

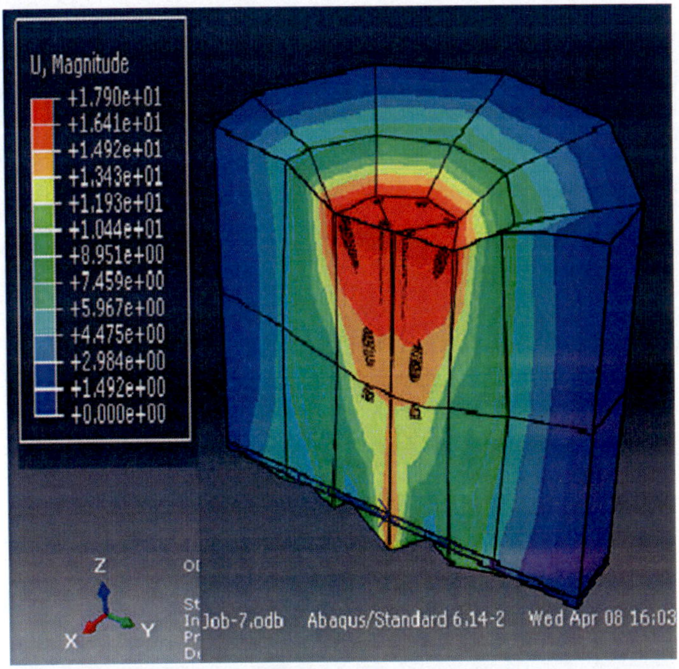

Fig. 8 Displacement contour of the PVC rod reinforced soil (1.5B)

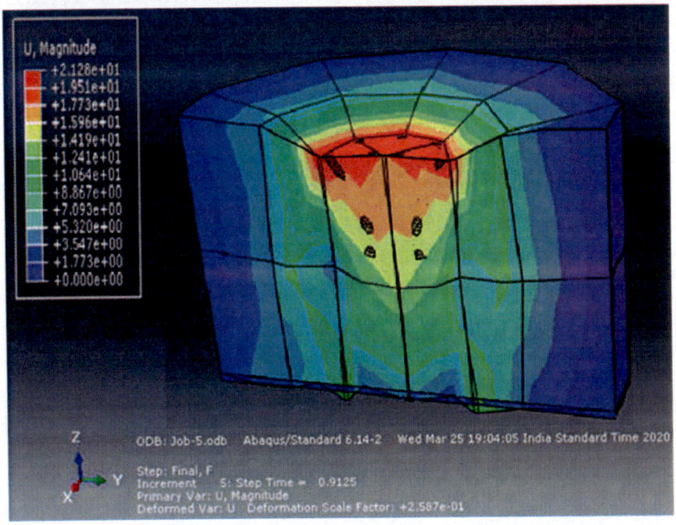

Fig. 9 Displacement contour of the PVC rod reinforced soil (2.0B)

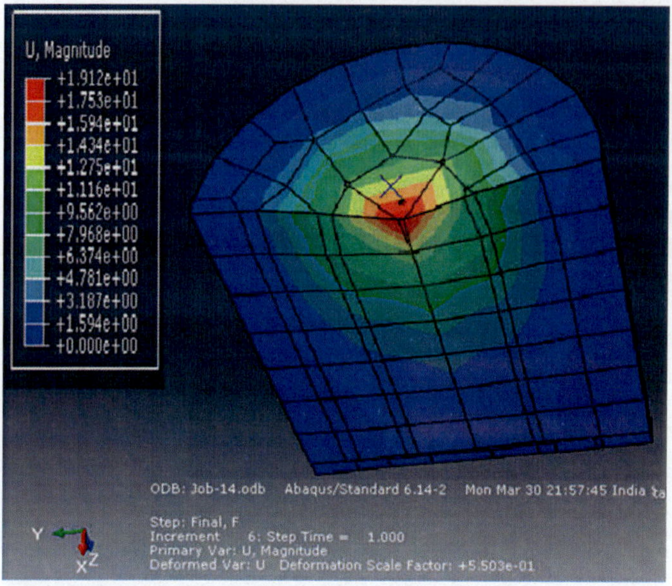

Fig. 10 Displacement contour of the PVC rod reinforced soil (2.5B)

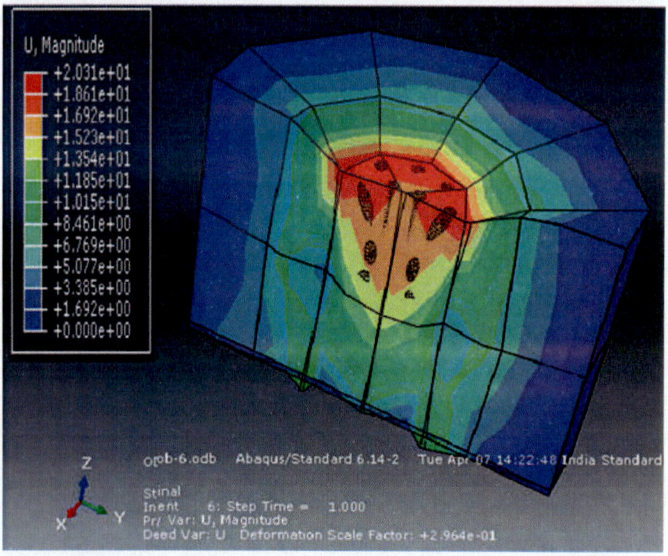

Fig. 11 Displacement contour of PVC rod reinforced soil (L = 1.5B, d = 8 mm)

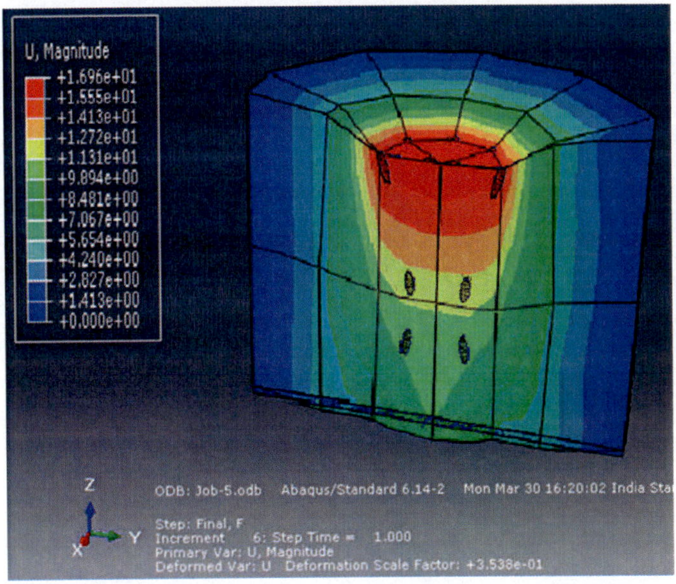

Fig. 12 Displacement contour of PVC rod reinforced soil (L = 2B, d = 8 mm)

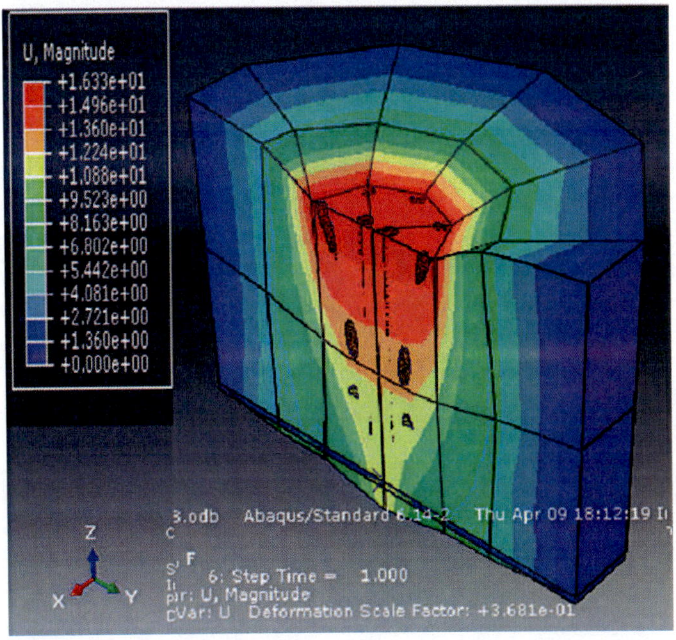

Fig. 13 Displacement contour PVC rod reinforced soil (L = 2.5B, d = 8 mm)

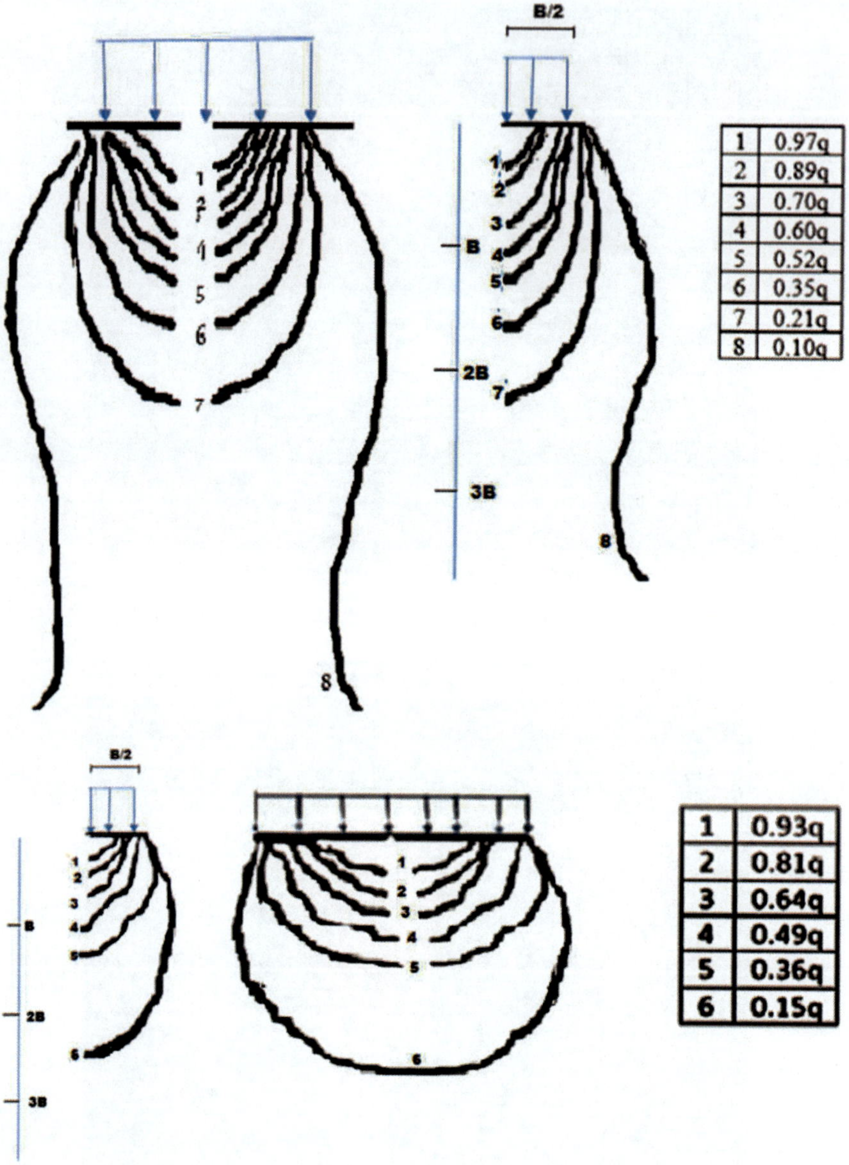

Fig. 14 Pressure bulb

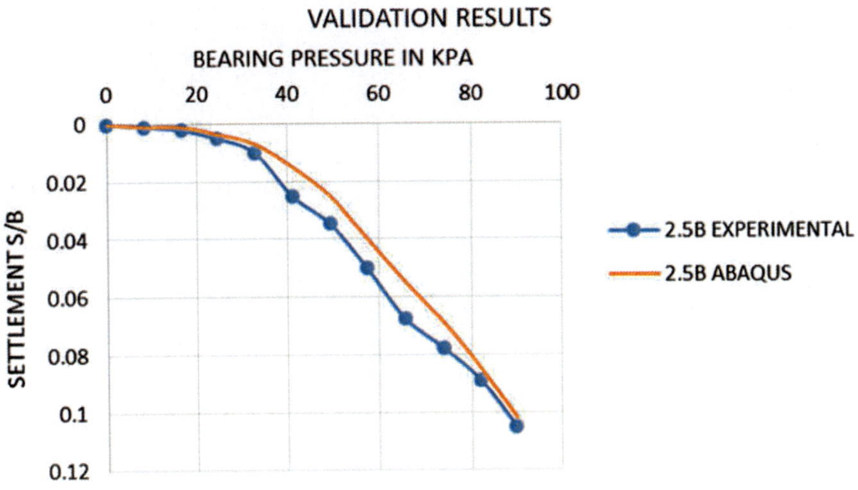

Fig. 15 Validation graph for reinforcement with PVC rods ABAQUS results and experimental results are compared

3.3 *Validation*

The numerical model is compared with the experimental values using stress–settlement curve for reinforcement with PVC rods. The results of the experimental program are analogous with the FEM analysis results. It is observed that the experimental and numerical simulation shows a similar trend (Fig. 15).

4 Conclusions

On evaluating the experimental program and numerical analysis results, the following conclusions about the response of footing placed over PVC rods–soil system are made:

1. PVC rod reinforcement in the vicinity of pressure bulb will enhance the strength and stiffness of soil.
2. Confinement of the soil can be achieved by placing the reinforcement with different length, diameter, spacing and extent of reinforcement leading to significant improvement in the load–settlement response.
3. As the length of the PVC rod in the vicinity of the pressure bulb increases, there is a considerable decrease in the deformation of the soil structure.
4. When the length increases considerably, there is no much effects on the deformation. Thus from the cost-effective point of view, the optimum length of PVC rod is 2.5B, where B is the base of footing.

5. FEM analysis with Mohr–Coulomb constitutive behavior for sand can be used to solve the bearing capacity problem of the footing placed over the PVC rod–soil system.
6. Settlement is reduced to 65.2% for 6 mm diameter PVC rod (two-array) compared to unreinforced soil.
7. Settlement is reduced to 67.1% for 6 mm diameter PVC rod (two-array) compared to unreinforced soil.
8. Various stress contours are plotted, and it is clear that when the reinforcement are added to the sand bed, at a particular depth the stress values get decreased.
9. Substantial improvement in the bearing capacity of the sand can be derived by reduced stress values. Thus it is proved that a combination of vertical and inclined reinforcement gives better results.

Reference

1. Srivastava, A., Chaitanya, R.: Load settlement response of footing placed over buried flexible pipe through a model plate load test. J. Geotech. Eng. ASCE **13**(2), 477–481 (2013)
2. Jha, J.N., Shukla, S.K.: Bearing capacity and settlement characteristics of sand subgrades with vertical reinforcement supporting a square footing. Int. J. Geosynth Ground Eng. **16**(1) (2015). Springer
3. Craig, H.B., Milind, V.K.:. Reinforcing sand with strips of reclaimed high density polyethylene. J. Geotech. Eng. ASCE 399–417 (1994)
4. Thushara, M., Johnson, S.A.: Experimental study on vertical and inclined reinforcement using PVC rods in sand. In: Proceedings of Indian Geotechnical Conference, Paper no: 83. Bengaluru (2018)

Strength and Deformation Characteristics of Subgrade Soil Stabilized with Plastic Covers

Monica Simon, A. Vismaya, P. K. Jayasree, and Leema Peter

1 Introduction

The fast growth of population and the resulting increase in construction activities led to the scarcity of land with desirable soil conditions. Nowadays, engineers were forced to use the problematic soils for construction purposes through various stabilization methods. Soil stabilization can be explained as the technique of improving the performance of soil by chemical, physical, or biological means, and thereby making it suitable for construction activities. This process can be achieved by mixing soil with several admixtures like lime, cement, calcium chloride, fly ash, and bituminous materials but they are highly expensive. The stabilization using lime and cement results in various environmental issues and are not adopted nowadays. Due to the increasing cost and harmful effects produced by the additives, alternative methods for soil stabilization are to be found.

Plastic becomes an integral part of our lives. It is produced on a gigantic scale all over the world and its global production exceeds 150 million tons per year [1]. The plastic wastes being nonbiodegradable are mainly disposed by burning and that releases toxic gases harmful to the health. Therefore, the plastic products should be disposed properly for better future. Recycling of plastics is a promising alternative for plastic waste management, but it can be done up to 3–4 times. The impact of using plastic waste in civil engineering constructions was analyzed by various researchers. Plastic waste was mixed with cement [2] to produce sturdy and flexible concrete slabs. In India, it has now become a rule for all road manufacturers to use plastic waste, along with bituminous mixes, for road construction. Reclaimed high-density polyethylene (HDPE) which resists tensile force was used to reinforce locally available sandy soil to enhance its engineering properties [3]. Research studies were conducted to check

M. Simon · A. Vismaya · P. K. Jayasree (✉) · L. Peter
College of Engineering Trivandrum, Trivandrum 695016, Kerala, India
e-mail: jayasreepk@cet.ac.in

C. N. V. Satyanarayana Reddy et al. (eds.), *Ground Improvement and Reinforced Soil Structures*, Lecture Notes in Civil Engineering 152,
https://doi.org/10.1007/978-981-16-1831-4_12

for the alternative of stabilizing soils using waste plastics in the form of bottles and bags. In order to find out the consolidation characteristics, experiments were performed on clayey soil and it was found that the plastic waste stabilized specimen obtains a lower initial void ratio [4]. Higher values of CBR were achieved upon adding plastic strips of particular size and amount to locally available sand at 4% plastic content [5].

Due to the harmful effects caused by the industrial waste, many developing countries were forced to use them in the road construction, and it is based on technical, economical, and ecological considerations. The characteristics of clayey soil were enhanced by the introduction of industrial wastes combined with lime and cement. The decrement of the friction angle was attained on the inclusion of polypropylene in the form of fibers to silty soil treated with lime and rice husk ash [6], while cohesive nature of the mixture boosted initially and then dropped with further addition of fiber, and the optimum value was obtained at 0.4% fiber content. The utilization of waste polyethylene material for soil stabilization can be considered as an eco-friendly solution for plastic waste management [7]. The effectiveness of HDPE plastic strips was studied by conducting experiments on both solid and perforated strip reinforced sandy soils and the results indicate that the greater improvement in the friction angle happens at 0.1% strip content, length of 15 mm and with perforations of 2 mm diameter [8]. It was also seen that the strength reduces considerably beyond the optimum length of plastic strips. The incorporation of plastic bottle strips reduces the compressibility of soil along with higher compressive strength [9]. The use of shredded plastic covers in weak soil has greater influence on strength compared to that of plastic in strip form. The inclusion of finely shredded plastic waste shows greater improvement in strength compared to that of coarsely shredded plastic waste [10].

In this study, the potential of plastic covers on the strength and deformation characteristics of subgrade soil is investigated. If unconventional materials are used to replace subgrade, it becomes difficult to predict the pavement performance [11]. Therefore, in the present study the utilization of plastic cover in soil reinforcement was analyzed by performing unconfined compressive strength (UCC) tests by the uniform mixing of plastic strips with the soil at different percentages. The UCC tests were also conducted by changing the thickness and aspect ratio (AR) of the plastic strips added.

2 Experimental Studies

2.1 Materials Used

Waste plastic covers used for the investigation were locally collected within the thickness ranges of 15, 30, and 45 μm. Wide width tensile strength test [12] as per ASTM D 4595 were carried out to find the tensile strength of plastic covers, and

Table 1 Properties of plastic waste

Thickness (μm)	Tensile stiffness (kN/m)
15	0.5
30	0.9
45	1.2

Table 2 Properties of soil

Property	Value
Specific gravity	2.56
Gravel (%)	1
Sand (%)	40
Silt (%)	32
Clay (%)	27
Liquid limit (%)	53
Plastic limit (%)	37
Plasticity index (%)	16
Shrinkage limit (%)	22
MDD (g/cc)	1.6
OMC (%)	20.5
UCS (kN/m^2)	49
CBR (%)	3

the results obtained are presented in Table 1. The soil was also collected locally which was found to be clayey in nature upon visual inspection. The geotechnical characteristics of soil are summarized in Table 2. The soil was classified as MH [13].

2.2 *Methodology*

The plastic covers were cut into strips of size 10 mm × 10 mm, aspect ratio (AR) = 1 and 20 mm × 10 mm (AR = 2) using scissors and measuring ruler shown in Figs. 1 and 2. At the initial stage, the UCC tests were performed by adding plastic strips (AR = 1) at different amounts ranging from 0.1% to 0.7%. The tests were done for plastic strips of thickness 15, 30, and 45 μm. In order to find the effect of aspect ratio tests were also done by changing the size of strips for all the thickness ranges.

Fig. 1 Plastic strips with AR = 1

Fig. 2 Plastic strips with AR = 2

3 Results and Discussion

3.1 Effect of Thickness and Aspect Ratio of Plastic Strips on Strength Characteristics

The stress–strain relationship of soil obtained by the inclusion of plastic strips of varying thicknesses 15, 30, and 45 μm are presented in Figs. 3, 4, and 5, respectively. From the graphs it is clear that introduction of plastic strips in soil increased the peak stress. The UCS of natural soil increases from 49 to 152 kPa for 0.5% plastic mixed soil. The UCS value is found to be decreasing as the percentage of plastic increased

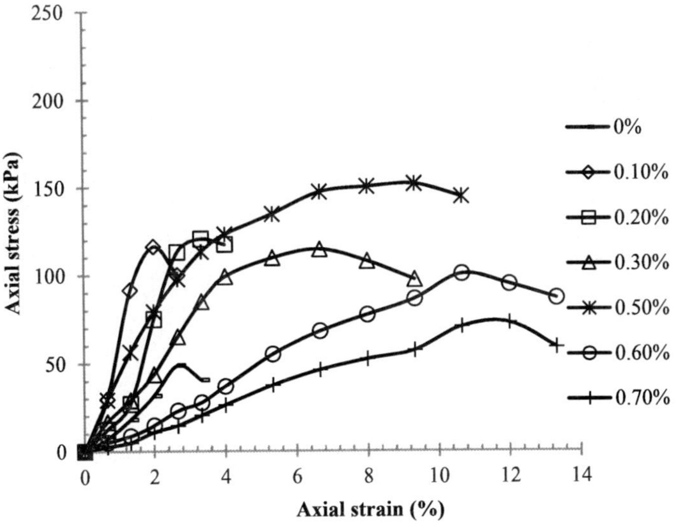

Fig. 3 UCC results of soil treated with plastic strips of thickness 15 μm (AR = 1)

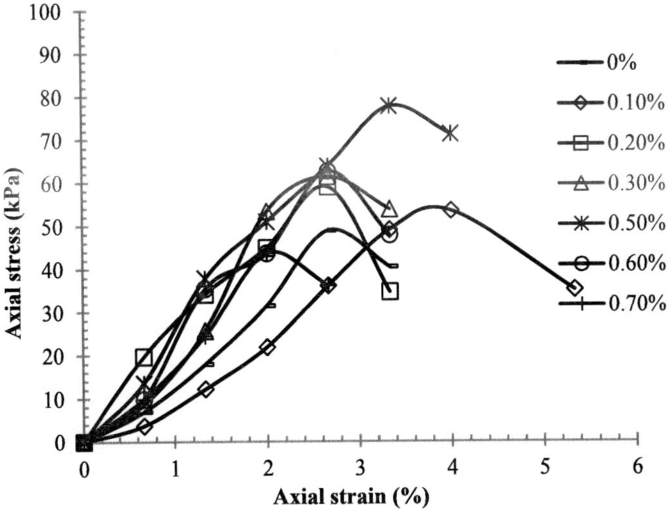

Fig. 4 UCC results of soil treated with plastic strips of thickness 30 μm (AR = 1)

beyond 0.5%. Similarly the maximum value of UCS obtained for plastic strips with thickness 30 and 45 μm are 78 and 92 kPa, respectively, at 0.5% plastic content. The strength increases up to an optimum percentage for all the thickness ranges and then decreases. It can also be observed from these figures that strain to reach peak stresses increases with plastic content.

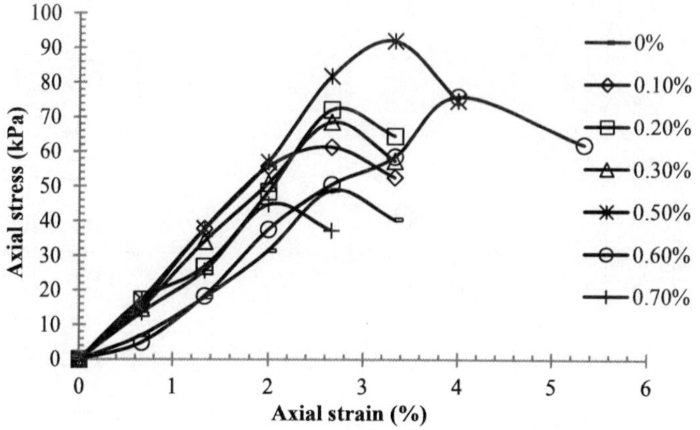

Fig. 5 UCC results of soil treated with plastic strips of thickness 45 μm (AR = 1)

It has been already examined that the plastic strips having a thickness of 15 μm (AR = 1) is having the peak value of compressive strength. This is three times more than that of an unreinforced soil. Further tests were also conducted by changing the size of strips to 20 mm × 10 mm (AR = 2) for all the thickness ranges. The results obtained with the change in aspect ratio of plastic strips are shown in Figs. 6 and 7. From the graphs shown it can be inferred that the optimum improvement in UCS is obtained at 0.5% for both the thickness ranges.

The increase in compressive strength may be as a result of increase in total contact area between the soil and the plastic strips. The addition of plastic strips consequently

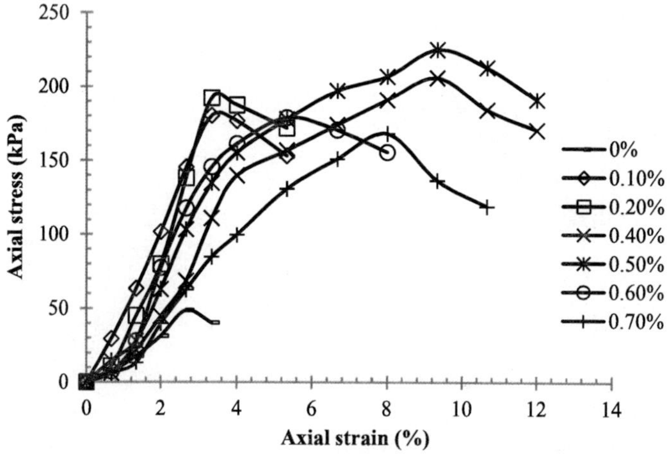

Fig. 6 UCC results of soil treated with plastic strips of thickness 15 μm (AR = 2)

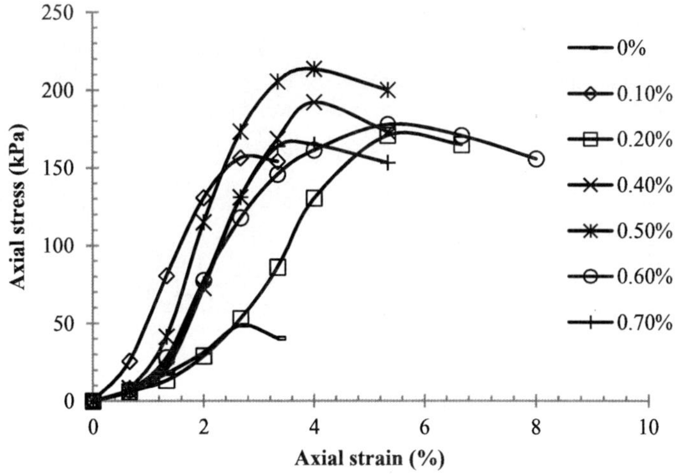

Fig. 7 UCC results of soil treated with plastic strips of thickness 30 μm (AR = 2)

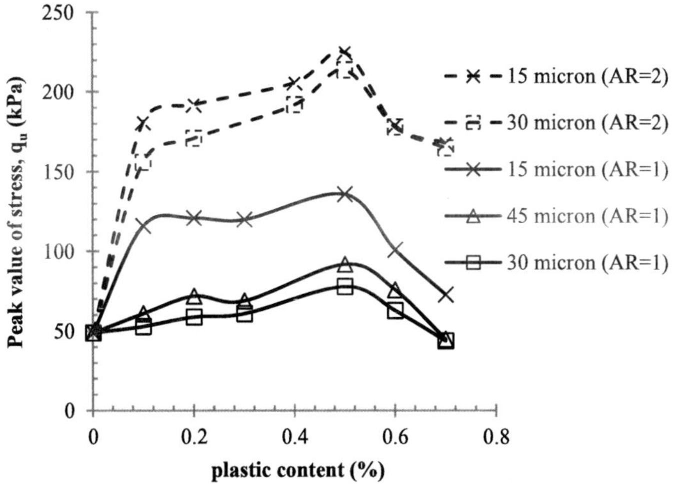

Fig. 8 Variation of q_u with different percentage of plastic

enhanced the friction between the soil and the plastic, which leads to tensile stresses in the plastic material which absorbs the more load above the normal soil's capacity.

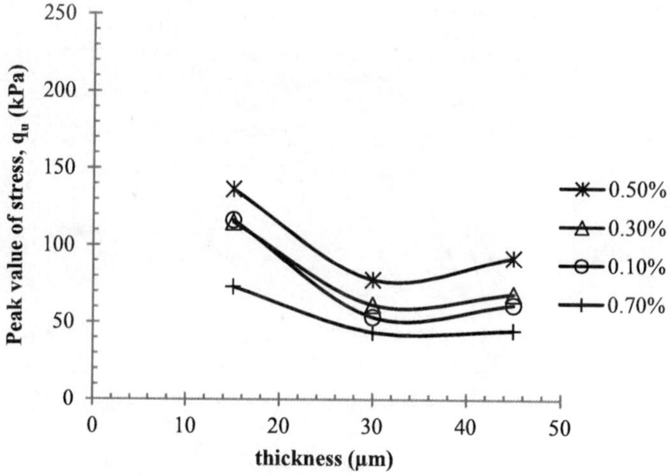

Fig. 9 Variation of q_u with thickness (AR = 1)

3.2 *Comparison of Results*

The variation in UCS with the change in plastic content was presented in Fig. 8. The maximum improvement in UCS can be seen by addition of plastic strips of thickness 15 μm for both the aspect ratios. The increase in thickness of plastic strips results in decrease in compressive strength. This may be because of decrease in friction between the soil and the plastic due to the decrease in contact between particles. When the aspect ratio of plastic strips goes on increasing, notable improvement in UCS can be seen, i.e., as the strip length increases keeping the width constant, higher strength was achieved. However, beyond the optimum length, there will be reduction in the compressive strength. This may be due to the increased contact between the plastic strips which leads to the decrease in soil–plastic interaction.

The variation in UCS with the change in thickness of plastic was shown in Fig. 9. The optimum percentage of plastic strips for soil is 0.5% for all the thickness ranges. Further increase in plastic content causes strength reduction. As the plastic content increases beyond 0.5%, UCS decreased due to the increased interaction between the plastic strips due to more overlapping of plastic and it results in reduced soil plastic interaction.

4 Conclusions

1. The maximum value of UCS is obtained by adding 0.5% plastic with aspect ratio 2 and thickness 15 μm.
2. The UCS value increases with the increase in aspect ratio of plastic strips.

3. The strain to reach peak stresses increases with plastic content.
4. The nature of soil changes from brittle to ductile during the inclusion of plastic strips.

References

1. Peddaiah, S., Burman, A., Sreedeep, S.: Experimental study on effect of waste plastic bottle strips in soil improvement. Geotech. Geolog. Eng. J. (2018). Springer
2. Thorneycroft, J., Ball, R.J.: Performance of structural concrete with recycled plastic waste as a partial replacement for sand. Construct. Build. Mater. (2017). https://doi.org/10.1016/j.conbuildmat.2017.11.127
3. Choudhary, A.K., Jha, J.N., Gill, K.S.: A study on CBR behavior of waste plastic strip reinforced soil. Emir. J. Eng. Res. **15**(1), 51–57 (2010)
4. Okoro, C., Vogtman, J., Yousif, A., Agnaou, M., Khoury, N.: Consolidation characteristics of soils stabilized with lime, coal combustion product and plastic waste. Geo-frontiers 2011, ASCE, 1202–1209 (2011)
5. Choudhary, A.K., Jha, J.N., Gill, K.S.: Utilization of plastic wastes for improving the sub-grades in flexible pavements. J. Paving Mater Pavement Anal. ASCE, 320–325 (2010)
6. Muntohar, A.S., Widianti, A., Hartono, E., Diana, W.: Engineering Properties of Silty Soil Stabilized with Lime and Rice Husk Ash and Reinforced with Waste Plastic Fiber. J. Mater. Civ. Eng. ASCE **25**(9), 1260–1270 (2013)
7. Ilies, N.M., Circu, A.P., Nagy, A.C., Ciubotaru, V.C., Kisfaludui-Bak, Z.: Comparative study on soil stabilization with polypropylene waste materials and binders. Procedia Eng. Elsevier **181**(2017), 444–451 (2017)
8. Kalumba, D., Chebet, F.C.: Utilization of polyethylene (plastic) shopping bags waste for soil improvement in sandy soils. In: Proceedings of the 18th International Conference on Soil Mechanics and Geotechnical Engineering, pp. 3223–3226 (2013)
9. Sivakumar Babu, G.L., Chouskey, S.K.: Analytical model for stress-strain response of plastic waste mixed soil. J. Hazard. Toxic Radioact. Waste **16**(3), 219–228 (2012)
10. Jegan, B.K.A., Salini, U.: Use of shredded waste plastic in stabilization of subgrade soil. In: Proceedings of the International Conference on Geotechnics for High Speed Corridors, pp. 524–526, Trivandrum (2019)
11. Lee, S.W., Fishman, K.L.: Waste products as highway materials in flexible pavement system. J. Transp. Eng **119**(3), 433–449 (1993)
12. ASTM: Standard test method for tensile properties of geotextiles by the wide-width strip method, D4595, West Conshohocken, PA (2011)
13. IS: 1498: Classification and Identification of Soils for General Engineering Purposes, Bureau of Indian Standards, New Delhi (1970)

Enhancement of Soil Stabilization by Electrokinetic Process

V. K. Stalin and S. Balasundari

1 Introduction

The increase in demand for land space all over the world due to rapid urbanization has increased the need to construct on soft soil grounds, which were considered unsuitable for construction, thereby causing excessive settlement and failure of the structure. To stabilize these soils there are many ground improvement techniques and stabilization techniques. Conventional remediation methods have been known successful in minimizing several damages; however, they are expensive, time-consuming, and may be difficult to implement in some existing structures. In this regard, the electrochemical or electrokinetic (EK) treatment method can be used as an alternative soil treatment method for remediation of those deficiencies.

Ranjitha and Manjari Blessing [5] have attempted to evaluate the use of electrokinetic stabilization (EKS) as an effective method to stabilize soft clayey soil. EKS of soft soil was done by using graphite electrodes as anode and cathode in the first method, and similarly, a combination of graphite and zinc electrodes as anode and cathode in the second method. The results show that the liquid limit was reduced up to 37.27 from 67.45%, plastic limit increased from 36.11 to 37.4%, plasticity index decreased from 31.34 to 13.73%, and free swelling index was decreased to 2.9% from 26%. The results showed that the graphite–graphite electrode combination was efficient when compared to graphite–zinc electrode combination. Samidurai et al. [3] studied the effects of EK treatment at different voltages at different intervals of time. When the voltage was increased the water collected at the cathode was also increased. The test results indicated that the compressive and the shear strength of the soil has considerably improved after the electrokinetic stabilization treatment of the soil. Hongtao Fu et al. [2] conducted laboratory experiments to determine the

V. K. Stalin · S. Balasundari (✉)
Department of Civil Engineering, College of Engineering, Anna University, Chennai 600025, India

C. N. V. Satyanarayana Reddy et al. (eds.), *Ground Improvement and Reinforced Soil Structures*, Lecture Notes in Civil Engineering 152,
https://doi.org/10.1007/978-981-16-1831-4_13

optimal voltage gradient of electrokinetic consolidation of Wenzhou clay slurry with a small amount of fine sand. The results showed that a steady current was achieved under relatively lower voltage gradients (0.5 V/cm), while higher voltage gradients (1 V/cm) reduced the drainage time. Voltage gradients greater than 1.0 V/cm had no significant effect on the drainage rate.

Wayne et al. [4] investigated the effects of high-voltage electric fields on the properties of disturbed marine sediment. The primary experimental data indicated that high-voltage electrokinetics increased the undrained shear strength of marine sediment up to 267% and the pullout resistance of steel plates up to 88%. Geeta and Rakaraddi [1] aimed to evaluate the use of electrokinetic stabilization as an effective method to strengthen soft clayey soil. After electrokinetic processing, the soil was tested for the shear strength test. Three different anode and cathode electrokinetic systems were employed in this study, under constant voltage of 30 V for periods of 3, 7, and 14 days. The chemicals sodium silicate and calcium chloride at different concentrations of 0, 1, and 1.5 mol/l were poured at anode and cathode, respectively. The combined effect of these processes together with various geochemical reactions alters the chemical composition of the soil porous medium and thereby alters the physicochemical properties of the soil. The experimental results suggest the potential of developing an electrokinetic treatment technique to stabilize the shear strength of expansive soils effectively and efficiently. Maheshwari and Chandrakanth [6] studied the influence of different arrangement of electrodes on the shear strength of the soil. The various parameters that have varied throughout the test are voltage (20 V, 30 V), spacing (10 cm, 15 cm), and different arrangement of electrodes such as three-anode–one-cathode (3A-1C), two-anode–two-cathode (2A-2C), one-anode–one-cathode (1A-1C), and three-cathode–one-anode (3C-1A). The variation in current and resistance was recorded and the results indicated that the shear strength of soil samples improved considerably after electrokinetic treatment. Before the start of the test, UCC strength was found to be 8.2 kN/m2. After electrokinetic treatment, the maximum UCC strength was found out in 3A-1C arrangement (28.64 kN/m2). In the case of a 2A-2C arrangement, the UCC strength increases from 8.2 to 22.57 kN/m^2.

In the present study, an attempt is made to carry out the electrokinetic remediation technique to improve and stabilize the soft clay and enhance the rate of dewatering of soft clay using geotextiles placed vertically. Also, the efficiency of dewatering using plate electrodes and strip electrodes is compared for varying voltage and initial moisture content of the soil.

2 Materials

Soil sample required for the experimental study was collected from Saidapet (13.0297°N, 80.2380°E), Chennai and tested for its index properties, as shown in Table 1. Based on the plasticity characteristics, the soil is classified as "CH" type (clay of high compressibility or high plasticity).

Table 1 Properties of soil

S. no	Properties	Value
1	Liquid limit, w_l (%)	60
2	Plastic limit, w_p (%)	30
3	Plasticity index, I_p (%)	30
4	Specific gravity, G	2.71
5	Free swell index (%)	100
6	Clay percentage (%)	71
7	Silt percentage (%)	22
8	Gravel percentage (%)	7
9	IS soil classification	CH

The geotextile adopted is a non-woven type made from high resistance staple fiber polypropylene with needle punching. Needle-punched non-woven geotextiles have far superior flow rates with less tendency to clog than heat-bonded non-woven geotextiles, and therefore ideal for drainage and stabilization. It is available in the market. The permeability of the geotextile was found to be 0.005 cm/s.

3 Methodology

3.1 Electrokinetic Cell Setup

To study the occurrence of electrokinetic phenomena in soils, an electrokinetic cell is designed and fabricated. The cell includes the components like tank open at the top, electrodes, voltmeter, ammeter, and AC to DC transformer. Figures 1 and 2 show the schematic representation of the cylindrical and rectangular electrokinetic cell.

3.2 Procedure for Dewatering of Remolded Clay Through Electrokinetic Cell

The following procedure was adopted in the EK cell for dewatering of remolded clay for varying parameters such as IMC, voltage, and electrode material.

1. To start with, the tanks (namely, rectangular and cylindrical tanks) are filled with equal volume of soil sample by hand remolding. The electrodes are fitted in the box with the help of the slits provided and the spacing between the electrodes is maintained a constant throughout the study.
2. At the cathode, provision is given to collect water during the process of passage of voltage across the soil sample. The electrodes are connected to the transformer

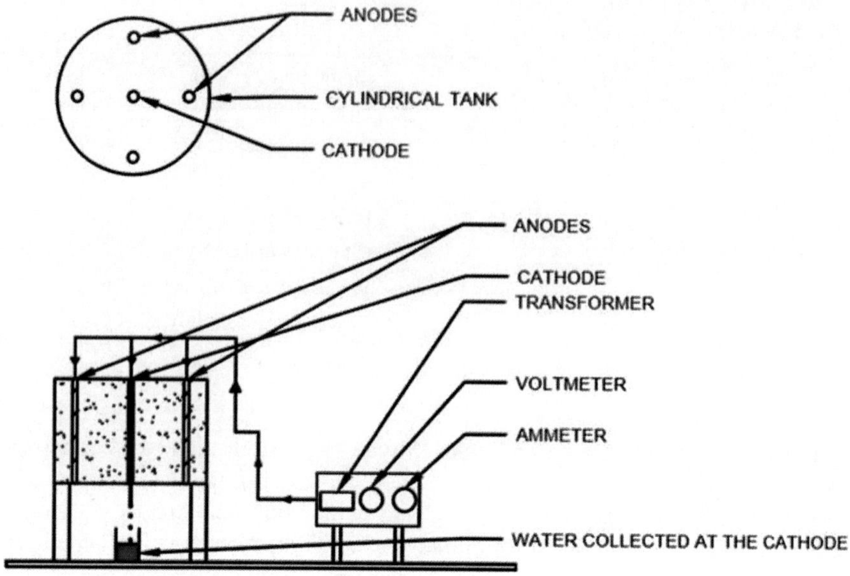

Fig. 1 Schematic representation of cylindrical electrokinetic cell (plan and sectional view)

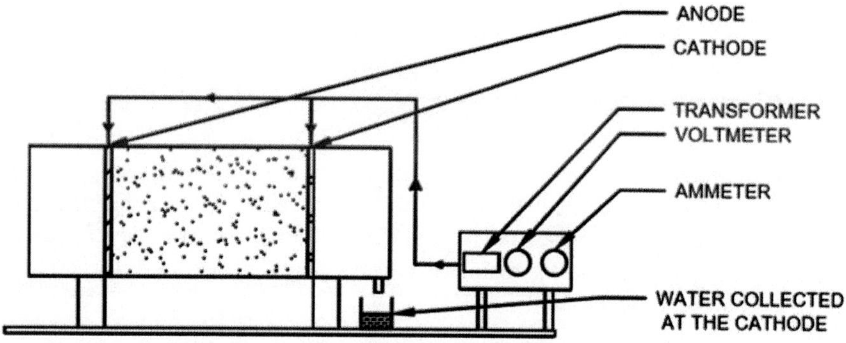

Fig. 2 Schematic representation of rectangular electrokinetic cell (sectional view)

setup which converts the AC to DC. On applying the required voltage across the electrodes for a specific time, the water is simultaneously collected at the cathode.

3. The volume of water collected is noted at a regular time interval till the sample reaches a steady-state attaining no more water coming out of sample at the cathodic point. After passing the current, at the end of the test, the soil sample is removed and is tested for its final moisture content, and vane shear strength test is performed in the dewatered soil.

4. The experiment is then carried out by placing the geotextile vertically in the soil between the anode and the cathode penetrating for the entire depth, and the electrokinetic dewatering process is performed.
5. The time-cumulative volume of water curves is compared for different voltages, initial moisture content, and different material of electrode (Plate electrode and strip electrode). The duration of the test depends upon the time taken for the water collected at the cathode to achieve a steady state.

4 Results and Discussion

The volume of water collected at the cathodic points at a regular time interval until the completion of the tests in EK cell such that there is no more withdrawal of water at the cathode are analyzed in the following section for varying voltage and initial moisture content and with the inclusion of geotextile.

4.1 Effect of Voltage

Voltage is one of the important parameters in the way that it solely decides the time to attain a particular rate of dewatering. Logically, one can say that applying higher voltage yields higher dewatering and it is proven in the case of laboratory studies, but upon the application of higher voltage for a longer time, results in the shrinking of soil mass at the interfaces of electrokinetic box. High voltage even though increases the rate of dewatering, however crack propagation at early stage is going to hamper the efficiency of dewatering. Especially at 90 V, irrespective of the initial moisture content, material and type of electrode, the soil mass shrinks to a greater extent which validates the above discussion. Hence in a real-time application, an optimum voltage has to be applied which achieves a higher rate of dewatering by spending lesser energy.

Figure 3 represents the volume of water collected at the cathode with time for 60 V using plate electrodes and strip electrodes. Similarly, graphs are obtained by applying 90 V. It can be inferred from the graphs that, for the same IMC, as the voltage increases, the time taken to attain a steady state decreases. From the comparison between the electrode types, the plate electrode yields more volume of water than the strip electrode for a given volume of soil taken and for a given voltage and IMC.

Similar trends could be found on experiments conducted for voltages of 60 V and 90 V with the inclusion of geotextile placed vertically in between the anode and cathode conducted for plate electrodes and strip electrodes and the graphs for the above are shown in Fig. 4. The time taken for the water to attain a steady rate for varying voltage and IMC is shown in Table 2.

Table 2 shows the increased time taken to achieve constant volume rate of water, and Table 3 shows the cumulative volume of water collected at the cathode with

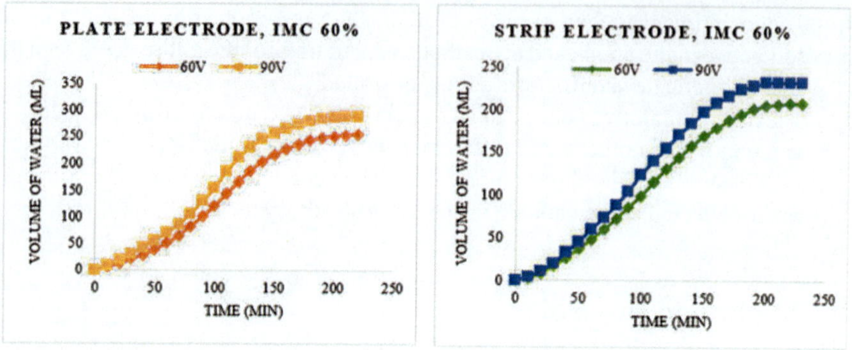

Fig. 3 The cumulative volume of water collected at the cathode with time for 60% IMC using plate electrodes and strip electrodes

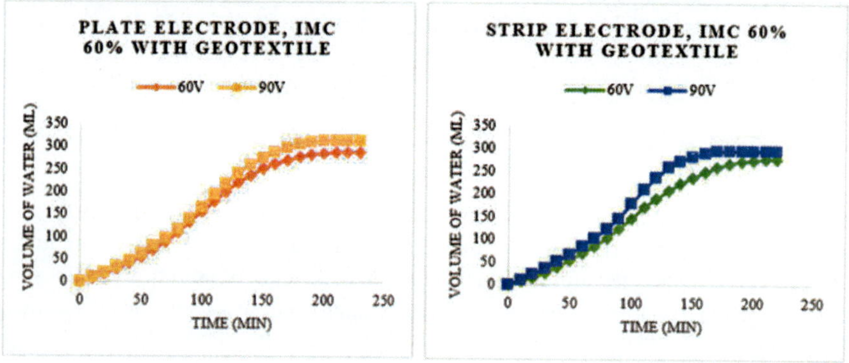

Fig. 4 The cumulative volume of water collected at the cathode with time for 60% IMC using plate electrodes and strip electrodes with the inclusion of geotextile

Table 2 Time lapse for achieving a steady volume of water for 60 and 70% IMCs of soil for varying voltage

Voltage (V)	Type of electrode	Time taken to attain a steady volume of water at the cathode (min)			
		60% IMC		70% IMC	
		Virgin clay	Clay with geotextile	Virgin clay	Clay with geotextile
60 V	Plate electrode	230	220	300	290
	Strip electrode	220	210	320	280
90 V	Plate electrode	220	210	270	250
	Strip electrode	210	170	310	260

Table 3 Cumulative volume of water collected for 60 and 70% IMCs of soil for varying voltage of 60 and 90 V

Voltage (V)	Type of electrode	The cumulative volume of water collected (ml)			
		60% IMC		70% IMC	
		Virgin clay	Clay with geotextile	Virgin clay	Clay with geotextile
60 V	Plate electrode	253	287	310	338
	Strip electrode	208	275	270	314
90 V	Plate electrode	289	313	330	367
	Strip electrode	233	295	300	341

the increasing voltage and initial moisture content. This can be attributed to the fact that with the increase in IMC of soil, the volume of water available in the soil to be collected at the cathode is also high. It can be observed that as the voltage increases, for the same IMC of soil, the time taken to achieve a steady rate of volume decreases because water drains more quickly at the cathode.

From Fig. 5, it is clear that the cumulative volume of water collected at cathode increases with increase in applied voltage for given initial moisture content. The cumulative volume of water collected upon the application of 90 V is found to increase by around 10% more than the volume of water collected on applying 60 V. Further, an increase of 8–16% of the volume of water was achieved by the inclusion of geotextile on placing vertically between the electrodes.

Table 4 shows the variation in the final moisture content of soil on varying the initial moisture content for varying voltages. The comparison is made for plate electrodes and strip electrodes for varying voltages of 60 V and 90 V in virgin clay and

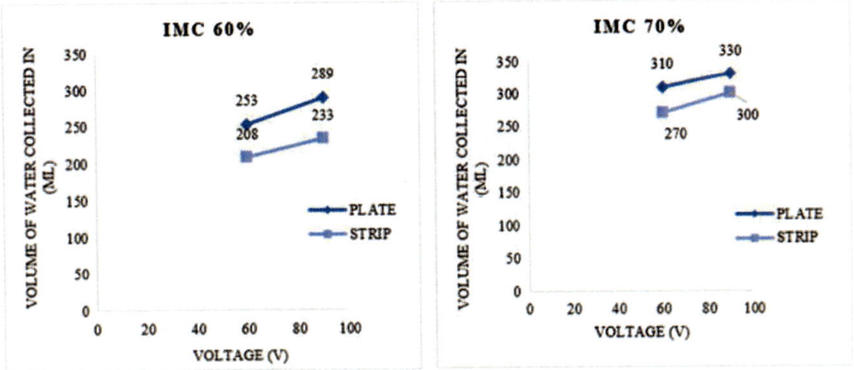

Fig. 5 Cumulative volume of water collected at the cathode for IMC of 60% for varying voltage

Table 4 Effect of voltage, IMC, and type of electrode on the FMC

Voltage (V)	Initial moisture content IMC (%)	Description	Final moisture content FMC (%)	
			Plate electrode	Strip electrode
60	60	Virgin clay	39.2	48
		Clay with geotextile	35.1	45.5
	70	Virgin clay	44.2	54.2
		Clay with geotextile	39.8	50.7
90	60	Virgin clay	36.8	46.8
		Clay with geotextile	32.1	43.1
	70	Virgin clay	40.1	51.5
		Clay with geotextile	37.7	47.1

with the inclusion of geotextile. It can be observed that the final moisture content on applying 90 V has dropped more when compared to that of applying 60 V and the placement of geotextile vertically between the electrodes has further increased the percentage reduction in water content. This decrease in the FMC of soil is more pronounced for 70% IMC than that of 60% IMC of soil for increasing voltage values. The current cannot flow effectively between the electrodes (through the soil mass) for lower IMCs of soil. This may be the reason why variation in FMC is higher for higher IMC of soil.

Also, from Table 5 it is evident that the percentage reduction in water content is more in the case of the rectangular tank using plate electrodes than cylindrical tank using strip electrodes for a given volume of soil and given initial moisture content with and without the inclusion of geotextile. The percentage reduction in moisture content using plate electrodes ranges from 34.67 to 46.14% and that of strip electrode ranges from 20 to 32.71%. This may be due to the increased surface area of the plate electrodes which are in contact with the soil than compared to the surface area of the strip electrodes. The presence of geotextile further increases

Table 5 Percentage reduction in water content

Voltage (V)	Initial moisture content IMC (%)	Description	Reduction in water content (%)	
			Plate electrode	Strip electrode
60	60	Virgin clay	34.67	20
		Clay with geotextile	41.5	24.17
	70	Virgin clay	36.86	22.57
		Clay with geotextile	43.14	27.57
90	60	Virgin clay	38.67	22
		Clay with geotextile	46.5	28.17
	70	Virgin clay	42.71	26.43
		Clay with geotextile	46.14	32.71

Table 6 Vane shear strength of soil with varying voltage

Voltage (V)	Initial moisture content IMC (%)	Description	Vane shear strength (kPa)	
			Plate electrode	Strip electrode
60	60	Virgin clay	12.5	10.51
		Clay with geotextile	13.78	11.4
	70	Virgin clay	7.68	6.42
		Clay with geotextile	8.7	7.8
90	60	Virgin clay	13.9	11.64
		Clay with geotextile	14.2	12.89
	70	Virgin clay	8.82	7.52
		Clay with geotextile	9.6	8.4

the efficiency of dewatering as they possess a higher permeability than soft clay; they form as drainage paths for dewatering to take place at a faster rate and more effectively thereby reducing the moisture content of the soft clay to a greater extent in a shorter duration, which results in earlier consolidation of the clay strata.

On the application of a voltage across the soil mass, the soil gets dried up at anode. The portion of the soil that has dried up is removed and the soil sample is subjected to vane shear strength test. Table 6 shows the variation of vane shear strength at the end of the EK test for varying voltage, initial moisture content of the soil, and type of electrode. Before the start of the test, the vane shear strength of soil was found to be around 11 and 5 kPa for initial moisture content of 60 and 70%, respectively. After the completion of the test as indicated by the constant cumulative volume of water, the sample was tested for vane shear strength and is tabulated below. A maximum vane shear strength of 14.2 and 12.89 kPa was obtained upon the application of 90 V in soil containing IMC of 60% using plate electrodes and strip electrodes, respectively, with the inclusion of geotextile. For soil with 70% IMC, the maximum vane shear strength was found to be 9.6 and 8.4 kPa using plate electrodes and strip electrodes, respectively, with the inclusion of geotextile on the application of 90 V between the electrodes.

4.2 *Effect of Initial Moisture Content (IMC)*

Initial moisture content (IMC) is one of the important factors that decides the range of all the other parameters to be involved in the EK-based dewatering process. In fact, it is IMC and amount of fine-grained particles responsible for the selection of EK dewatering process instead of the open well dewatering process. Despite the soil's resistance, ions move through the soil with the help of water present in it. Practically, the IMC level of the site decides the implementation of the entire process of electrokinetics itself. During the testing, it is observed that when IMC is higher,

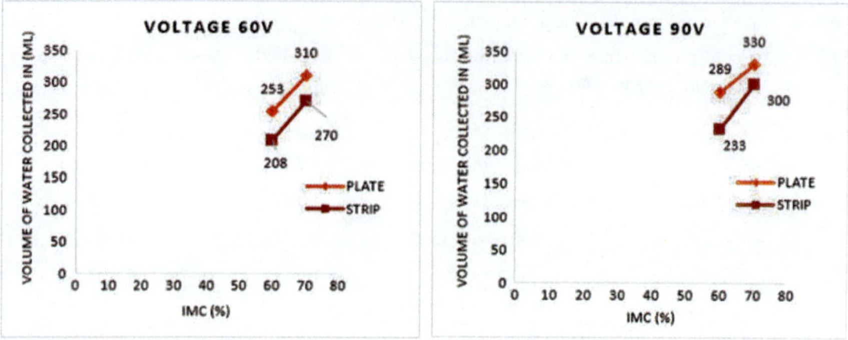

Fig. 6 Cumulative volume of water collected at the cathode for voltage 60 V and for varying IMC

the rate of dewatering achieved is also higher. When IMC is high there is an effective transfer of current; hence, water molecules are easily drawn toward cathodes.

It is usual practice to continue the process of dewatering using electro-osmotic principle only in case of ground with more clay content that too in a slurry state (as indicated by its high IMC even higher than liquid limit consistency of the in situ soils). Otherwise, the well-point system is sufficient enough to dewater in the case when the soil is having more sand content. In this investigation, an attempt is made to study the soft clay with varying IMCs of 60 and 70%. The IMC of soil decides the voltage to be applied and the time required for the EK process.

Figure 6 shows that the cumulative volume of water collected at cathode increases with the increase in the initial moisture content of the soil for an applied voltage. Hence soils with IMCs well above the liquid limit are only fairly treated by other dewatering techniques and then this process can be adopted. Now focusing toward the lower range limit of the IMC, at lower IMC, the soil–water discontinuity occurs due to the formation of microcracks at lesser time interval compared to higher IMC of soil resulting in loss of contact with the electrodes. Due to volumetric shrinkage of remolded clay soil that has been observed in a short span of time at higher voltages soon after the start of passing current also leads to loss of contact between the soil and electrode.

5 Conclusions

From the results obtained from the experiments, the effect of voltage and IMC on the properties of soft clay are analyzed and the following conclusions are drawn.

1. As the voltage increases, the volume of water collected at the cathode also increases. The increase on applying 90 V is 10% more than on applying 60 V. With the increase in voltage, there is a reduction in FMC of soil and the vane shear strength increases irrespective of the IMC of soil.

2. On the application of current through the soil mass between the electrodes, the soil gets dried up to such an extent that cracks are formed throughout the soil mass which hamper the effectiveness of electro-osmotic process. The cracks developed are a function of voltage applied.
3. The introduction of geotextile increases the rate of dewatering and there is an early attainment of steady state before propagation of cracks. A further increase of 8–16% of volume of water was achieved by inclusion of geotextile placed vertically between the electrodes.
4. The use of plate electrodes were considered to be more efficient than strip electrodes. The percentage reduction in moisture content using plate electrodes ranges from 34.67 to 46.14% and that of strip electrode ranges from 20 to 32.71%.
5. As the IMC of the soil increases, the time taken for the volume of water to attain a steady state also increases. This is because of more amount of water present in the soil to get collected at cathode on application of DC.

References

1. Geeta Megur, A., Rakaraddi, P.G.: A Study on stabilization of soil by electro Kinetic method. Int. J. Res. Eng. Technol. **3**, 104–109 (2014)
2. Hongtao, Fu., Yuan, L., Wang, J., Cai, Y., Xiuqing, Hu., Geng, X.: Influence of high voltage gradients on electrokinetic dewatering for wenzhou clay slurry improvement. Soil Mech. Found. Eng. **55**, 400–406 (2018)
3. Jeevanandan, K., Ragul, R., Samidurai, V.: Soil stabilization by electrokinetic technique. Int. J. Adv. Res. Civil Struct. Environ. Infrastruct Eng. Dev. **3**, 278–281 (2017)
4. Julie Shang, Q., Wayne Dunlap, A.: Improvement of soft clays by High-Voltage Electrokinetics. J. Geotech. Eng. **122**(4), 274–280 (1996)
5. Maheshwari Bisanal, G., Chandrakanth, S.M.: Study on stabilization of soil by electrokinetic method. Int. J. Eng. Technol. Sci. Res. **4**, 103–107 (2017)
6. Manjari Blessing, B.V., Ranjitha, K.: Soil stabilization by electrokinetic method. Int. J. Sci. Res. **6**, 1316–1320 (2015)

Effect of Modification on Coir Fiber in Durability and Shear Parameters in Flyash Soil Mixture

S. Athira and G. S. Manju

1 Introduction

Soil is a construction material which has been used from time immortal. Thonnakkal soil is building challenges to civil engineers in improving its properties based upon the need because of its poor mechanical properties. Hence this soil was adopted for the study. From the start of the industrial and economic revolution the most important issue in ahead of the industries is the disposal of the industrial waste, and it is a major issue within the current situation. The solution to the above problem is to use these industrial wastes to a most level for varied functions, like road construction, highways and embankments. Therefore, flyash was used in the study and its optimum dosage to soil was found out. A lot of work has been studied and done on the strength deformation behavior and characteristics of fiber strengthened soil, and it has been obtained that addition of fiber in soil improves the overall engineering properties of soil. Analyzing the addition of treated fiber in soil is less attempted. Natural fibers are lot more advantageous over synthetic due to their handling ability, availability, low cost, and since they are biodegradable will not cause disposal problem [1]. The effectiveness of fiber depends on the strength of fiber and its interaction at normal stress with the soil. Among the natural fibers, coir fiber is used in the study which has highest lignin content of all. The tensile strength and durability of coir fibers can be altered by changing the cellulose packing and physical changes in it. The hydroxyl groups (OH) present in cellulose and their hydrogen bond formation is what makes coir fiber amenable to chemical modification. Quick precipitation method was chosen, where the structure of coir fiber was altered using $CaCl_2$ and NaOH solution. Therefore, the aim of the research is to identify the applicability of proposed modification on coir fiber by conducting tensile strength, water absorption

S. Athira (✉) · G. S. Manju
College of Engineering Trivandrum, Trivandrum 695016, Kerala, India

C. N. V. Satyanarayana Reddy et al. (eds.), *Ground Improvement and Reinforced Soil Structures*, Lecture Notes in Civil Engineering 152,
https://doi.org/10.1007/978-981-16-1831-4_14

and durability of individual fiber, analyzing the unconfined compressive strength of soil–flyash mix using modified fibers, and thereby improving the soil properties of clay using naturally available material and waste material as additives.

2 Literature Review

Natural fiber and its application in geotechnical engineering have attracted researchers since many years. Coir is a natural biodegradable fiber with high amount of lignin content of around 40–45%, cellulose content of 32–43% and hemi-cellulose content of 0.15–0.25% [2]. The coir fiber has more individual fiber tensile strength compared to other natural fiber. This individual strength of fiber will be mobilized and works together with the soil. Hence coir as reinforcement helps in improving the strength at an early stage and it provides good stability to soil because of its high initial strength [3]. Addition of randomly distributed fibers to soil in improving geotechnical properties of soil has been found to be beneficial. Test results from the graph indicated that unconfined compression test and indirect tensile test were higher at 1% addition of fiber, which is clear from the graph showing higher slope at 1% fiber, indicating more ductile behavior in soil–lime mixture [4]. More fiber content dominated the fiber–fiber interaction, thereby forming lumps, and full contact with soil particles is not possible. Coir fiber of 1–1.5% helped in reducing swelling and compressibility of soil to a large extent [5]. Natural fibers are easily amenable to modifications done chemically due to hydroxyl groups present throughout it and its capability in forming hydrogen bond with the new particles. Many researchers have tried lot of surface coating and chemical treatment on fibers to improve its tensile strength and durability. The effect of gum resin coating on sisal fiber in shear and compressive strength on soil–fiber mix was investigated. Initially, the UCC value was low but upon drying it improved. Thus, gum resin could improve the shear property and potential durability of fiber–soil mix [6]. A study was analyzed on the comparison between CCl_4 and NaOH-treated fiber on consolidated undrained test. The results showed that the peak deviator stress can be significantly improved by the presence of treated fiber and this improvement, cohesion, friction angle was highest for the one treated with CCl_4 [7]. Behavior of NaOH-treated coir fiber in clay-pond ash mixture stabilized with cement was investigated. The rate of reduction in peak strength and stiffness was observed by the addition of treated coir fiber, and the behavior of the composite changed to ductile nature which initially showed brittle behavior [8].

Nanotechnology is a recent approach of fiber modification by impregnating with nanoparticles onto the surface and pores, thereby introducing a new function onto the fiber surface and improving their mechanical properties. The effect of nanomodification in coir fiber through quick precipitation method and its effect on properties of soil are less attempted. The effect of reinforcement by nanomodification method in water hyacinth (WH) fibers by ferric hydroxide was studied. The water absorption capacity in treated fiber reduced as the surface changed to hydrophobic nature and also because it was less exposed to water surface. The treated fiber compared to unmodified WH fiber increased the tensile strength of around 1.25 times. This improvement in tensile strength of modified fibers reflected in the increase of unconfined compressive strength of soil–fiber mix [9]. The effect of nanomodification on coir fiber through quick precipitation method in limed marine clay improved the tensile strength by 63 and 33% for $Al(OH)_3$ and $Fe(OH)_3$ modified fibers. The shear strength, cohesion intercept, frictional angle and durability increased by the modification of fiber [10].

3 Experimental Studies

3.1 Materials

The soil used in the study is collected from a depth of more than 5 m at Thonnakkal region at Trivandrum district. The soil properties are tabulated in Table 1. Flyash used in the study was collected from Tuticorin thermal power plant, India. The obtained Class F flyash was one with high siliceous content and low calcareous content according to IS 3812–1. The chemical composition of flyash is shown in Table 2. Coir fiber was collected from Neyyattinkara Coir Cluster, Trivandrum.

Table 1 Properties of soil

Property	Value
Color	Pinkish white
Natural moisture content (%)	20
Specific gravity	2.5
Liquid limit (%)	46
Plastic limit (%)	25
Plasticity index (%)	21
Sand (%)	48
Silt (%)	14
Clay (%)	38
Maximum dry density (g/cc)	1.8
Optimum moisture content (%)	16.67
Unconfined compressive strength (kN/m^2)	28.49
California bearing ratio	2.68

Table 2 Properties of flyash

Chemical composition	Component (wt %)
$SiO_2 + Al_2O_3 + Fe_2O_3$	92.11
SiO_2	61.53
MgO	0.63
SO_3	0.82
Na_2O	0.39
Total chlorides	0.008
LOI	0.95
Mean particle size	24 μm
Fineness-specific surface (Blaine)	365 m^2/kg

3.2 Nano Modification of Coir Fiber

The process is carried out through quick precipitation method in which $Ca(OH)_2$ nanoparticles were loaded onto the surface of fibers and into the fiber pores. The procedure was carried out at ambient conditions of temperature and pressure. The procedure is as follows. Initially, 50 g of coir fibers was submerged into 500 ml aqueous solution of 0.5 M $CaCl_2$ for 24 h in order to uniformly fill the pores and the fiber surface with $CaCl_2$ solution. Then the submerged coir fibers were separated and incubated in a different beaker containing 500 ml of 0.5 M sodium hydroxide solution. After that it was kept in ambient conditions for another 24 h. Here $Ca(OH)_2$ quickly precipitates as nanoparticles on the fiber surfaces and on its pores. Finally, the coir fibers were separated from sodium hydroxide solution and the unwanted residues such as NaCl and NaOH were removed washing with distilled water. At last, fibers are dried at ambient room temperature. The chemical reaction undergoing during the process is represented through a chemical equation given below and the entire process of modification is depicted in Fig. 1

$$CaCl_2 + 2NaOH \rightarrow Ca(OH)_2 + 2NaCl \quad (1)$$

3.3 Testing Program

3.3.1 SEM Analysis

The surfaces of chemically modified and unmodified coir fibers were examined with scanning electron microscope. For proper conduction the samples were sputtered by ejecting gold atoms. The properly sputtered samples were then positioned inside the SEM chamber for obtaining micrographs. The micrograph is obtained at 1000 × magnification and the images are then digitally analyzed.

3.3.2 Water Absorption

The water absorption was carried out by drying the fiber and immersing in distilled water in accordance with Saha et al. [11] at room temperature. The portion of fiber was taken out from distilled water at fixed time interval. It was wiped with a cloth carefully to remove the excess water and weighed to get the water absorption. This process is done for longer immersion time period until a steady value of water absorption is obtained.

3.3.3 Tensile Test

Coir fibers were randomly selected and tension test was done as per ASTM D-3379–75: 1989 in Universal Testing Machine at constant rate of extension of 5 mm/min. Samples of uniform shape and average thickness 0.15 mm were tested at gage length of 20 ± 5 mm. The tab was prepared according to ASTM D-3379–75 [12] as shown in Fig. 1. Strands of fiber were subjected to tensile test and average breaking load was taken, expressed in Newtons (N).

3.3.4 Alternate Wetting and Drying Study

To study durability on untreated and modified coir fiber, alternate wetting and drying processes were carried out, as per the condition mentioned in Sumi et al. [13]. The fiber samples were carefully weighed accurately and immersed in alkaline medium (pH 10.0) and neutral medium (pH 7.0) separately at room temperature (27 ± 1 °C) for seven days. For complete immersion of fiber in solution during the test period, fiber to solution ratio (by weight) was maintained 1:10 in a tray. Afterwards fibers were removed and placed on steel container to drain out excess water for 30 m and exposed to sunlight for seven days. 1 cycle was considered as 7 days of wetting and 7 days of drying and the tensile test of each set of fibers was determined at the end of each cycle. At the end of the durability test period, the fibers were taken out and

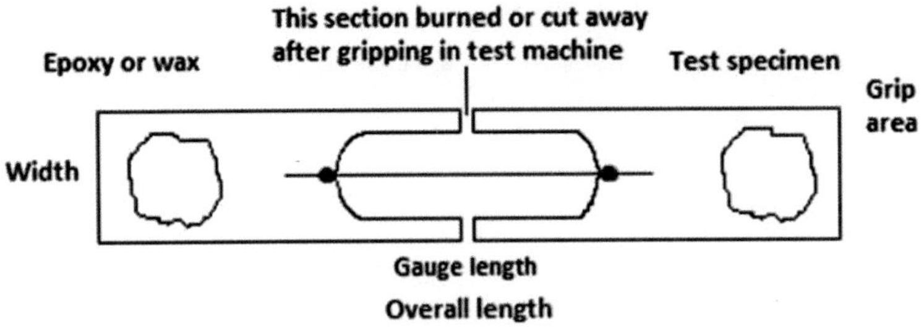

Fig. 1 Set up for holding individual fiber strand for tensile test

unconfined compressive strength of sample reinforced with these fibers (at optimum percentage alone) was found out.

3.3.5 Unconfined Compressive Strength

UCC was done to find out the optimum flyash content, and then different percentages of untreated and modified fiber (0.5, 1, and 1.5%) were added to soil-optimum flyash mix to study the effect of modification of coir fiber in the mix. The samples are prepared at OMC and compacted to maximum dry density (MDD) obtained from compaction test [15]. The samples were wrapped with plastic sheet and kept for curing for 7 and 28 days prior to testing.

4 Results and Discussion

4.1 SEM Analysis

Figure 2a, b illustrates the SEM image of the fiber surface of unmodified and modified coir fiber, respectively. It is seen that clusters of the proposed materials, in the range of 110 nm, changed the morphology into a relatively rougher surface.

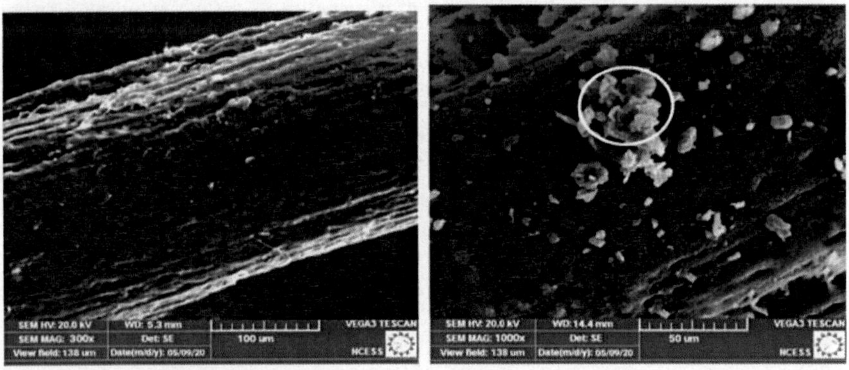

(a) (b)

Fig. 2 Surface structure of coir fiber **a** unmodified **b** $Ca(OH)_2$ modified

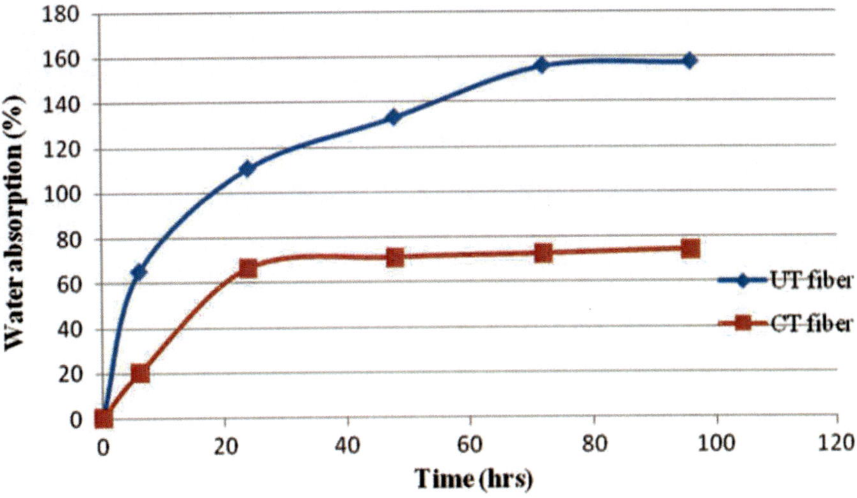

Fig. 3 Percentage variation in water absorption of fibers with time

4.2 *Water Absorption*

Treatment on fiber helped to reduce the hydrophilicity of the fiber. Figure 3 shows variation in water absorption with time. In chemical treatment, the lignin and water dissolvable constituents were initially removed to some extent and dissociable groups were partly blocked by the treatment which might have led to a reduction in swelling tendencies. There was 160% water absorption in untreated fiber which was reduced to 74% upon treatment.

4.3 *Tensile Test and Elongation*

As per IS 235–1989 code, tensile test of individual fiber is expressed in terms of its breaking strength, i.e., the maximum force supported by a specimen until rupture in a tensile test [14]. It is usually expressed in Newtons (N). The average tensile load of untreated was 6.58 N and when it was chemically treated it increased to 8.05 N, i.e., 22.4% increase. Figure 4 shows the load extension graph of treated and untreated fiber. The extension of fiber was also found to increase. This is due to the nanoparticle formation within the fiber and this might increase the surface area of fiber acting as a single unit during tensile test.

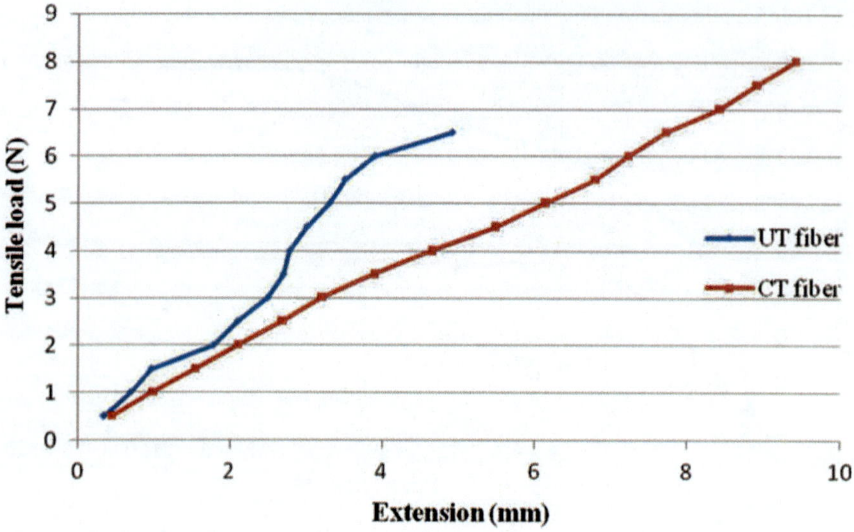

Fig. 4 Load extension curve of fibers

4.4 Effect of Treated Fiber on Unconfined Compression Test

Addition of flyash showed a great increase in unconfined compressive strength with a brittle failure. Figure 5 shows the behavior of stress–strain curve with the addition of flyash. It showed a distinct failure stress at a strain of about 4.5–6.5% after which they collapsed. 10% flyash showed maximum unconfined compressive strength which was

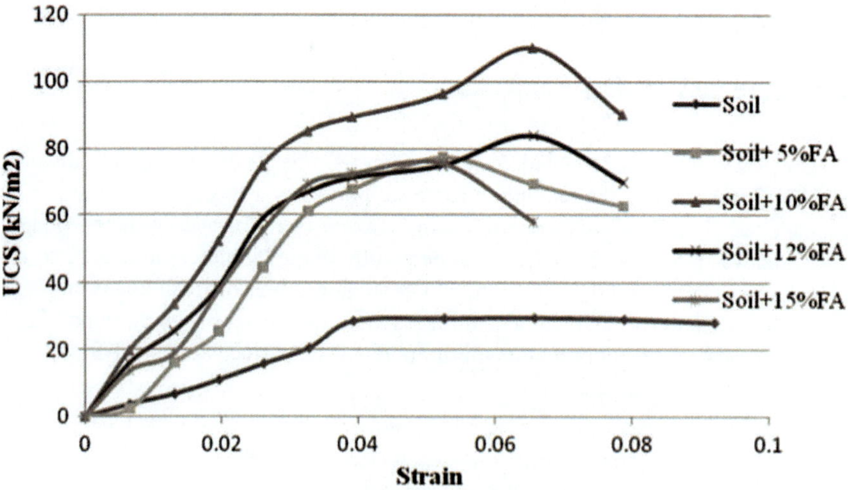

Fig. 5 Stress versus strain curves for various flyash content

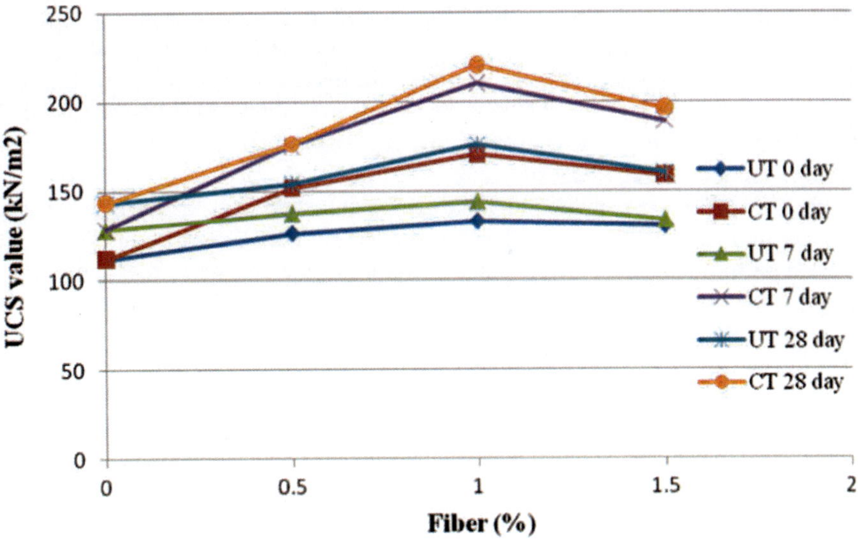

Fig. 6 Variation in UCS value upon addition of treated and untreated fiber

109.74 kN/m^2. Hence that was taken as optimum. After that different percentages of fiber were added to soil-optimum flyash mix. The variation in UCC value upon addition of untreated (UT) and chemically treated fiber (CT) for 0, 7 and 28 days is shown in Fig. 6. Fiber-reinforced soil samples exhibited a ductile behavior more at 1% fiber content.

If above 1% fiber is added, the UCS value is reduced, which might be due to large content of fiber adheres themselves and its contact and bond with soil becomes less. Addition of chemically treated fiber showed a higher unconfined compressive strength improvement in soil. The maximum value was observed at 28 days curing at 1% fiber content. This increase is because of improved tensile strength of fiber upon treatment that improved the overall strength of soil. Upon curing the UCS value also increased, which might be due to the crystallization of nanoparticles within the surface and cellulosic pores of fiber, and the surface of fiber changed the morphology to a rougher surface, as evident in SEM analysis, which led to more interfacial adhesion with soil and frictional resistance to force application improvement, and consequently, the strength of soil–flyash–fiber mixture increased.

4.5 Effect of Treatment on Durability of Fiber and UCC Value

The fiber was subjected to accelerated degradation study and the tensile load was noted at the end of each cycle. Percentage reduction in tensile load at neutral and alkaline medium is expressed in Fig. 7. At the end of fifth cycle there was 27.38% decrease for untreated fiber at neutral medium and 30.39% decrease at alkaline medium, whereas upon treatment the rate of decrease was less, which was 8.11% under neutral medium and 12.52% under alkaline medium. The 1% of degraded fiber samples (optimum percentage from UCC test) obtained at the end of fifth cycle of durability study was added to soil to study the effect of decrease in UCC value. The values are tabulated in Table 3. The untreated fiber in alternate wetting and drying cycle test had faster loss in tensile load compared to treated fiber. It is because more water penetrates during wetting cycle and more released during drying cycle that causes hydration–dehydration. This behavior can be extended for a longer immersion period of time. Under chemical treatment the surface morphology changed and it reduced the dissolution of lignin, especially in alkaline medium. The enhanced tensile strength may be due to the crystallization of the $Ca(OH)_2$ particle into the cellulosic pores and available capillaries present within the fiber. Alkaline medium resulted in faster loss of strength, which can be due to the capacity of alkali to remove the waxy and oily coating of fiber and exposing them to moisture intake. Treatment on fiber showed sufficient durability and they were able to resist the adverse pH condition compared to untreated fiber. Thus from the results it is clear that $Ca(OH)_2$ impregnation into the fiber was found to be effective in its potentiality for long-term use.

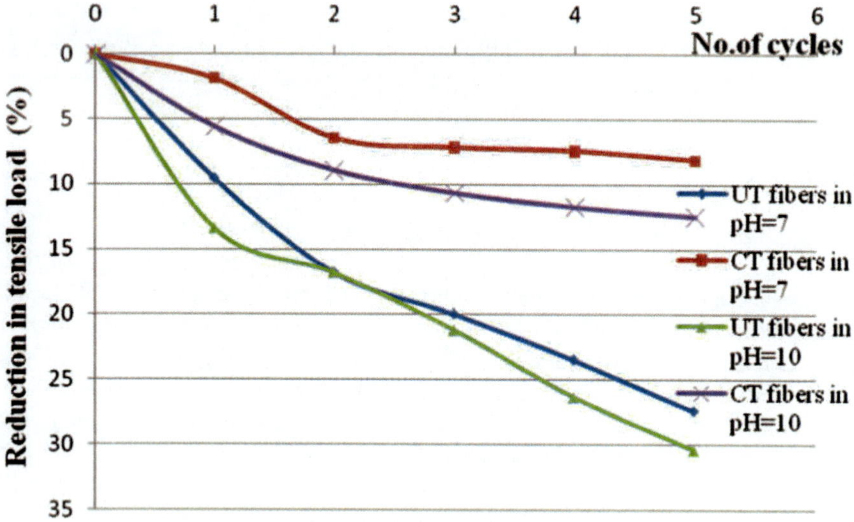

Fig. 7 Percentage reduction in tensile load with cycle for fibers in neutral and alkaline medium

Table 3 Tensile load of fiber and UCC of soil sample by the addition of fibers subjected to alternate wetting and drying cycles

Fiber type	Untreated fiber		Chemically treated fiber	
	Neutral medium	Alkaline medium	Neutral medium	Alkaline medium
Tensile load of fiber at the end of fifth cycle (N)	4.778	4.58	7.397	7.042
% decrease	27.38	30.39	8.11	12.5
UCC value (kN/m^2) (Soil + 10%flyash + 1%fiber)	120.11	113.98	162.01	158.9
% decrease	10	14	2	4

5 Conclusions

1. $Ca(OH)_2$ nanoparticle impregnation was tested for improving the durability or extending the life of coir fiber under adverse condition. Suitably pretreated coir fiber by the above process showed much retention of its tensile strength after different cycles under neutral and alkaline medium.
2. Percentage reduction in water absorption up to about 74% from 160% could be attained with this treatment.
3. Modification of fiber did not reduce the mechanical property of fiber. There was 22.4% increase in tensile load upon treatment.
4. The morphology of fiber changed to a rougher surface; thereby it provided better adhesion and interlocking ability with the soil.
5. The optimum flyash was obtained to be 10%, hence the combination of 10% flyash and 1% treated fiber showed the maximum unconfined compressive value. Its effect of curing was also studied which showed improvement in UCS value.
6. Addition of degraded modified fiber obtained at the end of durability test showed less decrease in UCC value compared to degraded untreated fiber.

References

1. Aswathy, R., Beena, K.S.: Emergence of nanomodified fibres as soil reinforcement—a review. In: Green Buildings and Sustainable Engineering, pp. 359–365. Springer, Singapore (2019)
2. John, M.J., Anandjiwala, R.D.: Recent developments in chemical modification and characterization of natural fiber-reinforced composites. Polym. Compos. **29**(2), 187–207(2008)
3. Anggraini, V., Asadi, A., Huat, B.B., Nahazanan, H.: Effects of coir fibers on tensile and compressive strength of lime treated soft soil. Measurement **59**, 372–381(2015)
4. Olgun, M.: Effects of polypropylene fiber inclusion on the strength and volume change characteristics of cement-fly ash stabilized clay soil. Geosyn. Int. **20**(4), 263–275 (2015)

5. Sivakumar Babu, G.L., Vasudevan, A.K., Sayida, M.K.: Use of coir fibers for improving the engineering properties of expansive soils. J. Nat. Fibers **5**(1), 61–75 (2008). Taylor and francis
6. Kafodya, I., Okonta, F.: Effect of fibre surface coating on the mechanical properties of natural fibre-reinforced soil. Int. J. Geotech. Eng. 1–11 (2018). Taylor and Francis
7. Khatri, V.N., Dutta, R.K., Venkataraman, G., Shrivastava, R.: Shear strength behaviour of clay reinforced with treated coir fibres. Periodica Polytech. Civil Eng. **60**(2), 135–143(2016)
8. Yadav, J.S., Tiwari, S.K.: Behaviour of cement stabilized treated coir fibre-reinforced clay-pond ash mixtures. J. Build. Eng. **8**, 131–140 (2016)
9. Bordoloi, S., Patwa, D., Hussain, R., Garg, A., Sreedeep, S.: Nano-Particle coated natural fiber impregnated soil as a sustainable reinforcement material. Int. J. Geomech. (2017). ASCE
10. Anggraini, V., Asadi, A., Farzadnia, N., Jahangirian, H., Huat, B.B.: Reinforcement benefits of nanomodified coir fiber in lime-treated marine clay. J. Mater. Civil Eng. **228**(6), 06016005(2016). ASCE
11. Saha, P., Manna, S., Sen, R., Roy, D., Adhikari, B.: Durability of lignocellulosic fi bers treated with vegetable oil–phenolic resin. Carbohyd. Polym. **87**(2), 1628–1636 (2012)
12. ASTM D 3379–75 (1989): Standard Test Method for Tensile Strength and Young's Modulus for High Modulus Single-Filament Materials. ASTM International, West Conshohocken, PA, USA
13. Sumi, S., Unnikrishnan, N., Mathew, L.: Durability studies of surface-modified coir geotextiles. Geotex. Geomembr. **46**(6), 699–706(2018)
14. IS 235–1989: Textile Fibers—Tensile Characteristics of Individual Fibers—Method for Determination
15. IS: 2720-Part 10: Indian Standard Methods of Tests for Soils—Determination of Unconfine Compressive strength, Bureau of Indian Standards, New Delhi (1991)

Dynamic Compaction of Sandy and Silty Soils Near Delhi for Liquefaction Mitigation

Amit Somwanshi, Sandeep Ghan, and Manoj Tipnis

1 Introduction

Ministry of Railways (MOR), Government of India has planned to construct a high axle load dedicated freight corridor (DFC) covering about 3325 km on two corridors: Eastern and Western Corridors. Certain stretches in the DFC alignment near Delhi indicate the presence of loose to medium dense sandy silt and silty sand deposits which are liquefying up to a depth of 8.0–11.0 m from the ground level. Figure 1 shows the location map of the DFC alignment near Delhi where subsoil was liquefying. In order to mitigate the risk of liquefaction, ground improvement by densification of the loose sandy and silty subsoil was necessary.

The current paper briefly describes the details of the ground improvement using dynamic compaction technique to create a suitable ground to support the railway embankment.

2 Subsoil Profile

The soil exploration of the present site was done at 500 m interval along the railway alignment. The boreholes were terminated at refusal strata. The strata encountered are generally of sandy silt of low plasticity (ML-CL) and silty sand (SM). The variation in SPT-N values with depth of boreholes performed in liquefaction susceptible stretches is presented in Fig. 2. Ground water table was encountered at about 1.5–7.1 m depth below ground level in the liquefaction susceptible stretches.

A. Somwanshi (✉) · S. Ghan · M. Tipnis
Transportation Infrastructure IC, EDRC (RREC), L & T Construction, Mumbai 400093, India
e-mail: amitsomwanshi@lntecc.com

C. N. V. Satyanarayana Reddy et al. (eds.), *Ground Improvement and Reinforced Soil Structures*, Lecture Notes in Civil Engineering 152,
https://doi.org/10.1007/978-981-16-1831-4_15

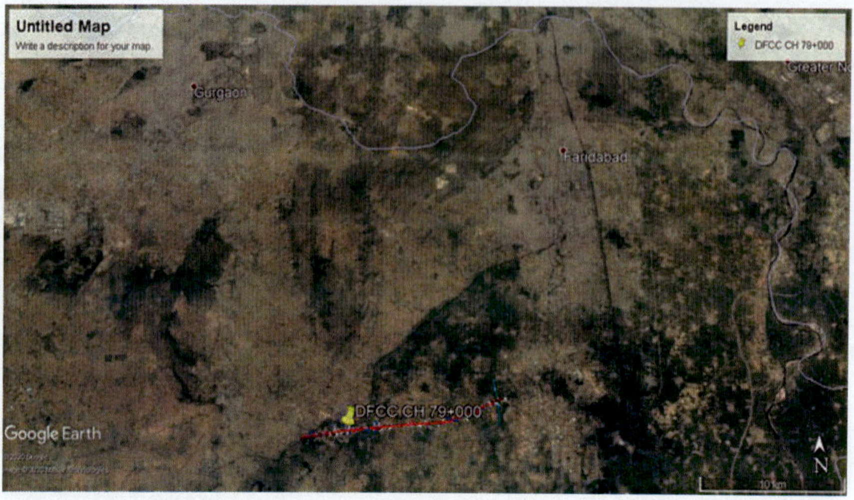

Fig. 1 Location map of DFCC alignment near Delhi

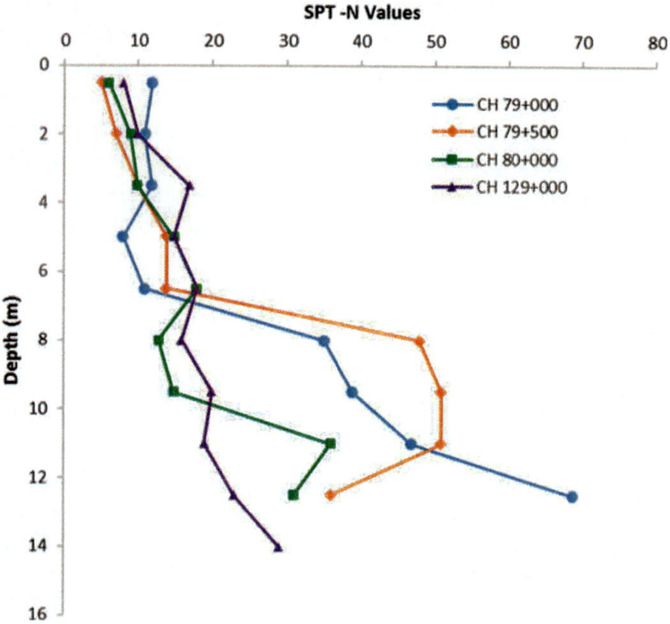

Fig. 2 SPT-N versus depth

As seen from Fig. 2, it is evident that the field SPT-N values ranges from 5 to 20, especially in the depth range of 0.5–11.0 m from the ground level at this certain stretches. This indicates the loose to medium dense nature of subsoil.

3 Liquefaction Susceptibility Analysis

3.1 Methodology and Design Parameters

Due to the presence of sandy and silty soils of low SPT-N values, shallow ground water table and level of ground shaking expected at the site due to an earthquake, it was concluded that the site had potential for liquefaction. Therefore, the liquefaction analysis was performed to determine the density of soils required to minimize the potential of liquefaction. These densities were then compared with densities of existing soils to determine the magnitude of liquefaction potential of the site and level of site improvement needed.

According to Fig. 1 and Annexure E of IS 1893 (Part1)-2016 [1] which shows the seismic zones and zone factors of important towns, the proposed site falls under Zone-IV having zone factor as 0.24. A design earthquake magnitude of 7.0 and peak ground acceleration (PGA) of 0.24 g were considered for the analysis based on earthquake histories and is recommended by IS 1893(Part1)-2016 [1].

The liquefaction potential calculations were carried out as per the procedure detailed in Annexure F of IS 1893 (Part1)-2016 [1]. A critical factor of safety of 1.1 was considered for the liquefaction analysis.

Two variables are required to evaluate liquefaction resistance of the soil. (1) The seismic demand on a soil and (2) the capacity of the soil to resist liquefaction. The first variable is related to the seismic load expressed as cyclic stress ratio (CSR). The second variable can be related to field tests, i.e., SPT, CPT and VST tests and is expressed as CRR. For the DFCC project this variable will be calculated using SPT tests. The factor of safety follows from FOS = CRR/CSR with a correction for the magnitude of the seismic load.

Design water table is considered at ground level except for embankment stretch at CH 129 + 000 where water table is measured at 7.1 m depth below the ground level during geotechnical investigation. The design water table is considered at 4.1 m depth below the ground level for the analysis of stretch CH 129 + 00.

3.2 Results

As per the detailed liquefaction analysis, the authors estimated that the soils up to a depth of 8.0–11.0 m from the ground level are susceptible to liquefaction in the event of the design earthquake at the certain stretches of embankment. The plots of

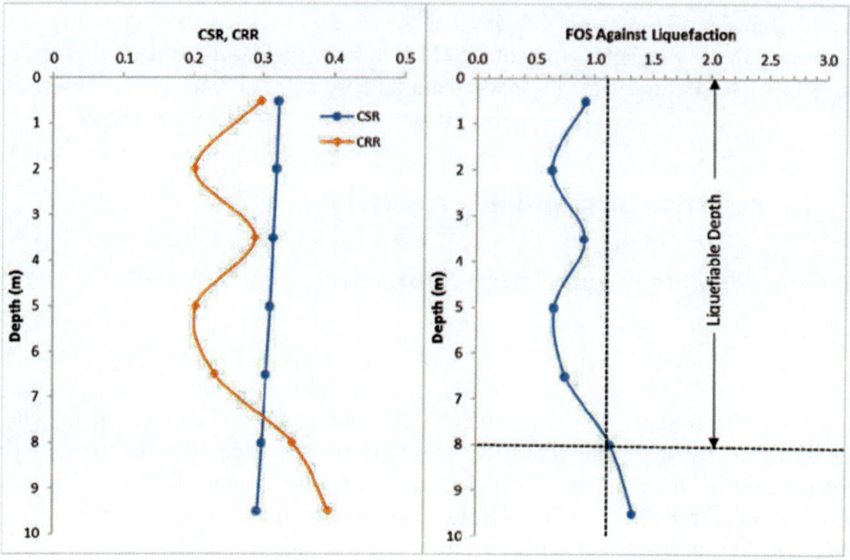

Fig. 3 Liquefaction susceptibility analysis (CH 79 + 000)

CSR, CRR and computed factor of safety against liquefaction for stretch CH 79 + 000 are presented in Fig. 3. The minimum required SPT-N values are calculated to achieve the required safety factor against liquefaction.

Densification of soils has been identified as a remedial measure against soil liquefaction due to earthquake shaking. The number of options for improving the ground were discussed, and finally, after verifying the site constraints, local conditions, material availability and the cost, it was decided to use dynamic compaction technique of ground improvement.

As per IRC 75:2015 [2], clause no. 5.2.8.1, the dynamic compaction method can be used for the densification process for sandy and silty soils to mitigate the potential risk of liquefaction.

4 Ground Improvement

4.1 Design Concept

In dynamic compaction technique, the soil is densified by using a heavy drop weight. The drop weight is lifted by a crane and repeatedly dropped onto the ground surface in a grid pattern. The basic principle behind the technique consists in the transmission of high energy waves through a compressible soil layer in order to improve at depth of its geotechnical properties.

Table 1 Required parameters of dynamic compaction

S. no	Required depth of improvement (m)	Weight of tamper (Tons)	Height of fall (m)
1	8.0	30.0	9.0
2	11.0	30.0	16.5

As per Mitchell [3], the depth of influence (D) up to which the ground is improved by using the tamper of weight (W) falling from the height (H) is given by:

$$D = n\sqrt{(WH)} \tag{1}$$

where D and H are in meters and W is in tons.

$n = 0.5$, modification factor taken as per Mitchell [3].

The weight of tamper (W) and height of fall (H) for the required depth of improvement (D) of 8.0 m and 11.0 m is calculated using Eq. 1 and given in Table 1.

4.2 *Site Execution*

The dynamic compaction was done in two phases, followed by a levelling phase. The energy was applied to the soil in phases on a grid pattern over the entire area in two phases. The proposed arrangement and spacing of dynamic compaction points is given in Fig. 4.

The depth of improvement generally depends on the total amount of energy applied to the soil, which is a function of the weight of the tamper and the drop height. At each point on a 4 m × 4 m grid, a 30-ton tamper was dropped repeatedly from a height of 9.0 m and 16.5 m for depth of improvement of 8.0 m and 11.0 m, respectively. Spacing of drop points is commonly selected to be 1.5–2.5 times the diameter or width of a tamper. Usually, 7 or more blows were applied at each point and craters of about 0.5–1.0 m depth were formed. The square tamper, 1.84 m high, 2.4 m in width and weighing 30 tons, was made of concrete with steel casing. The in-place soils below depth of craters are compacted due to vibrations and dissipation of excessive pore pressures generated during compaction. Figures 5 and 6 show the dynamic compaction in progress at the grid points and resulting crater.

The resources used for the execution of dynamic compaction.

(a) Friction crane—200 tons capacity
(b) Concrete tamper—30 tons weight
(c) Sling and D-shackle.

After the dynamic compaction, the area was levelled and compacted with an 8-ton vibratory roller. A minimum lag time of one week was given between each subsequent pass to allow the excess pore water pressures to dissipate.

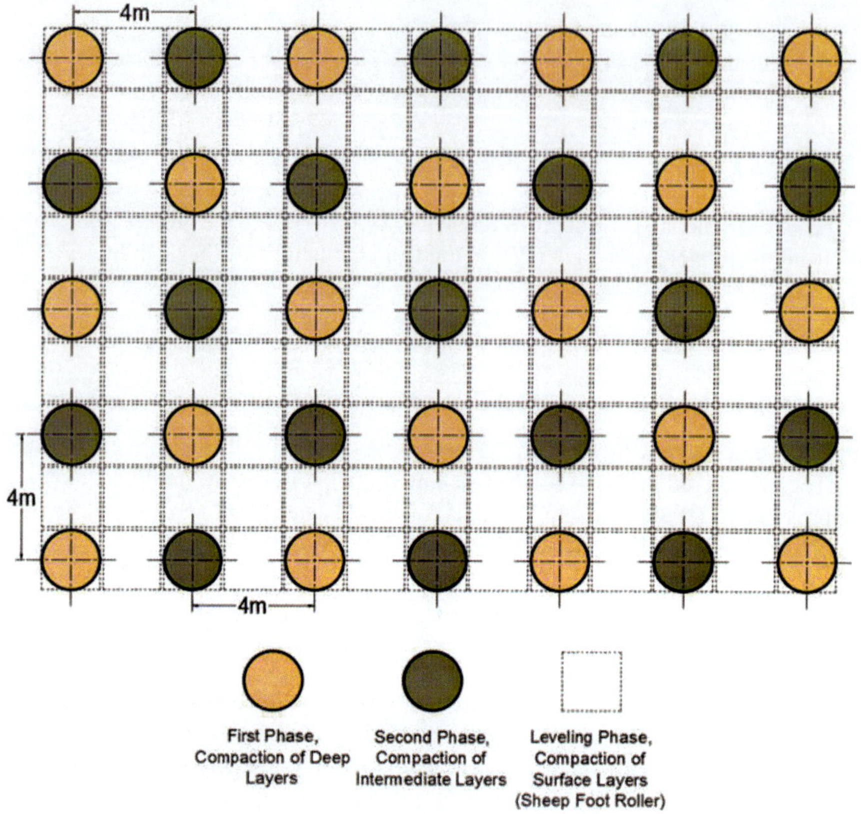

Fig. 4 Arrangement of dynamic compaction grid

5 Field Testing and Improvement

Following the area-wide dynamic compaction, standard penetration tests were conducted at predetermined locations. Figures 7, 8 and 9 present a typical comparison of SPT-N values in a borehole before and after dynamic compaction. It is observed from Figs. 7, 8 and 9 that the SPT-N values show a substantial improvement after area-wide dynamic compaction because of the interlocking of the soil grains as a result of densification.

Figures 7, 8 and 9 also show N-values required to mitigate the liquefaction potential. As can be seen from Figs. 7, 8 and 9, most of the N-values measured post improvement are greater than those required to mitigate the liquefaction potential.

Fig. 5 Dynamic compaction at the site in progress

Fig. 6 Dynamic compaction at the site in progress

6 Conclusions

Liquefaction analyses performed for a railway embankment site near Delhi are presented. The liquefaction analysis showed that the existing soils at the site had significant liquefaction potential. The site soils were densified using dynamic compaction technique. Based on the results of this investigation, the following conclusions can be drawn.

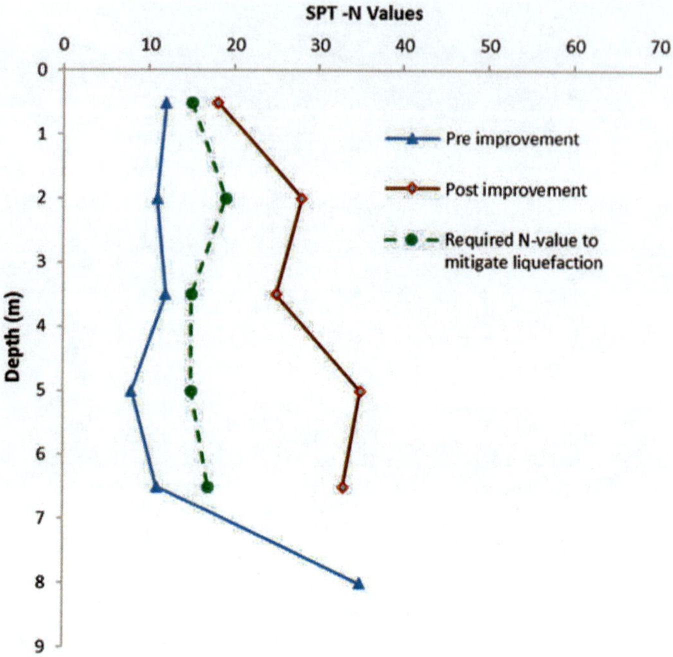

Fig. 7 SPT data before and after compaction (CH 79 + 000)

Fig. 8 SPT data before and after compaction (CH 79 + 500)

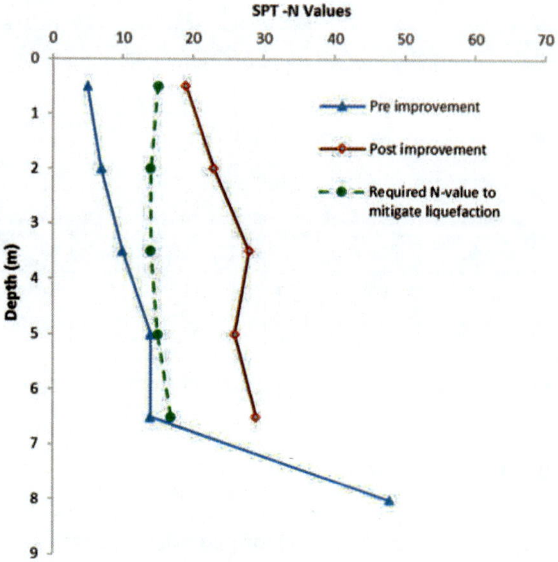

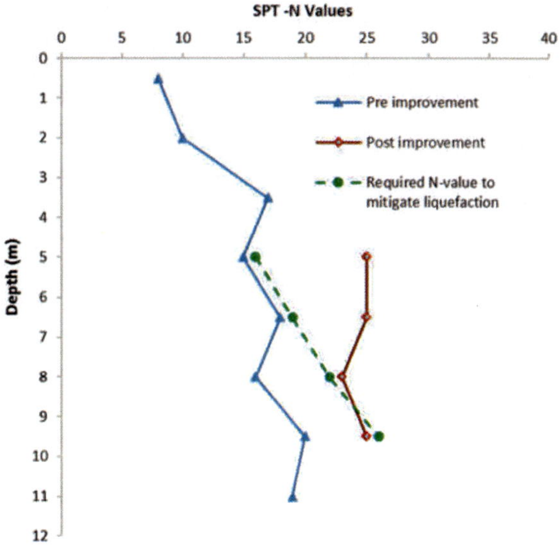

Fig. 9 SPT data before and after compaction (CH 129 + 000)

1. The dynamic compaction was effectively used to compact the sandy and silty subsoils up to 11.0 m depth below the ground level.
2. The compaction achieved was adequate to mitigate the liquefaction potential of the subsoils at the site.
3. In seismic zones with liquefiable soils, dynamic compaction technique provides technically sound and cost-effective solution.

References

1. IS 1893-Part 1: Methods of test for soils—Criteria for earthquake resistance design of structures, Bureau of Indian Standards, New Delhi (2016)
2. IRC 75–2015: Guidelines for the design of high embankments
3. Mitchell, J.K.: Soil improvement—state of the art report. In: 10th ICSMFE Proceedings, 4, Stockholm, pp. 509–565 (1981)

Lateral Displacements of Soft Ground Under Embankment Loading

M. Bhanuchitra, V. Padmavathi, and M. R. Madhav

1 Introduction

Embankments were used in earlier times to impound and divert water. The construction of embankments is playing a vital role in the development of communication networks such as highways and railways. Due to necessity embankments are often constructed even on soft soils. Hence the design and construction of embankments over soft soils have become a major issue due to low shear strength and high compressibility of soft ground. The estimation of settlements and lateral displacements of the embankment for good serviceability is the main design concern for embankments on soft soil. Rapid construction of the embankment generates large horizontal stresses in soft ground, which in turn leads to lateral soil movement under the toe of the embankment. The heaving of soil adjacent to the toe of the embankment may cause excessive deformations of the adjacent structures. It is essential to predict settlements and lateral displacements of the ground beneath an embankment and a reasonable criterion on the limits of lateral soil displacements incorporated in the design. Generally when the soft ground is subjected to embankment loading plastic flow or squeezing of soil occurs near the top of the ground. During construction, due to the fast rate of load application and low permeability of clays, an undrained response prevails which leads to the development of lateral displacements. The failure of soft ground under embankment is obviously related to the settlements and lateral displacements of the embankment. Matsuo and Kawamura [4] proposed a plot (Fig. 1) between the ratio of lateral displacement to vertical settlement and vertical settlement to predict the failure of an embankment. The rate of increase in settlements and lateral displacements [7] are functions of undrained shear parameters of the ground. Loganathan et al. [3] study numerically the deformations of soft subsoil during embankment loading and consolidation stages using finite element and finite difference methods.

M. Bhanuchitra (✉) · V. Padmavathi · M. R. Madhav
Department of Civil Engineering, JNT University, Hyderabad 500085, India

C. N. V. Satyanarayana Reddy et al. (eds.), *Ground Improvement and Reinforced Soil Structures*, Lecture Notes in Civil Engineering 152,
https://doi.org/10.1007/978-981-16-1831-4_16

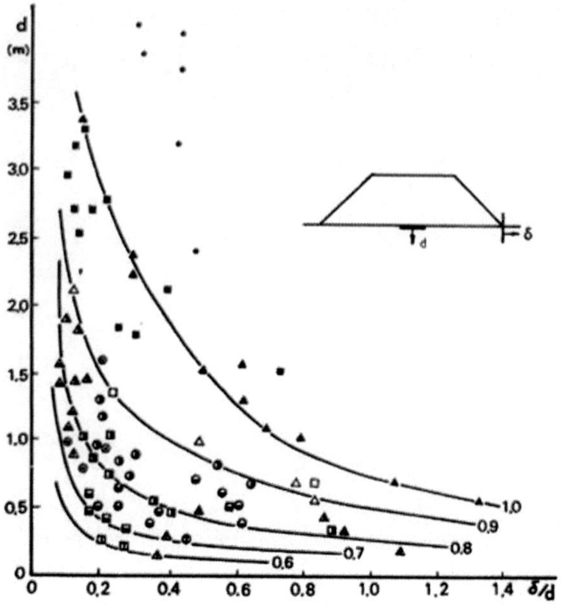

Fig. 1 (δ/d–d) Diagram for prediction of failure

The factors influencing lateral displacement induced by embankment loading are (a) Magnitude and loading rate of the embankment loading; (b) Undrained shear strength (c_u) of the subsoil; (c) Deformation characteristics of the soft subsoil. The provision of a stiff layer at the base embankment would minimize the deformations. Many experimental and numerical studies are available to estimate the effect of reinforcement on deformations of subsoil under the embankment. Chai et al. [1] compared embankment behavior with and without reinforcement with respect to the factor of safety (FOS) and concluded that the factor of safety increases and the lateral displacements reduce with the provision of reinforcement. Huang et al. [2] suggested that the increase of embankment soil stiffness results in the reduction of lateral displacements at the toe of the embankment. Rankine et al. [5] developed equations to determine lateral displacements at the toe and at a short distance from the toe of the embankment based on the results of numerical analysis. Safadoust et al. [6] concluded that the increase of reinforcement stiffness does not lead to a reduction in the settlement compared to the reduction in horizontal displacement. Various techniques are available to control deformations and for the rapid construction of embankments by implementing different methods. Placing a stiff reinforced granular bed (RGB) over soft soil is one of the simplest techniques of ground improvement to minimize the lateral deformations of soft soil.

The present study is concentrated on the lateral displacements during the construction of embankment on a reinforced granular bed treated as a two-layer system consisting of a stiff (upper) layer over soft ground. The objective of this paper is to estimate the effect of stiffness of reinforced granular bed (RGB) on the performance of embankment founded on soft soil through the prediction of lateral displacements for different modular ratios.

2 Methodology

A finite element analysis is carried out to investigate the effect of reinforced granular bed on the deformations of subsoil due to incremental or stage-wise loading of the embankment. PLAXIS 2D software is used and the model is set up in-plane strain condition with 15-noded triangular 1551 number of elements. The mesh generated is medium to course refined to finer in the high-stress regions. As the embankment is symmetric, only a half-section of the embankment is considered in the analysis.

To eliminate possible boundary effects on the results, the foundation soil is extended to a distance more than ten times horizontally and five times the embankment width vertically. Mohr–Coulomb model is used to simulate soil behavior. The soil parameters considered in the analysis are shown in Table 1.

Embankment founded on homogeneous soft soil is analyzed first and then a two-layer system with 2.0 m thick RGB over soft soil. The embankment is 6.0 m in height with a base width of 31.0 m with side slopes of 2H: 1 V. The sequence of construction is modeled by 0.5 m thick fill placed for each raise such that the total height of the embankment is reached in 12 increments. RGB at the top of the soft ground is freely draining while the soft clay is treated as in an undrained state. The geometry and PLAXIS model of embankment with homogeneous soil and of two layer system are shown in Figs. 2a, b and 3.

The settlements and lateral deformations are predicted for various deformation modular ratios, E_1/E_2, where E_1 and E_2 are the deformation moduli of the stiff RGB and soft layer respectively. Deformations are estimated for modular ratios of 1, 5, 20, 50 and 100 for homogeneous and two-layer systems. The modular ratio, $E_1/E_2 = 1$ represents homogeneous soil and $E_1/E_2 > 1$ a stiff foundation bed over a soft layer.

Table 1 Soil parameters

Parameter	Embankment soil	Layer 1-RGB	Layer-2 (Soft clay)
Material Model	Mohr–Coulomb (MC)	Mohr–Coulomb (MC)	Mohr–Coulomb (MC)
γ_b (kN/m^3)	17	20	15
Cohesion kPa	5	3	10
φ (°)	32	38	0
Deformation Modulus, EMPa	20.0	25.0–50.0	5.0
Poisson's ratio, ν	0.3	0.3	0.35

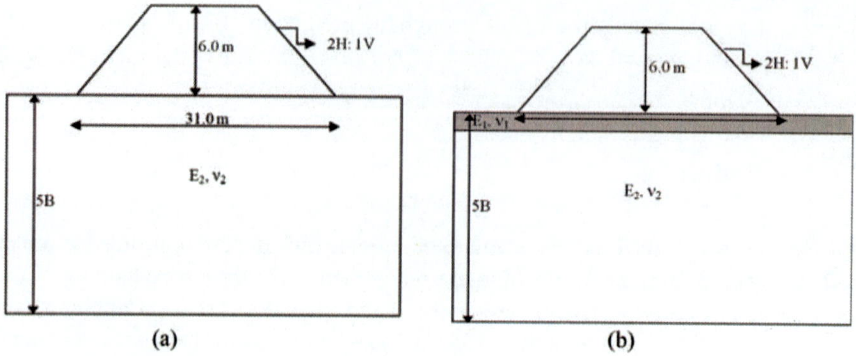

Fig. 2 Geometry of the embankment **a** homogeneous soil and **b** two-layer system

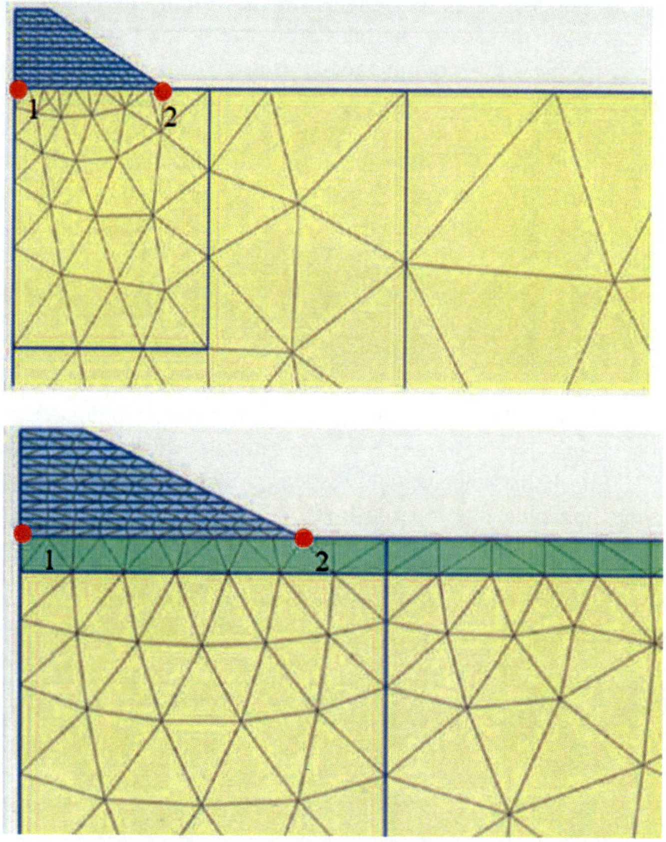

Fig. 3 PLAXIS model of embankment

3 Results and Discussion

3.1 Settlements

The variations of surface settlements beneath the center of embankment with an increase in height of fill for a homogeneous and two-layer system of different modular ratios are shown in Fig. 4. Near linear increase of settlements at the end of the construction are observed for the thickness of embankment fill till about 4.0 m and non-linear increase for thickness greater than 4.0 m. Figure 5 presents the surface settlement profiles beneath the embankment at the end of construction. The predicted surface settlements decrease marginally from 175 to 165 mm for the modular ratio

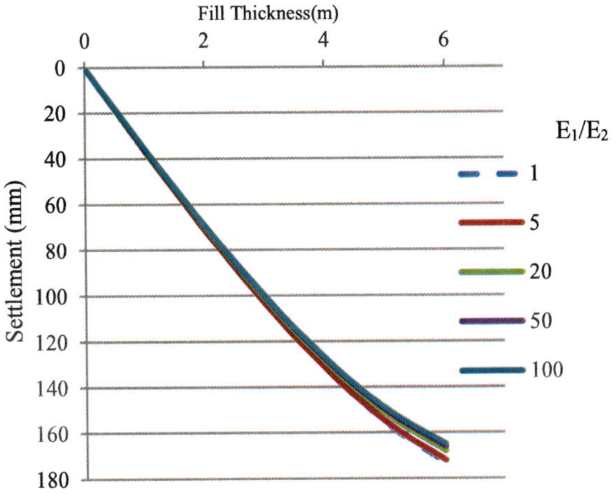

Fig. 4 Variation of settlements with height of fill

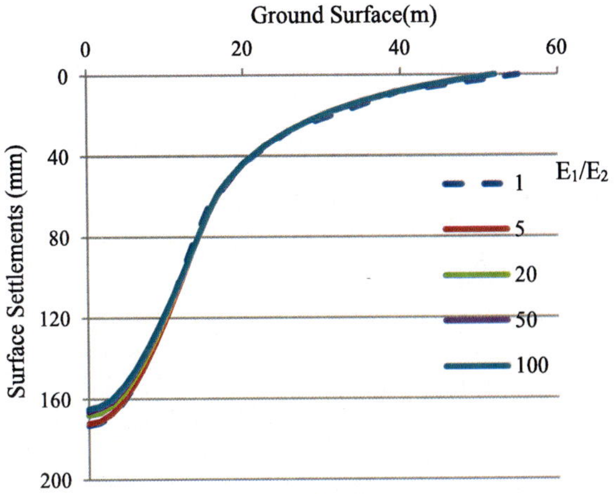

Fig. 5 Surface settlement profiles at the end of construction

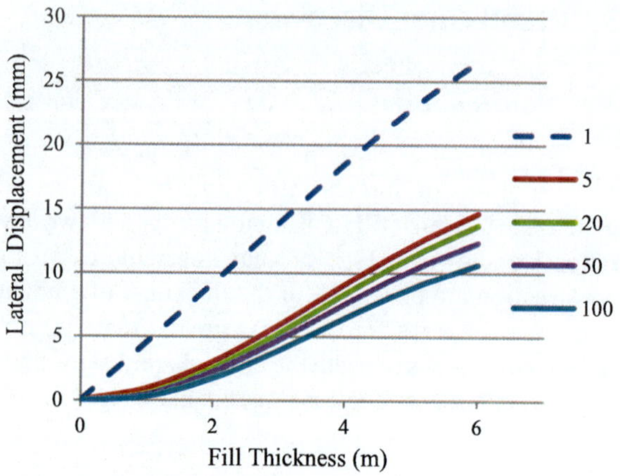

Fig. 6 Lateral displacements at the toe of embankment with incremental load

increasing from 1 to 100. The settlements of the two-layer system are not much different compared to those of homogeneous layer.

3.2 Lateral Displacements

The variation of lateral displacements at the toe of the embankment, at point 2 (Fig. 3) with a height of fill for homogeneous and two-layered soil cases are shown in Fig. 6. For the modular ratio $E_1/E_2 = 1$ (homogeneous soil), the lateral displacement at the ground surface at point 2 is 27 mm. and The displacements at the toe decrease from 14 to 11 mm with an increase of modular ratio from 5 to 100. Figure 7 presents the lateral displacement profile under the toe of the embankment at the end of construction. With the provision of a stiff layer over the soft soil, the lateral displacements reduce over the full depth of the soft layer. At a depth of 2 m that is at the top of the soft clay layer, the lateral displacements reduce marginally from 30 to 27 mm with the increase of modular ratio from 5 to 100.

3.3 Construction Control Chart

Matsuo's stability chart plots the ratio of lateral displacement by settlement at the center as abscissa and settlement beneath the center of the embankment as ordinate as in Fig. 8. Failure is indicated by the points moving away from the center and towards the top. The converse is true, i.e., increased stability or factor of safety is indicated if the points move to the left and downwards from the numerical analysis Results from the numerical analysis reported here for the embankment indicate the displacements curves moving towards the left in the Matsuo chart. The provision of a

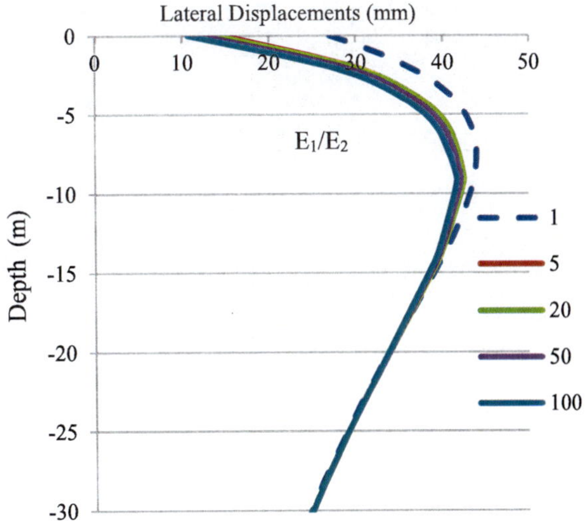

Fig. 7 Lateral displacements under the toe of the embankment

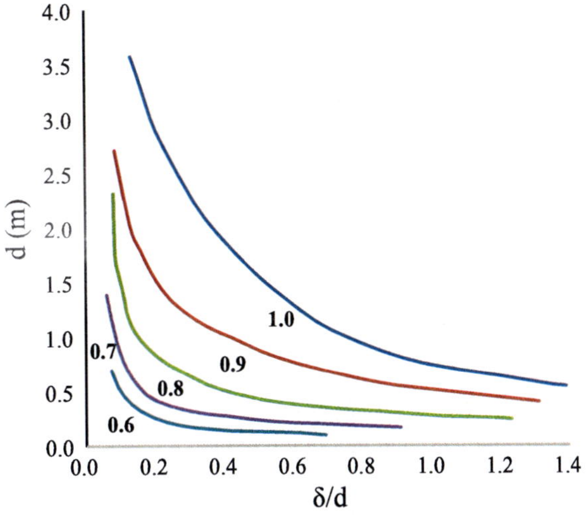

Fig. 8 Matsuo's stability

reinforced granular bed over the soft clay decreases settlements marginally but lateral displacements by a larger amount. Consequently, the ratio of lateral displacement to settlement decreases with the increase of modular ratio (E_1/E_2) from 1 to 100. The curves shift towards to the left from right as shown in Fig. 9. The embankment is on the safer side and the factor of safety more than 1.67 as per the Matsuo plot.

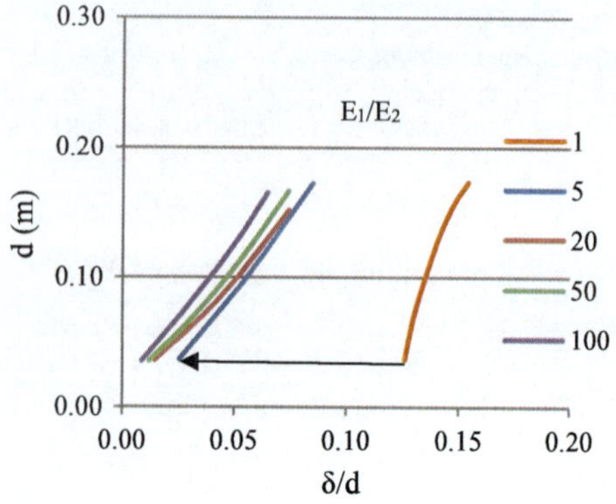

Fig. 9 Matsuo plot for proposed section

4 Conclusions

The influence of stiff granular layer over soft layer on deformations beneath an embankment is estimated for different modular ratios by finite element analysis using the PLAXIS program. Conclusions that can be drawn from this study are:

1. Provision of reinforced granular bed over the soft clay does not contribute to the reduction in settlements.
2. The lateral displacements at the toe of the embankment get reduced up to 60% with the placement of stiff layer over soft layer compared to those without granular bed and lead to the prediction of a higher factor of safety.

References

1. Chai, J.C., Miura, N., Shen, S.L.: Performance of embankments with and without reinforcement on soft subsoil. Can. Geotech. J. **39**(8), 838–848 (2002)
2. Huang, W., Fityus, S., Bishop, D., Smith, D., Sheng, D.: Finite-Element parametric study of the consolidation behavior of a trial embankment on soft clay. Int. J. Geomech. **6**(5), 328–341 (2006)
3. Loganathan, N., Balasubramaniam, A.S., Bergado, D.T.: Deformation analysis of embankments. J. Geotech. Eng. **119**(8), 1185–1206 (1998)
4. Matsuo, M., Kawamura, K.: Diagram for construction control of Embankment on soft ground. Soils Found. **17**(3), 37–52 (1977)
5. Rankine, B., Sivakugan, N.: Short term lateral deformation induced in an elastic medium through the application of rigid embankment loading. In: Proceedings 10th Australia New Zealand on Geomechanics, vol. 1, pp. 214–219 (2007)
6. Safadoust, J., Amiri, S.N., Esamaeily, A.: Numerical analysis of reinforced embankment over foundation. J. Eng. Sci. Technol. Rev. **6**(3), 153–159 (2013)
7. Tavenas, F., Mieussens, C., Bourges, F.: Lateral displacements in clay foundations under embankments. Can. Geotech. J. **16**, 532–550 (1979)

Influence of Eco-Sand Drains on the Performance of Consolidation Characteristics Founded on Soft Clay Deposits

Gowtham Padmanabhan, T. Subitha, and K. S. Kishore

1 Introduction

In the world, major infrastructure development projects get affected due to expansive soils or ultra-soft soil, due to its high compressibility, high moisture content, poor permeability and very low bearing capacity [7, 16, 20]. There are several ground improvement techniques such as Sand drains, Sand compaction piles (SCP), Sand wick drains and Prefabricated vertical drains are available to remediate the problems associated with the soft clay [4, 6]. Sand drains are one of the traditional ground improvement techniques and have been used extensively to accelerate the consolidation process of soft clay subsoil by preloading. This method possesses many advantages. Firstly, it accelerates the consolidation process by reducing the drainage path in a radial direction and also it reinforces the soft clay layer to provide a better bearing capacity of the overall foundation Atkinson and Eldred [2], Long [17], Kirmani [16].

The consolidation of the soil is the process of expelling some volume of water from the pores between the solid particles. Hence, the rate of consolidation is controlled by the compressibility, permeability and maximum length of the drainage path. In this study, the effectiveness of Eco – sand drains were evaluated by modeling its behavior in three gang consolidometer apparatus available in SNS Engineering, Coimbatore. Sand drains are constructed by driving down casings or hollow mandrels into the soil. The holes are then filled with sand, only after casings are removed. When a surcharge is applied at the ground surface, the pore water pressure in the clay will increase, and there will be drainage in the vertical and horizontal directions. The horizontal drainage is induced by the sand drains. Hence, the process of dissipation of excess

G. Padmanabhan (✉)
Geotechnical Engineering Division, Government College of Technology, Tamilnadu, India
e-mail: gowtham@eq.iitr.ac.in

T. Subitha · K. S. Kishore
Department of Civil Engineering, SNS College of Engineering, Tamilnadu, India

C. N. V. Satyanarayana Reddy et al. (eds.), *Ground Improvement and Reinforced Soil Structures*, Lecture Notes in Civil Engineering 152,
https://doi.org/10.1007/978-981-16-1831-4_17

pore water pressure created by the loading and hence the settlement is accelerated Naga and Bouazza [1], Radhakrishnan [19], Deng et al. [5], Nogami [18], Bo et al. [3].

The Eco-sand was collected from ACC Cements, Madukkarai Cement Plant, Coimbatore. Eco-sand is a very fine particle, a by-product from cement manufacture which can be used to increase the efficiency of the soil. The generation of Eco-sand is approximately 500 Tons per day and is disposed of as waste. As sustainability in the geotechnical engineering field is the need of the hour, by utilizing the eco – sand instead of natural river sand in the construction of sand drains, sustainability is attained to a greater extent.

2 Materials

2.1 Soil Sample

The representative soil sample is collected in the SNS college of engineering campus. The locations are 11.0178 N and 76.9380 E. The properties of the soft soil are tabulated in Table 1 [8–15].

Table 1 Properties of soft clay

S. no.	Properties		Results
1	Initial moisture content		13.80%
2	Specific gravity		2.7
3	Dry sieve analysis	% of Gravel	2.4%
		% of Sand	25.3%
		% of Silt & Clay	72.3%
4	Free swell index		55%
5	Liquid limit (w_L)		58%
6	Plastic limit (w_p)		29%
7	Shrinkage limit (w_s)		14%
8	Flow index (I_f)		16.68%
9	Plasticity index (I_p)		29%
10	Soil classification		CH
11	Optimum moisture content		21.6%
12	Maximum dry density		1.682 g/cc

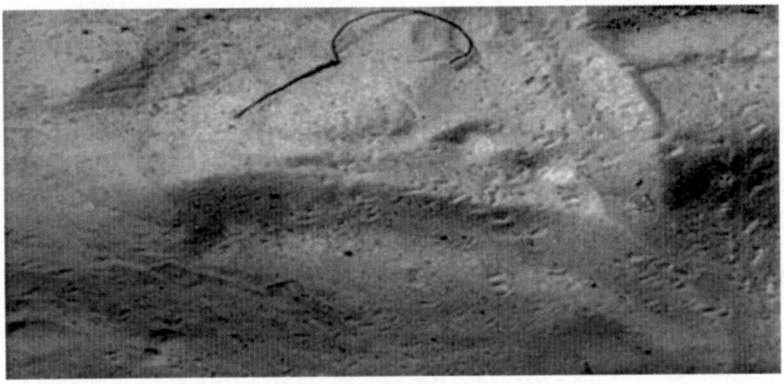

Fig. 1 Eco-sand collected from ACC Cements, Coimbatore

Table 2 Properties of Eco-sand

S. no.	Properties		Results
1	Initial moisture content		4.50%
2	Specific gravity		2.66
3	Dry sieve analysis	% of Gravel	0%
		% of Sand	96.3%
		% of Silt & Clay	3.7%
4	Soil classification		SP
5	Permeability (m/day)		0.252

2.2 *Eco-Sand*

The Eco-sand was collected from ACC Cements, Madukkarai Cement Plant, Coimbatore. The Eco-sand is shown in Fig. 1. The properties of the eco-sand are tabulated in Table 2.

3 Experimental Works

3.1 *Experimental Setup*

The Three gang consolidometer test was conducted as per IS 2720 (Part–15), 1986, using a soil sample of 20 mm thickness and 60 mm diameter.

The area replacement ratio formula used in the study was discussed below,

Area replacement ratio (%) = Area of the improvement zone/Total area of the specimen.

The ratio will vary according to the number of eco-sand drains installed in the test setup. Higher the number of drains; the higher the area replacement ratio.

3.2 Installation of Drains

The Eco-Sand Drains (ESDs) are installed through the mandrels. Circular mandrels are used for the ESDs. The mandrels are installed with utmost care without disturbing the soils to a greater extent. The mandrels are fabricated according to the various sizes and shapes so as to ease the installation. Experimental procedure.

1. For Consolidation testing, it is generally desirable that the applied pressure at any loading stage be double than the preceding stage. The test may therefore be continued using a loading sequence that would successively apply stress of 0.1, 0.2, 0.5, 1.0, 2.0, 4.0, and 8.0 kg/cm2 on the soil specimen.
2. For each loading increment, after application of load, readings of the dial gauge shall be taken using a time sequence such as 0, 0.25, 1, 2.25, 4, 6.25, 9, 16, 25, 36, 49, 64, 81, 100 min and 24 h. These time sequences facilitate plotting of thickness or change of thickness of specimen against the square root of time or against log time.
3. The data concerning dial readings with time for each pressure increment for both loading and unloading stages shall be recorded.
4. The data obtained after specimen assembly concerning the final wet weight of the specimen and the dry weight shall be recorded.

4 Results and Discussion

The main aim of the present study is to identify the optimum configuration of the drains. The results were obtained by carrying out the consolidation test in the three gang consolidometer apparatus strictly following the Indian codal provisions. The dimensions of the eco-sand drains used in the experimental study were listed in Table 3.

Three parameters are considered for predicting the optimum dimensions of the vertical drains

- Coefficient of Consolidation (C_v)
- Compression Index (C_c)
- Time taken for 90% Consolidation

Table 3 Dimensions of Eco-sand drains

S. no.	Replacement ratio (%)	Diameter (mm)	Depth (mm)
1	4	12.5	18
2	6	15	18
3	8	17.5	18
4	10	20	18

The comparison charts are prepared by plotting the e versus log P curve for virgin soil and for all the configurations and the time taken for 90% consolidation was calculated based upon the square root of the time plot method. 1 kg/cm^2 pressure condition is considered for plotting the square root of the time plot graph (Figs. 2, 3, 4 and 5).

The consolidation performance Eco-sand drains initially increases with an increase in dimensions, once it reaches its optimum configuration, the performance of both the vertical drains starts decreasing with an increase in dimensions due to the smear and transition zone effects.

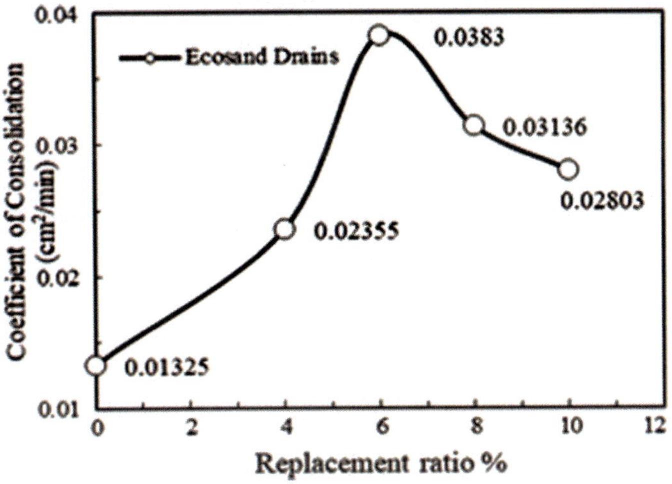

Fig. 2 Variation of co-efficient of consolidation with respect to replacement ratio

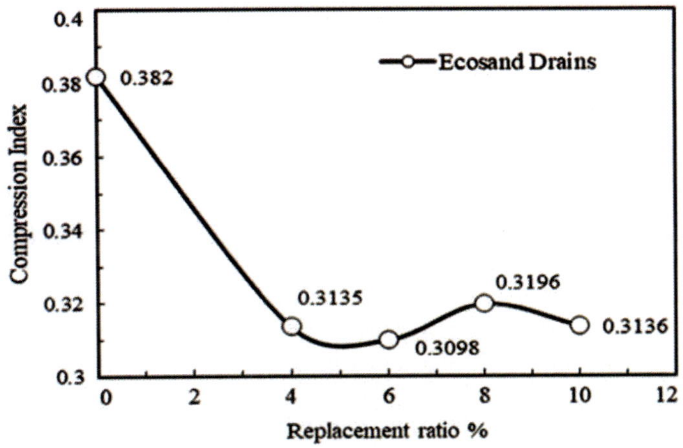

Fig. 3 Variation of compression index with respect to replacement ratio

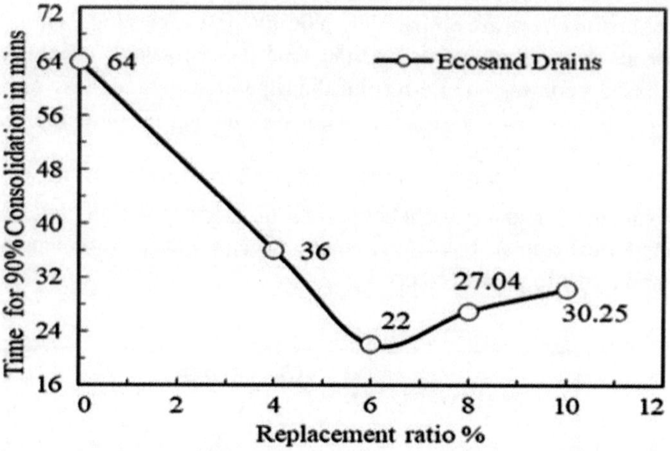

Fig. 4 Time taken to attain 90% consolidation with respect to replacement ratio

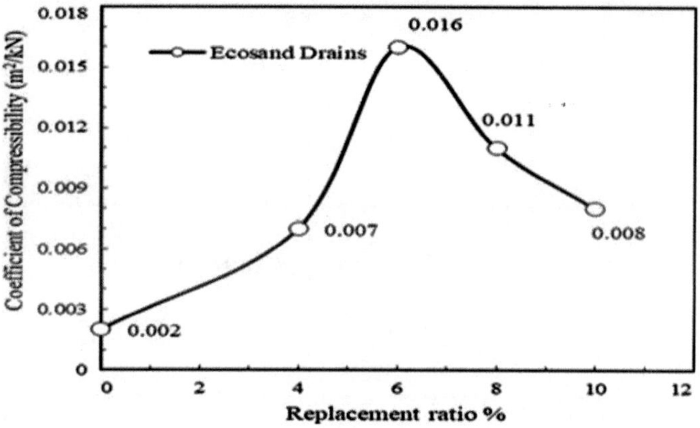

Fig. 5 Variation of co-efficient of compressibility with respect to replacement ratio

5 Conclusions

1. The installation of vertical drains improves the engineering properties and the performance of highly compressible soils to a greater extent and proves to be a promising ground improvement measure.
2. All the results substantiate that the consolidation characteristics of the highly compressible soils increase to a greater extent due to the installation of vertical drains.
3. The decrease in soil compressibility with a reduction of void ratio has a significant impact on consolidation behavior.

4. The performance of 6% area replacement ratio of eco-sand drains was found optimum in enhancing the consolidation characteristics and increases the load-bearing capacity of the soft clay deposits.
5. The Eco-sand was used instead of natural river sand in Sand Drains in this study, shows potential in improving the performance of high compressible soils under lower cost, and this study provides a solution for the disposal problem of Eco-sand and promotes sustainability.

References

1. Abuel-Naga, H.M., Bouazza, A.: Equivalent diameter of prefabricated vertical drain. Geotext. Geomembr. **27**(3), 227–231 (2009)
2. Atkinson, M.S., Eldred, P.J.L.: Consolidation of soil using vertical drains. Géotechnique **31**(1), 33–43 (1981)
3. Bo, M.W., Arulrajah, A., Horpibulsuk, S., Leong, M.: Laboratory measurements of factors affecting discharge capacity of prefabricated vertical drain materials. J. Soils Found. **56**, 129–137 (2017). (Elsevier)
4. Chai, J.C., Shen, J.S.L., Liu, M.D.: Predicting the performance of embankments on PVD improved subsoils. J. Comput. Geotechn. **93**, 222–231 (2018). (Elsevier)
5. Deng, Y.B., Liu, G.B., Liu, M.M., Lie, K.H.: Consolidation behaviour of soft deposits considering the variation of prefabricated vertical drain discharge capacity. J. Comput. Geotech. **62**, 310–316 (2014). (Elsevier)
6. Hansbo, S.: Consolidation of clay by band-shaped prefabricated vertical drains. Ground Eng. **12**(5), 16–18 (1979)
7. Indraratna, B., Zhong, R., Rujikiatkamjorn, C.: An analytical model of PVD–assisted soft ground consolidation. In: Proceedings of "Advances in Transportation Geotechnics 3, (ICTG 2016), vol. 143, pp. 1376–1383 (2016)
8. IS: 1498: Classification and Identification of Soils, Bureau of Indian Standards, New Delhi (1970)
9. IS: 2720-Part 2: Methods of test for soils–Determination of Water Content, Bureau of Indian Standards, New Delhi (1973)
10. IS: 2720-Part 3: Methods of test for soils–Determination of Specific Gravity, Bureau of Indian Standards, New Delhi (1980)
11. IS: 2720-Part 4: Methods of test for soils–Grain Size Analysis, Bureau of Indian Standards, New Delhi (1985)
12. IS: 2720-Part 5: Methods of test for soils–Determination of Liquid and Plastic Limit, Bureau of Indian Standards, New Delhi (1985)
13. IS: 2720-Part 7: Methods of test for soils–Determination of Water Content-Dry Density Relation, Bureau of Indian Standards, New Delhi (1980)
14. IS: 2720-Part 10: Methods of test for soils–Determination of Unconfined Compressive Strength, Bureau of Indian Standards, New Delhi (1991)
15. IS: 2720-Part 16: Methods of test for soils–Laboratory Determination of CBR, Bureau of Indian Standards, New Delhi (1987)
16. Kirmani, S.M.H.: Consolidation of soil for foundation by using Sand Drains. IEP–SAC J. 2004–2005 (2005)
17. Long, D.O.C.: Soil improvement by vertical drains, Ph.D. Thesis, University of Illinois at Urbana-Champaign (1991)
18. Nogami, T.: Consolidation settlement with sand drains–analytical and numerical approaches. J. Geotech. Geoenvironmental Eng. Am. Soc. Civil Eng. (ASCE) **129**(9), 838–848 (2015)

19. Radhakrishnan. G.: Study of consolidation accelerated by Sand Drains. In: Proceedings of "Indian Geotechnical Conference (IGC)–2010", GEOtrend, IIT Bombay (2010)
20. Shadab, G.: Ground Improvement using preloading with prefabricated vertical drains: a case study. In: Proceedings of "Indian Geotechnical Conference (IGC)–2013", IIT Roorkee (2013)

Effect of Fiber Reinforcement on the Strength of Geopolymerised Soil: An Experimental Investigation and Numerical Modeling

Mazhar Syed, Yash V. Kasat, Gaurav B. Sarda, and Anasua GuhaRay

1 Introduction

The dual nature of black cotton soil (BCS) exhibits low volumetric stability upon moisture imbalance, leading to cause destruction of lightweight structures founded on them. Convention cementitious binders like lime and cement are the most utilizable binders for strengthening the geoengineering characteristics of expansive soils and efficiently reduce the problems associated with swelling and shrinkage. The production of these binders, however, contributes 7–8% of annual greenhouse gases around the world. The global demand for portland cement-based (PC) binders has increased and can reach up to 200% by the end of 2050 [1]. Usage of low carbon emission binders such as Alkali Activated Binder (AAB) is an alternative cementitious product for effectively reinforcing the soil. Typically, AAB is a long-chain polymeric sodium aluminosilicate compound, synthesized from alumina and silica-rich precursors through geopolymerisation [2]. AAB helps to eliminate the demand for Portland cement binders by overcoming the issue of disposing of fly ash and slag, thereby saving the related costs during landfills. In terms of energy, costs, and environmental impacts, AAB is recorded to produce almost 80% less CO_2 compared to Portland cement. In comparison, AAB's global warming potential is 70% lesser than the PC-based binders [3]. AAB also possessed early strength gain with low hydration heat, the higher resistance to acid and sulfate attack, excellent durability, improved freeze-thaw resistance, superior workability, and binding properties relative to hardened cement binders [4]. A substantial amount of progress was recorded over the last decades in applying geopolymerisation in ground improvement [5–7]. Although the aluminosilicate precursor-based AAB-treated soils efficiently enhance the compressive shear strength, but it exhibits weak tensile and flexure resistance. The existence

M. Syed · Y. V. Kasat · G. B. Sarda · A. GuhaRay (✉)
BITS-Pilani Hyderabad Campus, Secunderabad 500078, Telangana, India
e-mail: guharay@hyderabad.bits-pilani.ac.in

C. N. V. Satyanarayana Reddy et al. (eds.), *Ground Improvement and Reinforced Soil Structures*, Lecture Notes in Civil Engineering 152,
https://doi.org/10.1007/978-981-16-1831-4_18

of shrinkage cracking is crucial in the summer season when this type of soil is encountered. The problem of shrinkage cracking can be effectively dealt with by including discrete fibers [8, 9]. Over the last few years, discrete fibers (natural/artificial) gained popularity in cemented soil stabilization owing to their higher tensile and durability [10, 11]. However, not many studies were conducted to compare the geomechanical behavior of both natural/artificial fiber-reinforced AAB soils. In the present study, an effort has been made to compare the geoengineering characteristics between polypropylene fiber (PF) and chemically modified coir fiber (TCF) reinforced AAB treated soil at different fly ash-slag proportions. To determine the effectiveness of both fiber-AAB-soil mixtures, a series of microstructural and geotechnical tests are conducted at different dosages of fibers. Based on the experimental results, a numerical analysis is performed using commercially available finite element software to correlate the flexural failure patterns for both PF and TCF reinforced soil beam.

2 Materials and Methods

2.1 Raw Materials

Black cotton soil (BCS) was collected from the Nalgonda region of Telangana state, while class F fly ash and slag were directly obtained from Ramagundam National Thermal Power plant and JSW Cement Ltd., Andhra Pradesh, respectively. BCS was classified as high plasticity clay (CH) according to the Unified Soil Classification System (USCS) and found to contain 78% clay. A constant length of 12 mm length of polypropylene and 25 mm length for coir fiber were obtained from Go-green product industries, Tamil Nadu. Coir fiber was chemically treated with 10 M of NaOH before inclusion as soil reinforcement. The different physicochemical and engineering properties of raw materials are provided in Table 1 as per ASTM codes.

Table 1 Properties of raw materials

Soil Properties	Values	Fiber	PF	CF
Specific gravity	2.65	Diameter (μm)	33	30
Free swell index (%)	88.30	Density (g/cc)	0.91	0.88
Moisture content (%)	24.0	Cellulose (%)	–	45
Liquid limit (%)	62.0	Hemicellulose (%)	–	1
Plasticity index (%)	38.0	Lignin (%)	–	48
Dry density (g/cc)	1.65	Ash (%)	–	3
Tensile strength (kPa)	25.8	Tensile strength (mPa)	330	94
Compression strength (kPa)	187	Elastic modulus (mPa)	350	330
Flexural strength (kPa)	100	Melting Point (^{0}C)	168	–

2.2 *Alkali Activated Binder (AAB)*

AAB was prepared by mixing sodium silicate, sodium hydroxide, and aluminosilicate precursors (fly ash and slag) with a mass ratio of 129:10.57: 400 [3, 12]. Both sodium silicate and sodium hydroxide chemicals were obtained from Hychem Chemicals Ltd. in the form of solution and pellets, respectively. Fly ash and slag proportions were varied in the AAB mixture to obtain an optimum mix.

2.3 *Sample Preparation*

BCS was uniformly mixed with 5% of AAB paste (total dry mass of soil) by varying fly ash and slag content with a constant 0.4 w/s ratio in the chemical binders. AAB blended soil was compacted in a container of (950 × 480 × 150) mm in 3 layers with a 9 kg steel rammer under a free fall of 310 mm. The compacted soil was covered with moist jute bags for up to 48 h to remove excess heat during the geopolymerization reaction. Prior to mixing either PF or TCF (0.5 and 1% by mass of soil) in the AAB-treated BCS, it was oven-dried. A series of microstructural and geotechnical tests were performed for both fiber-AAB composite BCS.

2.4 *Geotechnical Characteristics*

A series of microstructure and geotechnical tests were performed on both untreated soil and fiber-reinforced AAB mixed soil. The influence upon the addition of PF and TCF in AAB treated soil was analyzed through unconfined compressive strength split tensile strength (STS) and flexure strength tests. UCS test on both untreated soil and fiber-reinforced AAB treated soils were conducted as per ASTM D2166 [13] standard by compacting them in a split cylindrical mold of 76 mm height with an inner diameter of 38 mm at their MDD-OMC values, respectively. A static strain rate of 1.25 mm/min was maintained throughout the test.

STS tests were conducted as per ASTM D3967 [14] on the Marshal Stability machine by attaching a loading strip of 12.5 mm on the load frame with a constant strain rate of 50.5 mm/min. Both AAB treated and fiber-reinforced soils were compacted in a cylindrical mold of 100 mm diameter and 80 mm thickness. The following formula calculates split tensile strength (S_t)

$$S_t = 2p/td\,\Pi \tag{1}$$

where

P = ultimate load at which failure of sample occurred (N),
t = thickness of specimen (mm),
d = diameter of the specimen (mm).

The flexure strength test was conducted using a three-point bending flexure machine as per ASTM D-1635 [15] standard by molding a soil beam of 280 mm × 70 mm × 70 mm into five layers with a 3 kg steel rammer under a free fall of 310 mm. Flexural strength tests (S_f) mPa calculated by using the formula as follow.

$$S_f = 3pl/2bh^2 \tag{2}$$

P = ultimate breaking load at which sample failure occurred (N),
l = length of support on beam specimen (mm),
b = width of the beam specimen (mm), h = depth of the beam specimen (mm).

2.5 *Numerical Modeling*

The soil beam was simulated using three-point bending flexure in commercially available finite element software ABAQUS. The basic soil, AAB, and fiber material properties (such as elastic modulus, Poisson ratio, and c-φ parameters) were used in the finite element model to track flexural resistance variations. In addition, the soil beam was assumed to be an elastic material, and rigid supports were provided. To demonstrate the flexural behavior of the soil beam, a load versus deflection chart was developed and compared with the experimental results.

3 Results and Discussion

3.1 *Fourier Transform Infrared Spectroscopy (FTIR)*

FTIR molecular bonding was analyzed using a JASCO FTIR 4200 setup with K.Br. Pellet arrangement under 4000-400 cm^{-1} spectral range. O-H stretching vibrations' transmittance peaks were found around 3600 cm^{-1} for both untreated and AAB-treated soils in Fig. 1. A slight reduction in the intensity of AAB and fiber-reinforced soils was found at 3600 cm^{-1}, followed by the O-H hydroxyl alcohol group at 3420 cm^{-1}. This marginal change in bonding may be due to clay particles' chemical weathering action [16, 17]. The broadband detected at 1630 cm^{-1}, and 1440 cm^{-1} correspond to C = C alkene and CH_2 bending vibrations in both fiber-reinforced soil. This carbonation reaction may be induced due to cellulose's active existence in the fiber alkaline matrix [12]. Moreover, a sharp band attributed to Si-O-Si's asymmetric stretching vibration around 1050 cm^{-1} in the clay particles. The combined mixture of aluminosilicate compounds and fiber alkalinity in the soil showed a broader peak of (Al)-O bending vibration at 790 cm^{-1} along with the Si-O stretching group around 510-490 cm^{-1}, respectively. Thus, the spectrum peaks from untreated soil, AAB,

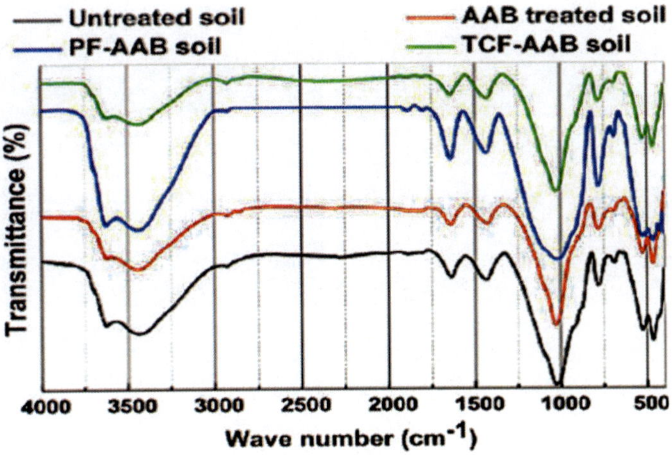

Fig. 1 FTIR spectroscopy of untreated and fiber AAB-treated soil

and fiber-reinforced AAB treated soil showed similar bonds with a chemical shift of about 10 cm^{-1}.

3.2 *Stereomicroscope and SEM Images*

The soil's physical features were captured at different magnifications using the stereomicroscope of SX27 Olympus with the least dimension of 20 μm. Surface micrographs were captured using a Thermo Scientific Apreo SEM provided by Field Electron-Ion Company. 20 kV excitation voltage with the help of the Gentle beam of electromagnetic lenses was adapted to screen the surface. Figure 2a–h displayed the stereomicropic and SEM images of untreated, AAB treated, and fiber-reinforced AAB soil, respectively. Untreated soil revealed a light brownish color region, which indicated the occurrence of the smectite group. Irregular aggregated particles in the form of vitreous texture were observed with minor surface cracks in Fig. 2a–b. The hardened AAB paste deposit by filling the pores between the clay particles was observed in Fig. 2c. Moreover, a series of spherical particles with different sizes were detected in Fig. 2d. The occurrences of irregular hallow spherical particles on the soil surfaces might be due to the presence of fly ash. Figure 2e–f showed the densely compacted microstructure of the PF-AAB soil matrix, which interlocked the particles by forming a spatial thread groove network [8]. The vigorous morphology changes were also noticed in TCF-AAB soil in Fig. 2g–h which acts as bridge surfaces. Thus, the combined fiber-AAB-soil mixtures aid in controlling the tensile cracking behavior with a higher linkage effect.

Fig. 2 Stereomicroscopic and SEM images of **a–b** Untreated soil, **c–d** AAB treated soil, **e–f** PF-AAB treated soil, **g–h** TCF reinforced AAB soil

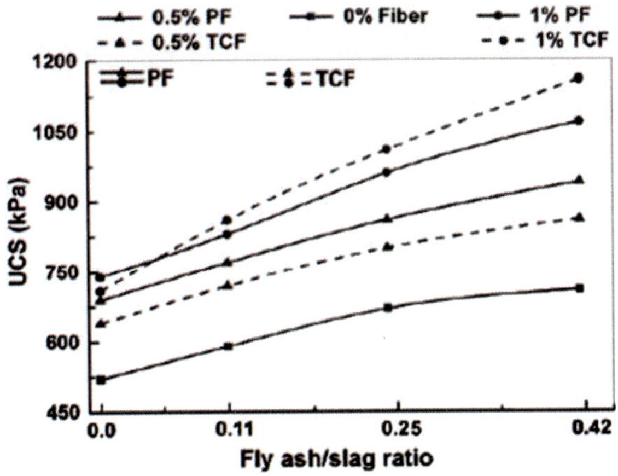

Fig. 3 Variation of UCS for PF and TCF reinforced AAB treated soil at different fiber dosages

3.3 Unconfined Compression Strength (UCS)

Figure 3 plotted UCS of two kinds of fiber-reinforced AAB soil at varying fibers and fly ash-slag ratio. The solid lines indicated the PF-AAB, and dotted lines represented the TCF-AAB-soil mixture. The figure revealed that the combined addition of fibers in the fly ash slag-based AAB soils had a significant effect on the rate of compressive shear strength gain. As the replacement of fly ash with slag content increased, the rate of geopolymerisation reaction and confinement bonding efficiency of soil-fiber was greatly enhanced. Moreover, PF-AAB UCS increased by over 15% compared to TCF-AAB at lower fiber dosages. The relative gain in PF's shear strength upon TCF reinforced AAB soil might be attributed to the development of higher interfacial surface roughness, which mobilized the friction during loading [9, 11]. It was also interesting to note that TCF's initial addition did not cause much impact on stiffness behavior. The combined effects of TCF (beyond 0.5%) and 70/30 fly ash/slag attained higher shear strength among all compositions. In addition, the marginal improvement was observed in the PF-AAB reinforced soil at a higher dosage (1%) in the soil. This might be due to smooth surface texture formation, which does not easily allow the soil to compact [8, 18]. Thus, the combined addition of fibers in AAB can efficiently enhance the compression shear strength with higher linkage effects.

3.4 Split Tensile Strength (STS)

Figure 4 showed the tensile strength of fiber-reinforced soil specimens with varying fly ash-slag and fiber dosages. It was found that, when the fiber dosage was constant, the ITS results for AAB soils were almost similar, regardless of the fiber type and fly ash/slag proportions. The gain in tensile strength followed a parallel pattern to that of UCS for both PF and TCF reinforced soils. Upon comparison, the PF-AAB

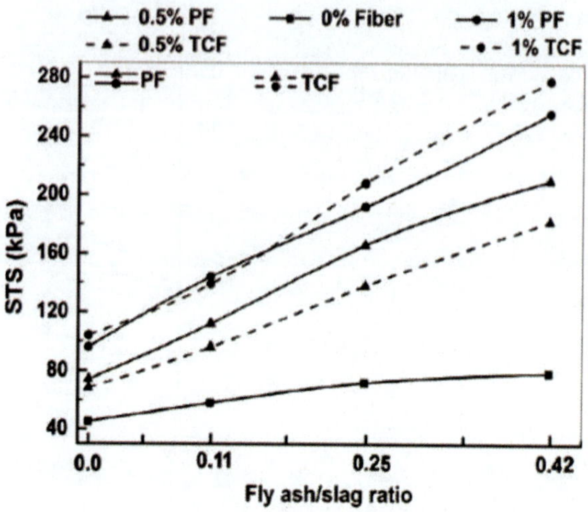

Fig. 4 Variation of STS for PF and TCF reinforced AAB treated soil at different fiber dosages

showed greater stretching and interlocking density over TCF-AAB soil mixtures. This beneficial effect of PF reinforcement might be attributed to its higher interfacial frictional resistance across the soil cementing matrix [10, 18]. Furthermore, PF had actively regulated the strain cracks along soil fiber's failure plane by producing a spatial thread groove network. Hence, the increase in slag dosages with fiber reinforcement in the AAB soil aided by strengthening the tensile resistance capacity and restricting the relative movement of fiber.

3.5 *Flexural Strength (S_f)*

Figure 5a plotted flexural strength of PF and TCF reinforced at two different percentages (0.5 and 1%) in the fly ash-slag-based AAB soils. It can be seen that both fibers enhanced the rate of increase in flexural resistance with an increase in slag and fiber dosages. The improvement was much more significant in PF-AAB soil relative to the TCF-AAB-soil mixture. The relative gain in flexural and shear strength in PF reinforced soil might be due to their higher contact area (confinement bonding) and frictional resistance [19, 20]. Moreover, slag-based AAB had the potential to reinforce the soils through active cementation compounds due to the rapid dissolution of pozzolanic precursors in the presence of reactive alumina and silica soil substances. Thus, the vigorous changes in soil fiber interlocking density and surface morphology enhanced the flexural resistance potentially. Figure 5b shows the schematic soil beam arrangement and load supports.

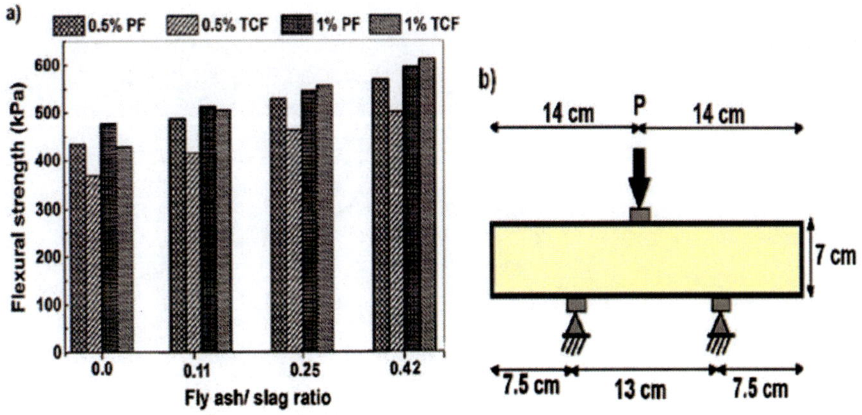

Fig. 5 **a** Variation of flexural strength for PF-TCF-AAB soil **b** Typical flexural soil beam set up

3.5.1 Finite Element Model

Figure 6 shows the three-point bending soil flexural beam model before the application of load. The meshed 2.5 mm × 2.5 mm soil beam was shown in Fig. 6a. As the load's application increased, the soil beam significantly deformed and failed at the midpoint on the beam's bottom side. A maximum 4.5 mm deflection for untreated soil occurred at the midpoint, which was shown in Fig. 6b. Similar studies are carried out for TCF-AAB and PF-AAB soil. Figure 7 showed the load-deflection curve obtained from the experimental and finite element analysis of untreated soil, TCF, and PF reinforced AAB soil beam. The results of the finite element analysis indicated a good correlation with that of the experimental results.

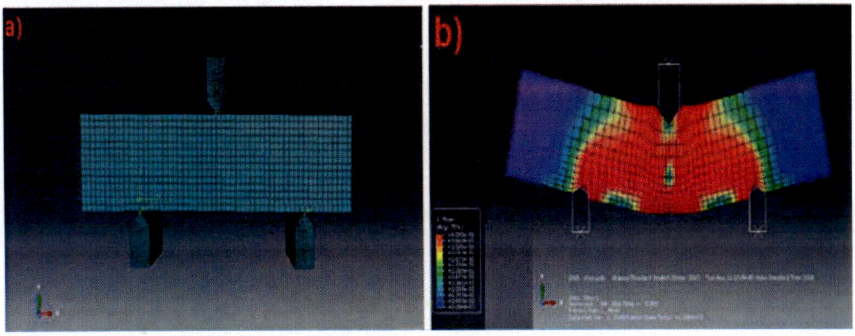

Fig. 6 **a** Load actuator soil beam model setup **b** Soil beam deflection at mid-span

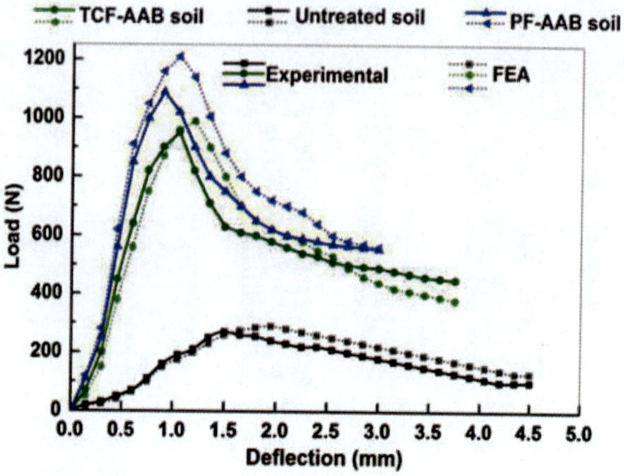

Fig. 7 Variation of flexural load and deflection curve for soil beam

4 Conclusions

The present experimental study investigated the effects of fiber reinforcement in alkali-activated binder (AAB) treated expansive soil. The following conclusions can be drawn from this analysis:

1. The PF-AAB soil showed higher strength performance than the TCF-AAB-soil mixture. UCS and STS of AAB-BCS increased by 48% and 38% for PF, and 39% and 33% for TCF reinforced soil when the fiber dosages were low.
2. Greater mobilization of friction and interfacial bonding helped PF to yield higher mechanical strength performance than TCF-AAB-soil, irrespective of dosage of fiber and slag in the AAB compound.
3. The combined addition of AAB and fiber had a significant effect on the rate of gain of flexural strength; these values notably increased with fiber dosages and fly ash/slag ratio.
4. Micrographs of fiber-AAB soils revealed new surface morphology, and it can act as a spatial thread-bridge network, which improves the particle holding and interlocking density around the fiber surfaces.
5. PF-AAB soil mixture attained the highest flexural strength upon TCF-AAB soil mixture. At higher fiber dosages, both PF and TCF play a substantial role in regulating flexure crack development.
6. The flexural failure pattern obtained from the finite-element model showed similar trends to that of the physical test failure in the laboratory.

References

1. Taylor, M., Tam, C., Gielen, D.: Energy efficiency and CO2 emissions from the global cement industry energy efficiency and CO2 emission reduction potentials and policies in the cement. IEA-WBCSD Workshop, pp. 4–5 (2006)
2. Davidovits, J.: Properties of geopolymer cements, Alkaline Cements and Concretes, Kiev, Ukraine, pp. 1–19 (1994). https://doi.org/10.1073/pnas.0811322106
3. Mazhar, S., Guharay, A.: Stabilization of expansive clay by fibre-reinforced alkali-activated binder: an experimental investigation and prediction modelling (2020). https://doi.org/10.1080/19386362.2020.1775358
4. Provis, J.L., van Deventer, J.S.J.: Alkali Activated Materials, 13th ed. Springer International Publishing, UK (2014). https://doi.org/10.1007/978-94-007-7672-2
5. Miao, S., Wei, C., Huang, X., Shen, Z., Wang, X., Luo, F.: Stabilization of highly expansive black cotton soils by means of geopolymerization. J. Mater. Civ. Eng. **29**, 04017170 (2017). https://doi.org/10.1061/(asce)mt.1943-5533.0002023
6. Komljenović, M., Baščarević, Z., Bradić, V.: Mechanical and microstructural properties of alkali-activated fly ash geopolymers. J. Hazard. Mater. **181**, 35–42 (2010). https://doi.org/10.1016/j.jhazmat.2010.04.064
7. Murmu, A.L., Dhole, N., Patel, A.: Stabilisation of black cotton soil for subgrade application using fly ash geopolymer. Road Mater. Pavement Design. **0**, 1–19 (2018). https://doi.org/10.1080/14680629.2018.1530131
8. Tang, C., Shi, B., Gao, W., Chen, F., Cai, Y.: Strength and mechanical behavior of short polypropylene fiber reinforced and cement stabilized clayey soil. Geotext. Geomembr. **25**, 194–202 (2007). https://doi.org/10.1016/j.geotexmem.2006.11.002
9. Consoli, N.C., Arcari Bassani, M.A., Festugato, L.: Effect of fiber-reinforcement on the strength of cemented soils. Geotext. Geomembr. (2010). https://doi.org/10.1016/j.geotexmem.2010.01.005
10. Sudhakaran, S.P., Sharma, A.K., Kolathayar, S.: Soil stabilization using bottom ash and areca fiber: experimental investigations and reliability analysis. J. Mater. Civ. Eng. **30**, 1–10 (2018). https://doi.org/10.1061/(ASCE)MT.1943-5533.0002326
11. Moghal, A.A.B., Chittoori, B.C.S., Basha, B.M., Al-Shamrani, M.A.: Target reliability approach to study the effect of fiber reinforcement on UCS Behavior of lime treated semiarid soil. J. Mater. Civ. Eng. **29**, 04017014 (2017). https://doi.org/10.1061/(asce)mt.1943-5533.0001835
12. Gupta, S., GuhaRay, A., Kar, A., Komaravolu, V.P.: Performance of alkali-activated binder-treated jute geotextile as reinforcement for subgrade stabilization. Int. J. Geotech. Eng. 1–15 (2018). https://doi.org/10.1080/19386362.2018.1464272
13. ASTM D 2166: Standard test method for unconfined compressive strength of cohesive soil (2006)
14. ASTM D 3967: Standard Test Method for Splitting Tensile Strength of Soil Specimens, ASTM D3967 West Conshohocken, PA (2016)
15. ASTM D1635: Standard Test Method for Flexural Strength of Soil-Cement Using Simple Beam with Third-Point Loading, ASTM D1635West Conshohocken, PA (2019)
16. Adhikari, B., Khattak, M.J., Adhikari, S.: Mechanical and durability characteristics of flyash-based soil-geopolymer mixtures for pavement base and subbase layers. Int. J. Pavement Eng. **0**, 1–20 (2019). https://doi.org/10.1080/10298436.2019.1668562
17. Mwaikambo, L.Y., Ansell, M.P.: Chemical modification of hemp, sisal, jute, and kapok fibers by alkalization. J. Appl. Polym. Sci. **84**, 2222–2234 (2002). https://doi.org/10.1002/app.10460

18. Pourakbar, A.: Application of alkali-activated agro-waste reinforced with wollastonite fibers in soil stabilization. J. Mater. Civ. Eng. **29**, 04016206 (2016). https://doi.org/10.1061/(asce)mt.1943-5533.0001735
19. Anggraini, V., Asadi, A., Farzadnia, N., Jahangirian, H., Huat, B.B.K.: Reinforcement benefits of nanomodified coir fiber in lime-treated marine clay. J. Mater. Civ. Eng. **28**, 1–8 (2016). https://doi.org/10.1061/(ASCE)MT.1943-5533.0001516
20. Correia, A.A.S., Venda Oliveira, P.J., Custódio, D.G.: Effect of polypropylene fibres on the compressive and tensile strength of a soft soil, artificially stabilised with binders. Geotext. Geomembr. **43**, 97–106 (2015). https://doi.org/10.1016/j.geotexmem.2014.11.008

Strength Behavior of Polypropylene Fiber Reinforced GGBS Based Geopolymer Clay Blends

T. V. Nagaraju, Ch. Sarada Sharmila, and A. S. P. V. Santhoshi

1 Introduction

In the last few decades, green materials and their applications in the civil engineering infrastructure have been gaining significant attention. Geopolymers are the emerging green materials comprised of silica and alumina for the development of geopolymerization products. In general, sodium silicate gel and sodium hydroxide or potassium hydroxide pellets are used to prepare the alkali-activator [1]. The alkali-activators were reacted with the Al and Si ions present in the precursor and form three-dimensional Si-Al chains with strong bonds. And also, the formation of Si–O-Al bonds leads to a dense geopolymer matrix with mild brittle behavior when exposed to thermal curing specimens [1, 5]. Fly ash-based geopolymers are commonly used blends because of fly ash consists of rich aluminum trioxide and silicon dioxide, which helps in the formation of N-A-S–H compounds [7]. Moreover, other precursors such as ground granulated blast furnace slag (GGBS), silica fume and rice husk ash (RHA) are also useful for the acceleration of condensation and geopolymerization process [9, 11]. Moreover, in geopolymer soils, the minerals present within the soil contributes pozzolanic reactions when Ca^{2+} ions in the pozzolanic material reacts with Si^{4+} and Al^{3+} ions in the clay form cementitious compounds of calcium silica hydrate (C-S–H), calcium alumina hydrate (C-A-H) and C-A-S–H [9]. Both geopolymerization products and cementitious products result in high stiffness and moderate brittle behavior.

At present, in geopolymers research, there is a shift towards the necessity or potential of fibers imbedded in the geopolymer blends in order to improve flexural resistance. This can be helpful for a wide range of geopolymer applications in civil infrastructure. The utilization of fibers embedded in cement concrete has been in

T. V. Nagaraju (✉) · Ch. S. Sharmila · A. S. P. V. Santhoshi
Department of Civil Engineering, S.R.K.R. Engineering College, Bhimavram 534204, India
e-mail: tvnraju@srkrec.edu.in

C. N. V. Satyanarayana Reddy et al. (eds.), *Ground Improvement and Reinforced Soil Structures*, Lecture Notes in Civil Engineering 152,
https://doi.org/10.1007/978-981-16-1831-4_19

vogue, and since many studies related to fibers reinforced soils also increased tremendously [2, 4, 10]. The strength characteristic of the fiber-reinforced soils depends on the type of fiber, amount of fiber and aspect ratio of fiber. The fiber inclusion in expansive clays shows significant enhancement in swell-shrink behavior, i.e. rate of heave and swelling pressure. Moreover, during swell-shrink behavior, the contact area of fibers exposed to soil blends allows greater internal friction between the soil and fibers. Further, increases the greater resistance against swelling and shrinkage. Friction resistance of fiber-reinforced soil mainly depends on the aspect ratio, that is when the length is more interlocking performance is high, and when the width is more there is an improvement in the contact area, thereby swelling resistance increases [8]. And also, the inclusion of fibers such as natural fibers or synthetic fibers shows significant enhancement in shear strength, permeability, compressibility and bearing capacity [4, 6, 8].

The objective of this paper is fiber embedded in the geopolymer clays is to date limited and the tensile behavior of the geopolymer blends reinforced with fibers is focused in this research work. Unconfined compressive strength and split tensile strength tests were conducted on the geopolymer soil blends with varying precursor and fiber content. SEM analysis was carried to characterize the GGBS, clay and treated geopolymer clays.

2 Experimental Investigation

2.1 Materials

The expansive clay used in this study is a locally available in Bhimavaram, India. The expansive clay consists of montmorillonite material and having high swelling and shrinkage behaviour. GGBS was used in this study as a precursor for effective condensation and geopolymerization process, which is collected at the local market in Vijayawada, India. Other raw materials used in this investigation were sodium hydroxide pellets and sodium silicate gel. Moreover, for determining split tensile strength, polypropylene fibers were used as a reinforcement material embedded in geopolymer soil blends. The index properties of expansive clay are shown in Table 1.

Table 1 Index properties of expansive clay

Properties	Liquid limit, %	Plastic limit, %	Soil Classification (IS)	Specific gravity	Fines, %	Free swell index, %
Value	86	24	CH	2.66	94	180

2.2 *Sample Preparation*

Geopolymer solution was prepared using 10 molar NaOH solution and Na_2SiO_3 gel, and the ratio of $Na_2SiO_3/NaOH$ is maintained as 2.5. Flat expansive clay used in this study was powdered into smaller sizes using mechanical tampers and then sieved into a 4.75 mm I.S. sieve. Pulverized soil sample was oven-dried for 24hours at a temperature of 110 °C. It was mixed with a prefixed amount of geopolymer solution and precursor content was mixed homogeneously and compacted in a standard cylindrical mould of size 38 mm diameter and 76 mm height using a manual static compactor.

2.3 *Methods*

Many researchers have suggested different approaches to evaluate mechanical behaviour of soils using unconfined compressive strength (UCS), direct tensile strength test, split tensile strength test (STS) and flexural strength test. This study was comprised of UCS and STS on geopolymer blended expansive clays. The specimens for the UCS and STS were prepared in the standard cylindrical moulds of dimensions 38 mm diameter and 76 mm height. UCS test was conducted on geopolymer soil blends by varying amount of precursor as 0, 5, 10, 15, 20 and 25% by dry of soil. The specimens were loading at a rate of 1 mm/min until the specimen failed.

For determining split tensile strength of geopolymer soil blends, cylindrical specimens of the same dimensions used in the UCS were prepared. The load is applied on the surface center line on each side of the support. A split tensile strength test was conducted for cylindrical soil specimens placed horizontally in between the compressed blocks in the compression test in accordance with the Kumar et al. [3].

Scanning electron microscopy (SEM) tests were carried out on the GGBS based geopolymer blends to know about surface texture and developed geopolymer products.

3 Results and Discussion

3.1 *Effect of GGBS Based Geopolymers on Unconfined Compressive Strength*

The stress–strain curves of the GGBS based geopolymer clay blends for different percentages of GGBS content in the blends are shown in Fig. 1. It shows that, for a given strain, maximum axial compression stress increases with increasing precursor content. UCS values (kg/cm^2) of GGBS based geopolymer clay blends were 8.1, 22.8, 24.71, 26.33, 30.45 and 32.24 when GGBS content varied as 0, 5, 10, 15, 20

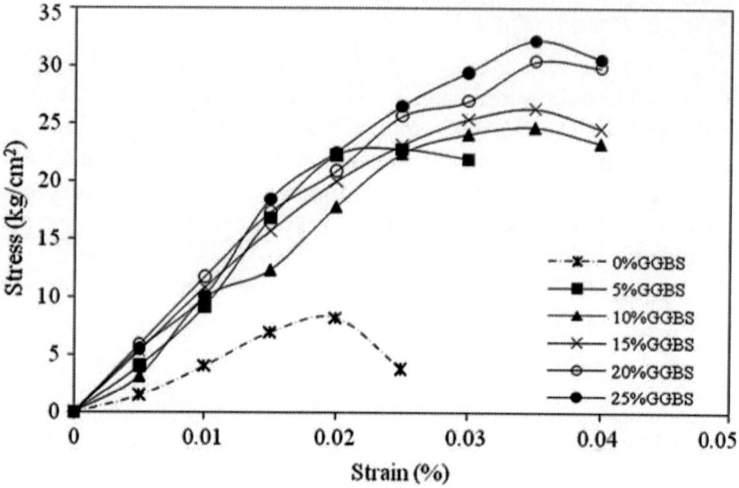

Fig. 1 Stress–strain behavior of untreated clay and GGBS based geopolymer clays

and 25% by dry weight of soil respectively. The results showed the enhancement of strength is due to the increasing precursor content, i.e. increasing Si/Al ratio.

3.2 *Effect of GGBS Based Geopolymers and Polypropylene Fibers on Split Tensile Strength*

Split tensile strength results of the GGBS based geopolymer mixtures, with and without polypropylene fiber inclusion are shown in Table 2. It can be seen that, for a given precursor content, the inclusion of polypropylene shows significant improvement in the tensile behavior. This effect is more evident in the higher percentage precursor 920% and 25% GGBS) blends. The fibers embedded in the dense geopolyme matrix contribute adhesion or internal friction between the fibers

Table 2 Effect of GGBS and polypropylene on the split tensile strength

GGBS content (%)	Split tensile strength (kg/cm^2) of geopolymer soil blends with varying polypropylene fiber content as					
	0%	0.2%	0.4%	0.6%	0.8%	1.0%
0	0.58	0.65	2.15	3.08	4.75	3.65
15	0.63	2.85	5.45	6.18	9.24	7.20
20	0.88	3.20	7.22	8.64	11.18	9.15
25	1.05	4.55	9.85	11.20	13.26	10.85

and geopolymer blends. The higher percentage of fines presence in the GGBS enhances the dispersion of fibers in the geopolymer matrix.

3.3 Micro-Analysis

Micrographs of GGBS-based geopolymers show that alkali-activators react with the precursor, when GGBS was activated with 10 M of NaOH. The microstructure of the GGBS geopolymer clay blends at higher GGBS content (20 and 25%) clearly shows micro-cracks and pockets on the surface do not effective on strength characteristics due to strong Si–O-Si bonds. Moreover, pockets on the surface are formed due to the higher concentration leaching of Al and Si oxides from GGBS by alkali dissolution (vide Fig. 2c and d).

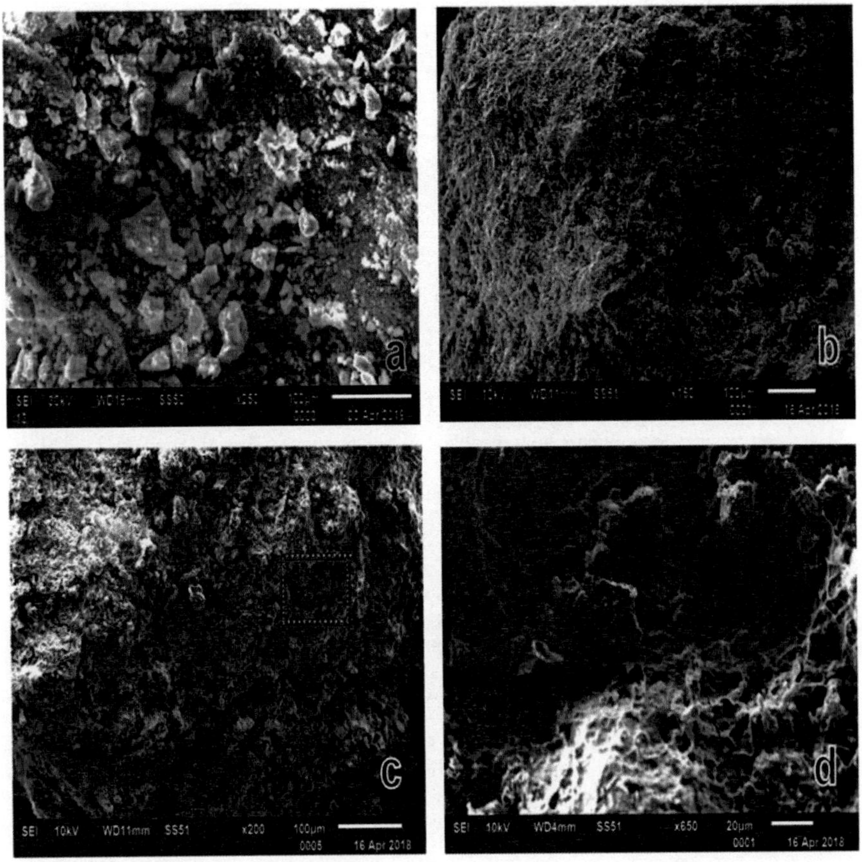

Fig. 2 SEM micrographs of GGBS geopolymer clay blends with varying GGBS content: **a** 0% GGBS, **b** 15% GGBS **c** 20% GGBS **d** 25% GGBS

Figure 2a and b, by comparison, it was clearly evident that a dense matrix was found when the precursor is added (vide Fig. 2b). The addition of precursor content to the geopolymer soils contributes to the effective condensation and geopolymerization process. Further, allows the formation of geopolymerization products (hydrates of Na-S-Al).

4 Conclusions

The utilization of geopolymer soils in civil engineering infrastructure presents a sustainable opportunity by bulk utilization of industrial by-products. Previous studies on strength characteristics of geopolymer soils show positive results. Moreover, studies on the flexural and tensile behavior of geopolymer soils were limited. This study provides the effect of fibers on the tensile behavior of expansive clays. The following conclusions were drawn from this work.

The UCS values of GGBS based geopolymers increased with increasing precursor content. Results also suggested that bulk utilization of GGBS is possible while using geopolymers (up to 25%).

Split tensile strength of the geopolymer clay blends were improved with increasing polypropylene fibers content up to 0.8%. SEM micrographs are helpful to explain the formation of geopolymer products.

Finally, it can be concluded that the inclusion of polypropylene fibers into the geopolymer clays can give scope for a wide range of applicability.

References

1. Davidovits, J.: Geopolymers and geopolymeric materials. J. Therm. Anal. **35**(2), 429–441 (1989)
2. Kumar, A., Gupta, D.: Behavior of cement-stabilized fiber-reinforced pond ash, rice husk ash–soil mixtures. Geotext. Geomembr. **44**(3), 466–474 (2016)
3. Kumar, A., Walia, B.S., Bajaj, A.: Influence of fly ash, lime, and polyester fibers on compaction and strength properties of expansive soil. J. Mater. Civ. Eng. **19**(3), 242–248 (2007)
4. Kumar, A., Walia, B.S., Mohan, J.: Compressive strength of fiber reinforced highly compressible clay. Constr. Build. Mater. **20**(10), 1063–1068 (2006)
5. Pan, Z., Sanjayan, J.G., Rangan, B.V.: Fracture properties of geopolymer paste and concrete. Mag. Concr. Res. **63**(10), 763–771 (2011)
6. Prabakar, J., Sridhar, R.S.: Effect of random inclusion of sisal fibre on strength behaviour of soil. Constr. Build. Mater. **16**(2), 123–131 (2002)
7. Phummiphan, I., Horpibulsuk, S., Sukmak, P., Chinkulkijniwat, A., Arulrajah, A., Shen, S.L.: Stabilisation of marginal lateritic soil using high calcium fly ash-based geopolymer. Road Mater. Pavement Des. **17**(4), 877–891 (2016)
8. Soltani, A., Deng, A., Taheri, A.: Swell–compression characteristics of a fiber–reinforced expansive soil. Geotext. Geomembr. **46**(2), 183–189 (2018)
9. Suksiripattanapong, C., Kua, T.A., Arulrajah, A., Maghool, F., Horpibulsuk, S.: Strength and microstructure properties of spent coffee grounds stabilized with rice husk ash and slag geopolymers. Constr. Build. Mater. **146**, 312–320 (2017)

10. Tang, C., Shi, B., Gao, W., Chen, F., Cai, Y.: Strength and mechanical behavior of short polypropylene fiber reinforced and cement stabilized clayey soil. Geotext. Geomembr. **25**(3), 194–202 (2007)
11. Zhang, M., Guo, H., El-Korchi, T., Zhang, G., Tao, M.: Experimental feasibility study of geopolymer as the next-generation soil stabilizer. Constr. Build. Mater. **47**, 1468–1478 (2013)

Effect of Marble Dust on Strength Characteristics of Rice Husk Stabilized Soil

Srikalpa Rajguru Mahapatra, Monoswi Manini Sahoo, and Ruparshree Ragini Sahoo

1 Introduction

Due to the peculiar cyclic swell shrink behavior of expansive soil always creates a problem for civil engineers in the construction area. The soil swells when it absorbs moisture and shrinks when the moisture goes out. It creates severely damage on lightly loaded structures Due to this movement of water such as foundations, pavements and linings and residential building. An Estimate was conducted regarding annual damage of civil engineering structures on expansive soil, which shows that, In UK£150 million, In US$1000 million, and many billion pounds worldwide. In India this occupied more than 20% of total soil [1, 2]. The presence of clay mineral montmorillonite widely responsible for this nature of the soil. Different methods are conducted to modify the soil for construction purpose, stabilization using industrial wastes is one of them. The Densification of expansive soil using Phosphogypsum (a waste product from phosphoric acid industry) and observed that different proportions of Phosphogypsum when mixed with soil reduces its plasticity and liquid limit thereby making the soil more workable [3–5]. The CBR value of the soil increased from 2 to 9% and the free swell of the soil reduced considerably when 40% Phosphogypsum was added. the mix could not be compacted properly. When the proportion of Phosphogypsum was increased beyond 40%. Rice husks are the shells produced during dehiscing operation of paddy, which varies from 20% [6] to 23% [1, 2, 7] by weight of the paddy. The rice husk is waste material and disposed of by dumping or burning in the boiler for processing paddy. About 20% of its weight as ash generate due to the burning of rice husk [6]. Silica is the main constituent of rice husk ash (RHA) and the quality (% of amorphous and unburnt carbon) depends upon the burning process [8, 9]. The RHA is defined as a pozzolanic material due to its high amorphous silica content [9]. Generally, In India, 100 million tons of paddy

S. R. Mahapatra (✉) · M. M. Sahoo · R. R. Sahoo
Civil Engineering Department, VSSUT, Burla 768017, India

C. N. V. Satyanarayana Reddy et al. (eds.), *Ground Improvement and Reinforced Soil Structures*, Lecture Notes in Civil Engineering 152,
https://doi.org/10.1007/978-981-16-1831-4_20

are produced and more than 4 million tons of RHA generating annually. To lack of cementations, properties. Rice husk ash cannot be used alone for the Modification of soil [4, 10]. So, it is used as a binder with Lime, cement, lime sludge, Calcium chloride, etc. [7, 11–15]. Marble dust are the wastes generated during marble cutting and polishing. The stabilizing effect of marble dust on engineering properties of expansive soil and has found varied success [16–18]. The main objective of this paper is to study the effect of Marble dust on UCS, Compaction, Soaked CBR, swelling pressure and durability characteristics of an expansive soil stabilized with the optimum percentage of Rice husk ash.

From the review of the available literature, it is observed that various efforts have been found to study the possible utilization of different industrial wastes for the stabilization of expansive soil. However, studies related to the stabilization of expansive soil using RHA are Limited particularly in conjunction with Marble dust that is why an attempt has been made in this experimental work to know the effect of Marble dust on expansive soil stabilized with the optimum percentage of Rice husk ash.

2 Experimental Work and Methodology

ln geotechnical constructions will reduce the problems faced by the rice mill for its disposal mostly because of its large-scale utilization of rice husk ash and its property closely related to the natural earth material. So, assessment to check the behavior of rice husk ash at different condition before its use as a construction material in Civil engineering structure. Even though adequate substitutes for full-scale field tests are not available; tests at a laboratory scale provide a measure to control many of the variable encountered in practice. Details of sample preparation, the material used, and the testing procedure conducted have been outlined below.

2.1 Expansive Soil

The soil sample used for this study is collected from the local area of Balasore. The overall geotechnical properties of soil are classified as Clay with high plasticity (CH) in the IS Soil Classification System. The properties of the soil used in the investigation are given in Table 1.

Table 1 Properties of the natural expansive soil

Soil properties	Value
Specific-gravity	2.47
Free swell index (%)	70
IS classification of soil	CH
Liquid-limit (%)	66.65
Plastic-limit (%)	32.25
Plasticity-index (%)	28.38
Maximum dry density(g/cc)	1.5
Optimum moisture content (%)	23.04
UCS(KN/m^2)	75.3
CBR (%) (unsoaked)	4.67
CBR (%) (soaked)	1.72

Table 2 Chemical composition of RHA and Marble dust

Constituent	Chemical composition of RHA (%)	Chemical composition of Marble dust (%)
SiO2	75.2	5.2
Al2O3	5.2	3.8
Fe2O3	1.02	0.9
CaO	1.4	32.1
MgO	1.75	17.7
Loss on ignition	15.43	40.3

2.2 *Rice Husk Ash (RHA)*

The RHA was collected from Ma TariniRice Mill at Balasore (Odisha). Well-burnt RHA passing through 425 μm aperture before use. The chemical composition RHA is given in Table 2.

2.3 *Marble Dust*

The marble dust was obtained from the marble cutting and polishing industry at Balasore (Odisha). The chemical composition of Marble dust is given in Table 3.

Table 3 Proportioning of Soil-RHA-Marble dust mixes

SI. no.	Proportioning of soil-RHA mixes		Proportioning of soil-RHA-Marble dust mixes			
	Soil (%)	RHA (%)	Mix no.	Soil (%)	RHA (%)	Marble dust (%)
1	100	0	M1	90.00	10.00	0.00
2	95	5	M2	85.00	10.00	5.00
3	90	10	M3	80.00	10.00	10.00
4	85	15	M4	75.00	10.00	15.00
5	80	20	M5	70.00	10.00	20.00
6	–	–	M6	65.00	10.00	25.00
7	–	–	M7	60.00	10.00	30.00

2.4 *Sample Preparation and Experimental Program*

First, the soil was air-dried and it was pulverized with the help of a wooden hammer. Then it was sieved with 425 μm I.S sieve. Before performing the various tests on materials, the proportioning of ingredients was ascertained. The guidance on proportioning was taken from the literature. The amount of RHA was varied from 5 to 20% by dry weight of soil in the step of 5% and Marble powder from percent to 5% by dry weight of soil. For the combination of both materials with soil 10% of RHA was mixed with 0–30% Marble powder for the laboratory study total of 5 mixes were prepared. For conducting different tests on Standard Proctor Compaction tests, UCS tests, stabilized expansive soil with Marble powder, the Marble powder was added from 0 to 25% at an increment of 5%.

3 Results and Discussion

The main aim of this project is to analyze the improvement in geotechnical properties like compaction characteristics, unconfined comprehensive strength tests, direct shear tests and CBR tests, etc. The variation in the properties of marble dust-treated rice husk ash is discussed in this section. Different types of samples/mixes were prepared for the test, described earlier in the desired proportion, weight, etc.

3.1 *Geotechnical Properties of Soil-RHA Mix*

RHA is a very fine pozzolanic, light, bulky, and highly porous material. This material is solidified while suspended in exhausted and collected in the rice mill industry. It is produced fluidized bed systems, a great furnace, and a suspension/entrained combustion reactor. Reactive of RHA is attributed to its high content of amorphous silica, and to its very large area governed by the porous structure of the particles. The properties of Soil-RHA mix are presented in Table 4.

Effect of RHA on Atterberg's limit of expansive soil

It is observed that the liquid limit of expansive soil alone was 66.65%. Adding various proportions of Rice Husk Ash has significant effects on the liquid limit of soil. The liquid limit decreased from 66.65% to 55.9% (Fig. 1) as Rice Husk Ash content is increased from 0 to 20%. Similarly, the plasticity index of soil decreased from 34.4% to 29.4% (Fig. 2) with the increase of Rice Husk Ash content in the soil.

Effect of RHA on compaction characteristics of soil

An increase in pèrcentage of RHA the MDD of soil goes decreasing (Figs. 3, 4), decreases to 1.4 g/cc from 1.5 g/cc when 20% RHA was added which is attributed to the lower value of specific gravity of rice husk has compared to soil. The OMC goes on increasing (Fig. 5) irrespective of the percent addition of RHA. The OMC increases to a value of 26.54% when 20% RHA was added. The increase in the optimum moisture content may be caused by the absorption of water by the RHA.

Effect of RHA on UCS of soil

RHA the UCS of soil goes on increasing up to 10% (Figs. 6, 7) By increasing the percentage, further addition of RHA, decreases the UCS of the soil. The UCS of soil increases 87.3 kN/m^2 from 75.3 kN/m^2 of native soil, when 10% RHA was added.

It is observed that cohesion and angle of internal friction of the soil initially increased as the RHA increased, peaking at an RHA content of 10%. (Figs. 8, 9) The angle of internal friction increases considerably from 31.88° to 34.61° for 10% RHA content, then decreased with higher content. The increased shear strength of the treated soil was attributed to the additional friction produced by the coarse particles of the RHA embedded on the shear plane.

Effect of RHA on CBR test of expansive soil

The CBR of soil goes on increasing up to 10% of addition of RHA (Figs. 10, 11), further addition of RHA, decreases the CBR of the soil. The CBR of soil increases to 3.24% from 1.72% of native soil, when 10% RHA was added.

3.2 Geotechnical Properties of Marble Dust Treated RHA

The strength of Rice husk ash generally improves with time due to pozzolanic reactions. Reactive silica and free lime contents are necessary for pozzolanic reactions to take place. strength of rice husk ash can significantly improve, When adding marble dust. The properties of Soil - RHA- marble dust mix are presented in Table 5.

Table 4 Summary of geotechnical properties of Soil-RHA mix

SI. no.	Mixes	Atterberg limit			Compaction test		UCS (kN/m^2)	DST		CBR (Soaked) (%)
		LL (%)	PL (%)	PI (%)	OMC (%)	MDD (g/cc)		Cohesion (kPa)	Angle (°)	
1	100% s	66.65	32.25	34.40	23.04	1.50	75.30	68.00	31.88	1.72
2	95%S + 5%R	62.15	30.20	31.95	23.51	1.48	83.50	89.00	32.86	2.55
3	90%S + 10%R	58.35	27.30	31.05	24.92	1.45	87.30	114.00	34.61	3.24
4	85%S + 15%R	57.20	26.70	30.50	25.38	1.43	80.50	103.00	30.20	2.97
5	80% S + 20%R	55.90	26.50	29.40	26.54	1.40	77.50	94.00	28.99	2.25

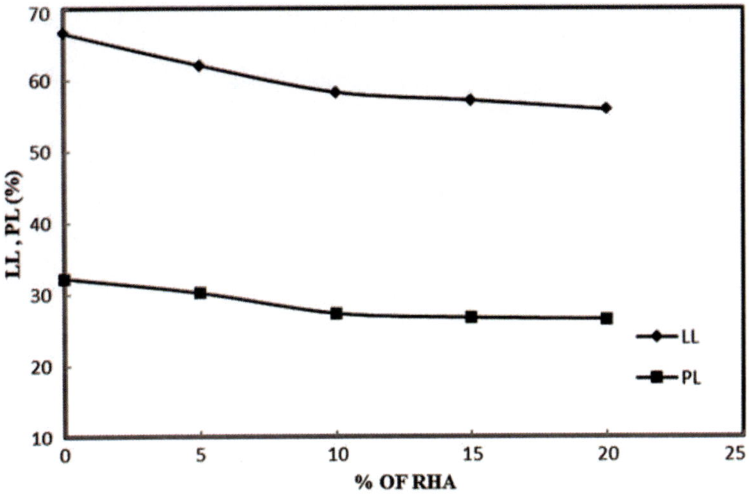

Fig. 1 Variation of LL and PL with RHA (%)

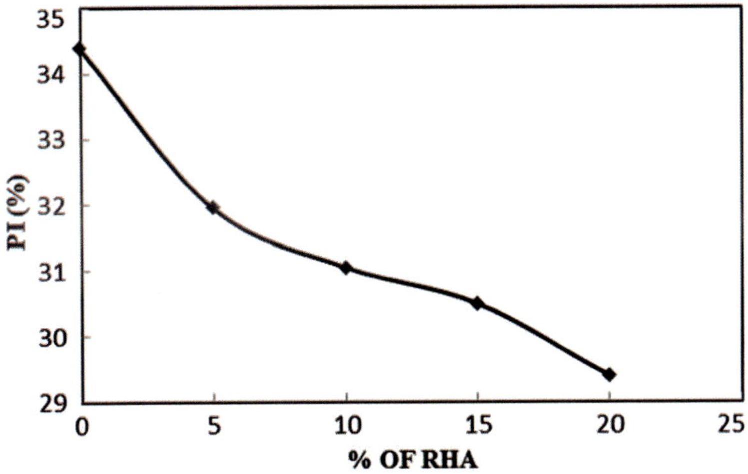

Fig. 2 Variation in PI with RHA (%)

Effect of RHA on Atterberg's limit of expansive soil

Marble Dust has significant effects on the liquid limit of the RHA stabilized soil; the liquid limit goes on decreases rapidly (Fig. 12) and the plastic limit of Soil alone was 34.4%. Marble Dust has substantial effects on the plastic limit of the RHA stabilized soil, the plastic limit goes on decreases rapidly (Fig. 12) and the Plasticity index was found to decreases from 31.05% to 26.77% (Fig. 13) due to adding various fractions.

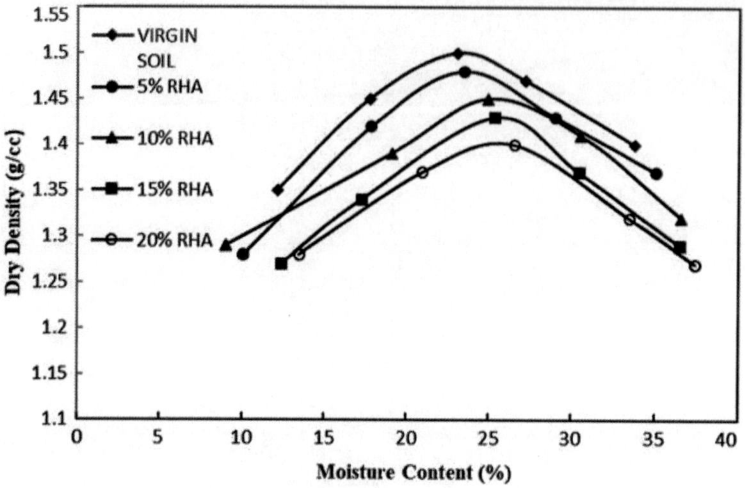

Fig. 3 Variation in compaction curve of soil with RHA (%)

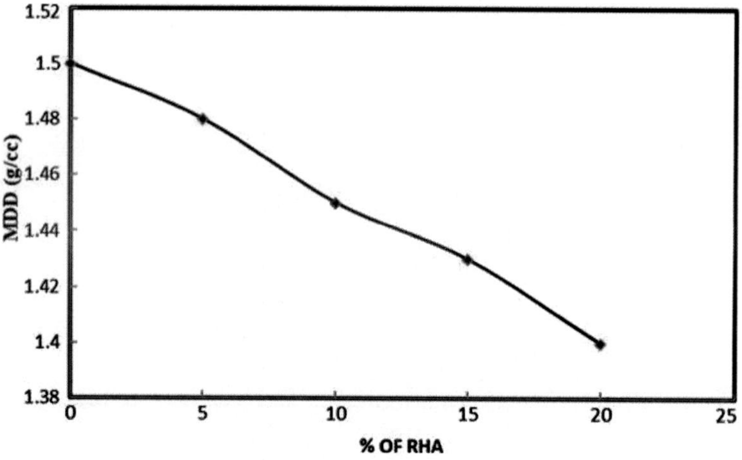

Fig. 4 Effect of RHA on MDD of soil

Effect of RHA on Compaction characteristics of soil

The MDD of soil goes on increasing (Fig. 14), increases to 1.69 g/cc from 1.48 g/cc when 25% Marble Dust was added with an increase in the percentage of Marble dust. Further adding of Marble Dust decreases the MDD of the soil goes on decreasing, decreases to 1.67 g/cc from 1.69 g/cc when 30% Marble dust was added. The MDD of the soil increases to 1.69 when 25% Marble dust was added to soil stabilized with 10% Rice husk ash. The reverse thing happens in case OMC. The OMC

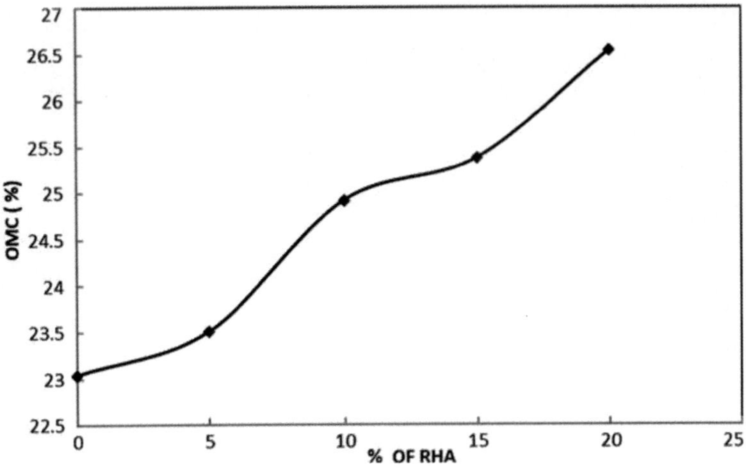

Fig. 5 Effect of RHA on OMC of soil

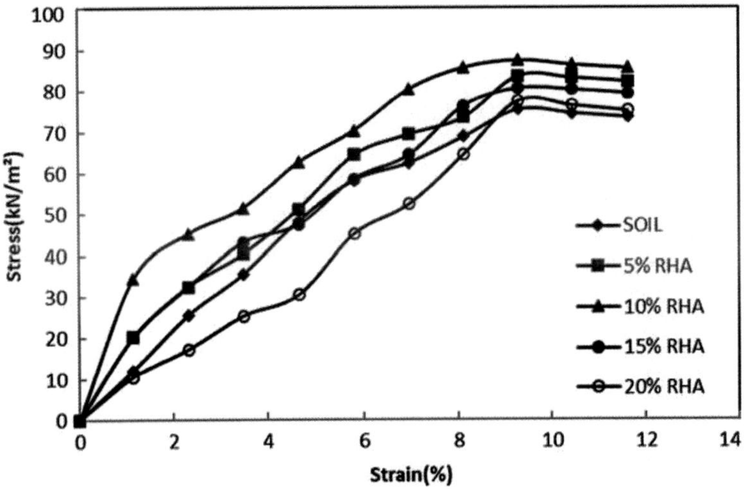

Fig. 6 Effect of RHA on the stress strain behavior of soil

decreases to 24.92% when 10% Marble dust was added to soil. The OMC goes on decreasing (Fig. 15), decreases to a value of 21.09% when 30% Marble dust was added to soil stabilized with 10% Rice husk ash, With increase in different percentage of Marble dust.

Effect of RHA on UCS of Soil

Addition of different percentage of Marble dust has considerable effects on the UCS of the RHA stabilized soil, it is observed that by the addition of 10% Marble dust

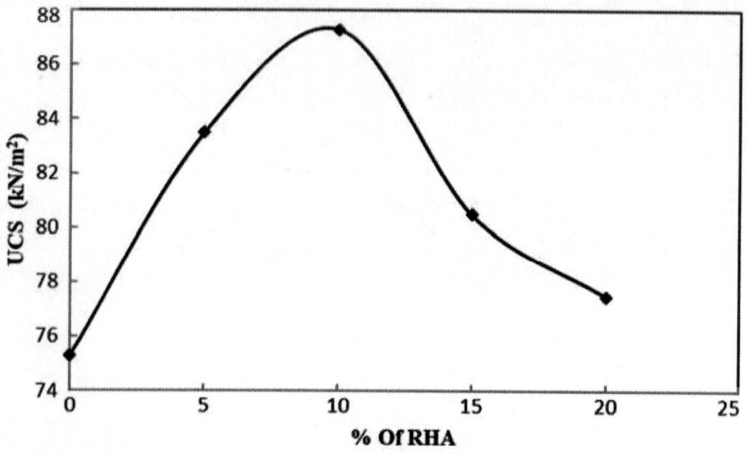

Fig. 7 Effect of RHA on UCS of soil

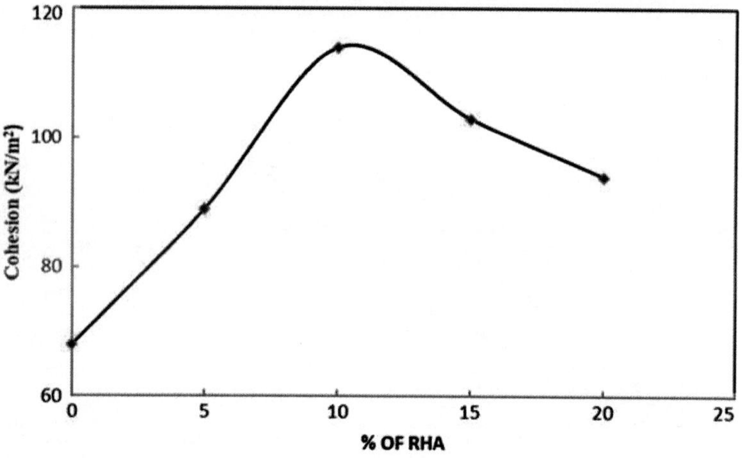

Fig. 8 Effect of RHA on cohesion of soil

the UCS of soil increases to 139.4 kN/m2. With an increase in the percentage of Marble dust the UCS of soil further goes on increasing (Figs. 16, 17), increases to 165.7 kN/m2 from 87.3 kN/m2 when 15% Marble dust was added further addition of Marble dust decreases the UCS of the soil. When 30% Marble dust was added to soil stabilized with 10% rice husk ash. The UCS of the soil decreases to 135.1 kN/m2. The UCS attains the highest value when the percentage of Marble dust is 15%.

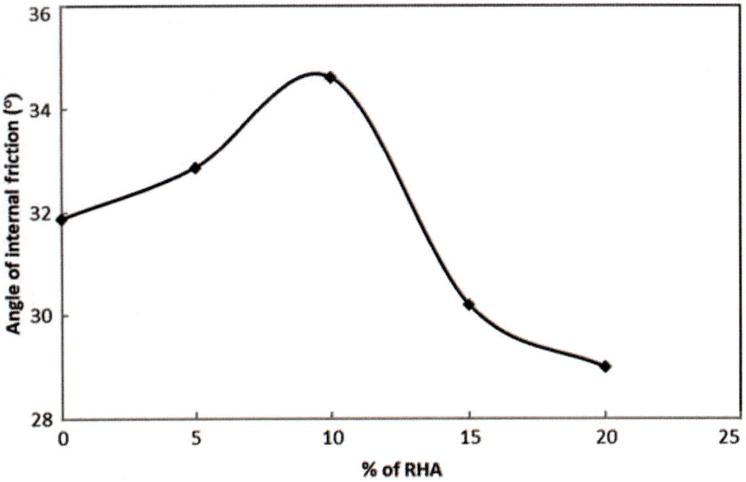

Fig. 9 Effect of RHA on angle of internal friction of soil

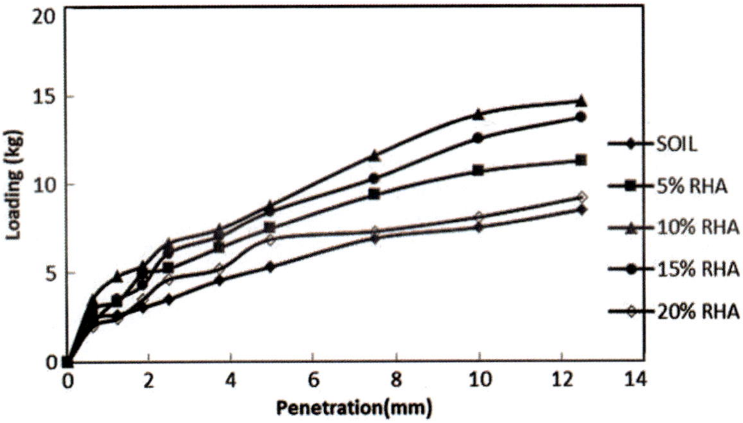

Fig. 10 Variation in CBR of soil with RHA (%)

Effect of RHA on direct shear test of soil

The angle of internal friction increased considerably to 43.23° from 34.61° when 15% Marble Dust content, then decreased with higher content (Fig. 18). The increased shear strength of the treated soil was attributed to the additional friction produced by the coarse particles of the Rice Husk Ash embedded on the shear plane. However, the cohesion underwent obvious change when the proportion of soil: RHA: MD ratio was 75:10:15, which corresponded to the peak shear strength (Fig. 19).

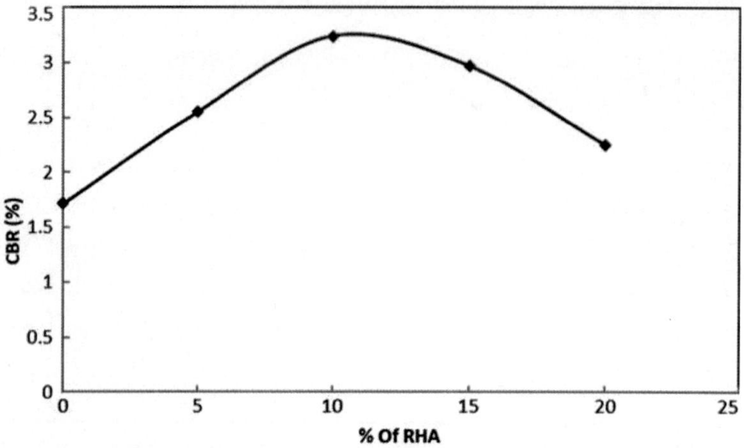

Fig. 11 Effect of RHA on CBR of soil

Effect of RHA on CBR test of expansive soil

Addition of different percentage of Marble dust has considerable effects on the Soaked CBR of the RHA stabilized expansive soil. With increase in the percentage of Marble dust the Soaked CBR of soil further goes on increasing (Fig. 20), increases to 5.15% from 3.24% when 20% Marble dust was added further addition of Marble dust decreases the Soaked CBR of the soil. The Soaked CBR of the soil decreases to 4.58% when 30% Marble dust was added to expansive soil stabilized with 10% rice husk ash (Fig. 21). The Soaked CBR attains the highest value when the percentage of Marble dust is 15%.

4 Conclusions

1. The optimum proportion of expansive soil: RHA was 90: 10 by weight. Based on direct shear strength and UCS test results,
2. The expansive soil stabilized with the optimum ratio of rice husk ash; the plasticity index of the soil reduces from 31.05% to 26.77% with the increase in the percentage of marble dust which was found to be promising on adding the various percentages of marble dust
3. The MDD starts to decrease and OMC starts to increase with the addition of RHA to expansive soil.
4. The cohesion and angle of internal friction of expansive soil both increased as the content of RHA increased, giving a peak value when the RHA content was 10%, while the cohesion showed no marked change.
5. The UCS of the expansive soil increased by 15.94% compared to untreated soil, with the addition of 10% RHA,

Table 5 Summary of geotechnical properties of Soil-RHA-Marble Dust Mix

Mixes	Marble dust (%)	Atterberg limit			Compaction test		UCS (kN/m^2)	DST		CBR (Soaked) (%)
		LL (%)	PL (%)	Pl (%)	OMC (%)	MDD (g/cc)		Cohesion (kPa)	Angle (°)	
M1	0	58.35	27.30	31.05	24.92	1.48	87.30	114.00	34.61	3.24
M2	5	56.75	26.35	30.40	24.28	1.51	117.70	121.00	36.51	3.64
M3	10	54.85	25.20	29.65	23.27	1.55	139.40	128.00	40.27	4.72
M4	15	53.64	24.64	29.00	22.67	1.58	165.70	143.00	43.23	5.15
M5	20	53.03	24.53	28.50	22.18	1.64	153.60	132.00	38.47	4.78
M6	25	52.35	23.75	27.85	21.80	1.69	148.20	118.00	36.90	4.66
M7	30	50.87	22.50	26.77	21.09	1.67	135.10	112.00	35.55	4.58

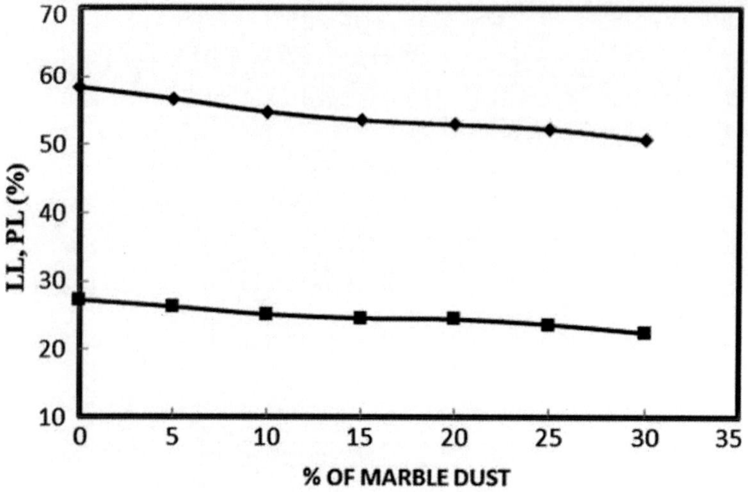

Fig. 12 Variation IN LL and PL for RHA stabilized soil with marble dust (%)

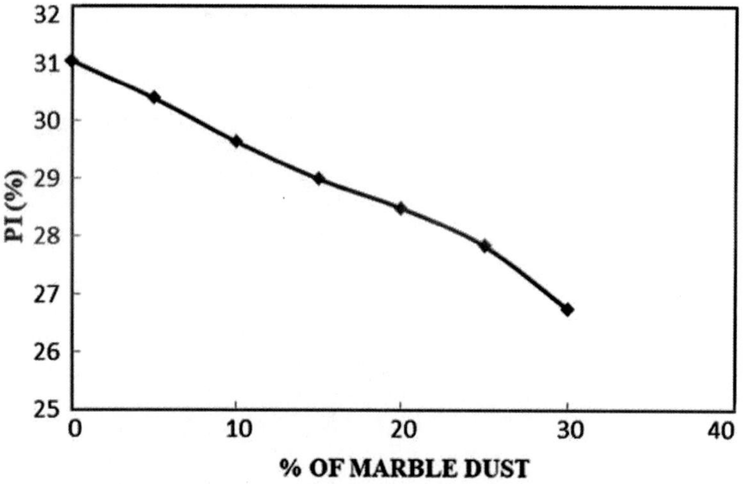

Fig. 13 Variation of PI for RHA stabilized soil with variation in Marble dust (%)

6. The MDD goes on increasing and OMC goes on decreasing up to 25% due to the addition of different ratio of marble dust to the expansive soil stabilized with the optimum ratio of rice husk ash. Both MDD and OMC go on decreasing further the addition of marble dust.
7. An addition of 15% MD corresponded to the peak the strength and cohesion of soil- RHA blends increased as the content of MD increased.

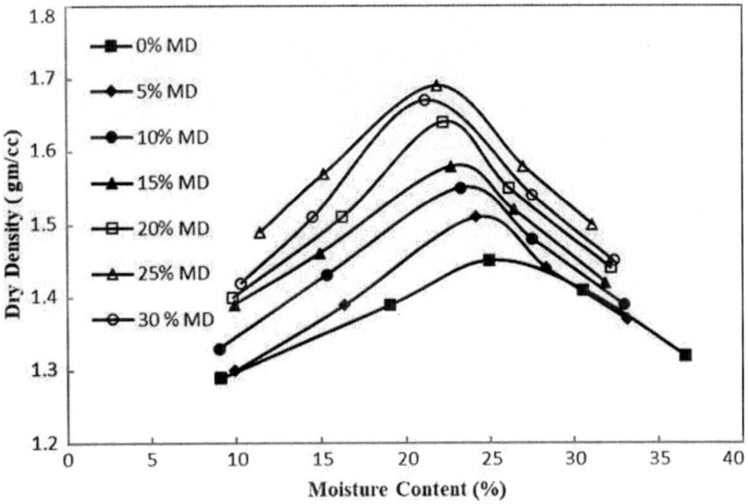

Fig. 14 Variation in compaction curves for RHA stabilized soil with Marble dust (%)

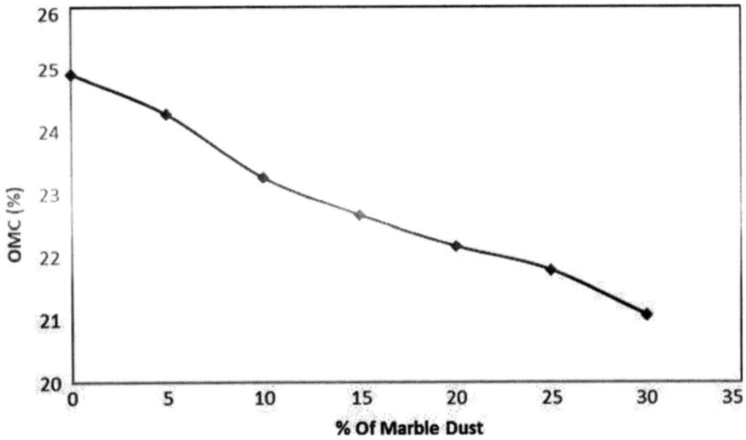

Fig. 15 Effect of Marble dust on OMC of RHA stabilized expansive soil

8. Adding more Marble dust reduces the UCS of the expansive soil. Unconfined compressive strength improved nearly by 120.05% of the virgin soil after adding 15% Marble dust.
9. Adding the various ratio of marble dust to the expansive soil stabilized with optimum ratio of rice husk ash, for soaking condition the CBR value increased till 15% marble dust, further adding of marble dust reduces the CBR value of the expansive soil. There is a 199.42% increase CBR value in soaking condition when related with natural soil.

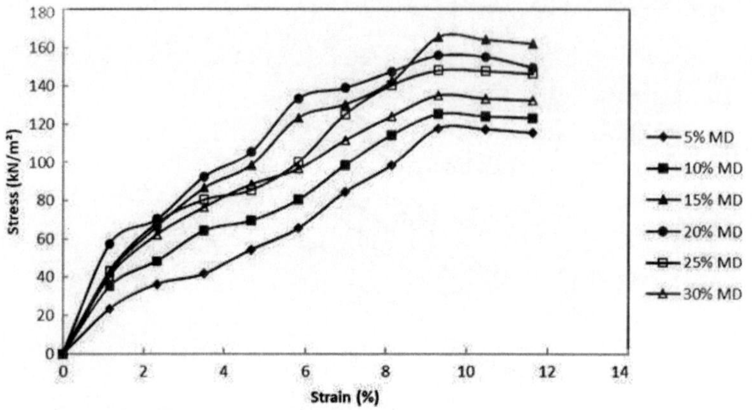

Fig. 16 Variation on the stress-strain behavior for RHA stabilized soil with Marble dust (%)

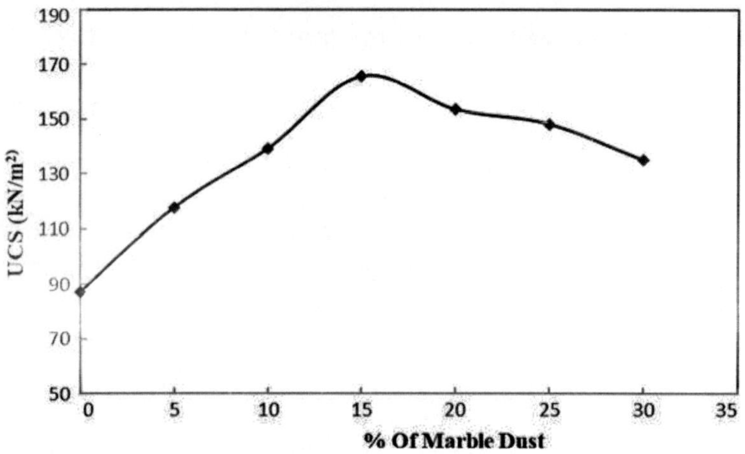

Fig. 17 Variation in UCS for RHA stabilized soil with Variation in Marble dust (%)

10. For better stabilization result, the optimal percentage of Soil: Rice husk ash: Marble dust was found to be 75: 10: 15.

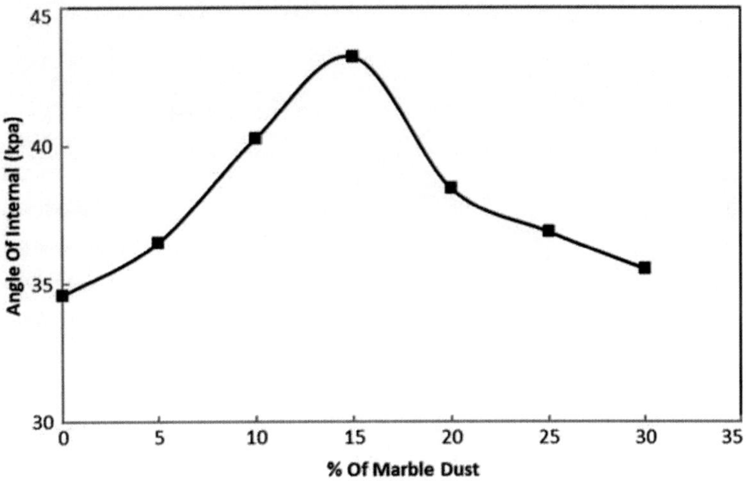

Fig. 18 Variation in Angle of internal friction for RHA stabilized soil with varying percentage of Marble dust

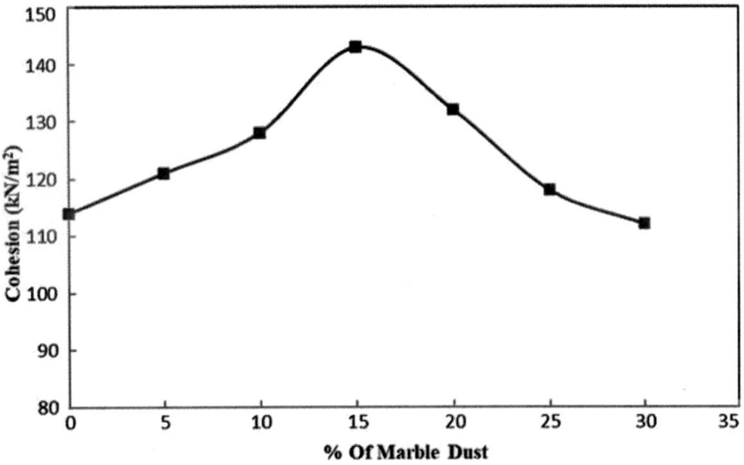

Fig. 19 Variation in Cohesion of RHA stabilized soil with variation in Marble dust (%)

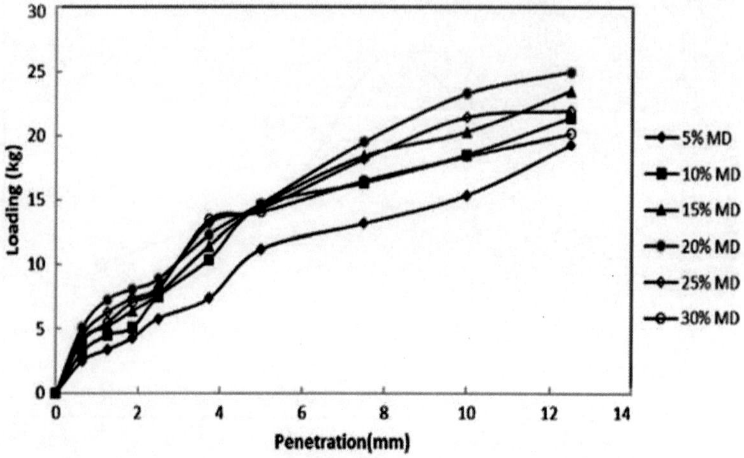

Fig. 20 Variation in CBR for RHA stabilized soil with variation in Marble dust (%)

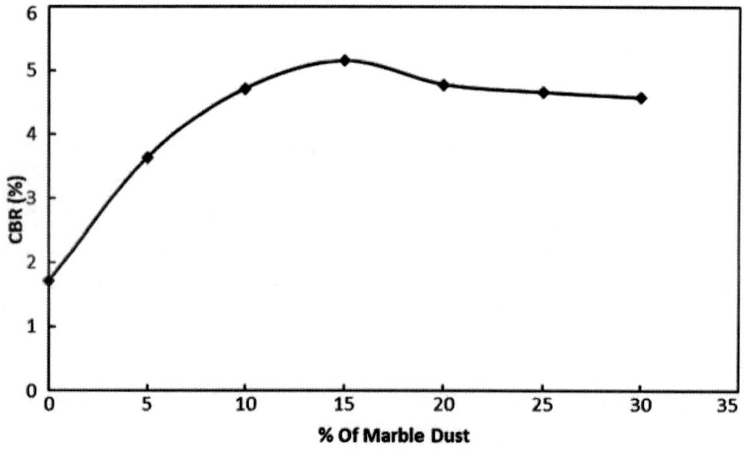

Fig. 21 Effect of Marble dust on soaked CBR of RHA stabilized expansive soil

References

1. Chen, F.H.: Foundations on expansive soils. Chen & Associates, Elsevier Publications, USA (1988)
2. Gourley, C.S., Newill, D., Shreiner, H.D.: Expansive soils: TRL's research strategy. In: Proceedings of 1st International Symposium on Engineering Characteristics of Arid Soils (1993)
3. Mishra, A.K., Mathur, R.: Use of phosphogypsum–an industrial by-product in stabilization of black cotton soils. J. Highw. Res. Bull. (2004). (Nov. No-70)
4. Nelson, J.D., Miller, D.J.: Expansive Soils: Problem and Practice in Foundation and Pavement Engineering. Wiley, New York (1992)

5. Ramakrishna, A.N., Pradeep Kumar, A.V.: Effect of moisture content on strength behaviour of BC soil-rice husk-lime mixes. Indian Highw. **36**(1), 49–58 (2008)
6. Mehta, P.K.: Concrete structure, properties and materials. Prentice- Hall, Englewood Cliffs, N.J (1986)
7. Della, V.P., Ku¨hn, I., Hotza, D.: Rice husk ash as an alternate source for active silica production. Mater. Lett. **57**, 818–821 (2002)
8. Nair, D.G., Jagadish, K.S., Fraaim, A.: Reactive pozzolanas from rice husk ash: An alternative to cement for rural housing. Cem. Concr. Res. **36**, 1062–1071 (2006)
9. Payá, J., Monzó, J., Borrachero, M.V., Mellado, A., Ordoñez, L.M.: Determination of amorphous silica in rice husk ash by rapid analytical method. Cem. Concr. Res. **31**, 212–231 (2001)
10. Haji Ali, F., Adnan, A., Choy, C.K.: Geotechnical properties of a chemically stabilized soil from Malaysia with rice husk ash as an additive. Geotech. Geol. Eng. **10**, 117–134 (1992)
11. Basha, A.M., Hashim, R., Muntohar, A.S.: Effect of cement- rice husk ash on the plasticity and compaction of soil. Electron. J. Geotech. Eng. **8**, Bundle A (2003)
12. Chandra, S., Kumar, S., Anand, R.K.: Soil stabilization with rice husk ash and lime sludge. Indian Highw. **33**(5), 87–97 (2005)
13. Muntohar, A.S., Hantoro, G.: Influence of rice husk ash and lime on engineering properties of a clayey subgrade. Electron. J. Geotech. Eng. **5**, 1–13 (2000)
14. Ramakrishna, A.N., Pradeep Kumar, A.V.: Stabilisation of black cotton soil using rice husk as hand cement. In: National Conference on Civil Engineering Meeting the Challenges of TOMORROW. GND Engineering College, Ludhiana, pp. 215–220 (2006)
15. Sharma, S.R., Phani Kumar, B.R., Rao B.V.: Engineering behaviour of a remolded expansive clay blended with lime, calcium chloride and Rice-husk ash. ASCE J. Mater. Civil Eng. **20**(8), 509–515 (2008)
16. Başer, O.: Stabilization of expansive soils using waste marble dust. Master of Science thesis submitted to Civil Engineering Department, Middle East, Technical University (2009)
17. Palaniappan, K.A., Stalin, V. K.: Utility effect of solid wastes in problematic soils. Int. J. Eng. Res. Ind. Appl. **2**(1), 313–321 (2009)
18. Swami, B.L.: Feasibility study of marble dust in highway sector. Highw. Res. Bull. **67**, 27–36 (2002). (December)

Swelling Characteristics of Fly Ash Based Geopolymer Expansive Clay Blends

T. V. Nagaraju and K. N. Mounika

1 Introduction

Expansive clay poses great damage to lightly loaded infrastructure such as pavements, embankments, retaining walls and buildings due to their drastic swell-shrink behavior [1]. The swell-shrink behavior of expansive clays causes due to the expanding lattice structure of montmorillonite minerals [1, 2]. Expansive clays are the most common deposits available in many parts of India, especially in the east coast [3].

There are many modification techniques are in practice to alter the properties of expansive clays viz. chemical amelioration, mechanical stabilization, under-reamed piles, stone columns, surcharge loading and sand drains [4, 5]. Among all the modification techniques, chemical amelioration of expansive clays is quite successful in pavement subgrades, embankment fills and foundation soil. In chemical amelioration, many chemical additives were emerged past and present. However, choosing sustainable eco-friendly material and their applications is much needed. Cement and lime are common additives to improve swell-shrink behavior [6]. These are particularly effective in increasing the strength characteristics due to the formation of clay flocks and cementitious products (C–S–H and C–A–H) [6, 7]. However, cement and lime are energy-intensive and emits CO_2 during the manufacturing process [8]. Other hand, there are many industrial by-products available to enhance the engineering behaviour of expansive clays. Industrial by-products having higher amount of Ca, Si and Al oxides contribute pozzolanic reaction, which after the pozzolanic reaction generates strong cementitious compounds. Fly ash, silica fume, ground granulated blast furnace slag (GGBS) and rice husk ash (RHA) were pozzolanic materials

T. V. Nagaraju (✉)
Department of Civil Engineering, SRKR Engineering College, Bhimavaram 534204, India
e-mail: tvnraju@srkrec.edu.in

K. N. Mounika
Department of Civil Engineering, JNTU Kakinada, Kakinada 533003, India

C. N. V. Satyanarayana Reddy et al. (eds.), *Ground Improvement and Reinforced Soil Structures*, Lecture Notes in Civil Engineering 152,
https://doi.org/10.1007/978-981-16-1831-4_21

used as a stabilizer shows significant improvement in the swell-shrink behavior of expansive clays [9, 10]. Swell-shrink properties of expansive clays such as rate of heave, swell potential, swelling pressure, compressibility and linear shrinkage were improved with pozzolanic materials due to the formation of flocculation of grains and cementitious compounds [10].

In recent years, geopolymers are using as a sustainable alternative for lime and cement. Geopolymers are prepared based on the inorganic alumino-silicate polymer. Geopolymer binders are depended on many factors which include molarity of alkali solution, precursor (mineral composition), the texture of materials (glassy phase or amorphous phase), Si/Al ratio, thermal exposure and fineness of materials. Several attempts [11–14] have been made to study the role of geopolymers in altering the properties of soils. The competitiveness of geopolymerization blends increased because of the improvement in strength, ductility, shrinkage strains and stiffness. During geopolymerization process, oligomers were formed due to the Si and Al oxides dissolution. Further, hardening of geopolymer gel contributes dense geopolymer products with the distribution of pores. Syed et al. [15] reported that volumetric instability of expansive clays can be improved by using geopolymers, they form a strong cementitious products. By comparison, clayey soils treated with geopolymers attain higher strength improvement but not more than cement stabilized clays. However, geopolymer blends show ductile behavior when compared with ordinary Portland cement blends [16].

This paper presents the swell-shrink behavior of expansive clays stabilized with fly ash-based geopolymers. One-dimensional swell consolidation tests were carried on the fly ash-based geopolymers with varying precursor content as 5, 10, 15, 20 and 25% by dry weight of clay. Furthermore, tested oedometer samples were analyzed for micro-structural behavior using scanning electron microscopy.

2 Experimental Investigation

2.1 *Materials*

Expansive clay used in this study, which is highly swelling clay in the coastal region of Andhra Pradesh, India. Table 1 shows the index properties of the expansive clay, while Fig. 4a shows the SEM micrograph of expansive clay. Fly ash was considered as a precursor in this study, which is collected from NTPC, Vijayawada, India. Table 2

Table 1 Properties of soil

Properties	Liquid limit, %	Plastic limit, %	Soil classification (IS)	Specific gravity	Fines, %	Free swell index, %
Value	85	23	CH	2.65	94	180

Table 2 Chemical composition of clay and fly ash

Material	SiO_2, %	Al_2O_3, %	Fe_2O_3	CaO, %	MgO, %
Clay	63.15	19.35	4.35	0.67	1.79
Fly ash	33.5	23.05	6.10	27.1	4.6

shows the chemical composition of both soil and fly ash. Alkali-activators for altering expansive clays were prepared using sodium hydroxide pallets and sodium silicate gel.

2.2 *Laboratory Tests*

Geopolymers were prepared by maintaining 10 M of NaOH solution and ratio Na_2SiO_3/NaOH as 2.5. geopolymer specimens were prepared at an initial density of 12kN/m^3. Moreover, in the tests on fly ash-based geopolymer clay blends, the soil was replaced by the required amount of precursor content. The geopolymer blend was compacted four layers of each 5 mm thickness using a manual static compactor. One-dimensional swell consolidation tests were conducted on fly ash-based geopolymer clay blends according to ASTM D4546-03. Rate of heave, swell potential, swelling pressure, rebound and linear shrinkage of geopolymer clay blends were determined. The rate of heave was monitored until the equilibrium heave reaches, generally undergoes swell by inundation for 4320 min (3 days). The swell potential was measured under an initial surcharge of 5 kPa when the sample was inundated with water. Swelling pressure (p_s) was determined with the help of e-logp curves with respect to their initial void ratio.

In addition to swell compressibility tests on geopolymer blends, micro-structural tests were conducted on the blends. Scanning electron microscopy (SEM) analysis was carried to know the behavior of texture and morphology of the geopolymer blends with the intension of relating these results to associated swell-shrink behaviour.

3 Results and Discussion

3.1 *Swell-Shrink Behaviour*

Figure 1 plots indicate that the variation of rate of heave with respect to the time interval when the geopolymer specimens inundated with water. The rate of heave decreased with increasing fly ash content in the fly ash-based geopolymer clay blends. The reduction in have is due to the fly ash microspheres counteracts the expanding silica lattice of montmorillonite mineral. Moreover, clay pockets were interlocked with strong geopolymer products.

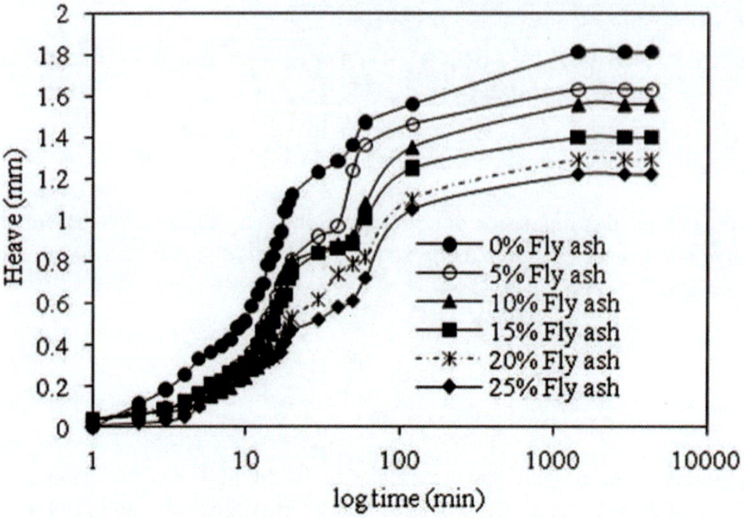

Fig. 1 Rate of heave with varying precursor content

Test data shows that with increasing fly ash content in the fly ash-based geopolymer clay blends rebound and linear shrinkage were improved. Significant reduction in rebound (axially) and shrinkage (laterally) was due to the resistance offered by the inter molecule geopolymer products (Table 3).

Figure 2 shows that fly ash-based geopolymers caused an appreciable decrease in swell potential (from 9.05 to 6.10) at the addition of 25% fly ash content. An improvement in swell potential is due to the flocculation of clays with alkali-activators.

Figure 3 plots show the e-logp curves of geopolymer-treated expansive clay and fly ash-based geopolymer-treated clays with varying fly ash content. Swelling pressure (p_s) increased with increasing fly ash content in the blends up to 15%. Thereafter, increasing fly ash content, swelling pressure could not be determined due to the dense geopolymer matrix.

Table 3 Test data

Property	Fly ash content					
	0%	5%	10%	15%	20%	25%
Rate of heave (mm)	1.81	1.63	1.56	1.40	1.29	1.22
Swell potential (%)	9.05	8.15	7.80	7.00	6.45	6.10
Swelling Pressure (kPa)	112	152	214	225	–	–
Rebound (mm)	0.43	0.26	0.20	0.14	0.11	0.06
Linear Shrinkage (%)	8	7	4	2	–	–

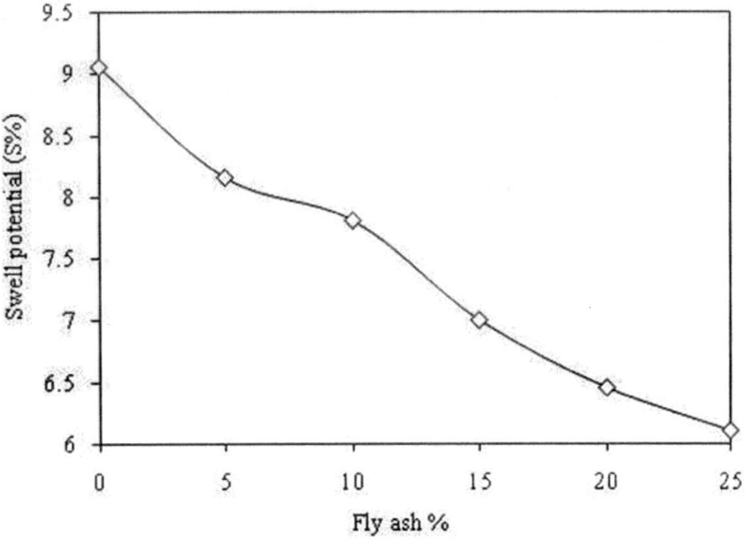

Fig. 2 Swell potential with varying precursor content

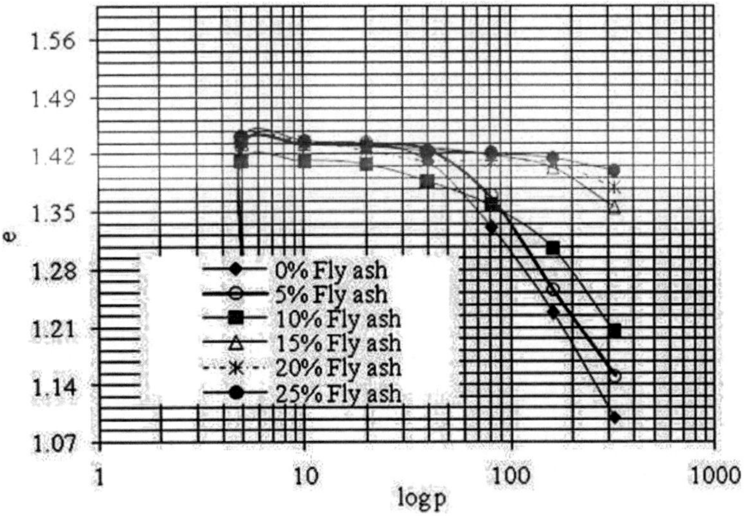

Fig. 3 e-logp curves with varying precursor content

3.2 *Micro-Structure Studies*

The micrographs of fly ash-based geopolymer blends were clearly evident that morphology and geopolymer products were improved with the increasing precursor content.

Figure 4c, d shows the geopolymer gel phases and interface between the geopolymer products and clay particles. Moreover, unreacted fly ash particles (spherical particles) were also found in the geopolymer blends at higher precursor content (255) addition (vide Fig. 4e). In Fig. 4d, morphology highlighted in box-shaped, clearly evident that fly ash particles (Al and Si oxides) were dissolved and exhibits compacted micro-structure. This indicates an effective geopolymerization and resulted in higher resistance against swelling and shrinkage.

4 Conclusions

The following broad conclusions are drawn from the above experimental study:

(1) Rate of heave, swell potential, rebound and linear shrinkage were decreased with increasing fly ash content in the fly ash-based geopolymer clays. Swell potential showing an improvement of 32.5% when fly ash content increased from 0% to 25%.

(2) Swelling pressure ($_{ps}$) of the fly ash-based geopolymers improved with the addition of fly ash content in the blends. However, determination of swelling pressure was not possible for blends having higher fly ash content 20% and 25%, due to the strong geopolymer matrix.

(3) SEM micrographs show that the geopolymer gel phases and compacted micro-structure resulted in higher resistance against both swelling (axially) and shrinkage (laterally).

(4) Swell-shrink behavior of expansive clays treated with fly ash-based geopolymers can be utilized as construction material in pavement sub-grades, backfill and cushion layer.

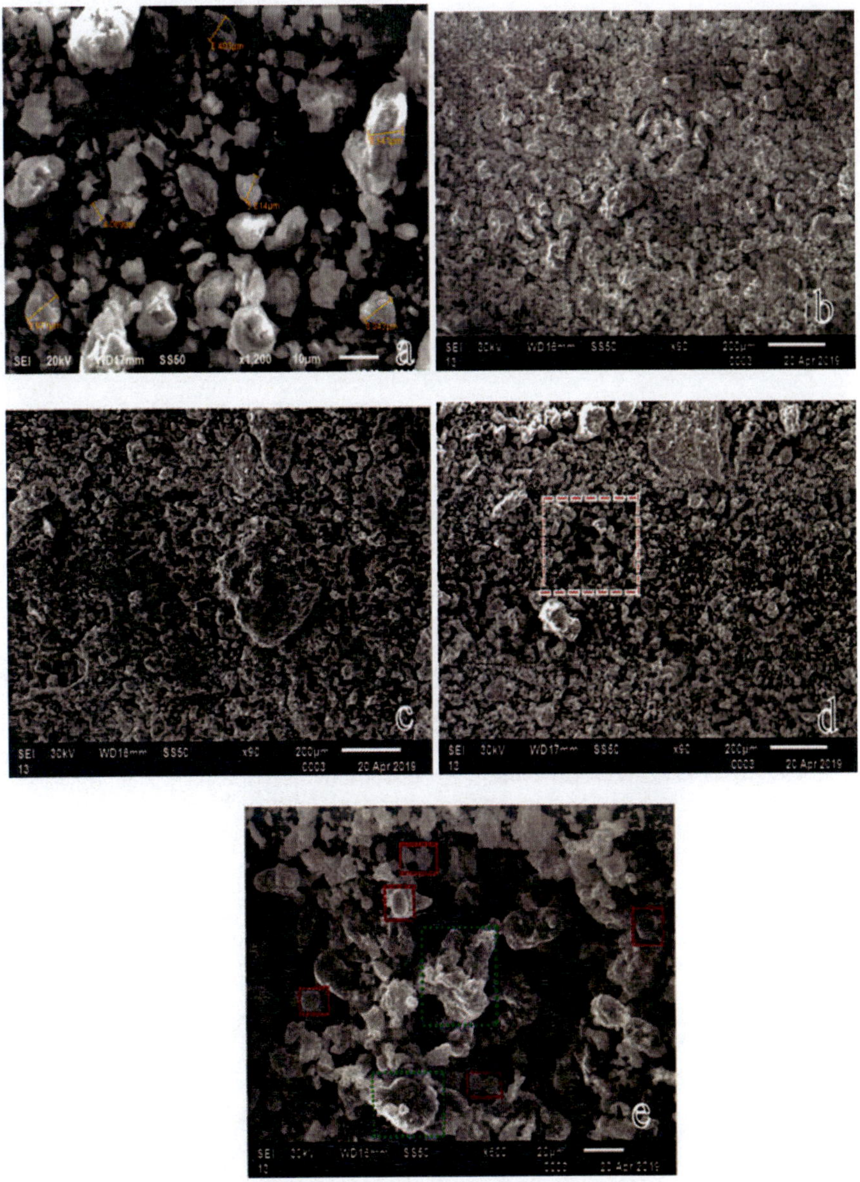

Fig. 4 SEM micrographs of clay treated with fly ash geopolymers with varying fly ash content **a** virgin expansive clay **b** 15% fly ash **c** 20% fly ash and **d** 25% fly ash

References

1. Chen, F.H.: Foundations on Expansive Soils. Elsevier Scientific Publishing Co., Amsterdam (1988)
2. Basma, A.A., Al-Homoud, A.S., Malkawi, A.I.H., Al-Bashabsheh, M.A.: Swelling-shrinkage behavior of natural expansive clays. Appl. Clay Sci. **11**(2–4), 211–227 (1996)
3. Nagaraju, T.V., Satyanarayana, P.V.V.: Geotechnical aspects of various constructions along the canal embankment using rice husk ash as stabilizer. In: Ground Improvement Techniques and Geosynthetics, pp. 143–150. Springer, Singapore (2019)
4. Kumar, K.S.R., Thyagaraj, T.: Comparison of lime treatment techniques for deep stabilization of expansive soils. Int. J. Geotechn. Eng., 1–19 (2020)
5. Welsh, J.P.: In situ testing for ground modification techniques. In: Use of In Situ Tests in Geotechnical Engineering, pp. 322–335. ASCE (1986)
6. Saride, S., Puppala, A.J., Chikyala, S.R.: Swell-shrink and strength behaviors of lime and cement stabilized expansive organic clays. Appl. Clay Sci. **85**, 39–45 (2013)
7. Sirivitmaitrie, C., Puppala, A.J., Chikyala, V., Saride, S., Hoyos, L.R.: Combined lime and cement treatment of expansive soils with low to medium soluble sulfate levels. In: GeoCongress 2008: Geosustainability and Geohazard Mitigation, pp. 646–653 (2008)
8. Gartner, E.: Industrially interesting approaches to "low-CO_2" cements. Cem. Concr. Res. **34**(9), 1489–1498 (2004)
9. Bose, B.: Geo engineering properties of expansive soil stabilized with fly ash. Electron. J. Geotech. Eng. **17**(1), 1339–1353 (2012)
10. Phanikumar, B.R., Nagaraju, T.V.: Swell and compressibility of GGBS–clay mixes in lumps and powders: effect of 4% lime. Indian Geotechn. J. **49**(2), 161–169 (2019)
11. Nagaraju, T.V., Prasad, C.D.: New prediction models for compressive strength of GGBS-based Geopolymer clays using swarm assisted. In: Advances in Computer Methods and Geomechanics: IACMAG Symposium 2019, Vol. 1, p. 367. Springer Nature (2020)
12. Onyelowe, K.C., Bui Van, D., Nguyen Van, M.: Swelling potential, shrinkage and durability of cemented and uncemented lateritic soils treated with CWC base geopolymer. Int. J. Geotechn. Eng., 1–16 (2018)
13. Singhi, B., Laskar, A.I., Ahmed, M.A.: Investigation on soil–geopolymer with slag, fly ash and their blending. Arabian J. Sci. Eng. **41**(2), 393–400 (2016)
14. Zhang, M., Guo, H., El-Korchi, T., Zhang, G., Tao, M.: Experimental feasibility study of geopolymer as the next-generation soil stabilizer. Constr. Build. Mater. **47**, 1468–1478 (2013)
15. Syed, M., GuhaRay, A., Agarwal, S., Kar, A.: Stabilization of expansive clays by combined effects of Geopolymerization and fiber reinforcement. J. Inst. Eng. (India): Series A, **101**(1), 163–178 (2020)
16. Ghadir, P., Ranjbar, N.: Clayey soil stabilization using geopolymer and Portland cement. Constr. Build. Mater. **188**, 361–371 (2018)

A Critical Review on Stabilisation of Expansive Soils with Compensating Materials

T. Ashok Kumar, T. Thyagaraj, and R. G. Robinson

1 Introduction

For ensuring the stability of the civil engineering structures, the sub-structures are constructed over different types of soils. However, some soils in their natural state may not be suitable for construction purposes, and such soils are identified as problematic. Moreover, these soils require special treatment before using it as a foundation material. Due to the detrimental volume change experienced by the expansive soils upon moisture intake, they are identified as problematic soils. Expansive soils are residual soils derived from the parent materials of igneous (basalt) and sedimentary (limestone, sandstone and shale) based rocks when subjected to chemical weathering at the high alkaline environment [1, 2]. In India, almost 20% of the land area is covered by expansive soil deposits and mostly found in the states of Maharashtra, Madhya Pradesh, Karnataka, Telangana, Andhra Pradesh, Tamil Nadu, Rajasthan and Gujarat [3, 4], Kumar et al. [5]. The unpredictable volume change of expansive soils may disturb the service life of the structures if not addressed and treated properly.

Several methods were developed and followed as mitigation measures. However, methods such as moisture control, pre-wetting of soil, surcharge loading, placement of compensating materials (removal and replacement technique), compaction control and chemical modification gained popularity [6], Nelson and Miller [7]. Among the mitigation methods, the removal and replace technique (placement of compensating) is one of the attractive solutions for lightly loaded structures due to material availability, field workability and economic considerations. Hence, a detailed review was carried to bring out the significance and limitations of the different compensating materials in mitigating the volume change of expansive soils.

T. Ashok Kumar (✉) · T. Thyagaraj · R. G. Robinson
Indian Institute of Technology Madras, Chennai 600036, India

C. N. V. Satyanarayana Reddy et al. (eds.), *Ground Improvement and Reinforced Soil Structures*, Lecture Notes in Civil Engineering 152,
https://doi.org/10.1007/978-981-16-1831-4_22

2 Different Types of Compensating Materials and Their Mechanism in Controlling the Volume Change of Expansive Soils

2.1 Gravel Material

Holtz [8] suggested soil replacement as a retrofitting technique for Mohawk and Welton canal, constructed over expansive soil. Lightly compacted gravel and sand were used as a compensating material. Since it is a remedial measure, the material was placed over the partially swollen sample, and the compensating material not only provides the surcharge effect but also it minimises the differential movement. However, the placement conditions such as density index and thickness of the compensating material were not detailed for expansive soils with varying degrees of soil expansion.

2.2 Sand Material

Satyanarayana [9] and Moussa et al. [10] suggested sand as a compensating material to control the volume change. It works on the principle that during monsoon, the wet sand consumes less volume, and hence it accommodates the excess volume change of underlying expansive soil. Similarly, during the shrinkage process of expansive soil, the excess voids are compensated by bulking nature of sand in a partially saturated condition. The swell potential of expansive soil stabilised with sand as cushion material showed maximum reduction when sand was placed at a lesser density index. A recent study by Kumar [11] also reported similar observations and showed that the swell reduction was mainly due to the reduction in the void ratio of the sand upon inundation. Earlier observations showed that sand cushion could be used for low and medium swelling soils; however, the placement conditions were not generalised. Besides, compensating materials like gravel and sand are not encouraged due to its higher hydraulic conductivity, and it may establish a water reservoir over the expansive soils.

2.3 Non-swelling Soil

Based on the field observations, Chen [6] reported that a structure resting on 5 feet (1.5 m) of granular soil (SP–SC) followed by expansive soil did not result in any distress. However, the mechanism behind the control of swell was not precise, i.e. whether the seepage water did not reach the expansive soil or it was because of the uniform heave of expansive soil that did not result in any distress of the structure.

Chen [6] suggested using non-swelling soils with a liquid limit of 30 to 50% and the finer fractions (<75 μm) ranging from 5 to 50%. Besides, he also suggested that material should be impervious, and the relative compaction of the compensating material should be in the range of 90–100% of standard Proctor density. The thickness of the non-swelling soil ranges from 3 to 5ft (about 0.9 to 1.5 m). Nelson and Miller [7] also supported the observations reported by Chen [6].

2.4 *Cohesive Non-swelling Soil (CNS)*

The continuous works of Katti [3] and his co-workers proposed cohesive non-swelling (CNS) soil as a compensating material and showed that the swelling was controlled due to the development of cohesive forces in the CNS material. This method is also extended to field applications for lightly loaded structures such as canal lining and pavement construction. Based on the laboratory and field investigations, Katti [3] proposed the specifications, field conditions and thickness of CNS material to minimise the heave of expansive soil during saturation. The term cohesive non-swelling could be used only when the compensating material meets the specifications proposed by Katti [3], and other materials as non-swelling soils. The CNS should be compacted to standard Proctor condition, and the measured swell pressure of the CNS should be less than 10 kPa. The thickness of the CNS material was decided based on the measured swell pressure. IS-9451: 1994 also suggested the specification proposed by Katti [3] for swell control in expansive soils in particular for canal applications. Recently, Kumar et al. [12] suggested CNS material as a compensating material in the retrofitting works of the distressed industrial building.

2.5 *Chemically Stabilised Soil (CSS)*

Practically, it is not easy to obtain the soils with the exact specification of CNS material in the vicinity of the construction site (location), and this makes the designer to choose alternate methods [3]. Later researchers tried to get compensating materials using industrial wastes like fly ash and rice husk ash by using directly or blending with cement or lime as a cushion [13, 14]. Few experiments were conducted to study the possibility of using native expansive clay as CNS material by mixing with suitable admixtures and called this material as chemically stabilised soil (CSS) cushion [3, 15].

Rao et al. [16] recommended the use of lime admixed fly ash as cushion material. They also suggested that the thickness of the cushion should be half the depth of the active zone for efficient functioning. Murthy and Praveen [17] suggested that the expansive clay blended with 0.5% $CaCl_2$ and 8% rice husk ash can also be used as an alternate to natural CNS material.

3 Cyclic Swell-Shrink Behaviour of CNS/CSS Stabilised Expansive

3.1 Soil

Several laboratory studies were performed to bring out the wet-dry effects on expansive soil stabilised with compensating materials. Rao [4] reported that CNS was effective only during the first cycle of wetting, and its efficacy was lost in the subsequent wetting cycle. A recent study by Kumar [11] also reported similar observations where CNS was used as a compensating material. Few studies were also carried out on natural, cement/lime stabilised murrum as a CNS material and subjected to wetting and drying cycles [13, 18]. Rao and Rao [19] performed swell-shrink studies to evaluate the suitability of cement and lime stabilised fly ash as cushion material. They found that the cement stabilised fly ash cushion was more effective than the lime stabilised fly ash cushion in controlling the swell-shrink movements.

4 Depth of Active Zone

Due to the cyclic process of precipitation and evapotranspiration, the water content fluctuates at the surface level of expansive soil deposit, and it decreases or increases with depth according to the environmental condition (summer and winter). Nelson et al. [20] defined the active zone as the zone of soil that contributes to the heave due to soil expansion at any particular time. The selection of the method of treatment to the expansive soils depends on the depth of the active zone and swell pressure. It is essential to evaluate the active zone for a given site because almost 60–80% of volume change occurs in the top 50% of active zone depth [21], Fityus et al. [22].

The depth of the active zone changes with time and also it depends on the soil profile, hydraulic conductivity and other environmental factors. In places, where the soil profile is stratified, the depth of the active zone can be determined by plotting w/w_L(w is the water content and w_L is the liquid limit) or plasticity index (I_P) versus depth [7]. Typically, in India, the depth of active zone was found to be 3.5 to 5 m in Madhya Pradesh, 1.5 m in Maharashtra and NIT Warrangal campus, Telangana [17, 23]. The maximum depth of the active zone was found in the place Colorado (USA), and it was obtained as 16 m [24].

4.1 Thickness of Soil Removal in the Active Zone

Ardani [25] has developed guidelines for the thickness of removal of soil, which was proposed and followed by the Colorado department of highways. Here, the thickness of removal is optimised based on the plasticity index of soil, and the same

Table 1 Thickness of removal of soil in the active zone based on plasticity index

Plasticity index (%)	Thickness of removal (ft)
10–20	2
20–30	3
30–40	4
40–50	5
>50	6

is summarised in Table 1. Further, the sub-excavated soil should be compensated with impervious materials.

As discussed in the above section, the process of removal of soil in the active zone itself controls the volume change before placing the compensating materials [21], Fityus et al. [22]. Therefore, it is more meaningful if the replacement of soil is decided based on the active zone, rather than depending on plasticity, the thickness of the clay layer and swell pressure. In general, it is complicated to evaluate the active zone profile for the given site, and this makes uncertainty in heave measurement [26].

5 Laboratory Constraints in Understanding the Behaviour of Expansive Soil Stabilised with Compensating Material

Swell-consolidation and constant volume methods are mostly used for the determination of swell pressure of layered soil (ES + CNS/CSS/Sand). Fredlund [27] reported that the swell pressure of expansive soil (ES) from the swell-consolidation method should be corrected for compressibility of filter paper, porous stone and apparatus. In a layered soil system, the compressibility of compensating material is entirely different from the expansive soil. So far, no studies were reported to show how to account for the corrections for layered soils, which pose different compressibility behaviour in saturated conditions. A recent study by Kumar [11] showed that the swell pressure of ES + CNS was found to be higher than the swell pressure of ES alone. The higher swell pressure might be due to the rigid behaviour of CNS material (comparatively with expansive soil) and sidewall friction, which offer more resistance against the applied pressure. Besides, it is also challenging to simulate environment-dependent active zone profile in laboratory conditions.

6 Summary and Conclusions

This paper explores the development of different compensating materials to overcome the detrimental volume change of expansive soil induced upon inundation. It is clear from the above review that the placement of compensating material is successful in controlling the volume change and still followed by the many practising engineers across the world. Further, the natural compensating materials are ineffectiveduring wet-dry cycles. Besides, it is also evident from the above discussion that mechanisms of compensating materials in controlling the volume change differ among the researchers, and it might be due to limited field trials. Therefore, better understandings of compensating materials could be established by performing many field trials with the support of instrumentation results.

References

1. Seiple, W.R.: Engineering geology of expansive soils. MSc Dissertation, University of Southern California, USA (1968)
2. Donaldson, G.W.: The occurrences of problems of heave and the factors affecting its nature. In: 2nd International Research and Engineering Conference on Expansive Clay Soils, A & M Press, Texas, 25–36 (1969)
3. Katti, R.K.: Search for solutions to problem in black cotton soil. Indian Geotechn. J. **9**, 1–88 (1978)
4. Rao, K.S.S.: Swell-shrink behaviour of expansive soils-geotechnical challenges. Indian Geotechn. J. **30**(1), 1–68 (2000)
5. Kumar, T.A., Raheena, M., Robinson, R.G., Thyagaraj, T.: A rapid method of determination of swell potential and swell pressure of expansive soils using constant rate of strain apparatus. Geotechn. Testing J. (2020)
6. Chen, F.H.: Foundations on Expansive Soils. Elsevier Scientific Publishing Company, New York (1975)
7. Nelson, J.D., Miller, D.J.: Expansive soils: Problems and Practise in Foundation and Pavement Engineering. Wiley (1992)
8. Holtz, W.G.: Expansive clays-properties and problems. Quat. Colorado School Mines **54**(4), 448–468 (1959)
9. Satyanarayana, B.: Behavior of expansive soil treated or cushioned with sand. In: 2nd International Conference on Expansive Soils, Texas, 308–316 (1969)
10. Moussa, A.A., Abd-el-Meguid, M.A., Okdah, S.M., Heikal, A.H.: Effect of sand cushion on swelling and swelling pressure of expansive silty clay. In: 11th International Proceedings on soil mechanics and foundation engineering, San Francisco (1985)
11. Kumar, T.A.: Development of rapid characterization methods and stabilisation of expansive soil. Ph.D. Dissertation, Indian Institute of Technology Madras, Chennai (2020)
12. Kumar, T.A., Robinson, R.G., Thyagaraj, T.: Distress of an industrial building constructed on an expansive soil—a case study from India. Proceedings of Institution of Civil Engineers—Forensic Engineering **171**(3), 121–126 (2018)
13. Sahoo, J., Pradhan, P., Rao, K.: Behavior of stabilised soil cushions under cyclic wetting and drying. Int. J. Geotech. Eng. **2**(2), 89–102 (2008)
14. Sivapullaiah, P.V., Rao, K.S.S., Gurumurthy, J.V.: Stabilisation of rice husk ash for use as cushion below foundations on expansive clay. Ground Improvem. **8**(4), 137–149 (2004)
15. Katti, R.K., Katti, A.R.: Behaviour of saturated expansive soil and control methods. Central Board of Irrigation and Power, New Delhi (1994)
16. Rao, M.R., Rao, A.S., Babu, R.D.: Efficacy of cement-stabilized fly Ash cushion in arresting heave of expansive soils. Geotechn. Geolog. Eng. **26**(2), 189–197 (2008)
17. Murthy, R.V., Praveen, G.V.: Use of chemically stabilised soil as cushion material below light weight structures founded on expansive soils. J. Mater. Civ. Eng. ASCE **20**(5), 392–400 (2008)
18. Sahoo, J.P., Pradhan, P.K.: Effect of lime stabilised soil cushion on strength behaviour of expansive soil. Geotech. Geol. Eng. **28**(6), 889–897 (2010)

19. Rao, A.S., Rao, M.R.: Behaviour of expansive soils under stabilized fly ash cushions during cyclic wetting and drying. Int. J. Geotech. Eng. **4**(1), 111–118 (2010)
20. Nelson, J.D., Overton,D and Durkee, D.: Depth of wetting and the Active Zone. Expansive clay soils and vegetative influence on shallow foundations, ASCE, 95–109 (2001)
21. Rao, R.R., Rahardjo, H., Fredlund, D.G.: Closed-form heave solutions for expansive soils. J. Geotechn. Eng. ASCE **114**(5), 573–588 (1988)
22. Fityus, S.G., Smith, D.W., Allman, M.A.: Expansive soil test site near Newcastle. J. Geotechn. Geoenviron. Eng. ASCE **130**(7), 686–695 (2004)
23. Soni, K.M.: Construction on expansive soils. Building and construction, NBM and CW, Athena Information solutions Pvt. Ltd, New Delhi (2009)
24. Holtz, W.G.: Volume change in expansive soils and control by lime treatment. Proceedings of the Second International Conference on Expansive soils, Texas, 157–174 (1969)
25. Ardani, A.: Expansive Soil Treatment Methods in Colorado. Colorado Department of Transportation, Colorado, US (1992)
26. Coduto, D.P.: Foundation Design: Principles and Practices. PearsonEducation, London, UK (2013)
27. Fredlund, D.G.: Consolidometertest procedural factors affecting swell properties. In: 2nd International Conference on Expansive Clay Soils, College Station, TX: Texas A & M Press, pp. 435–456 (1969)
28. IS 9451: Guidelines for lining of canals in expansive clay. Bureau of Indian Standards, New Delhi (1994)

Swell-Shrink Behaviour of Lime Pile and Lime Slurry-Treated Expansive Soil

K. S. R. Kumar and T. Thyagaraj

1 Introduction

Expansive soils occur all over the world and are found in countries such as Australia, Africa, China, Canada, India, Saudi Arabia, South Africa and USA [1–3]. These soils are very much suitable for agriculture and are commonly called as black cotton soil, and occupy 20% of the total land in India [1, 3–5]. These soils undergo swelling and shrinkage due to changes in seasonal moisture content, as a result, it causes severe distress to the structures constructed over them [6–8]. Several researchers all over the world have studied the swell-shrink phenomenon of expansive soils. Studies carried out by [9–12, 3] showed that the maximum swell occurs during the 2nd wetting cycle and attains equilibrium after 3–4 wetting cycles. The effect of aging on swell-shrink behavior of an expansive soil was studied by [13]. Their study showed that the aging effect is predominant during the 1st wetting cycle which is due to particle rearrangements and the formation of bonds. Whereas with an increase in swell-shrink cycles the effect of aging slowly nullified and the behaviour of aged specimens almost becomes equal to specimens that were not cured.

Studies by [14, 15], Rao and Shivanada [16–18] found that the pozzolanic reactions formed between free lime and clay particles were lost with the increase in number of swell-shrink cycles. Investigation by [19] showed that the extent of swelling on lime and cement-treated expansive soil mostly depends on the montmorillonite content present in the soil. The investigation also showed that the cement treatment is more effective in reducing the swell-shrink behaviour when compared to lime-treated expansive soil. Literature review shows that previous research studies on the swell-shrink behaviour of lime stabilized expansive soils by direct mixing

K. S. R. Kumar (✉) · T. Thyagaraj
Department of Civil Engineering, IIT Madras, Chennai 600036, India

T. Thyagaraj
e-mail: ttraj@iitm.ac.in

C. N. V. Satyanarayana Reddy et al. (eds.), *Ground Improvement and Reinforced Soil Structures*, Lecture Notes in Civil Engineering 152,
https://doi.org/10.1007/978-981-16-1831-4_23

with lime is effective. But in the field, lime pile and lime slurry techniques were adopted for the expansive soils extending greater depths. The swell-shrink mechanism involved by direct mixing with lime and through permeation (lime pile (LP) and lime slurry (LS)) is entirely different. Most of the previous studies were carried out by direct mixing of stabilizer with soil and very few studies have been carried out on swell-shrink behaviour of expansive soil stabilized through permeation. Hence, in the present study, an attempt is made to study the swell-shrink behaviour of expansive soil through laboratory tests using LS and LP techniques and the two results are compared.

2 Materials

An expansive soil collected from Trichy, Tamil Nadu, India and commercially available $Ca(OH)_2$ were used in the present investigation. The procedures for the determination of physico-chemical properties, grain size distribution, index properties and standard Proctor compaction characteristics were described in detail in [6]. The properties of the Trichy expansive soil used for the present laboratory testing are shown in Table 1.

Table 1 Properties of expansive soil (after [6])

Property	Value
pH	8.52
Pore water salinity (mg/l)	321
Specific gravity (Gs)	2.75
Free swell index (FSI) (%)	270
Initial Consumption of Lime (ICL) (%)	3.5
Atterberg limits	
Liquid limit (%)	92
Plastic limit (%)	22
Shrinkage limit (%)	8
Grain size distribution (%)	
Sand	33
Silt	18
Clay	49
Unified soil classification symbol	CH
Standard Proctor compaction characteristics	
Maximum dry unit weight (kN/m^3)	16.8
Optimum moisture content (OMC) (%)	20

3 Compaction of Soil and Sampling Procedure

LS and LP treatments were carried out in test moulds of diameter 385 mm and height 400 mm. The required quantity of expansive soil was calculated based on OMC (20%), dry unit weight (16.8 kN/m^3) and statically compacted to a thickness of 300 mm in three layers in test moulds for LS and LP treatments using a hydraulic jack. Central holes were made at the centre of the test moulds by statically pushing a 75 mm mild steel pipe. After making the central holes, one test mould was allowed to dry under the direct sunlight for the development of desiccation cracks and LS was poured into the central hole of desiccated soil. Installation of LP was done by filling the central hole with the lime powder and by adding water in three different stages. After the complete permeation of LS and installation of LP in the test moulds, the moulds were allowed to cure for 30 days. Swell-shrink studies were carried out on undisturbed soil specimens collected at a radial distance of 1.5d and depths of 0–90 mm and 200–290 mm for LS and LP-treated soils. The detailed procedure of compaction, implementation and sampling programme for LS and LP test moulds were clearly explained in the [6].

4 Results and Discussion

Figure 1 compares the time-swell plots of LS-treated specimen collected at a radial distance of 1.5D and depth of 0–45 mm during different wetting cycles. A nominal swell potential of 0.04% can be observed during the 1st wetting cycle. The swell potential increased to 1.18% and 5.45% during the 2nd and 5th wetting cycles (Fig. 1). The initial decrease in the swell potential of the LS-treated specimen during the 1st wetting cycle may be due to the formation of strong pozzolanic bonds between the clay and lime slurry (Rao and Thyagaraj 2003; Thyagaraj and Suresh 2012), [6]. Whereas, the increase in the swell potential of LS-treated specimens with an increase in wetting cycles may be due to partial breakage of these pozzolanic bonds [14, 15, 6, 16–18, 20].

The time-swell plots during different wetting cycles of LP-treated specimens sampled at a radial distance of 1.5D and 0–45 mm depth are plotted in Fig. 2. The swell potential of LP-treated specimen was compared with untreated specimen compacted to the same dry density and water content (Fig. 2). From Fig. 2, it can be clearly observed that the second swell potential was the maximum for both LP and untreated specimens in comparison with the swell potential during other wetting cycles. The equilibrium was achieved after 3rd wetting cycle in both LP and untreated specimens. The swell potential of LP and untreated specimen compacted to the same dry density and water content at the 5th wetting cycle was 13.88% and 14.64% (Fig. 2).

Figure 3 compares the time-swell plots of the LS and LP-treated soil specimens sampled at 1.5D radial distance and 200–245 mm depth, respectively. The swell

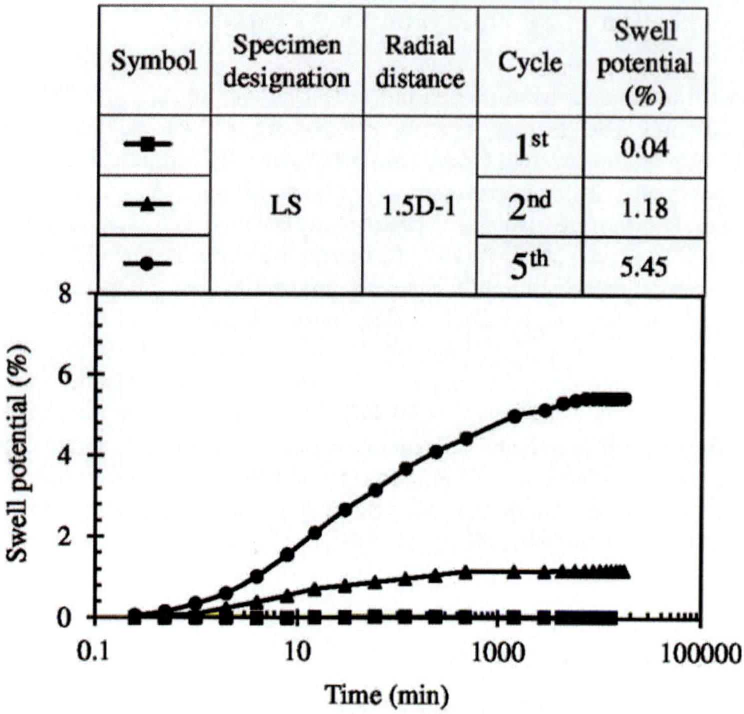

Fig. 1 Swell potential of LS-treated specimens sampled at 1.5D and 0–45 mm depth

potential of LS-treated specimens increased from 3.58% during the 1st wetting cycle to 16.12% and 20.67% during 2nd and 5th wetting cycles (Fig. 3). In the case of LP-treated specimens the maximum swell potential is observed during the second wetting cycle (22.21%) and attained equilibrium after 3rd wetting cycle, and exhibits a swell potential of 19.11% during 5th wetting cycle. The weaker pozzolanic bonds that formed in LS-treated specimen collected at a depth of 200–245 mm might have lost with an increase in wetting and drying cycles and thus increased the swell potential of LS-treated specimen (Fig. 3). This clearly shows that LS treatment depends on the extent of cracks that was developed with depth. Whereas, the LP-treated specimen collected at a radial distance of 1.5D at depths of 0–45 mm and 200–245 mm did not show any reduction in swell potential with an increase in wetting and drying cycle and almost acts like an untreated specimen (Figs. 2 and 3). This is due to less permeation of lime into highly impervious expansive soil [6, 21].

Symbol	Specimen designation	Radial distance	Cycle	Swell potential (%)
- □ -	Untreated	-	1st	3.71
—■—	Treated	1.5D-1		2.81
- △ -	Untreated	-	2nd	16.10
—△—	Treated	1.5D-1		15.17
- ⊖ -	Untreated	-	5th	14.64
—●—	Treated	1.5D-1		13.88

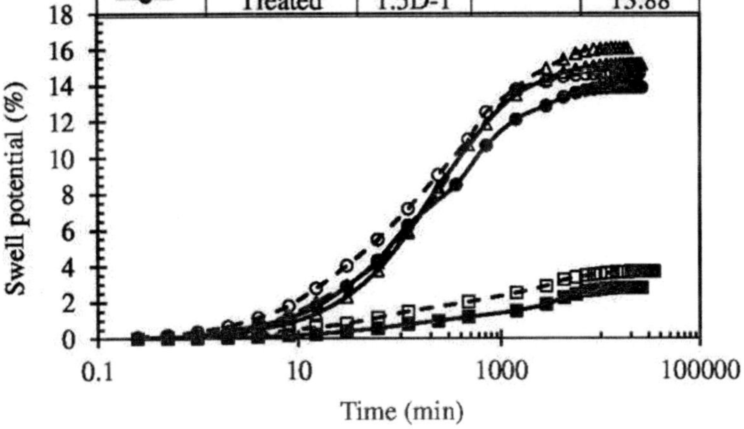

Fig. 2 Swell potential of untreated and LP-treated specimens sampled at .5D and 0–45 mm depth

5 Conclusions

The swell-shrink behaviour of LS and LP-treated expansive soil in a compacted state was studied in the present investigation and the following conclusions were drawn from the swell-shrink studies.

1. The LS-treated specimen collected at a radial distance of 1.5D and depth of 0–45 mm showed better performance in reducing the swell potential when compared to the specimen collected at a radial distance of 1.5D and depth of 200–245 mm. This indicates that LS treatment mainly depends on the extent and depth of cracks that was developed in the expansive soil.
2. The LP-treated specimens sampled at a radial distance of 1.5D and a depth of 0–45 mm and 200–245 mm did not show improvement and almost behaves like untreated specimen compacted to the same density and water content. This shows that lime from LP did not permeate to greater radial distance into highly impervious expansive soil.

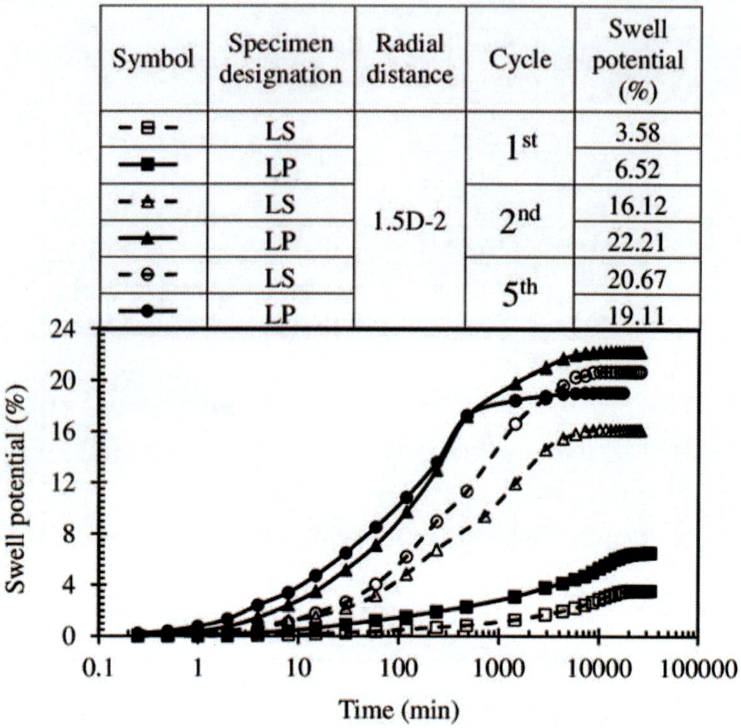

Fig. 3 Swell potential of LS and LP-treated specimens sampled at .5D and 200–245 mm depth

References

1. Ashok Kumar, T., Robinson, R.G., Thyagaraj, T.: Distress of an industrial building constructed on an expansive soil: a case study from India. Proc. Inst. Civil Eng. Forensic Eng. **171**(3), 121–126 (2018)
2. Chen, F.H.: Foundations on expansive soils. Elsevier, Amsterdam and New York (1988)
3. Rao, K.S.S.: Swell-shrink behavior of expansive soils: geo-technical challenges. Indian Geotechn. J. **30**(1), 1–69 (2000)
4. Katti, R.K.: Search for solutions to problems in black cotton soils. First IGS Ann. Lect. Indian Geotechn. J. **9**, 1–80 (1978)
5. Patil, N.G., Tiwary, P., Pal, D.K., Bhattacharyya, T., Sarkar, D., Mandal, C., Chandran, P., Ray, S.K., Prasad, J., Lokhande, M., Dongre, V.: Soil water retention characteristics of black soils of India and pedotransfer functions using different approaches. J. Irrigation Drainage Eng. ASCE **139**(4), 313–324 (2013)
6. Kumar, K.S.R., Thyagaraj, T.: Comparison of lime treatment techniques for deep stabilization of expansive soils. Int. J. Geotechn. Eng. (2020). https://doi.org/10.1080/19386362.2020.1775359
7. Thyagaraj, T., Zodinsanga, S.: Laboratory investigations on in situ stabilization of an expansive soil by lime precipitation technique. J. Mater. Civ. Eng. ASCE **27**(7), 06014028 (2015)
8. Thyagaraj, T., Samuel, Z., Kumar, K.S.R.: Relative efficiencies of electrolytes in stabilization of an expansive soil. Int. J. Geotech. Eng. **10**(2), 107–113 (2016)

9. Popescu, M.: Behaviour of expansive soils with crumb structures. Proceedings, 4th International Conference on Expansive soils, Vol. 1, Denver, CO, pp. 158–171 (1980)
10. Osipov, V., Bik, N.N., Rumjantseva, N.A.: Cyclic swelling of clays. Applied Clay Sci. Amsterdam Netherlands **2**(7), 363–374 (1987)
11. Dif, A.F., Bluemel, W.F.: Expansive soils under cyclic drying and wetting. Geotechn. Testing J. GTJODJ **14**(1), 96–102 (1991)
12. Day, R.W.: Swell—Shrink behaviour of expansive compacted clay. J. Geotechn. Eng. **120**(3):618–623 (1994)
13. Rao, K.S., Tripathy, S.: Effect of aging on swelling and swell–shrink behaviour of a compacted expansive soil. ASTM Geotech. Test J. **26**(1), 36–46 (2003)
14. Rao, S.M., Reddy, B.V.V., Muttharam, M.: Effect of cyclic wetting and drying on the index properties of a lime-stabilised expansive soil. Ground Improvem. **5**(3), 107–110 (2001)
15. Rao, S.M., Reddy, B.V.V., Muttharam, M.: The impact of cyclic wetting and drying on the swelling behaviour of stabilized expansive soils. Eng. Geol. **60**, 223–233 (2001)
16. Rao, S.M., Shivananda, P.: Swelling behaviour of lime stabilized specimens subjected to wetting-drying cycles. In: Proceedings of Chemo-Mechanical Coupling in Clays: from Nano-Scale to Engineering Applications (Di Maio C, Hueckel T and Loret B (Eds)). Swets and Zeitlinger, Lisse, the Netherlands, pp. 95–104 (2002)
17. Guney, Y., Sari, D., Cetin, M., Tuncan, M.: Impact of cyclic wetting–drying on swelling behavior of lime-stabilized soil. Build. Environ. **42**, 681–688 (2007)
18. Khattab, S.A.A., Al-Mukhtar, M., Fleureau, J.M.: Long-term stability characteristics of a lime-treated plastic soil. J. Mater. Civ. Eng. **19**(4), 358–366 (2007)
19. Chittoori, B.C.S., Puppala, A.J., Pedarla, A.: Addressing clay mineralogy effects on performance of chemically stabilized expansive soils subjected to seasonal wetting and drying. J. Geotech. Geoenviron. Eng. **144**(1), 04017097 (2018)
20. Stoltz, G., Cuisinier, O., Masrouri, F.: Weathering of a lime-treated clayey soil by drying and wetting cycles. Eng. Geol. **181**, 281–289 (2014)
21. Rao, S.M., Venkataswamy, B.: Lime pile treatment of black cotton soils. Ground Improv. Inst. Civ. Eng. London **6**(2), 85–93 (2002)

Effect of Lignosulfonate on Strength and Deformation Behavior of Swelling Soil

Richa Mudliar and S. Rajesh

1 Introduction

In arid and semi-arid regions, the expansive soil undergoes swelling and shrinkage due to adverse climatic conditions. The volumetric deformation of structures over these soils can damage buildings, pipes, roads, and other structures. Expansive soil mainly consists of reactive montmorillonite minerals that have open sites for the attraction of positive charge are neutralized by the absorption of mono- or divalent cations. For achieving the desirable geotechnical properties, stabilization is an economical and long-lasting method. Traditional stabilizers have been proved to be effective but may pose danger to the environment (like, increase in soil pH, brittle failure, contaminated groundwater) possibly due to the formation of ettringite and thaumasite [1]. This made researchers and field engineers focus on environment-friendly stabilizers.

The use of non-traditional stabilizers is becoming popular because of the non-toxic and environmentally friendly nature like lignosulfonate, enzymes, polymers, tree resins. Lignosulfonate (LS) is produced as the by-product in the pulp industry and therefore is an economical and readily available polymer [2]. Lignosulfonate has been proven beneficial for expansive soils [3, 4]. The optimum dosage of lignosulfonate may need to be decided based on tests like swell, permeability, consolidation, unconfined compressive strength, and CBR tests. [5] performed wetting-drying cycles on highly plastic clays treated with lignosulfonate for direct shear and unconsolidated undrained triaxial tests. It was reported that with the increase in wet-dry cycles, shear

R. Mudliar (✉) · S. Rajesh
Department of Civil Engineering, IIT Kanpur, Kanpur 208016, India
e-mail: vstricha@iitk.ac.in

S. Rajesh
e-mail: hsrajesh@iitk.ac.in

C. N. V. Satyanarayana Reddy et al. (eds.), *Ground Improvement and Reinforced Soil Structures*, Lecture Notes in Civil Engineering 152,
https://doi.org/10.1007/978-981-16-1831-4_24

strength parameters (cohesion and angle of internal friction) do not change significantly but at the end of the 4th cycles, the cohesion has increased by 30 kPa proving the efficiency of chemical in increasing strength. However, literature including the effect of LS on UCS has not established the details other than the optimum amount of chemical which is fixed based on basic geotechnical properties test s [6]. Hence, there is a need to establish a beneficial aspect of adding LS to the expansive soils. The objective of the present study is to investigate the efficacy of lignosulfonate for the stabilization of expansive soil. The engineering properties of expansive soil with different dosages of LS were evaluated. From the test results, the optimum dosage of the chemical needed for enhancing the strength and deformation of expansive soil was quantified.

2 Materials Used

The artificial swelling soil used in the present study was prepared by mixing 20% sodium bentonite and 80% fly-ash (by weight). The mix proportion was decided based on the swelling and permeability property of the mix [6, 7]. The soil mix comprises more than 60% of particles that are less than 2 mm in diameter. The liquid limit (LL) and plasticity limit (PL) of the bentonite are 574% and 66%, respectively. The PI of bentonite indicates that the bentonite is a very highly plastic material and has high swelling potential. The fly-ash(type C) used in the present study was collected from Panki thermal power plant. Table 1 provides a summary of the index and engineering properties of the soil tested following the ASTM standards.

Table 1 Summary of index and engineering properties of swelling soil

Property	ASTM code used for evaluation	Soil (20B80FA)
Liquid limit (%)	[8]	108
Plastic limit (%)	[9]	40
Shrinkage limit (%)	[10]	20
Maximum dry density (gm/cc)	[11]	1.2
Optimum moisture content (%)		29.22
Swell (%)	[12]	9.9% (OMC −2%); 2.8% (OMC +2%)
Compression index	[13]	0.473 (OMC −2%); 0.172 (OMC +2%)
Unconfined compressive strength (kPa)	[14]	416.8 (OMC −2%) 315.28(OMC +2%)

3 Methodology

The artificial swelling is preparedafter oven-drying of constituent soils at 105 °C to constant mass and sieving through 1 mm.

Free-swell index test is performed according to IS:2720 (Part 40) 1977. The oven-dried soil passing through 425 micron and chemical is mixed in 1:4 proportion. Test is performed for 0, 1, 2, 3% chemical mix. Two cylinders are taken one is filled with kerosene and the other with water (3 trials). Soil is added gradually in a cylinder filled with water. The time given for each subsequent addition of 0.1 gms is 5 min. Relative change with respect to volume attained in kerosene is calculated to find the free swell index.

For tests to be performed according to placement conditions, the calculated quantity of Ca-lignosulfonate was mixed with water to prepare the LS solution of the chosen dosage. The soil and the prepared LS solution were mixed thoroughly and compacted to the required density in a consolidation ring to obtain a sample size of 60 mm diameter and 20 mm thickness for conducting the swell-consolidation test. After compaction, the specimen along with the consolidation ring was wrapped in a plastic sheet and kept in constant-water content condition by cutting off the air supply for 24 h. After the moisture equilibrium period, water was allowed to flow through the soil under a seating pressure of 5 kPa [12]. Due to the intake of moisture, soil specimen tends to swell which was measured using the dial gauge. The measurements were taken until the values become constant.

Cylindrical soil samples of 38 mm in diameter and76 mm in height were used for the unconfined compressive strength (UCS) test. The chemical dosages adopted for the UCS test are 0, 1, 2, 3% LS. The samples were sealed in a polythene sheet and cured under constant water content conditions for 7, 14, 28 days. After curing, the specimen was sheared at 0.210 mm/min until failure [14].

4 Results and Discussion

4.1 Effect of Lignosulfonate on Plasticity

According to the cone-penetration test, the liquid limit is defined as the water content corresponding to the depth of penetration of 20 mm. Figure 1 shows the variation of the liquid limit of soil at various dosages of LS chemical. A significant reduction in the liquid limit was noticed up to the dosage of 3% LS but beyond which negligible change in liquid limit can be observed. The plastic limit of soil at different dosages of the chemical was found to be almost constant, hence the plasticity index decrease with the increase in the chemical dosage. Figure 2 shows the plasticity chart with the position of the treated soils. It can be noticed that soil classification has changed from CH (with LL = 108%, PI = 68%) to MH and ML depending on the dosage of chemicals, which shows the significance of LS in reducing the plasticity nature of the soil.

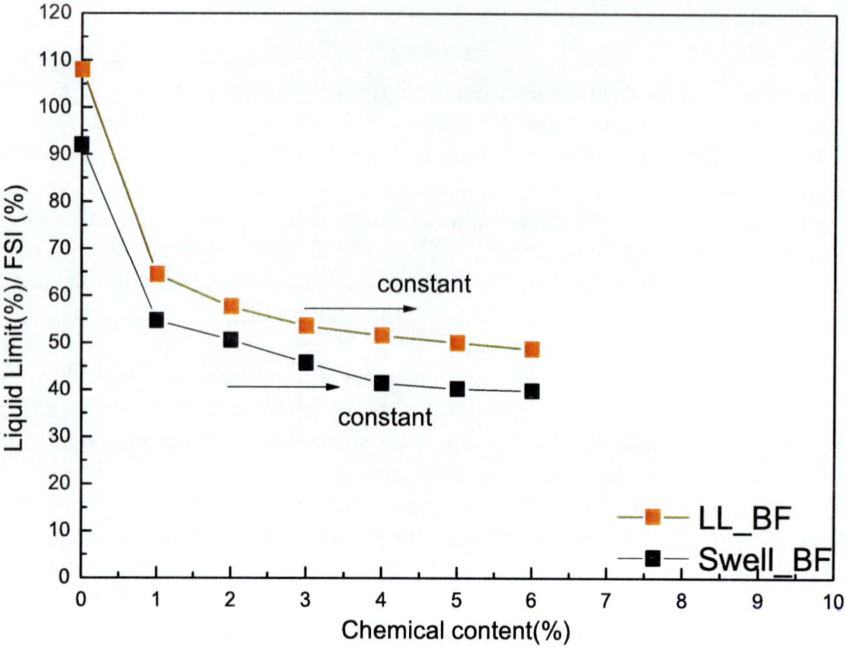

Fig. 1 Variation of liquid limit and FSI of soil with LS content

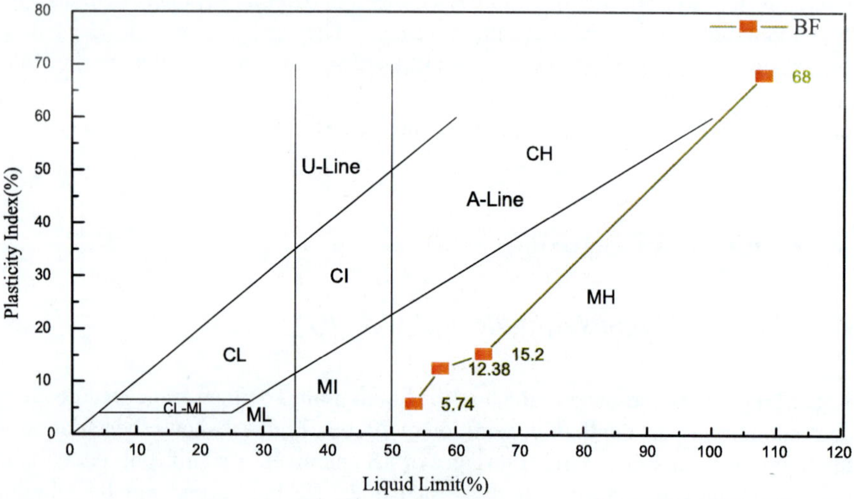

Fig. 2 Soil classification with variation in LS content

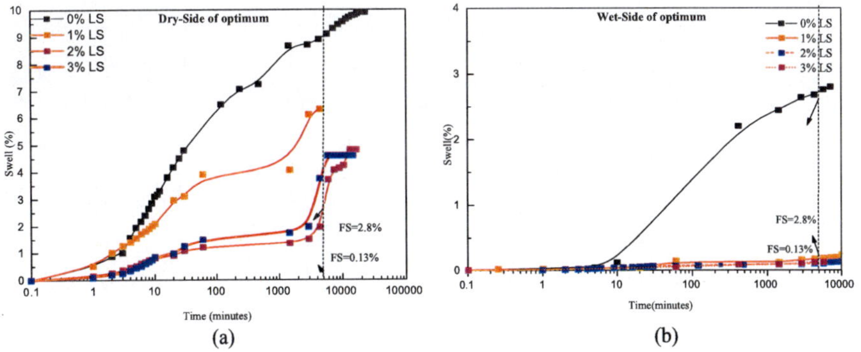

Fig. 3 Variation of percentage swell of untreated and treated soils compacted at **a** dry and **b** wet side of optimum with time

4.2 Effect of Lignosulfonate on Swelling Property

The free swell index (FSI) of the soil is the index when soil is allowed to freely swell with water relative to the original volume occupied by the soil in non-polar liquid like kerosene. The variation of the FSIof soil with different dosages of LS is also shown in Fig. 1. It can be noticed that FSI follows the same trend as the liquid limit, which implies, swelling nature of the expansive soil can be controlled with the addition of LS but the extent of reduction in swelling depends on the dosage of chemicals. The influence of placement condition (wet/dry side of optimum) and the surcharge loading on the swelling characteristics of soils can be obtained from the 1D swell-consolidation test and the results are shown in Fig. 3. It can be noticed that the percent swell is relatively higher for the dry side of the optimum (OMC-2%) than the wet side of the optimum (OMC + 2%), irrespective of the chemical dosage. The swelling is relatively less in the wet side of optimum could be due to the presence of dispersed structure as compared to the dry side of optimum where the flocculated structure is dominant. Moreover, with the increase in the chemical dosage, the percent swell has decreased significantly, irrespective of the placement condition. At 2% of the chemical, a significant reduction in the percent swell can be noticed at the dry side of optimum.

4.3 Effect of Lignosulfonate on Consolidation Characteristics

Figure 4 shows the consolidation characteristics of untreated and treated soils. It also shows the swelling pressure of the soil. The swell pressure (Ps) is the pressure required to bring the void ratio of the soil to the initial state of compaction. As expected, the swelling pressure of untreated soil is higher at the dry side of optimum

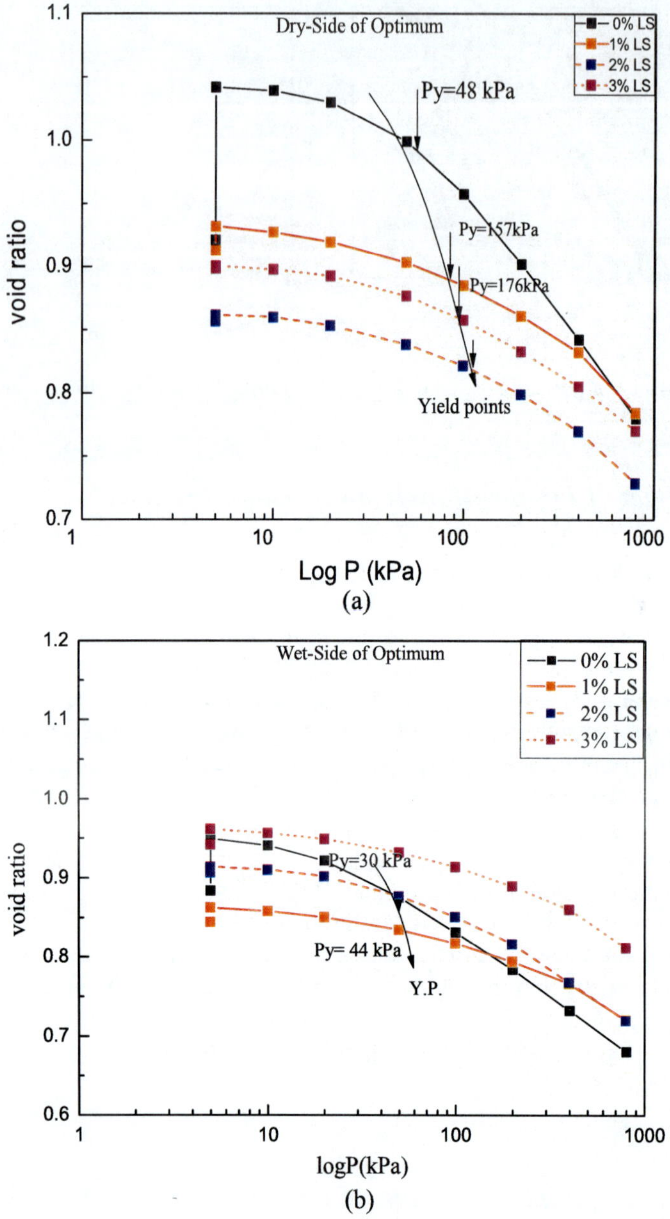

Fig. 4 Consolidation characteristics of untreated and treated soil at **a** dry and **b** wetside of optimum

Table 2 Swell-Consolidation properties of soil

Chemical % in soil	Dry side of optimum		Wet side of optimum	
	Ps (kPa)	Cc	Ps (kPa)	Cc
0	161	0.473	30	0.172
1	30	0.127	14	0.084
2	13	0.116	10	0.16
3	10	0.104	10	0.129

(1.61 kPa) when compared to the wet side of optimum (0.31 kPa). With the addition of LS, the swell pressure of soil has been drastically reduced. The maximum reduction in the swell pressure at the dry side of optimum (91.92%) was noticed for 3% of LS while at the wet side of optimum maximum reduction was noticed at 3% of LS when compared to untreated soil. The compression index (Cc), which is the slope of the virgin compression curve was found to decrease with the increase in the chemical dosage. Table 2 provides a summary of the swell pressure and the compression index of the untreated and treated soil at various placement conditions. The yield stress or the pre-consolidation pressure can be determined from the 1D consolidation test. It can be noticed that pre-consolidation pressure increases with the increase in the chemical dosage. The maximum increase in the pre-consolidation pressure at the wet and dry side of optimum was found to be 25% and 72.72% at LS percentage of 3%.

4.4 *Effect of Lignosulfonate on Strength Characteristics of Soil*

Figure 5a, b shows the variation of axial stress with axial strain for soil samples treated with different dosages of LS under unconfined condition. The results are shown for samples cured for 28 days. The UCS of treated soil was found to be relatively higher than the untreated soil but the enhancement in the strength depends on the dosage of the chemical. The amount of chemical which has given the least increase in strength is termed as the critical amount of chemical, while the amount of chemical which satisfies the minimum criteria is termed as the optimum amount of chemical. At the LS dosage of 1%, the maximum increase in the UCS of soil compacted at the dry and the wet side of optimum was found to be 50 kPa and 200 kPa respectively. Similarly at a 2% LS dosage increase is 40 kPa at dry side of the optimum and 100 kPa at the wet side of the optimum. Interestingly, at a 3% LS dosage, a minimum increase in the UCS was observed at both dry and wet sides of optimum. Therefore, the critical and optimum amount of LS chemical was found to be 3% and 2% respectively. The toughness is measured as the area under the stress-strain curve. The toughness of soil was found to increase significantly at the wet side of optimum than the dry side of optimum but the decrease in the toughness depends on the dosage of the chemical. The wet side of optimum has 43.86% more toughness compared to the dry side of optimum after 28 days curing period for 1% LS. While with subsequent addition, it is found to increase for the dry side of optimum by 400 kPa up to 3% LS and 200 kPa at wet-side of optimum.

The effect of the curing period on the stress-strain characteristics of the soil was evaluated by conducting the UCS test on soil samples mixed with a critical dosage of

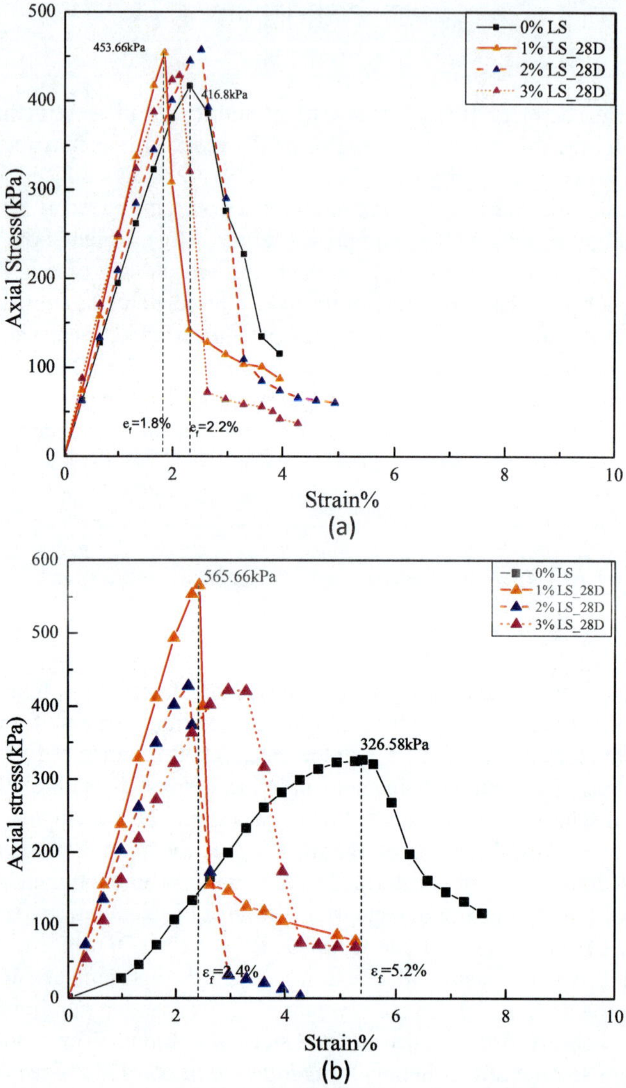

Fig. 5 Stress versus strain curve for soil specimen compacted at **a** dry-side and **b** wet-side of optimum with varying amount of lignosulfonate for 28 days curing **c** dry-side and **d** wet-side of optimumwitha critical amount of chemical for different curing periods

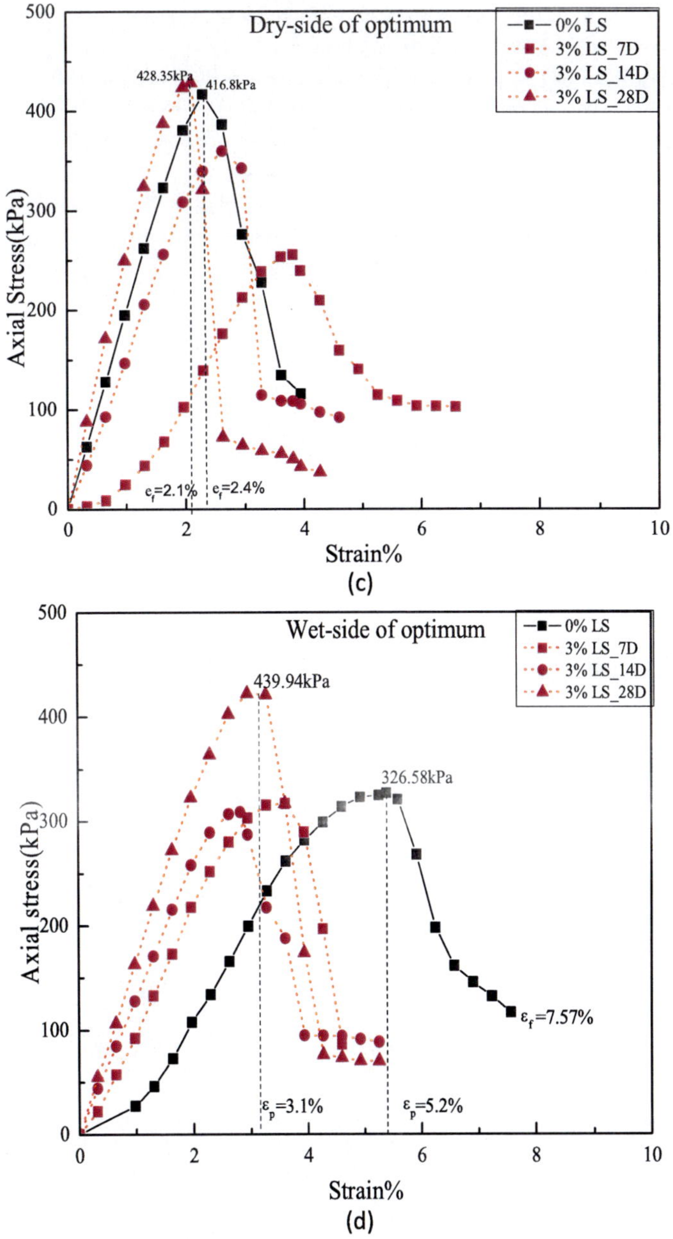

Fig. 5 (continued)

chemical (3% LS) at the end of 7,14, and 28 days curing period. Figure 5c, d shows the stress-strain curve of soil samples compacted at the dry and wet side of optimum. After the addition of LS, toughness decreases from 937.28 kPa to 730.69 kPa for 1% LS as the peak is achieved in smaller strain values at the dry side of the optimum while a decrease is 300 kPa at the wet side of optimum for 3% LS, the toughness increases for the dry side of optimum by 200 kPa while for the wet side of optimum decreases by 50%. For each of the samples, three identical specimens were tested. The unconfined compressive strength of soil obtained from the three trials was found to be almost the same. The plots for stress-strain used here belongs to the values of the lowest peak strength.

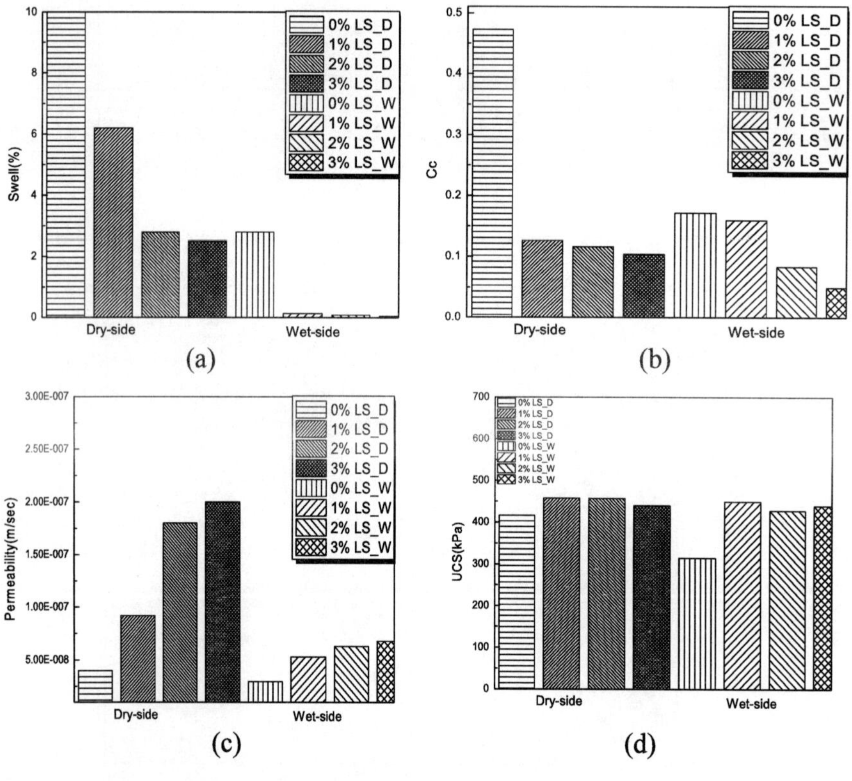

Fig. 6 Optimum dosage of lignosulfonate considering **a** percent swell, **b** compression in dex, **c** permeability, and **d** UCS

4.5 Optimum Dosage of Lignosulfonate

Figure 6 shows the optimum dosage of lignosulfonate for effective stabilization of expansive soil in terms of various engineering properties. The permissible values of each of the engineering properties are also shown in Figure. The optimum dosage of LS in terms of percentage swell is found to be 2% (Fig. 6a). The percentage swell is found to be constant beyond 2% LS for both dry and wetside of optimums. Interestingly, with the addition of LS to the swelling soil, the compression index of treated soil was found to be less than the permissible value, irrespective of the dosage of chemical and placement conditions (Fig. 6b). A minimum dosage of 1% LS is sufficient for soil stabilization considering the compression index. For 2% and 3% LS, the decrease in compression index is approaching constant close to 0.15 at the dry and wet side of optimum. The coefficient of permeability of soil was found to increase with the increase in chemical amount, irrespective of the placement condition (Fig. 6c). A significant increase in the permeability of the order of one was observed when the dosage of chemical increases from zero to 3%. As the UCS (Fig. 6d) of the treated soil is significantly higher (>50 kPa) than the untreated soil, the optimum dosage of chemical in terms of UCS is 1% LS. It can be observed for 1% LS in the wetside of the optimum significant increase is 100 kPa, while in the dry side of optimum the increase is 30 kPa, and approx. 50 kPa increase for 2% and 3% LS at dry side of optimum.

5 Conclusions

The efficacy of using lignosulfonate in stabilizing the swelling soil was assessed in this study. The index and the engineering properties of untreated and treated soils were evaluated. Based on the experimental results, the following conclusions are made:

1. Treatment of the high plasticity clay with the LS, results in an increase in the optimum water content and a decrease in maximum dry unit weight with change in soil type from highly plastic (CH) to medium (MI) and low plastic (ML) silt.
2. Free-swell index decreased from 90% (high swelling) to 40% (medium-low swelling) at 2% and 3% LS. While the swell% under compaction state is found to reduce from 161 kPa to 31 kPa at dry side of optimum and 30 kPa to 10 kPa at the wet side of optimum for 1% LS and 10 kPa for 2% and 3% LS addition.
3. From the unconfined compressive test, it can be observed that with increasing the LS-percentage from 0 to 3, the UCS increase from 400 kPa to 460 kPa (i.e., the relative maximum increment was up to 15%) at the dry side of optimum and 328 to 565 kPa (72% increase) at wet-side of optimum. The major contribution

to this increase in UCS is attributed to the formation of coat around the particles which prevent water to come in contact with the smectite group in the soil.

4. It can be concluded that 2% LS addition to soil changes the soil type from CH to MI decreasing swell-index from 90 to 40% bringing the soil type to silty. The Swell% has been decreased from 9.9% to 4% more than 50% at the dry side of optimum and negligible swelling at the wet side of optimum. The swell% is observed to follow a constant trend after an increase beyond 2% LS. The compression index for 2% LS addition is 50% from 0.473 to 0.127 at dry-side of the optimum and wet side of optimum has Cc value of 0.15. UCS has been found to be 50 kPa more than untreated at dry-side of optimum and 100 kPa after 28 days of curing. Therefore the addition of 2% lignosulfonate to the soil is found to be optimum dosage according to swell%, permeability, compression index and strength enhancement.

References

1. Puppala, A.J., Hanchanloet, S.: Evaluation of a new chemical (SA-44/LS-40) treatment method on strength and resilient properties of a cohesive soil. Transport Research Board, 78th Annual Meeting, 10–14 January, Washington, DC, USA 990389 (1999)
2. Gandini, A., Belgacem, M.: Lignin as components of macromolecular materials. In: Monomers, Polymers, and Composites from Renewable Resources. Elsevier, Oxford, UK, pp. 243–273 (2008)
3. Alazigha, D.P.: The efficacy of lignosulfonate in controlling the swell potential of expansive soil and its stabilization mechanisms (2015)
4. Alazigha, D.P., Indraratna, B., Vinod, J.S., Ezeajugh, L.E.: The swelling behaviour of lignosulfonate-treated expansive soil. Proc. Inst. Civ. Eng. Gr. Improv. **169**(3), 182–193 (2016)
5. Ta'negonbadi, B., Noorzad, R.: Stabilization of clayey soil using lignosulfonate. Transp. Geotech. **12**, 45–55 (2017)
6. Alazigha, D.P., Vinod, J.S., Indraratna, B., Heitor, A.: Potential use of lignosulfonate for expansive soil stabilisation. Environ. Geotech. **6**, 480–488 (2017)
7. Younus, M.M., Sreedeep, S.: Evaluation of bentonite-fly ash mix for its application in landfill liners. J. Test. Eval. **40**(3), 357–362 (2012)
8. Mollamahmutoğlu, M., Yilmaz, Y.: Potential use of fly ash and bentonite mixture as liner or cover at waste disposal areas. Environ. Geol. **40**(11–12), 1316–1324 (2001)
9. BS-1377-part-2: Methods of test for Soils for civil engineering purposes, Part 2: Classification tests
10. ASTM D4318: Standard Test Methods for Liquid Limit, Plastic Limit, and Plasticity Index of Soils. ASTM International, West Conshohocken, PA, pp. 1–14 (2005)
11. Analysis, P., Constants, S., Limit, P., Method, W., Statements, B.: Standard test method for shrinkage factors of soils by the mercury method **1**(4), 3–6 (1999)
12. ASTM D698, "Test Method for Laboratory Compaction Characteristics of Soils Using Standard Effort (12.400 ft-lbf/ ft 3 (600 kN-m/ m 3)).ASTM International, West Conshohocken, PA., pp. 1–13 (2012)

13. ASTM D4546-14. Pycnometer, G. Purposes, S. Barrel, T. Soil, P. Limit, and V. Ovens: Standard Test Methods for One-Dimensional Swell or Collapse of Cohesive Soils 1. ASTM International, West Conshohocken, PA, pp. 1–9 (2011)
14. ASTM D2435: One-Dimensional Consolidation Properties of Soils. ASTM International, West Conshohocken, PA. vol. 11, pp. 1–10 (2011)

Strength and Compressibility of Kaolinite Clay Stabilized with Lime Sludge and Fly Ash

S. G. Burra, P. K. Kolay, V. K. Puri, and S. Kumar

1 Introduction

With the growing demand for energy, most of the industrial processes end up with generating by-products. Most of these by-products are transferred to a landfill, which increases the cost of transportation as well as any environmental regulations depending on the type of by-product generated from the industry. Using these by-products in construction, to stabilize the soil and to improve its geotechnical aspects, has been gaining more interest with regards to promoting a sustainable environment. Unlike high-rise structures, the ability to use these by-products in pavement construction, will lead to consumption of most of the material and prove economical with respect to construction costs. Following are few studies that used by-products from industries to treat expansive or problematic soils to reduce the settlement or volume change potential.

Savaş [1] studied dispersive soil stabilized with lime and natural zeolite, his findings demonstrated an improvement in the effect of dispersivity, swell properties and compressibility. Pozzolanic reaction between the soil, lime and natural zeolite reduced the compressibility of the dispersive soil. Fattah et al. [2] have found that soft soil stabilized with lime and silica fume improved the Atterberg limits and reduced

S. G. Burra · P. K. Kolay (✉) · V. K. Puri · S. Kumar
School of Civil, Environmental and Infrastructure Engineering, 1230 Lincoln Drive, MC 6603, Carbondale, IL 62901, USA
e-mail: pkolay@siu.edu

S. G. Burra
e-mail: burra@siu.edu

V. K. Puri
e-mail: puri@engr.siu.edu

S. Kumar
e-mail: kumars@ce.siu.edu

C. N. V. Satyanarayana Reddy et al. (eds.), *Ground Improvement and Reinforced Soil Structures*, Lecture Notes in Civil Engineering 152,
https://doi.org/10.1007/978-981-16-1831-4_25

the compressibility index. The effect was more pronounced when lime and silica fume were used together, than using them individually. Amiralian et al. [3] have found that the minimum percentage of lime usage, to have a significant effect on the compressibility index should be more than 2%. Reduction in compressibility index with an increase in lime content is due to the Pozzolanic reactions between lime and soil. The compression index, coefficient of compressibility, and the coefficient of volume change decreased with an increase in fly ash and dolochar content for expansive soil [4]. Eberemu [5] has found that with the increase in rice husk ash content, the plasticity index decreased. The coefficient of consolidation increases at higher consolidation pressure. Pre-consolidation pressure increased, and compression index decreased with an increase in rice husk ash percentage up to 16%. However, at higher molding water content, the pre-consolidation pressure and compression index reduced. Phanikumar and Singla [6] found that heave swelling potential and vertical swelling pressure decreased, with an increase in fiber content for a given fiber length. However, this was valid up to a fiber length of 0.25%. At 0.30% fiber content, the swell potential and vertical swelling pressure increased. Lekan and Ojo [7] found that lateritic soils stabilized with tire ash can be appropriate for recovering low-lying marginal land for foundations. Compression index and swell index were reduced, and pre-consolidation pressure was increased with an increase in tire ash content up to 8%. The magnitude of primary consolidation in the normally consolidated state increases, while in the overconsolidated state it decreases [8]. Geopolymer cements (GPC) are manufactured under alkali-activated conditions and the properties of these cements are better than ordinary Portland cement. Soils stabilized with GPC's can manage consolidation settlement, resist high temperatures, and resist salts, sulphate, corrosion and acid attacks [9]. Rice husk ash has been proved more effective in reducing the consolidation settlement of the soil in comparison with stone dust and fly ash [10]. The oil-contaminated soil demonstrated a decrease in unconfined compressive strength and void ratio and an increase in the coefficient of consolidation and volume compressibility [11]. The compression index, expansive index, coefficient of consolidation and recompression index decreased for the expansive soil treated with an increase in lime and bagasse ash content [12]. For structures built on clay modified by fly ash, the amount of settlement decreases and the rate of settlement increases, reducing the time required for reaching the final settlement [13]. The improvement in consolidation characteristics of lateritic soil modified with cement addition, is due to the formation of a solid microstructure. The optimum content of cement was found to be 6%, to have a better mechanical performance [14]. Pre-consolidation pressure increased with lime and fly ash additions, leading the soil to form an over-consolidated soil. Also, the compression index was reduced with lime and fly ash-modified soils [15].

Based on the literature review, most of the researchers used the by-products available in their vicinity to stabilize soil and reduce settlement. An attempt has been made in this study to stabilize commercially available EPK clay to reduce the compression index by adding industrial by-products. The by-products used are sourced from the vicinity, like fly ash from a coal-fired thermal power plant and lime sludge from a water treatment plant. The findings can be applied to pavement construction, in stabilizing a weak subgrade.

2 Materials

2.1 EPK Clay

The EPK clay used in this study also called as kaolinite, is a commercially available material supplied by Edgar minerals, FL, USA. It is classified as silt with high plasticity (MH), according to the Unified Soil Classification System (USCS). The specific gravity was determined using a gas pycnometer and the value came out to be 2.61.

2.2 Lime Sludge

Lime Sludge (LS) used in the present study is a by-product from treating water. It is collected from a water treatment plant in Southern Illinois (Saline Valley Conservancy District water Treatment plant, Saline County, IL). The water is treated using various chemical processes, and suspends a softening residual matter high in concentrations of calcium oxide (CaO). The sample is like a slurry, it has been air dried, crushed, mixed homogeneously and sieved through 2 mm sieve prior to use in experiments. The percentage of fine sand, silt and clay size fraction are determined using a laser particle analyzer, and the values came out to be 12.5%, 76.5%, and 11.0%, respectively. The specific gravity was determined using a gas pycnometer, and the value came out to be 2.47.

2.3 Fly Ash

The Fly Ash (FA) used in the present study is from a coal-fired thermal power plant located at Southern Illinois University Carbondale (SIUC). The fly ash is classified as Class C fly ash as per ASTM C618-19 [16]. The specific gravity determined using a gas pycnometer was found to be 2.76.

3 Methods

All the testing procedures is this study are according to the respective American Society of Testing and Materials (ASTM) standards.

3.1 Physical Properties Test

Specific Gravity Test. The specific gravity for all the EPK clay mixes with fly ash and lime sludge were conducted using an equipment called Gas pycnometer, manufactured by Quantachrome Corporation, FL, USA. Depending on the type of material to be tested, the test set up has three different types of sleeve sizes (small, medium, and large). Prior to testing, the respective sleeve size has to be calibrated, and the representative sample of known weight has to be taken in the sleeve. The sleeve is loaded into the equipment, and the instructions are followed according to the equipment manual to determine the specific gravity. The specific gravity is calculated, taking an average of 5 runs, making the readings accurate and reliable.

3.2 Engineering Properties Test

Standard Proctor Test. The standard Proctor test was conducted on the proposed mixes i.e. EPK clay, EPK clay with lime sludge and EPK clay with lime sludge and fly ash. Lime sludge was used in varying percentages of 2%, 4%, 6% and 8% by dry weight of the clay and lime sludge and fly ash together were used in varying percentages 5%, 10%, 15% and 20% by dry weight of the clay.

The standard Proctor tests for the proposed mixes were carried out using a Harvard miniature compaction device, having a mold size of 33 mm in diameter and 71 mm in height. The compaction hammer used, had a diameter of 12.7 mm and weight (compaction effort) of 20 lb (9.07 kg). The samples were compacted in 5 equal layers and 5 blows per layer.

Unconfined Compressive Strength (UCS) Test. The unconfined strength samples were prepared in a standard UCS mold, having a height to diameter ratio of 2:1. The samples were compacted in 3 layers and 25 blows per layer, corresponding to its optimum moisture content and dry density determined from the standard Proctor tests. The samples were extruded then, wrapped in plastic film and placed in a plastic bag for a curing period of 7 and 14 days in a water bath having temperature maintained at 25 °C. The samples for 0 days curing, were tested right after the extrusion from the mold. The compacted samples were loaded in a UCS test machine with a strain rate of 1.27 in/min (32.26 mm/min) until the load reading drops or the strain value exceeds 15%. The stress-strain graphs were plotted based on the load and deformation readings and the peak value of the graph is taken as the unconfined compressive strength.

Consolidation Test. The consolidation tests for the proposed mixes were carried out in an automated consolidation device, manufactured by GeoComp Corporation, Boston, USA. The samples were compacted in the consolidation ring based on the standard Proctor moisture-density relationship. The top and bottom of the sample is covered with a filter paper followed by porous stones. The consolidation cell is loaded in the test set up and the load is zeroed. Then, the cell is inundated with water and the load is zeroed again before starting the test. The test runs automatically once it is initiated, with a specified sequence of loading and unloading. The data is then downloaded and necessary calculations are done.

4 Results and Discussion

4.1 Physical Properties Test

Specific Gravity Test. The specific gravity test results for EPK clay mixed with different percentages of lime sludge, different percentages of lime sludge and fly ash are presented in Table 1. From Table 1, it can be observed that the specific gravity values increased with an increase in the percentage of lime sludge from 2%–8%, lime sludge and fly ash from 5%–20%.

Table 1 Specific gravity values for EPK clay, EPK clay with different percentages of lime sludge (LS), EPK clay with different percentages of lime sludge (LS) and fly ash (FA)

Mix details	Specific gravity
EPK	2.5699
EPK + 2% LS	2.5595
EPK + 4% LS	2.5612
EPK + 6% LS	2.5655
EPK + 8% LS	2.5674
EPK + 5% LS + 5% FA	2.5842
EPK + 10% LS + 10% FA	2.5969
EPK + 15% LS + 15% FA	2.6014
EPK + 20% LS + 20% FA	2.6020

4.2 Engineering Properties Test

Standard Proctor Test. The test results for the standard Proctor test for EPK clay mixed with lime sludge is presented in Fig. 1. For the virgin EPK clay, the optimum moisture content and the dry unit weight from Fig. 1 is 32.17% and 12.98 kN/m^3, respectively. However, with an increase in the percentage of lime sludge from 2% to 6%, there was an increase in dry unit weight from 12.75 kN/m^3 to 13.15 kN/m^3. Further, at 8% lime sludge the dry unit weight had a slight decrease to 13.03 kN/m^3. The moisture content, on the whole, decreased from 36% to 31% with an increase in lime sludge percentage from 2% to 8%.

The test results of the standard Proctor test for EPK clay mixed with lime sludge and fly ash are presented in Fig. 2. From Fig. 2, it can be observed that, with an increase in fly ash and lime sludge percentage from 5% to 20%, the dry unit weight increased from 12.92 kN/m^3 to 13.16 kN/m$^{3.}$ Also, with the increase in lime sludge

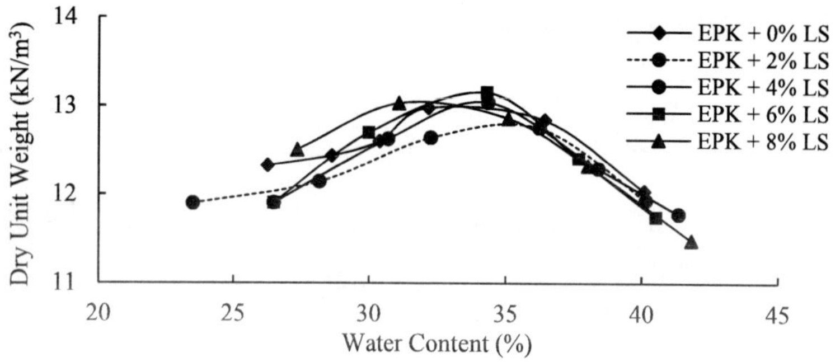

Fig. 1 Compaction curve for EPK clay with different percentages of LS

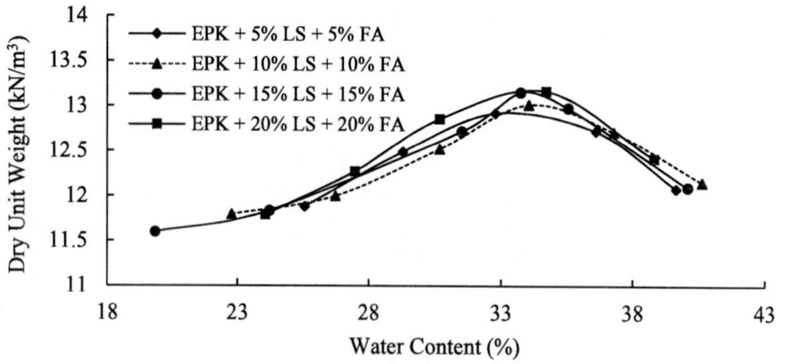

Fig. 2 Compaction curve for EPK clay with different percentages of LS and FA

and fly ash content from 5%–10%, the optimum moisture content (OMC) increased from 32.77% to 34.03%. Then, at 15% the OMC decreased to 33.72% and increased to 34.69% at 20% lime sludge and fly ash content.

Unconfined Compressive Strength (UCS) Test. The test results of the unconfined compressive strength (UCS) mixed with different percentages of lime sludge are presented in Fig. 3. From Fig. 3, it can be seen that the UCS value increases with an increase of lime sludge content from 2% to 8% for a curing period of 7 days. However, the UCS value increased with the increase in lime sludge content from 2% to 6% and had a slight decrease at 8% for a curing period of 0 and 14 days.

The test results of the UCS mixed with different percentages of lime sludge and fly ash are presented in Fig. 4. From Fig. 4, it can be observed that, for a curing period of 0 days, the UCS value increased with the increase in lime sludge and fly

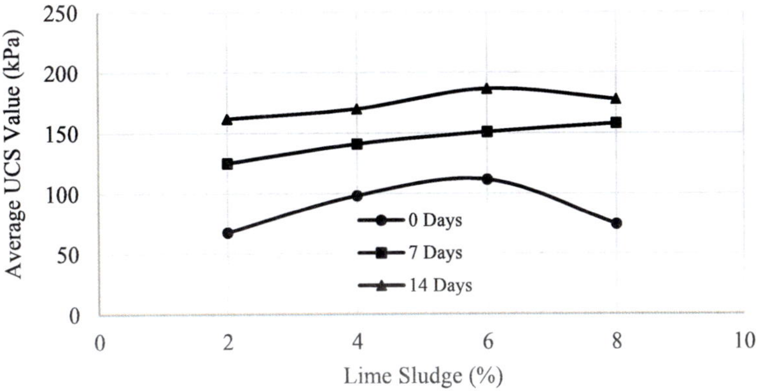

Fig. 3 UCS values of EPK with different percentage of LS at 0, 7, and 14 days curing

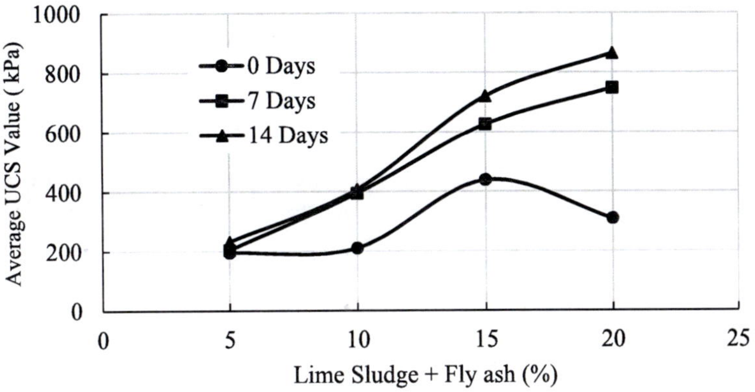

Fig. 4 UCS values of EPK with different percentage of LS and FA at 0, 7 and 14 days curing

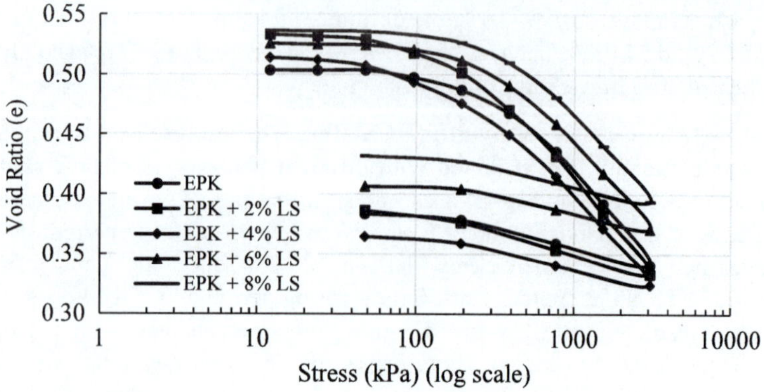

Fig. 5 Consolidation test for EPK clay with different percentages of Lime sludge

Table 2 Compression Index (C_c) of EPK clay with different percentages of Lime sludge

Mix details	C_c
EPK + 0% LS	0.1453
EPK + 2% LS	0.1603
EPK + 4% LS	0.1443
EPK + 6% LS	0.1370
EPK + 8% LS	0.1353

ash content from 5% to 15% and decreased by 20%. However, for a curing period of 7 and 14 days, the UCS value increasedwith the increase in lime sludge and fly ash percentage from 5% to 20%.

Consolidation Test. The consolidation test results for EPK clay stabilized with lime sludge are presented in Fig. 5 and Table 2. From Table 2, it can be observed that, the compression index value increased at 2% lime sludge content. As the percentage of lime sludge increased, the compression index values decreased gradually in comparison with the EPK clay.

The consolidation test results of EPK clay stabilized with lime sludge and fly ash are presented in Fig. 6 and Table 3. From Table 3, it can be observed that with the increase in lime sludge and fly ash content, the compression index value decreased at 5% addition, then had a slight increase for 10% and 15% and decreased for 20% addition in comparison to the original EPK clay.

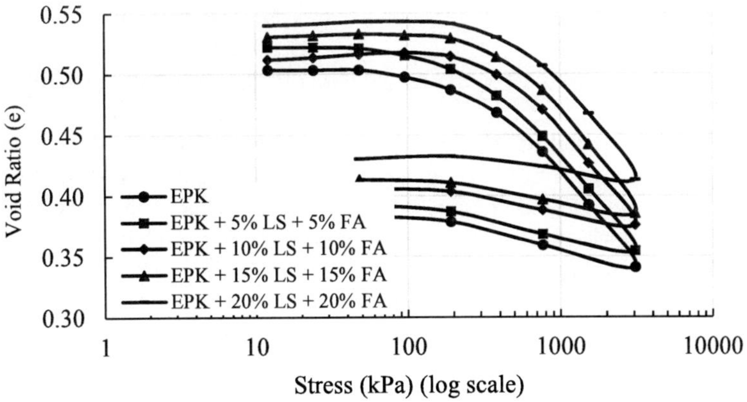

Fig. 6 Consolidation test for EPK clay with different percentages of Lime sludge and Fly ash

Table 3 Compression Index (C_c) of EPK clay with different percentages of Lime sludge and Fly ash

	C_c
EPK + 0% LS + 0% FA	0.1453
EPK + 5% LS + 5% FA	0.1443
EPK + 10% LS + 10% FA	0.1467
EPK + 15% LS + 15% FA	0.1487
EPK + 20% LS + 20% FA	0.1318

5 Conclusions

The aim of the study was to stabilize the EPK clay with lime sludge, the combination of lime sludge and fly ash, and to observe the compressibility behavior. The following conclusions were drawn based on the results.

1. The specific gravity values of EPK clay mixed with lime sludge, lime sludge and fly ash mixture demonstrated an increase in value with an increase in the percentage of lime sludge, lime sludge and fly ash addition.
2. The dry unit weight for EPK clay mixed with different percentages of lime sludge had an increase from 2% to 6% and had a slight decrease for 8%. Overall, the OMC decreased from 36% to 31% with an increase in lime sludge content from 2% to 8%.
3. The dry unit weight for EPK clay mixed with different percentages of lime sludge and fly ash mixture had an increase from 5% to 20%. Overall, the OMC increased from 5% to 10%, decreased at 15% and then increased by 20%.
4. The UCS strength of EPK clay mixed with lime sludge, demonstrated an increase in value for a curing period of 7 days with the increase in lime sludge content from 2% to 8%. For a curing period of 0 and 14 days, the UCS value increased

with an increase in lime sludge content from 2% to 6% and had a slight decrease of 8%.

5. The UCS strength of EPK clay mixed with different percentages of lime sludge and fly ash had an increase in value for a curing period of 7 and 14 days from 5% to 20% in addition. However, for a curing period of 0 days, the UCS value increased from 5% to 15% and decreased by 20% addition of lime sludge and fly ash. Curing played an effective role in strength gain and the UCS strength of EPK clay with lime sludge and fly ash was 4 times more in comparison with EPK clay with lime sludge.
6. The compression index (C_c) values of EPK clay mixed with different percentages of lime sludge had a slight increase at 2% addition and decreased thereafter in comparison with EPK clay.
7. The compression index values of EPK clay mixed with different percentages of lime sludge and fly ash did not follow a decreasing trend, but had a slight increase for 10% and 15% addition and decreased for 20% addition. At the higher percentage of both lime sludge, lime sludge and fly ash, the compression index value reduced in comparison with EPK clay.

References

1. Savaş, H.: Consolidation and swell characteristics of dispersive soils stabilized with lime and natural zeolite. Sci. Eng. Compos. Mater. **23**(6), 589–598 (2016)
2. Fattah, M.Y., Al-Saidi, A.A., Jaber, M.M.: Consolidation properties of compacted soft soil stabilized with Lime-Silica fume mix. Int. J. Sci. Eng. Res. ISSN, 2229–5518 (2014)
3. Amiralian, S., Chegenizadeh, A., Nikraz, H.: Laboratory investigation on the effect of lime on compressibility of soil. In: Proceedings of the International Conference on Civil and Architectural applications (ICCAA'2012), pp. 89–93. Planetary Scientific Research Center (2012)
4. Mohanty, S.K., Pradhan, P.K., Mohanty, C.R.: Consolidation and drainage characteristics of expansive soil stabilized with fly ash and dolochar. Geotech. Geol. Eng. **34**(5), 1435–1451 (2016)
5. Eberemu, A.O.: Consolidation properties of compacted lateritic soil treated with rice husk ash. Geomaterials **1**(3), 70 (2011)
6. Phanikumar, B.R., Singla, R.: Swell-consolidation characteristics of fiber-reinforced expansive soils. Soils Found. **56**(1), 138–143 (2016)
7. Lekan, A.O., Ojo, T.A.: Consolidation properties of compacted lateritic soil stabilized with tire ash. J. Eng. Manuf. Tech., 36–44 (2013)
8. Salehi, M., Sivakugan, N.: Effects of lime-clay modification on the consolidation behavior of the dredged mud. J. Waterway Port Coastal Ocean Eng. **135**(6), 251–258 (2009)
9. Onyelowe, K., Van, D.B.: Structural analysis of consolidation settlement behavior of soil treated with alternative cementing materials for foundation purposes. Environ. Technol. Innov. **11**, 125–141 (2018)
10. Jain, A., Puri, N.: 1-Dimensional consolidation characteristics of clay stabilized with major industrial wastes of Haryana. In: Proceedings of Indian Geotechnical Conference, pp. 22–24 (2013)
11. Ijimdiya, T.S.: The effects of oil contamination on the consolidation properties of lateritic soil. Dev. Appl. Oceanic Eng. (DAOE) **2**(2), 53–59 (2013)

12. Manikandan, A.T., Moganraj, M.: Consolidation and rebound characteristics of expansive soil by using lime and bagasse ash. Int. J. Res. Eng. Technol. **3**(4), 403–411 (2014)
13. Phanikumar, B.R., Sharma, R.S.: Volume change behavior of fly ash-stabilized clays. J. Mater. Civ. Eng. **19**(1), 67–74 (2007)
14. Mengue, E., Mroueh, H., Lancelot, L., Eko, R.M.: Physicochemical and consolidation properties of compacted lateritic soil treated with cement. Soils Found. **57**(1), 60–79 (2017)
15. Sureban, V.: Consolidation characteristics of fly ash and lime treated black cotton soil. In: Proceedings of International Conference on recent Trends in Transportation, Environmental and Civil Engineering, pp. 49–52 (2011)
16. ASTM C618–19, Standard specification for coal fly ash and raw or calcined natural pozzolan for use in concrete, ASTM International, West Conshohocken, PA (2019)

Slope Stability Analysis of Stacker–Reclaimer Embankment Under the Influence of Adjacent Stockpile

Riddhi M. Gupta, Hardik Gajjar, and Jaykumar Shukla

1 Introduction

The infrastructure sector is one of the fastest developing sectors in developing countries. Due to economic growth and progressive increase in population, readily available soil with reasonable shear strength and bearing capacity is scare. For the past few decades several ground improvement techniques have been very well explored and implemented [1, 2]. The selection of any ground improvement technique is mainly governed by soil type, resource availability, its environmental impact and economic viability. Based on available previous study Ground improvement by bottom feed stone column explicitly for loose sandy silt/soft clay type soil is considered as one of the most proven methods [3]. Over the period of time several analytical as well as empirical has been develop to understand the settlement as well as stability behavior of bottom feed stone column [4, 5].

The present study has been carried out in order to understand the influence of 20 m iron ore stockpile on stacker–reclaimer (S-R) berm. Figure 1 for site reference and arrangement of iron ore stoke pile with stacker–reclaimer berm facility [6].

For the material handling work, the stacker–reclaimer crane runs on the rails provided on the berm. To anticipate the vertical load from the stacker–reclaimer facilities, a pile foundation system has been provided [12]. It should be noted that the influence of pile on stability is not considred for present case. Moreover, several underground utilities such as fire pipes and drains were also passing underground the stacker–reclaimer berm. Based on available soil data and preliminary calculations it was found that there may excessive displacement and instability occur if week soft soil strata were not modified.

R. M. Gupta (✉) · H. Gajjar · J. Shukla
Geodynamics, Vadodara, India
e-mail: riddhigupta95@gmail.com

C. N. V. Satyanarayana Reddy et al. (eds.), *Ground Improvement and Reinforced Soil Structures*, Lecture Notes in Civil Engineering 152,
https://doi.org/10.1007/978-981-16-1831-4_26

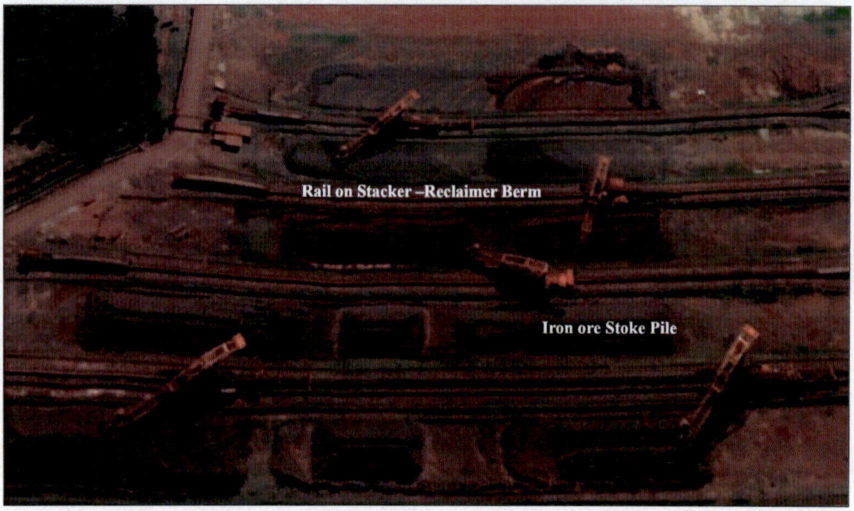

Fig. 1 Reference site picture with iron ore stockpile and stacker–reclaimer berm facility

Local stability of stacker–reclaimer (S-R) berm without the influence of stockpile was carried out and it was not of major concern. However, while calculating the stability considering the stockpile influence, deep-seated failure was observed beneath the (S-R) berm. Based on the soil profile and previous ground improvement measure of the same site bottom-feed vibro stone column is selected as the ground improvement technique for the stability of the proposed stacker–reclaimer berm. In the present case, the height of the stockpile is divided into equal "h" m thick layer and the influence of its sequential filling on the adjacent stacker–reclaimer berm is studied. Stage construction methodology has been adopted for stability analysis. Stockpile was discretized into eight stages (each of equal height).

Parametric study has been carried out by varying the friction angle and grid spacing of stone column to understand its influence on stability. For each stage of stockpile, filling comparison is made between improved and unimproved ground condition. The results are summarized in various charts.

2 Ground Characterization

The general ground profile was prepared based on available soil borehole detail data. From the soil description and standard penetration test (SPT) 'N' value, it was concluded that the soil strata was majorly governed by interbedded layers of sandy silt and clayey silt type of soil. Moreover, from SPT value it can be classified as weak soil. The value of soil cohesion observed was <30 kPa for unimproved ground conditions.

Table 1 Ground characterization

Sr. no.	Soil description	Depth (m)	Average SPT 'N' value
1	Sandy clay	0–2	4
2	Silty sand	2–4	3
3	Soft clay	4–7.45	1
5	Gravelly sand	7.45–9.45	1
6	Sandy clay	9.45–11	1
7	Clayey sand	11–13	6
8	Sandy clay	13–16	4
9	Silty sand	16–23.45	7
10	Sandy silt	23.45–28.95	15
11	Hard sandy silt	28.95–34.4	50

Table 2 Material property used for limit equilibrium analysis

Material description	Bulk density (γ), kN/m^3	Angle of internal friction (φ)	Cohesion (C), kPa
Fill layer	18	30	0
S-R berm	20	35	05
Stone column	18	40	0
Iron ore stockpile	28	40	0

The material model used for stacker–reclaimer berm, stockpile and soil layer are the Elasto-Plastic Mohr Column model. The stone column was modeled as a material with friction property only.

Berm soil is also modeled as cohesion-less deposit. Moreover, for the stockpile, which is an iron ore deposit only friction angle is estimated. The generalised site SPT vales are provided in Table 1. The material property for the berm, stockpile and stone column considered are given in Table 2.

2.1 2-Dimensional LEM Model Validation

A two layer slope has been taken from literature [7] and analyzed to validate the methodology adopted in using the GEO-SLOPE International Ltd. (Slope/w) software. The parameter used for the stability analysis with respect to embankment fill and foundation material is given in Table 3. The Morgenstern-Price method have been chosen in Slope/w for the limit equilibrium analysis.

Table 3 Soil property used in validation model [7]

Material parameter	Unit	Sand	Clay	Embankment fill	Stone column	Equivalent parameter
Saturated unit weight	kN/m^3	18	16	18	17	16.2
Cohesion	kPa	0	20	0	0	16
Friction angle	Degree	30	0	32	38	8.9

Comparison is made between finite difference method (FDM) and Limit Equilibrium method based on Itasca International Inc. [11] and GEO-SLOPE International Ltd [13] Software respectively. The location of the critical slip surface observed by these two methods is also observed. Only factor of safety is discussed for validation purpose. The height of the embankment is taken as 5.0 m. The slope of the embankment with respect to horizontal is 26.56° (0.5 V:1 H).

The model Dimensions and configuration adopted for validation are shown in Fig. 2a. Figure 2b gives the factor of safety value obtain with LEM-based analysis with GEOSLOPE software.

The factor of safety value obtained was 1.557 for equivalent ground conditions which are very close to the reported FOS 1.6 [7], which validates the methodology of analysis.

3 Numerical Modeling

In the present study, the geotechnical design is mainly concerned with the stability of S-R berm under the influence of adjacent iron ore stockpile. Two-dimensional Slope stability analysis is carried out with the help of GEOSLOPE software (GEO-SLOPE International Ltd.). Limit equilibrium analysis based on the method of slices was selected to perform the stability check. Here also the Morgenstern-Price method has been chosen in Slope/w for the limit equilibrium analysis. The slope of the S-R berm was considered as 35° with respect to the ground surface.

Among the various simplified geometric models such as unit cell, longitudinal gravel trench model, cylindrical gravel ring model, three-dimensional model, etc., here homogenization method is selected for modeling. The Complex 3-Dimensional field problem is converted into the equivalent 2-D plain strain problem with homogenization or equivalent homogeneous soil model. In this method, the stone column and soft soil are replaced by an equivalent homogeneous soil with improved shear strength parameters [8, 9]. The area treated with a stone column is replaced by equivalent homogeneous soil. For conservative analysis, the water table was assumed at the ground surface. In addition to this surcharge load with the uniform intensity of 20 kPa was applied on the (S-R) berm.

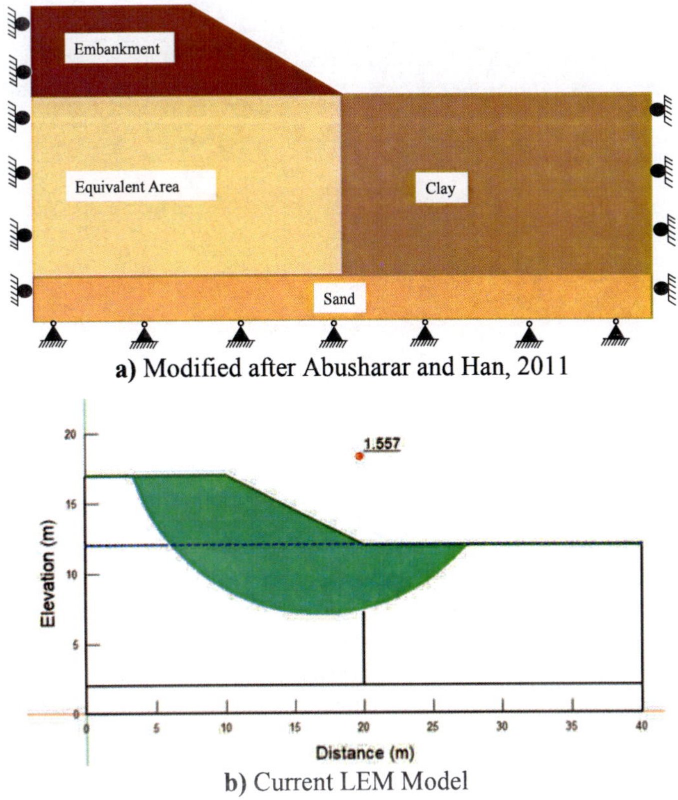

Fig. 2 Model validation using limit equilibrium approach. **a** Geometry used for model validation. **b** Present LEM model

The ratio of amount of soft soil replaced by stone column is called area replacement ratio (a_s). The area replacement ratio plays a significant role in analyzing the stone column to improve the ground condition. By varying the area replacement ratio (a_s) and improved material parameter, various combinations can be analysis at the design stage. The combinations of different stone column diameter, spacing and length can be made without making a change in the geometry of the model. When the problem has a complex geometry, this method is more suitable in order to simplify modeling with the stone column. The weighted average of soil and stone column parameters taken according to area ratio (a_s). The equivalent soil parameters are computed as follows [1].

$$c_{eq} = c_c a_s + c_s(1 - a_s) \tag{1}$$

$$\varphi_{eq} = \tan^{-1}(a_s \tan\varphi_c + (1 - a_s)\tan\varphi_s) \tag{2}$$

Table 4 Detail of various configuration explored

Case number	Series	Angle of internal friction (°)	Grid spacing (m)	Column diameter (m)
Case 1	A1	38	1.5	1.0
	B1	38	1.7	1.0
	C1	38	2	1.0
Case 2	A2	40	1.5	1.0
	B2	40	1.7	1.0
	C2	40	2	1.0
Case 3	A3	42	1.5	1.0
	B3	42	1.7	1.0
	C3	42	2	1.0

$$\gamma_{\text{eq}} = \gamma_c a_s + \gamma_s (1 - a_s) \quad (3)$$

Here, a_s is the area replacement ratio.

Area replacement ratio (a_s) was calculated for different grid spacing and friction angle of stone column, with constant column diameter as given in Table 4. The material parameters are derived based on Eqs. (1)–(3). The square grid pattern was considered for further calculation. To capture the trend of the stone column parameters (i.e. angle of internal friction, diameter and grid spacing) following different case were studied.

It should be noted that physical phenomenon such as radial drainage towards the column, stress-concentration on soil and stone column cannot be reproduced by homogenization method. Though, after calibrating the parameters of the equivalent homogeneous soil, the overall response of the system on a large scale may be correctly predicted [8–10].

4 Result Analysis

The influence on the value of FOS corresponding to the stage construction of the stoke pile has been examined for (S-R) berm facility. Comparison is made between unimproved and improved ground conditions considering the effect of varying stone column spacing and angle of internal friction of stone material. The diameter of the column was kept constant and a sequential filling of stockpile was made. In case 1 the angle of internal friction was kept as 38° while the spacing of the stone column varies as 1.5, 1.7 and 2.0 m respectively (Fig. 3). Similarly, for case 2 with a varying spacing of 1.5, 1.7 and 2.0 m, the column friction angle was kept as 40° (Fig. 4). And for case 3 with a varying spacing of 1.5, 1.7 and 2.0 m, the friction angle was 42° (Fig. 5). The factor of safety for (S-R) berm was found 1.635 for unimproved

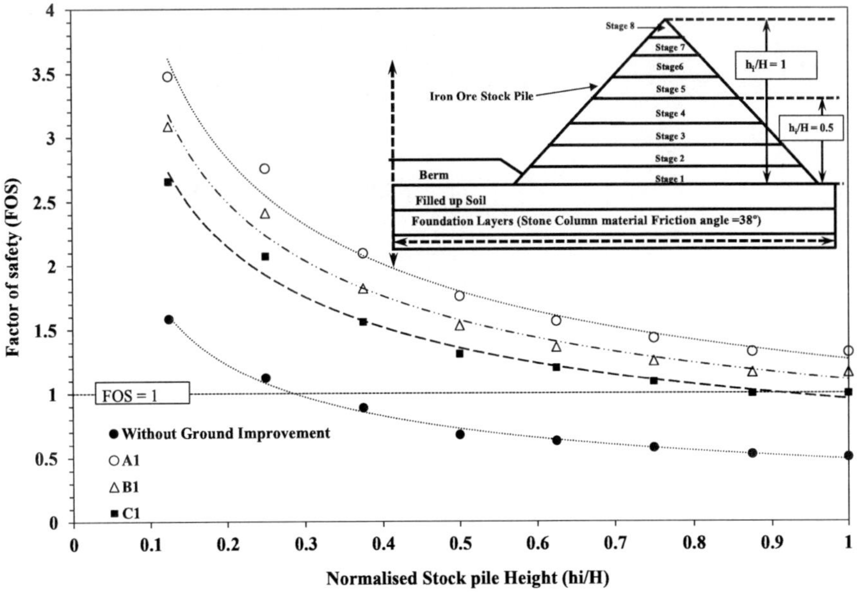

Fig. 3 Influence of stockpile height and column spacing for stone column friction angle 38°

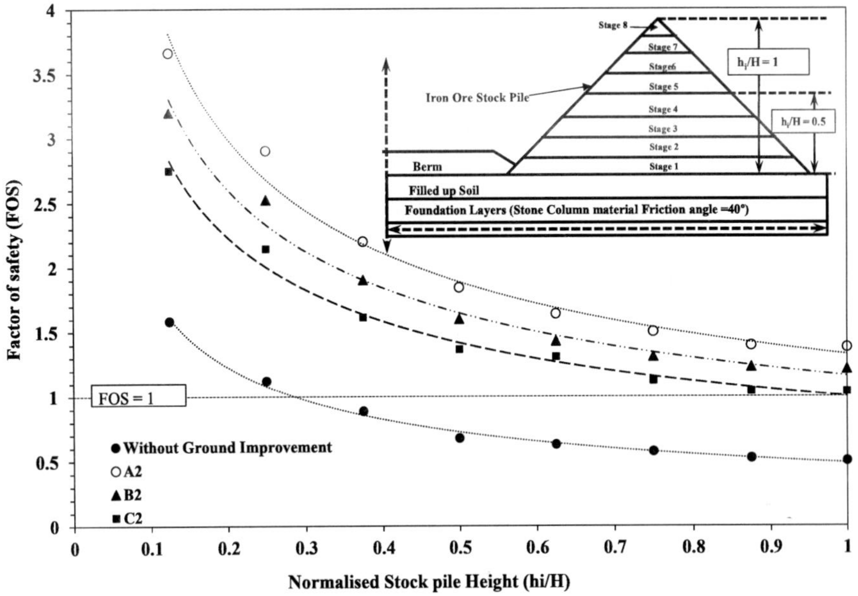

Fig. 4 Influence of stockpile height and column spacing for stone column friction angle 40°

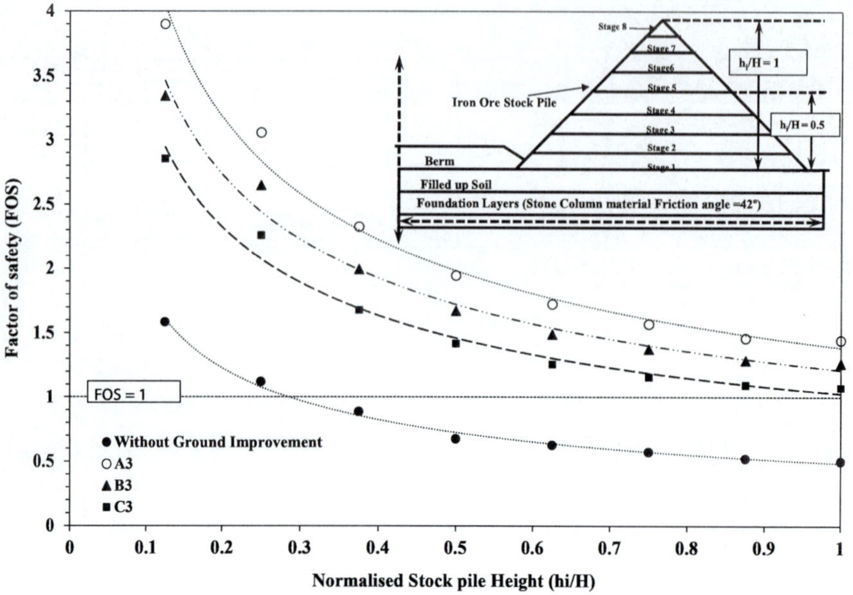

Fig. 5 Influence of Stockpile height and column spacing for stone column friction angle 42°

ground condition. The same rages from 1.87 to 1.814 with a stone column for the improved ground condition for (S-R) berm facility. This analysis was made without taking into account the stockpile influence.

The result of the effect of friction angle of the stone column and varying column spacing on (S-R) Berm under the influence of stockpile is as follows.

From the above results for the unimproved ground condition, it is to be noted that the berm is nearly stable up to the "2 h". However, with further increase in height of stacking from "3 h" to "8 h", the (S-R) berm along with stockpile becomes unstable and leads to deep-seated failure due to soft compressible soil. When analysis made considering ground improvement with stone column keeping friction angle of stone as 38° and column spacing of 1.5 m the FOS observed ranges between 3.477 and 1.321 for Stages 1–8, respectively.

Similarly, for column spacing of 1.7 m FOS observed was 3.086–1.164 and for 2.0 m FOS observed was 2.656 and 0.998 for Stage 1–8.

The percentage variation in spacing of 22 and 43% results in percentage variation in FOS of 12% to 24% respectively. Nearly 50% variation is observed between area replacement ration and FOS. As by increasing the spacing the area replacement ration decreases which directly affects the stability of the berm. The soil strength parameter gets improve with a reduction in stone column spacing, as more amount of stone replaces the poor soil. Similar results were observed for stone friction angle of 40 and 42. Moreover, it should be noted that as the friction angle of stone material increases i.e. from 38 to 42, the factor of safety for similar spacing increases (Figs. 3, 4 and 5). For Case C1 the FOS for the stage "7 h" and "8 h" was found less than unity, while for similar stage FOS found more than unity with the same spacing for Case C2 and C3. The increased friction angle of stone material dramatically increases the overall stability.

5 Conclusions

In this study, it was observed that even though the (S-R) berm without any means of ground improvement was found stable, shows deep-seated failure with an increase in the height of adjacent stacking. It is to be noted that in lieu of detailed analysis and modeling it would lead to the instability and fatal failure of the berm along with the stockpile.

In order to optimize the spacing and minimize the cost parametric study plays a vital role. Further stone column analysis with equivalent area approach saves considerable time in modeling and gives reasonably accurate results. It can be concluded that an increase in friction angle of stone material leads to greater stability against deep-seated failure. Moreover, as the spacing of the stone column increases, the area replacement ratio decreases lead to comparatively less improvement in the shear strength parameter of the equivalent area. This ultimately leads to a decrease in the factor of safety of the stone column improved ground. The higher area ratio results in an increased factor of safety, which ultimately leads to greater stability.

References

1. Christoulas, S., Giannaros, C., Tsiambaos, G.: Stabilization of embankment foundations by using stone columns. Geotech. Geol. Eng. **15**, 247–258 (1997a). https://doi.org/10.1007/bf00880828
2. Ambily, A.P., Gandhi, S.R.: Behavior of stone columns based on experimental and FEM analysis. J. Geotech. Geoenviron. Eng. **133**(4), 405–415 (2007). https://doi.org/10.1061/(ASCE)1090-0241(2007)133:4(405)
3. Dheerendra Babu, M.R., Nayak, S., Shivashankar, R.: A critical review of construction, analysis and behaviour of stone columns. Geotech. Geol. Eng. **31**(1), 1–22 (2013). https://doi.org/10.1007/s10706-012-9555-9
4. Prieb, H.J.: The design of vibro replacement. Ground Eng. **28**(10), 31–37 (1995). https://doi.org/10.1016/0148-9062(96)80092-1
5. McCabe, B.A., Nimmons, G.J., Egan, D.: A review of field performance of stone columns in soft soils. Proc. Inst. Civil Eng.: Geotech. Eng. **162**(6), 323–334 (2009). https://doi.org/10.1680/geng.2009.162.6.323
6. Shukla, J., Sengupta, S.: Geotechnical challenges and opportunities in large infrastructure projects—learning from failures. December, 213–223 (2019). https://doi.org/10.1007/978-981-13-0505-4_19
7. Abusharar, S.W., Han, J.: Two-dimensional deep-seated slope stability analysis of embankments over stone column-improved soft clay. Eng. Geol. **120**, 103–110 (2011a). https://doi.org/10.1016/j.enggeo.2011.04.002
8. Castro, J.: Modeling stone columns. Materials **10** (2017a). https://doi.org/10.3390/ma10070782

9. Castro, J.: Groups of encased stone columns: Influence of column length and arrangement. Geotext. Geomembr. **45**(2), 68–80 (2017b). https://doi.org/10.1016/j.geotexmem.2016.12.001
10. Ng, K.S., Ann, S., Tan, H.: Parametric study on the settlement improvement factor of stone column groups. ESTEEM Acad. J. **10**, 55–65 (2014). https://doi.org/10.13140/2.1.5092.8328
11. Itasca Consulting Group, Inc.: FLAC/Slope User's Guide, Version 5.0. 84 p. (2006)
12. Gupta R.M., Shukla J.C., Joshi N.H. (2021) A study of piled raft foundation. In: Patel S., Solanki C.H., Reddy K.R., Shukla S.K. (eds) Proceedings of the Indian Geotechnical Conference 2019. Lect. Notes in Civil Eng, vol **133**. Springer, Singapore. https://doi.org/10.1007/978-981-33-6346-5_48
13. GEO-SLOPE International Ltd 1200, 700 - 6th Ave SW, Calgary, AB, Canada T2P 0T8

Improvement of Bearing Capacity of Stone Columns: An Analytical Study

Manita Das and Ashim Kanti Dey

1 Introduction

The most familiar characteristic of the soft soil is that it undergoes an excessive settlement on the application of surcharge load. This is due to its low shear strength and high compressibility, which increases the importance of soil improvement techniques for the construction of soft soil. Amongst the various well-established soil improvement methods, the use of stone columns is proved to be a good alternative to a pile foundation as it is cost-effective, and environmentally friendly [1]. In the case of soft soil, Golakiya et al. [2] also preferred the use of stone columns *to* the use of pile due to the prohibitive costs for its large length and negative drag force. Moreover, this method shows excellent performance in the improvement of load-carrying capacity, minimization of the foundation settlement, and, accelerating the consolidation settlements by shortening the drainage pathway [2]. Bergado et al. [3] carried out a full-scale load test and reported q_u of soft soil was increased by three to four times with the installation of stone columns [3]. Rao and Reddy [4] reported that the end bearing does not have any effect on q_u for column length greater than ten times of column diameter [4]. Several researchers tried to obtain the load-carrying capacity (q_u) of stone columns through numerical and analytical methods using different concepts. For example, some used unit cell concept [5–9], that is, a single column surrounded by six stone columns; some [10–14] used the behaviour of homogeneous soil with improved soil properties; some [15–17] usduced numerical model; some [18, 19] used 3-D finite element technique, etc.

M. Das (✉)
Civil Engineering Department, ITER, SOA Deemed to Be University, Bhubaneswar 751030, Odisha, India

A. K. Dey
Civil Engineering Department, NIT, Silchar 788010, Assam, India

C. N. V. Satyanarayana Reddy et al. (eds.), *Ground Improvement and Reinforced Soil Structures*, Lecture Notes in Civil Engineering 152,
https://doi.org/10.1007/978-981-16-1831-4_27

It is now well understood that the stone column or granular pile achieves its bearing capacity from the shear strength of the surrounding soft soil. Upon loading, the stone column bulges and pushes the soft soil which in turn tries to resist by imparting radial stress based on its shear strength. For very soft clay (c_u < 12.5 kPa) the radial stress is very low and thus columns bulge or expand excessively, which is the main reason behind the failure of the stone columns [20]. Again, due to bulging, there is a settlement of the ground surface resulting in the limitation of its application. The expected settlement will be around 40–50% of the total settlement of untreated ground as per the Greenwood chart [21]. Thus, the construction of residential buildings over stone columns may become unsafe. Researchers have tried to reduce the bulging effect in order to reduce the settlement. Notable works include geosynthetic encasement [22–28], placement of sand bed reinforced with geogrid [29], jacketing the stone columns with tubular wire mesh [30], etc. All the techniques have their inherent merits and demerits.

The present study shows the application of a compacted soil-cement bed (SCB) over the stone column to reduce the bulging effect of the stone columns. Numerical analysis based on Plaxis-2D was conducted on both ordinary stone column (OSC)s and stone columns underlying SCB. It is observed that the use of SCB reduces the bulging effect and increases the load-carrying capacity of stone columns by more than two times. The degree of improvement depends on the thickness of the soil-cement bed and an optimum value is obtained. A limited number of small-scale laboratory tests were also conducted to validate the analytical results.

2 Numerical Modelling

Numerical modeling for this study was performed by using Plaxis2D software. Many researchers obtained satisfactory results using this software. For example, Marto et al. [31] conducted the analytical study using the software Plaxis 2D for the prediction of bearing capacity of geogrid reinforced stone columns [31]: Phutthananon et al. [32] developed a numerical model on the behavior of conventional deep cement mixing and T-shaped column constructed in an embankment [32], etc. In the present study, a group of three stone columns of diameter 20 cm under a circular footing was taken and an axisymmetrical analysis was carried out. Figure 1 shows a sectional elevation. The Mohr-Coulomb failure criterion was considered to govern the stage of failure of the clay, soil-cement bed, and stone column materials. All the material properties that were applied to the numerical models are listed in Table 1. The values of Poisson's ratio (ν) for both the column materials and clay were taken from the data available in the literature [33]. The other parameters were obtained from the experiments conducted in the laboratory as mentioned in the subsequent paragraphs.

In this finite element model, fifteen nodded triangular elements were considered. All the analyses were carried out with the application of loading in increments. Figure 1 presents the boundary conditions considered which are vertical roller for lateral boundaries and fixed for a bottom boundary. The other dimensions are: tank

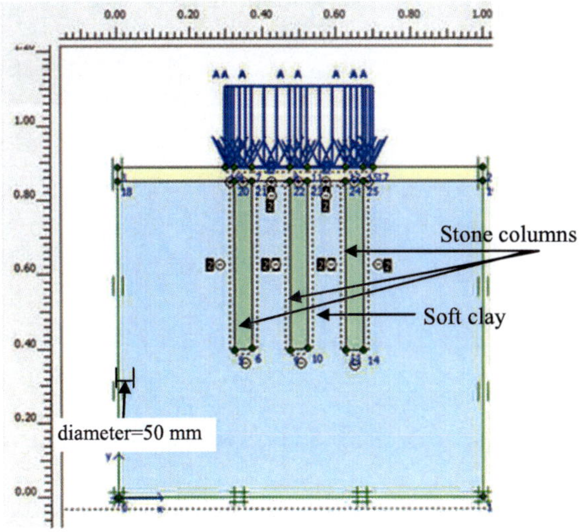

Fig. 1 Sectional elevation of three stone columns for numerical modeling

Table 1 Properties of clay, stone and soil-cement utilised in the PLAXIS-2D analysis

Material	Constitutive model	γ (kN/m^3)	E (MPa)	ν	c (kPa)	ϕ (°)
Clay	Mohr-Coulomb	15	15	0.35	10	0.1
Stone	Mohr-Coulomb	17.0	55	0.3	0	44
SCB	Mohr-Coulomb	15.2	91	0.21	45.4	42

Note μ = Poisson's ratio; γ = unit weight; E = elasticity modulus; c = cohesion; ϕ = friction angle

size 1 m × 1 m × 1 m; stone column diameter, d = 50 mm, and length, l = 40 cm. The depth (t) of the SCB was considered to be variable like, t = 0.5D, 0.75D and 1D (where D = diameter of footing). After the application of load, the settlement of stone columns was recorded and plotted against the bearing pressure as shown in Fig. 2. One additional analysis on ordinary soft clay was also performed for a comparison purpose. It is observed that for the ordinary soil, the bearing pressure increases with the settlement. For a single stone column without any soil-cement bed, the bearing pressure increases with settlement and the increase is nearly two times the value for normal soil. With the introduction of the soil-cement bed, the bearing pressure increases many fold. However, a sudden failure is observed when the settlement is more than 10 mm. That means, up to 10 mm settlement, the application of stone columns underlying SCBis completely safe. From the graph, the q_u of the soft soil was noticed as 7.7 and 32 kPa corresponding to the settlement of 10 and 50 mm respectively. With the use of single stone columns, the q_u of soft soils can be increased by 148 and 72% corresponding to the settlement of 10 and 50 mm respectively. The q_u of soft soil can be improved by 1900 and 470% with the utilisation of stone columns underlying soil-cement bed corresponding to the settlement of 10 and 50 mm respectively. For the group stone columns, the load-carrying capacity

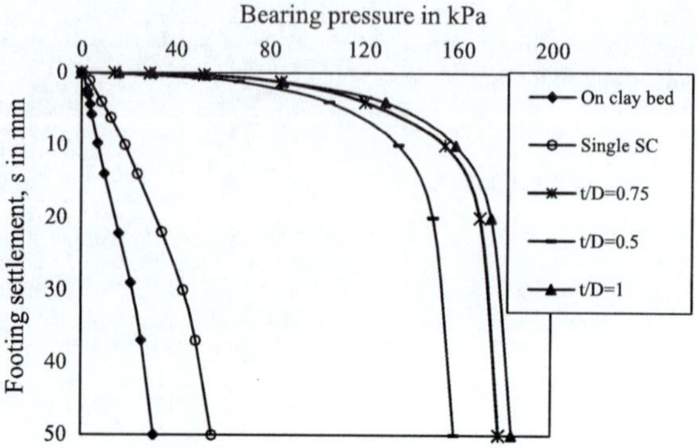

Fig. 2 Bearing pressure-settlement response of SCB over a single stone column

of the soft soil is increased by 523 and 249% with the use of group stone columns corresponding to the settlement of 10 and 50 mm respectively. For the group stone columns underlying SCB, the bearing pressure was increased by 2536 and 572% corresponding to the above-mentioned settlements as shown in Fig. 3. It is also observed from Figs. 2 and 3 that with an increase in t from 0.5D to 1D, the bearing pressure also increases, however, after $t = 0.75\text{D}$, the increment in q_u was marginal. So, the optimum thickness was considered as 0.75D. The improvement in bearing capacity can also be defined by using a term Improvement Factor (I.F.) which is stated as the ratio of bearing pressure acting on the reinforced clay bed to that of the unreinforced soft soil bed at the same value of footing settlement (s/D %). Figure 4 shows I.F. versus non-dimensional settlement (s/D in %) graph for all the combinations of reinforcements which shows a distinct improvement in q_u for both single and group stone columns with the placement of SCB. The I.F. versus s/D (in %) graph of clay bed reinforced with only SCB (without stone column) is also shown in the figure for comparision purpose. The maximum increment was noticed corresponding to the settlement of 2–3% of the diameter of footing. At this point, the

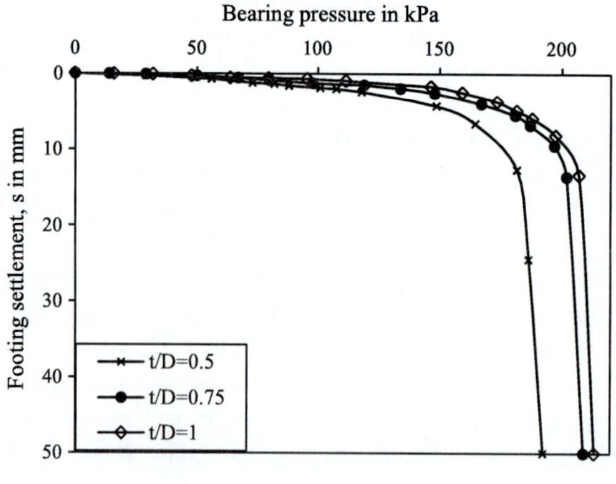

Fig. 3 Bearing pressure-settlement response of SCB over a group of 3 stone column

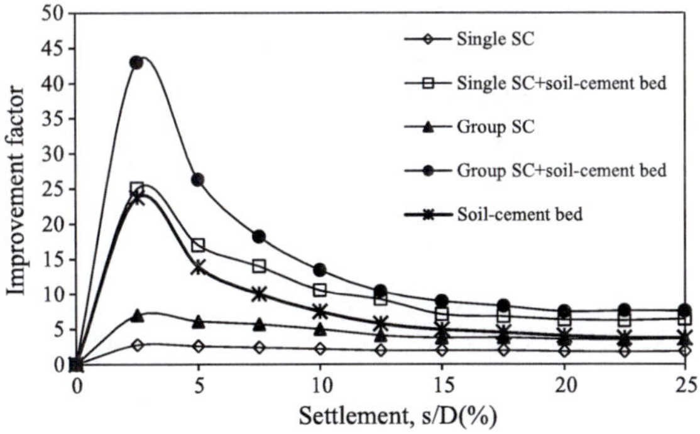

Fig. 4 Variation of improvement factor or stress concentration factor with single and group stone columns with and without SCB

values of improvement factors are 2.75, 7, 24.5, 25, 45 for clay bed with the single stone column, group stone column, only SCB, single stone column underlying SCB, and group stone column underlying SCB respectively. Therefore, it can be concluded that the maximum IF is obtained with the placement of SCB over the stone columns at a settlement of 2–3% of the diameter of footing. The deformed mesh of the single and group stone columns without SCB obtained from the analysis are shown in Figs. 5a and b whereas Fig. 5c and d show the deformed mesh of single and group stone columns underlying SCB respectively. The deformations due to loading are also clearly observed from these figures. That means maximum improvement can be achieved by combining the group of stone columns with SCB.

3 Experimental Investigation

3.1 Material Used

To compare the numerical results obtained from PLAXIS-2D, three experiments (one with clay bed alone, one with a group of OSC without SCB, one with a group of stone columns underlying SCB) were conducted in the laboratory. Each experiment was conducted in a steel tank of size 1 m × 1 m × 1 m. The dimensions of a stone column were diameter $d = 50$ mm and length, $l = 400$ mm and thickness of SCB, $t = 0.75D = 15$ cm (optimum value).

For constructing the clay bed, the tank was wrapped with a polythene sheet to minimize side friction. Clay was collected from a nearby paddy field, pulverized, thoroughly mixed with water, put in the tank in layers and uniformly compacted. The compacting effort was achieved after a number of trials so that the density of clay would be 15 kN/m^3. In all the tests the water content was kept constant as 33% corresponding to the undrained cohesion of 10 kPa. For the tests on clay bed and OSC, the thickness of the clay bed was 90 cm, and for the test on stone columns

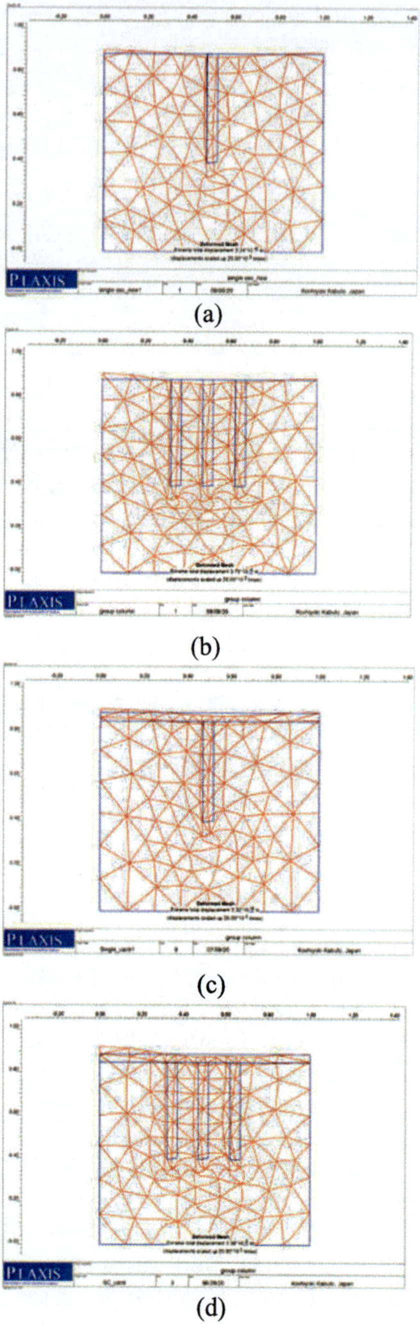

Fig. 5 Deformed mesh of **a** single stone column, **b** group stone columns, **c** single stone column underlying SCB, and **d** group stone columns underlying SCB

Table 2 Properties of clay

Properties	Value
Liquid limit (%)	42.4
Plastic limit (%)	20.6
Plasticity index (%)	21.8
Specific gravity	2.6
Bulk unit weight at 34% water content (kN/m^3)	15.8
Undrained shear strength (kPa)	10
USCS classification system	CL

underlying SCB, the thickness of the clay bed was 75 cm. The geotechnical properties of clay are presented in Table 2.

After constructing the clay bed, three boreholes were dug using an auger of 50 mm diameter up to the required depth for the construction of stone columns. Then the boreholes were cleaned with repeated insertion of the auger. Before filling the holes with stones, the diameter and length of the hole were checked with an open-ended steel pipe having inner and outer diameters of 48.5 and 50 mm, respectively.

The holes were filled by the stones of size 2–6 mm with uniform compaction in each layer. For compacting the stones in each layer, one tamping rod of 10 mm diameter was used. The number of blows was adjusted after a number of trials to achieve a density of 17 kN/m^3. It is very difficult to maintain a uniform relative density of stones placed in the stone column. A number of trials were made with a different number of blows for each of five layers and a plot was made between a number of blows and the relative density of the stone. The number of blows corresponding to 70% relative density was adopted for the present experiments.

The spacing between the columns was kept as 2.5 times the diameter of the stone column.

The soil-cement stabilized bed was placed over the columns in one of the tests. The SCB was prepared by thoroughly mixing locally available c−ϕ soil, reddish in colour, with ordinary Portland cement, and water in a proportion of 1part of cement with 6 parts of soil. All the materials were mixed and placed over the columns by compacting the mixture uniformly in three layers. Each layer was compacted by using a square hammer with uniform compaction energy to achieve a density of 15.2 kN/m^3.

The shear strength parameters of the three materials were obtained at the compacted density in the laboratory and are shown in Table 1. The parameters c and ϕ were obtained by direct shear test; Poisson's ratio was obtained from unconfined compressive strength (UCS) test and Young's modulus was obtained from Four-point bending test.

3.2 Loading Arrangement

One steel plate of 20 cm diameter and 10 mm thickness was used as a footing for all the tests. The diameter of the loading plate was chosen so that all three columns would be covered by the plate. Figure 6 shows the arrangement of stone columns used in the experimental setup. After placing the footing at its specific position, that is, centre of the bed, a uniformly distributed load was applied on it through the hydraulic jack of 10 t capacity. The experimental setup is shown in Fig. 7.

The applied loads were measured using a load cell, which was connected to a loading frame as shown in Fig. 7. To estimate the settlement that occurred due to the applied load, two LVDTs of capacity 50 and 25 mm were located at diametrically opposite ends of the footing. The LVDTs and load cell were connected to a data acquisition system to store the data in a computer. Two dial gauges were also fixed on the opposite diagonal for validation of the data obtained from LVDTs.

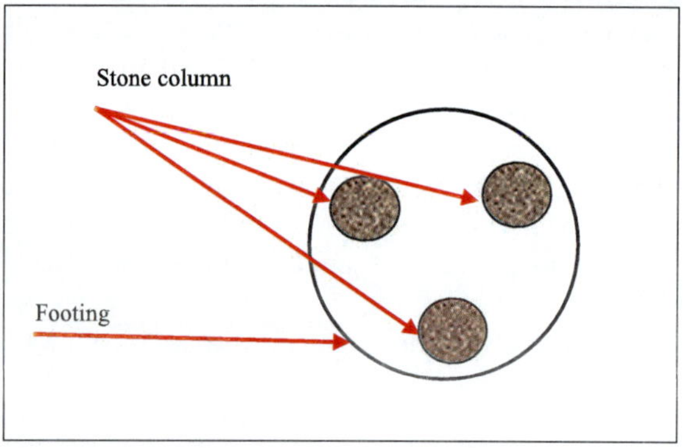

Fig. 6 Arrangement of three stone columns

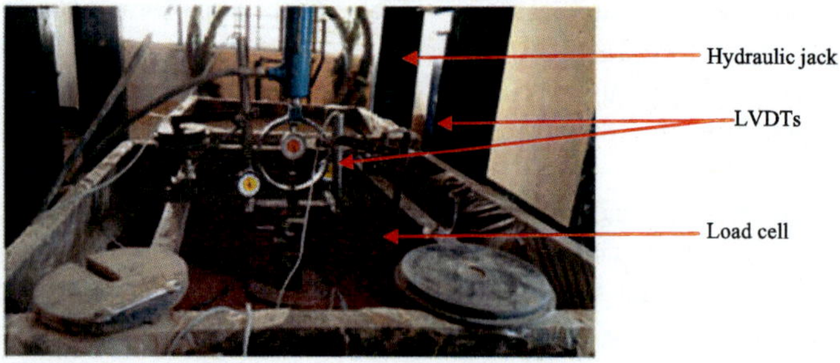

Fig. 7 Experimental setup

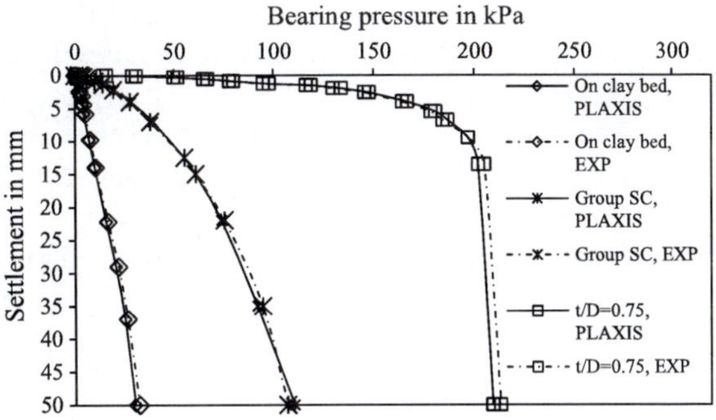

Fig. 8 Comparison of results of FEM and experimental model tests

3.3 *Test Procedure*

The procedure of load test involves the application of uniformly distributed load on footing and measurement of settlements of the plate. In all the tests, the load was applied in equal increments and was maintained until the settlement became <0.02 mm/h [34]. The loading was continued till 50 mm settlement or failure whichever occurred earlier.

4 Comparison of Numerical Results with Experimental Results

The results of the finite-element analysis are compared with the results obtained from the experimental model test results. The results were compared in terms of bearing pressure versus settlement response as shown in Fig. 8 for unreinforced clay bed, clay bed reinforced with group OSC, and clay bed reinforced with group columns underlying SCB. The figure presents a very good match between the numerically obtained results and experimentally obtained responses which reflects the accuracy of the FEM results.

5 Conclusions

The study reported herein presents the response of a soil-cement bed on the improvement in the bearing capacity of a single and a group of floating stone columns in the soft clayey ground. The following concluding remarks are obtained from the present study:

1. With the use of single stone columns, the bearing capacity of soft soils can be raised by 148 and 72% corresponding to the settlement of 10 and 50 mm respectively.
2. The bearing capacity of soft soil can be increased by 1900 and 470% with the construction of single stone columns underlying SCB corresponding to the settlement of 10 and 50 mm, respectively.
3. By using the group of three stone columns, the bearing capacity of soft soil can be increased by 523 and 249% with the use of group stone columns corresponding to the settlement of 10 and 50 mm, respectively.
4. The bearing capacity of soft soil can be increased by 2536 and 572% corresponding to the settlement of 10 and 50 mm respectively with the use of group stone columns underlying SCB.
5. Based on the maximum improvement percentage of bearing capacity, the optimum thickness of SCB was obtained as 0.75 times of footing diameter.

Thus from this study, it is clear that the bearing capacity of soft soil can be increased manifold with the installation of stone columns underlying a soil-cement bed. The technique also reduces the bulging of stone columns and thereby minimizes settlement of footing.

References

1. Shien, N.K.: Numerical study of floating stone column. A Thesis Submitted for the Degree of Doctor of Philosophy Department of Civil And Environmental Engineering, National University of Singapore (2013)
2. Golakiya, H.D., Lad, M.D.: Ground improvement by using stone columns. Int. J. Emerg. Technol. Innov. Res. **2**(11), 133–144. ISSN: 2349-5162, www.jetir.org
3. Bergado, D.T., Rantucci, G., Widodo, S.: Full scale load tests on granular piles and sand drains in the soft Bangkok clay. In: Proceedings International Conference on In Situ Soil and Rock Reinforcement, Paris, p.p 111–118 (1984)
4. Rao, S.N., Reddy, K.M.: Load transfer in stone column in soft marine clay. In: Proceedings, Indian Geotechnical Conference, Madras, India, pp. 403–406 (1996)
5. Priebe, H.J.: Evaluation of the settlement reduction of a foundation improved by vibro–replacement. Bautechnik **5**, 160–162 (1976)
6. Aboshi, H., Ichimoto, E., Enoki, M., Harada, K.: The composer: a method to improve characteristics of soft clays by inclusion of large diameter sand columns. In: Proceedings of International Conference on Soil Reinforcement: Reinforced Earth and Other Techniques, pp. 211–216. Champssur-Marne, Paris (1979)
7. Balaam, N., Booker, J.: Effect of stone column yield on settlement of rigid foundations in stabilized clay. Int. J. Numer. Anal. Methods Geomech. **9**(4), 331–351 (1985). https://doi.org/10.1002/nag.1610090404
8. Van Impe, W.F., Madhav, M.R.: Analysis and settlement of dilating stone column reinforced soil. Österreichische Ingenieur und Architekten-Zeitschrift **137**(3),114–121 (1992)
9. Ambily, A.P., Gandhi, S.R.: Behavior of stone columns based on experimental and FEM analysis. J. Geotech. Geoenviron. Eng. **133**(4), 405–415 (2007)
10. Schweiger, H.F., Pande, G.N.: Numerical analysis of stone column supported foundations. Comput. Geotech. **2**(6), 347–372 (1986)

11. Lee, J.S., Pande, G.N.: Analysis of stone-column reinforced foundations. Int. J. Numer. Anal. Meth. Geomech. **22**(12), 1001–1020 (1998)
12. Wang, W.G., Leung, C.F., Ichikawa, Y.: A simplified homogenization method for composite soils. J. Comput. Geotech. **29**, 477–500 (2002)
13. Jellali, B., Bouassida, M., Buhan, P.D.: Stability analysis of an embankment resting upon a column-reinforced soil. Int. J. Numer. Anal. Meth. Geomech. **35**, 1243–1256 (2011). https://doi.org/10.1002/nag.954
14. Abdelkrim, M., Buhan, P.D.: An eleastoplastic homogenization procedure for predicting the settlement of a foundation on a soil reinforced by columns. Eur. J. Mech. A/Solids **26**, 736–757 (2007)
15. Das, M., & Dey, A. K. Determination of bearing capacity of stone column with application of neuro-fuzzy system. KSCE Journal of Civil Engineering, 22(5), 1677–1683 (2018)
16. Das, M., Dey, A.K.: Prediction of bearing capacity of stone columns placedin soft clay using ANN model. Geotech. Geol. Eng (2018). https://doi.org/10.1007/s10706-017-0436-0
17. Das, M., Dey, A.K.: Prediction of bearing capacity of stone columns placed in soft clay using SVR model. Arab. J. Sci. Eng. 1–11 (2018)
18. El Kamash, W., Han, J.: Displacements of column-supported embankments over soft clay after widening considering soil consolidation and column layout: numerical analysis. Soils Found. **54**(6), 1054–1069 (2014)
19. Killeen, M.M., McCabe, B.A.: Settlement performance of pad footings on soft clay supported by stone columns: a numerical study. Soils Found. **54**(4), 760–776 (2014)
20. Ranjan, G., Rao, A.S.R.: Basic and applied soil mechanics. New Age International, Daryaganj, New Delhi, India (2007)
21. Bowles, J.E.: Foundation Analysis and Design, 4th edn., McGraw-Hill, Singapore
22. Van Impe, W.F.: Soil improvement techniques and their evolution. Balkema, Rotterdam (1989)
23. Murugesan, S., Rajagopal, K.: Geosynthetic encased stone columns: numerical evaluation. J Geotext. Geomembr. **24**(6), 349–358 (2006)
24. Lo, S.R., Zhang, R., Mak, J.: Geosynthetic-encased stone column in soft clay: a numerical study. Geotext. Geomembr. **28**, 292e302 (2010)
25. Dash, S.K., Bora, M.C.: Improved performance of soft clay foundations using stone columns and geocell-sand mattress. Geotext. Geomembr. **41**, 26–35 (2013)
26. Castro, J.: Groups of encased stone columns: influence of column length and Arrangement. Geotext. Geomembr. **45**, 68–80 (2017). http://dx.doi.org/10.1016/j.geotexmem.2016.12.001
27. Cengiz, C., Güler, E.: Seismic behavior of geosynthetic encased columns and ordinary stone columns. Geotext. Geomembr. **46**, 40–51 (2018)
28. Chen, J.F., Wang, X.T., Xue, J.F., Zeng, Y., Feng, S.Z.: Uniaxial compression behavior of geotextile encased stone columns. Geotext. Geomembr. **46**, 277–283 (2018)
29. Deb, K., Samadhiya, N.K., Namdeo, J.B.: Laboratory model studies on unreinforced and geogrid-reinforced sand bed over stone column-improved soft clay. Geotext. Geomembr. **29**(2), 190–196 (2011)
30. Black, J.A., Sivakumar, V., Madhav, M.R., Hamill, G.A.: Reinforced stone columns in weak deposits: laboratory model study. J. Geotech. Geoenviron. Eng. **133**(9), 1154–1161 (2007). https://doi.org/10.1061/(ASCE)1090-0241(2007)133:9(1154)
31. Marto, A., Moradi, R., Helmi, F., Latifi, N., Oghabi, M.: Performance analysis of reinforced stone columns using finite element method. Electron. J. Geotech. Eng. **18**, 315–323 (2013)
32. Phutthananon, C., Jongpradist, P., Jongpradist, P., Dias, D., Baroth, J.: Parametric analysis and optimization of T-shaped and conventional deep cement mixing column-supported embankments. Comput. Geotech. **122**, (2020)
33. Bowles, J.E.: Foundation analysis and design, 5th edn. McGraw-Hill, New York (1997)
34. IS (Indian Standard): Method of load test on soils. New Delhi, India, IS: 1888-1982 (1983)

Applicability Analysis of Stone Column Against Liquefaction Under Repeated Dynamic Events

Aravind Murali, M. D. Godson, Ganesh Kumar Shanmugam, and Nanthakumar Subramani

1 Introduction

Soil is a complex component that shows a variety of problematic behavior under dynamic loading conditions. One such most common complex phenomenon is liquefaction, occurring after an earthquake in loose sandy deposits. During shaking, the saturated sand completely loses its shear strength affecting the stability of foundation structures. Some of the evidence includes Niigata, Japan (1964) and Alaska (1964), the Tangshan, China (1976), the Kocaeli, Turkey (1999), the Wenchuan, China (2008), and the Chilean earthquake (2010). The greater the magnitude of approaching seismic load, the greater the generation of excess pore pressure, and the higher will be the soil deformation. During this event, the excess pore water pressure generated will densify the soil bed in the post liquefaction stage [1]. The occurrence of soil densification when subjected to earthquake load also improves resistance to liquefaction in the subsequent earthquake. But still, mitigation of pore water pressure generation under repeated shaking will require attention to improve the soil resistance against liquefaction.

Stone columns or gravel drains have proven to be an appropriate ground improvement technique to mitigate the effects of liquefaction [2–5]. The granular pile acts as a porous medium in dissipating the excess pore pressure generated during an earthquake, thereby reducing the soil's susceptibility to liquefaction [6]. The effectiveness of these drains ultimately depends on the permeability of the material under consideration. Detailed experimental studies have been carried out on a fully saturated silty sand bed to view its response to liquefy under earthquake loads and evaluate gravel drains' effectiveness in controlling liquefaction [7]. However, studies on the effect

A. Murali · M. D. Godson · N. Subramani
SRM Institute of Science and Technology, Kattankulathur 603203, Tamil Nadu, India

G. K. Shanmugam (✉)
CSIR—Central Building Research Institute, Roorkee 247667, Uttrakhand, India

C. N. V. Satyanarayana Reddy et al. (eds.), *Ground Improvement and Reinforced Soil Structures*, Lecture Notes in Civil Engineering 152,
https://doi.org/10.1007/978-981-16-1831-4_28

of re-liquefaction resistance with stone column improvement are not available. The most suitable examples to reveal the impacts of re-liquefaction caused by the after-shocks of a seismic shaking are the Pacific Coast of Tohoku Earthquake in Japan (2011) [8]. There is a common idea among the researchers that densification of the soil stratum susceptible to liquefy will induce resistance against liquefaction in the subsequent shaking [9–11]. This paved the way to take up densification and drainage as a tool to control the effects of soil re-liquefaction [12, 13].

This study evaluated the stone column's performance in addressing the effects of liquefaction and re-liquefaction experimentally under repeated incremental acceleration loading. The experiments were carried out in a tank, placed over a uniaxial shaking table. The sand bed of 40% relative density was prepared and subjected to repeated acceleration loading with increments, leaving time intervals to dissipate the pre-generated pore water pressure between each testing. A comparative analysis was conducted with stone columns installed ground and the virgin sand bed. The liquefaction assessment parameters such as pore water pressure, the dissipation rate of excess pore water pressure, foundation settlement, and soil displacement were monitored and analyzed.

The experimental studies revealed that in the case of untreated deposits, the potential to re-liquefy in the subsequent incremental accelerations might be due to the effect of non-uniform soil densification with depth which induces generation of pore water pressure in the subsequent loading. From the comparative analysis, it can be concluded that the installation of stone columns increased the rate of pore water pressure dissipation and improves the seismic response of liquefiable deposits, even at repeated shaking events.

2 Materials and Methodology

2.1 Soil Used for Sample Preparation

The soil for the study was obtained from the Solani river bed in Roorkee and subjected to basic characterization, which revealed that the greyish brown-toned ground falls under the category of poorly graded sand with the Coefficient of uniformity (C_u) and Coefficient of gradation (C_c) as 2.634 and 1.147, respectively. The soil properties are listed in Table 1. The soil contained higher amounts of finer particles, as observed from the gradation data in Table 1.

2.2 Tank Specifications

The laboratory scaled model tests were performed using a rigid Perspex rectangular tank, having a dimension of 1.4 m × 1.0 m × 1.0 m, mounted over a uniaxial shaking

Table 1 Soil index properties

Sl. no.	Parameters		Value	Unit
1	Type of soil		Poorly graded sand	
2	Specific gravity (G)		2.65	No unit
3	Minimum density (γ min)		1.4011	g/cc
4	Maximum density (γ max)		1.6644	g/cc
5	40% RD	Density (γ)	1.494	g/cc
		Permeability (k)	0.00811	cm/s
		Cohesion (c)	0	kPa
		Angle of internal Friction (φ)	32	°
6	Youngs modulus (E)		12,000	kPa
7	Sand	Coarse (4.75–2 mm)	0.034	%
		Medium (2–0.425 mm)	9.77	%
		Fine (0.425–0.075 mm)	88.25	%
8	Silt and clay (<0.075 mm)		1.946	%

Fig. 1 Perspex glass tank placed over shaking table

table, as shown in Fig. 1. To minimize boundary effects during shaking, a 50 mm thick Polyethylene foam is attached on both sides of the container.

2.3 Sample Preparation

The method of soil bed preparation plays a crucial role in representing the response of in situ soil towards liquefaction [14–17]. Depending on the method of soil packing, the initial stress condition to initiate liquefaction may broadly vary under defined stress cycles for a sample of the same gradation and density. Considering the above, wet sedimentation method was adopted in this study for preparing saturated ground deposit.

For testing, a 600 mm thick, saturated sand bed was prepared inside the tank. The quantity of sand and water required to attain a fill density of 40% was estimated and measured. To achieve uniformity throughout the tank's depth, the sample was filled gradually by dividing the fill volume into three equal parts of the required water and soil. The sand was drizzled down into the container at a pre-calculated height (IS 2720) through a conical hopper arrangement having an inverted solid cone with a 60° angle attached at the end, which resulted in achieving a uniformly spread soil bed [18].

During sample preparation, glass tube piezometers and strain-based pore pressure transducers were used for monitoring pore pressure response. The instruments were placed centrally along the fill volume at 0.2 and 0.4 m height from the tank's base to monitor the excess pore water pressure developed during seismic loading.

2.4 Design and Installation of Gravel Drains

The stone columns were designed based on the guidelines given in IS 15284 Part 1 [19]. The design criteria and the properties of stone chips used in constructing the gravel drain are listed in Table 2.

Initially, the arrangement of stone columns is marked on the ground bed appropriately with proper dimensions. Then a hollow PVC pipe of outer diameter equivalent to the stone column diameter was driven inside the prepared sand bed, and subsequently, the sand inside the PVC pipe was removed and replaced with stone aggregates compacted in three layers to achieve the required density. The entire assembly was then left undisturbed for 24 h.

Table 2 Design criteria and properties of stone columns

Sl. no.	Parameters	Description	Units
1	No. of stone columns	3	No units
2	Area replacement ratio (ARR)	5	%
3	Pattern of installation	Triangular	No units
4	Diameter of stone column	160	mm
5	Type of stones	Granite chips	No units
6	Classification	Uniform and angular	No units
7	Gradation	2–10	mm
8	Unit weight	16 ± 0.2	kN/m^3
9	Relative density	73	%

2.5 Scale Down Model of Foundation

To monitor foundation response under dynamic shaking, a scaled-down shallow footing model was designed and used following dynamic similitude laws as given in Eq. (1) [20].

$$N_{EI} = N_k \times N_L^3 \tag{1}$$

where,

N_{EI} = Scale factor for flexural rigidity

N_K = Scale factor for stiffness

N_L = Scale factor for linear dimensions

A scale down factor (n) equal to 10 was used to model shallow footing. The foundation was modeled for 115 mm length, 115 mm wide, and 30 mm thick using steel material having a modulus of elasticity of 200 GPa. For all the test series, the foundation model was installed at 30 mm depth inside the prepared ground centrally for evaluating its settlement during seismic shaking.

2.6 Testing Conditions

Based on the peak ground acceleration (PGA) of several severe earthquakes, an acceleration of 0.3 and 0.4 g at 5 Hz frequency simulating high and very high-intensity shaking was selected and applied to the ground. The shaking was carried out for 200 cycles lasting for 40s for both unreinforced and stone column reinforced beds. To assess the liquefiable sand bed's re-liquefaction characteristics, the incremental acceleration was given after 24 h, allowing the complete dissipation of EPWP generated from the previous acceleration loading. The generated pore-pressure, soil displacement, and foundation settlement were estimated and compared using piezometers and strain-based transducers for both unreinforced bed and stone column bed.

3 Results and Discussion

3.1 Effect of Pore Pressure Generation on Liquefaction Potential

The shaking table tests were carried out on the soil bed packed with 40% relative density for both stone column (SC) reinforced soil bed and unreinforced bed to assess stone columns' efficiency in mitigating liquefaction. An incremental acceleration of

Fig. 2 Unreinforced bed—40% RD—initial condition

Fig. 3 SC reinforced bed—40% RD—initial condition

0.3 and 0.4 g were given to the prepared sand beds. The generated excess pore pressure and the corresponding pore pressure ratio for each incremental loading are measured and compared for the unreinforced and stone column reinforced bed conditions. The initial bed conditions before shaking are shown in Figs. 2 and 3, respectively.

The prepared bed condition was left undisturbed for 24 h before subjecting it to an acceleration load of 0.3 g initially. The excess pore water pressure generated during the shaking is measured using glass tube piezometers and strain-based transducers to understand the variation of generated pore pressure along the depth of fill. The pore water pressure distribution for the initial 0.3 g acceleration loading is shown in Fig. 4a and b.

The generation of EPWP was observed by sand boils in the case of the unreinforced condition, whereas in the case of stone column reinforced bed the column dissipates the generated pore-water pressures. The generated pore water during shaking is shown in Fig. 5a and b, respectively.

Based on the results observed, it can be concluded that the generated excess pore water pressure measured by the transducers were found to be maximum in the bottom and minimum in the top for both unreinforced and reinforced bed condition. This effect may be due to the effect of overburden loading. Provision of gravel drains minimize generation of pore water pressure during shaking and dissipated generated excess pore water pressures mitigating the occurrence of liquefaction. For the untreated soil, the top and bottom transducers recorded a peak pore pressure of 1.35 and 2.76 kPa, which was 25.2 and 17.4% higher than the stone column reinforced bed.

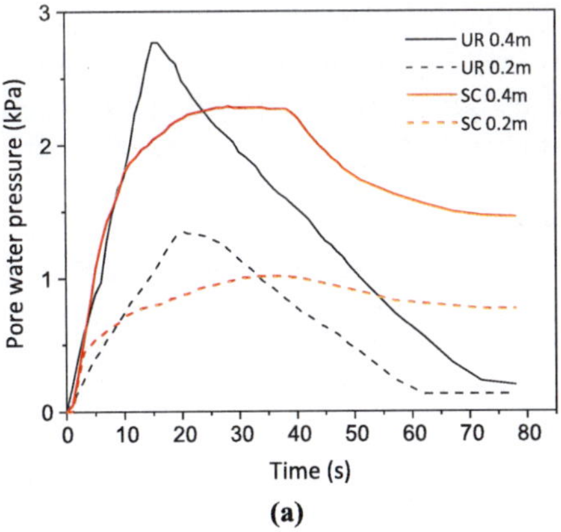

Fig. 4 **a** Porewater pressure generation—0.4 and 0.2 m from top—0.3 g. **b** Porewater pressure generation—0.4 and 0.2 m from top—0.4 g

Fig. 5 **a** Sand boils depicting generation of EPWP in unreinforced bed. **b** Water seeping out initially from the gravel drains in SC reinforced bed

The time is taken by the soil bed to liquefy, as recorded by the top and bottom transducers, were the 20 and 15 s, which was prolonged to 30 and 35s, respectively, in the case of stone column reinforced bed. This proves that stone columns have improved permeability characteristics that mitigate generation and improves dissipation rates of generated EPWP and delays the occurrence of liquefaction. This observation was in line with earlier studies [21–23].

After letting enough time for the dissipation of EPWP generated during 0.3 g acceleration loading, an incremental acceleration of 0.4 g was applied to the same bed condition to assess the effect of re-liquefaction. The bottom transducers recorded higher EPWP, similar to in the case of 0.3 g acceleration for the unreinforced condition (Fig. 5b). The peak pore water pressure in the case of unreinforced condition was 1.23 and 2.79 kPa recorded in the top and bottom transducers, respectively, causing re-liquefaction of the soil bed. However, only 0.5 and 1.31 kPa excess pore pressure was observed in the case of a stone column reinforced bed. The reduction was about 65 and 53% for the top and bottom conditions. Also, the time to reach the peak pore pressure in the case of a stone column reinforced bed sensed by the bottom piezometer was found to be 27 s, which is almost 2.5 times greater than that seen in pristine bed condition.

When 0.3 and 0.4 g are compared, there is a gradual increase in the generation of EPWP as the acceleration loading increases in the unreinforced condition. Contradicting this statement, in 0.4 g, the bed reinforced with gravel drains reported a deficient generation of pore pressure than in the case of 0.3 g. This may be due to the intensive rearrangement of particles, causing the soil bed to densify at higher rates expelling all the water in between the particles at the post liquefaction stage of 0.3 g loading. This combined effects of densification and pore pressure generation after 0.3 g resulted in a lesser generation of pore pressure in 0.4 g acceleration loading. Hence it may be concluded that stone columns effectively mitigate liquefaction and re-liquefaction effects due to intense earthquakes.

3.2 *Effect of Pore Pressure Ratio (r_u)*

The pore pressure ratio is the ratio between the pore water pressure (U) and the overburden pressure ($\sigma_{vo'}$), which can indicate a soil's susceptibility to liquefaction [24]. Figure 6 represents the pore pressure ratio variation with time for the unreinforced and SC reinforced bed condition for 0.3 g and 0.4 g acceleration loading sensed by the bottom transducer. It can be seen that in both the loading conditions, the unreinforced bed liquefies, owing to a pore pressure ratio of about 0.66 and 0.67, respectively [23]. In the case of a stone column reinforced bed, under 0.3 g, the pore pressure ratio of 0.57 was observed due to initial ground preparation. However, the installed stone

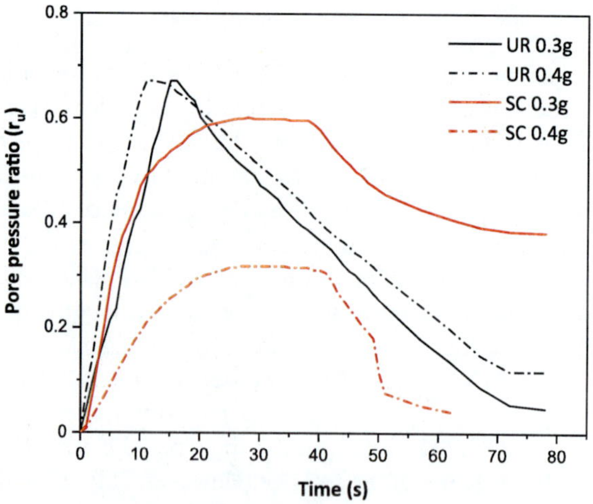

Fig. 6 Porewater pressure ratio—0.3 g and 0.4 g—40 cm from top

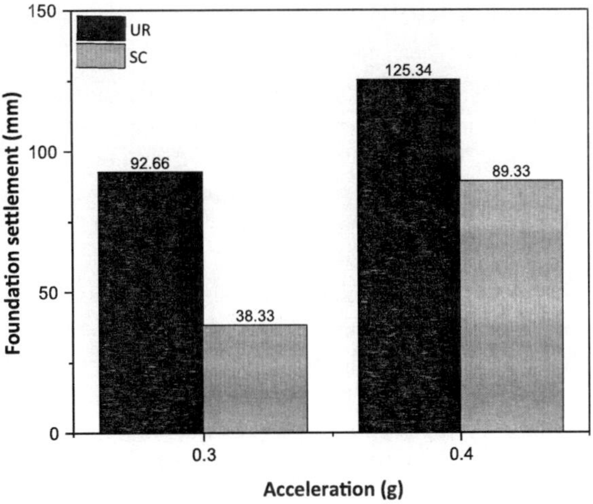

Fig. 7 Foundation settlement

columns delay the generation of peak pore pressure ratio and effectively dissipates generated EPWP. Under 0.4 g loading, the peak pore pressure ratio further reduced to 0.31 and improves reliquefaction resistance of the prepared ground. The reduction may be due to the effects of soil densification due to successive acceleration and the higher dissipation rates of stone columns, as discussed in the earlier sections. This proves the stone columns' efficiency to mitigate liquefaction when subjected to the seismic load of higher peak ground accelerations (PGA).

3.3 Soil Displacement and Foundation Settlement

Figure 7 represents the foundation settlement versus the incremental acceleration loading condition compared between unreinforced stone columns reinforced sand beds.

From the above graph, it is visible that when there is an increase in the acceleration loading intensity, the foundation settlement has also increased considerably. In the unreinforced condition, the maximum foundation settlement observed after giving 0.3 and 0.4 g incremental accelerations were found to be 93 and 125 mm, which was reduced to 38 mm and 89 mm with a percentage reduction of about 59% and 29% respectively in the case of stone column reinforced ground bed. This concluded that the gravel drains installation has considerably reduced the foundation settlement implying a higher bearing capacity of the stone column reinforced ground.

4 Conclusions

In this study, the efficiency of the stone column in mitigating liquefaction and re-liquefaction was experimentally evaluated. An incremental repeated acceleration loading of 0.3 and 0.4 g was applied to a tank placed on a uni-axial shaking table filled with Solani River sand compacted at a relative density of 40%. The generation of excess pore water pressure was higher and faster in the bottom layers than the top when subjected to 0.3 and 0.4 g loading, respectively, in the unreinforced condition. The foundation settlement proportionally increased as the successive acceleration loading increased, attaining a value of 93 mm and 125 mm, respectively. To reinforce the liquefiable ground and minimize the EPWP generation and foundation settlement, stone columns with a 5% area replacement ratio were installed. At 0.3 g acceleration loading, the peak pore pressure generation was reduced by 17.39% in the bottom layers with stone column improvement.

Further, in the successive 0.4 g acceleration loading, the stone columns reinforced bed proved to be more effective similar to 0.3 g load application due to the combined effect of densification and drainage characteristics. The pore pressure ratio reduction is 0.32, inferring that the bed did not liquefy at repeated shaking, thus proving gravel drains more efficient to mitigate liquefaction even at higher successive acceleration loading. It can be concluded that the stone column reinforced bed improves liquefaction resistance, delays generation of pore water pressure, and improve the safety of the foundation even at repeated shaking events.

References

1. Thevanayagam, S., Martin, G.R., Shenthan, T., Liang, J.: Post-liquefaction pore pressure dissipation and densification in silty soils. International Conferences on Recent Advances in Geotechnical Earthquake Engineering and Soil Dynamics. 23 (2001)
2. Adalier, K., Elgamal, A.: Mitigation of liquefaction and associated ground deformations by stone columns. Eng. Geol. **72**(3–4), 275–291 (2004)
3. Bayati, H., Bagheripour, M.H.: Shaking table study on liquefaction behaviour of different saturated sands reinforced by stone columns. Mar. Georesour. Geotechnol. **37**(7), 801–815 (2019)
4. Seed, H.B., Booker, J.R.: Stabilization of potentially liquefiable sand deposits using gravel drains. J. Geotech. Geoenviron. Eng. **103**(ASCE 13050) (1977)
5. Baez, J.I., Martin, G.R.: Quantitative evaluation of stone column techniques for earthquake liquefaction mitigation. In: Proceedings of the 10th World Conference on Earthquake Engineering, Madrid, Spain, vol. 3, pp. 1477–1483 (1992)
6. Sasaki, Y., Taniguchi, E.: Shaking table tests on gravel drains to prevent liquefaction of sand deposits. Soils Found. **22**(3), 1–14 (1982)
7. Huang, C., Sui, Z., Wang, L., Liu, K.: Mitigation of soil liquefaction using stone columns: an experimental investigation. Mar. Georesour. Geotechnol. **34**(3), 244–251 (2016)
8. Huang, Y., Yu, M.: Review of soil liquefaction characteristics during major earthquakes of the twenty-first century. Nat. Hazards **65**(3), 2375–2384 (2013)
9. Krishna, A.M., Madhav, M.R., Latha, G.M.: Liquefaction mitigation of ground treated with granular piles: densification effect. ISET J. Earthq. Technol. **43**(4), 105–120 (2006)

10. Mitchell, J.K.: Mitigation of liquefaction potential of silty sands. In: From Research to Practice in Geotechnical Engineering, pp. 433–451 (2008)
11. Salem, Z.B., Frikha, W., Bouassida, M.: Effects of densification and stiffening on liquefaction risk of reinforced soil by stone columns. J. Geotech. Geoenviron. Eng. **143**(10), 06017014 (2017)
12. Ye, B., Yokawa, H., Kondo, T., Yashima, A., Zhang, F., Yamada, N.: Investigation on stiffness recovery of liquefied sandy ground after liquefaction using shaking-table tests. In: Soil and Rock Behavior and Modeling, pp. 482–489 (2006)
13. Ye, B., Ye, G., Zhang, F., Yashima, A.: Experiment and numerical simulation of repeated liquefaction-consolidation of sand. Soils Found. **47**(3), 547–558 (2007)
14. Mulilis, J.P., Arulanandan, K., Mitchell, J.K., Chan, C.K., Seed, H.B.: Effects of sample preparation on sand liquefaction. J. Geotech. Eng. Div. **103**(2), 91–108 (1977)
15. Maheshwari, B.K., Singh, H.P., Saran, S.: Effects of reinforcement on liquefaction resistance of Solani sand. J. Geotech. Geoenviron. Eng. **138**(7), 831–840 (2012)
16. Varghese, R.M., Latha, G.M.: Shaking table tests to investigate the influence of various factors on the liquefaction resistance of sands. Nat. Hazards **73**(3), 1337–1351 (2014)
17. Banerjee, R., Konai, S., Sengupta, A., Deb, K.: Shake table tests and numerical modeling of liquefaction of Kasai river sand. Geotech. Geol. Eng. **35**(4), 1327–1340 (2017)
18. IS 2720 (Part 14): Methods of Test for Soil Part 14 Determination of Density Index (Relative Density) of Cohessionless Soils. Bureau of Indian Standard, New Delhi (1983)
19. IS 15284 (Part 1): Design and Construction for Ground Improvement-Guidelines. Bureau of Indian Standard, New Delhi (2003)
20. Moncarz, P.D., Krawinkler, H.: Theory and application of experimental model analysis in earthquake engineering, vol. 50. Stanford University, California (1981)
21. El-Sekelly, W., Abdoun, T., Dobry, R.: Liquefaction resistance of a silty sand deposit subjected to pre-shaking followed by extensive liquefaction. J. Geotech. Geo-environ. Eng. **142**(4), 04015101 (2016)
22. Darby, K.M., Boulanger, R.W., DeJong, J.T.: Volumetric strains from inverse analysis of pore pressure transducer arrays in centrifuge models. Geotechnical Earthquake Engineering and Soil Dynamics V: Liquefaction Triggering, Consequences, and Mitigation, pp. 626–636. American Society of Civil Engineers, Reston, VA (2018)
23. Padmanabhan, G., Shanmugam, G.K.: Reliquefaction assessment studies on saturated sand deposits under repeated acceleration loading using 1-g shaking table experiments. J. Earthq. Eng. 1–23 (2020)
24. Seed, B., & Lee, K.L.: Liquefaction of saturated sands during cyclic loading. J. Soil Mech. Found. Div. **92**(ASCE# 4972 Proceeding) (1966)

Application of Geosynthetics for Geotechnical Challenges

A. L. Ksheeraja and N. Srilatha

1 Introduction

Construction over soft soil has always been a challenge for geotechnical engineers. The low shear strength, high compressibility and low bearing potential characterize soft soils and these factors make the embankment construction difficult on those soils. Weak subsoil may pose serious problems if not treated correctly. The nature of the soil was lateritic and was observed to be incapable to satisfy the bearing pressure requirement for operations of the cranes which will be employed for the erection of heavy columns. Hence, there was a need to build a heavy-duty working platform or a hardstand area for the purpose to be solved.

A hardstand area is a paved area or place where machinery and equipment are stationed to work for the project which is being executed. The paving is usually designed to be thicker and more durable to support the weight of heavy vehicles such as large cranes or heavy trucks. To obtain the required hardstand area, a suitable ground improvement technique must be adopted as a solution for improving the bearing capacity and controlling the settlements at that particular location.

Depending upon the subsoil strata and bearing capacity requirement ground improvement using the cellular mattress and Mechanically Stabilized Layers are incorporated at different locations as optimal foundation solution. Geogrids used in stabilization of soil layers known as Mechanically Stabilized Layers and the geocell mattress is proposed as the solution for stiffening granular layers to increase load distribution, control total settlements and minimize differential settlements over low strength and variable foundation soils.

A. L. Ksheeraja
Indian Institute of Technology, Hyderabad, India

N. Srilatha (✉)
Ramaiah Institute of Technology, Bangalore 560054, Karnataka, India
e-mail: sreelathaN@msrit.edu

C. N. V. Satyanarayana Reddy et al. (eds.), *Ground Improvement and Reinforced Soil Structures*, Lecture Notes in Civil Engineering 152,
https://doi.org/10.1007/978-981-16-1831-4_29

Ong et al. [1] discussed the use of geocell mattresses with mechanically stabilized layer (MSL) for the construction of a heavy-duty working platform for an offshore facilities fabrication yard over the soft soil. The geocell mattress with a thickness of 1 m was constructed using stiff geogrids to form a honeycomb structure and filled with aggregates. Subsequently, the MSL with a thickness of 2 m was constructed on top of the geocell mattress to form a stiff and stable working platform by maximizing the pressure distribution of applied loading onto the soft foundation soils effectively. This paper also discussed on design and construction of the geocell mattress combined with mechanical stabilized layers and presented the results from a large-scale plate bearing test conducted to verify the performance of the foundation system constructed. The result from the plate bearing test showed minimal settlement when the maximum load applied was 60 t/m^2. This indicates the geocell mattress with MSL had been constructed successfully to meet its design requirements.

Dobie et al. [2] proposed guidelines for the design, construction, operation and maintenance of working platforms and these are very important for many projects to support cranes or piling rigs. Finite element analysis was performed to check the mechanical behaviour of working platforms. The results from the parametric analysis were validated by full-scale testing subjected to bearing capacity failure. From the results, it is observed that there is a significant increase in bearing capacity with the inclusion of geogrid resulting in stabilization of the granular material. Ooi et al. [3] discussed the use of different types of geogrid applications in soft ground in Malaysia.

Rakowski and Kawalec [4] discussed on mechanical stabilization of granular material provided by geogrid. By including one or more layers of geogrid in a granular layer, a composite material with better properties and performance are created and this is often described as a Mechanically Stabilized Layer (MSL). Figure 1 shows the interlocking mechanism, which restrains the movement of aggregate particles within the geogrid aperture, is identified as the lateral confinement effect that can be mobilized from a stiff geogrid. Through the interlocking mechanism and lateral confinement, the aggregate layer can be stabilized without excessive deformation of the surface.

Fig. 1 Interlocking mechanism of stiff geogrid providing lateral confinement

The cellular foundation mattress is an open top, continuous and honeycombed structure formed from a series of interlocking cells [5]. The use of a cellular foundation mattress or stratum was proposed as it maximizes the pressure distribution of applied load onto the low bearing capacity of the soft foundation soils. It also forms a firm, stiff, stable working platform enabling safe construction plant travels and enables faster construction as compared to RC foundation construction.

Latha [6] presented the methods of design available for geocell-supported embankments. Two of the earlier methods are considered in this paper and a third method is proposed and compared with earlier methods. In the first method called the slip line method, plastic bearing failure of the soil was assumed and the additional resistance due to geocell layer is calculated using a non-symmetric slip line field in the soft foundation soil. In the second method based on slope stability analysis, a general-purpose slope stability program was used to design the geocell mattress of required strength for the embankment. In the third method, geocell reinforcement is designed based on the plane strain finite element analysis of embankments. The geocell layer is modelled as an equivalent composite layer with modified strength and stiffness values. The strength and dimensions of a geocell layer are estimated for the required bearing capacity or permissible deformations. These three design methods are compared through a design example. It is observed that the design method based on finite element simulations is most comprehensive because it addresses the issue of permissible deformations and also gives complete stress, deformation and strain behaviour of the embankment under given loading conditions.

The majority of the researchers have successfully predicted the ultimate bearing capacity of shallow footing resting on the horizontal surface of the ground Terzaghi and Peck [7], Meyerhof [8], Hansen [9], Vesic [10] and the predictions are based on the laboratory investigation of undisturbed and disturbed soil samples. In the present study, conventional methods have been used to determine the bearing capacity of a stratified soil deposit. Ground improvement scheme was proposed to enhance the bearing capacity of the soil, limit post-construction long term settlements to tolerable limits for the given loading conditions of the proposed foundation and is validated through the numerical studies.

2 Materials and Methods

Field tests were conducted to find the properties of the soil stratum at the location of the project site. Table 1 shows the properties of stratified soil deposit considered for the present study. SBC of the existing stratified deposit of soil was calculated by considering the weighted average of shear strength parameters up to the influence zone of the footing (5 m) respectively. The angle of internal friction of the soil deposit is 0° and the cohesion of the soil as 24 kPa. The ultimate bearing capacity of the original soil stratum as per IS code 6403-1981 was calculated as 165.62 kPa and this is the optimum value of bearing capacity compared to the bearing capacity calculated using other conventional methods [11].

Table 1 Properties of stratified soil deposit

Soil description	Thickness of the layer (m)	C (kPa)	ϕ (deg)	ϒ (kN/m^3)	Observed standard penetration number
Clayey soil	0–4	12	0	15	4
Clayey silt	4–8.5	36	0	15	8
Sandy soil	8.5–10.7	0	34	18	50

3 Finite Element Analysis

3D Finite Element Analysis was performed instead of 2D analysis due to the rectangular shape of the loaded area considered. Separate models were developed for the layered deposit of the original soil stratum and proposed reinforced soil stratum but were identical in all respects apart from the geometry of the MSL, cellular mattress and geometry of loaded area. Based on the available data, the footing area of size 13.5 m × 5.5 m × 10.7 m was considered for the analysis and the crane contact area at the location is 13.5 m × 5.5 m and the depth of soil profile is 10.7 m. Trials were carried out to select a suitable ground improvement scheme to be adopted for the proposed project to meet the requirements and to satisfy the permissible settlements. Ground improvement scheme adopted for the particular area consists of three layers of MSL using three layers of triaxial geogrids underlain by 0.65 m thick cellular mattress respectively. The MSL was compacted in equal lifts of three, each of thickness 300 mm and above that, a concrete layer of 0.2 m thickness was provided. The water table was located at 0.2 m below the ground level. The geometry of the original soil stratum and reinforced soil stratum models are shown in Fig. 2 and Fig. 3, respectively.

The FEA mesh is shown for the original soil stratum and reinforced soil stratum in Fig. 4 and Fig. 5 respectively. The medium-mesh factor was selected with a mesh

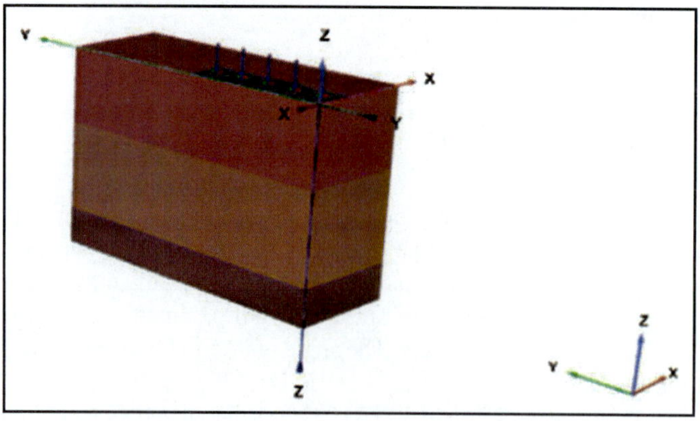

Fig. 2 Geometry model of soil stratum

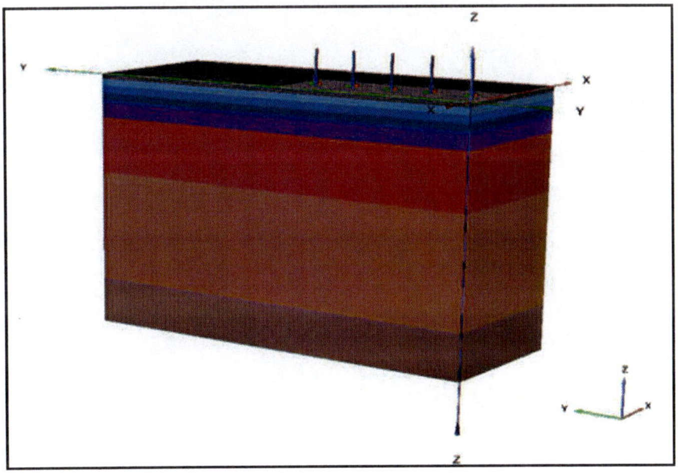

Fig. 3 Geometry model of reinforced soil stratum

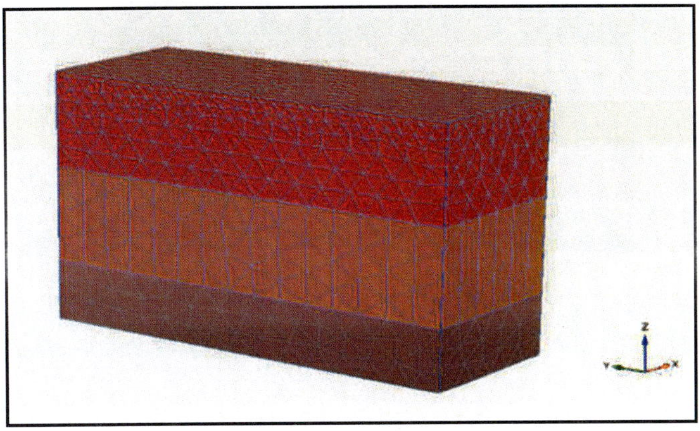

Fig. 4 FEA mesh of original soil stratum

factor of 1.0. In both cases, the symmetry of the problem was used to model one-quarter of the loaded area and geometry with two of the vertical mesh boundaries forming planes of symmetry along the centerlines of the loaded areas. The loading was applied to the rectangular area of 2.75 m × 6.75 m which is a quarter of the overall area. The fixed base of the mesh was placed at 10.7 m below ground level coincide with the bedrock at the proposed location of the hard stand area. Clayey soil (coloured red) shown in Fig. 5b was adopted to 4 m depth overlying clayey silt (coloured yellow) to 8.5 m depth overlying dense sandy soil (coloured brown) to 10.7 m depth where hard stratum was encountered. To get reinforced soil stratum, the natural ground was assumed to be excavated to 1.65 m and replaced with geo cellular mattress and MSL respectively. The Layout of reinforced soil stratum of thickness

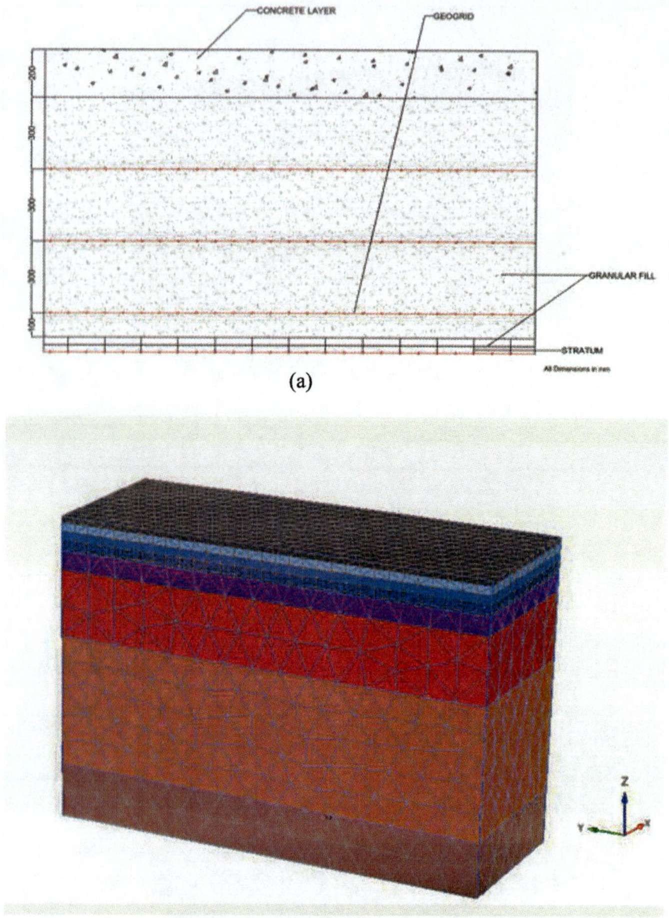

(a)

(b)

Fig. 5 **a** Layout of reinforced soil stratum of thickness 1.65 m includes MSL, geogrid and cellular mattress. **b** FEA mesh of reinforced soil stratum

1.65 m includes MSL, geogrid and cellular mattress adopted as shown in Fig. 5a. The reinforced soil stratum namely a 1.0 m thick MSL (coloured different shades of blue in Fig. 3) and three layers of Triaxial geogrid overlying 0.65 m thick stratum (coloured purple). Both the MSL and mattress were assumed to be composed of the granular fill material. The top layer was composed of reinforced cement concrete of thickness 0.2 m respectively.

The in situ layers were modelled using linear elastic perfectly plastic Mohr-Coulomb model with the input parameters of soil stratum shown in Table 2. The reinforced concrete pavement surface was modelled as a linear elastic material and Table 3 shows the assumed properties of MSL, cellular mattress and reinforced

Table 2 Input Properties of soil used in the analysis

Property	Soft silt	Stiff silt	Dense sand
Drainage	Undrained	Undrained	Drained
Elasticity (MPa)	6	19	35
Poisson's ratio	0.2	0.2	0.2
Angle of internal friction	0	0	34
Bulk unit weight (kN/m^3)	17	17	19
Cohesion (kPa)	16	48	0.1

Table 3 Assumed physical and mechanical properties of MSL and cellular mattress

	MSL	Cellular mattress	Reinforced cement concrete
Elasticity (kN/m^2)	50,000	80,000	3 × 107
Poisson's ratio	0.2	0.2	0.15
Angle of internal friction	0	0	–
Unit weight (kN/m^3)	17	17	24

cement concrete, respectively. The required safe-bearing pressure of the area was assumed as 220 kN/m^2.

Figure 6 shows the total displacement of the modelled original soil stratum. The total displacement of the soil stratum from Fig. 6 is 3.46 m. Figure 7 shows the predicted displacements of the reinforced soil stratum. The predicted total displacement of improved ground is observed to be 58.56 mm.

Figure 8 shows the deformed mesh of the original soil stratum showing clearly the failure pattern of the ground. Figure 9 shows the deformed mesh of the improved ground after subjecting to the surface load.

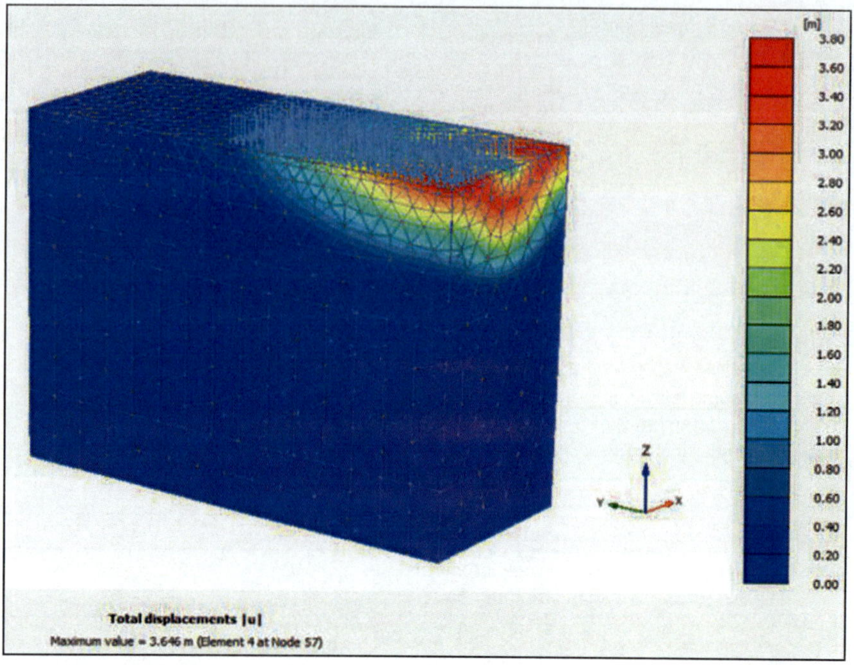

Fig. 6 Total displacements of original soil stratum

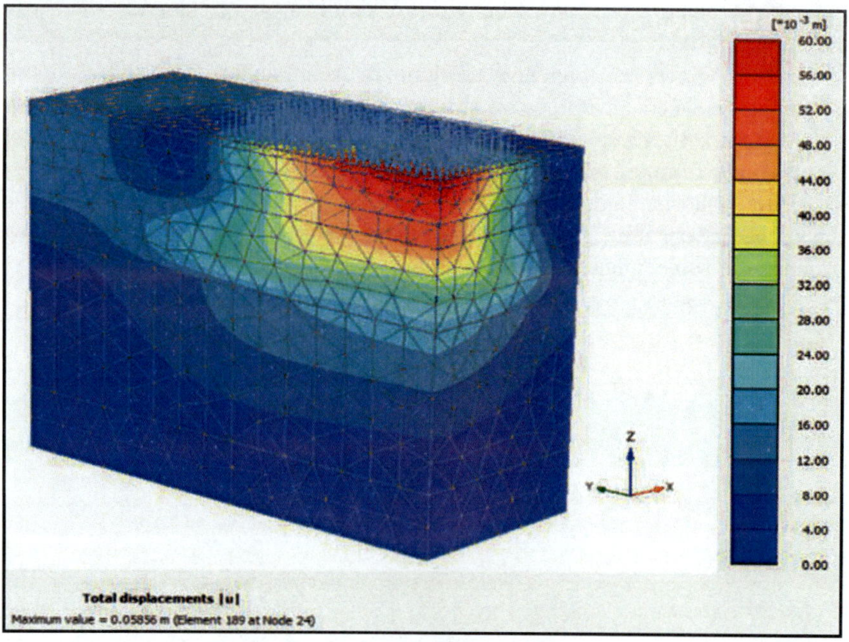

Fig. 7 Total predicted displacements of reinforced soil stratum

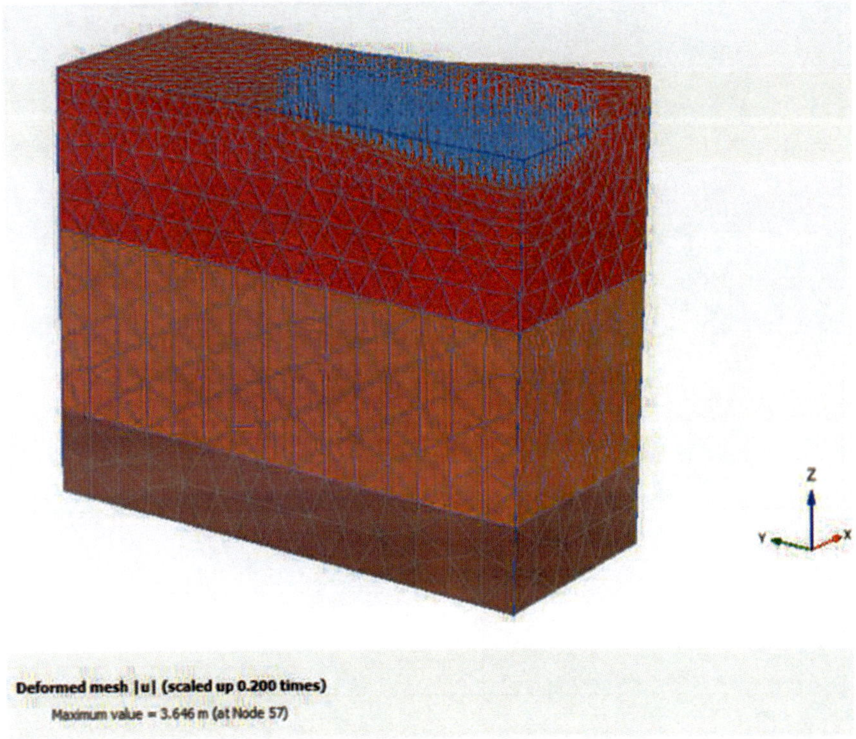

Fig. 8 Deformed mesh of original soil stratum

4 Conclusions

The present study presented the numerical results of load tests on rectangular footing supported on a geogrid mechanical stabilized layer overlying geo cellular mattress. The bearing capacity of the natural soil stratum was calculated using conventional methods. Based on the findings from the present investigation, the following conclusions are drawn:

1. Provision of cellular mattress in the overlying granular fill improves the load-carrying capacity and settlements are observed to be reduced substantially.
2. A maximum settlement of 58.56 mm was predicted in the area which is less than the allowable settlement of 75 mm.
3. Therefore, the predicted settlement satisfies the specified settlement required in the present case.

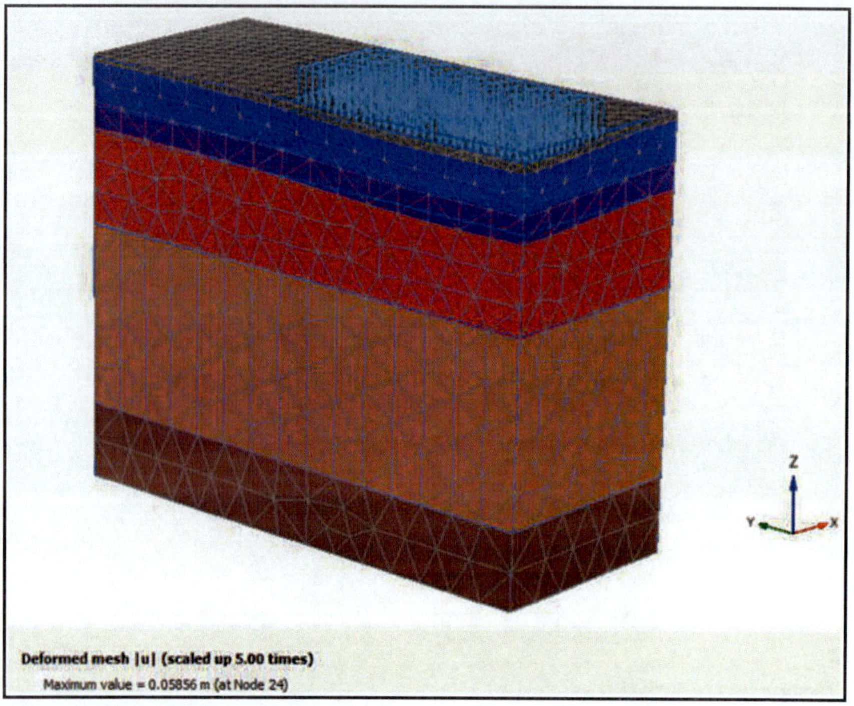

Fig. 9 Deformed mesh of reinforced soil stratum

References

1. Ong, R., Song, W.K. Minh, T.N., Thinh, P.X.: Geocell mattress for heavy duty working platform over soft soil. Geotech. Hanoi **2011**(2011), 153–158
2. Dobie, M.J.D., Lees, A.S., Buckley, J., Bhavsar, R.: Working platforms for tracked plant–BR 470 guideline and a revised approach to stabilisation design with multiaxial hexagonal geogrids. In: Proceedings of the 13th Australia New Zealand Conference on Geomechanics (2019)
3. Ooi, T.A., Tee, C.H., Chan, C.B., Ong, R.: A review of geogrid worling platform in soft ground in Malaysia. In: Proceedings of the 18th International Conference on Soil Mechanics and Geotechnical Engineering, pp. 2969–2972 (2013)
4. Rakowski, Z., Kawalec, J.: Mechanically stabilized layers in road construction. In: XXVII Baltic Road Conference, Ryga, Sierpień (2009)
5. BSI: BS8006-1:2010, Code of practice for strengthened/reinforced soils and other fills. British Standard Institution (2010)
6. Latha, G.M.: Design of geocell reinforcement for supporting embankments on soft ground. Geomech. Eng. **3**(2), 117–130 (2011)
7. Terzaghi, K., Peck, R.B.: Soil Mechanics in Engineering Practice, 1st edn. Wiley, New York (1948)
8. Meyerhof, G.G.: An investigation for the foundations of a bridge on dense sand. In: Proceedings of the 3rd International Conference on Soil Mechanics and Foundation Engineering, vol. 2, pp. 66–70 (1953)
9. Hansen, J.B.: A general formula for bearing capacity. Danish Geotechnical Institute, Copenhagen, Denmark, Bulletin, no. 11, pp. 38– 46 (1970)

10. Vesic, A.S.: Bearing capacity of shallow foundations. In: Winterkorn, H.F., Fang, H.Y. (eds.) Foundation Engineering Handbook, Van Nostrand Reinhold, New York, USA pp. 121–147(1975)
11. Bowles, L.E.: Foundation Analysis and Design. Mcgraw-Hill (1996)

Stabilization of Artificial Soils Using Waste Marble Dust

K. M. N. Saquib Wani, B. A. Mir, and Ishfaq Rashid Sheikh

1 Introduction

With enormous load on civil engineers especially geotechnical engineers to use available weak soil deposits for construction purposes, improving the properties of weak soils for a feasible construction be it roads, buildings, dams, retaining walls, backfill materials etc. becomes a necessity. Researchers have always tried to improve weak soil by stabilizing them [14, 15]. Soils can be stabilized or improved by a variety of techniques available at hand like chemical stabilization, inclusion and confinement, physical methods like freeze-thaw, vibration etc. [1, 18, 19, 21]. Due to the uneconomical aspects or the adverse effect on the environment, most of these techniques do not adhere to sustainable practices and as such find no use in the present day world apart from in some extreme cases where there is no alternative. Researchers have now focused on more environment friendly and sustainable practices of soil stabilization. The harmful gases released during the production of chemicals or additives like cement, grout materials, epoxy's etc. have hindered the use of such materials to a great extent paving way for more environment friendly practices like geo-cell inclusion, biological processes etc. [22–29]. Thus, an alternative to such materials is a dire need, the substance that proves beneficial in both reducing carbon emissions and side by side enhances the properties in comparison to conventional materials. One such material is the waste produced from marble manufacturing units. This has high levels of lime present in it and acts as a pozzolanic material if added to the soil. Reactions similar to cement addition take place between water and clay particles thereby increasing the strength [2].

Previously, researchers used waste marble fines to stabilize lateritic soils and recommended it as a low-cost soil stabilizer and well suited for road construction

K. M. N. S. Wani (✉) · B. A. Mir · I. R. Sheikh
National Institute of Technology, Union Territory of J&K, Srinagar, India
e-mail: sakibwani_02phd17@nitsri.ac.in

C. N. V. Satyanarayana Reddy et al. (eds.), *Ground Improvement and Reinforced Soil Structures*, Lecture Notes in Civil Engineering 152,
https://doi.org/10.1007/978-981-16-1831-4_30

[2]. Various other forms of waste like granite waste, marble dust, ceramic wastes etc. have been utilized efficiently by researchers and have given good results based on strength characteristics [3, 24, 31]. Researchers evaluated the potential of WMD in red tropical soils and concluded that maximum strength and CBR values were noticed at 7–10 days curing and 8% WMD [20]. An increase in mechanical properties of soil with the incorporation of WMD suggesting it to be a sustainable material for solid waste management was also proposed [30]. Effects of waste marble dust and polypropylene fiber contents on mechanical properties of gypsum stabilized earthen showed promising results [5].

In the present study, an attempt has been made to reinforce artificial soils made in the laboratory by a different mix of clay (kaolinite) and dredged soil. The main idea is to correlate the application of this technique with other soil types that have similar properties. WMD produced from marble manufacturing units proves very detrimental in terms of waste as well as environment including health hazards. This study aims to stabilize artificial soils by incorporating different percentages of WMD and studying the UCS as well as penetration resistance (CBR) characteristics.

2 Materials and Testing Methodology

2.1 *Artificial Soil*

Laboratory prepared artificial soil consisted of 60% dredged soil and 40% kaolinite clay. 50 kg of soil in batches was mixed manually by a shaker in the laboratory to attain a uniform mix. The soil was then air-dried and stored for further use. Researchers have previously worked with artificial soil by varying sand and clay percentages in varying proportions [4]. The properties of the mixed soil are given in Table 1. The new soil is classified as silt with intermediate compressibility with a sufficient amount of clay (42%). Based on strength properties, we can conclude that the soil is not feasible for any construction material unless stabilized or improved. The particle size distribution curve of the soil is shown in Fig. 1.

2.2 *Waste Marble Dust (WMD)*

WMD was collected from Nowgam area of Srinagar district as many marble supplying units are present in the vicinity. The marble dust was fine in nature with very less coarse particles. WMD passing 300μ sieve was used in the study. Similar fine materials were also used by previous researchers in different stabilization techniques [2, 17]. Waste material was collected in bulk in plastic bags and transported to the lab where only 300μ sieve passing material was used and air-dried to remove

Table 1 Properties of artificial soil (60% dredged soil + 40% kaolinite)

Properties	Description (value)
Specific gravity of soil, G	2.5
Specific gravity of WMD, G	2.7–2.9
% Finer than 75 μm	90
Sand (%)	10
Silt (%)	48
Clay (%)	42
Liquid limit (%)	37
Plastic limit (%)	23
Plasticity index (%)	14
Classification	MI
Maximum dry unit weight (kN/m^3)	17
OMC (%)	19
UCS at OMC, q_u (kPa)	65
CBR, Un-soaked (%)	6
CBR, soaked @ 94 h (%)	3

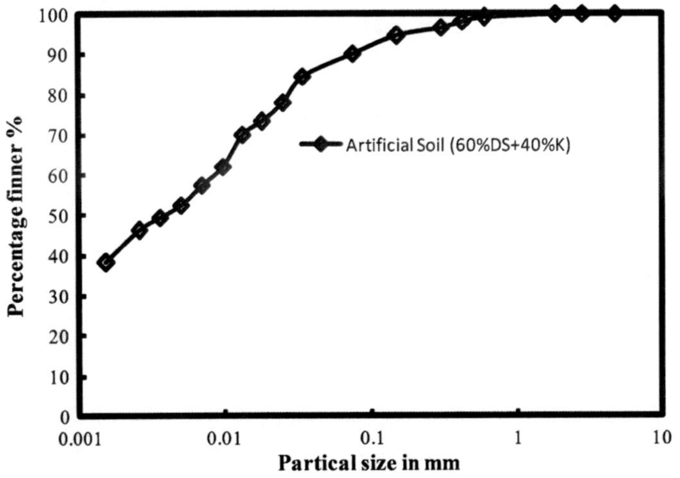

Fig. 1 Particle size distribution curve

any moisture. Figure 2 shows the XRD graph of WMD confirming the presence of lime and other constituents.

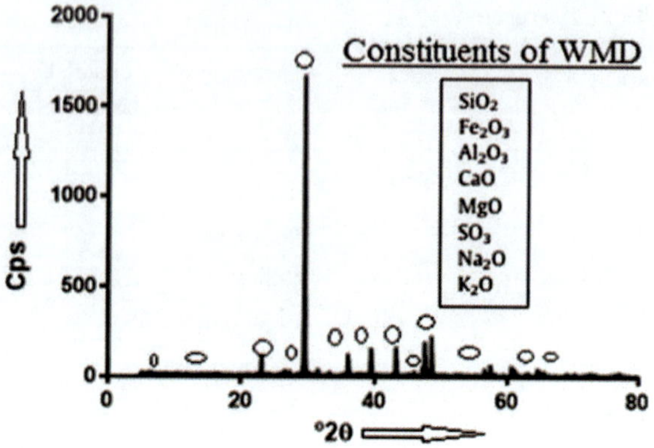

Fig. 2 XRD analysis of WMD

2.3 *Testing Methodology*

The preliminary set of testing consisted of basic geotechnical laboratory testing, which included tests on remolded samples as shown in Table 1. Tests like PSD, specific gravity, Atterberg limit tests UCS, CBR etc. [6–12] were performed to determine the engineering properties of the soil. The final set of testing included UCS and CBR. WMD in different percentages by weight of dry soil was added to the soil and UCS as well as CBR tests were performed. UCS samples were cured for 7 days in desiccators whereas CBR testing was performed on both un-soaked and soaked samples, soaked samples were submerged in water for 96 h with two 2.5 kg surcharge loads over the samples as per standard codal provisions [8].

3 Results and Discussions

3.1 *Effect of WMD on the UCS of Stabilized Artificial Soil*

The addition of WMD increases the strength of treated samples in comparison to the control sample. The UCS values increased from 65 kPa in the untreated state to 490 kPa at 14% WMD addition and 7 days curing. The strength gain from immediate tests to cured sample tests showed a significant gain in strength as can be seen in Fig. 3 whereby strength gain is significant up to 14% WMD. At 21% WMD strength, the gain is not sufficient although it tends to increase owing to the brittle nature of specimens, which tend to fail vertically thereby reducing the strength.

The main reason for the increase in strength of UCS samples is attributed to a series of pozzolanic reactions between lime-rich WMD and the silica- and alumina-rich soil sample, in the presence of water, forming calcium-silicate hydrates and calcium-aluminate hydrates, respectively.

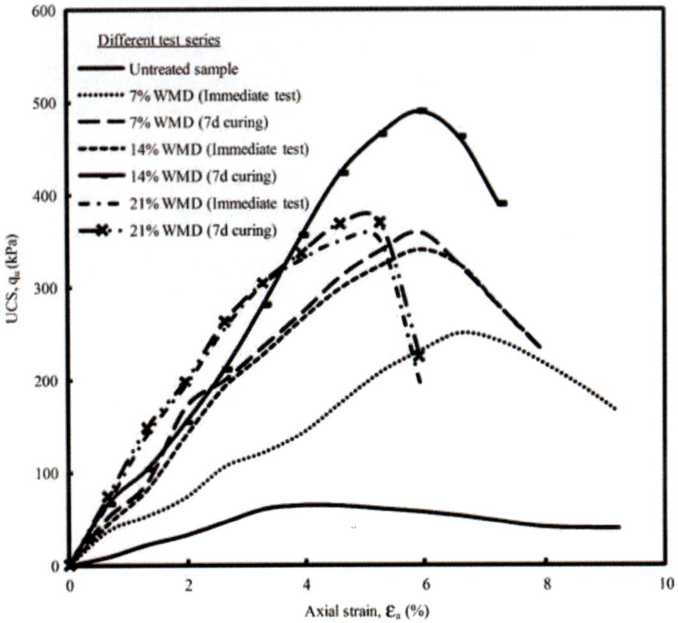

Fig. 3 UCS graphs (stress vs. strain) for treated soil samples

$$\underset{\text{(Calcium-silicate-hydrate)}}{Ca^{2+} + 2OH^{-} + Sio_2 \rightarrow Cao \cdot Sio_2 \cdot H_2O;} \tag{1}$$

$$\underset{\text{(Calcium-aluminate-hydrate)}}{Ca^{2+} + 2OH^{-} + Al_2O_3 \rightarrow Cao \cdot Al_2O_3 \cdot H_2O\,.} \tag{2}$$

Similar observations were also reported by other researchers although the samples they used were different but definitely followed the same trend. At 7 days curing and 8% WMD strength, gain was maximum as reported by Okagbue and Onyeobi [20]. Similarly, UCS of 255 kPa was reported by researchers after 28 days of curing using WMD [30]. The problem often encountered with such stabilizing techniques is the transition from ductile to brittle behavior, which must be taken care of. It can be seen from Fig. 3 that the failure at 21% WMD is abrupt. Hence in this study, we recommend the use of 14% as optimum WMD content.

3.2 *Effect of WMD on the CBR of Stabilized Artificial Soil*

CBR values play an important role in designing the pavement. Sub-grade, sub-base and base course thickness for pavements depend on the CBR values of the material. IRC-37 [13] gives different guidelines for pavement construction. CBR values in this study show promising results as the un-soaked CBR enhanced from an initial value of 6–18% at 21% WMD addition whereas the soaked value increases from 3 to 12% for the same conditions (Fig. 4a–c). The enhancement in CBR is significant and attributed to the formation of CSH and CAH (Eqs. 1 and 2). The fact that soaked CBR values show a higher increase is because of the curing effect in water, which solidifies the bonding between grains. Many other researchers reported the similar finding of increased CBR values although the materials used were different but can be positively taken as a reference in this study [16, 20, 30].

4 Conclusions

Based on the experimental results, the following conclusions can be drawn from the present study.

1. Artificial soils prepared in the laboratory by incorporating different proportions of silt and clay proved beneficial enough to correlate the properties of weak soils.
2. UCS values increased significantly owing to the formation of cementitious compounds. An increase in UCS value from 65 to 490 kPa at 14% WMD content and 7 days curing were noticeable.
3. CBR values increased for both un-soaked as well as soaked conditions proving it to be beneficial for pavement construction. The increase in CBR by about 200% in un-soaked and 400% in soaked condition is attributed to the pozzolanic reaction within the soil matrix.
4. The inclusion of WMD can prove beneficial in restoring the ecological balance of the area and reduce the health hazards like choking and lung diseases, only if WMD is utilized in bulk.

5 Future Scope of Work

This study aimed at putting forth the benefits of using WMD, which is considered a hazardous waste affecting both the groundwater and ecosystem and creates health problems. Permeability, shear strength, tri-axial and plate load testing can pave way for a more sustainable practice in terms of soil improvement.

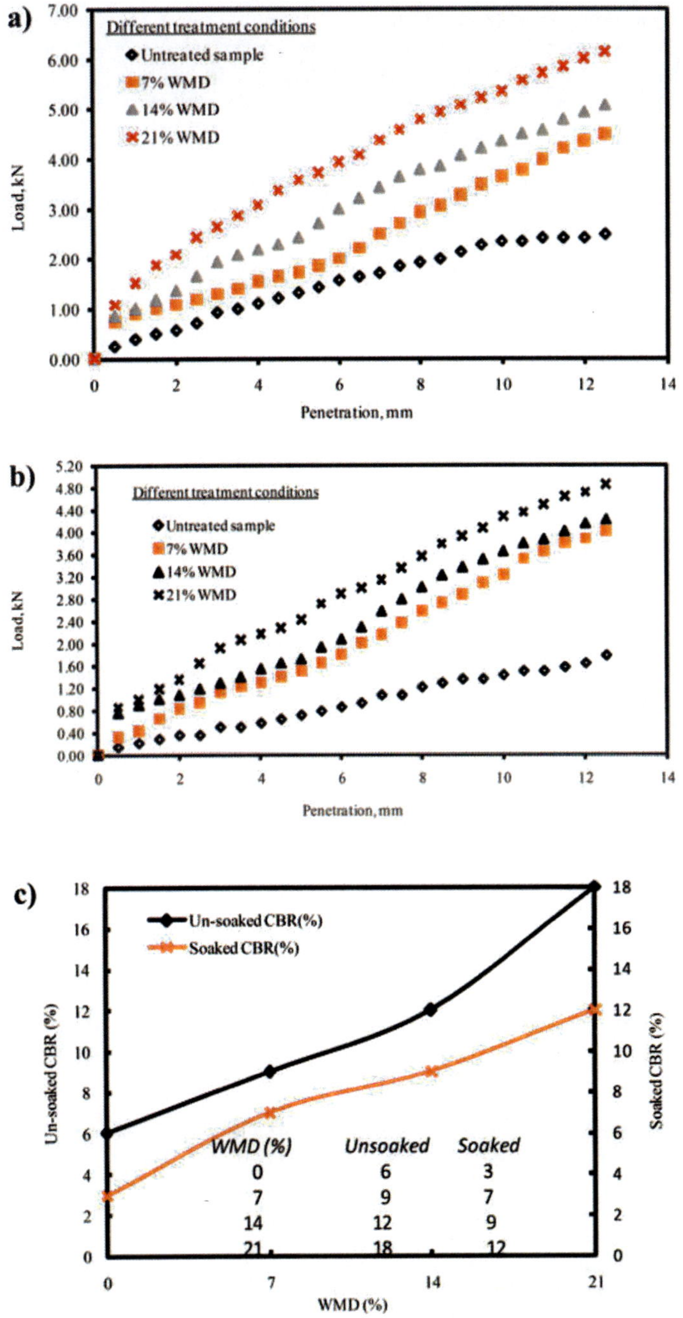

Fig. 4 CBR values for **a** Un-soaked samples; **b** Soaked samples; **c** Variation in CBR values

Acknowledgments The authors would like to thank the Department of Civil Engineering and Administration, National Institute of Technology, Srinagar for their support. Thanks are due to the faculty and support staff of Geotechnical Engineering Division. The authors would like to credit Mr. Mohd Ismael and Mr. Ab. Aziz for their help during testing and curing of the samples.

References

1. Adamson, L.G., Chilingar, G.V., Beeson, C.M., Armstrong, R.A.: Electrokinetic dewatering, consolidation and stabilization of soils. Eng. Geol. **1**(4), 291–304 (1966). https://doi.org/10.1016/0013-7952(66)90011-1
2. Akinwumi, I.I., Booth, C.A.: Experimental insights of using waste marble fines to modify the geotechnical properties of a lateritic soil. J. Environ. Eng. Landsc. Manag. **23**(2), 121–128 (2015). https://doi.org/10.3846/16486897.2014.1002843
3. Al-Bared M.A.M., Marto A.: Evaluating the compaction behaviour of soft marine clay stabilized with two sizes of recycled crushed tiles. In: Pradhan B. (eds.) GCEC 2017. Lecture Notes in Civil Engineering, vol. 9. Springer, Singapore (2019). https://doi.org/10.1007/978-981-10-8016-6_90
4. Al-Jeznawi, D., Sanchez, M., Al-Taie, A.J., Zielinski, M.: Experimental studies on curling development of artificial soils. J. Rock Mech. Geotech. Eng. **11**(6), 1264–1273 (2019). https://doi.org/10.1016/j.jrmge.2019.02.006
5. Balkis, A.P.: The effects of waste marble dust and polypropylene fiber contents on mechanical properties of gypsum stabilized earthen. Constr. Build. Mater. **134**, 556–562 (2017). https://doi.org/10.1016/j.conbuildmat.2016.12.172
6. Bureau of Indian Standards IS 1498: Classification and Identification of Soils for General Engineering Purposes. Bureau of Indian Standards, New Delhi (1970)
7. Bureau of Indian Standards IS 2720-Part 10: Laboratory Tests Procedure for Unconfined Compressive Strength Test. Bureau of Indian Standards, New Delhi (1986)
8. Bureau of Indian Standards IS 2720-part 16: Laboratory Determination of CBR. Bureau of Indian Standards, New Delhi (1979)
9. Bureau of Indian Standards IS 2720-part 3(1): Determination of Specific Gravity of Finegrained Soils. Bureau of Indian Standards, New Delhi (1980)
10. Bureau of Indian Standards IS 2720-part 4: Determination of Grain Size Distribution. Bureau of Indian Standards, New Delhi (1985)
11. Bureau of Indian Standards IS 2720-part 5: Determination of Atterberg Limits. Bureau of Indian Standards, New Delhi (1985)
12. Bureau of Indian Standards IS 2720-part 7: Determination of Water Content-Dry Density Relation Using Light Compaction. Bureau of Indian Standards, New Delhi (1980)
13. Congress, I.R.: Guidelines for the design of flexible pavements. Indian Code of Practice, IRC, 37 (2001)
14. Daraei, A., Sherwani, A.F.H., Faraj, R.H., et al.: Stabilization of problematic soil by utilizing cementitious materials. Innov. Infrastruct. Sol. **4**, 33 (2019). https://doi.org/10.1007/s41062-019-0220-5
15. Das, B.M., Khing, K.H., Shin, E.C.: Stabilization of weak clay with strong sand and geogrid at sand-clay interface. Transp. Res. Rec. **1611**(1), 55–62 (1998). https://doi.org/10.3141/1611-07
16. Fırat, S., Yılmaz, G., Cömert, A.T., Sümer, M.: Utilization of marble dust, fly ash and waste sand (Silt-Quartz) in road subbase filling materials. KSCE J. Civil Eng. **16**(7), 1143–1151 (2012). https://doi.org/10.1007/s12205-012-1526-4
17. Karaşahin, M., Terzi, S.: Evaluation of marble waste dust in the mixture of asphaltic concrete. Constr. Build. Mater. **21**(3), 616–620 (2007). https://doi.org/10.1016/j.conbuildmat.2005.12.001

18. Maurya, S., Sharma, A.K., Jain, P.K., Kumar, R.: Review on stabilization of soil using coir fiber. Int. J. Eng. Res. **4**(6), 296–299 (2015)
19. Mokhtari, M., Kalantari, B.: Soft soil stabilization using stone columns—a review. Electron. J. Geotech. Eng. **17**, 1459–1466 (2012)
20. Okagbue, C.O., Onyeobi, T.U.S.: Potential of marble dust to stabilise red tropical soils for road construction. Eng. Geol. **53**(3–4), 371–380 (1999). https://doi.org/10.1016/S0013-7952(99)00036-8
21. Rafalko, S.D., Filz, G.M., Brandon, T.L., Mitchell, J.K.: Rapid chemical stabilization of soft clay soils. Transp. Res. Rec. **2026**(1), 39–46 (2007). https://doi.org/10.3141/2026-05
22. Sheikh, I.R., Shah, M.Y.: Experimental investigation on the reuse of reclaimed asphalt pavement over weak subgrade. Transp. Infrastruct. Geotech. (2020). https://doi.org/10.1007/s40515-020-00115-w
23. Sheikh, I.R., Shah, M.Y.: Experimental study on geocell reinforced base over dredged soil using static plate load test. Int. J. Pavement Res. Technol. **13**, 286–295 (2020). https://doi.org/10.1007/s42947-020-0238-2
24. Wani, K.M.N.S., Mir, B.A.: Stabilization of weak dredged soils by employing waste boulder crusher dust: a laboratory study. Geotech. Geol. Eng. (2020). https://doi.org/10.1007/s10706-020-01472-6
25. Wani, K.M.N.S., Mir, B.A.: Unconfined compressive strength testing of bio-cemented weak soils: a comparative upscale laboratory testing. Arab. J. Sci. Eng. (2020). https://doi.org/10.1007/s13369-020-04647-8
26. Wani, K.M.N.S., Mir, B.A.: A laboratory-scale study on the bio-cementation potential of distinct river sediments infused with microbes. Transp. Infrastruct. Geotech. (2020). https://doi.org/10.1007/s40515-020-00131-w
27. Wani, K.M.N.S., Mir, B.A.: An experimental study on the bio-cementation and bio-clogging effect of bacteria in improving weak dredged soils. Geotech. Geol. Eng. (2020). https://doi.org/10.1007/s10706-020-01494-0
28. Wani, K.M.N.S., Mir, B.A.: Influence of microbial geo-technology in the stabilization of dredged soils. Int. J. Geotech. Eng. (2019). https://doi.org/10.1080/19386362.2019.1643099
29. Wani, K.M.N.S., Mir, B.A.: Microbial geo-technology in ground improvement techniques: a comprehensive review. Innov. Infrastruct. Sol. **5**, 82 (2020). https://doi.org/10.1007/s41062-020-00335-6
30. Yilmaz, F., Yurdakul, M.: Evaluation of marble dust for soil stabilization. Acta Phys. Pol., A **132**(3), 710–711 (2017). https://doi.org/10.12693/APhysPolA.132.710
31. Zainuddin, N., Mohd Yunus, N.Z., Al-Bared, M.A.M., Marto, A., Harahap, I.S.H., Rashid, A.S.A.: Measuring the engineering properties of marine clay treated with disposed granite waste. Measurement **131**, 50–60 (2019). https://doi.org/10.1016/j.measurement.2018.08.053

Compressibility Characteristics of Guar Gum-Treated Expansive Soil

Kopparthi Venkata Vydehi and Arif Ali Baig Moghal

1 Introduction

Expansive soil causes a lot of distresses to the structures built over it due to swelling and shrinkage under saturated or partially saturated conditions. Chemical stabilization is often preferred to deal with expansive soils compared to other stabilization methods [20]. Cement, lime and fly ash are used as the most common additives to improve the geotechnical properties of expansive soils [10, 13, 12]. Recent studies admit that the utilization of these additives contributes to CO_2 gas emission into the environment and can trigger groundwater contamination affecting the growth of vegetation due to alteration of soil pH [1, 21]. To overcome these limitations, the use of bioengineering solutions as an alternate treatment strategy is proposed to improve the geotechnical properties viz., shear strength, permeability and compressibility [7, 11, 8, 16, 17]. Nugent et al. [19] carried out one-dimensional consolidation tests on kaolinite treated with guar gum. Results indicated that the compressibility was reduced at lower dosages due to the hydrogen bond between GG and soil. Experimental investigations on silty clay indicated that a reduction of 4% collapse potential was observed with 2% guar gum [7, 9].

Guar gum originates from the guar bean, a vegetable plant. It is a water-soluble polysaccharide composed of D-galactose (36.6%) and D-mannose (63.1%) monomer units. India, the largest producer of guar gum accounts for 80% of world production annually. The present study deals with the investigation of compressibility characteristics of the locally available black cotton soil treated with guar gum.

K. V. Vydehi (✉) · A. A. B. Moghal
Department of Civil Engineering, National Institute of Technology Warangal, Warangal 506004, Telangana, India
e-mail: vydehi56@student.nitw.ac.in; kvvydehi252@gmail.com

A. A. B. Moghal
e-mail: baig@nitw.ac.in; reach2arif@gmail.com

C. N. V. Satyanarayana Reddy et al. (eds.), *Ground Improvement and Reinforced Soil Structures*, Lecture Notes in Civil Engineering 152,
https://doi.org/10.1007/978-981-16-1831-4_31

Table 1 Physical properties of BC soil

Property	Value	Standard adopted
Specific gravity	2.66	ASTM D854 [6]
Liquid limit (%)	47	ASTM D4318 [4]
Plastic limit (%)	22	ASTM D4318 [4]
OMC (%)	13.47	ASTM D698 [5]
MDD (kN/m^3)	17.95	ASTM D698 [5]
Soil classification (USCS)	CL	ASTM D2487 [2]

Note OMC—Optimum moisture content; MDD—Maximum dry density; USCS—Unified soil classification system

2 Materials

2.1 Soil

The black cotton (BC) soil used in the present study was collected at a depth of 3 ft after removing the top loose soil near the National Institute of Technology, Warangal Campus. The collected soil was air-dried, pulverized and preserved for further experiments. The physical properties were determined in accordance with relevant ASTM standards and are presented in Table 1. According to USCS classification, the soil has been classified as lean clay.

2.2 Guar Gum

The guar gum used for the study was sourced from SRL Chemicals Pvt. Ltd., Mumbai, Maharashtra. The gum powder is almost white to pale yellowish in color (Fig. 1).

3 Methodology

The compressibility characteristics were studied by carrying out a series of one-dimensional consolidation tests on biopolymer-treated soil at varying dosages (0.5, 1.0, 1.5, 2.0 and 2.5% by dry weight of soil). The soil was mixed with various dosages of guar gum and compacted at targeted fixed dry density (17.95 kN/m^3) in a consolidation ring of 6 cm diameter and 2 cm height. Samples were allowed to fully saturate at a seating pressure of 6.25 kPa and then loaded under vertical pressures of 12.5, 25, 50, 100, 200, 400 and 800 kPa (maintaining a load increment ratio of unity) and each load was kept for a period of 24 h as per ASTM D4186M-12e1 [3, 14, 15]. Compression index values were determined from the void ratio and consolidation pressure plots for each load increment.

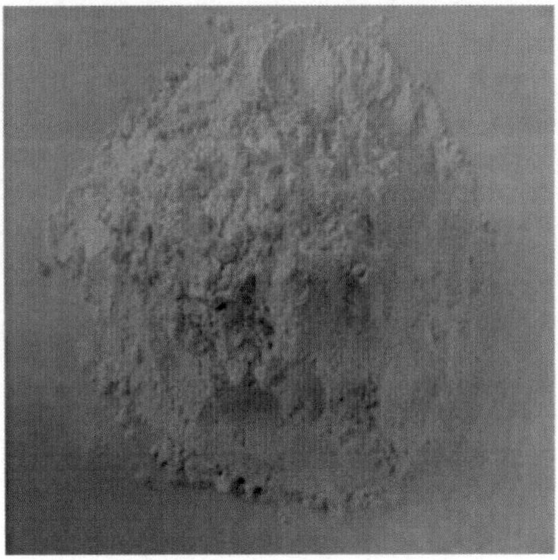

Fig. 1 Guar gum powder

4 Results and Discussions

4.1 Physical Properties

Figure 2 shows the variation of liquid limit and plasticity index of soil treated with guar gum. It is observed that with the increase in dosage of GG, both the liquid limit and plasticity index values increased. The liquid limit of soil increased by 2

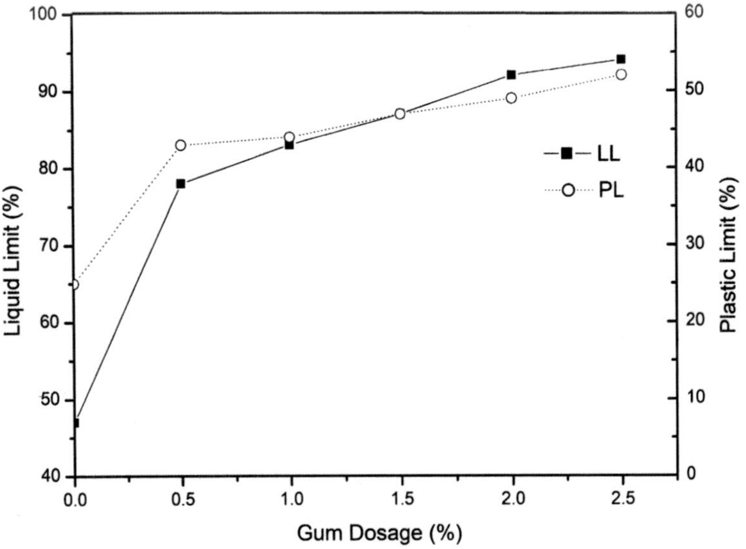

Fig. 2 Variation of liquid limit and plasticity index with guar gum

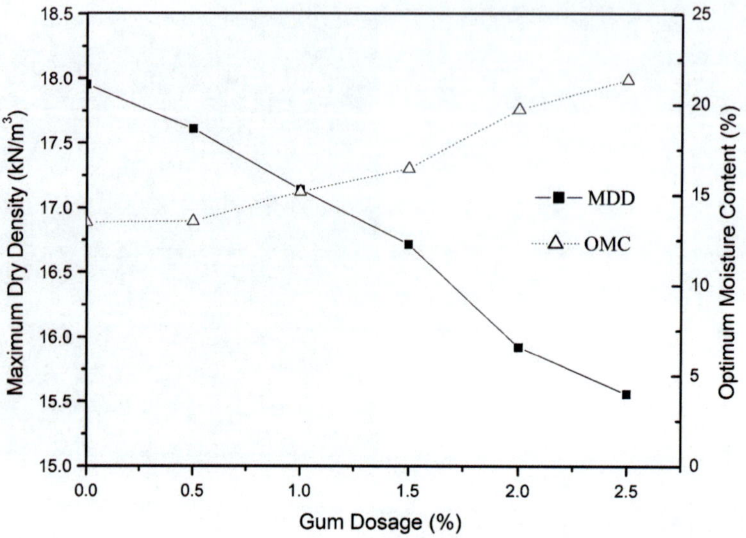

Fig. 3 Variation of MDD and OMC with guar gum

manifolds with 2.5% GG compared to untreated soil and it is attributed to the increase in viscosity of solution [18].

Figure 3 shows the variation of maximum dry density (MDD) and corresponding optimum moisture content (OMC) of GG-treated soil with varying dosages. From Fig. 3, it is observed that the MDD value reduced by 13.33% and the OMC value increased by 58.6 with 2.5% GG addition. This is attributed to the replacement of soil grains with guar gum strands leading to a reduction in mass of the treated soil. Further, the presence of a higher quantity of GG absorbs more water resulting in the reduction of soil–biopolymer mass. The maximum dry density value of the soil–biopolymer mix depends upon the type of the biopolymer, the viscosity of the solution, gradation of the soil and fines in the soil [7].

4.2 *Compressibility Characteristics*

The compression curves (e vs. log (p)) of GG-treated soil at varying dosages are shown in Fig. 4. The curves follow a similar trend irrespective of dosage of gum.

The coefficient of compression index (C_c) value of soil at different gum dosages is presented in Fig. 5. With 0.5% GG, the gum absorbs more water due to its hydrophilic nature, resulting in increased swelling and compression. With 1% dosage, the gum binds the soil particles facilitating crosslinking of the individual soil grains. It is attributed to the development of a hydrogen bond between the soil particles and the hydroxyl group (OH^-) of GG starts at this dosage and hence the compression values are reduced. However, beyond 1%, any further addition of gum (up to 2.5%) causes

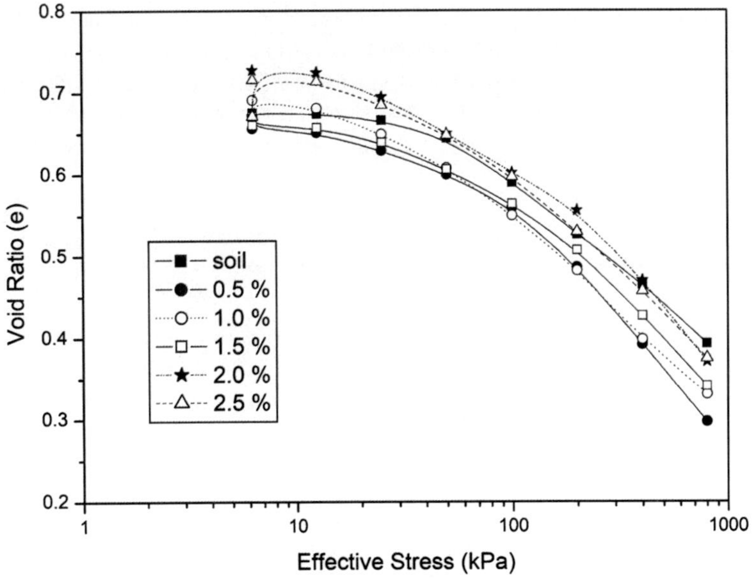

Fig. 4 Void ratio versus effective stress (log scale) of guar gum-treated soil

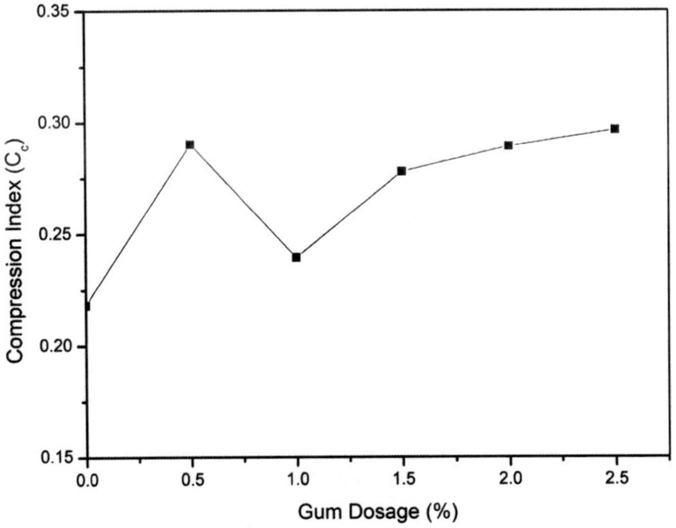

Fig. 5 Variation of compression index of guar gum-treated soil

an increase in compression due to the development of repulsive forces between negatively charged clay particles and hydroxyl group (OH^-) of guar gum [19]. Accordingly, the Cc values of biopolymer-treated soil (at all selected dosages) are observed to be more than the untreated case. Hence it is corroborated that the addition of guar gum delays the consolidation process resulting in increased compression values [8].

5 Conclusions

In the present study, the effect of guar gum on compressibility characteristics of expansive soil has been studied. One-dimensional consolidation tests were performed at various dosages of guar gum. The conclusions of the present investigation based on experimental results are listed as follows:

- The increase in liquid limit and plasticity index values is attributed to an increase in the solution viscosity, especially at higher gum dosages.
- The hydrophilic nature of gum causes absorption of more water, leading to an increase in OMC and a corresponding reduction in the MDD values of the soil–biopolymer mixtures.
- The compressibility values increased at higher gum dosage, which is attributed to the increase in repulsion between negatively charged clay particles and the hydroxyl group (OH^-) of guar gum.
- Cross-linking of soil particles and the development of hydrogen bonds are the key factors affecting compressibility characteristics of soil–biopolymer mix.

References

1. Andrew, R.M.: Global CO_2 emissions from cement production, 1928–2018. Earth Syst. Sci. Data **11**, 1675–1710 (2019)
2. ASTM D2487-17e1: Standard practice for classification of soils for engineering purposes (Unified soil classification system). In:ASTM International, West Conshohocken, PA, USA (2017)
3. ASTM D4186M-12e1: Standard test methods for one dimensional consolidation properties of saturated cohesive soils using controlled strain loading. In: ASTM International, West Conshohocken, PA (2012)
4. ASTM D4318-17e1: Standard test methods for liquid limit, plastic limit and plasticity index of soils. In: ASTM International, West Conshohocken, PA, USA (2017)
5. ASTM D698-12e2: Standard test methods for laboratory compaction characteristics of soil using standard effort (12 400 ft-lbf/ft3 (600 kN-m/m3)). In: ASTM International, West Conshohocken, PA (2012)
6. ASTM D854-14: Standard test methods for specific gravity of soil solids by water pycnometer. In: ASTM International, West Conshohocken, PA, USA (2014)
7. Ayeldeen, M.K., Negm, A.M., El Sawwaf, M.A.: Evaluating the physical characteristics of biopolymer/sand mixtures. Arab. J. Geosci. **9**(371), 1–13 (2016)

8. Cabalar, A.F., Awraheem, M.H., Khalaf, M.M.: Geotechnical properties of a low-plasticity clay with biopolymer. J. Mater. Civil Eng. **30**(8): 04018170-10 (2018)
9. Moghal, A.A.B., Vydehi, K.V.: State-of-the-art review on efficacy of xanthan gum and guar gum inclusion on the engineering behavior of soils. Innovative Infrastructure Solutions. 6(2), 108 (2021)
10. Kaniraj, S.R., Havanagi, V.G.: Behavior of cement-stabilized fiber-reinforced fly ash-soil mixtures. J. Geotech. Geoenviron. Eng. **127**(7), 574–584 (2001)
11. Latifi, N., Horpibulsuk, S., Meehan, C.L., Majid, M.Z.A., Tahir, M.M., Mohamad, E.T.: Improvement of problematic soils with biopolymer-an environmentally friendly stabilizer. J. Mater. Civil Eng. **29**(2): 04016204-11 (2017)
12. Moghal, A.A.B.: State-of-the-art review on the role of fly ashes in geotechnical and geoenvironmental applications. J. Mater. Civil Eng. **29**(8), 04017072-1–04017072-14 (2017)
13. Moghal, A.A.B., Sivapullaiah, P.V.: Role of lime leachability on the geotechnical behavior of fly ashes. Int. J. Geotech. Eng. **6**(1), 43–51 (2012)
14. Moghal, A.A.B., Obaid, A.A.K., Al-Refeai, T.O.: Effect of accelerated loading on the compressibility characteristics of lime-treated semiarid soils. J. Mater. Civil Eng., ASCE **26**, 1009–1016 (2014)
15. Moghal, A.A.B., Obaid, A.A.K., Al-Refeai, T.O., Al-Shamrani, M.A.: Compressibility and durability characteristics of lime treated expansive semiarid soils. J. Test. Eval. **43**(2), 255–263 (2015)
16. Moghal, A.A.B., Lateef, M.A., Mohammed, S.A.S., Lemboye, K., Chittoori, B.C.S., Almajed, A.: Efficacy of Enzymatically induced calcium carbonate precipitation in the retention of heavy metal ions. Sustainability **12**, 7019 (2020a)
17. Moghal, A.A.B., Lateef, M.A., Mohammed, S.A.S., Ahmad, M., Usman, A.R.A., Almajed, A.: Heavy metal immobilization studies and enhancement in geotechnical properties of cohesive soils by EICP technique. Appl. Sci. **10**, 7568 (2020b)
18. Nugent, R.A., Zhang, G., Gambrell, R.P.: Effect of exopolymers on the liquid limit of clays and its engineering implications. J. Transp. Res. Board (2101), 34–43 (2009)
19. Nugent, R.A., Zhang, G., Gambrell, R.P.: The effect of exopolymers on the compressibility of clays. Geofrontiers (GSP) 3935–3944. ASCE (2011)
20. Sivapullaiah, P.V., Baig, M.A.A.: Lime leachability and CBR of class F fly ashes. In: The 14th Asian Regional Conference on Soil Mechanics and Geotechnical Engineering, vol. 1, pp. 44–49, Hong Kong, China (2011)
21. Soldo, A., Miletić, M., Auad, M.L.: Biopolymers as a sustainable solution for the enhancement of soil mechanical properties. Nat.: Scie. Rep. **10**, Article number: 267 (2020)

Influence of Microstructural and Geotechnical Behavior of Expansive Soil Using Ultra-Fine Slag and Calcium Chloride

R. Suresh and V. Murugaiyan

1 Introduction

The expansive soil problems were first investigated by the U.S. Bureau of Reclamation (USBR) in 1938 [1]. Expansive soils occur above the water table; they swell with an increase in moisture content and shrink with a decrease in moisture content [2, 3]. These types of soils are known worldwide for their volume change behavior due to moisture fluctuation because of their intrinsic mineralogical behavior . Expansive soils are in arid and semi-arid regions in the world such as Australia, Canada, China, India, South Africa, and the United States. India has an extensive track of expansive soils known as black cotton soil covers about 20% of the total land area [4, 5]. The swelling phenomenon is attributed predominantly to the presence of montmorillonite clay minerals in the soil. The montmorillonite is hydrated aluminum silicates with three-layered lattice structures in which intraparticle space when occupied by layers of water molecules gives rise to swelling and related phenomena. The change in volume in montmorillonite clay can exert sufficient stress on a building, sidewalk, driveways, basement floors, pipelines, and foundations to cause damages. While it is true that swelling soils are probably responsible for most of the cracking and movement of lightly loaded structures, other aspects of foundation movement cannot and should not be ignored [6]. Since the expansive soils are found worldwide, the challenges to the Civil Engineers in one felt around the globe. If not adequately treated, expansive soils may act as natural hazards resulting in damages to structures [7–10].

R. Suresh (✉) · V. Murugaiyan
Department of Civil Engineering, Pondicherry Engineering College, Puducherry, India
e-mail: rangasuresh307@pec.edu

C. N. V. Satyanarayana Reddy et al. (eds.), *Ground Improvement and Reinforced Soil Structures*, Lecture Notes in Civil Engineering 152,
https://doi.org/10.1007/978-981-16-1831-4_32

2 Materials and Methods

2.1 Materials

Expansive Soil

The expansive clay soil is collected from Ariyankuppam, which is located in Puducherry, India. The soil is collected in a dry condition at a depth of 1 m below the ground level and preserved in the laboratory. Identification of the index and engineering properties of expansive soils are shown in Table 1.

Ultra-fine Slag

Alccofine is an ultra-fine slag; it performs superior to all other mineral admixtures used in India. It is a micro-fine material of particle size much finer than other hydraulic materials like cement, lime, fly ash etc. manufactured by Ambuja Cement Private Limited in India. Alccofine-1203 and Alccofine-1101 are two types with low calcium silicate and high calcium silicate respectively. Alccofine-1203 has a lime content of 34% and an average particle size of 4 μ (range 0–17 μ). Alccofine has almost the same binding characteristics as silica fume. It controls high reactivity because of controlled granulation, and it also improves workability by reducing the water demand. Chemical composition and physical properties are tested by alccofine micro materials, pissurlem, Goa. Alccofine-1203 properties are given in Table 2.

Calcium Chloride

Commercially grade $CaCl_2$ has the composition of 50% $CaCl_2$ and 50% $MgCl_2$ (Figs. 1, 2 and 3).

Table 1 Physical properties of soil

Properties of soil	Results
Specific gravity	2.65
Liquid limit (W_L)	74.48%
Plastic limit (W_P)	26.45%
Shrinkage limit (W_S)	18.5%
Free swell index (FSI)	90%
Water absorption (W_A)	68.92%
Cation exchange capacity (CEC) meq/100 g	50.96
Unified soil classification (USCS)	CH
OMC (%)	20.19
MDD (kN/m^3)	16.55
UCS (kPa)	104
Swell potential (%)	36.86
Swell pressure (kPa)	256

Table 2 Physical and chemical properties of Alccofine-1203

Properties	Results
Physical properties	
Particle size distribution (mm)	
D10	1.5
D50	4.3
D90	9.0
Specific gravity (g/cc)	2.88
Bulk density (kg/m^3)	6
Chemical properties	
SiO_2	35.6%
Al_2O_3	21.4%
Fe_2O_3	1.3%
CaO	33.6%
SO_3	0.12%
MgO	7.98%

3 Results and Discussion

See Fig. 4.

3.1 Index and Compaction Characteristics

The influence of alccofine and $CaCl_2$ on Atterberg limits (liquid limit, plastic limit, and shrinkage limit) of expansive soil is shown in Fig. 1. Results show that liquid limit decreases and plastic limit increases; hence the difference between liquid limit and plastic limit is the plasticity index. Plasticity index is reduced from 48.03 to 16.45% when the soil is blended with 8% alccofine + $CaCl_2$ 1%.

The compaction characteristics of untreated and treated soils are shown in Fig. 1. The results of compaction show that the maximum dry density increases from 16.55 to 17.76 kN/m^3 and optimum moisture content reduces from 20.19 to 16.10% with an increase in 8% alccofine and 1% $CaCl_2$ binder; that is, for sample which shows maximum strength.

3.2 Unconfined Compression Strength

Unconfined compressive strength (UCS) tests were conducted with alccofine-1203 and $CaCl_2$ was added independently and blended into the expansive soil samples.

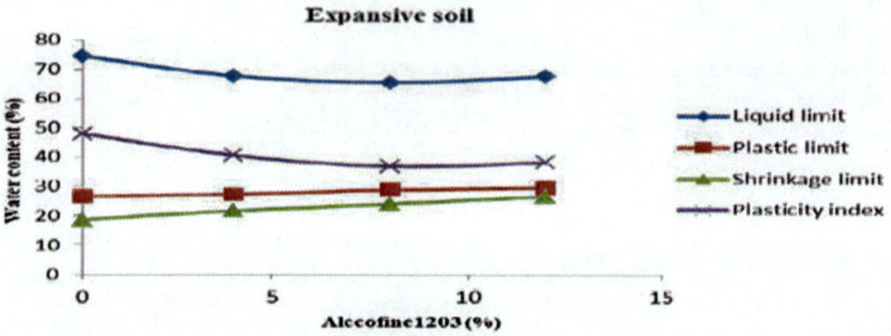

(a) Soil + Alccofine

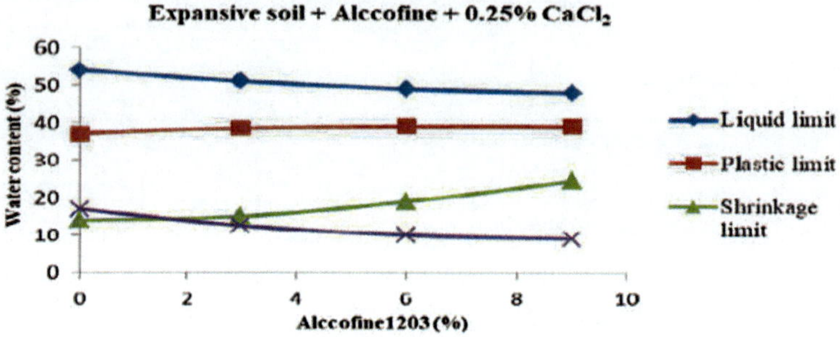

(b) Soil + Alccofine + 0.5% $CaCl_2$

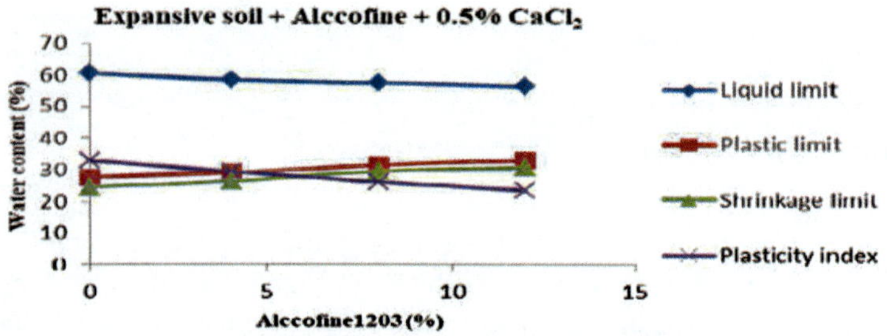

(c) Soil + Alccofine + 1% $CaCl_2$

Fig. 1 Atterberg limits: **a** Soil + Alccofine; **b** Soil + Alccofine + 0.5% $CaCl_2$; **c** Soil + Alccofine + 1% $CaCl_2$

UCS tests were performed on both intrinsic soil and chemically treated soil. The UCS value for intrinsic soil is 104 kPa. The percentage of alccofine (4, 8, and 12%) and $CaCl_2$ (0.5, 1.0, 1.5, and 2.0%) was added by dry weight of the soil. The UCS values are shown in Fig. 1. Optimum increase was noticed at 8% alccofine and 1% $CaCl_2$. The UCS strength was increased from 104 to 1270 kPa. Beyond 8% of alccofine with 1% $CaCl_2$ resulted in a slight decrease in UCS values.

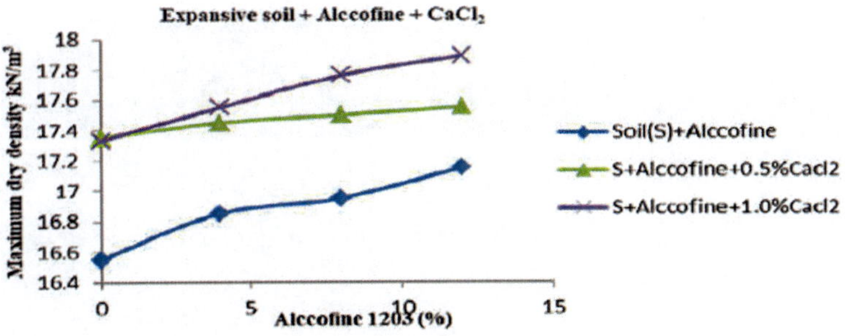

(a) Maximum dry density

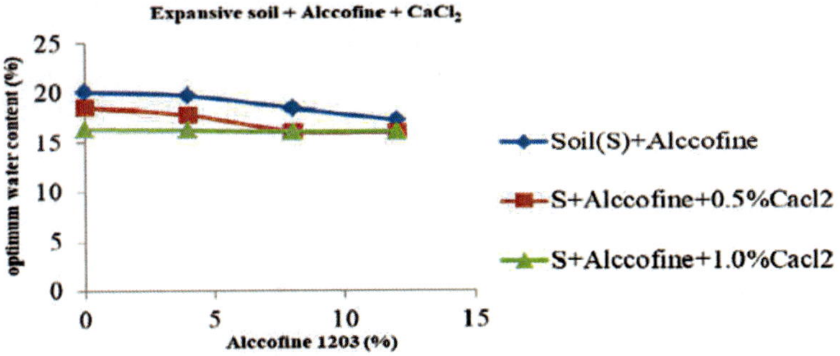

(b) Optimum moisture content

Fig. 2 **a** Maximum dry density; **b** Optimum moisture content

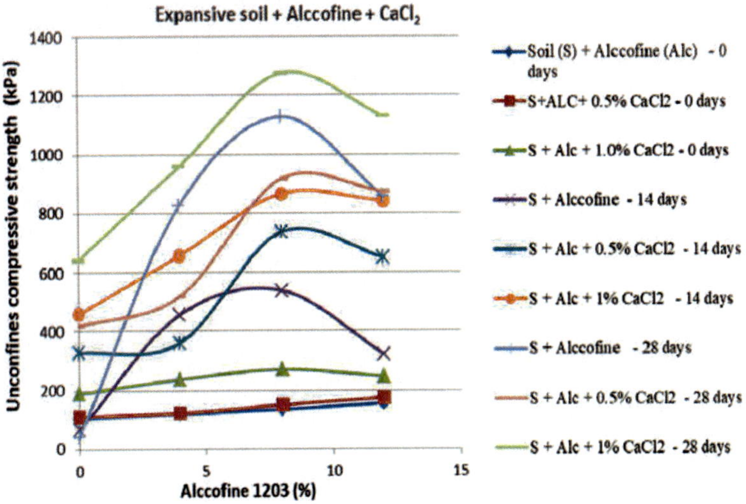

Fig. 3 Variation of UCS with Soil + Alccofine + $CaCl_2$

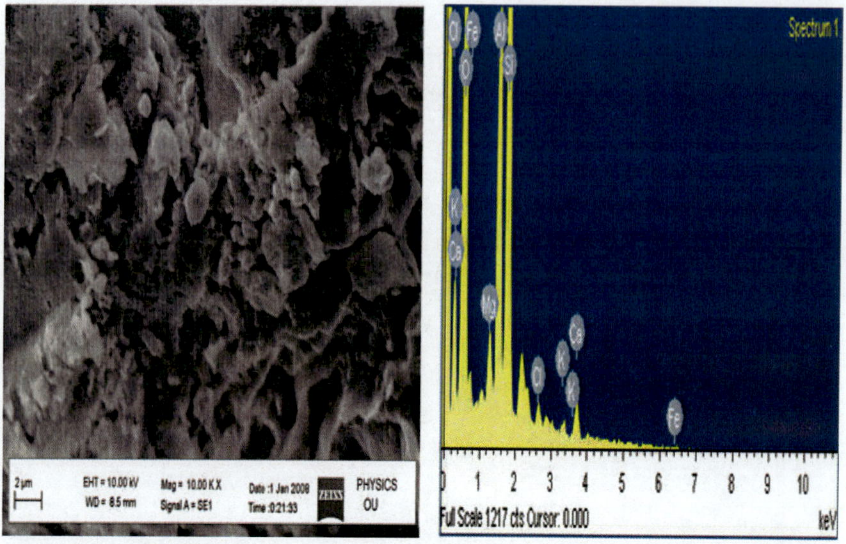

(a) SEM with EDAX analysis for untreated soil 0 days

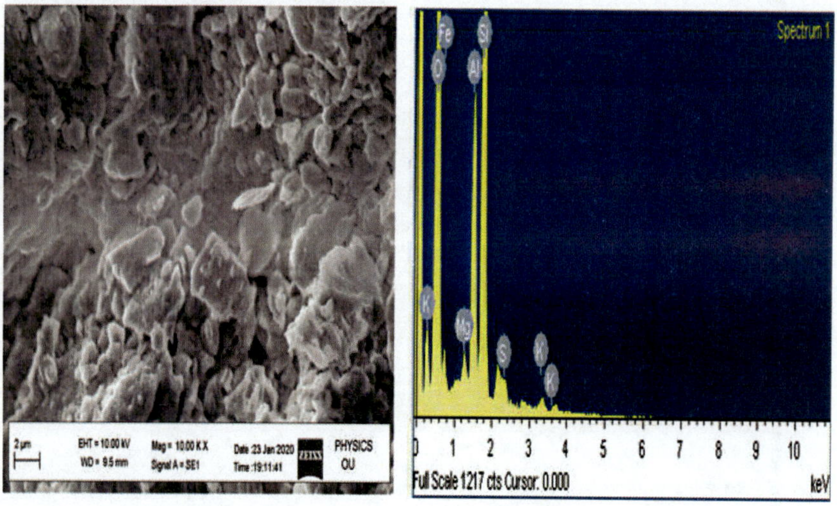

(b) SEM with EDAX analysis for treated soil (soil + 8%alccofine+1%$CaCl_2$) for 28 days of curing.

Fig. 4 **a** SEM with EDAX analysis for soil; **b** SEM with EDAX analysis for soil + 8% Alccofine-1203 + 1.0% $CaCl_2$ 28 days

3.3 SEM Analysis

SEM and EDAX spectrum analysis for clay soil, alccofine-1203, and clay soil + alccofine-1203 8% + $CaCl_2$ 1% are shown in Fig. 4a and b. These studies are conducted in order to observe the individually and changes in the soil is blended with admixture of 0 days. Eminent peaks Fe, Si, Al are observed in Fig. 4a and Fe, Si, and Al are observed in clay soil. Figure 4b is observed eminent peaks are Fe, Au, Si, O, Mg, Ca, and Al. The test was performed mainly for the identification of the various cementations compounds on the soil stabilized with 8% alccofine + $CaCl_2$ 1% binder; that is, for sample which shows maximum strength.

3.4 XRD Analysis

X-ray diffraction peaks identify for clay soil, alccofine-1203, and clay soil + alccofine-1203 8% + $CaCl_2$ 1%. The most important peaks traced were related to CH, which were identified at $2\Theta = 26°$ to 36° as can be seen from Fig. 5a and b; the addition of alccofine-1203 and $CaCl_2$ in the soil causes CH-related peaks to appear at the aforementioned 2Θ. It has carried out to confirm the formation of new minerals which can play a significant role of strength improvement behavior, calcium stabilized for soil admixture.

4 Conclusions

1. The addition of alccofine-1203 and $CaCl_2$ to the soil decreased the liquid limit and plasticity index while increasing the shrinkage limit. It is found that the addition of binders causes flocculation of clay. Particles and increases the number of coarse particles which help in reducing the Atterberg limits.
2. The optimum moisture content (OMC) was found to decrease from 20.19 to 16.10% while the maximum dry density (MDD) increases from 16.55 to 17.76 kN/m^3 with binding content.
3. Unconfined compressive strength (UCS) tests were conducted with alccofine-1203 and $CaCl_2$ was added independently and blended into the expansive soil samples. UCS tests were performed on both intrinsic soil and chemically treated soil. The UCS value for intrinsic soil is 104 kPa. The percentage of alccofine (4, 8, and 12%) and $CaCl_2$ (0, 0.5, and 1.0%) was added by dry weight of the soil. An optimum increase was noticed at 8% alccofine and 1% $CaCl_2$. The UCS strength was increased from 104 to 1270 kPa. Beyond 8% of alccofine with 1% $CaCl_2$ resulted in a slight decreased in UCS values.
4. SEM, XRD studies confirm the formation of reaction products such as Ca, Mg, and Si contributes to strength significantly. In XRD, the addition of

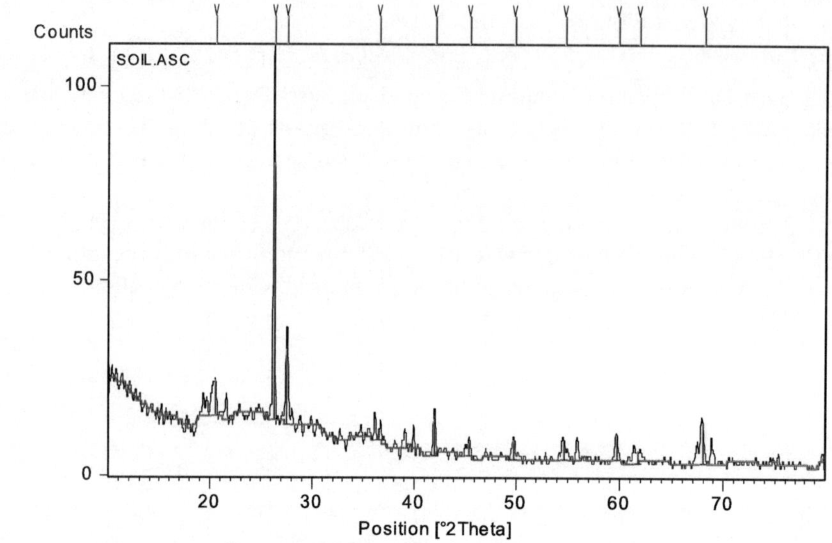

(a) Expansive soil

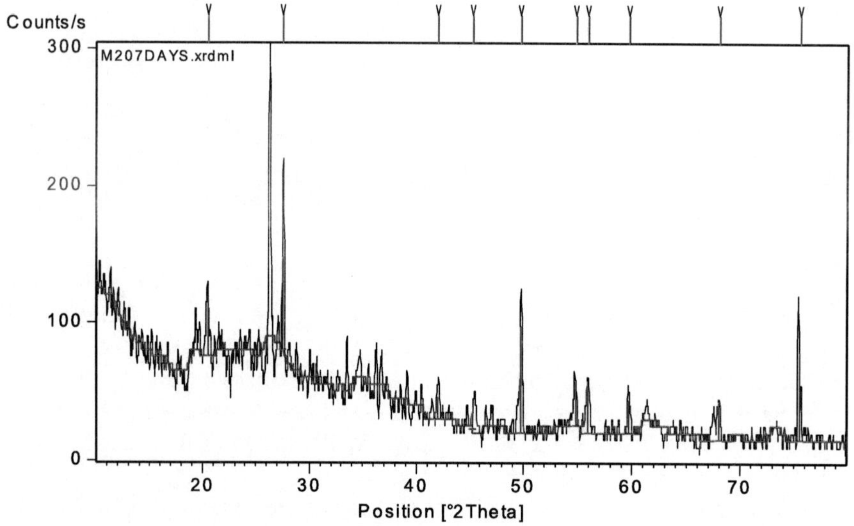

(b) Soil+8%Alccofine+ 1.0%$CaCl_2$

Fig. 5 XRD analysis for **a** soil; **b** soil + 8% Alccofine + 1.0% $CaCl_2$

alccofine-1203 and $CaCl_2$ in the soil causes CH-related peaks to appear at the aforementioned 2Θ.

Acknowledgments This research is financially supported by the Council of Scientific and Industrial Research (CSIR), New Delhi, India (Acknowledgement number 141268/2K18/1; File number 08/518(0001)/2019 EMR-1).

References

1. Chen, F.H.: Foundations on Expansive Soils. Elsevier, Amsterdam, The Netharland (1975)
2. Sivapullaiah, P.V., Subba Rao, K.S., Gurumurthy, J.V.: Stabilization of rice husk ash for use as cushion below foundation on expansive soils. Ground Improv. **8**(4), 137–149 (2004)
3. Ramanamurty, V., Hari Krishna, P.: Stabilisation of expansive clay bed using calcium chloride solution. Ground Improv. **10**(1), 39–46 (2006)
4. Ramanamurty, V., Praveen, G.V.: Use of chemically stabilized soil as cushion material below light weight structures founded on expansive soils. J. Mater. Civil Eng. (ASCE) **20**(5), 392–400 (2008)
5. Lal, R.: Encyclopedia of Soil Science, vol. 1. CRC Press, Boca Raton (2007)
6. Al Rawas, A.A.: Microfabric and mineralogical studies on the stabilization of an expansive soil using cement by-pass dust and some type of slags. Canadian Geotech. J. **39**(5), 1150–1167 (2002)
7. Al Rawas, A.A., Taha, R., Nelson, J.D., Al-shab, T.B., Al-siyabi, H.: A comparative evaluation of various addition used in the stabilization of expansive soils. ASTM Geotech. Test. J. **25**(2), 199–209 (2002)
8. Phanikumar, B.R.: A study of swelling characteristics and granular pile-anchor foundation system in expansive soils. Ph.D. thesis, Jawaharlal Nehru Technological University, Hyderabad, India (1997)
9. Behnood, A.: Soil and clay stabilization with calcium- and non-calcium-based additives: a state-of-the-art review of challenges, approaches and techniques. Transp. Geotech. **7**, 14–32 (2018). https://doi.org/10.1016/j.trgeo.2018.08.002
10. Rao, A.S.: A study of swelling characteristics and behavior of expansive soils. Ph.D. thesis, Kakatiya University, Warangal, India (1984)

Development of New Grout Mix Using Kota Stone Slurry for Sustainable Development

Neha Shrivastava and Hemlata Kumawat

1 Introduction

Cement is commonly used as a binding material in grouting. On the other side, the stone industries generate a large volume of solid waste, which has financial and environmental concerns, so, the sustainable development demands the utilization of solid waste in various engineering applications. This study focuses on the reuse of stone waste in cement grout mix as partial replacement.

The stone waste used as partial replacement of cement in this study is Kota stone slurry, which is largely available in Kota, Rajasthan. The silica and limestone are the major constituents of Kota stone. Kota stone is quarried at Kota district, Rajasthan, India in the fine-grained form of limestone. There are many mines are located in Ramganjmandi town and in the Kota district. Limestone comes in category of carbonate sedimentary rocks, which is generally made of the skeletal fragments of marine organisms like molluscs, foraminifera and coral. The crystal forms of calcium carbonate ($CaCO_3$) like aragonite and calcite are the major minerals of the Kota stone waste.

Limestone is approximately 10% of the sedimentary rocks. The reason behind the erosion of limestone is the karst landscape in which the limestone becomes soluble in water and weal acids. Most of the cave systems all around the world are through limestone rock beds. Limestone has various uses: as an important constituent of concrete (Portland cement), in base/sub-base of roads as an aggregate, in building materials, as a chemical feedstock for the production of lime, as filler material in

N. Shrivastava (✉)
Department of Civil Engineering, Malaviya National Institute of Technology, Jaipur 302017, India
e-mail: neha.ce@mnit.ac.in

H. Kumawat
Department of Civil Engineering, Swami Keshvanand Institute of Technology, Management and Gramothan, Jaipur, 302017, India

C. N. V. Satyanarayana Reddy et al. (eds.), *Ground Improvement and Reinforced Soil Structures*, Lecture Notes in Civil Engineering 152,
https://doi.org/10.1007/978-981-16-1831-4_33

products such as toothpaste and paints, as a popular decorative addition to rock gardens or as a soil conditioner.

The Kota stone powder waste disposed of the industries causes an adverse impact on the surrounding environment. The desired smooth shape of Kota stone blocks is achieved by cutting them into smaller blocks. In the cutting process of Kota stone blocks into finer blocks, about 25% of the original stone mass is lost. The major objective of this study is to evaluate the durability and strength parameters of the grout mix containing Kota stone slurry. The experimental program includes preparing grout mixes with Kota stone slurry partially replaced with cement.

2 Literature Review

The main aim of Hanley and Pavía [4] is to investigate the relationship between strength and workability of natural hydraulic lime mortar. This investigation is useful to find out the optimum use of lime in buildings. In this study, several mortar mixes with different water contents were prepared. The workability, flexural strength and workability were tested for all the mortar mix.

Patil et al. [10] investigated the partial replacement of cement by fly ash in concrete mix and also as admixture. In this research, the case study for the Deep Nagar thermal power plant of Jalgaon District in MS is performed. The concrete mixtures were prepared for fly ash addition from 5 to 25% and tested to find compressive strength. The test results indicated that the substitution of cement by fly ash leads to loss in compressive strength and also slows down its hardening. This investigation provides an environmentally favorable fly ash disposal method for Deep Nagar. In this study, M20 grade concrete mix was utilized as per IS 456-2000. In this study, the concrete includes the proportion of cement: fine aggregate: coarse aggregate as 1:1.5:3 and water–cement ratio of 0.5.

Beemamol et al. [1] examined the durability and strength of masonry mortar in which fine aggregates were replaced 50 and 100% by ceramic waste (locally called tailing sand). The industrial waste (fly ash) was also used as replacing material up to 25%. The workability of mortar was tested for 1:3 binders to sand ratio and 1:4 water to binder ratio. The experiment program includes Flexure Strength Test, Rapid Chloride Permeability Test, Strength Test. On the basis of test results, it was concluded that that the durability of mix increases due to the replacement of fine aggregates by ceramic waste but the effect on strength was insignificant.

Braga et al. [2] studied the replacement of cement content in mortars by crushed concrete aggregates. The number of standard tests was performed in this study to determine the binding properties of concrete fines in mortar. The experiment program includes the following tests—Compressive Strength Test, Bulk Density, Adhesive Strength Test, Consistency Test, Flexure Strength Test, Water Retentivity Test, Permeability Test, Dimensional Instability Test. The mortar mix designs were prepared with water–cement ratio 1:3, 1:4 and 1:5. The test results were also compared with reference mortars with fine wastes.

Melichar et al. [8] explored the basic mechanical and physical parameters of cement–polymer blended materials. The influence of admixtures on the aggregates was also observed. Further, it was observed that micro silica represents 5–7.5% of cement weight in terms of thermal resistance. A polymer–cement paste of the proportion 3:7:1 is prepared by replacement of basic material with polymer. The tests performed to evaluate strength are Flexure Strength Test, Compressive Strength Test and Bulk Density test.

Rai et al. [11] evaluated the transverse strength and compressive strength of a 1:3 ratio mortar mix in which cement was replaced by low calcium fly ash as 15, 20, 25, 33%, and natural sand was replaced by quarry dust as 20, 50 and 100%. The test results indicated that due to pozzolanic activity and efficient micro filling ability, the combined fly ash and quarry dust provided excellent performance in mortar mix.

Thaker [15] developed a correlation between a very low workable mix and a high workable mix. The tests performed in this study were marsh cone test and mini flow table test. Marsh cone test is applicable for cement paste having W/C ratio beyond 0.50 whereas the same flow diameter was achieved in mini flow table test above 0.50 W/C ratio. By modifying cement paste cone volume, it is possible to evaluate the flow behavior of cement paste up to W/C ratio 0.62.

Singh and Bansal [13] investigated the replacement of cement or sand in concrete mix by a certain percentage of silica fumes. The test results indicate that silica fume improves the mechanical properties of concrete as it contains pozzolanic material silica in non-crystalline form. Three concrete mixes were designed as proportions 1:3, 1:4 and 1:6 and tested for compressive strength. The experimental results indicated that the addition of silica fume in concrete mix improves the compressive strength.

Satyapriya [12] evaluated the optimum percentage of partial replacement of cement in mortar by rice husk ash. The following tests were conducted in this study—initial setting time, normal consistency, bulk density test, final setting time, compressive strength test, water absorption test, density test. The test results concluded that the optimum percentage of rice husk ash as partial replacement of cement in mortar mix is 0–20%.

Panda and Parhi [9] investigated the mix design process of high-strength concrete with 60 N/mm^2 specified characteristic compressive strength as per IS-10262:1982. In this study, the coarse aggregates in concrete mix were partially replaced by marble chips. Super-plasticizers were also used in high-strength concrete design. This study concluded that high strength in concrete mix can be achieved by utilizing good quality ingredients, by proper mixing and using low water—cement ratio and by proper casting and testing and by following calculations of mix design. This study also concluded that by addition of super-plasticizers in ordinary concrete improves its properties like durability, strength etc.

Sun et al. [14] investigated the properties of high-performance cement mortar (HPCM) and its optimum formulation. The grouting of cement mortar was also studied in this research. The results indicate that expansion admixture, polycarboxylate super plasticizer and accelerators have different impacts on the strength, workability and the shrinkage in HPCM, and the workability of HPCM is improved by adding these three admixtures. This study also concluded that the grouting rate of more than 90% of the semi flexible material provided better pavement performance.

3 Waste Material

There are several Kota stone mines are located in the district Kota of Rajasthan. The places where the Kota stone are quarried in Rajasthan are Kota, Ramgajmandi, Chechat, Jhalawar, Modak, Kherabad etc. Two types of wastes are generated: first is quarry/sawing/cutting from in situ stone site and the second is polishing waste from construction sites. The raw stone blocks are cut either into slabs or tiles of various thicknesses (generally 2 or 4 cm), by using diamond blades. During this processing of stone, the blades of the cutter needed a shower of water for cooling and to absorb dust. The waste water that comes through the stone processing is very alkaline in nature and affects the polishing of slabs when reusing in processing. The volume of this wastewater is very large so it is stored in large pits where suspended particles settle down and the slurry is collected in trucks for disposal. This wastewater stone slurry contains a high amount of stone powder in it. So, this waste Kota stone slurry is used in this present study in grout mix. The types of stone waste and generation process are presented in Figs. 1 and 2.

Chemically, Kota stone is siliceous calcium carbonate rock. The chemical composition is given in Table 1.

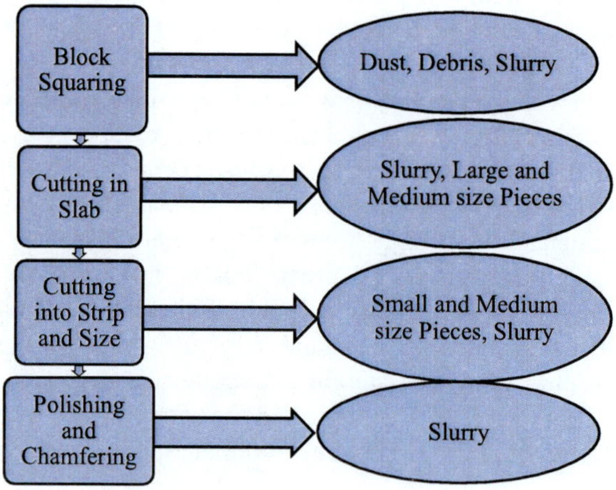

Fig. 1 Generation of Kota stone waste

Fig. 2 Forms of Kota stone waste

Table 1 Chemical formation of Kota stone

Chemical	Percentage (%)
Silica (SiO_2)	20–25
Lime (CaO)	38–42
Alumina (Al_2O_3)	2–4
Magnesium oxide	4.13
Potassium oxide	0.40
Ferrous oxide	0.85
Titanium dioxide	0.05
Sodium oxide	1.21
Loss on ignition	30–32

4 Experimental Investigation and Results

4.1 Tests Performed on Kota Stone Waste

Sieve analysis (IS 2720-part IV): Kota stone waste in powder slurry form is collected from a dumping field in Kota. Sieve analysis test is conducted on its sample. The test results are presented as particle size distribution curve as given in Fig. 3. The waste can be classified as poorly graded sand (SP) based on the shape of the curve.

Specific gravity by pycnometer (IS 2720-part III): The average specific gravity of the three samples of Kota stone waste is determined as **2.15**.

Liquid limit by cone penetrometer (IS 2720-part V): The cone penetration test results on the Kota stone waste samples are presented in Fig. 4. Liquid limit (L.L.) of the waste is determined as **43.4%**.

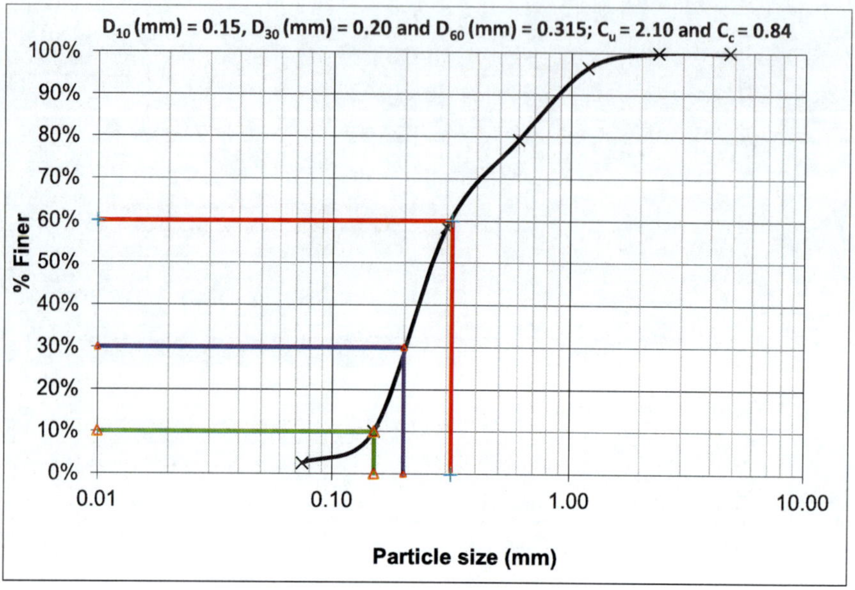

Fig. 3 Particle size distribution curve of Kota stone waste

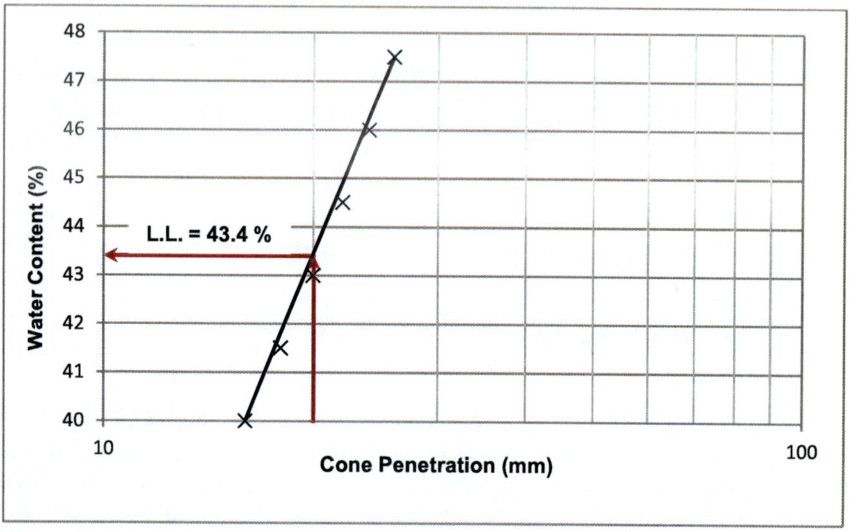

Fig. 4 Plot of cone penetration test

4.2 Tests Performed on the Mortar

Sample preparation: Five different mix samples were tested by replacing the cement content with Kota stone slurry in various percentages. Water cement ratios in various mixes are according to proportion (1:6). Table 2 presents the details of the sample preparation.

Compressive strength test results: Compressive strength test on cubes of size 50 mm was conducted at the age of 7 and 28 days as per Indian standards [3]. The desired strength for cubes was taken as 9 MPa. The test results of all the mixes are shown in Fig. 5. 28 days compressive strength of all the mixes is lying above 7.5 MPa or above so the mixes can be grade as MM 7.5 (clause 6.1, IS 2250 [3]). As per Table 2, water–cement ratio increases with Mix. No. 1–5, accordingly compressive strength should fall from mix 1 to mix 5. On the other hand, the replacement of cement with

Table 2 Different proportions of materials by weight

Mix number	Percentage replacement	Cement (kg)	Kota stone slurry (kg)	Sand (kg)	Water (kg)	Water cement ratio
1	0	1.00	0.00	8.7	0.8	0.8
2	5	0.95	0.05	8.7	0.8	0.83
3	10	0.90	0.10	8.7	0.8	0.89
4	15	0.85	0.15	8.7	0.8	0.94
5	20	0.80	0.20	8.7	0.8	1

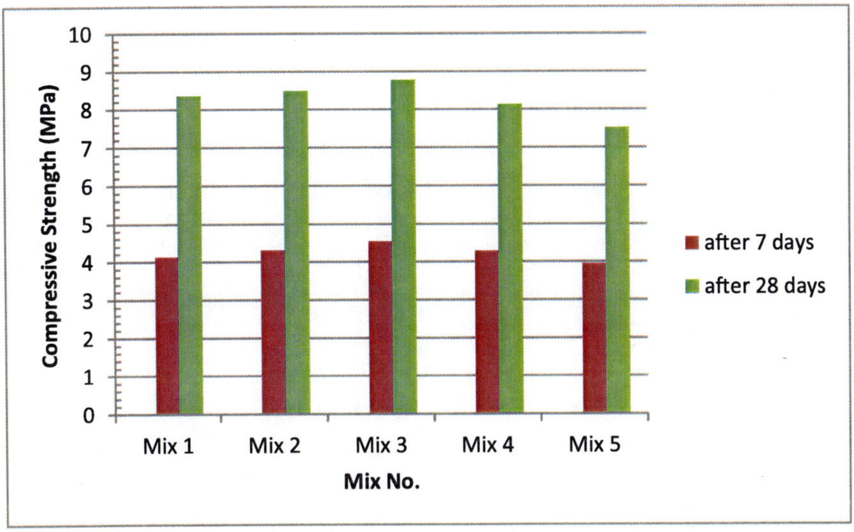

Fig. 5 Compressive strength comparison after 7 and 28 days

Kota stone slurry may impart some additional strength as that compressive strength does not fall. The optimum value has been obtained for mix 3, which gives the value of 8.78 MPa at the age of 28 days.

5 Conclusions

(1) Test results show that the compressive strength of the mortar was increased up to 10% replacement of cement by Kota stone slurry, after that it was decreased. So it is adequate to replace cement up to 10% without any loss in strength. Thus, Kota stone slurry can be a perfect alternative for partial replacement of cement.

(2) Kota stone slurry as a substitute for cement will not only increase the strength but also it will decrease the waste, which is produced every year in abundance. The use of Kota stone slurry will reduce the quantity of CO_2 produced in the manufacturing process of cement, which is very harmful to nature. This emission of carbon dioxide can be reduced by sustainable use of the waste as a binding material in the mortar.

(3) The present study focused on the investigation of workability and preliminary compressive strength of the mixtures prepared with Kota stone slurry replacements. As a future recommendation, the durability can also be checked by performing microscopic studies.

References

1. Beemamol, U.S., Nizad, A., Nazeer, M.: Investigations on cement mortar using ceramic tailing sand as fine aggregate. Am. J. Eng. Res. (AJER) **3**, 28–33 (2013)
2. Braga, M., de Brito, J., Veiga, R.: Reduction of the cement content in mortars made with fine concrete aggregates. Mater. Struct. **47**(1–2), 171–182 (2014)
3. Bureau of Indian Standards (BIS): Indian standard code of practice for preparation and use of masonry mortars. IS 2250 (1981)
4. Hanley, R., Pavía, S.: A study of the workability of natural hydraulic lime mortars and its influence on strength. Mater. Struct. **41**(2), 373–381 (2008)
5. IS: 2720-Part 3: Methods of test for soil–determination of specific gravity (1980)
6. IS: 2720-Part 4: Methods of test for soils–grain size analysis (1985)
7. IS: 2720-Part 5: Determination of liquid limit, plastic limit and shrinkage limit (1985)
8. Melichar, T., Bydžovský, J., & Černý, V.: Research and development of lightweight repair mortars with focus on their resistance to high temperatures. Int. J. Civ. Environ. Eng. **8**(7), 799–802 (2014)
9. Panda, D., Parhi, P.K.: Analysis on mix design of High Strength Concrete (HSC) using IS: 10262-1982 and partial replacement of waste marble aggregates as coarse aggregate. Int. J. Sci. Res. Public. **7**(6), 2250–3153 (2017)
10. Patil, S. L., Kale, J. N., & Suman, S.: Fly ash concrete: a technical analysis for compressive strength. Int. J. Adv. Eng. Res. Studies (IJAERS) **2**(1), 128–129, 2249–8974 (2012)

11. Rai, B., Kumar, S., Satish, K.: Effect of fly ash on mortar mixes with quarry dust as fine aggregate. In: Advances in Materials Science and Engineering (2014)
12. Satyapriya: Study of performance of cement mortar by partial replacement of cement using rice husk ash. Int. Res. J. Eng. Technol. (IRJET) **4**(6), 2395-0072 (2017)
13. Singh, H., Bansal, S.: Effect of silica fume on the strength of cement mortar. Int. J. Res. Eng. Technol. **4**(2) (2015)
14. Sun, Y., Cheng, Y., Ding, M., Yuan, X., Wang, J.: Research on properties of high-performance cement mortar for semiflexible pavement. In: Advances in Materials Science and Engineering (2018)
15. Thaker, P., & Arora, N. K.: Selection of test method to quantify workability of cement paste and mortar for very low workable to high workable mix. Int. J. Eng. Sci. Res. Technol. (IJESRT) **4**(12), 2277–9655 (2015)

Experimental Study on Waste Tire Chips-Reinforced Sand Using Cyclic Plate Load Test

Sanket Rawat, Ravi Kant Mittal, and Ashirbad Satapathy

1 Introduction

Waste tire management has put forward a major challenge to environmentalists all around the world. Waste tires are disposed and converted to the form of tire-derived fuel to be consumed further in cement kilns, power, and paper plants. However, the rise in air pollution due to this practice has become a serious environmental concern [1]. A major percentage of waste tires (40%) produced in the United States was used in the generation of tire-derived fuels, while other applications included manufacturing of ground rubber products (26%), landfills (13%), roads (6%), retread (7%) etc. However, as per the U.S. Scrap Tire Management Summary—2017 [2], the use of waste tires in civil engineering applications has witnessed a significant rise as it increased by almost 97% from 2013 to 2017 (327.78 thousand tons in 2013 to 646.78 thousand tons in 2017), while its use as tire-derived fuel reduced by 13% from 2013 to 2017 [2]. The use of tire wastes in geotechnical engineering is very promising as it can help in the sustainable consumption of large quantities of wastes [3]. The utilization of various forms of waste tires has also been found to be very effective in the case of shallow foundations due to the observed substantial improvement in bearing capacity (up to 11 times) at different settlement ratios under static conditions [4, 5]. However, the existing laboratory studies (using cyclic triaxial tests, resonant column tests) on granulated rubber or powdered rubber reinforced sand have suggested a negative effect on the stiffness of the reinforced soil mass due to the rubber inclusion under dynamic loading. The main limitation of these tests was the size of the representative reinforced soil sample (approximately 70 mm diameter) and, therefore, the mean size of the tire reinforcement was restricted to <4.75 mm

S. Rawat · R. K. Mittal · A. Satapathy (✉)
Department of Civil Engineering, BITS-Pilani, Pilani Campus, Pilani 333031, Rajasthan, India
e-mail: satapathy.shrbd@gmail.com

C. N. V. Satyanarayana Reddy et al. (eds.), *Ground Improvement and Reinforced Soil Structures*, Lecture Notes in Civil Engineering 152,
https://doi.org/10.1007/978-981-16-1831-4_34

in most of the cases to avoid the boundary effects [6–8]. This reduction in stiffness may be overcome if larger interaction between soil and reinforcement can be established. Therefore, the use of larger size of tire reinforcement (tire chips/tire shreds) needs further investigation under cyclic/dynamic loading. Also, the application of waste tire chips as reinforcement in the case of machine foundations needs to be established where dynamic load applies repetitively over a very large period, though its magnitude is small. In designing a machine foundation, the coefficient of elastic uniform compression of soil (C_u) is the most important parameter. Nevertheless, this has not been comprehensively investigated.

Moghaddas Tafreshi et al. [9] studied the cyclic response of tire chip (mean size 14 mm) reinforced sand spread in a different number of layers below the foundation and evaluated the optimum thickness and the number of reinforced layers [9]. It was observed that C_u decreases with the inclusion of rubber sand mixture, while the damping ratio was found to increase. However, in their study, the rubber -sand mixture was spread in layers of varying thickness starting from 0.2D (D = diameter of footing) distance below the ground level. The present study, on the other hand, evaluates the effect of sand tire chip mixture if spread continuously below the square foundation under cyclic loading conditions as the increase in bearing capacity and settlement reduction are found substantial under both static concentric and eccentric inclined loading conditions for this type of arrangement [4, 5, 10].

2 Test Materials

2.1 Tire Chips

Waste tire chips of size 12–14 mm and specific gravity = 1.08 are used in the present study. The tire piece mixture was free from any metal or steel. Figure 1 shows the picture of the tire chips used.

Fig. 1 Tire chips used in the experimental study

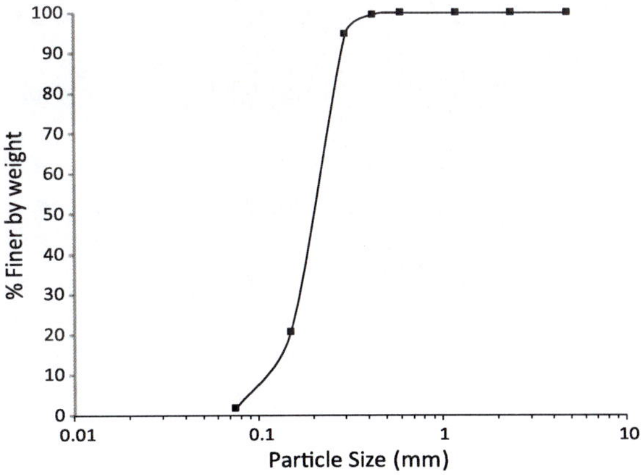

Fig. 2 Grain size distribution curve of local sand

2.2 Soil

Sandy soil used in the experimental program was procured from a local site. The soil is classified as poorly graded sand as per the Unified soil classification system. The grain size distribution curve is shown in Fig. 2. The average grain size of the sand particles is 0.2 mm, coefficient of uniformity = 1.86, and specific gravity of the sand is 2.65 assessed as per ASTM D2487 [11] and ASTM D854 [12] respectively. Maximum and minimum dry density of the sand calculated as per ASTM D4254 [13] are 1510 kg/m^3 and 1360 kg/m^3, respectively. Poison's ratio of the sand is assessed to be 0.30.

3 Experimental Program

3.1 Test Equipment

Model cyclic plate load tests were performed on square footing resting on reinforced or unreinforced sand bed having relative (or equivalent relative) density of 30 or 70% depending upon the testing requirements. Model footing was machined to the precise size of 75 mm × 75 mm and was made of mild steel plate of thickness 12.5 mm. Bottom of the footing was made irregular to simulate the field conditions. The size of the tank was decided as per the zone of influence for the size of the footing [14]. Tank of cross-section 500 mm × 500 mm and height 500 mm was used. The deflection of the sides of the tanks was measured during some preliminary tests, which came out to be negligible and hence ensured the rigidity of the tank.

3.2 Preparation of Sand Bed Below the Footing

First, unreinforced sand was placed in the tank to the desired height in layers of 75 mm thickness. Each layer of unreinforced sand was loosely filled to achieve a relative density of 30%. Thereafter, the sand reinforced with the desired percentage of waste tire chips was placed manually in layers and compacted to achieve 70% equivalent relative density. The dry density of sand (ρ_d) corresponding to 30 and 70% equivalent relative density values was determined as 1.39 g/cm^3 and 1.44 g/cm^3 respectively [4, 5, 10]. These dry density (ρ_d) values were adhered to while placing the sand layers of required weight in the desired depth. 70% equivalent relative density for the reinforced sand was mainly kept to assess the effect of both compaction and inclusion of tire chips on loosely packed sand. The main purpose behind studying the effect of both compaction and reinforcement is that if loosely packed poorly graded sand is to be improved in the field, it can be mixed with the recommended amount of reinforcement and then correspondingly compacted as well to get into a denser state. The mixing of sand and tire chip was done manually. Quantity of the tire chips by weight (depending upon the desired equivalent relative density, 70% in the present case) was added in different proportions in the top reinforced zone in equal layers. Mass of the mixture corresponding to a depth was placed and compacted with a wooden rammer.

3.3 Loading System

The top of reinforced or unreinforced sand was properly leveled, and the square footing was placed in position for testing. The load application was carried out by a hand-operated jack system with 50 kN capacity, supported against a relatively strong reaction beam. The load was increased to a certain desired value and then was brought down to zero. This sequential loading and unloading made it possible to separate the elastic (recoverable component) and plastic (non-recoverable component) settlement. Two dial gauges were mounted on the footing, and the average of the two readings was taken as the settlement of the footing. Settlement at every increment of loading and unloading was recorded when readings of dial gauges became reasonably constant and then it was utilized for obtaining pressure-settlement curves. The schematic diagram of the test setup is shown in Fig. 3.

3.4 Testing Details

A total of six tests on model square footing were conducted under cyclic loading conditions to study the effect of tire chip content. Each test was repeated up to three times to ensure the reproducibility of the obtained test data. The testing details are

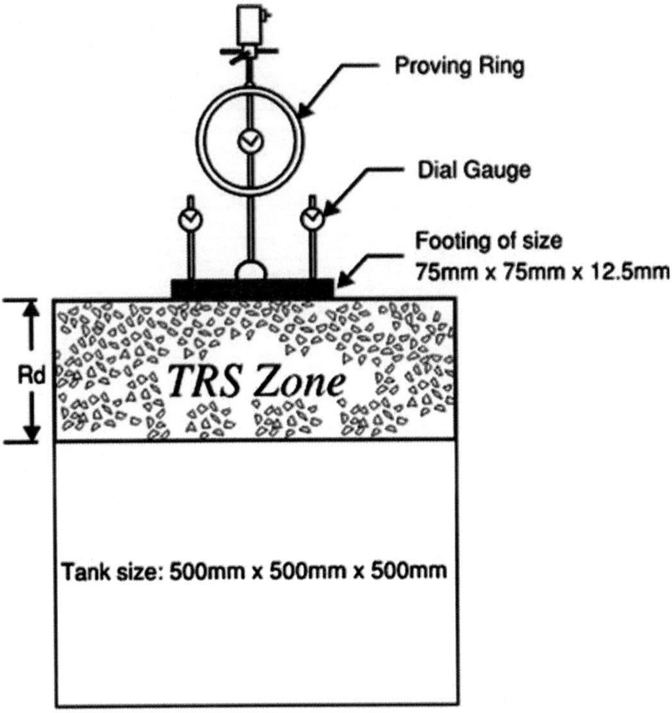

Fig. 3 Schematic diagram of the test setup

mentioned in Table 1. Mittal and Gill [4] recommended the depth of the reinforced zone as 1B (B = width of the footing) to obtain the optimum performance under static loading conditions from both strength and economy perspective. Therefore, in the present study, the depth of reinforced zone is maintained as 1B.

The unreinforced sample test series name is taken as UR30, which implies 'Unreinforced Sand at 30% relative density', while the reinforced sample test series name

Table 1 Details of the performed cyclic plate load tests

Test series	Type of zone below foundation	Relative density (D_r)	Percentage of tire chip reinforcement	Depth of reinforced zone (R_d)	Number of tests performed
UR30	Unreinforced	30%	–	–	1
TRS70A1B05	Reinforced	Equivalent relative density = 70%	5	1B	5
TRS70A1B10			10	1B	
TRS70A1B20			20	1B	
TRS70A1B30			30	1B	
TRS70A1B40			40	1B	

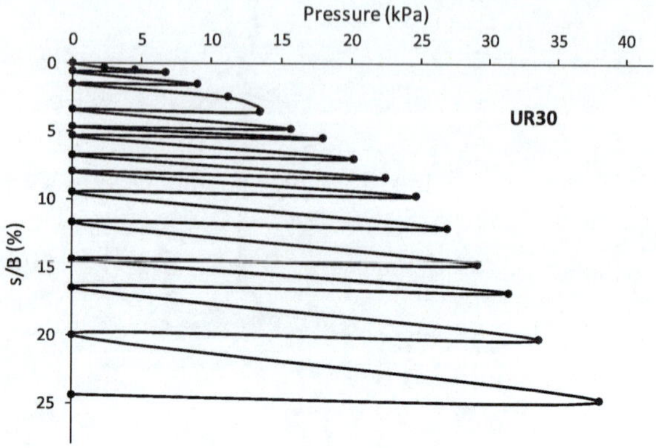

Fig. 4 Pressure settlement curve of unreinforced sand ($D_r = 30\%$)

is considered in the form TRS70A1B10 implying 'Tire-chip Reinforced Sand sample at 70% equivalent relative density reinforced in top 1B layer by the addition of tire chips equal to 10% by weight of sand in that layer'. It is to be noted that relative density of all the unreinforced layers below the reinforced layers was maintained at 30%.

4 Results and Observations

The present section describes the effect of cyclic loading on the pressure-settlement response of unreinforced sand and tire chips-reinforced sand. A brief discussion on the optimum amount of reinforcement and the pattern of variation of C_u is also presented. The settlement values are presented in a non-dimensional form (settlement/Width, s/B%) to utilize them in field applications avoiding possible scale effects.

4.1 Unreinforced Sand

Figure 4 depicts the pressure-s/B (%) curve of the unreinforced sand at 30% relative density.

4.2 Reinforced Sand

Pressure-s/B (%) curves of reinforced sand are shown in Fig. 5 for different tire chips content. It can be roughly observed from Fig. 5 that the pressure-settlement behavior of reinforced sand improves on an increase in the tire chip content even after the addition of a very small quantity i.e. 5% by weight. Moreover, the pressure was found to be significantly increased at 30% tire chip content ratio at all strain levels.

The results are shown only till the mixture with 30% tire chip content as the addition of tire chips beyond this content (i.e. 40%) led to major mixing difficulties. The compaction of the mixture to attain the desired equivalent relative density at

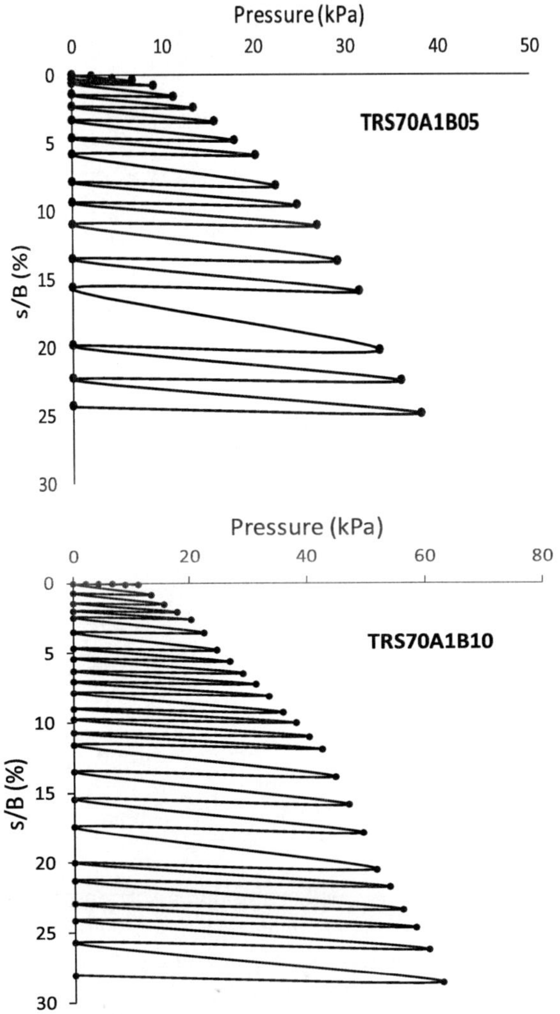

Fig. 5 Pressure settlement curve of reinforced sand

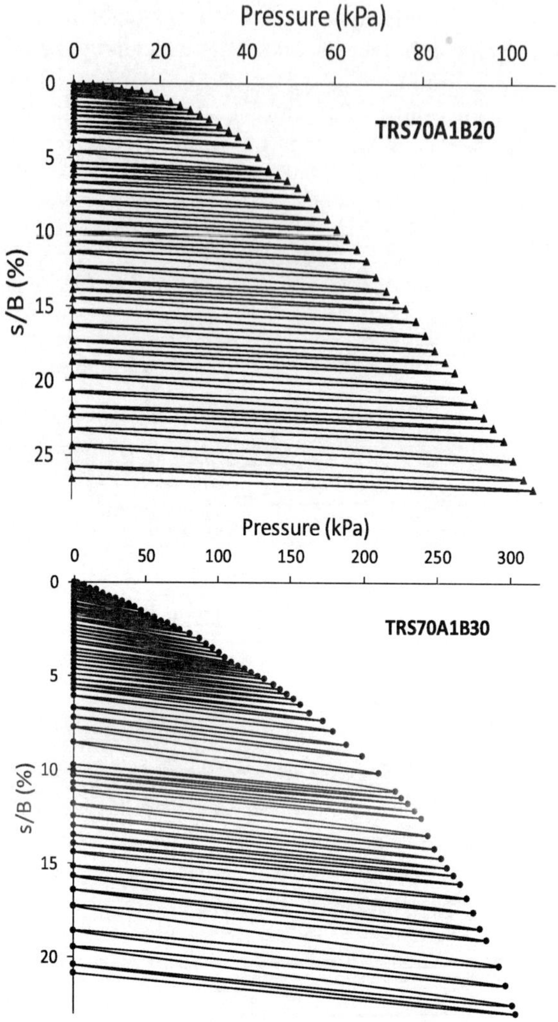

Fig. 5 (continued)

40% tire chip content was very challenging due to the absorption of compactive energy by the high amount of rubber content. Therefore, this content of tire chips is not recommended for field applications. Figure 6 shows the sample at 40% tire chip content and it can be clearly seen that the soil is merely visible from the top and the complete top area is almost occupied by the tire chips due to poor mixing.

Fig. 6 Tire chip reinforced sand (Sample: TRS70A1B40)

5 Discussion

5.1 Bearing Capacity Ratio

Figure 7 represents the bearing capacity ratio (BCR) obtained at different tire chip contents. BCR is defined as the ratio of the bearing capacity of reinforced sand to the bearing capacity of unreinforced sand at a given strain level. BCR values are reported at both low strain level (2–5%) and high strain level (10–20%) in the present study. It can be clearly observed that BCR increases on increase in the tire chip content at

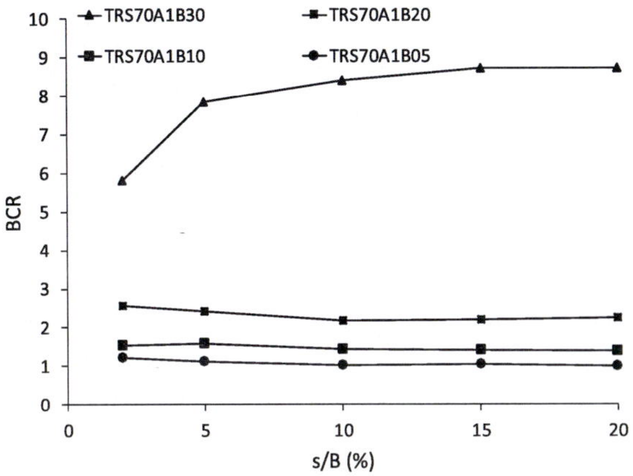

Fig. 7 Comparison of BCR at different tire chip contents

all strain levels. The rise in BCR was not significant at very low tire chip content (i.e. at 5%), however, BCR increased to approximately 2.5 times when the tire chip content was increased to 20%. Furthermore, the most significant increase in BCR was noticed at 30% tire chip content where approximately 8.5 times improvement in BCR was observed. Also, at higher strain level, the BCR showed similar or higher improvement.

5.2 *Coefficient of Elastic Uniform Compression and Shear Modulus*

Coefficient of elastic uniform compression, C_u, can be obtained by the slope of the pressure–elastic settlement curve. For the same, loading-unloading curves of the cyclic plate load tests (Figs. 4 and 5) were analyzed to compute the elastic settlement at each load cycle and subsequently, the pressure–elastic settlement curves were drawn. Figure 8 shows the best-fit curves of pressure v/s elastic settlement for all the sample types. It can be clearly observed that with an increase in the tire chip content, the slope of the pressure–elastic settlement curve increases.

Moreover, it is evident that the C_u is directly proportional to the shear modulus of the soil as given in equation 1 [15] and hence, the relative increase in the shear modulus (G) would be the same (as the other factors in equation 1, A = area of footing, ν = Poison's ratio, have no or negligible effect of the reinforcement content or type and remain almost same). Figure 9 is a clearer representation of the increase in the shear modulus. Shear modulus ratio can be defined as the ratio of the shear modulus of the reinforced sand to that of unreinforced sand (G in the present case

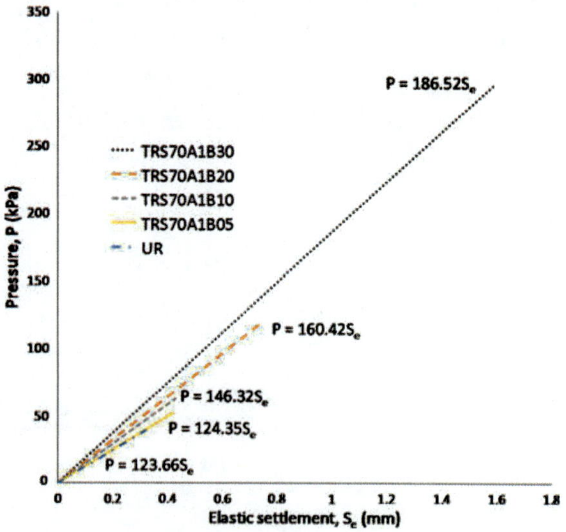

Fig. 8 Pressure versus elastic settlement curve at different tire chip contents

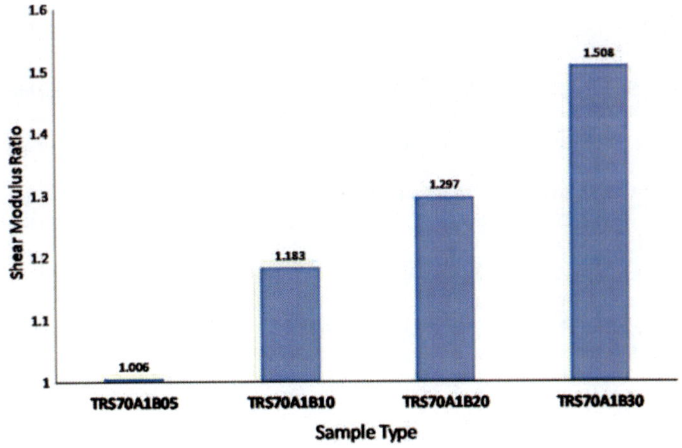

Fig. 9 Comparison of Shear modulus at different tire chip contents

for unreinforced sand is 2872.63 kPa). It can be clearly observed (Fig. 9) that tire chip addition can lead to an increase in G of up to 1.5 times. Therefore, it can be suggested that with an increase in the tire chip content, the shear modulus increases. This may be due to the establishment of a greater contact area, hence a greater friction being mobilized between the tire chip and the soil grains, which in turn leads to the increase in the stiffness of the mixture. The reduction in shear modulus found in case of granulated or powdered rubber soil mixture was possibly due to the similar size of both rubber and soil grain, which does not allow the establishment of proper inter-particle friction or bonding.

$$G = \frac{c_u\left(1 - \nu^2\right)\sqrt{A}}{2.26 \times (1 + \nu)} \tag{1}$$

6 Conclusions

The present study is mainly aimed towards the analysis of the effect of shear modulus on an increase in the content of tire chips. Contrary to the existing results on granulated or powdered rubber–sand mixture, it has been observed that the addition of tire chips can lead to an increase in the coefficient of elastic uniform compression or the shear modulus. An increase of 1.01–1.51 times was observed on varying the tire chip content from 5 to 30%. This result can, therefore, be utilized in future studies related to the sand–tire chip mixture. However, further evaluation of the accurate rise in the shear modulus is recommended using dynamic block vibration tests on the field.

References

1. Ayrilmis, N., Buyuksari, U., Avci, E.: Utilization of waste tire rubber in manufacture of oriented strandboard. Waste Manag. **29**(9), 2553–2557 (2009)
2. U.S. Scrap Tire Management Summary—2017, U.S. Tire Manufacturing Association, Washington (2018). https://www.ustires.org/scrap-tire-markets. Accessed 28 Nov 2018
3. Reddy, S.B., Krishna, A.M., Reddy, K.R.: Sustainable utilization of scrap tire derived geomaterials for geotechnical applications. Indian Geotech. J. **2017** (2017). https://doi.org/10.1007/s40098-017-0273-3
4. Mittal, R.K., Gill, G.: Sustainable application of waste tire chips and geogrid for improving load carrying capacity of granular soils. J. Clean. Prod. **200**, 542–551 (2018)
5. Gill, G., Mittal, R.K.: Use of waste tire-chips in shallow footings subjected to eccentric loading—an experimental study. Constr. Build. Mater. **199**, 335–348 (2019)
6. Feng, Z.Y., Sutter, K.G.: Dynamic properties of granulated rubber-sand mixtures. Geotech. Test. J. **23**(3), 338–344 (2000)
7. Hyodo, M., Yamada, S., Orense, R., Okamoto, M., Hazarika, H.: Undrained cyclic shear properties of tire chip-sand mixtures. In: Hazarika, H., Yasuhara K. (eds.) Proceedings of the International Workshop on Scrap Tire Derived Geomaterials—Opportunities and Challenges, Yokosuka, Japan, pp. 187–196 (2007)
8. Ehsani, M., Shariatmadari, N., Mirhosseini, S.M.: Shear modulus and damping ratio of sand-granulated rubber mixtures. J. Central South Univ. **22**(8), 3159–3167 (2015)
9. Moghaddas Tafreshi, S.N., Darabi, N.J., Dawson, A.R.: Cyclic loading response of footing on multilayered rubber-soil mixture. In: Geomechanics and Engineering, vol. 14, no. 2, pp. 115–129. Techno Press (2018)
10. Mittal, R.K., Gill, G.: Pressure settlement behaviour of strip footing resting on tire-chip reinforced sand. Int. J. Geotech. Eng. (2017). https://doi.org/10.1080/19386362.2017.1408195
11. ASTM (American Society of Testing and Materials): Standard practice for classification of soils for engineering purposes (Unified soil classification system). In: ASTM D2487-17, ASTM International, West Conshohocken, PA (2017)
12. ASTM (American Society of Testing and Materials): Standard test methods for specific gravity of soil solids by water pycnometer. In: ASTM D854-14, ASTM International, West Conshohocken, PA (2014)
13. ASTM (American Society of Testing and Materials): Standard test methods for minimum index density and unit weight of soils and calculation of relative density. In: ASTM D4254-16, ASTM International, West Conshohocken, PA (2016)
14. Harikumar, M., Sankar, N., Chandrakaran, S.: Behaviour of model footing resting on sand bed reinforced with multidirectional reinforcing elements. Geotext. Geomembr. **44**, 568–578 (2016)
15. Barkan, D.D.: Dynamics of Bases and Foundations, p. 23. McGraw-Hill, New York (1962)

Influence of Crumb Rubber and Polypropylene Fibre on the Behaviour of Cemented Black Cotton Soil

Vineeth Reddy Karnati, Abhishek Koukuntla, and Sowjanya Karnati

1 Introduction

The Indian Black Cotton (BC) soils are well known for their expansive behaviour and appear predominantly in black colour, which is abundantly used for cotton crops. These soils cover about one-fifth of the total land area in India [1]. With the ever-growing population and the infrastructure needs of the society, it becomes almost impossible to carry out construction activities only on the limited good soils that offer adequate strength and reliability [2]. Carrying out the construction activities on these BC soils results in severe damage to structures, which can be easily observed in the walls. Several researchers across the globe studied the volume change behaviour of the expansive soils and provided various ground improvement techniques to mitigate the excessive swell-shrink behaviour [3–7]. Besides swell-shrink behaviour, the BC soils also possess very low strength and the research has been carried out in improving the strength of the BC soil by chemical modification [4, 7–10], by the inclusion of various kinds of natural [3, 11, 12] and artificial fibres [1, 8, 11, 13– 15].

V. R. Karnati (✉)
Manglam Consultancy Services Hyderabad, Hyderabad 500035, Telangana, India

A. Koukuntla
Department of Civil Engineering, Anurag Group of Institutions, Hyderabad 500038, Telangana, India

S. Karnati
Department of Civil Engineering, SV University College of Engineering, Tirupati 517502, Andhra Pradesh, India

C. N. V. Satyanarayana Reddy et al. (eds.), *Ground Improvement and Reinforced Soil Structures*, Lecture Notes in Civil Engineering 152,
https://doi.org/10.1007/978-981-16-1831-4_35

The ground improvement techniques not only help in modifying the existing poor soils but also providing the basis for the utilization of a lot of wastes and industrial by-products that require a lot of inputs in terms of cost and space [16]. Research has been initiated in recent years in utilizing this geotechnical engineering arena for effective utilization of plastic wastes [17, 18] and rubber wastes [19–21]. The utilization of rubber wastes in the ground improvement applications offers not only in effective disposal of wastes but also additional advantages of being the lightweight material [22, 23] and high-damping efficiency in earthquake engineering applications [24–27]. Inclusion of tyre material in the clayey soil improves the load-carrying capacity of soil [28]. The amount of improvement in the strength, quantity of material and durability are some important factors to be considered in soil–rubber mixtures [29, 30].

Several studies have shown that there is an optimum lime content up to which the increase in strength is marginal and later which it decreases [31–33]. For further improvement in lime stabilized soil, additional treatment can be done by reinforcing the material with some non-reactive materials [10, 34]. Several studies were carried out on lime–fibre mixtures and rubber mixed with cemented soils. In this study, an attempt is made to check the effect of blended polypropylene fibres and crumb rubber on the compaction and strength behaviour of lime cemented black cotton soil.

2 Materials

2.1 Soil

The BC soil sample used in the current study was collected from Alampur town, Gadwal District of Telangana, India. Based on the physical observation, it was found to be black cotton soil. The soil was tested for its basic properties as per IS:2720 [35–39] and are presented in Table 1 and the grain size analysis is presented in Fig. 1.

2.2 Lime

The negatively charged surface of the fine clayey particles absorbs the free calcium ions from the lime resulting in the alteration of the mineralogy of the particles. At higher concentrations of Ca^{2+}, isomorphous substitution takes place resulting in increased strength of the soil because of pozzolanic reactions and formation of cementitious products. Lime addition to the highly reactive clay minerals not only increases the strength but also reduces the swell-shrink behaviour as the mineralogy itself gets altered. Commercially available hydraulic lime was used for the study.

Table 1 Soil properties

Sl. no.	Property	Value
1.	Specific gravity, G	2.73
2.	*Grain size analysis*	
	Gravel (%)	1.05
	Sand (%)	33.13
	Fines (%)	65.28
3.	*Atterberg's limits*	
	Liquid Limit, LL (%)	68.3
	Plastic Limit, PL (%)	19.29
	Plasticity Index, PI (%)	49.01
4.	Indian standard soil classification	CH
5.	Differential free swell index (%)	65%
6.	*Compaction characteristics*	
	Optimum Moisture Content, OMC (%)	19.5
	Maximum Dry Density, MDD (%)	1.41
7	Unconfined compressive strength (kPa)	53.5

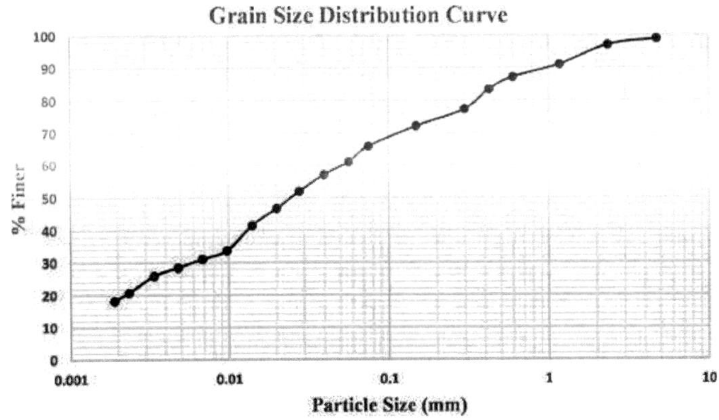

Fig. 1 Grain size distribution of Soil Sample

2.2.1 Polypropylene Fibres

Polypropylene, a monomer of petroleum-based polymer, propylene, is a 100% synthetic fibre that is transformed from 85% polypropylene. Polypropylene is thermoplastic, white, mechanically rugged material, and is resistant to many chemical solvents, acids and bases. The polypropylene to be used in this study was obtained from the local market.

2.3 *Crumb Rubber*

Crumb rubber is a granular form of rubber recycled from truck scrap tyres and automotive. The tyre cord (fluff) and the steel are removed from the tyre rubber during the recycling process, leaving a granular consistency material. The crumb rubber was procured from the locally available market and used for the study. The gradation of tyre crumbles was found equivalent to poorly graded sand (SP) as per I.S. Classification. Mixing crumb rubber with the clayey soil is similar to the replacement of some portion of clayey soil with granular material. In the virgin soil and the cemented soil, the bond between the clayey particle and the crumb rubber could be studied with the help of microstructural analyses, which was out of the scope of the current study.

3 Experimental Methodology

In this study, blended polypropylene fibres and crumb rubber are used to reinforce the lime cemented BC soil. The methodology adopted for this study is explained below.

3.1 *Crumb Rubber Reinforcement*

In the current study, the BC soil was mixed with varying percentages of crumb rubber i.e., 0, 0.5, 1, 2, 5 and 10% by weight of soil. The OMC and MDD of the respective specimens were respectively calculated and the specimens were prepared for testing.

3.2 *Lime Treatment*

The BC soil was mixed with the optimum lime content, which is found based on the initial lime consumption value found from pH test as per ASTM D6276 [40] and the Atterberg limits as per IS:2720 [36]. The compaction characteristics of the lime cemented BC soil are found by carrying out the light compaction test as per IS:2720 [37]. The unconfined compressive strength (UCS) value of the cemented soil was found as per IS:2720 [38] by preparing the specimens at OMC and MDD obtained and curing them in desiccators for 7, 14 and 28 days respectively.

3.3 Lime + Crumb Rubber Reinforcement

The crumb rubber was added along with 4% lime into the BC soil at varying percentages of 0.5, 1, 2, 5 and 10% by weight of soil. The OMC and MDD of all the mixtures were respectively calculated and specimens were prepared for testing. The UCS values of the specimens were found after 7, 14 and 28 days of curing in desiccator.

3.4 Lime + Polypropylene + Crumb Rubber Reinforcement

The lime cemented BC soil was further reinforced with short discrete polypropylene fibres of 12 mm length. An optimum of 0.25% is chosen as fibre content based on the available literature [9, 14, 34]. The OMC and MDD of reinforced lime cemented soil were found and the specimens were prepared for UCS testing at 7, 14 and 28 days of curing in desiccator.

Furthermore, the lime cemented BC soil was mixed with varying percentages of crumb rubber at 0, 0.5, 1, 2, 5 and 10% by weight of the BC soil plus 0.25% polypropylene respectively. The OMC and MDD of all the mixtures were found and the specimens were prepared for UCS testing at 7, 14 and 28 days curing in desiccator.

4 Results and Discussion

The results of various tests conducted as discussed in the methodology were described in the following sub-sections. In the subsequent sections, BC stands for black cotton soil, L stands for hydraulic lime, PP stands for polypropylene and CR stands for crumb rubber.

4.1 Lime Treatment

The initial lime consumption value was found using the pH testing on the lime mixed soil and the Atterberg's limit tests whose results are shown in Figs. 2 and 3 respectively. Based on the results of these tests, it was identified that 4% hydraulic lime could be sufficient to modify the behaviour of BC soil. At this lime content, the soil obtained a pH value of about 12.4, which is suggesting that long-term stabilization reactions could be achieved and also the soil is behaving almost non-plastic (PI = 6%). The UCS of the lime-modified soil was found to be 263.9 kPa after 28 days curing, which is about 490% of that of the virgin soil.

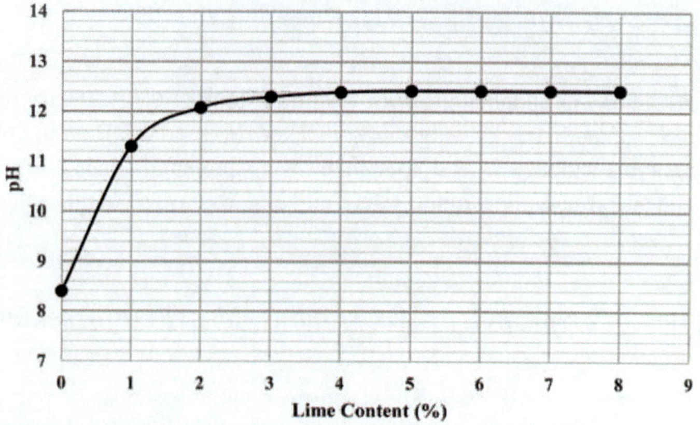

Fig. 2 Variation of pH of soil with different hydraulic lime contents

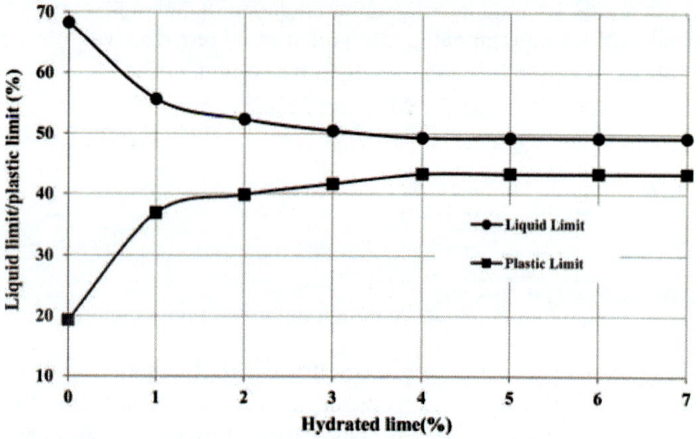

Fig. 3 Variation of Atterberg's limits with different hydraulic lime contents

4.2 Crumb Rubber Reinforcement

The results of the UCS test on the BC soil mixed with varying percentages of crumb rubber are presented in Figs. 4 and 5. Based on the test results, it was observed that the UCS value of the BC soil increased with the addition of the CR up to 5% and then decreased with the maximum value being 149.1 kPa, which is about 280% of that of the original soil. The initial modulus of elasticity was not found to be varied much except in case of 1% CR addition.

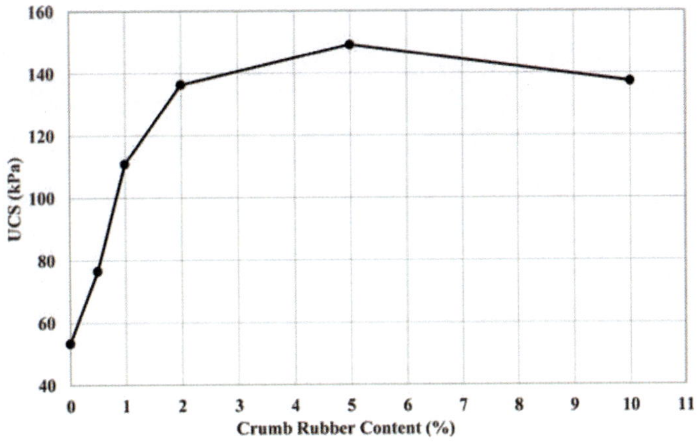

Fig. 4 UCS values of BC + CR mixtures

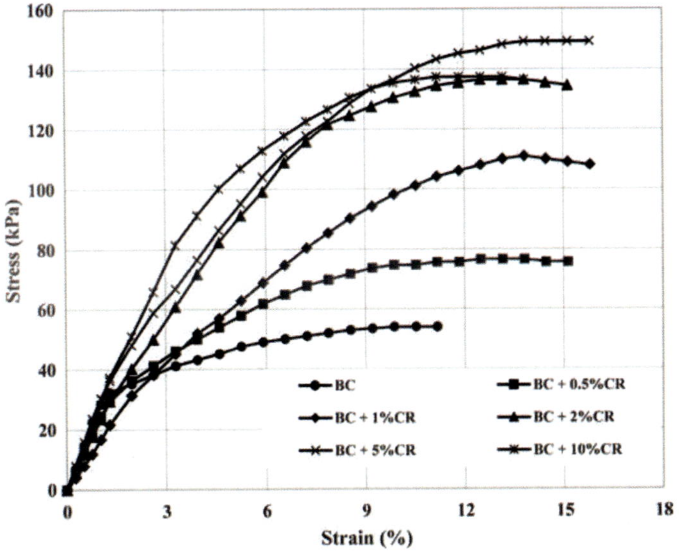

Fig. 5 Stress versus strain curves for BC + CR mixtures

4.3 *Lime + Crumb Rubber Reinforcement*

The UCS test results of the lime-modified BC soil mixed with varying percentages of CR are presented in Fig. 6 and the average stress versus strain variations of different specimens after 28 days of curing are presented in Fig. 7. Examination of test results provides a basic understanding of the improvement possible with the rubber crumbles added together with lime. The strength of the BC + 4%L + CR samples is shown a

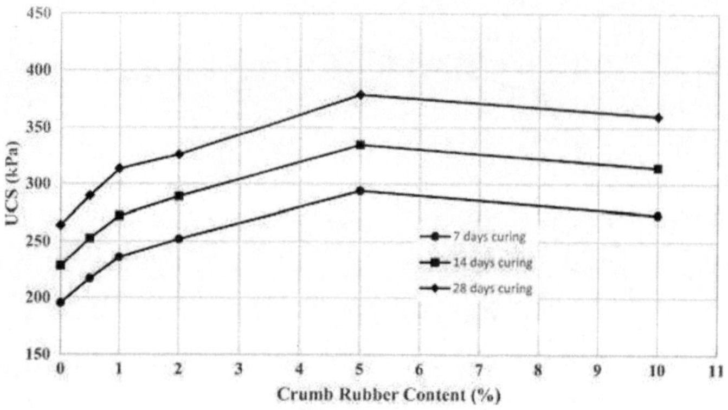

Fig. 6 UCS values of BC + 4%L + CR mixtures

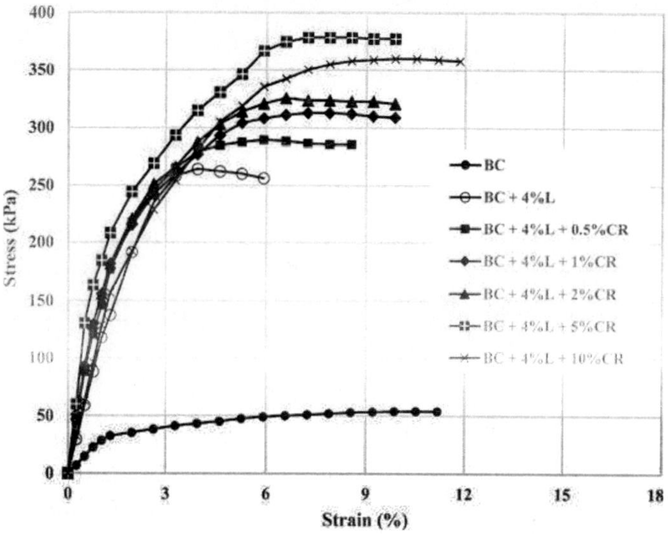

Fig. 7 Stress versus strain curves for BC + 4%L + CR mixtures

similar trend as in the case only BC + CR mixtures giving the maximum strength at 5% CR addition. An increase in the UCS value, after 28 days curing, of about 700 and 140% was observed compared to that of virgin BC soil and 28 days strength of BC + 4%L mixture respectively.

4.4 Lime + Polypropylene + Crumb Rubber Reinforcement

Furthermore, a little strength increase was observed when the PP and CR were blended together with the lime and BC soil, which can be seen from Fig. 8. The stress–strain variation of these specimens after 28 days of curing are presented in Fig. 9, which shows the improvement in the initial modulus of elasticity with PP reinforcement in BC + 4%L mixture and further similar increase with CR blending. 5%CR addition was found to optimum value giving the highest UCS value. This value corresponds to about 920, 188 and 130% times that of the Virgin BC soil, BC + 4%L and BC + 4%L + 0.25%PP mixtures after 28 days curing.

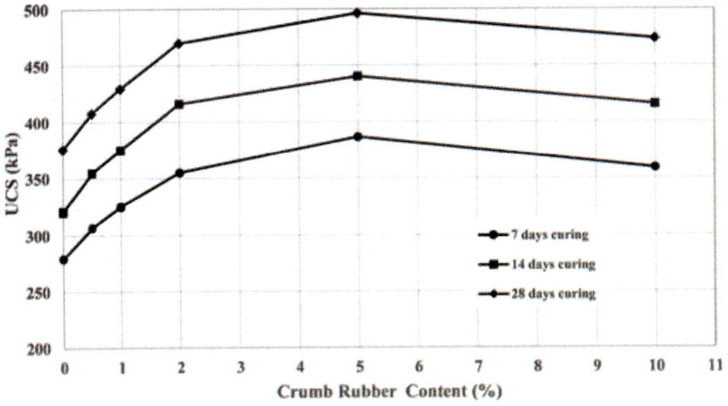

Fig. 8 UCS values of BC + 4%L + 0.25%PP + CR mixtures

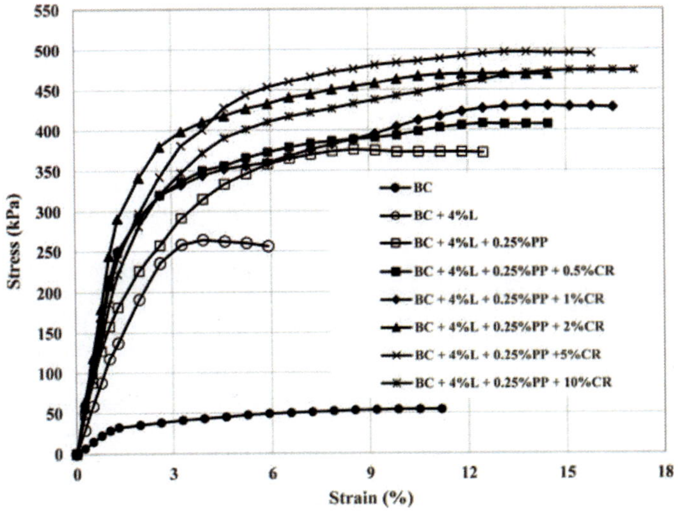

Fig. 9 Stress versus strain curves for BC + 4%L + 0.25%PP + CR mixtures

5 Conclusions

Tyre being a hazardous industrial waste requires a lot of available land for its disposal. Geotechnical engineering applications being a suitable option for bulk use of these materials, it is very important to study their applicability. Hence, a series of tests were performed to study the effect of the randomly distributed CR and discrete short PP fibres on the strength of the uncemented and lime cemented soil. The addition of CR to the virgin soil and the lime-modified soil resulted in a noticeable increment in the UCS values. Increasing CR content increased the strength up to 5% and then got reduced, which shows that 5% CR can be the optimum amount to be adopted to use for soil stabilization. Blended PP and CR addition also shows a similar trend. The bridging effect of the fibres could impede the growth of tension cracks. Addition of 4% hydraulic lime-modified the highly plastic BC soil to almost non-plastic soil thereby offering sufficient bond strength to the fibres that contributed to the strength of PP reinforced lime cemented soil. 5%CR addition resulted in the strength of about 2.8 times the virgin BC soil. In the lime-modified soil, 5%CR addition resulted in the strength of about 1.4 and 7 times the strength of lime-modified and virgin BC soil respectively. Furthermore, the addition of 0.25%PP + 5%CR to the lime-modified soil showed the strength of about 1.3, 1.9 and 9 times the strength of PP reinforced lime-modified soil, lime-modified and virgin BC soil. It was concluded that 5% CR can be adopted as an optimum additive for BC soil improvement provided the other characteristics like swell-shrink behaviour, durability behaviour and microstructure etc. also were studied.

References

1. Kumar, A., Walia, B.S., Bajaj, A.: Influence of fly ash, lime, and polyester fibers on compaction and strength properties of expansive soil. J. Mater. Civ. Eng. **19**(3), 242–248 (2007)
2. Karnati, V.R., Munaga, T., Gonavaram, K.K., Amitava, B.: Study on strength and leaching behavior of biogeochemical cemented sand. Geomicrobiol J. **37**(7), 670–681 (2020)
3. Jayasree, P.K., Balan, K., Peter, L., Nisha, K.K.: Volume change behavior of expansive soil stabilized with coir waste. J. Mater. Civil Eng. **27**(6), 04014195 (2015)
4. Kampala, A., Horpibulsuk, S.: Engineering properties of silty clay stabilized with calcium carbide residue. J. Mater. Civ. Eng. **25**(5), 632–644 (2013)
5. Malekzadeh, M., Bilsel, H.: Swell and compressibility of fiber reinforced expansive soils. Int. J. Adv. Technol. Civil Eng. **1**(2), 42–45 (2012)
6. Moghal, A.A.B., Chittoori, B.C., Basha, B.M., Al-Mahbashi, A.M.: Effect of polypropylene fibre reinforcement on the consolidation, swell and shrinkage behaviour of lime-blended expansive soil. Int. J. Geotech. Eng. **12**(5), 462–471 (2018)
7. Rogers, C.D.F., Glendinning, S., Roff, T.E.J.: Lime modification of clay soils for construction expediency. In: Proceedings of the Institution of Civil Engineers-Geotechnical Engineering, vol. 125, no. 4, pp. 242–249 (1997)
8. Muntohar, A.S.: Influence of plastic waste fibers on the strength of lime-rice husk ash stabilized clay soil. Civil Eng. Dimens. **11**(1), 32–40 (2009)

9. Tang, C., Shi, B., Gao, W., Chen, F., Cai, Y.: Strength and mechanical behavior of short polypropylene fiber reinforced and cement stabilized clayey soil. Geotext. Geomembr. **25**(3), 194–202 (2007)
10. Yadav, J.S., Tiwari, S.K.: Effect of inclusion of crumb rubber on the unconfined compressive strength and wet-dry durability of cement stabilized clayey soil. J. Build. Mater. Struct. **3**(2), 68–84 (2016)
11. Chauhan, M.S., Mittal, S., Mohanty, B.: Performance evaluation of silty sand subgrade reinforced with fly ash and fibre. Geotext. Geomembr. **26**(5), 429–435 (2008)
12. Devi, R.: Effect of inclusion of coir fibers of varying length on unconfined compressive strength of expansive clay. Int. Res. J. Eng. Technol. **3**(6), 2097–2100 (2016)
13. Anil, U., Prasad, S.D., Prasad, D.S.V., Raju, G.P.: A study on stabilization of expansive soil using tile waste and Recron-3S fibres. SSRG Int. J. Civil Eng. **5**(8), 12–16 (2018)
14. Jiang, H., Cai, Y., Liu, J.: Engineering properties of soils reinforced by short discrete polypropylene fiber. J. Mater. Civ. Eng. **22**(12), 1315–1322 (2010)
15. Priya, C.S., Archana, S., Albert, A.B., Deeraj, A.D.: Stabilization of clayey soil using polypropylene fiber. Int. Res. J. Eng. Technol. (IRJET) **4**(4), 1252–1255 (2017)
16. Suluguru, A.K.: Experimental investigations on potential of brick-based building-derived materials for geotechnical applications. Innov. Infrastruct. Sol. **5**(85), 1–11 (2020)
17. Ru, A.K., Shafin, M.M., Madhavan, R., Rajkumar, R.: Waste plastic fiber reinforced soil. Thesis submitted to Velammal Engineering College, Chennai (2015)
18. Vasudevan, R., Sekar, A.R.C., Sundarakannan, B., Velkennedy, R.: A technique to dispose waste plastics in an ecofriendly way—application in construction of flexible pavements. Constr. Build. Mater. **28**(1), 311–320 (2012)
19. Hambirao, G.S., Rakaraddi, P.G.: Soil stabilization using waste shredded rubber tyre chips. IOSR J. Mech. Civil Eng. **11**(1), 20–27 (2014)
20. Kokila, M.L., Bhavithra, G., Hemapriya, V., Iniya, C., Madhunigga, P.: Experimental investigation on soil stabilisation using rubber crumbs on expansive soil. World J. Res. Rev. **4**(4) (2017)
21. Patil, U., Valdes, J.R., Evans, T.M.: Swell mitigation with granulated tire rubber. J. Mater. Civ. Eng. **23**(5), 721–727 (2011)
22. Eid, H.T., Stark, T.D., Evans, W.D., Sherry, P.E.: Municipal solid waste slope failure. I: Waste and foundation soil properties. J. Geotech. Geoenviron. Eng. **126**(5), 397–407(2000)
23. Eldin, N.N., Senouci, A.B.: Use of scrap tires in road construction. J. Constr. Eng. Manag. **118**(3), 561–576 (1992)
24. Dhanya, J.S., Boominathan, A., Banerjee, S.: Response of soil-tyre mixture subjected to cyclic loading. In: Proceedings of the 16th World Conference on Earthquake Engineering, Chili, vol. 1662 (2017)
25. Madhusudhan, B.R., Boominathan, A., Banerjee, S.: Static and large-strain dynamic properties of sand–rubber tire shred mixtures. J. Mater. Civil Eng. **29**(10), 04017165 (2017)
26. Nakhaei, A., Marandi, S.M., Kermani, S.S., Bagheripour, M.H.: Dynamic properties of granular soils mixed with granulated rubber. Soil Dyn. Earthq. Eng. **43**, 124–132 (2012)
27. Vinod, J.S., Sheikh, M.N., Mashiri, S.: Cyclic behaviour of scrap-tyre soil mixtures. In: Frontiers in Geotechnical Engineering, pp. 303–311. Springer, Singapore (2019)
28. Cetin, H., Fener, M., Gunaydin, O.: Geotechnical properties of tire-cohesive clayey soil mixtures as a fill material. Eng. Geol. **88**(1–2), 110–120 (2006)
29. Priyadarshee, A., Kumar, A., Gupta, D., Pushkarna, P.: Compaction and strength behavior of tire crumbles–fly ash mixed with clay. J. Mater. Civil Eng. **30**(4), 04018033 (2018)
30. Youwai, S., Bergado, D.T.: Strength and deformation characteristics of shredded rubber tire sand mixtures. Can. Geotech. J. **40**(2), 254–264 (2003)
31. Bell, F.G.: Lime stabilization of clay minerals and soils. Eng. Geol. **42**(4), 223–237 (1996)
32. Dash, S.K., Hussain, M.: Lime stabilization of soils: reappraisal. J. Mater. Civ. Eng. **24**(6), 707–714 (2012)
33. Diamond, S., Kinter, E.B.: Mechanisms of soil-lime stabilization. Highway Res. Rec. **92**, 83–102 (1965)

34. Tang, C.S., Shi, B., Cui, Y.J., Liu, C., Gu, K.: Desiccation cracking behavior of polypropylene fiber–reinforced clayey soil. Can. Geotech. J. **49**(9), 1088–1101 (2012)
35. IS 2720-Part 3-1: Methods of Test for Soils—Determination of Specific Gravity, Fine Grained Soils. Bureau of Indian Standards, New Delhi (1980)
36. IS 2720-Part 5: Methods of Test for Soils—Determination of Liquid Limit and Plastic Limit Of Soils. Bureau of Indian Standards, New Delhi (1985)
37. IS 2720-Part 7: Methods of Test for Soils—Determination of Water Content—Dry Density Relation Using Light Compaction. Bureau of Indian Standards, New Delhi (1983)
38. IS 2720-Part 10: Methods of Test for Soils—Determination of Unconfined Compressive Strength. Bureau of Indian Standards, New Delhi (1991)—RA 1995
39. IS 2720-Part 40: Methods of Test for Soils—Determination of Free Swell Index of Soils. Bureau of Indian Standards, New Delhi (1977)
40. ASTM D6276-19: Standard test method for using ph to estimate the soil-lime proportion requirement for soil stabilization. In: ASTM International, West Conshohocken, PA (2019). www.astm.org

Importance of Verticality of PVD in Consolidation Settlement of Prefabricated Vertical Drain-Improved Soft Soils

Rai Bahadur Reang, Sujit Kumar Pal, and Sanjay Paul

1 Introduction

Highly compressible soft soils possess high natural moisture content, even reach close to the liquid limit, low hydraulic conductivity, and low undrained shear strength. If proper measures are not taken to improve those soils, the structures constructed may face stability problems. However, expensive pile foundations are adopted in some cases. Now-a-days, the application of prefabricated vertical drains (PVDs) in the area of ground improvement has proven to be the fastest-growing method among all other techniques. Thus, it plays a vital role in geotechnical engineering constructions. It includes large-scale infrastructures such as highways, railways, ports, and airports [1, 9, 10, 12]. The time required for consolidation of the soil reduces to few months with PVDs and preloading. However, the performance or efficiency of PVDs is dependent on many factors, such as the type and discharge capacity of the PVDs, the permeability of the soil, preloading pressure and its period, type of the soil, buckling or deformation of the PVD, etc. Among all other factors, the PVD's verticality is also an influencing factor on which consolidation settlements mostly depend. In practice, sometimes, the verticality of PVDs may not be maintained due to defective machinery measures or improper monitoring arrangements during installation. Consequently, adequate functioning of the PVDs is hindered to some extent, and hence the construction completion period also may be delayed, followed by unwanted high maintenance costs.

In this paper, an attempt is made through experimental studies on PVD-treated soft soils to determine the adverse effects of verticality of PVDs on consolidation settlements of the silty-clay soil. It is hoped that this research study will help the researchers and academicians working on PVDs to some extent.

R. B. Reang (✉) · S. K. Pal · S. Paul
National Institute of Technology Agartala, Agartala, Tripura, India

C. N. V. Satyanarayana Reddy et al. (eds.), *Ground Improvement and Reinforced Soil Structures*, Lecture Notes in Civil Engineering 152,
https://doi.org/10.1007/978-981-16-1831-4_36

2 Objectives of the Present Study

The present study's objectives are to determine the adverse effects of the verticality of PVD on consolidation settlements of silty-clay soil and to find out the impact of the time. Curves exhibiting consolidation deviation results (due to tilting of the PVD) are presented in this study. Hence knowledge about the significance of the verticality of PVDs is achieved from the curves.

3 Experimental Programs

3.1 Materials Used in This Study

The soil is classified as silty clay (CI) of medium plasticity and of grey-white in color. The soil was collected from Ananda Nagar, Agartala, Tripura state, India. Geotechnical properties of the soil are given in Table 1. The prefabricated vertical

Table 1 Properties of the soil used

Parameters	Test method	Results
Specific gravity	IS: 2720 (Part 3) [2]	2.59
Natural moisture content (%)	IS: 2720 (Part 2) [3]	82.23
Liquid limit (%)	IS: 2720 (Part 5) [5]	52.20
Plastic limit (%)	IS: 2720 (Part 5) [5]	28.71
Plasticity index (%)	IS: 2720 (Part 5) [5]	16.49
Grain size	IS: 2720 (Part 4) [4]	
Sand (%)		2.82
Silt (%)		44.72
Clay (%)		52.46
Coefficient of consolidation (c_v) (m^2/s)	IS: 2720 (Part 15) [7]	5.65×10^{-8}
Permeability (m/s)	IS: 2720 (Part 17) [8]	
Standard Proctor Test	IS: 2720 (Part 7) [6]	
Optimum moisture content (OMC) (%)		23.65
Maximum dry density (MDD) (g/cm^3)		1.58

Table 2 Properties of the PVD used

Parameters	Test method	Results
Core material	ASTM D 276	PP
Width (mm)	ASTM D 3774	100.00
Thickness (mm)	ASTM D 5199	4.00
Width-width tensile strength (kN/m)	ASTM D 4595	≥7.00
Tensile strength (kN/m)	ASTM D 4595	≥2.50
	EN ISO 10319	
Elongation at 0.5 kN (%)	ASTM D 4595	≤10
	EN ISO 10319	
Discharge capacity, 300 kPa ($i = 1$) (cm^3/s)	ASTM D 4716	≥100
Permeability (cm/s)	ASTM D 4491	$\geq 1.4 \times 10^{-4}$
Permittivity (s^{-1})	ASTM D 4491	0.44
Apparent opening size, O_{90} (μ)	EN ISO 12956	<85

drain was collected from Meccaferi Pvt. Ltd., India. Its breadth is 100 mm, and the width is 4 mm, so the equivalent diameter of the PVD is 66 mm. Its properties are shown in Table 2.

3.2 Laboratory Model and Preparation of Soil Paste

In this study, a physical model made of 6 mm thick steel of 452 mm internal diameter and 600 mm internal high is used. A sketch of the model is shown in Fig. 1. Vertical pressures are applied with the help of a lever arm and its system. Soil is mixed with water, and paste is prepared at the water content of 47% (near the liquid limit) in a tray, and later the soil bed is prepared in stages inside the model along with a PVD.

3.3 Test Procedure

In this study, PVDs are preinserted, i.e., no mandrel is used to install the PVD eliminating soil disturbance. Two cases had been studied: (i) vertical PVD and (ii) tilted PVD. In both cases, the magnitudes of vertical pressures and time are kept the same. The ray diagram of this arrangement is shown in Fig. 1.

(i) **Vertical PVD.** In this case, a PVD is, first, placed centrally inside the model, making an angle of 90° (φ) with the model base. Then soil paste is poured inside it in three stages, keeping the soil's undrained shear strength as 20 kPa. A polythene paper is used to minimize friction between the soil and the model wall surface. After preparing the soil bed, a sand layer of thickness 2 cm is

Fig. 1 Tilting of a PVD of height 'H' at an angle 'θ' with the vertical line, making an angle 'φ' with the baseline and a horizontal displacement 'x' at top surface

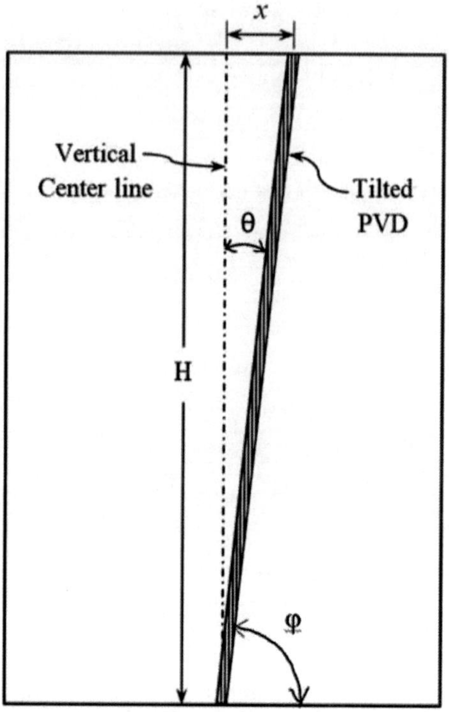

formed, and a seating plate is placed through which a vertical pressure of 10 kPa is applied with the help of a loading bar and kept for 7 days. After that, an incremental vertical surcharge pressure of 10 kPa is applied at an interval of 7 days up to 63 days, and the corresponding consolidation settlements are recorded with the help of dial gauges.

(ii) **Tilted PVD**. In this case, a PVD is held making an angle of 85°55′ (φ) at the bottom center with the model base, i.e., at an angle of 4°45′ (θ) with the vertical line. It is so obtained by tilting the PVD by a small amount of 5 cm (x) length from the model's center point. After forming the sand layer, the same test procedures are followed, just as in the first case. On the other hand, if an 8 m long PVD is mistakenly dislocated horizontally during the installation due to defective machinery measures or improper monitoring arrangements, an angle 4°45′ (θ) is formed with the vertical line making 66 cm (x) displacement at the ground surface. Taking into account this example, the tilting angle of 4°45′ (θ) was considered in the laboratory study to resemble the same case.

4 Results and Discussions

4.1 Consolidation of Vertical PVD Treated Soil

In this case, there is no dislocation of the PVD at the top surface, i.e., the PVD is vertically straight, hence $x = 0$. When the soil is treated with the PVD by keeping it vertical, the consolidation settlement rate is very high within 35 days of superimposing the load. At the same time, the total vertical pressure reaches 50 kPa, a steep curve is observed within this period that implies large settlements of the soil. The reason is the quick horizontal movement of excess pore water within days. After 35 days, the consolidation settlement started decreasing and gives the result as 27.80 mm. The reason is clogging in opening pores of the PVD by soil particles and buckling at the top portion, which significantly affects the discharge capacity. Crimping, folding, bending, kinking, or buckling may reduce the discharge capacity, hence settlement, significantly or even completely [11]. The magnitude of clogging never decreases but increases with time. Due to this phenomenon, the consolidation rate gradually decreases after this point and almost becomes constant, showing the settlement result as 29.40 mm after 63 days, even though the total pressure now is 90 kPa. It is presumed that the settlement value could have been more if the incremental loading rate is more than 10 kPa.

4.2 Consolidation of Tilted PVD-Treated Soil

Here, the PVD is dislocated at the top surface from the center point by 5 cm, i.e., $x = 5$ cm. After imposing the vertical pressure, the consolidation settlement rate is very high at the beginning period between 1 and 21 days, and a very steep curve can be seen (Fig. 3). But, after this point, this rate gradually starts decreasing, and the settlement becomes 22.60 mm after 21 days. It may be attributed to the extreme clogging and buckling at the top portion of the PVD. Figure 2 shows the distribution of action of vertical force on two faces (assumed); front face (F) and back face (B). In this case, since the PVD is tilted, the PVD's front face encounters more pressure as the force is acting vertically due to which clogging expedites in this face compared to another face (backside). However, fewer folds in the buckling portion are observed than in the first case (vertical PVD-treated soil). These are the reasons the tilted PVD-treated soil exhibits less consolidation settlement than vertical PVD-treated soil. After 63 days, the consolidation result is observed as 24.60 mm.

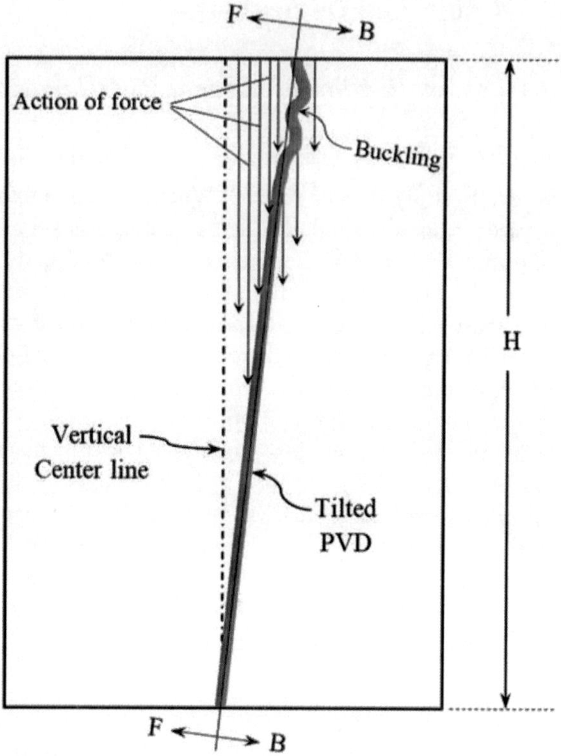

Fig. 2 Action of vertical force on the front face (F) and back face (B) of the tilted PVD

4.3 Comparison of the Results

Consolidation settlement results are compared and are shown in Fig. 3. From this figure, it can be seen that untreated soil has the lowest consolidation settlements,

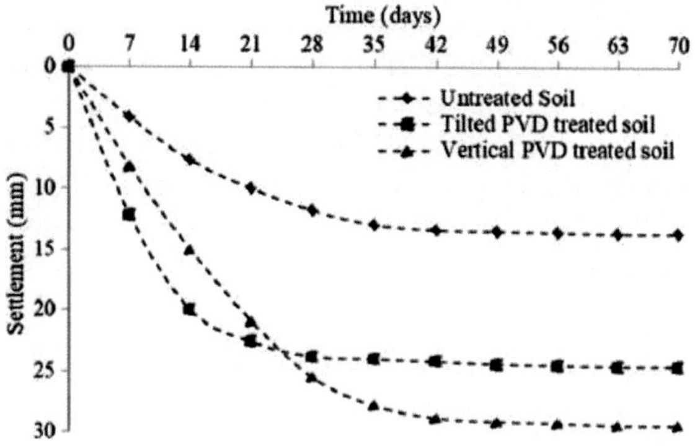

Fig. 3 Comparison of the consolidation settlement results

and more consolidation settlements are observed in the case of vertical PVD-treated soil. But, at the beginning stage, a comparative steep curve is observed (between 1 and 24 days) in tilted PVD-treated soil. It gradually decreases and tends to be almost constant after 42 days producing 29.40 and 24.60 mm settlements for vertically PVD-treated soil after 63 days. Therefore, the deviation in the result is 4.80 mm. In other words, if the verticality of the PVD is changed, there must be some deviations in the consolidation results. This deviation even may be more significant if the vertical pressure is changed to a higher magnitude.

5 Conclusions

Based on the study made above, the following points may be obtained:

1. Consolidation settlement of untreated soil increases if the soil is treated with PVDs.
2. Tilted PVD expedites the settlement of the soil in the beginning periods only; it does not give the ultimate expected results.
3. Tilting of the PVD significantly affects the consolidation settlement. So, it may be concluded that the verticality of PVDs is also an important issue, and great care should be taken care of during installation in fields. Keeping it as right-angled with the horizontal datum line gives significant results.

References

1. Indraratna, B., Rujikiatkamjorn, C., Ewers, B., Adams, M.: Class a prediction of the behavior of soft estuarine soil foundation stabilized by short vertical drains beneath a rail track. J. Geotech. Geoenviron. Eng. ASCE **136**(5), 686–696 (2010)
2. IS: 2720-Part 3/sec.1: Indian Standard Methods of Test for Soils: Determination of Specific Gravity of Fine-Grained Soils. BIS, New Delhi (1980)
3. IS: 2720-Part 2: Indian Standard Methods of Test for Soils: Determination of Moisture Content Using Rapid Moisture Meter. BIS, New Delhi (1980)
4. IS: 2720-Part 4: Indian Standard Methods of Test for Soils: Grain Size Analysis. BIS, New Delhi (1985)
5. IS: 2720-Part 5: Indian Standard Methods of Test for Soils: Determination of Atterberg's Limit. BIS, New Delhi (1985)
6. IS: 2720-Part 7: Indian Standard Methods of Test for Soils: Light/Standard Proctor Compaction Test of Soil. BIS, New Delhi (1980)
7. IS: 2720-Part 15: Indian Standard Methods of Test for Soils: Determination of Consolidation Properties. BIS, New Delhi (1986)
8. IS: 2720-Part 17: Indian Standard Methods of Test for Soils: Laboratory Determination of Permeability. BIS, New Delhi (1986)
9. Karim, M.R., Lo, S.C.: Estimation of the hydraulic conductivity of soils improved with vertical drains. Comput. Geotech. **63**, 299–305 (2015)
10. Karunaratne, G.P.: Prefabricated and electrical vertical drains for consolidation of soft clay. Geotext. Geomembr. **29**(4), 391–401 (2011)

11. Tran-Nguyen, H.H., Edil, T.B., Schneider, J.A.: Effect of deformation of prefabricated vertical drains on discharge capacity. Geosynthetics Int. **17**(6), 431–442 (2010)
12. Watabe, Y., Shinsha, H., Yoneya, H., Ko, C.: Description of partial sandy layers of dredged clay deposit using penetration resistance in installation of prefabricated vertical drains. Soils Found. **54**(5), 1006–1017 (2014)

Effect of Addition of Scrap Tire Chips in Stabilization of Clayey Sand

C. N. V. Satyanarayana Reddy, G. Tabitha, S. Srikanth Reddy, and K. Chandranna

1 Introduction

Cohesive soils are invariably used in the construction of embankments for canal bunds, earth dams and highway and railway bridge approaches. Impervious non-swelling clays are preferred in canal bunds and inner shells of the earth dams whereas clayey sands and clayey gravels are preferred in highway/railway embankments for strength. In bridge approaches, as the required height of embankments is more, the usage of clays is permitted except that the top 1–2 m portion is filled with soils of higher strength. As the stability of embankment depends on shear strength of material to prevent slope failures, certain countries [15] specify minimum CBR values for use of soils in embankments of varying heights. So, it is essential to ensure adequate shear strength for cohesive soils to avoid slip failures in high embankments. So, in the present study, an attempt is made to improve the CBR of clayey sand to make it suitable for use in the construction of moderate to high embankments by stabilizing with the addition of shredded scrap tire chips of 10 and 20 mm size in proportions of 5, 10, 15 and 20% by weight of soil.

Studies of stabilization of soils using scrap tire chips have been conducted by researchers across the world to find useful and effective means of disposal of scrap tires and to improve the properties of soil and aggregate [3, 6, 12, 16]. Granular soils are reported to have significant benefits through stabilization by the addition of tire chips. Tire chips are added to soil in sizes varying from 30 to 110 mm and it is

C. N. V. Satyanarayana Reddy (✉) · G. Tabitha
Andhra University College of Engineering, Visakhapatnam 530003, Andhra Pradesh, India
e-mail: prof.cnvsreddy@andhrauniversity.edu.in

S. Srikanth Reddy
S.R.K.R. Engineering College, Bhimavaram 534201, India

K. Chandranna
Vignan's Institute of Information Technology, Duvvada 530049, Visakhapatnam, India

C. N. V. Satyanarayana Reddy et al. (eds.), *Ground Improvement and Reinforced Soil Structures*, Lecture Notes in Civil Engineering 152,
https://doi.org/10.1007/978-981-16-1831-4_37

reported that the angle internal friction improved significantly and CBR of granular soils increased by 2–3 folds. However, it is reported that clays do not have significant stabilizing action by the addition of tire chips from studies conducted using tire chips of sizes above 30 mm [14].

Many researchers have carried out studies on the influence of addition of tire shreds/chips in clay soils. It is observed that the addition of different proportions of tire chips (5–50%) significantly improved the CBR and unconfined compressive strength and reduced compressibility and permeability of clay [1, 2, 5]. Partial replacement of black cotton soil/clay soil with tire chips (4.75–2.00 mm) has resulted in a decrease in weight of fill as well as swell pressure. Hence soil stabilized with tire chips can be used as a lightweight fill material for retaining walls and embankments constructed on relatively weak soils resulting in lower earth and bearing pressures [3, 5–7]. Tire shreds can be effectively used for landfill applications subjected to extreme acidic conditions and are highly durable [11].

2 Details of Study

2.1 Properties of Soil

The clayey sand used in the study is procured from P.M. Palem, Visakhapatnam and tested for engineering properties in a laboratory by performing tests as per relevant codes of practice, IS 2720. The engineering properties of soil determined from the tests are presented in Table 1.

The soil is classified as clayey sand based on gradation and plasticity characteristics as per IS 1498 [8]. The soil has fines in excess of 40% and poor drainage and low CBR value in soaked condition. The soil has insignificant swelling as the free swell index is 20%. The clayey sand is suitable for the construction of highway embankments of heights up to 3–9 m but not suitable for embankments of height exceeding 9 m based on CBR value as per TRH-09 [15].

2.2 Properties of Tire Chips

Tire chips of size 10 mm × 10 mm and 20 mm × 20 mm are cut from scrap tires for use in the study. The tire chips have a specific gravity of 1.05, bulk density of 0.64 g/cc and water absorption of 2.2%.

Table 1 Engineering properties of soil

Property	Value
Specific gravity	2.68
Grain size distribution	
a. Gravel (%) b. Sand (%) c. Fines (%) (i) Silt (%) (ii) Clay (%)	0 58 42 28 14
Plasticity characteristics	
a. Liquid limit (%) b. Plastic limit (%) c. Plasticity index (%)	37 22 15
IS classification symbol	SC (clayey sand)
Compaction characteristics	
a. Maximum dry density (g/cc) b. Optimum moisture content (%)	1.85 13.4
Coefficient of permeability (cm/s)	1.12×10^{-6}
Shear parameters	
a. Angle of internal friction b. Cohesion (KN/m^2)	24° 42
California bearing ratio (%)	3.2
Free swell index (%)	20

3 Studies on Soil-Tire Chip Mixes

3.1 Compaction Characteristics

The clayey sand under study is mixed with shredded tire chips of 10 and 20 mm size in proportions of 5–20% by weight of soil, in increments of 5%. The compaction characteristics of the soil-tire chip mixes are determined from IS heavy compaction tests and the compaction characteristics obtained from the tests are presented in Table 2.

3.2 Strength Characteristics

The shear parameters of soil-tire chip mixes are determined by testing specimens of 50 mm diameter and 100 mm length prepared by compacting at OMCs and respective MDDs in Triaxial test apparatus. The shear parameters are also determined on saturated specimens in undrained conditions. The shear parameters are determined

Table 2 Compaction characteristics of tire chips stabilized soil

Size of tire chips	Details of mix	Compaction characteristic	
		OMC (%)	MDD (g/cc)
10 mm	Soil + 0% tire chips	13.4	1.85
	Soil + 5% tire chips	12.9	1.83
	Soil + 10% tire chips	11.8	1.80
	Soil + 15% tire chips	11.4	1.74
	Soil + 20% tire chips	10.8	1.72
20 mm	Soil + 0% tire chips	13.4	1.85
	Soil + 5% tire chips	12.6	1.82
	Soil + 10% tire chips	11.2	1.78
	Soil + 15% tire chips	10.9	1.72
	Soil + 20% tire chips	10.4	1.70

by plotting Mohr's circles and by plotting the failure envelopes. The shear parameters of soil-tire chip mix specimens determined from the tests are given in Table 3.

In addition to shear parameters, the CBR values of soil-tire mixes are important in highway embankment application. Hence, CBR tests [9] are performed on the soil-tire chip mix specimens prepared at OMCs and respective MDDs after soaking for 96 h. CBR values are determined by plotting load-penetration curves and applying zero correction (wherever it is required) as per IS 2720 part 16. The results of CBR tests are given in Table 4.

3.3 *Permeability of Soil-Tire Chip Mixes*

The permeabilities of clayey sand and tire chips mixed clayey sand specimens prepared at OMCs and respective MDDs are tested after saturation in variable head permeability test apparatus. The test is performed as per IS 2720 part 17 [10]. The results of permeability tests are presented in Table 5.

Table 3 Shear parameters of tire chips stabilized soil

Size of tire chips	Soil-tire chips mix	Shear parameters			
		OMC-MDD compacted state		Saturated state	
		Angle of internal friction	Cohesion (KN/m^2)	Angle of internal friction	Cohesion (KN/m^2)
10 mm	Soil + 0% tire chips	24°	42	21°	35
	Soil + 5% tire chips	26°	38	23°	32
	Soil +10% tire chips	28°	34	26°	30
	Soil +15% tire chips	30°	32	28°	28
	Soil +20% tire chips	32°	29	30°	25
20 mm	Soil + 0% tire chips	24°	42	21°	35
	Soil + 5% tire chips	25°	40	22°	33
	Soil +10% tire chips	26°	36	24°	32
	Soil +15% tire chips	28°	33	29°	30
	Soil +20% tire chips	30°	30	26°	27

Table 4 CBR of tire chips stabilized soil

Size of tire chips	Mix details	Soaked CBR (%)
10 mm	Soil + 0% tire chips	3.2
	Soil + 5% tire chips	4.2
	Soil + 10% tire chips	4.9
	Soil + 15% tire chips	5.3
	Soil + 20% tire chips	4.4
20 mm	Soil + 0% tire chips	3.2
	Soil + 5% tire chips	3.6
	Soil + 10% tire chips	4.2
	Soil + 15% tire chips	5.1
	Soil + 20% tire chips	4.2

Table 5 Permeability of tire chips stabilized soil

Size of tire chips	Mix details	Permeability (cm/s)
10 mm	Soil + 0% tire chips	1.10×10^{-6}
	Soil + 5% tire chips	1.12×10^{-5}
	Soil + 10% tire chips	4.30×10^{-5}
	Soil + 15% tire chips	8.40×10^{-5}
	Soil + 20% tire chips	1.12×10^{-4}
20 mm	Soil + 0% tire chips	1.10×10^{-6}
	Soil + 5% tire chips	2.92×10^{-5}
	Soil + 10% tire chips	6.34×10^{-5}
	Soil + 15% tire chips	9.34×10^{-5}
	Soil + 20% tire chips	1.32×10^{-4}

3.4 Coefficient of Elastic Uniform Compression of Tire Chips Stabilized Soil

The effect of addition of tire chips on the coefficient of elastic uniform compression (C_u) is evaluated from small-scale cyclic load tests conducted specimens prepared in CBR Mould by compacting in IS heavy compaction condition after soaking for 96 h. The load tests are conducted in a self-straining loading frame using the CBR plunger as loading plate. Loading is done up to anticipated safe bearing capacity (150 kPa) in increments of 10 kPa and load at each stage is maintained till settlement reached equilibrium. The load at each stage is released and the plate is allowed to rebound and then loaded to the next higher load. Based on the dial gauge readings of loading and unloading stages, elastic settlements are determined and plots are made between applied load intensities and corresponding elastic settlements as shown in Figs. 1 and 2. The values of C_u determined as the slopes of liner portions of the plots of different specimens from the tests are presented in Table 6.

The values of C_u are determined for foundations of base areas equal to or more than 10 m^2 using Barkan's equation [4] and are presented in Table 6. The values are applicable for circular vibrating base as circular load plate is used in cyclic load tests.

As the values of C_u are influenced by the shape of footing [13] in soils, the values are to be modified for shape effect for use with vibrating bases of square shape.

4 Discussion

The compaction characteristics in Table 2 indicate that both OMC and MDD values of soil-tire chip mixes decreased with an increase in addition of tire chips. This is attributed to lower water absorption and lower density of tire chips compared to soil. At a given percentage addition of tire chips, 10 mm tire chips resulted in relatively

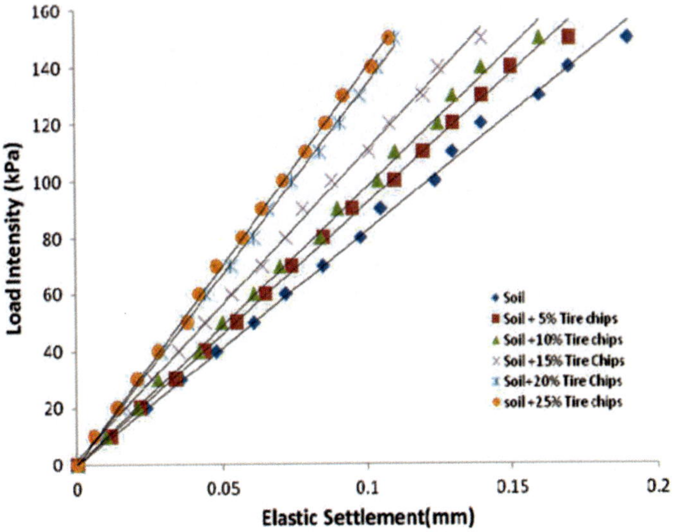

Fig. 1 Load–elastic settlement plots of clayey sand stabilized with scrap tire chips of 10 mm size

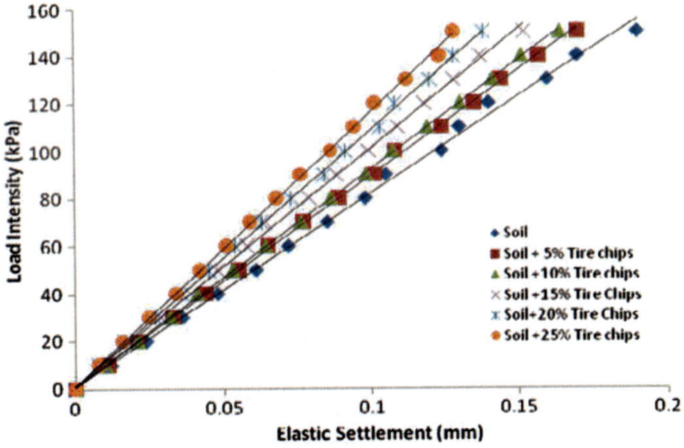

Fig. 2 Load–elastic settlement plots of clayey sand stabilized with scrap tire chips of 20 mm size

higher MDD compared to 20 mm tire chips due to better mixing of tire chips due to smaller size.

The results presented in Table 3 indicate that the angle of internal friction of tire chips stabilized soil increased and cohesive strength decreased with an increase in the percentage of tire chips. The increased angles of internal friction are due to the inclusion of coarse-grained tire chips, which are relatively tougher compared to soil. A decrease in cohesion is resulting from increased heterogeneity of tire chips mixed soil with increased tire chips addition. At a given percentage, the addition of tire

Table 6 Values of coefficient of elastic uniform compression of clayey sand and scrap tire chips stabilized clayey sand

Tire chips content by weight of soil (%)	Size of tire chips: 10 mm		Size of tire chips: 20 mm	
	$(C_u)_p$ in kN/m^3	$(C_u)_f$ in kN/m^3	$(C_u)_p$ in kN/m^3	$(Cu)_f$ in kN/m^3
0	8.2×10^5	1.15×10^4	8.2×10^5	1.15×10^4
5	9.1×10^5	1.28×10^4	9.0×10^5	1.26×10^4
10	9.5×10^5	1.33×10^4	9.4×10^5	1.32×10^4
15	11.3×10^5	1.58×10^4	10.2×10^5	1.43×10^4
20	13.5×10^5	1.89×10^4	11.1×10^5	1.55×10^4
25	14.2×10^5	1.99×10^4	11.9×10^5	1.67×10^4

chips, 10 mm size tire chips, resulted in slightly higher values of angle of internal friction over 20 mm size, which is attributed to a relatively better distribution of tire chips in soil due to smaller size and better mobilization of friction with sand fraction of soil. The permeability of soil is observed to increase with an increase in the addition of tire chips (Table 4) with a relatively higher increase for 20 mm size tire chips over 10 mm tire chips at a given percentage addition of tire chips. This is due to the possible higher void ratio of stabilized soil due to mixing of coarser tire chips. Tire chips of 20 mm size in the soil make the compaction less effective and result in higher voids.

From Table 5, it can be seen that the soaked CBR value of soil increased and attained maximum value at 15% addition of tire chips, with relatively higher CBR for 10 mm size tire chips addition compared to 20 mm size tire chips addition. Hence, the optimum percentage of tire chips for the stabilization of clayey sand under study is 15% for both 10 and 20 mm size tire chips. However, 10 mm size tire chips are to be preferred to stabilize the soil as 20 mm size tire chips will have a relatively more chance for punching under the loads and result in lesser shear strength.

The study indicates the addition of 15% of tire chips helps in improving the soaked CBR by 65% and with a soaked CBR value of 5.3%, it is suitable for constructing embankments with heights of 9–15 m. Furthermore, with increased permeability, its performance will be better as it is transformed to semi-pervious soil from nearly impervious soil.

From Table 6, it can be seen from the results of small-scale cyclic load tests that the values of C_u of clayey sand increased with an increase in tire chips content, with 10 mm size tire chips resulting in slightly higher values. The results reveal that the addition of 15–20% of tire chips of 10 mm size in clayey sand increased C_u by 35–65% whereas the addition of 15–20% of tire chips of 20 mm size has resulted in an increase of C_u by 25–35% only.

5 Conclusions

The following conclusions are drawn from the studies conducted on stabilization of cohesive soil under study with 10 and 20 m sized scrap tire chips.

1. Optimum percentage of tire chips for stabilizing clayey sand under study is 15% by weight of soil.
2. 10 mm size tire chips exhibited a better stabilizing effect in clayey sand under study compared to 20 mm size tire chips.
3. The percentage improvement in CBR value of clayey sand under study at optimum percentage addition of 10 mm size tire chips is 65%.
4. The clay soil under study stabilized by the addition of 15% of 10 mm size tire chips (by weight of soil) is suitable for use in the construction of embankments of 9–15 m as the CBR value is 5.3%.
5. The permeability of clayey sand stabilized with optimum content (15%) of 10 mm size tire chips improved by about 75 times the permeability of untreated soil.
6. The coefficient of elastic uniform compression of clayey sand under study increased by 35–65% by stabilizing with addition of 10 mm size scrap tire chips.

Hence, the stabilization of clayey sand with 10 mm size scrap tire chips may be considered for use as construction material in highway embankments of higher heights and also as an improved foundation bed for supporting machine foundations.

References

1. Ahmed, I., Lovell, C.W.: Rubber soils as light weight geomaterials, light weight artificial and waste materials for embankments over soft soils. In: Transportation Research Record, vol. 1422, pp. 61–70. National Academy Press, Washington, DC (1993)
2. Al-Tabbaa, A., Blackwell, O., Porter, S.A.: An investigation into the geotechnical properties of soil-tire mixtures. Environ. Technol. **18**(8), 860–885 (2010)
3. Srivastava, A., Pandey, S., Rana, J.: Use of shredded tire waste in improving geotechnical properties of expansive black cotton soil. Int. J. Geomech. Geoeng. **9**(4), 303–311 (2014)
4. Barkan, D.D.: Dynamics of Bases and Foundations. MC Graw Hill Co., New York (1962)
5. Edil, T.B.: Mechanical properties and mass behaviour of shredded tire—soil mixtures. In: Proceedings of International Workshop on Lightweight Geo-materials, 26–27 March 2002, pp. 17–32, Tokyo, Japan (2002)
6. Edil, T.B., Bosscher, P.J.: Engineering properties of tire chips and soil mixtures. Geotech. Test. J. GTJODJ **17**(4), 453–464 (1994)
7. Cetin, H., Fener, M., Gunaydin, O.: Geotechnical properties of tire-cohesive clayey soil mixtures as a fill material. Eng. Geol. **88**, 110–120 (2006)
8. IS 1498: Classification and Identification of Soils for General Engineering Purposes. Bureau of Indian Standards, New Delhi (1970)
9. IS 2720 Part 16: Method of Test for Soils: Laboratory Determination of CBR. Bureau of Indian Standards, New Delhi (1987)

10. IS 2720 Part 17: Method of Test for Soils: Laboratory Determination of Permeability. Bureau of Indian Standards, New Delhi (1986)
11. Reddy, K.R., Saicheck, R.E.: Characterisation and performance assessment of shredded scrap tires as leachate drainage material in landfills. In: Proceedings of Fourteenth International Conference on Solid Waste Technology and Management, Philadelpihia (1998)
12. Satyanarayana Reddy, C.N.V., Durga Rani, K.: Potential of shredded scrap tyres in flexible pavement construction. J. Indian Highways **41**(10), 19–24. Indian Roads Congress (2013)
13. Satyanarayana Reddy, C.N.V., Usha Rani, G.V.: Influence of shape of footing on coefficient of elastic uniform compression of foundation soils. Indian Geotech. J. **50**(4), 664–669, India (2020)
14. Tatlisoz, N., Benson, C., Edil, T.: Effect of fines on mechanical properties of soil-tire chip mixtures. In: Wasemiller, M.A., Hoddinott, K.B. (eds.) Testing Soil Mixed with Waste or Recycled Materials, ASTM STP 1275. American Society for Testing and Materials (1997)
15. TRH 9: Construction of Road Embankments, Technical Recommendations for Highways, pp. 1–42, South Africa (1982)
16. Yoon, S., Prezzi, M., Siddiki, N., Kim, B.: Construction of a test embankment using a sand–tire shred mixture as fill material. Waste Manag. **26**, 1033–1044 (2006)

Ground Modification Techniques for Deep Soft Soils Sites in Goa Region

Roshani K. Majik and Purnanand P. Savoikar

1 Introduction

In recent years, the increasing population has led to rapid urbanization and the demand for construction activities like highways, buildings and other infrastructures has increased. Space constraints and high cost of land for construction purposes have made engineers to utilize the land having weak soils, where the site conditions are not ideal. This is where the geotechnical engineer has to take the charge of improving the soil conditions in order to carry out construction on appropriate stable ground for different civil engineering projects. In the current scenario, the role of ground modification techniques has become a crucial task for various construction projects. Soils in the coastal belts like marine clay and soft alluvial soils are of weak composition, highly compressible, having very low permeability soft soils generally ranging from silts to clay at various locations and depths. To carry out construction activities in such soils, having low shear strength, it is necessary to stabilize the ground before constructing any infrastructure to prevent differential and total settlement and also to prevent liquefaction of the soil.

In the present study, six sites were selected in Goa and ground modification schemes have been designed for each site at Karmali, Margao, Siolim, Merces Vasco, Vaddem Vasco and Gogol Margao. Geotechnical soil investigations were carried out at these locations in Goa to evaluate the soil strata. Borelog data and soil samples were collected.

R. K. Majik · P. P. Savoikar (✉)
Goa Engineering College, Farmagudi 403401, Goa, India

C. N. V. Satyanarayana Reddy et al. (eds.), *Ground Improvement and Reinforced Soil Structures*, Lecture Notes in Civil Engineering 152,
https://doi.org/10.1007/978-981-16-1831-4_38

2 Literature Survey

The work of installing soil–cement columns besides a metro tunnel station to stabilize the soft soils, deep soil mixing using triple shaft method was proposed by Chen et al.[1]. It was observed that triple shaft caused unacceptable displacement. Thus, the field tests were conducted in two phases, Phase I was conducted using single shaft deep soil mixing and in Phase II, multiple shaft deep mixing was used to know the factors affecting the deep soil mixing process. Finally, it was concluded that the installation process could be modified without causing much disturbance by using a higher water/cement ratio at a lower mixing speed.

A detailed study of stone columns, their installation technique, designing of stone columns, equipment requirements and their failure mechanisms was reported by Golakiya and Lad [2]. In this paper, a case study is taken up and the stone columns are designed based on bore log data.

A thorough study on different ground modification techniques, their uses in different types of soils, the efficiency of each technique, also how economically feasible they are in the present scenario, was reported by Mishra [3].

An overview of the experimental, theoretical and numerical developments of soft ground modification using Prefabricated Vertical Drains (PVDs) along with natural fiber drains combined with surcharge and vacuum preloading was reported where it was found to be efficient and cost-effective technique to accelerate consolidation [4].

Soil modification technique using chemical stabilization is the most successful technique as reported by Nagaraju [5], which presents the potential of geo-polymer technology in soil stabilization that it could be used as an alternative to traditional stabilizers. Geo-polymer stabilization enhances soil properties and has high strength and low cost.

A case study was reported about a case study on ground improvement using stone columns installation followed by testing the integrity of stone columns using plate load test [6]. From the studies, it was concluded that the bearing capacity of the sub-soil strata was improved after opting for stone column technique and was found to be efficient and economical.

Design of soft ground improvement using sand compaction pile (SCP) was reported around the abutment of Railway Bridge [7]. The soil was found to be thick soft clay underlain by medium dense to dense sand layers. In order to control the lateral deformation of abutment, soft clay layer was required to be improved. It was concluded that using very low area replacement ratios (0.06), lateral displacement of abutment can be controlled if SCP is done under and around abutment.

The use of microbial induced carbonate precipitation (MICP) in soft clays was carried out to find out the efficiency of MICP in clay using soft clay specimen samples and mixing it with a solution containing Sporosarcina pasteurii bacteria (solution with different concentrations of nutrient salts) [8]. Mixing all these together resulted in the formation of calcium carbonate and thus simplifying the soft clays. It was concluded that it is feasible to use microbial-induced carbonate precipitation to increase the strength of soft clays.

3 Ground Modification Techniques

Ground modification techniques for shallow weak soils include methods like removal and replacement of weak soil, use of stone columns, dewatering techniques, use of sand compaction piles, sand drains or prefabricated vertical drains, etc. These methods are selected based on types of soils encountered, groundwater table position and type of structure proposed at the site. The new methods like microbial-induced carbonate precipitation (MICP) that has been recently adopted at some sites indicated that certain microorganisms can change the mechanical properties of the soil. This is mainly suitable for sand because sand has a very high permeability and also a large grain size. But experiments in clay have shown that it could also be used in clay as a binder. In this method, microorganisms are used for increasing the bond between the soil particles, which increases the shear strength. This method can only be used for very shallow depths. Bacteria like Microaerophilic bacteria, Anaerobic fermenting bacteria, Anaerobic respiring bacteria, etc. are used.

Another latest method of shallow ground improvement is the use of colloidal silica. Colloidal silica is used to stabilize the ground to mitigate liquefaction by injecting it into the soil. Colloidal silica is a suspension of silica particles in a liquid phase, which are capable of forming hydrogen bonds. Silica particles can be converted to gel by adjusting the pH concentration of the solution. Colloidal silica acts as moisture absorbent when induced into the soil and hence prevents liquefaction of the soil. It can also be used as high temperature binders. It is a very cost-efficient technique. Colloidal silica possesses properties like low viscosity and is non-toxic, hence it will not contribute in polluting the groundwater resources.

The various ground improvement techniques adopted for deep soft soil sites include methods like vibro compaction, dynamic compaction, blast densification, compaction grouting, jet grouting, permeation grouting, deep soil mixing, ground heating, ground freezing, etc. which are adopted based on the type of the soil, ground conditions and type of the structure proposed on the site.

4 Case Studies of Soft Soil Sites in Goa

4.1 Introduction

In the present paper, a detailed account of ground modification techniques for deep soft soils encountered in Goa is presented. Six sites were selected in Goa at the following places; Karmali, Margao, Siolim, Merces Vasco, Vaddem Vasco-da-gama and at Gogol Margao where soft soils were observed.

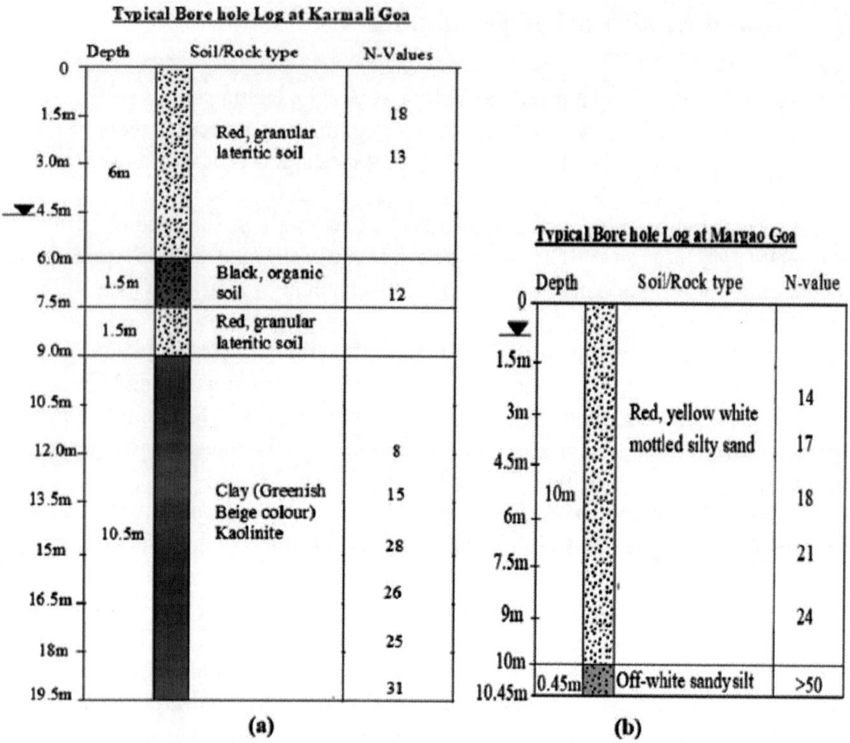

Fig. 1 Typical bore logs at **a** Karmali Goa and **b** near KRCL Margao Goa site

4.2 Ground Modification for Soft Soil at Karmali (Site I)

4.2.1 Site Conditions

This site is located at Karmali, adjacent to Carambolim Lake. A G + 1 RCC structure is proposed to be built at this site over an existing platform, which was built previously. The borehole was drilled up to a depth of 19.5 m from the ground level. Borehole log is shown in Fig. 1a.

4.2.2 Ground Modification Recommended

As per the site information obtained from the borelogs and laboratory testing, saturated marine clay is encountered at deep depths. The structure to be built is a low-rise building hence the soil does not have to carry heavy loads. Hence shallow ground modification is recommended. To stabilize the soil beneath the footings of the G + 1 RCC structure, excavation work using an excavator is to be carried out and the soil should be excavated up to a depth of 3.5 m. A layer of hard lateritic soil of 60 cm should be placed and well compacted below each footing. A plain cement concrete layer of 20 cm thick of grade 1:3:6 is recommended. All the columns should be connected to the plinth beams once the footings are complete. Taking into consideration the fact that the second row of columns is close to the bank of the lake and that very soft clay layer exits below, also that the water in the Lake rises during the monsoons, it is recommended to provide a cut-off wall near the second row of the columns on sloping pitched portion. This cut-off wall can be constructed by extending the plinth beams connecting the column by about 0.8 m and providing a vertical small RCC wall, which goes into the soil existing in the sloping portion by a meter. This will provide stability to the column near the bank and will prevent the movement of soil from sloping portion into the lake. This would be the most economical ground strengthening technique, which could be used to increase the bearing capacity of that area for low-rise buildings.

4.3 Ground Modification for Soft Soil at Margao (Site II)

4.3.1 Site Conditions

This site is located at Margao, near the Railway Station. A mid-rise building is proposed to be built at this site. The soil encountered here is soft silty sand. The borehole was terminated at a depth of 10.45 m. Hard rock stratum was not encountered beneath. The investigation consisted of one test borehole, which was drilled up to 10.45 m depth. The location of borehole was selected in consultation with the client. Standard penetration tests were performed in the borings in conjunction with samples. The top layer encountered is red, yellow-white mottled silty sand up to a depth of 10 m from the existing ground level. SPT values for the same varied from 14 to 24. This layer was followed by off-white sandy silt up to a depth of 10.45 m. SPT potential of this layer was more than 50. The water table was found at a depth of 1.0 m. A typical borehole log is shown in Fig. 1b.

4.3.2 Ground Modification Recommended

As per the geotechnical soil data obtained from the bore logs and laboratory testing, silty sand was found at this site up to a depth of 10.0 m from the existing ground level. No hard rock or very hard strata was available. Shallow ground modification is recommended to be implemented in order to enhance the soil properties. Since the soil found here is sandy in nature ground modification using Sand compaction piles (SCP's) is the best-suited method to be used. Since the water table was found at a depth of 1.0 m, it is recommended to carry out dewatering using a dewatering pumping system. It was recommended to use non-vibratory sand compaction piles as the vibratory equipment causes vibrations and noise pollution, which disturbs the surrounding environment.

Similar studies were carried out by Bicalho et al. [9], which were based on two case histories densification of loose sands using sand compaction piles. Two cases involving construction of a 10-storey building and a 6-storey apartment building. The results showed the feasibility of installing sand compaction piles for densifying loose soils. The successful installation of piles also resulted in an increase in the penetration resistance of cohesion-less soils. It was concluded that these piles produce a more homogeneous density, reducing future differential settlement. Harada et al. [10] reported various case studies, which were taken up to evaluate the effectiveness of the soil performance by installation of sand compaction piles and also discussed about the recent developments made in installing SCP's. The studies carried out also focused on the development of the non-vibratory sand compaction pile method over the vibratory method. It was concluded that the effectiveness of sand compaction piles was verified and proved through the cases taken up for evaluation.

4.4 Ground Modification for Soft Soil at Dando Siolim (Site III)

4.4.1 Site Conditions

This site is located at Dando Siolim—Goa. A low-rise building is proposed to be built at this site. Completely weathered rock was seen till the drilled depth of 9.60–10.12 m. The calculated safe bearing capacity is found to be 15 t/m^2 at a depth of 2.0 m where the footing is proposed to be built.

4.4.2 Ground Modification Recommended

As the structure to be built is a low-rise building, no heavy loads are to be taken up by the soil. The ground modification should be done for a depth of not more than 3 m. Isolated footings are proposed to be designed for the structure at 2.0 m depth. To strengthen the soil beneath, it is recommended to improve the soil by removing the soft layers of the soil for a depth of about 3.5 m and replacing it with well-compacted and well-graded soil. As the water table was found at a depth of 1.10 m, extensive dewatering should be carried out before implementing the ground modification technique. It is recommended to do ground improvement by removing the top 3.2 m soil, which is clayey sandy silt from the existing ground level for the entire foundation area (building plan +2 m all around) and laying a layer soil, which has requisite properties like lateritic soil for a depth of 50 cm and ramming it properly. A layer of rubble soling 20 cm thick should be laid over this layer and should be well compacted. On top of rubble soling, 30 cm of lateritic soil should be laid and rammed. Over this, rubble soiling of 20 cm should be laid and well compacted. The footings should be constructed over this. This incremental replacement of the layers will definitely result in improving the bearing capacity of soil. To check whether the bearing capacity of the soil has increased, it is suggested to conduct a plate load test over the improved soil to ascertain the improved SBC.

Gabr [11] reported that using replacement method instead of deep ground modification methods for lightweight buildings over layers of soft soils is an economical solution. In the studies carried out, the soft soil replaced with sand and gravel to support lightweight building on shallow foundation is examined by using centrifuge test and numerical modeling. From the studies conducted, it was concluded that the removal of unsuitable soil and replacing it with well-compacted soil can reduce the cost and settlement considerably.

4.5 *Ground Modification for Deep Soft Soil at Merces Vasco (Site IV)*

4.5.1 Site Conditions

This site is located at Merces, Vaddem Vasco. A high-rise residential building is proposed to be built at this site. From the typical borehole log of 40 m, the top layer of the soil observed in the boreholes is dark brown clayey sandy silt of about 3.0 m thick. This layer was followed by a layer of greyish silty clay mixed with fine sand from a depth of 3–8 m. SPT potential varied from 2 to 7 putting it in the soft to medium stiff category. The next layer found was greyish silty clay and the thickness of this layer was found to be 7.50 m. The SPT potential of the same varied from 1 to 3 putting it from very soft to soft category. This layer was followed by reddish-brown lateritic soil, which consisted of completely weathered lateritic rock fragments up to a depth of 40 m. Groundwater was encountered at a depth of 2.50–3.45 m below the existing ground level. It was proposed to build a raft foundation for the structure at a depth of 1.5 m. The safe bearing capacity at site was found to be 150kN/m^2. A typical borehole log is shown in Fig. 2b.

4.5.2 Ground Modification Recommended

As per the soil investigation data obtained from the bore logs and clayey, sandy silt was found at this site. No hard rock or very hard strata was available even at the depths of 40 m. As soil is very soft, the susceptibility to undergo settlement under heavy loads is very high, which might result in failure of the structure. In order to create a firm base for the structure to rest on the soil, it is necessary to opt for ground modification techniques. As the structure is a high-rise residential building, it is necessary to carry out ground improvement for a depth of about minimum 15 m, which will incorporate the area of the pressure bulb of the raft foundation making the soil stable to take up heavy loads. A combination of ground improvement techniques

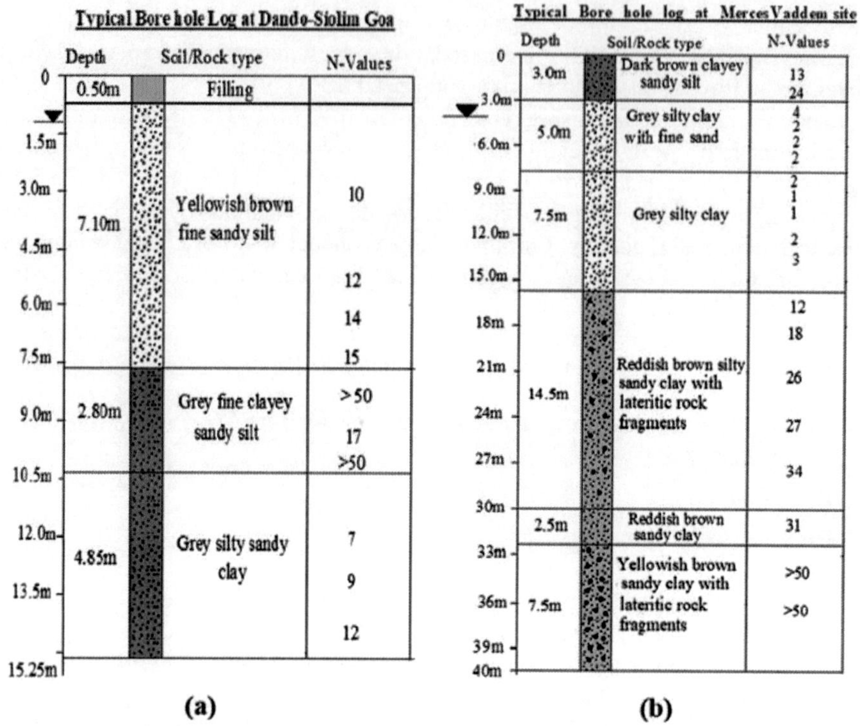

Fig. 2 Typical bore logs at **a** Dando Siolim and **b** Merces Vaddem Vasco site

of deep soil mixing supplementing vibro stone columns could be an effective solution for improving the ground conditions. By evaluating the sub-surface stratum, it is suggested to opt for deep soil mixing for a length of 8–15 m of the bore log. Over this, the vibro stone columns should be placed from a depth of about 1.5–8 m from the existing ground level. The designed raft foundation should be placed over this layer of stone columns at 1.5 m depth. Hence, the ground improvement is achieved. After completing ground improvement as above, plate load test can be done to ensure the effectiveness of ground improvement.

Similar method was studied, designed and executed by Shao and Kinley [12] where ground improvement technique for a building resting on alluvial deposits in San Diego, California, USA. The soil strata consisted of alluvial deposits. There were layers of saturated sand present in the soil stratum, which were liquefiable. Hence, a combined method of deep soil mixing and stone columns was proposed and implemented, which would also control settlement and reduce the risk of liquefaction.

4.6 Ground Modification for Deep Soft Soil at Vaddem Vasco (Site V)

4.6.1 Site Conditions

This site is located Vaddem Vasco-da-gama, Vasco. A high-rise commercial building is proposed to be built at this site. The soil encountered at this site at shallow depths is very soft marine clay up to 14.0 m and hard rock is encountered at a depth of 25 m. Very hard amphibolite rock with inclined joints was encountered at a depth of 23.5 m from the existing ground level. Groundwater was encountered at a depth of 0.6 m below ground level, which may vary with seasons. Borehole log at this site is shown in Fig. 3.

4.6.2 Ground Modification Recommended

Geotechnical observations show that very soft marine clayey silt is encountered at this particular site approximately at a depth of about 14 m and very hard amphibolite rock at 23.5 m. Since rock was found at considerable depth, thus, it is proposed to build a bored cast in situ pile foundation to take up the heavy loads and also to control differential settlement. Before commencing the ground improvement work, the extensive dewatering should be carried out, and water table was found at a depth of 1.10 m.

Bored cast in situ piles are used at similar sites for improving the ground conditions to take up heavy loads. Anirudhan [13] reported that there have been some defects observed in the bored cast in situ piles, which were constructed using rotary drilling rig in South Chennai. It was reported that after the concreting was completed and the temporary casing was withdrawn, it was noticed that there is an issue of sinking of concrete by 2.0 m in 30–40 min after concreting. High strain dynamic tests were conducted on the piles where concreting defects were observed, it was reported that pile capacities were decreased from designed 285 tons to 100 tons. From the thorough study of the case, it was concluded that the defects occurred due to the inappropriate construction practices and also due to improper identification of the soil stratum. The defect was rectified by installing more number of piles in that area.

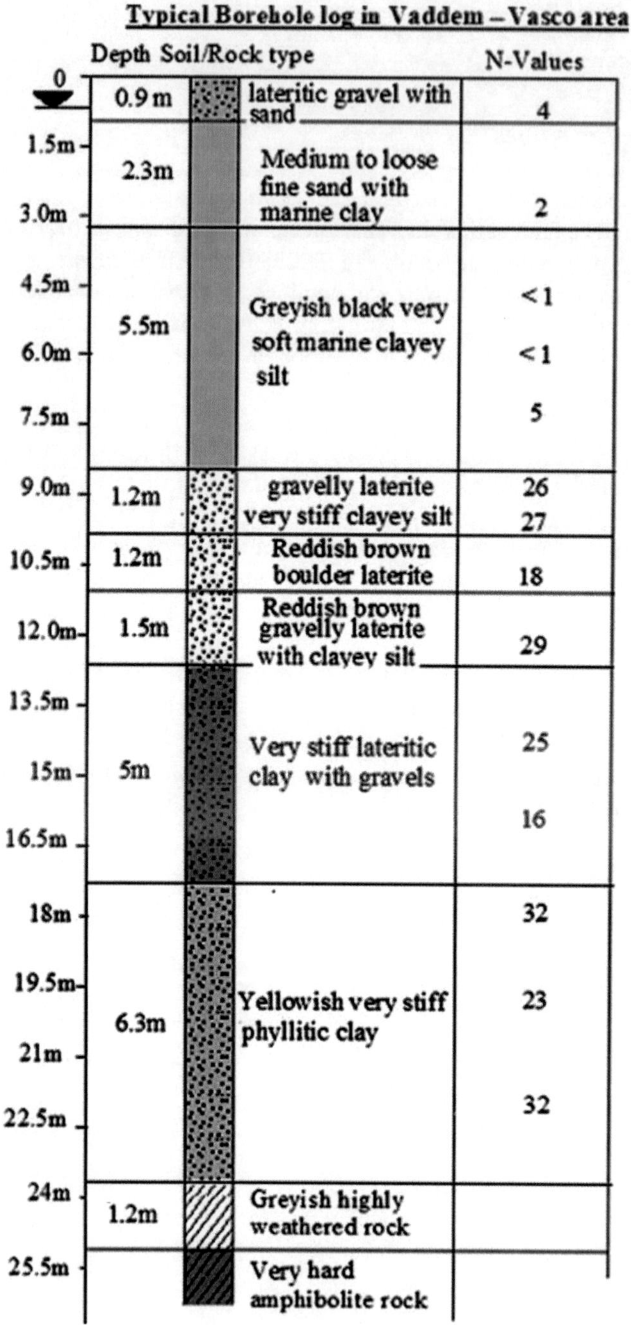

Fig. 3 Typical Borelogs at Vaddem Vasco area

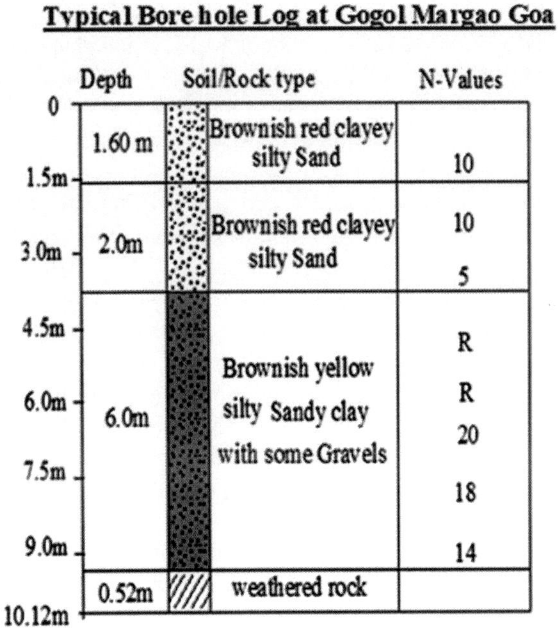

Fig. 4 Typical bore logs at **a** Vaddem—Vasco and **b** Gogol Margao site

4.7 Ground Modification for Deep Soft Soil at Gogol Margao (Site VI)

4.7.1 Site Conditions

This site is located at Gogol, Margao—Goa. A mid-rise building is proposed to be built at this site. Completely weathered rock was seen till the drilled depth of 9.60–10.12 m. From the typical borehole log of 10.12 m shown in Fig. 4, the first layer of the soil observed in the borehole is brownish-red clayey silty sand up to a depth of 3.8 m approximately. The SPT potential of the same varied from 5 to 10. The next layer observed was brownish-yellowish silty sandy clay with some gravels from a depth of 3.8–9.8 m. This layer was followed by a highly weathered rock from a depth of 9.6–10.12 m.

4.7.2 Ground Modification Recommended

The footing is proposed to be built at a depth of 2.00 m from the existing ground level. The existing site has a very low-bearing capacity and very high settlement characteristics, hence it is recommended to carry out ground improvement. Since rock was found at considerable depth thus, to control differential settlement and also to increase the bearing capacity of the soil, a hybrid composite ground modification method was proposed to be executed at this site, which was a combination of vibro stone columns and dynamic consolidation. Providing stone columns in the entire area at the site could be an uneconomical solution, thus, it was proposed to concentrate the stone columns only under the footings to control differential settlement at a depth up to which hard stratum was encountered. The stone columns were supplemented with dynamic consolidation to densify the remaining area surrounding the site. After the completion of ground improvement as above, plate load test can be done to ensure the effectiveness of ground improvement. Alternatively, stone column capacity testing can be done. These ground improvement techniques reduce the excessive settlements by densification.

A similar method was studied, designed and executed by Lopez et al. [14], a composite ground improvement program consisting of vibro replacement columns and dynamic compaction. Three concerns were identified at this site, which were liquefaction susceptibility, low-bearing capacity and differential settlement. It was proposed to install the stone columns concentrated under the footing to tackle the differential settlement of the building. Dynamic compaction was proposed to be performed over the entire area even including the area where stone columns were installed. From the study conducted, it was concluded that the combination of methods resulted in 40% cost reduction, and also the silts and clays experienced a surprisingly high level of improvement.

5 Conclusions

Ground modification techniques recommended for different sites depend upon the type of soil encountered like silt, clay, sand etc., position of groundwater table and depth of the soft soil layers. In addition to standard methods like vibro compaction, dynamic compaction, deep soil mixing, stone columns, sand drains etc. the newer techniques like microbial-induced carbonate precipitation (MICP) and colloidal silica are the latest emerging methods being studied. In the present study, shallow ground modification schemes are designed for soft soils encountered at Karmali, Margao and Siolim as low-rise and mid-rise buildings are proposed to be built, which do not have to carry significantly very heavy loads. In the present study, deep ground modification schemes are designed for deep soft soils at Merces Vasco, Vaddem Vasco and Gogol Margao as high-rise buildings are proposed to be constructed and these soils have to carry very heavy column loads.

References

1. Chen, J., Zhang, L., Zhang, J., Zhu, Y., Wang, J.: Field tests, modification, and application of deep soil mixing method in soft clay. J. Geotech. Geoenviron. Eng. ASCE (2013)
2. Golakiya, H.D., Lad, M.D.: Ground improvement by using stone columns. J. Emerg. Technol. Innov. Res. **2**(11) (2015)

3. Mishra, B.A.: Study on ground improvement techniques and its applications. Int. J. Innov. Res. Sci. Eng. Technol. **5**(1) (2016)
4. Indraratna, B., Baral, P., Rujikiatkamjorn, C., Nguyen, T.T.: Soft Ground Improvement—Theoretical, Experimental, Numerical and Field Studies, Indian Geotechnical Conference 2018, Indian Institute of Science Bengaluru—Karnataka (2018)
5. Nagaraju, V.T.: Potential of Geopolymer Technology Towards Ground Improvement, ACSGE Proceedings, BITS Pilani (2018)
6. Karthikeyan, A., Sahu, S.K.: A Case Study on Ground Improvement Technique Using Stone Column for Secondary Clarifier Resting on Loose Sand, ISSMGE, Symposium on Laterites and Lateritic Soils, Goa (2019)
7. Alam, M.J., Islam, M., Islam, M.S.: Development of soft soil improvement scheme for lateral stability of railway bridge abutment. In: Proceedings of 9th IACMAG Symposium , Gandhinagar (2019)
8. Xiao, J.Z., Wei, Y.Q., Cai, H., Wang, Z.W., Yang, T., Wang, Q.H., Wu, S.F.: Microbial-Induced carbonate precipitation for strengthening soft clay. Adv. Mater. Sci. Eng. Article ID 8140724 (2020)
9. Bicalho, K.V., Castello, R.R., d'Andréa, R.: The densification of loose sand using compaction piles. In: 5th international conference on case histories in geotechnical engineering, New York (2004)
10. Harada, K., Tsuboi, H., Tanaka, Y., Takehara, Y., Fukada, H.: Cases histories and recent development of the sand compaction pile method as a countermeasure against liquefaction. In: Proceedings of the 5th international conference on case histories in geotechnical engineering, New York (2004)
11. Gabr, A.K. :The uncertainties of using replacement soil in controlling settlement. J. Am. Sci. **8**(12) (2012)
12. Shao, L., Kinley, J.: Vibro replacement and soil mixing ground improvements at a shopping mall site in San Diego, California, USA. In: Proceedings of the 5th international conferences on recent advances in geotechnical earthquake engineering and soil dynamics and symposium in honour of professor I. M. Idriss (2010)
13. Anirudhan, I.V.: Construction of bored cast-in-situ piles using rotary drilling procedure—a case study. In: Proceedings of Indian Geotechnical Cinference, IGC2009, Guntur, India (2009)
14. López, R.A., Mathew, J.F., Rollins, K., Davis, S.W., Batchko, Z.: Composite ground modification system: vibro-replacement and dynamic compaction, Salt Lake County Detention Center, Utah. In: Fourth International Conference on Case Histories in Geotechnical Engineering, St. Louis, Missouri (1998)

Strength Characteristics of Kuttanad Soil Stabilized with a Biopolymer Guar Gum

Regi P. Mohan and P. Adarsh

1 Introduction

Clay soils are regarded as complex soils due to the undesirable characteristics, it possess low-bearing strength and high shrinkage or expansive characteristics. All these make the soil prone to failure even at low loads and make the structure constructed above this unstable. Going for a deep foundation is often not cost-effective. An alternative to this is soil stabilization to alter its undesirable characteristics. Soil stabilization is unavoidable in many geotechnical engineering applications such as road or embankment construction, slope stabilization, erosion control, foundation, etc. wherever undesirable characteristics of soils are encountered. Soil stabilization can be achieved by mechanical, physical, and chemical methods out of which chemical soil stabilization is most commonly preferred [1]. In many of the above situations where ground improvement is necessary, it may not be viable to use any of the methods alone. The situations may demand the usage of admixtures along with other methods or sometimes chemical soil stabilization alone to enhance the engineering characteristics of soil under consideration to the preferable value [2]. Stabilizing agents enhance the strength characteristics of weak subgrade soils resulting in reduced construction cost and improved performance. Significant amounts of the world's carbon dioxide (5–7%) and nitrogen oxide gas emissions occur during the production of conventional additives like cement, lime, etc. The production of these also accounts for around 15% of the total energy consumption in the global business sector [3]. This causes a high carbon footprint or high energy demand, leading to global warming. Hence the search for alternate additives in broader research areas resulted in the application of nanotechnology or biotechnology in soil stabilization. Bioenzymes, nanomaterials, etc. that can be grouped as nontraditional additives are dominant in recent years in the field of clay soil stabilization due to its advantages

R. P. Mohan · P. Adarsh (✉)
Sree Buddha College of Engineering, Alappuzha, Kerala, India

C. N. V. Satyanarayana Reddy et al. (eds.), *Ground Improvement and Reinforced Soil Structures*, Lecture Notes in Civil Engineering 152,
https://doi.org/10.1007/978-981-16-1831-4_39

from a sustainable point of view. These additives react faster, cause less energy demand in production and fewer carbon emissions. Hence they are eco-friendly. But the effects of these new generation additives on the properties of soil are still being researched.

Biopolymers are polymers that are produced by living organisms. These are stable, carbon-neutral, renewable, and hence environmentally friendly. The ultimate products of biopolymer decomposition are carbon dioxide, water, and ammonia [4]. Biopolymers have a broad spectrum of applications in agribusiness, biotechnology, processed food industry, chemical industry, power sector, environmental protection, and remediation [2]. Even from the 1940 s, the utilization of biopolymers as a soil strengthening agent for soil was under consideration. Commonly used biopolymers for soil modification were guar gum, xanthan gum, chitosan, sodium alginate, etc. Different researches have been conducted in the past several years to understand the suitability of various biopolymers in improving soil properties and studies revealed that biopolymers are capable of enhancing soil properties considerably [5–8]. Some studies revealed that the unconfined compressive strength of soil enhanced more than 1.5 times with the biopolymer addition [9–11]. Toxicity studies on some of these biopolymers done by the World Health Organization (1987) found that they do not cause health hazards [2]. Hence biopolymer modification of soil can be considered as an ecologically sound technology in the field of ground improvement [2]. Based on a review of previous studies, it is noted that, despite the beneficial properties of biopolymers, the lack of adequate research to understand their impact on soil behavior and the lack of standard methods for laboratory testing have prevented the use of biopolymers to become widely accepted for soil modification [12].

Kuttanad soils are soft highly organic clay or silt deposits with higher compressibility and lower strength characteristics found in the Kuttanad areas of the Alappuzha district in Kerala, India. Kuttanad comprises of Vembanad Lake and surrounding marshy land. These soils are unstable and had caused a lot of failures to the structures built over them. Hence studies are being conducted for the effective stabilization of Kuttanad clays by imparting desired characteristics using stabilizing agents for improved performance.

A detailed summary of the laboratory test program conducted for finding the influence of a nontraditional environmentally friendly soil stabilizer biopolymer, guar gum on strength characteristics of Kuttanad soil such as compaction, UC strength, and CBR strength is presented in this paper.

2 Materials and Testing

2.1 Material Characterization

Soil. Clay soil studied was collected from the Champamkulam area of the Kuttanad region in the Alappuzha district of Kerala. The soil was collected at 1.5 m depth

Table 1 Untreated soil characterization

Property	Value
Clay content (%)	30
Specific gravity	2.66
Liquid limit (%)	65
Plastic limit (%)	31
Plasticity index (%)	34
IS classification	CH
Maximum dry density (kN/m^3)	15.2
Optimum moisture content (%)	21.2
Unconfined compressive strength (UC) (kN/m^2)	93
Soaked CBR strength (%)	2.9
Unsoaked CBR strength (%)	4.0

from ground level. It is dark grey. Untreated soil characteristics were determined according to IS specification and are given in Table 1.

According to IS classification, collected soil is classified as inorganic clay of high plasticity from the plasticity chart. From Table 1, it can be inferred that the UC strength of base soil is moderately hard and is having low CBR strength.

Biopolymer. The biopolymer used in the study is guar gum (GG) or guaran powder. Guar gum is dry cream-cultured powder and was procured commercially. It is cold-water-soluble. It is a galactomannan polysaccharide extracted from guar beans, which are non-ionic. The main chemical constituent of guar gum is galactomannan, which amounts to 75–85%. The main characteristic of guar gum is its ability to form viscous colloidal solutions when mixed with water. The chemical structure of the biopolymer is given in Fig. 1. Guar gum is obtained from seeds of the guar plant, which is found in India. The guar seeds are mechanically dehusked, hydrated, milled, and screened according to their use [2].

Guar gum has a mannose: galactose ratio of approximately 2:1 [13]. The backbone in guar gum is a linear chain of β 1, 4—linked mannose remains on which galactose residues are 1, 6—linked at every second mannose, creating short side branches [14].

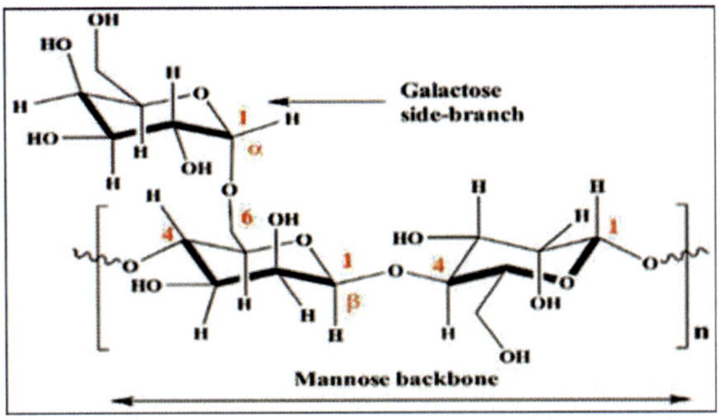

Fig. 1 Chemical structure of guar gum [1]

These are neutral polysaccharides with numerous hydroxyl groups. These hydroxyl groups are distributed in both main chains and side chains and form hydrogen bonding with each other, which renders higher viscosity [15]. In polar solvents, guar gum swells on dispersal and form strong hydrogen bonding while in non-polar solvents, it forms weak hydrogen bonding [16].

In the soil–biopolymer mixture, a portion of the biopolymer enters the voids between soil particles while some portion sticks to the soil surface. Chemisorptions, hydrogen bonding, and physical adsorption by Vander Waals bonds are possible at the soil–biopolymer interfaces. The ionic or electrostatic bonds that are the primary bonds have the maximum bond energy and Vander Waals forces, which cause physical adsorption to develop the weakest bonds [17]. As a result of bonding due to chemical reactions, a biopolymer encloses soil aggregates and interlinks to form a viscous and elastic membrane structure. These physicochemical reactions may usually take a few days. The modified structure causes an enhancement in the engineering characteristics of soil [10].

2.2 *Specimen Preparation and Testing*

Untreated soil is mixed with different proportions of biopolymer guar gum powder and soil samples were prepared according to IS specification. Guar gum was added in varying contents of 0.5, 1.0, 1.5, 2.0, and 2.5% by dry weight of soil taken for sample preparation to obtain soil–guar gum mixtures. Guar gum powder is mixed gently with water corresponding to OMC to avoid the formation of clumps and a homogenous solution is prepared. The soil was then mixed with this guar gum solution. Samples with varying contents of guar gum were prepared at maximum dry density (MDD) and optimum moisture content (OMC) of untreated soil and were stored in moistened conditions to minimize the loss of moisture. Prepared samples were cured without loss of moisture from 7 to 28 days before testing. All tests were conducted according to IS 2720 specification.

3 Results and Discussion

3.1 *Compaction Parameters*

For studying the influence of curing period on strength characteristics, standard proctor compaction tests according to IS 2720 Part 7: 1980 were conducted to determine the influence of biopolymer guar gum on MDD and OMC of treated samples after 7 days of curing period. Soil–guar gum mixtures were prepared at OMC and cured airtight in polythene bags for 7 days before the compaction tests were conducted. The compaction curves were obtained for various soil–guar gum mixtures and the influence of guar gum on OMC and MDD levels is shown in Fig. 2.

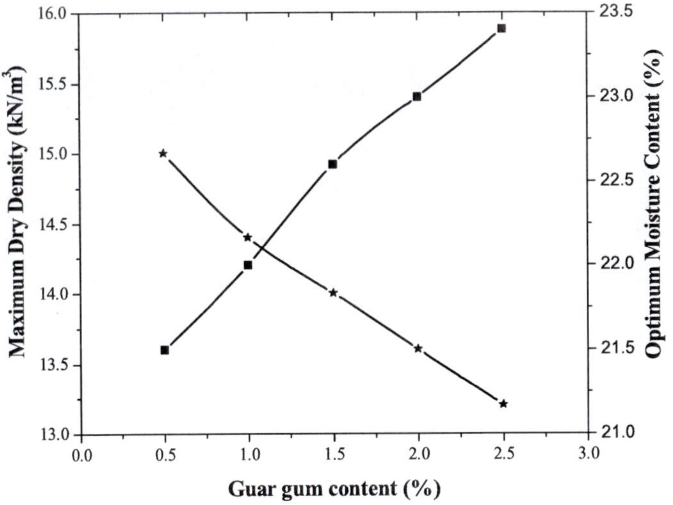

Fig. 2 Variation in compaction parameters with the addition of biopolymer guar gum after 7 days of curing

From Fig. 2, it can be seen that MDD decreases, and OMC increases with an increase in guar gum content. As the guar gum content increases from 0.5% to 2.5%, the MDD value reduces from 15.0 kN/m^3 to 13.2 kN/m^3, and the corresponding OMC increases from 21.5% to 23.4%. It follows the same trend as the previous studies [19, 20]. As the amount of biopolymer guar gum increases, being a lightweight material, these biopolymers absorb water and form suspension with high viscosity and fill the spaces between the soil grains [18]. Due to the low weight of the fine particles along with the high viscosity nature of guar gum, it inhibits particle interaction and moves away from each other thereby increasing the overall void volume and results in decreasing the dry density of biopolymer-treated soil. OMC increase is due to the increase in absorbed water utilized by the increased guar gum content. By increasing the solution concentration of biopolymer, the viscosity will increase, which will ultimately lead to more reduction in MDD.

3.2 *Effect on Unconfined Compressive Strength*

UCS tests were performed according to IS 2720 Part 10: 1991 under a constant strain rate of 1.2 mm/min on all combinations of soil–biopolymer mixture at different curing periods of 7, 14, and 28 days. For each guar gum–soil mixture, three similar cylindrical samples of 38 mm diameter and 76 mm height were prepared for UCS testing after 7–28 days of curing. These identical specimens of each mix were tested and axial stress vs strain curve is drawn. The peak axial stress values for each mix are taken from axial stress vs strain curves. The optimum content of guar gum is the

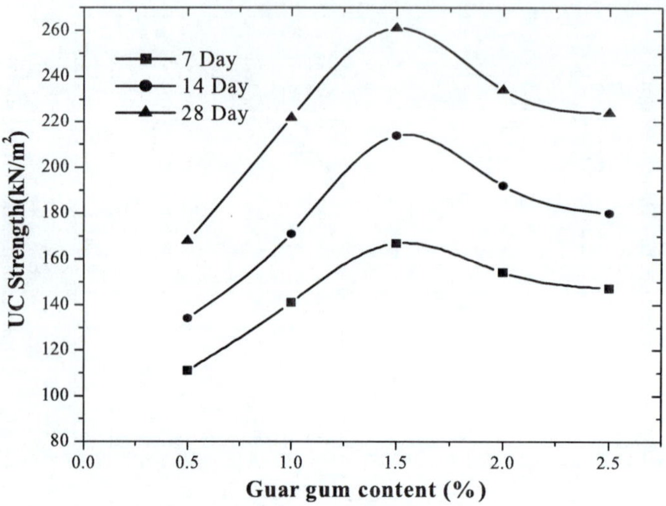

Fig. 3 Variation in UC strength with the addition of biopolymer guar gum for different curing periods

content that gave maximum UCS value at all curing periods. UCS values of different dosages of biopolymer samples for curing periods of 7, 14, and 28 days are given in Fig. 3.

From Fig. 3, it can be seen that, as the biopolymer guar gum content increases, the unconfined compressive strength increases until 1.5% biopolymer content and after that, the trend gets reversed. This is the trend at all curing periods. Hence 1.5% guar gum content can be considered as the optimum content of guar gum for maximum strength development. The UC strength at 1.5% is 261kN/m^2 at 28 days of curing. Quantitatively, the UC strength improved by 1.8 times the untreated clay soil strength. The higher strength enhancement with less aggregation and fewer voids is attributed to the cation bridging and hydrogen bonding between the electrically charged fine particles in the biopolymer and clay [20]. But beyond the optimal dosage of 1.5%, the reduction in UC strength is due to the higher viscosity with an increase in biopolymer content, which results in a lack of bonding between clay–guar gum–water mixtures.

It is also noted that the UC strength value increases with an increase in the curing period from 7 to 28 days in the case of soil–guar gum mixtures. The increase in strength with curing may be due to less aggregation and fewer voids.

3.3 *Effect on California Bearing Ratio Strength*

CBR strength testing was conducted according to IS 2720 Part 16: 1987 on all combinations of soil–biopolymer mixture at different curing periods of 7 and 28 days under both soaked and unsoaked conditions. For soaked CBR strength testing, prepared

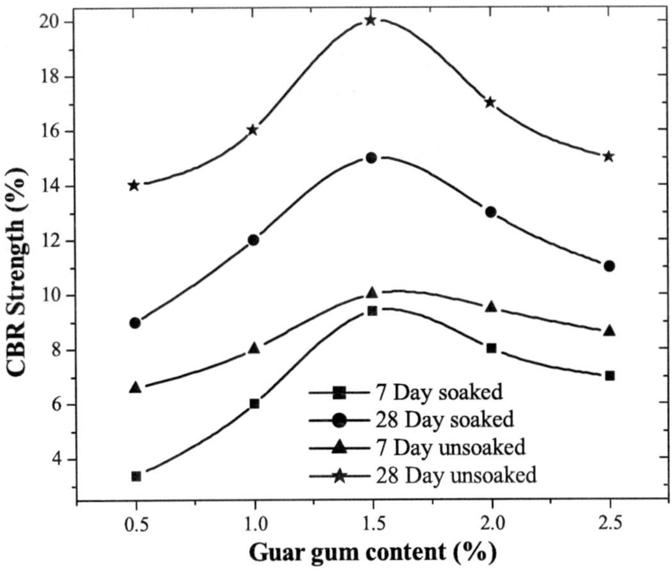

Fig. 4 Variation in CBR strength with the addition of biopolymer guar gum for different curing periods

samples in CBR mold were soaked in water for 72 h before conducting the test after covered curing. CBR strength values of soil–guar gum mixtures for curing periods of 7 and 28 days are given in Fig. 4.

From Fig. 4, it can be seen that, as the biopolymer content as well as curing period increases, CBR strength increases until optimum biopolymer content (1.5%) and after that CBR strength slightly reduces. This is the trend at all curing periods. From the results, it can be seen that after 28 days of curing, CBR strength of optimum biopolymer content-added samples increased considerably i.e. by about 5 times the respective strength values of the untreated sample under both soaked and unsoaked conditions.

4 Conclusions

This paper highlights the study on the influence of biopolymer guar gum as well as curing period on strength characteristics of weak Kuttanad clay soil through a set of laboratory experiments including compaction test, unconfined compressive strength test, and California bearing ratio strength test. The main findings are:

1. According to IS classification, soil collected for the study from Champamkulam of the Kuttanad region is clay of high compressibility (CH).

2. Untreated soil is having a high liquid limit, low UC strength, and CBR strength characteristics; which makes it unsuitable for load-bearing constructions.
3. With the addition of a biopolymer, the compaction characteristics of soil decrease.
4. The soil stabilized with 1.5% guar gum is the best soil–guar gum combination, which exhibits maximum UC strength.
5. The UC strength and CBR strength of guar gum stabilized soil increased with the increase in the curing period and also with the increase in additive dosages.
6. With guar gum stabilization, the UC strength of soil improved by 1.8 times, and the CBR strength of soil improved by 4 times than untreated soil.

Hence from the above findings, it can be concluded that the application of biopolymer guar gum will be a cost-effective and efficient method in the stabilization of weak subgrade soils, although more studies are needed on the durability characteristics of biopolymer stabilized soil under different environmental conditions.

References

1. Dehghan, H., Tabarsa, A., Latifi, N., Bagheri, Y.: Use of xanthan and guar gums in soil strengthening. Clean Technol. Environ. Policy **21**, 155–165 (2019)
2. Biju, M.S., Arnepalli, D.N.: Biopolymer modified soil: prospects of a promising green technology. In: Stalin, V.K., Muttharam, M. (eds.), Geotechnical Characterisation and Geoenvironmental Engineering IGC 2016 Volume 1, Lecture Notes in Civil Engineering, 16, pp. 163–169. Springer Nature Singapore Pte Ltd. (2019). https://doi.org/10.1007/978-981-13-0899-4_20
3. Gutiérrez, A.S., Eras, J.J.C., Gaviria, C.A., Caneghem, J.V., Vandecasteele, C.: Improved selection of the functional unit in environmental impact assessment of cement. J. Clean. Prod. **168**, 463–473 (2017)
4. Karlsson, S., Albertsson, A.: Biodegradable polymers and environmental interaction. Polym. Eng. Sci. **38**(8), 1251–1253 (1998)
5. Karimi, S.: A study of geotechnical applications of biopolymer treated soil with emphasis on silt, Ph.D. thesis, Civil Engineering Department, University of Southern California, Los Angeles (1997)
6. Khachatoorian, R., Ioana, G.P., Chang-Chin, K., Yen, T.F.: Biopolymer plugging effect: laboratory pressurized pumping flow studies. J. Petrol. Sci. Eng. **38**(1–2), 13–21 (2003)
7. Bouazza, A., Gates, W.P., Ranjith, P.G.: Hydraulic conductivity of biopolymer-treated silty sand. Geotechnique **59**, 71–72 (2009)
8. Taytak, B., Pulat, H.F., Yukselen, A.Y.: Improvement of engineering properties of soils by biopolymer additives. In: 3rd International Conference On New Developments In Soil Mechanics And Geotechnical Engineering, pp. 851–856, North Cyprus (2012)
9. Chang, I., Cho, G.C.: Strengthening of Korean residual soil with β-1,3/1,6-glucan biopolymer. Constr. Build. Mat. **30**, 30–35 (2012)
10. Liu, J., Bin, S., Hongtao, J., Huang, H., Wang, G., Kamai, H.: Research on the stabilization treatment of clay slope topsoil by organic polymer soil stabilizer. Eng. Geol. **117**, 114–120 (2011)
11. Chang, I., Jooyoung, I., Cho, G.C.: Introduction of microbial biopolymers in soil treatment for future environmentally-friendly and sustainable geotechnical engineering. Sustainability **8**(3), 1–23 (2016)

12. Latifi, N., Rashid, A.S.A., Siddiqua, S., Majid, M.Z.A.: Strength measurement and textural characteristics of tropical residual soil stabilised with liquid polymer. Measurement **91**, 46–54 (2016)
13. Pegg, A.M.: The application of natural hydrocolloids to foods and beverages. Natural Food Additives, Ingredients and Flavourings, 1st edn. Wood head Publishing (2012)
14. Ayeldeen, M., Negm, A., El Sawwaf, M.: Evaluating the physical characteristics of biopolymer/soil mixtures. Arab. J. Geosci. **9**, 371 (2016)
15. Wang, A., Wang, W.: Gum-g-Copolymers: synthesis, properties, and applications. polysaccharide based graft copolymers, chapter 5, 1st edn, pp. 149–203. Springer Heidelberg (2013)
16. Mudgil, D., Barak, S., Khatkar, B.S.: Guar gum: processing, properties and food applications—a review. J. Food Sci. Technol. **51**(3), 409–418 (2014). https://doi.org/10.1007/s13197-011-0522-x
17. Khatami, H.R., O'Kelly, B.C.: Improving mechanical properties of sand using biopolymers. J. Geotech. Geoenv. Eng. **139**(8), 1402–1406 (2013). ASCE
18. Kullayappa, G., Kumar, S.P.P.: Studies on strength characteristics of soil with Guar gum. Int. J. Sci. Res. **7**(6), 302–308 (2018)
19. Chang, I., Im, J., Prasidhi, A.K., Cho, G.C.: Effects of Xanthan gum biopolymer on soil strengthening. Constr. Build. Mater. **74**, 65–72 (2015)
20. Ayeldeen, M., Negm, A., El Sawwaf, M., Kitazume, M.: Enhancing, mechanical behaviors of collapsible soil using two biopolymers. J. Rock Mech. Geotech. Eng. **9**(2), 329–339 (2017)

Confined Reinforcement Barrier System for Mitigating the Generated Ground Motions in Liquefiable Soils

S. P. Vijay Kumar and S. Ganesh Kumar

1 Introduction

Many catastrophic damages have been caused to the civil infrastructures due to soil liquefaction. Failure due to liquefaction is commonly observed in saturated cohesionless soil deposits. Additionally, when buildings are located near to epicenter, the effect of combined horizontal and vertical ground acceleration responses created vertical deformation and failure of structure [1]. Indicatively, in saturated soil deposits, as the generated excess pore water pressure increases, it reaches a unity where the soil surrounding the structure tends to flow as a liquid apparently leading to the failure of structure. Even if the excess pore pressure developed below the center of the structure tends to be smaller than that away from the structure, the stability against combined horizontal and vertical acceleration should also be considered to ensure the safety of structures located in saturated ground deposits. Studies also show that the impact of generated ground motions is found to be higher in near-fault regions, higher magnitude seismic events and in shorter period events relative to the horizontal component [2].

Stone columns (SC) have been extensively used as drainage members in saturated loose soil deposits, which provide drainage and densification to the ground and mitigate the liquefaction effect and the associated ground deformations. However, the attenuation of incoming seismic vibrations in the saturated ground is least considered in stone column reinforcement. On the contrary, an alternative reinforcement system called confined isolation barrier reinforcement was used in this study to attenuate the ground motions generated during dynamic loading. The isolation barrier was

S. P. Vijay Kumar (✉)
IIT(ISM) Dhanbad, Dhanbad 826004, Jharkhand, India
e-mail: spvijay.18mt0390@agp.iitism.ac.in

S. Ganesh Kumar
CSIR-CBRI Roorkee, Roorkee 247667, Uttarakhand, India

C. N. V. Satyanarayana Reddy et al. (eds.), *Ground Improvement and Reinforced Soil Structures*, Lecture Notes in Civil Engineering 152,
https://doi.org/10.1007/978-981-16-1831-4_40

made with commercially available PU foam, which was widely accepted especially in shaking table tests as an absorbing boundary. The PUF-confined barrier system when installed in liquefiable soil deposits found to effectively absorb both the incoming horizontal ground accelerations and the vertical uplift motion due to the generation of excess pore pressure. The present research aims at comparing the sustainability of conventionally used stone column reinforcement system and PUF-confined barrier reinforcement system in liquefiable soils when subjected to repeated earthquake events. The performance of the barrier system was compared with conventional stone column reinforcement of 2.5% area ratio and tested under repeated acceleration loading conditions for liquefaction mitigation. Using uniaxial shaking table experiments, the performance of prepared saturated model ground without reinforcement, with stone column reinforcement and with developed PUF-confined barrier reinforcement systems is studied and presented.

2 Scope and Objective

The primary objective of this study is to compare the performance of the developed confined barrier reinforcement system that primarily uses an energy absorption mechanism to absorb incoming ground motions and the conventionally adapted stone column reinforcement system, which improves densification and drainage mechanism to stabilize the liquefiable ground when subjected to dynamic loading.

Stone columns are an array of crushed stone pillars. Reinforcement with gravel drains stiffens and increases the bearing capacity of soil deposit, thereby reducing liquefaction risk [3]. Considering their drainage and densification characteristics, they are employed in loose saturated ground deposits to increase the bearing capacity of soil as they offer easier dissipation of excess pore water pressure. However, the clogging effect of stone columns decreases their efficiency under multiple shaking events. The confined barrier reinforcement was prepared from polyurethane material, which is highly known for its shock-absorbing capacity and damping properties. PU foam infill trenches had a good vibration isolation capacity and proved to be very effective in damping out ground-borne vibrations [4]. In order to prevent the intrusion of finer sand particles, geotextile was wrapped around the PUF barrier and then installed.

For experimental studies, poorly graded sand was selected to model liquefiable ground having 40% relative density. The developed confined barrier reinforcement and stone column reinforcement were installed in the prepared model ground and tested under sub-sequential sinusoidal acceleration loading conditions in order to evaluate their isolation performance when subjected to repeated action of soil-liquefaction effect.

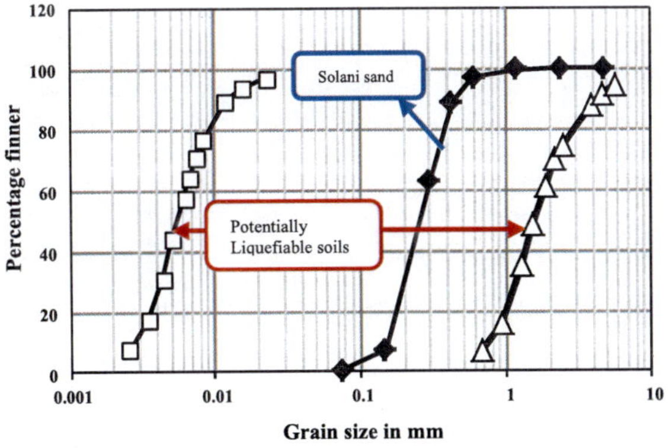

Fig. 1 Grain size distribution curve for Solani river sand

3 Methodology

3.1 Soil Selected for the Study

Soil sample for experimental studies was collected from the Solani river bed in Roorkee, Uttarakhand. Laboratory experiments such as specific gravity, sieve analysis, relative density and permeability tests were carried out to determine the soil properties and the soil is classified as poorly graded sand (SP) as per Indian standard classification [IS 2720 Part 4—1985]. The gradation curve for the soil is shown in Fig. 1. It can be seen that the gradation falls within the liquefaction susceptibility as proposed by Tsuchida [5]. For testing, the model tank of dimension 1.4 m × 1 m × 1 m was used for ground preparation.

3.2 Preparation of Un-Reinforced Model Ground

The saturated ground model was prepared using the wet sedimentation technique. The liquefaction response of soil highly depends on the method of sample preparation [6]. In the wet sedimentation method, the desired relative density for soil is achieved by first adding the water required for complete saturation of soil initially followed by sand pouring inside the tank. The sand was poured through a conical hopper having 60° inverted cone at the bottom from a calculated height, which was estimated from the height of fall experiments. The height of pouring was calculated as 110 mm for achieving 40% relative density. For experimental testing, ground bed having 40% density was prepared and tested with and without soil reinforcement under repeated acceleration loading conditions.

The saturated soil ground bed having 600 mm height was prepared in three layers each of 200 mm height for achieving uniformity in sand bed preparation. For monitoring pore water pressure generated during shaking, two piezometers were fixed at

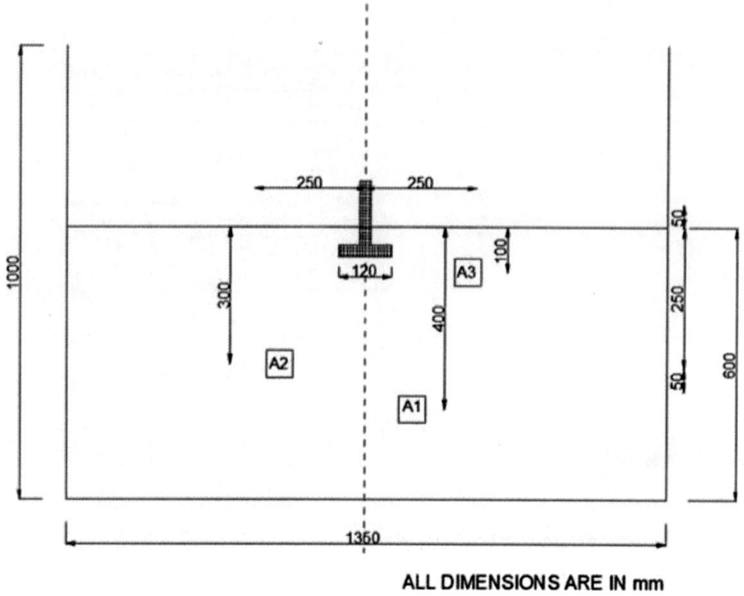

Fig. 2 Unreinforced model ground—Sectional view

a height of 200 and 400 mm from the bottom of the tank respectively. For estimating acceleration response, three accelerometers were placed at 100, 300 and 400 mm depth, respectively, from the top ground surface. For measuring foundation displacements, a scaled shallow footing model was embedded inside the prepared ground bed. Figure 2 showing the sectional view of the prepared model ground with accelerometer and piezometer connection details. Then, the prepared unreinforced ground bed was subjected to repeated acceleration loading. After completion of each test, the acceleration response and pore pressure variation were monitored and total foundation settlement was measured. The application of subsequent acceleration loading was carried out only after complete dissipation of generated pore water pressure from previous acceleration loading.

3.3 Stone Column Reinforcement

The stone column reinforcement was designed for an area replacement ratio of 2.5%. The column network was designed in a square pattern with four stone columns surrounding the foundation area each having diameter (D) 110 mm and height 600 mm. The center to center spacing of each column was calculated as 2.5D. Stone columns were installed in the ground model by displacement method [IS 15284 (Part1): 2003]. Coarse aggregates passing through 10 mm and retaining on 4.5 mm IS sieve were used for column construction.

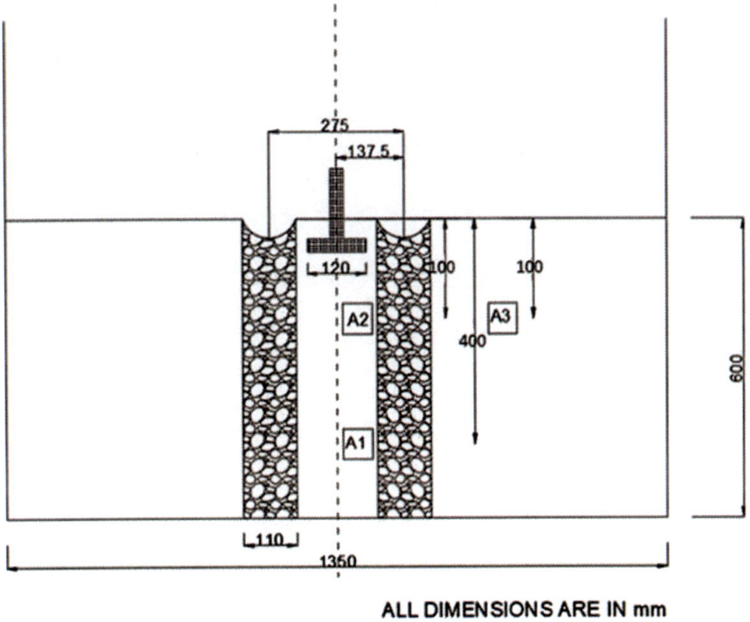

Fig. 3 Model ground reinforced with four stone columns—Sectional view

For stone column construction, a casing pipe having an outer diameter equivalent to diameter of stone column was driven to the entire depth of the soil bed initially. The soil inside the casing was removed using an auger arrangement. Then, the stones were filled as layers of 150 mm depth each. After filling each layer, the stones were compacted 25 blows using a hammer for achieving 76% equivalent to field density. After the installation of stone columns, scaled foundation model was embedded at the center of the model ground to measure the settlement. For measuring acceleration response with drainage reinforcement, three accelerometers are placed inside the model tank i.e. A1 being at 400 mm depth, A2 at 100 mm depth and A3 being outside the barrier area at 100 mm depth from top ground surface respectively. Using two piezometers, the generated excess pore water pressure during acceleration loading was measured. The stone column reinforced model ground was then subjected to repeated acceleration loading. Figure 3 shows the complete details of the prepared model ground reinforced with four stone columns in sectional view.

3.4 Confined Barrier Reinforcement

Polyurethane foam (PUF) was selected to design the confined isolation barrier considering its high resilience, abrasion resistance and vibration absorption characteristics. The liquid PU foam procured had a density of 1.1 g/cc, a damping ratio of about 0.08

and a volume expansion factor of 30 times the original volume. The foam barrier was developed by mixing resin and hardener in a ratio of 1:1 inside the inverted T-shaped mold. To prevent soil intrusion during dynamic loading, the surface of the barrier system was covered with geotextile material. Since the application of acceleration loading in saturated soils can generate combined horizontal and vertical acceleration amplitudes, the confinement barrier model was developed similar to the inverted T-shaped barrier system. This shape offers effective anchorage and also to absorb vertical ground accelerations generated due to shaking.

Two inverted T-shaped anchored barrier models were prepared for the study with dimensions of Heel: 200 × 500 × 50 mm, Toe: 100 × 500 × 50 mm and Stem: 50 × 500 × 300 mm. The isolation barriers were placed to confine the foundation area by maintaining 500 mm c/c spacing between the stems. After achieving a depth of 300 mm in ground preparation, the developed barrier model was placed over the ground and ground preparation was continued. The positioning of isolation barrier designed such that the stem portion absorbs horizontal ground motions and heel-toe portions prevent the vertical uplift due to excess pore water pressure during dynamic loading. Figure 4 shows the complete details of the PUF-confined barrier reinforcement system installed inside the ground deposit. For measuring acceleration response, three accelerometers were placed inside the model ground, A1 and A2 being at 400 and 100 mm depth from the top ground surface inside confinement whereas A3 being at 100 mm depth from ground surface outside the confinement and two piezometers were used to measure the generated excess pore pressure. The

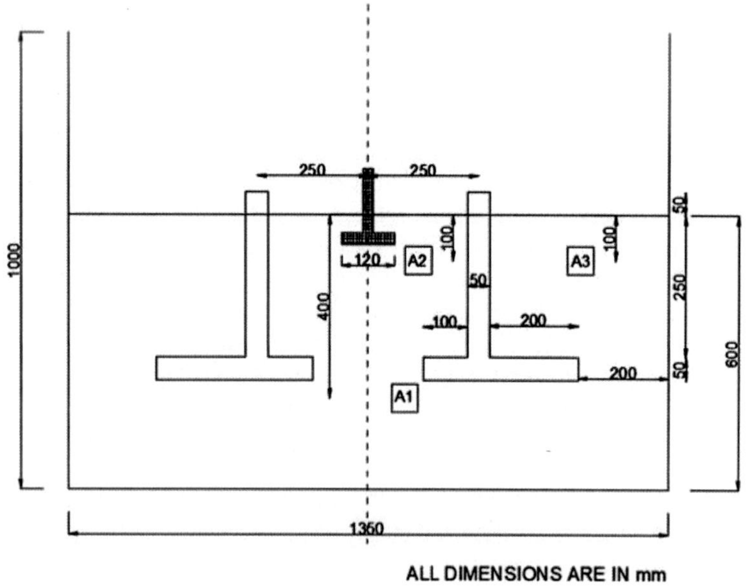

Fig. 4 PUF-confined barrier reinforcement system—Sectional view

reinforced model ground was then subjected to repeated acceleration loading and its acceleration response, pore pressure response and foundation settlement were observed and compared.

4 Experimental Testing Details

For experimental testing, repeated incremental sinusoidal acceleration loading having intensities 0.1, 0.2 and 0.3 g was applied to both unreinforced and reinforced model ground. The selected acceleration intensities tend to simulate the repetition of medium to severe earthquake conditions pertaining to the ground model [Resulting an intensity of VII to IX, as per Modified Mercalli Intensity Scale [7]. Also, the selection of repeated shaking is similar to the observed repeated foreshock and main shock events observed during earthquake loading. Hence repeated incremental acceleration loading was selected and applied to the unreinforced and reinforced model ground. The effects of foundation settlement, acceleration response and pore pressure response in model ground with and without reinforcement systems are evaluated. The efficiencies of confined barrier reinforcement and stone column reinforcement under the repeated occurrence of liquefaction effect in the saturated model ground are compared.

5 Results and Discussion

5.1 Acceleration Response

The obtained acceleration response of the prepared 40% density model ground under repeated sinusoidal acceleration loading conditions with and without provision of reinforcements with respect to time is converted into frequency domain having 5 Hz base frequency by performing fast Fourier transform. The resulting peak Fourier amplitude reduction observed near the foundation model embedded in model ground with stone column reinforcement and confined barrier reinforcement are shown in Fig. 5 to evaluate their effectiveness in mitigation of liquefaction induced ground motion due to the provided repeated incremental acceleration loading.

It was evident from the figure that the peak Fourier amplitude observed with stone column reinforcement condition had an average reduction of 69% and with PUF-confined barrier reinforcement, an average reduction of about 85% was observed as compared to that of unreinforced model ground. This shows that the confined barrier reinforcement was found to be more effective in controlling the generated ground motions and also efficient under repeated acceleration effect when compared to stone column reinforcement. This was mainly due to the damping characteristics of PU foam material, which attenuates the incoming motions into the structure.

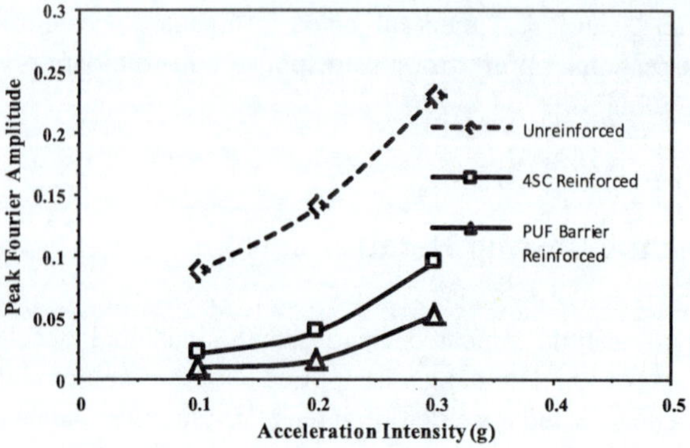

Fig. 5 Peak Fourier amplitude reduction observed with stone column reinforcement and confined barrier reinforcement

5.2 Amplitude Attenuation Ratio (A_R)

Amplitude attenuation ratio (A_R) is defined as the ratio of peak acceleration amplitude recorded with the provision of reinforcement to that without reinforcement under particular ground acceleration [8]. Figure 6 shows the variation of amplitude attenuation curve under repeated acceleration loading conditions with stone column reinforcement and confined barrier reinforcement.

$$A_R = \frac{\text{Peak acceleration amplitude of ground with reinforcement}}{\text{Peak ground acceleration amplitude of ground without reinforcement}}. \quad (1)$$

The variation pattern observed in both cases confirms the fact that liquefaction resistance of soil in a consolidated state is less as compared to its virgin

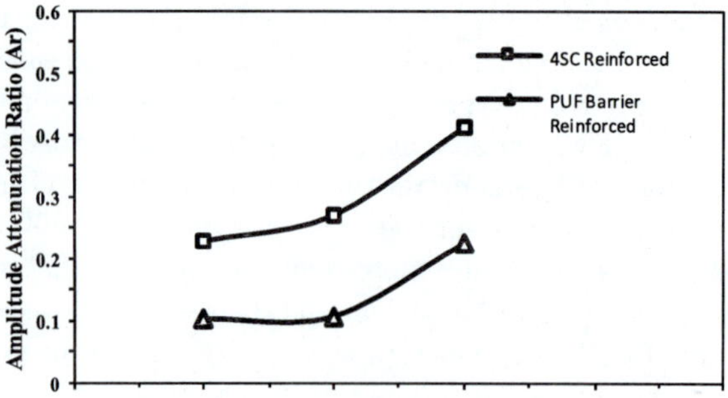

Fig. 6 Variation of Amplitude attenuation ratio curve observed under

state [9]. It is also evident that the efficiency of reinforcement systems tends to decay under repeated un-drained loading conditions. From the figure, it can be seen that the de-amplification capacity of PUF-confined barrier reinforcement under repeated acceleration loading conditions is comparatively higher than stone column reinforcement.

5.3 *Effect of Pore Water Pressure Ratio (r_u)*

The pore water pressure ratio (r_u) is defined as the ratio of excess pore water pressure U_{excess} to effective overburden pressure σ'_{vo}. The soil is considered to be liquefied when r_u value reaches unity. Due to the repeated un-drained shaking, the generation of pore water pressures from bottom to top makes soil at shallow depth more susceptible to liquefaction. Figure 7 shows the estimated peak pore pressure ratio recorded in top piezometer under repeated acceleration loading with and without reinforcement.

$$r_u = \frac{U_{excess}}{\sigma'_{vo}} \tag{2}$$

From the figure, it is evident that the provision of reinforcement within the liquefiable deposit improves the seismic resistance by minimizing the generation of pore water pressure. Comparatively, stone column improvement shows a better performance than PU foam barrier system. It is also verified from the figure that the provision of PU foam barrier also performs better in minimizing the generation of pore water pressure than untreated ground. The reduction in pore pressure ratio is found to be 54, 52 and 22% for PU foam-treated ground and 84, 79 and 73% for stone column-treated ground under repeated loading conditions. The enhanced reduction

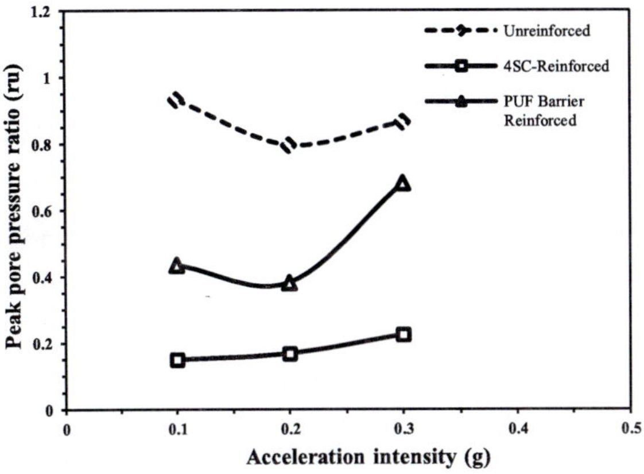

Fig. 7 Variation of peak pore pressure ratio observed under sub-sequential acceleration loading condition

in pore pressure ratio in stone column-treated ground may be due to the densification induced inside the ground bed during the stone column installation and due to drainage mechanism, which dissipates the pore water pressure generation more effectively.

5.4 Foundation Settlement

The foundation settlements during incremental sequential amplitude accelerations for both reinforced and unreinforced soil deposits are also evaluated and presented in Table 1.

It can be seen that the performance of stone column reinforcement was exceedingly well during initial accelerations. But, due to repeated shaking conditions, their efficiency gradually decreased due to clogging effect of stone columns. On the contrary, due to the enhanced absorption and confinement offered by PUF, the confined barrier was found more reliable during repeated loading conditions. At higher accelerations, slight deformation was observed in the barrier due to continuous generation of pore water pressures. However, in both the treated conditions, installation of reinforcement member reduces foundation settlement on average by about 45% in case of PU foam-treated ground and by about 48% for stone column-treated ground.

6 Conclusions

Based on the conducted shaking table experiments and obtained results, the following conclusions were made.

1. The confined barrier reinforcement performs well in mitigating incoming ground motions to the structures through its absorption and damping characteristics even under repeated acceleration loading conditions. However, in the absence of drainage characteristics, the performance reduces at high acceleration loading conditions.
2. Provision of stone column reinforcement improves soil densification and drainage characteristics in liquefiable deposits. However, under repeated acceleration loading condtions, the possibility of clogging due to intrusion of finer particles reduces its efficiency against liquefaction mitigation.
3. Selection of proper ground improvement techniques plays a major role in improving the seismic response of liquefiable deposits. Improvement in densification can improve seismic resistance but under repeated loading conditions, provision of drainage also contributes to improving resistance against liquefaction.

Table 1 Foundation settlement of ground model under sub-sequential acceleration loading

Acceleration intensity (g)	Foundation settlement (cm)				
	Unreinforced	Four stone column-reinforced	PUF-confined barrier reinforced	%Reduction with 4-SC reinforcement	%Reduction with PUF Barrier reinforcement
0.1	4.2	2.2	3.2	47.62	23.81
0.2	9.8	4.9	4.6	50.00	53.06
0.3	14.4	7.8	6.1	45.83	57.64

References

1. Papazoglou, A.J., Elnashai, A.S.: Earthquake Engineering and Structural Dynamics, vol. 25, pp. 1109–1137 (1996)
2. Bozorgnia, Y., Campbell, K.W.: The vertical-to-horizontal response spectral ratio and tentative Procedures for developing simplified V/H and vertical design spectra. J. Earthq. Eng. **8**(2), 175–207 (2004)
3. Baez, J.I., Martin, G.R.: Advances in the design of vibro systems for the improvement of liquefaction resistance. In: Symposium of Rround Improvement. Vancouver Geotechnical Society, Vancouver, BC (1993)
4. Tulika, B., Deepankar, C.: Efficiency of open and infill trenches in mitigating ground-borne vibrations. J. Geotech. Geoenviron. Eng. **144**(8)
5. Tsuchida, H.: Evaluation of liquefaction potential of sandy deposits and measures against liquefaction induced damage. In: Proceedings of the Annual Seminar of the Port and Harbour Research Institute (3–1)–(3–33) (1970)
6. Banerjee, R., Konai, S., Sengupta, A., Deb, K.: Shake table tests and numerical modelling of liquefaction of Kasai River Sand. Geotech. Geol. Eng. **35**(4), 1327–1340 (2017)
7. Robertson, P.K., Campanella, R.G.: Liquefaction potential of sands using the CPT. J. Geotech. Eng. **111**(3), 384–403 (1985)
8. Gao, M., Tian, S.P., Wang, Y.: Isolation of ground vibration induced by high speed railway by DXWIB: field investigation. Soil Dyn. Earthq. Eng. **131**(2020), 106039 (2019)
9. Wang et al.: Effect of previous cyclic shearing on liquefaction resistance of mississippi river valley silt. Part J. Mater. Civil Eng. **25**(10) (2012)

Performance of Vertically Confined Shallow Foundation on Reinforced Sand Under Concentric Loading

B. Kirtimayee and Narendra Kumar Samadhiya

1 Introduction

Bearing capacity and settlement are the two most challenging criteria considered during the design of foundation. In this regard, several methods for soil improvement have been applied to enhance the soil characteristics. Out of the different available techniques, soil confinement is the suitably applicable method accepted in the geotechnical field ensuring a safe bearing capacity. Many investigations have been made to study the effect of soil confinement technique to explore more in the geotechnical field. Rajgopal et al. [12] studied the influence of geocell confinement on the strength and stiffness behavior of granular soils performing a large number of triaxial compression tests. The effect of soil confinement on the ultimate bearing capacity of square footing under eccentric-inclined loading was investigated by Singh et al. [14]. The work of different researchers has shown that confiners made up of various materials like unplasticized polyvinyl chloride cylinders (Upvc), semiflexible vertical reinforcement, mild steel casing, plastic hollow cylinder, timber box [1, 5, 6, 9, 11] etc. can be used to improve the bearing capacity and reducing settlement. In the above work, researchers have recommended some parameters like confiner height, width and diameter of the confiner for obtaining optimum results. Eid et al. [4] carried out both physical and numerical modeling on the behavior of shallow foundation resting on laterally confined sand surrounded by sheet-pile walls to support excavation sides of sand underlain by a rock bed. Fattah et al. [8] reported that the use of vertically bounded walls for soil confinement mitigates the settlement from 5 to 160% depending on wall depth and distance from the footing. Some authors have investigated on use of skirts to confine the soil resulting in a significant improvement

B. Kirtimayee (✉) · N. K. Samadhiya
Department of Civil Engineering, Indian Institute of Technology Roorkee, Roorkee 247667, India

N. K. Samadhiya
e-mail: nksamfce@iitr.ac.in

C. N. V. Satyanarayana Reddy et al. (eds.), *Ground Improvement and Reinforced Soil Structures*, Lecture Notes in Civil Engineering 152,
https://doi.org/10.1007/978-981-16-1831-4_41

in the ultimate bearing capacity [2, 7, 10, 13]. Azzam and El Wakil [3] reported the behavior of circular footing on confined granular subgrade adjacent to the slope. Thakur and Dutta [15] carried out an experimental and numerical analysis to determine the ultimate bearing capacity of unskirted, singly skirted and doubly skirted hexagonal footing on sand varying its D_{10} values as 0.14, 0.45, 1.45 respectively. In the prospect of gaps, in the present work, both biaxial geogrids have been used as the confiner and horizontal reinforcement underneath the footing. The aim of this study is to find the improvement in the ultimate bearing capacity of square footing resting on horizontally reinforced sand confined laterally by a geogrid confiner under concentric vertical load.

2 Experimental Details

The model tests were conducted in an iron tank with internal dimensions of 1.0 m × 1.0 m × 0.8 m. To avoid lateral deformation, the tank walls were braced with iron section outside. A square footing of side B (20 cm) has been used in the present work. All the tests were carried out on poorly graded medium dense sand with uniformity coefficient (C_u) = 2.62, coefficient of curvature (C_c) = 1.13, specific gravity = 2.64, natural dry unit weight of sand = 15.24 kN/m^3, relative density of sand = 50%, maximum and minimum void ratio = 0.87 and 0.55, respectively, angle of shearing resistance = 32.6°. The physical properties of biaxial geogrids used for both lateral confiner and horizontal reinforcement are given in Table 1.

2.1 Test Program

Tests were performed in three series. In series A, only footing was used under vertical loading. In series B footing with confiner, in series C footing with confiner of given dimensions along with horizontal reinforcements inside were used underneath the

Table 1 Properties of Geogrid

Paramete	Value
Aperture size of Geogrid (mm)	25 x 25
Material	HDPE
Peak load (MD)	1.2 kN
Peak load (CMD)	2.5 kN
Stiffness modulus (kN/m)	2000
Thickness (mm)	1.57
Junction efficiency (%)	94
Sample area (sq.mm)	300

Table 2 Details of Laboratory model tests

Test series	Foundation configuration	Test parameters		Number of tests
		Variable	Constant	
A	Only footing			1
B	Footing with confiner	D/B = 1.0, 1.5, 2.0	d/B = 1	3
C	Footing with confiner and reinforcement	D/B = 1, 1.5, 2, N = 1–4, Y/B = 0.25, 0.5, 0.75, 1, L/B = 1, 1.5, 2.0	d/B = 1, y/B = 0.1B	12

footing. Parameters such as d, D, y, L and Y were normalized with respect to width of the footing (B). The top layer reinforcement layer (y) is kept at 0.1B (constant) below the base of the footing for all the tests. The consecutive horizontal reinforcement layer spacing is varied as 0.25B, 0.5B,0.75B and 1.0B. Variation in D/B is kept as 1.0, 1.5 and 2.0 for a constant depth of d/B = 1.0. The details of laboratory model tests are presented in Table 2.

3 Methodology

Sand bed was prepared for the model tests maintaining a uniform relative density of 50% throughout. In order to achieve the desired density, the height of fall was fixed by several trials in the test tank prior to the test. Lateral confiner was placed on the sand bed in the middle of the tank by using a plumb bob after the sand was filled to a required height, followed by the placement of horizontal reinforcement inside the confiner maintaining the certain spacing as mentioned. After the combined entity of confiner and horizontal reinforcements were placed in their desired position rest part of the tank was filled in the same procedure. Top surface of the sand bed was leveled by a wooden plate and footing was placed middle of the tank maintains the concentricity with the confiner. Four dial gauges with an accuracy of 0.01 mm placed on the four corners of the footing average of which provide the settlement of footing. Load was applied through a manually operated hydraulic jack associated with a precalibrated proving ring of capacity 30 kN measures the transferred load. Bearing capacity in all the model tests has been considered for 25 mm settlement. Schematic diagram for the experimental program is given in Fig. 1.

3.1 Numerical Analysis

Finite element software PLAXIS 3D has been used to verify the laboratory model test results. The nonlinerity of sand was modeled using the hardening soil model,

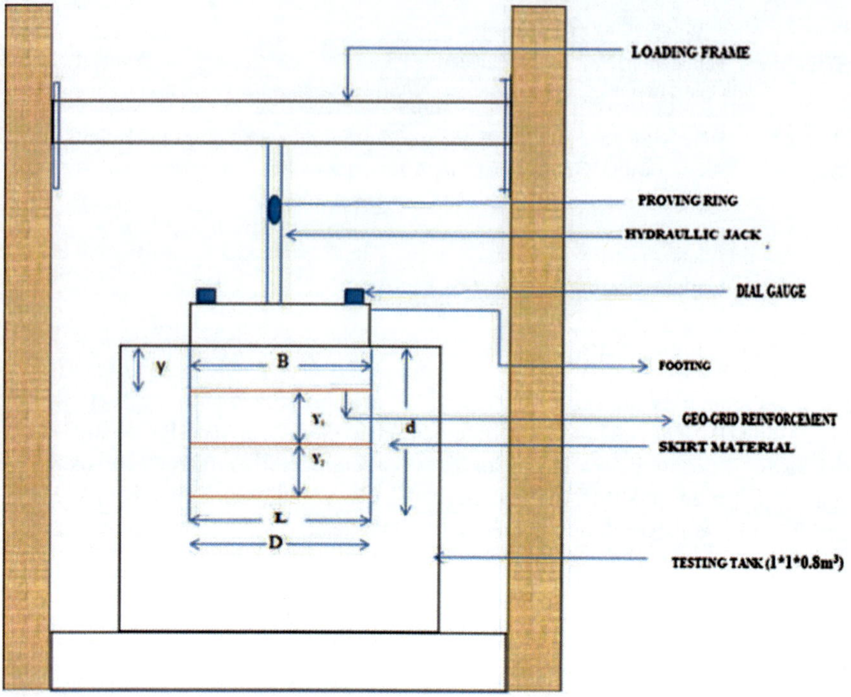

Fig. 1 Schematic diagram of the setup

an elastoplastic second-order hyperbolic isotropic hardening model. Input parameters (E_{50}, E_{oed}, E_{ur}) for sand were estimated using co-relation between the angle of shearing resistance and relative density. Numerical values for some basic parameters have been taken from the manual. The footing was modeled as a plate element and the elastic geogrid element was used to model both the confiner and horizontal reinforcements. A prescribed displacement of 25 mm was applied and load corresponding to 25 mm settlement was considered as ultimate load, which was compared with the ultimate load obtained from experimental results. Table 3 provides the values considered for the analysis (Figs. 2, 3 and 4).

4 Results and Discussion

Several laboratory tests, as detailed in Table 2, were carried out on model laterally confined footing resting on geogrid-reinforced medium dense sand. The results obtained from PLAXIS 3D have also been validated with the experimental results (Figs. 9, 10, 11 and 12). The improvement in bearing capacity of square footing with confiner and geogrid reinforced sand is represented using a non-dimensional factor, called bearing capacity ratio (BCR). BCR is defined as the ratio of footing ultimate

Table 3 Material properties

Parameter	Sand	Footing	Geogrid
Secant modulus of stiffness (E_{50})(kN/m^2)	18,600	–	–
Tangent oedometer stiffness (E_{oed}) (kN/m^2)	10,000	–	–
Unloading/reloading stiffness (E_{ur}) (kN/m^2)	55,800	–	–
Angle of internal friction	32.6	–	–
Unit weight (kN/m^3)	15.21	78.5	–
Thickness of footing (m)		0.025	
$E_1 = E_2$ value of footing (kN/m^2)	–	13.4e6	–
$G_1 = G_2$ value of footing (kN/m^2)		5.36e6	–
Poisson's ratio		0.25	
Stiffness modulus for Geogrid (kN/m)	–	–	2000

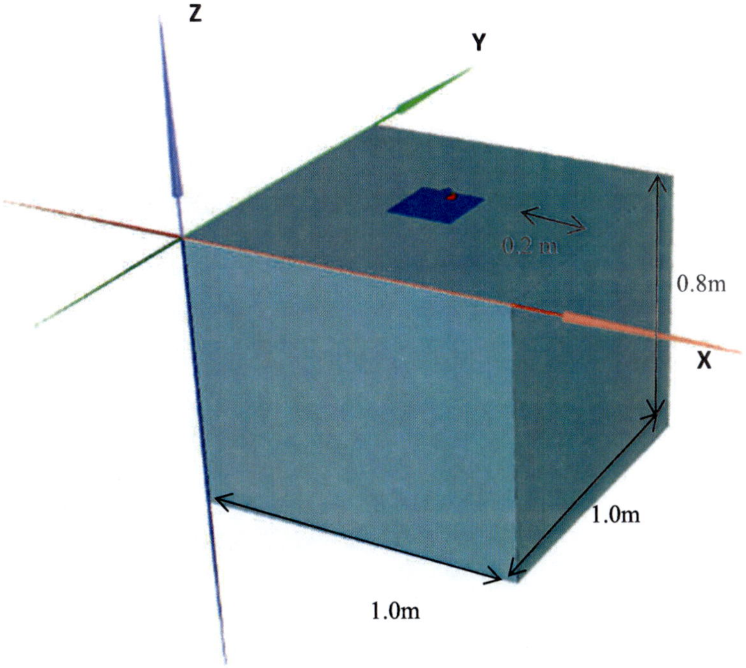

Fig. 2 Geometry modeled in Plaxis

bearing load intensity for reinforced sand ($q_{reinforced}$) to footing ultimate bearing load intensity for unreinforced ($q_{unreinforced}$) sand. Figure 5, 6, 7 and 8 illustrate the load intensities—settlement curves providing ultimate bearing capacity corresponding to 25 mm settlement. Figure 5 shows the load intensity–settlement curve for footing with only confiners in which depth of the confiner is kept 20.0 cm (d/B = 1) constant

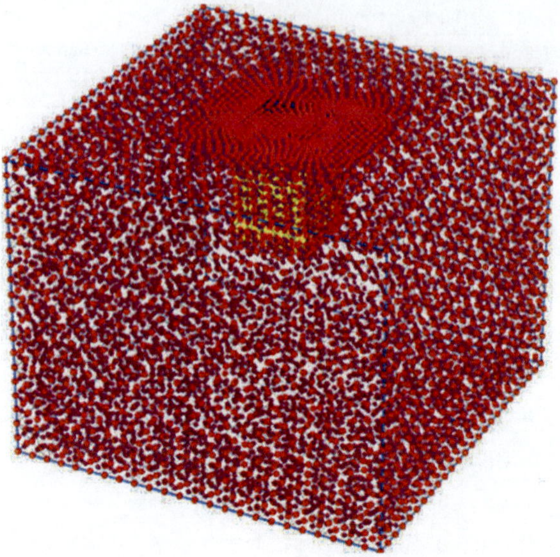

Fig. 3 Connectivity plot for (Footing + Confiner + Reinforcement) Configuration

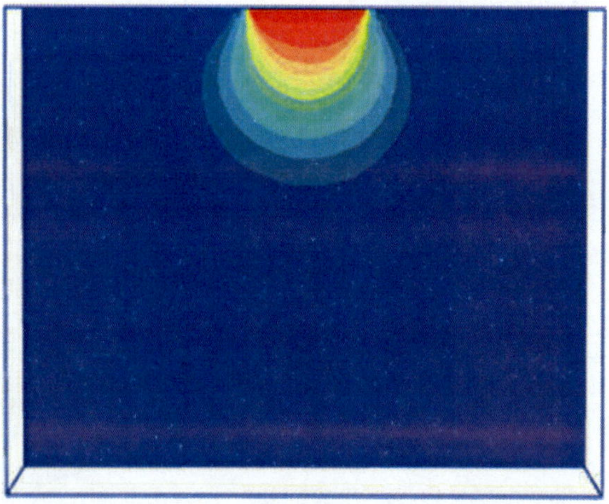

Fig. 4 vertical c/s for surface footing (Total displacement)

whereas top surface dimension of the confiner is varied as 20 cm × 20 cm (D/B = 1), 30 cm × 30 cm (D/B = 1.5), 40 cm × 40 cm (D/B = 2). Figures 6, 7 and 8 explain the variation of load intensity for footing with lateral confiner and horizontal reinforcements with a variation in spacing of reinforcement as 0.25B, 0.5B, 1.5B and 2.0B. Overall, it can be concluded that the use of both lateral confiner and reinforcement underneath the footing shows a noticeable effect, which is enumerated below individually. Load intensities and bearing capacity ratios for all the combinations are given in Table 4. Failure pattern for different arrangements of reinforcement and confiner observed in plaxis has been discussed below in Figs. 13, 14, 15 and 16.

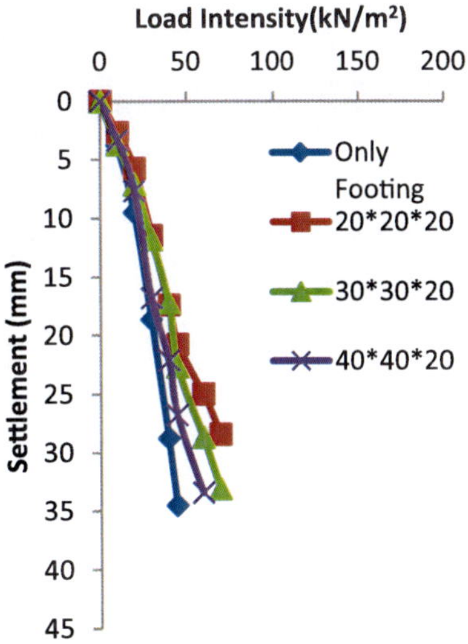

Fig. 5 Footing + Confiner (D/B = 1, 1.5, 2)

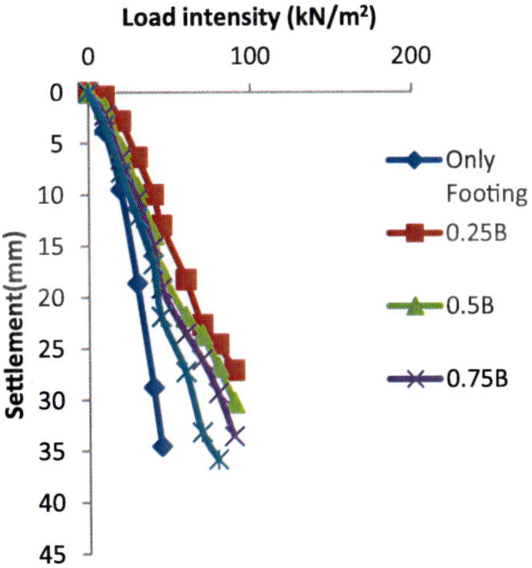

Fig. 6 Footing + Confiner (20*20*20) + Reinforcement

4.1 *Effect of Confiner*

It can be observed from Fig. 5 that with an increase in top surface dimension of the confiner, the load intensity starts decreasing as 59.5 kN/m^2(D/B = 1, d/B = 1), 50.3 kN/m^2 (D/B = 1.5, d/B = 1), 42.7 kN/m^2 (D/B = 2, d/B = 1) respectively. The corresponding BCRs are found to be 1.65, 1.4, 1.18, respectively, also showing a decremental nature substantiating the fact that with an increase in top surface dimension of the confiner, the degree of confinement decreases and more amount of sand spills out laterally due to little higher aperture size of geogrid. From Table 4,

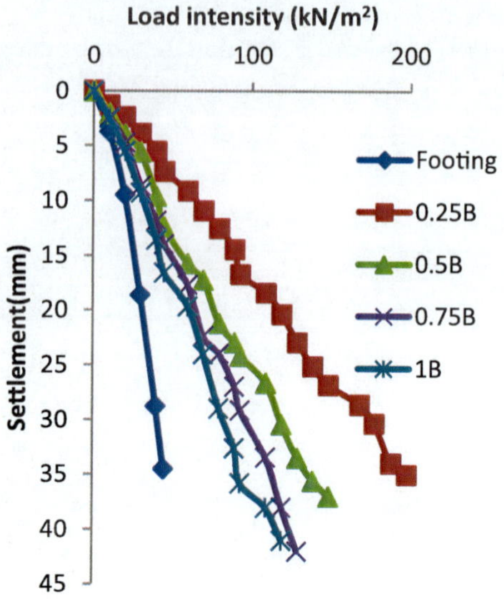

Fig. 7 Footing + Confiner (30*30*20) + Reinforcement

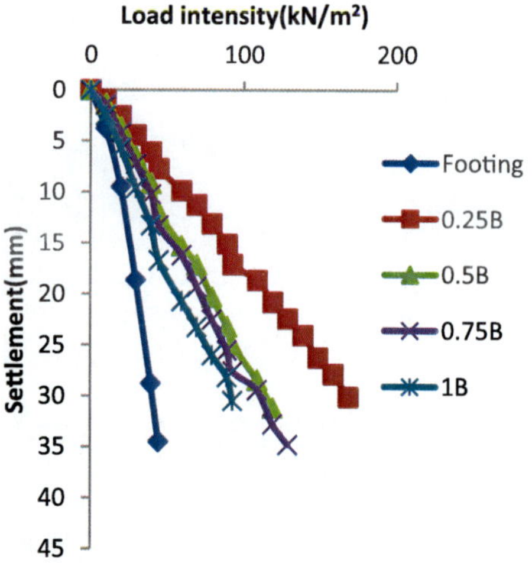

Fig. 8 Footing + Confiner (40*40*20) + Reinforcement

it can be observed that even though the load intensity decreased with increment in D/B value still it shows a noticeable increment of 1.65-fold increase for the confiner (D/B = 1, d/B = 1) compared to the surface footing (35.9 kN/m^2).

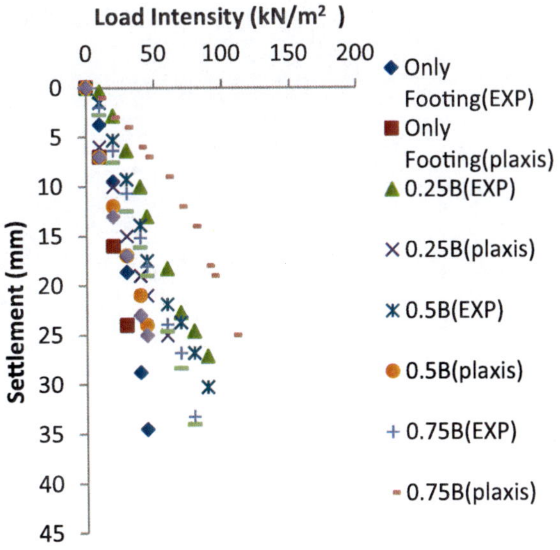

Fig. 9 Footing + Confiner (20*20*20) + Reinforement

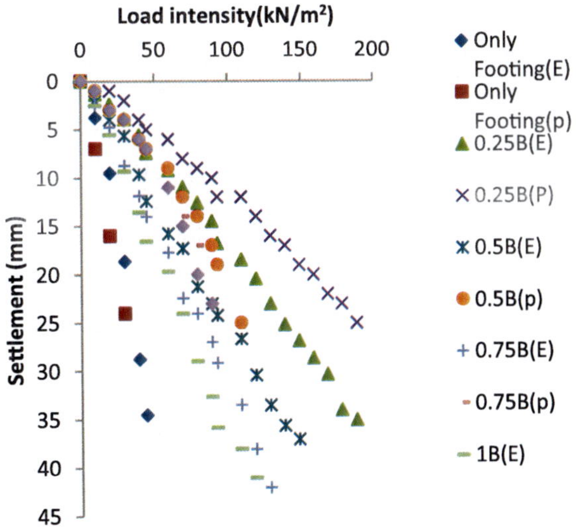

Fig. 10 Footing + Confiner (30*30*20) + Reinforement

4.2 *Effect of Horizontal Reinforcement and Spacing Inside Confiner*

The effect of horizontal reinforcement inside the confiner has been shown in Fig. 6, 7 and 8. Load intensities corresponding to 25 mm settlement for the 0.25B spacing were found to be 81kN/m^2 (D/B = 1), 138.2 kN/m^2 (D/B = 1.5), 143.1 kN/m^2 (D/B = 2), which shows a significant improvement in load carrying of the square footing.

Fig. 11 Footing + Confiner (40*40*20) + Reinforement

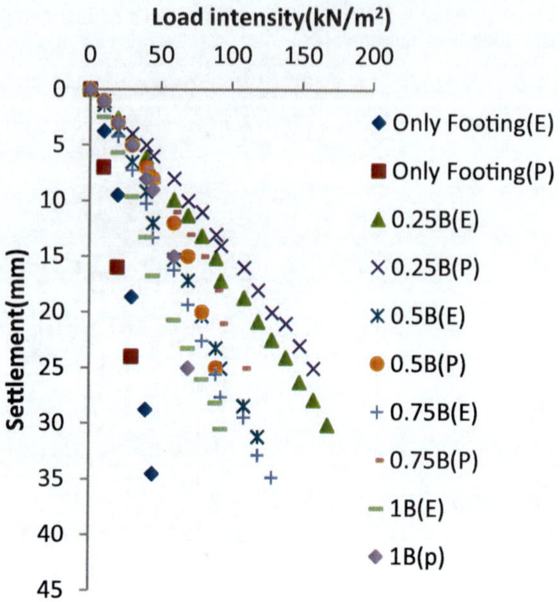

Fig. 12 Footing + Confiner (D/B = (1, 1.5, 2)

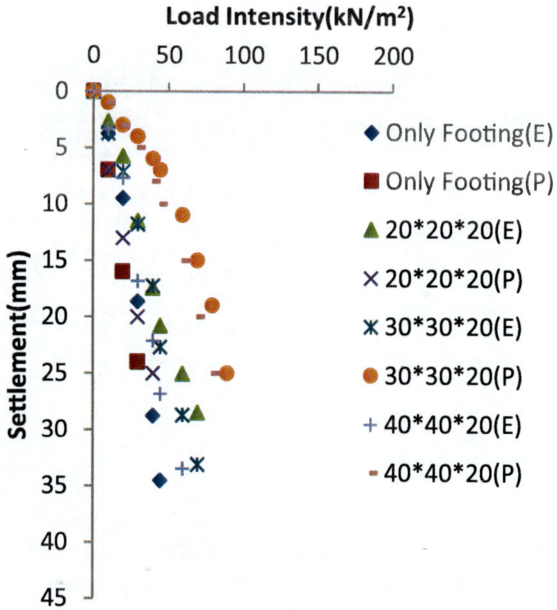

Table 4 Comparison between load intensities and BCR

Footing Configuration	Confiner Dimension (cm)	Horizontal Geogrid Spacing	Load Intensity (kN/m^2) (Experimental)	Load Intensity (kN/m^2) (Plaxis)	BCR (Experimental)	BCR (Plaxis)
Only footing		–	35.9	28.4	–	–
Footing + confiner	20*20*20	–	59.5	37.5	1.65	1.32
	30*30*20	–	50.3	76.75	1.4	2.71
	40*40*20	–	42.7	59	1.18	2.08
Footing + confiner + reinforcement	20*20*20 (D/B = 1, d/B = 1)	0.25B	81	55.7	2.25	1.96
		0.5B	73.6	27.95	2.05	1.04
		0.75B	65.1	100.4	1.81	3.53
		1.0B	53.2	43.4	1.68	1.52
	30*30*20 (D/B = 1.5,d/B = 1)	0.25B	138.2	179	3.84	6.3
		0.5B	98.2	106.5	2.73	3.75
		0.75B	82.7	76	2.3	2.67
		1B	71.4	82.5	1.98	2.90
	40*40*20 (D/B = 2, d/B = 1)	0.25B	143.1	158.75	3.98	5.59
		0.5B	92.9	94	2.58	3.3
		0.75B	87.5	91.5	2.43	3.22
		1B	75.8	72.25	2.11	2.54

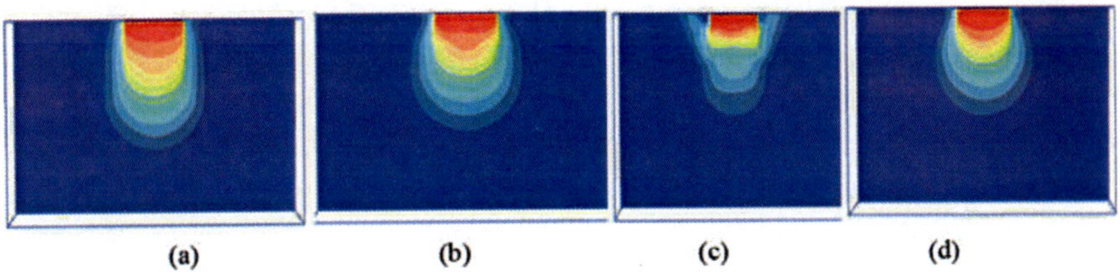

Fig. 13 Failure pattern for Footing + Confiner (D/B = 1, d/B = 1) + Reinforcement (spacing @ **a** 0.25B, **b** 0.5B, **c** 0.75B, **d** 1B)

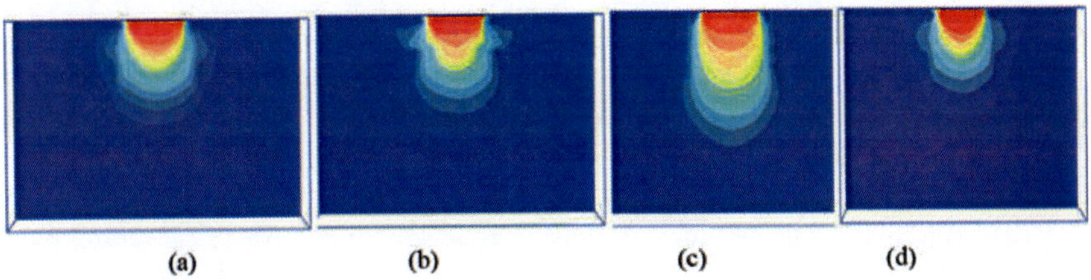

Fig. 14 Failure pattern for Footing + Confiner (D/B = 1.5, d/B = 1) + Reinforcement (spacing @ **a** 0.25B, **b** 0.5B, **c** 0.75B, **d** 1B)

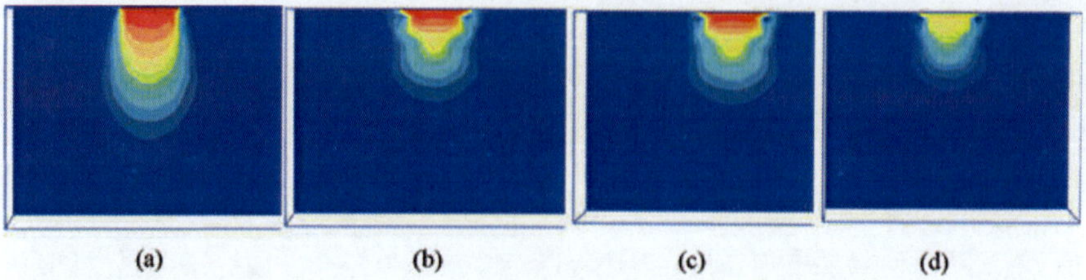

Fig. 15 Failure pattern for Footing + Confiner (D/B = 2, d/B = 1) + Reinforcement (spacing @ **a** 0.25B, **b** 0.5B, **c** 0.75B, **d** 1B)

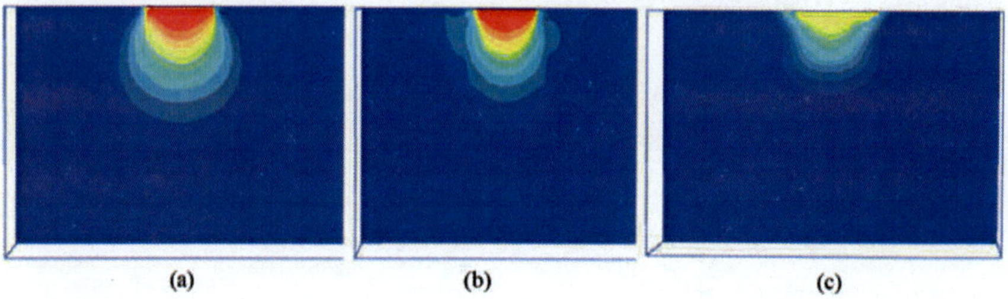

Fig. 16 Failure pattern for Footing + Confiner **a** D/B = 1, d/B = 1; **b** D/B = 1.5, d/B = 1; **c** D/B = 2, d/B = 1

This can be explained as horizontal reinforcements placed at closer spacing like 0.25B, it increases interlocking effect consequencing an increase in shearing resistance underneath the footing due to which lateral displacement of sand is restrained. From Table 4, it is found that as the spacing increases between the reinforcements, the load-bearing capacity starts reducing and mostly stays closer for 0.75 and 1.0B spacing.

4.3 Length of Horizontal Reinforcements

Observing Table 4, it is found that for series B with an increase in D/B value, the load intensities have been found to be decreased whereas for series C, the load intensity value shows a significant increment. This noticeable effect is ascertained not only due to the spacing of reinforcements but also complimented by the length and number of geogrids placed inside the confiner. As the D/B value increases, it simultaneously increases the L/B value, which provides a good anchorage for reinforcements. In the present, the work length of reinforcement equal to twice of footing width is considered for giving the optimum value of bearing capacity.

4.4 *Failure Pattern*

A cross-section of the total vertical displacement at 25 mm settlement under concentric vertical loading is shown in Figs. 13, 14, 15 and 16. It is seen that the failure mechanism for minimum spacing of 0.25B has shown a slow gradual vertical settlement, which is found to be more prominent with the increment of spacing. Figure 16 shows the failure pattern for footing with only confiner, which clears the fact that with an increase in D/B value, the effect of confiner reduces consequencing a rapid settlement.

5 Conclusions

Based on the laboratory investigations, the study brings forth the following conclusions that are closely supported by the PLAXIS results

1. The use of both confiner and horizontal reinforcements shows significant improvement in load intensity as compared to only confiner. With an increase in top surface dimension of confiner (d/B = 2) and minimum spacing of 0.25B gives optimum value in load intensity i.e. 143.1 kN/m^2, which is validated closely with plaxis results of 158.75kN/m^2.
2. In case of confiner, better performance is observed at (D/B = 1, d/B = 1), which is found to be providing lesser load intensity with increment in D/B value.
3. An increase in spacing value between reinforcements reduces down the load intensity and bearing capacity ratio together.
4. Aperture size and placement of confiner–reinforcement configuration play a vital role in the analysis.
5. From the failure pattern, it can be observed that with an increase in D/B value, the failure plane spreads more on both sides of the footing.
6. Both numerical and experimental results show very close validation though an unconventional value for 0.75B spacing of horizontal reinforcement for confiner dimension (D/B = d/B = 1) is noticed due to some execution error.

References

1. Amarasinghe, M.P., De Silva, L.I.N., Gallage, C.: The effect of lateral confinement on the settlement characteristics of shallow foundations on sand. Int. J. GEOMATE **15**(51), 258–265 (2018)
2. Al-Aghbari, M.Y., Dutta, R.K.: Performance of square footing with structural skirt resting on sand. Geomech. Geoeng. **3**(4), 271–277 (2008)
3. Azzam, W.R., El Wakil, A.Z.: Experimental and numerical studies of circular footing resting on confined granular subgrade adjacent to slope. Int. J. Geomech. **16**(1), 1–15 (2016)

4. Eid, H.T., Alansari, O.A., Odeh, A.M., Nasr, M.N., Sadek, H.A.: Comparative study on the behavior of square foundations resting on confined sand. Can. Geotech. J. **46**(4), 438–453 (2009)
5. Elsaied, A.E., Saleh, N.M., Elmashad, M.E.: Behavior of circular footing resting on laterally confined granular reinforced soil. HBRC J. **11**(2), 240–245 (2015)
6. El Sawaaf, M., Nazer, A.: Behavior of circular footings resting on confined granular sand. J. Geotech. Geoenviron. Eng. **131**(3), 359–366 (2005)
7. El Wakil, A.Z.: Bearing capacity of skirt circular footing on sand. Alex. Eng. J. **52**(3), 359–364 (2013)
8. Fattah, M.Y., Shlash, K.T., Mohammed, H.A.: Experimental study on the behavior of strip footing on sandy soil bounded by a wall. Arab. J. Geosci. **8**, 4779–4790 (2015)
9. Jha, J.N.: Effect of vertical reinforcement on bearing capacity of footing on sand. Indian Geotech. J. **37**(1), 64–78 (2007)
10. Joseph, M.A.S.A.: Behaviour of skirted footing resting on sea sand. Int. J. Sci. Res. Eng. Trends **4**(3), 470–475 (2018)
11. Krishna, A., Viswanath, B., Keshav, N.: Performance of square footing resting on laterally confined sand. Int. J. Res. Eng. Technol. **3**(6), 110–114 (2014)
12. Rajagopal, K., Krishnaswamy, N.R., Latha, G.M.: Behavior of sand confined with single and multiple geocells. Geotext. Geomembr. **17**(3), 171–184 (1999)
13. Renaningsih, Satria, I.F., Susanto, A., Listyawan, A.B.: Method to increase ultimate bearing capacity of skirted circular footing. In: AIP Conference Procedeeings 1855, 020013-1-9 (2017)
14. Singh, V.K., Prasad, A., Aggarwal, R.K.: Effect of soil confinement on ultimate bearing capacity of square footing under eccentric-inclined load. Electron. J. Geotech. Eng. **12**, 1–14 (2007)
15. Thakur, A., Dutta, R.K.: Experimental and numerical studies of skirted hexagonal footings on three sands. SN Appl. Sci. **2**(3), 1–11 (2020)

A Study on the Performance of Marine Clay Stabilized with Waste Materials

A. V. S. A. Srujana, M. Hemanadh, K. Ravi Kumar, S. Preethi, and Manchikanti Srinivas

1 Introduction

For many years mankind was wondering at the instability of earth materials, especially marine clay. Several investigations were made on the marine clay to improve its properties. The properties of saturated marine clay differ significantly from moist soil and dry soil. Marine clay is micro-crystalline in nature, and clay minerals like kaolinite, chlorite, and illite and non-clay minerals like quartz and feldspar are present in the soil. Construction on such ground may result in undesirable settlements due to the soft nature of the soil. Researchers from various fields have focused on solving environmental problems due to the production of wastes. Gidley and Sack [6] suggested methods of utilization of wastes, such as fly ash, iron slag, waste rock, mine tailings, and sludge in construction. In this paper three different additives are used to study the properties of treated marine clay. The additives used are GGBS, fly ash, sewage sludge, and lime. The reason for using the combination of sewage sludge and lime is lime reduces the decomposition nature of sewage sludge, when it is mixed with marine clay. GGBS consists of a substantial proportion of a glassy phase with a substantial content of Ca, Si, Al, and Mg-based compounds. Sewage sludge is one of the major unused waste materials, and if it is used directly to combine with the marine clay then the organic content in sewage sludge may disturb the clay structure. To reduce the organic content, lime is mixed with sludge in a proportionate manner.

Yi et al. [22] investigated the stabilization efficacy of alkali-activated ground-granulated blast furnace slag (GGBS) on marine soft clay. The results indicated that Na_2CO_3-GGBS mix had no stabilization efficacy for this marine soft clay. Moreover, NaOH-GGBS-stabilized clay yielded the highest UCS values at 7, 28, and 90 days; however, the UCS values decreased from 90 to 180 days because of microcracking.

A. V. S. A. Srujana · M. Hemanadh · K. Ravi Kumar · S. Preethi · S. Manchikanti (✉)
Department of Civil Engineering, Gayatri Vidya Parishad College of Engineering (Autonomous), Visakhapatnam 530048, India

C. N. V. Satyanarayana Reddy et al. (eds.), *Ground Improvement and Reinforced Soil Structures*, Lecture Notes in Civil Engineering 152,
https://doi.org/10.1007/978-981-16-1831-4_42

The performance of GGBS-fly ash-modified soft soils was also evaluated [4, 17, 18, 21] using compaction and California bearing ratio (CBR) tests.

Based on the results, it appears that soft soil can be effectively stabilized with the addition of fly ash–GGBS mixtures. It is observed that the plasticity index decreased and unconfined compressive strength (UCS) increased upon addition of slag to the soil. With the age of curing the UCS values increased indicating the chemical reaction between free lime and soil.

Buva and Wayal [2], Ouyang et al. [14], Mousa et al. [13], Monzó et al. [12], Lin et al. [10], Li and Lin [9] studied the properties of the incinerator fly ash derived from MSW incineration process, which indicated that fly ash is a potential source of jet-grouting admixture for soil improvement. The use of fly ash as an admixture in the stabilization of a soft marine clay resulted in stabilized samples with an improved strength, more than 75 times that of the untreated clay, and reduced both the plasticity and compression indices by about 69% and 23%, respectively. It is also noticed that the liquid limit of the marine clay has decreased by 23.88% on addition of 25% fly ash and it has further decreased by 26.30% when 9% lime is added, and the plastic limit of the marine clay improved by 6.93% on addition of 25% fly ash and it further improved by 20.46% when 9% lime is added. It is also found that the MDD of the marine clay has improved by 14.83% on addition of 25% fly ash and it has improved by 7.74% when 9% lime is added.

Quicklime and hydrated lime were used [1, 7, 11], to activate ground-granulated blast furnace slag (GGBS) for stabilization of marine soft clay. The UCS results demonstrated that hydrated lime-activated GGBS achieved slightly higher 90-day UCS in stabilized clay than quicklime-activated GGBS. Chandran et al. [3] investigated the efficacy of lime solution with different concentrations when added to the soft soil samples for stabilization and cured with water for 7, 14, 21, 28, and 35 days. From the test results, it was found that the unconfined compressive strength increased up to a curing period of 28 days and thereafter there is no appreciable effect. Stabilization of marine clay with cement, lime, and fly ash and bagasse ash has been studied [5, 15, 19] for strength improvement and plasticity behavior. Specimens were prepared by mixing with varying proportions of lime, cement, and fly ash with clayey soil separately. The unconfined compression test (UCS) and Atterberg's limits of the soil with different percentages of additives were determined separately after curing specimens for 7 days. The 7-day peak strength of soil-lime specimens was found at 7.5% lime content, although addition of 6% and 12% of lime was suitable.

The 7-day strength of specimens mix with fly ash was found significantly more than that of specimen without fly ash. For Wando marine clay improved [20] with fly ash, and the plasticity index reduced by about 18–26% with an increase of fly ash content. As the amount of fly ash increased, the composite soil resisted the compressive loading better and consequently showed lesser compressibility. Phanikumar and Nagaraju [16], Jamaludin et al. [8] conducted a laboratory investigation study to understand the effect of fly ash on the consolidation characteristics of compacted clay. The A-7-5 clay was stabilized with fly ash at different contents such as 0%, 5%, 10%, 15%, and 20%. The A-7-5 clay is stabilized with fly ash which causes a decrease in the maximum dry density (MDD) and an increase in the optimum moisture content (OMC) because fly ash requires more water during pozzolanic action. And it was observed that the A-7-5 clay became rough or granular and increased the rate of consolidation. It was found that the compression index (C_c) values decreased up to 10% of fly ash content.

2 Materials

2.1 Marine Clay

The marine clay used in this study is collected from Visakhapatnam Port. The index and engineering properties of marine clay are determined as per IS codes of practice. Samples are taken separately in a box to determine the initial moisture contents. The samples were tested for physical properties and the results are: specific gravity = 2.68, sand size(%) = 31, fines(%) = 69, field moisture content(%) = 72, liquid limit(%) = 68, plastic limit(%) = 21, plasticity index(%) = 47, maximum dry density(g/cc) = 1.50, optimum moisture content(%) = 26.4, Indian standard soil classification = CH, unconfined compressive strength (kN/m^2) = 41.11, California bearing ratio(%) (unsoaked and soaked) = 2.02 and 1.12, respectively.

2.2 Fly Ash

Fly ash was obtained from NTPC, Visakhapatnam. The chemical properties of the ash are for the most part influenced by the chemical content of the coal burned. Fly ash is generally gray in color and in powder form. The fly ash used in this study consists of silicon dioxide (SiO_2) = 43.6%, aluminum oxide (Al_2O_3) = 16.90%, ferric oxide (Fe_2O_3) = 5.80%, calcium oxide (CaO) = 24.30%, magnesium oxide (MgO) = 4.80%, titanium oxide(TiO_2), potassium oxide (K_2O) = 1.30%, and sulfur trioxide (SO_3) = 3.30%.

2.3 Lime

Lime mainly consisting of calcium oxide (CaO) = 74.23%, loss of ignition (LOI) = 24.35%, phosphorous oxide (P_2O_5) = 0.08%, magnesium oxide (MgO) = 0.74%, calcium sulfate (Ca_2SO_4) = 0.12%, ferric oxide (Fe_2O_3) = 0.17%, aluminum oxide (Al_2O_3) = 0.11%, and silica (SiO_2) = 0.20% was used in this study.

2.4 *Ground-Granulated Blast Furnace Slag*

It is a by-product of iron and steel industry. This slag contains calcium, magnesium, manganese, and aluminum silicates in various combinations. When powder form of GGBS is used in cement manufacturing industries, the grinding operations are costlier. For this reason, slag is stacked near iron industries, thus leading to massive consumption of area. Hence the investigation was undertaken to check the potential usage of granulated blast furnace slag to improve the geotechnical properties of marine clay. The GGBS used in this study has calcium oxide (CaO) = (30–50%), silica (SiO_2) = (28–38%), aluminum oxide (Al_2O_3) = (8–24%), and magnesium oxide = (1–18%) by weight.

2.5 *Sewage Sludge*

Sewage sludge is a product of wastewater treatment and it is a diluted suspension of solids. It consisted of silicon (Si) = 36.7%, iron (Fe) = 16.3%, aluminum (Al) = 14.7%, calcium(Ca) = 13.1%, phosphorous(P) = 11.8%, potassium (K) = 3.66%, titanium(Ti) = 1.00%, zinc (Zn) = 0.81%, magnesium(Mg) = 0.77%, sulfur (S) = 0.36%, manganese (Mn) = 0.16%, copper (Cu) = 0.13%, and nitrogen(N) = 0.17%. A constant amount of sewage sludge is added to the soil in combination with varying percentages of lime.

3 Laboratory Experimentation

The effectiveness of GGBS, fly ash, lime, and sewage sludge as stabilizers was studied using several laboratory tests. First, all the tests were conducted on marine clay alone, and later the tests were conducted on the marine clay mixed with 10% GGBS by weight of soil. Later, the percentage of GGBS was increased to 20%, 30%, and 40%. And the same increment was used in the case of fly ash too. In the case of sewage sludge and lime combination, the sludge was kept at a constant of 10% and lime was varied as 5%, 10%, 15%, and 20% by weight of marine clay, respectively. The tests were conducted in the laboratory to investigate the change in the characteristics of marine clay when compared with untreated marine clay. All the tests were conducted as per the procedures laid down in various parts of IS: 2720.

4 Results and Discussions

4.1 *Effect of GGBS, Fly Ash, Sewage Sludge, and Lime on Atterberg's Limits*

Liquid limit reduced from 68% to 59%, 57%, 51%, and 39% and the plastic limit increased from 21% to 23.5, 24.5, 26, and 28% when soil is replaced by GGBS in percentages of 10, 20, 30, and 40, respectively. Eventually the plasticity index decreased as shown in Fig. 1, and the same trend was observed for fly ash. The liquid limit decreased from 68% to 59, 53, 45, 40% and the plastic limit increased in the case of fly ash as shown in Fig. 2.

In the case of combination of sewage sludge and lime when mixed with the soil, the liquid limit decreased from 68% to 64%, 60%, 54%, 50% and the plastic limit increased from 21% to 28%, 31%, 34%, and 36% when the replacement percentages of lime are 5%, 10%, 15%, and 20% (keeping the sewage sludge of 10% as constant), respectively. Eventually the plasticity index decreased from 47% to 40%, 37%, 29%, and 26% as represented in Fig. 3.

4.2 *Compaction Test*

First, with increase in replacement with GGBS, the OMC decreased from 24% to 18%, 16% and 14%, and the MDD increased from 1.50 g/cc to 1.82 g/cc, 1.90 g/cc and 2.04 g/cc with replacements of 10%, 20%, 30%, and 40%, respectively. On further

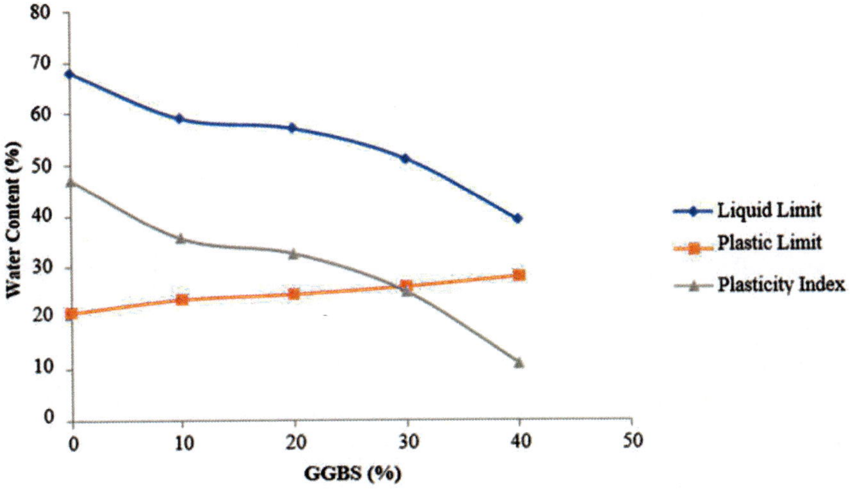

Fig. 1 Variation of Atterberg's limits of GGBS-treated clay

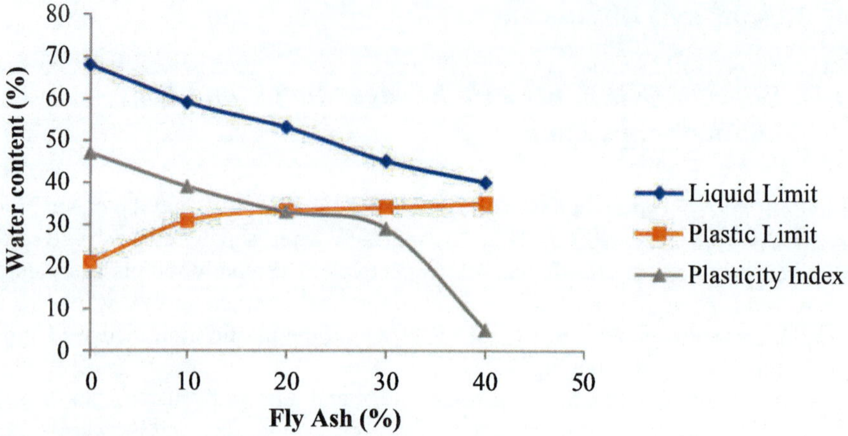

Fig. 2 Variation of Atterberg's limits of fly ash-treated clay

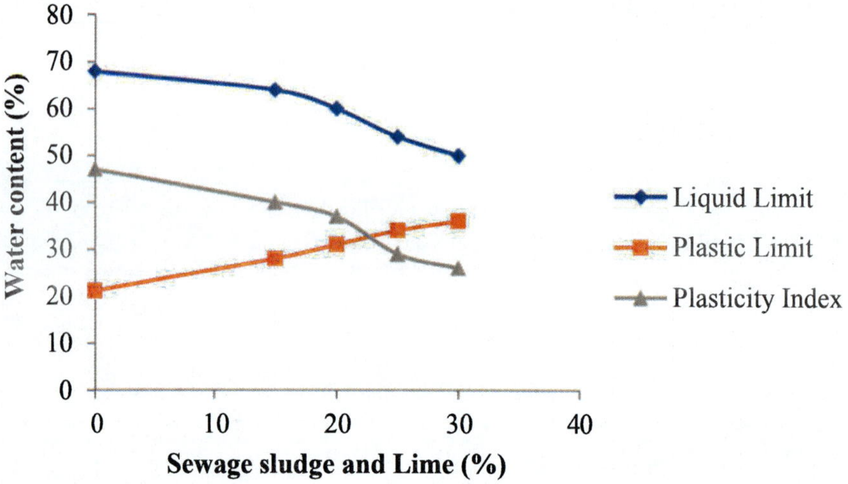

Fig. 3 Variation of Atterberg's limits of sewage sludge + lime + clay mixes

increase in GGBS by 10% the OMC decreased to 12% and the MDD decreased to 1.79 g/cc as shown in Fig. 4. However, Fig. 5 shows the compaction test results for clay–fly ash mixes.

First, with increase in replacement with fly ash the OMC (%) decreased from 24% to 20%, 18% and 16% and the MDD increased from 1.50 g/cc to 1.90 g/cc, 1.95 g/cc and 2.05 g/cc with replacements of 10%, 20%, 30%, and 40%, respectively. On further increase in fly ash by 10% the OMC decreased to 14% and the MDD decreased to 2 g/cc as shown in Fig. 5.

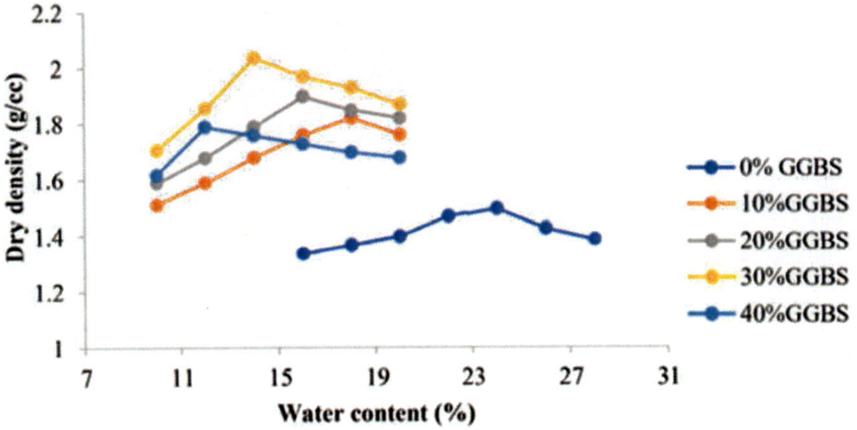

Fig. 4 Compaction curves of clay + GGBS mixes

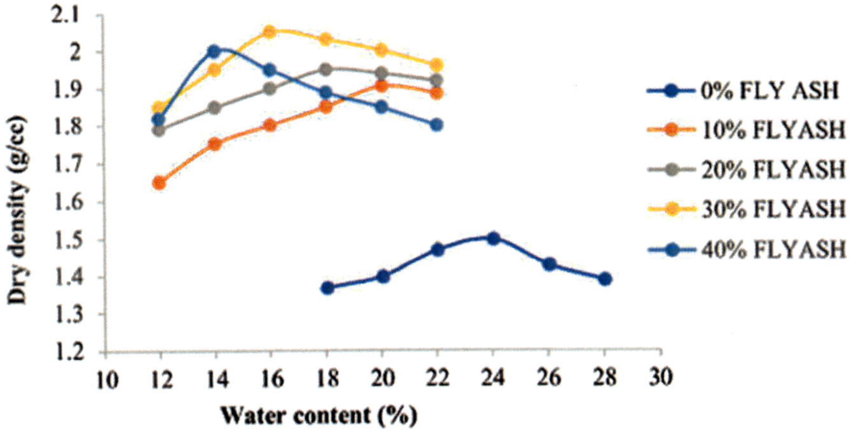

Fig. 5 Compaction curves of clay + flyash mixes

When sewage sludge and lime combination is used as an additive, the OMC increased from 24% to 26%, 28%, 30%, 32% and the corresponding MDD increased from 1.50 g/cc to 1.95 g/cc, 2.12 g/cc, 2.24 g/cc, 2.30 g/cc with the replacements of 5%, 10%, 15%, and 20% of lime (keeping sewage sludge of 10% as constant) (Figs. 6 and 7).

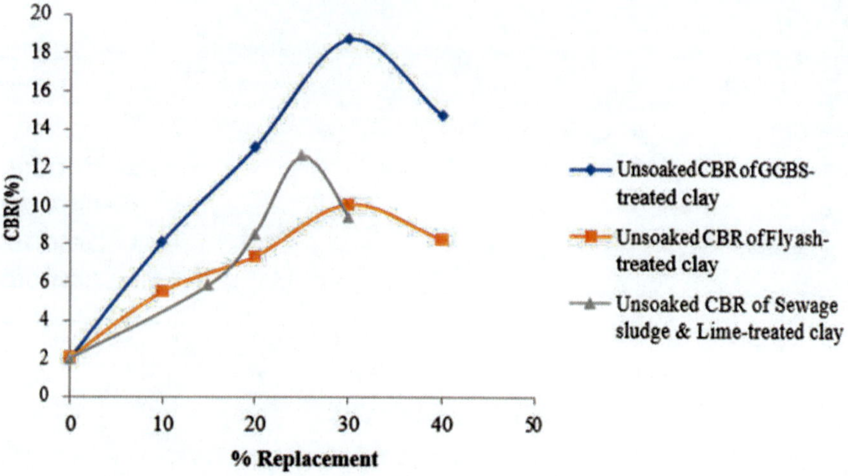

Fig. 6 Unsoaked CBR results for different additives

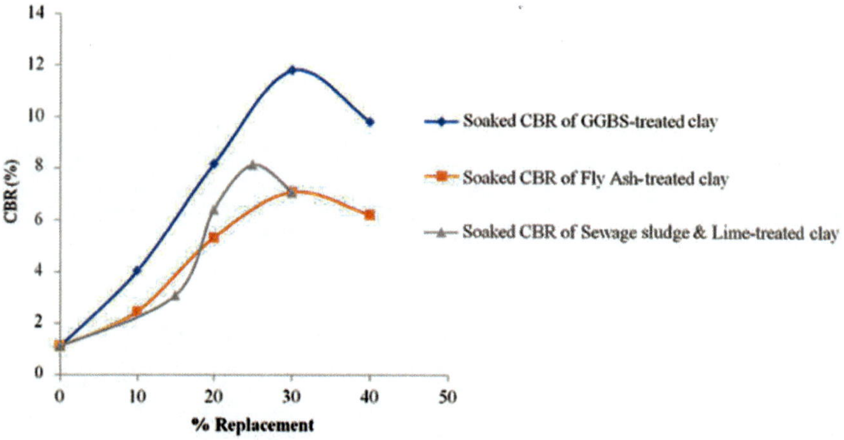

Fig. 7 Soaked CBR results for different additives

4.3 California Bearing Ratio (CBR) Test

Unsoaked and soaked CBR values were found from the tests. The CBR values of the soil in soaked and unsoaked conditions are 1.12% and 2.02%, respectively. The soaked CBR values increased to 4.04%, 8.15%, and 11.80% with the replacements of 10%, 20%, and 30% of GGBS, and the CBR values decreased in both soaked and unsoaked conditions with a replacement of 40% GGBS. The soaked CBR values increased to 2.45%, 5.31%, and 7.04% with the replacement of 10%, 20%, and 30% of fly ash and the CBR values decreased in both the soaked and unsoaked conditions

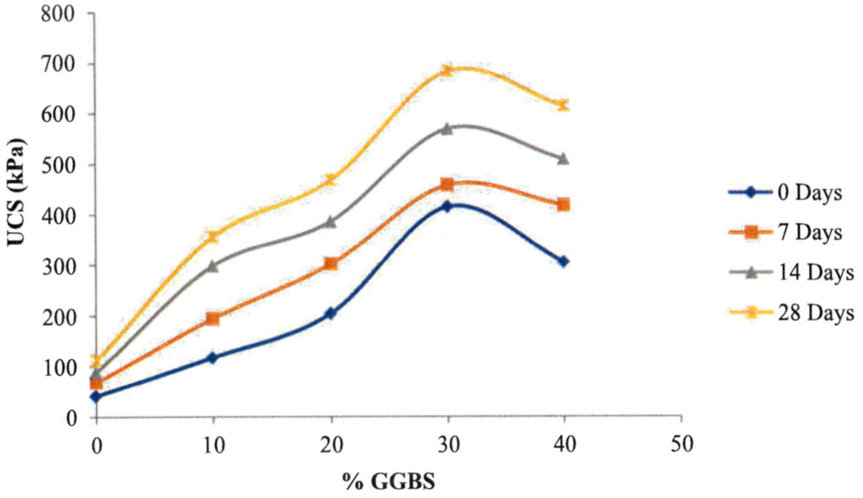

Fig. 8 Variation of UCS values for different curing periods for clay-GGBS mixes

with the replacement of 40% fly ash. The soaked CBR values increased to 3.09%, 6.40%, and 8.12% with the replacement of 5%, 10%, and 15% of lime (keeping 10% of sludge as constant) and the CBR values decreased in both the soaked and unsoaked conditions of 20% replacement of lime.

4.4 Unconfined Compressive Strength (UCS) Test

The UCS values increased with increase in the percentages of additives for curing durations of 0, 7, 14, and 28 days. For GGBS-mixed soil, the increase in UCS values is shown in Fig. 8. The UCS values decreased for 40% replacement of GGBS after 0, 7, 14, and 28 days of curing. The UCS values decreased at 40% replacement of GGBS after 0, 7, 14, and 28 days of curing.

In the case of fly ash, there is an increase in UCS values up to 30% of soil replaced with fly ash and the UCS values decreased at 40% replacement of fly ash for 0, 7, 14, and 28 days of (curing) testing as shown in Fig. 9.

A similar trend was observed in the case of clay–lime–sewage sludge mixes (keeping 10% sewage sludge as constant), as shown in Fig. 10.

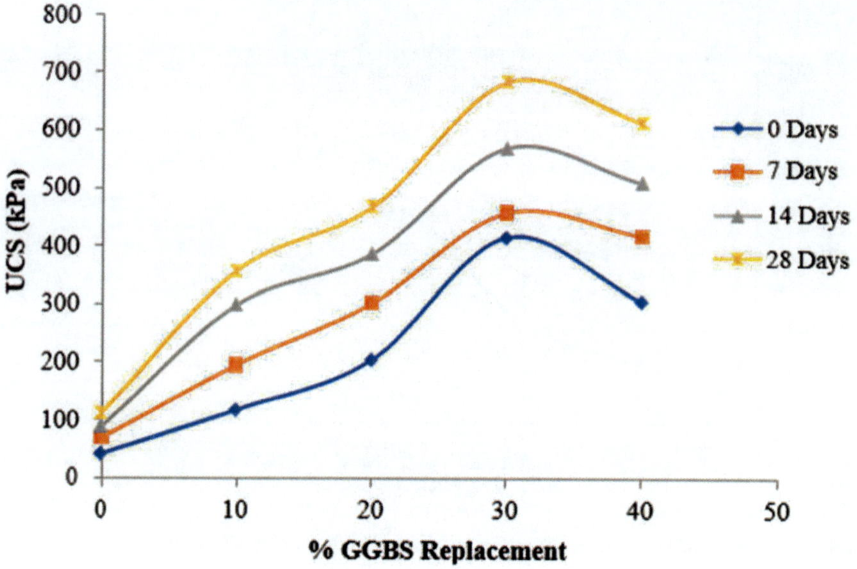

Fig. 9 Variation of UCS values for different curing periods for clay–fly ash mixes

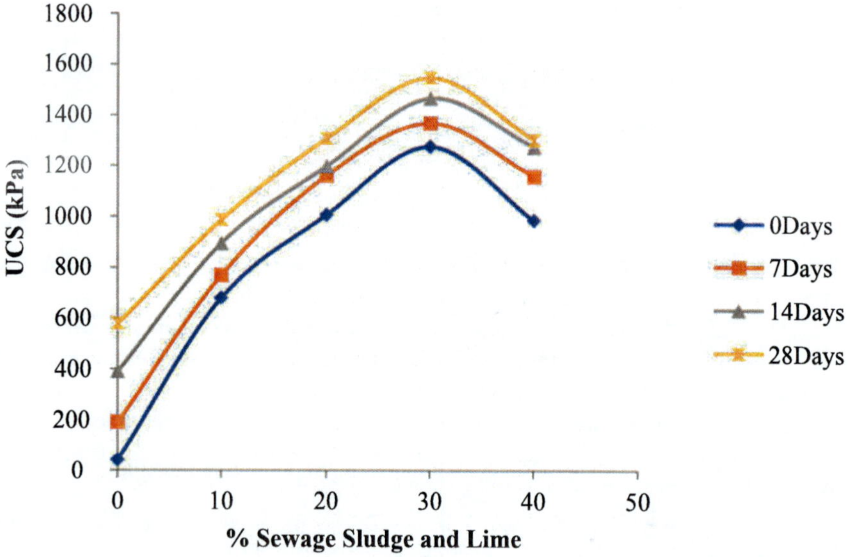

Fig. 10 Variation of UCS values for different curing periods for clay–sewage sludge–lime mixes

5 Conclusions

On the basis of experimental investigation, the following conclusions are drawn:

1. With the addition of any type of admixture used in this study and by the replacement of clay in 10, 20, 30, and 40 percentages, the liquid limit and the plasticity index values gradually decreased. When clay is treated with GGBS, the liquid limit and plasticity index values decreased appreciably by 42.6% and 76.5%, respectively, when compared with that of the untreated clay.
2. When clay was replaced with various admixtures in 10, 20, 30, and 40 percentages, the plastic limit values increased. Amongst all the additives used, the performance of sewage sludge–lime combination was the best, which resulted in an appreciable increase of 71% in the plastic limit value when compared with that of the untreated clay.
3. When clay was replaced with GGBS in percentages of 10, 20, 30, and 40, it was observed that with decrease in OMC values, the MDD values increased correspondingly. The same trend was observed when fly ash was used in the same percentages as mentioned above. But in the case of clay treated with sewage sludge–lime combination, an increase in MDD values was observed with a corresponding increase in OMC values. Hence, the sewage sludge–lime combination is not a suitable additive to improve marine clay.
4. The highest shear strength value of clay treated with GGBS is 340 kPa for 30% replacement. The value of 350 kPa was obtained for 30% replacement using fly ash. Similarly, the value of 770 kPa was obtained for replacement of 10% sewage sludge + 15% lime.
5. The California bearing ratio (CBR) value of clay treated with GGBS is 11.80% in soaked condition for 30% replacement by GGBS. Similarly, a value of 7.04% was obtained for 30% replacement using fly ash in soaked condition. In addition, a value of 8.12% was obtained for 30% replacement using 10% sewage sludge + 15% lime in soaked condition, whereas for untreated clay the CBR value is 1.12% in soaked condition. The CBR value of 30% GGBS as replacement is more than the CBR value of 30% fly ash as replacement. The cost of bituminous road constructed using fly ash-treated marine clay is more economical than using GGBS. Therefore, it may be concluded that GGBS and fly ash are effective stabilizers. According to the performance aspect, GGBS is better, but fly ash is better from an economic standpoint.

References

1. Balasubramaniam, A.S., Bergado, D.T., Buensucoso Jr., B.R., Yan, W.C.: Strength and deformation characteristic of lime treated soft clay. J. Geotech. Eng. **20**, 49–65 (1989)
2. Buva, V., Wayal, A.S.: Use of fly ash and lime for stabilization of soft soil. Electron. J. Geotech. Eng. 21.18, 6235–6246 (2016)

3. Chandran, S., Rajani, G, Padmakumar, P.: Stabilization of clayey soil using lime solution. In: 10th National Conference on Technological Trends (NCTT09), 1, 13–19 (2009)
4. Dhanalakshmi, P., Chandrakaran, S, Stabilisation of soft clay using ground granulated blast furnace slag and lime. In: Proceedings of 6th Young Geotechnical Engineers Conference, 1, 156–161 (2017)
5. Gandhi, K.S.: Expansive soil stabilization using bagasse ash. Int. J. Eng. Res. Technol. **1**(5), 1–3 (2012)
6. Gidley, J.S., Sack, W.A.: Environmental aspects of waste utilization in construction. J. Environ. Eng. **110**(6), 1117–1133 (1984)
7. Ibrahaim, M.A., Al-Kiki Khawla, A. K., Al-Juari, S., Khattab, A.A.: Strength, durability and hydraulic properties of clayey soil stabilized with lime and industrial waste lime. Al-Rafidain Eng. J. **16**(1), 102–116 (2008)
8. Jamaludin, N., Mohd.Yunus, N. Z., Jusoh, S. N., Pakir, F., Ayub, A., Zainuddin, N. E., Hezmi, M. A.: Nordiana Mashros, potential and future: utilization of waste material on strength characteristics of marine clay. In: IOP Conference Series: Materials Science and Engineering, pp. 1–8 (2019)
9. Chen, Li, Lin, Deng-Fong: Stabilization treatment of soft subgrade soil by sewage sludge ash and cement. J. Hazard. Mater. **162**(1), 321–327 (2018)
10. Lin, D.F., Lin, K.L., Luo, H.L.: A comparison between sludge ash and fly ash on the improvement in soft soil. J. Air Waste Manage. Assoc. **57**(1), 59–64 (2007)
11. Gharib, Mehdi, Saba, Hamidraza: Experimental investigation of impact of adding lime on Atterberg's limit in golestan province soils. Int. Res J. Appl. Basic Sci **3**(4), 796–800 (2012)
12. Monzó, J., Payá, J., Borrachero, M.V., Peris-Mora, E.: Mechanical behavior of mortars containing sewage sludge ash (SSA) and Portland cements with different tricalcium aluminate content. Cement Concrete Res. **29**(1), 87–94 (1999)
13. Attom, M., Mortula, M.Md, Munjed, A.: Shear strength stabilization using burned sludge ash, International Journal of Advances in Mechanical and Civil. Int. J. Adv. Mech. Civil Eng. **4**(3), 42–45 (2017)
14. Ouyang, C.F., Xu, Z.L., Lan, W.Z:. Study of the treatment and reuse of municipal sewage treatment plant sludge. Proceedings on 8th Annual Meeting Sanitary Engineering, 1, 221–232 (1998)
15. Pandian, N.S., Krishna, K.C.: California bearing ratio behavior of cement-stabilized fly ash-soil mixes. J. Test. Eval. **30**(6), 1–5 (2002)
16. Phanikumar, B.R., Nagaraju, T.V.: Swell and compressibility of GGBS–clay mixes in lumps and powders. Indian Geotech. J. **49**, 161–169 (2019)
17. Preetham, H.K., Nayak, S., Surya, E.V.: Experimental investigation on the stabilization of soft clay using granulated blast furnace slag. In: IOP Conference Series on Materials Science and Engineering, 1, 1–6 (2019)
18. Ramesh, B., Hima Bindu, P.: An experimental study on stabilization of clayey soil by using granulated blast furnace slag. Int. J. Trend Sci. Res. Dev. ISSN: 2456-6470, **3**(5) (2019)
19. Sen, P., Mukesh, M. D.: Evaluation of strength characteristics of clayey soil by adding soil stabilizing additives. Int. J. Earth Sci. Eng. **4**,1060–1061 (2011
20. Suneel, M., Jeonggeun Kwon, J.-C., WookJeon, C.: Long-term consolidation and strength behavior of marine clay improved with fly ash. J. Marine Geores. Geotechnol **28**, 105–114 (2010)
21. Tarun, K.R., LaxmikantYadu, S.K.C.: Strength characteristics of fly ash stabilized soil embankment and stability analysis using numerical modelling. In: Proceedings on Indian Geotechnical Conference, GeoNEst, 1, 14–16 (2017)
22. Yaolin, Y., Gu, L., Liu, S., Puppala, A.J.: Carbide slag–activated ground granulated blast furnace slag for soft clay stabilization. Can. Geotech. J. **52**, 656–663 (2015)

Influence of Biopolymer Treatment on Suction Characteristics of Bhavnagar Expansive Soil

Jahnavi Parmar, Vikas Yadav, and Saloni Pandya

1 Introduction

Long-term stability attributes of black cotton soils (expansive soils) depend upon the tendency of the soil to swell (increase in volume) and shrink (decrease in volume). Black cotton soil in partially saturated condition ($S_r < 100\%$) experiences acute shrinkage and swelling on account of timely change in the water content (w) of soil mass due to climatic variations. Approximately 20% of continental crust in India comprises shrinkage–swelling (expansive) soils. Expansive soils are encountered in numerous parts of Gujarat, Andhra Pradesh, Uttar Pradesh, Karnataka, western Madhya Pradesh and Maharashtra [31]. The vital problem related to expansive soils was noticed to be related to induced volumetric deformation, which causes terrible upheaving and enormous cracks in pavements and embankments; damage to the floor slabs; concrete and steel plinths failure; basement buckling due to increased lateral stresses; damages to retaining walls; deterioration of water pipelines etc.

Several different methods have been adopted for stabilization of expansive soils, which include physical, chemical and polymers methods. Physical methods comprise compaction methods and drainage provision; however, this method is found to be less popular [20] owing to meagre agreement between effectiveness of method and cost. Stability treatment by physical method was found to decrease over time and would provide impermanent stabilization. Chemical soil stabilization using lime, fly ash and cement are harmful, add toxicity and are not degradable [20, 30]. Both physical and chemical treatments were reported to be unsustainable [20, 30]. Biopolymer treatment has gained popularity as sustainable method for expansive soil stabilization. Biopolymer stabilization treatment was reported as efficient, sustainable and durable alternative to conventional techniques [5, 8, 14, 15, 20, 30].

J. Parmar (✉) · V. Yadav · S. Pandya
Civil Engineering, Navrachana University, Vadodara 391410, Gujarat, India

C. N. V. Satyanarayana Reddy et al. (eds.), *Ground Improvement and Reinforced Soil Structures*, Lecture Notes in Civil Engineering 152,
https://doi.org/10.1007/978-981-16-1831-4_43

Geological spread of such soils is found in the active zone/vadose zone of hyper arid, arid and semi-arid regions of the world [22] and hence is under unsaturated condition. Formation of expansive soils is attributed to the leaching and chemical weathering of basaltic rocks due to lower degree of precipitation, high evaporation owing to high temperature and fluctuating wetting and drying cycles. Detrimental volumetric changes caused by the swelling–shrinkage phenomena of expansive soils would be governed by soil–water chemistry (suction), salinity of soil, clay mineralogy and plasticity, extent of ground water table, extent of water content variation over space and time, vegetation cover, permeability, temperature, soil profile etc. [9, 21]. The degree of expansiveness of the unsaturated soil was found to be crucially governed by amount of negative pore water pressure (suction) present in the unsaturated expansive soils [22]. Amount of suction governs the macro and micro soil fabric arrangement of the expansive soil. Matric and osmotic suctions govern the crystalline and osmotic swelling in the expansive soils (Likos 2004) [24]. Hence, it is imperative to evaluate the influence of biopolymer stabilization treatment on the suction characteristics of treated shrinkage–swelling soils. The available literature on stabilization of expansive soil remains concentrated on the strength and swelling characteristics of expansive soils. Influence of biopolymer treatment on the suction characteristics of expansive soils remains unexplored. In the present study, an attempt has been made to evaluate the influence of biopolymer treatment on suction (total, matric and osmotic), strength, swelling and salinity characteristics of salt-enriched Bhavnagar black cotton soil.

2 Material Properties and Sample Collection

The soil for the present study was collected from the village Madhya, Bhavnagar district, Gujarat, India. The site was located near the salt production works (latitude: 21° 51′ 6″ and longitude: 72° 6′ 2″). The topographical map showing the soil collection site is shown in Fig. 1. Soil was collected at a depth of 0.5 m below the ground level. There were no construction works in and around the area and was purely dedicated for salt production works. Bhavnagar is located along the coast line of Gulf of Khambhat or Gulf of Cambay in the Arabian Sea. The geological origin of Bhavnagar soil was reported to have volcanic origin [19]. The volcanic outpourings and distribution of lava led to denudation, which led to the formation of a basaltic topography. Salinity in the soil has been developed due to gradual withdrawal of sea from the surrounding areas of Bhavnagar city near Gulf of Khambhat. The Gulf encompasses an area of high tides and is portrayed by the command of strong tidal currents.

The geotechnical properties and grain size distribution of Bhavnagar expansive soil are listed in Table 1. The soil has been classified as CH as per Indian Standard classification system [12]. Bhavnagar expansive soil exhibited high electrical conductivity indicating the extreme salinity of the soil (Table 1).

Fig. 1 Topographical map of Gujarat showing the Bhavnagar expansive soil collection site

Table 1 Geotechnical properties of Bhavnagar soil

Test	Bhavnagar soil
Gravel (%)	0.01
Sand (%)	0.33
Silt (%)	49.53
Clay (%)	50.13
OMC (%)	21.42
MDD (g/cc)	1.66
Free swell index (FSI) (%)	110
Liquid limit (%)	66.87
Plastic limit (%)	62.5
Shrinkage limit (%)	9.33
Specific gravity (%)	2.64
Electrical conductivity (μS/cm)	10900
Organic content (%)	12.63

2.1 Biopolymers

Biopolymers are natural polymers produced by living organisms. They are polymeric biomolecules derived from plant matter. Biopolymers contain monomeric units that form larger structures. Each monomeric unit is chemically bonded with each other. Properties such as renewability, sustainability, non-toxic, degradable, carbon neutral etc. embolden to be used in stabilization and treatment of expansive soil [8, 15, 20, 30].

2.1.1 Xanthan Gum

Xanthan gum is a microbial polysaccharide produced by the bacteria Xanthitalics Campestris by the process of fermenting glucose, sucrose or any other carbohydrate sources. It is polysaccharide which has strong sorption characteristics and creates microstructural interactions with clay forming a viscous hydrocolloid when added to water [20].

This biopolymer is used widely in various applications such as in the cosmetic industry, food industry, agriculture, civil engineering, pharmaceutical and petrochemical industries and in other sectors as a thickening agent, stabilizer or emulsifier, and when combined with other gums it can act as a gelling agent [30]. In the current study, the xanthan gum procured from National Chemicals (cost: Rs. 1265 per 100 g) was employed.

2.1.2 Guar Gum

In semi-arid regions guar has been used to replenish the soil. Guar gum or cluster bean is a polysaccharide composed mainly of the sugars, galactose and mannose. Guar gums made from guar beans have thickening and stabilizing properties [30]. In the current study, the guar gum procured from National Chemicals (cost: Rs. 250 per 100 g) was used.

2.1.3 Chitosan

A polysaccharide obtained from chitin is the second most abundant polysaccharide in the world. Chitosan has been used in medical applications such as antimicrobial and therapeutic biomaterials as it is biocompatible and non-toxic [5]. It has also been used as chelating agent as it possesses ability to bind with fats cholesterol, proteins and metal ions. Chitosan is obtained from waste of shrimp shells [8]. In the current study, the chitosan procured from National Chemicals (cost: Rs. 9840 per 100 g) was used.

3 Experimental Program

3.1 Total, Matric and Osmotic Suction

For evaluating suction characteristics of Bhavnagar expansive soil, in-contact and non-contact filter paper tests [1] were employed. The filter paper method covers large range of suction (10 kPa to 100 MPa) and has been described as the most efficient, quick, accurate and reliable method for suction measurements [2–4, 17, 23, 25, 27, 28].

For evaluating total suction (ψ) of expansive soil non-contact filter paper method was used and for evaluating matric suction (ψ_m) in-contact filter paper method was used. Osmotic suction (ψ_ϕ) is obtained by accessing difference in total suction and matric suction values. Whatman 42 filter paper was used during the experimental program. The oven dried soil was passed from 4.75 mm sieve. Suction evaluation of untreated soil was carried out of 1.58 g/cc (95% of MDD) and varying water contents (10%, 15%, 20% and 25%). For treated soil, the amount of biopolymer by weight was calculated and was added to the soil. Biopolymer-treated soil specimens were prepared at 1.58 g/cc dry density and 10% water content. For biopolymer treatment, percentages by weight of xanthan gum and guar gum used were 0.5%, 1% and 1.5% and percentages by weight of chitosan used were 0.12% and 0.16%. Percentage variation range was selected based on optimum percentage of biopolymer treatment report by several researchers [5, 8, 15, 20, 30]. Specimen treated with guar gum, xanthan gum and chitosan were kept under curing for 7 days, 7 days and 1 day, respectively [14, 15, 30].

Moist tamping method [23] was chosen as the sample preparation technique for both untreated and non-treated soils. The sequential sample preparation procedure and filter paper test methodology adopted for the present study has been shown in Fig. 2. After the completion of curing period two identical soil cakes were prepared. Whatman 42 filter paper of diameter 55 mm was sandwiched between two Whatman filter of diameter 65 mm and these three filter papers were placed between the prepared two soil cakes to measure the matric suction (in-contact filter paper method). The cakes were covered with insulating tape to prevent for any loss of moisture and then were kept in glass jar, which was also insulated using electrical tape. Whatman filter paper of diameter 55 mm was suspended above the soil cakes to measure total suction (non-contact filter paper method). The jar was sealed and kept for equilibrium process for 7 days in biological incubator. After the 7-day equilibration process, the weight of wet filter paper was measured and then dry weight of filter paper was measured after oven drying for 24 h. Table 2 presents the obtained values of suction (total, matric and osmotic) for different percentage of three biopolymers.

Fig. 2 Sequential sample preparation and filter paper test methodology

Table 2 Suction values at different percentages of biopolymer treated soil

	Biopolymer (%)	Ψ (kPa)	Ψ_m (kPa)	Ψ_ϕ (kPa)
Without treatment	0	2866.81	2483.59	383.22
Chitosan	0.12	2528.541	2353.47	175.071
	0.16	2483.59	2439.439	44.151
	0.20	2815.853	2765.795	50.058
Xanthan gum	0.5	2353.512	2190.53	162.94
	1	2971.52	2620.89	350.63
Guar gum	0.5	3888.92	3369.06	519.86
	1	2918.7	2866.81	51.89
	1.5	2716.62	2620.89	95.73

3.1.1 Need of Curing

Curing would enable biopolymer to build bond with the soil particles and assimilate alterations in the structural alignment of soil fabric. Curing would induce ductility among soil particles leading to higher strength and resistance to collapse. From the literature the curing period was found to be different for different biopolymers. Guar gum and xanthan gum required minimum of 7 days of curing, while chitosan needed only 1 day [14, 15, 30]. In the present study, curing was done in an isolated place where the change in temperature was observed to be negligible.

3.2 *Swell Pressure*

Swell pressure test was conducted on untreated and biopolymer treated soils at the obtained optimum percentage of each biopolymer. The test was performed using conventional oedometer setup in accordance to IS 2720 Part-XLI [13]. In this test, the as-compacted specimen prepared in the oedometer ring was placed in the consolidation setup and water was allowed to penetrate the soil specimen. Under seating load of 5 kPa, specimen was allowed to swell until the volume change became constant. Subsequently, the specimen was loaded sequentially from 10 kPa, 20 kPa, 50 kPa, 100 kPa, 200 kPa, 400 kPa and 800 kPa. The pressure increment, at which soil specimen reached its initial void ratio (e_0), was determined as swelling pressure of soil from e vs log σ_v'. The values acquired of swell pressure for untreated and treated Bhavnagar expansive soils are given in Table 3.

3.3 *Electrical Conductivity*

Qualification of the salinity in the soil was obtained by evaluating electrical conductivity of the Bhavnagar saline expansive soil. Electrical conductivity of untreated and biopolymer treated soil was evaluated at optimum percentages by conductivity meter as per IS 14767 [10]. The values obtained for electrical conductivity of soil are shown in Table 3. In this test, 20 g of oven dried soil was added to 40 ml distilled water in a closed beaker and was placed horizontally in shaking machine for 30 min

Table 3 Values for swell pressure, electrical conductivity and unconfined compressive strength at optimum percentages of biopolymer treated soil

	Optimum polymer percentage obtained (%)	Swell pressure (kPa)	Electrical conductivity (μs/cm)	Unconfined compressive strength (kPa)
Without treatment	0	53	10900	185
Chitosan	0.12	37	819	148
Xanthan gum	0.5	40	524	218
Guar gum	1.5	58	1605	279

to obtain a homogeneous solution. Electrode of conductivity meter was placed in the soil water solution to measure the electrical conductivity until the reading reached a constant value. The values obtained for electrical conductivity of soil are shown in Table 3.

3.4 Unconfined Compressive Strength

Unconfined compressive strength was conducted on untreated and biopolymer treated soil at optimum percentages of each biopolymer as per IS 2720 Part-10 [11]. The unconfined compression strength values obtained for the expansive soil are presented in Table 3.

4 Results and Discussions

4.1 Suction Characteristics of Bhavnagar Expansive Soil

Suction or water potential designates the propensity of water to transport from one place to other owing to surface tension, capillary action and osmosis phenomena. Suction is the measurable indication of the potential energy among pore water within the soil skeleton and the free water (hydraulic datum). It changes with the alteration in solute concentration termed as osmotic potential (osmotic suction) and due to alteration in cohesive/adhesive forces imbibing the water molecules to the solid surfaces designated as matric potential (matric suction).

The total suction, matric suction and osmotic suction of the untreated expansive soil at different gravimetric water contents (10, 15, 20 and 25%) are presented in Fig. 3. It was observed that the value of total suction, matric suction and osmotic suction reduced with increase in water content and degree of saturation. As depicted in Fig. 3, the total suction, matric suction and osmotic suction of the Bhavnagar saline expansive soil were found to be decreasing with the increase in water content. Increment in the as-compacted water content/degree of saturation would lead to enhanced bulk water proportion within the soil particles of unsaturated expansive soil [23]. Increase in the bulk water proportion would lead to progressive dilution of the solute concentration and gradual dissolution of the air–water interface causing reduction in osmotic potential and matric potential of the soil, respectively. Hence, the overall total suction of the soil would reduce with increase in the as-compacted water content.

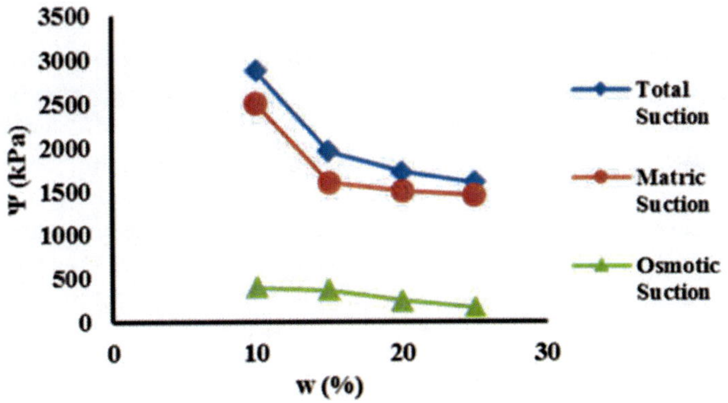

Fig. 3 Variation of total suction, matric suction and osmotic suction at different water content of untreated Bhavnagar expansive soil

4.2 Influence of Biopolymer Treatment on Suction, Swelling and Strength Characteristics of Saline Bhavnagar Expansive Soil

4.2.1 Treatment by Addition of Chitosan

Biopolymer treatment by chitosan was carried out by adding varying percentage of chitosan by weight to the Bhavnagar expansive soil. The percentage variation of chitosan was ranged between 0.12, 0.14 and 0.16%. The total suction, matric suction and osmotic suction characteristics at different percentages of chitosan have been presented in Table 2. It was observed that the optimum percentage of chitosan was obtained to be 0.12%. As compared to untreated soil, reduction in the total suction, matric suction and osmotic suction at optimum percentage of chitosan biopolymer treatment was obtained to be 338 kPa, 130 kPa and 208 kPa, respectively (Table 4). Chitosan treatment would lead to the formation of a sol–gel around soil particles [8]. It results in the development of an inorganic colloidal suspension (sol) and gelation of the sol in a continuous liquid phase (gel) to form a three-dimensional network structure. This would cause generation of thick gel around clay particles. The formation

Table 4 Comparison of suction values at optimum percentages for different biopolymer treatment of Bhavnagar saline expansive soil

	Optimum polymer percentage obtained (%)	Ψ (kPa)	Ψ_m (kPa)	Ψ_ϕ (kPa)
Without treatment	0	2866.81	2483.59	383.22
Chitosan	0.12	2528.541	2353.47	175.07
Xanthan gum	0.5	2353.47	2190.53	162.94
Guar gum	1.5	2716.2	2620.89	95.72

of thick sol–gel could lead to the development of enhanced true cohesion between the soil particles. As a result, soil particles would be held together by enhanced attraction which would lead to restricted development of air–water interface at lower water content in unsaturated soil. Considering the Toroidal approximation [16] the matric suction of soil would depend upon the surface tension, filling angle (θ) and spherical radii (r_1 and r_2) of the air–water interface. The creation of sol–gel could led to increase in the filling angle (θ), which would in turn lead to increase in the spherical radii (r_1 and r_2). This would lead to the restricted development of the air–water interface and hence would cause reduction in the matric suction of the soil. Osmotic suction was observed to reduce drastically by 208.15 kPa as compared to untreated soil (Table 2). It has been reported that chitosan enhances the plant growth by reducing osmotic potential in saline conditions [26]. It was found to be analogous to the electrical conductivity results of the chitosan-treated Bhavnagar expansive soil (Table 3). Creation of sol–gel could restrict the osmosis process within the soil skeleton due to solidification of the salt solutes, which could have led to decrease in the electrical conductivity, leading to significant decrease in the osmotic suction of the soil.

Swelling pressure of Bhavnagar expansive soil was evaluated for untreated and chitosan treated expansive soil at obtained optimum chitosan percentage (Table 3). It was observed that the addition of chitosan led to decrease in the swelling pressure of treated soil. Reduction in the total suction due to chitosan treatment led to decreased capillary force between the soil particles [22]; hence, the unbalanced negative charge would reduce, thus causing reduction in the affinity for water to create equilibrium. This might cause reduction in the swelling pressure of soil. However, chitosan treatment led to decrease in the unconfined compressive strength of the soil.

4.2.2 Treatment by Addition of Xanthan Gum

Xanthan gum treatment was carried out by adding varying percentage of xanthan gum by weight to the Bhavnagar black cotton soil. The percentage deviation of xanthan gum was varied from 0.5% to 1%. The total suction, matric suction and osmotic suction characteristics at different percentages of xanthan gum have been presented in Table 2. It was observed that the optimum percentage of xanthan gum was obtained to be 0.5%. As compared to untreated soil, reduction in the total suction, matric suction and osmotic suction at optimum percentage of xanthan gum biopolymer treatment was obtained to be 513 kPa, 293 kPa and 220 kPa, respectively (Table 4).

Inclusion of xanthan gum in the black cotton soil could lead to sorptive and microstructural changes within the soil skeleton. Xanthan gum monomers would form hydrogen or electrostatic bonding with clay particles. Xanthan monomers tend to link with clayey particles owing to hydrogen bonding and cation bridging between the carboxyl group (–COOH) and the hydroxyl (–OH) groups of xanthan gum and negatively charged soil surfaces [15]. As a result, a cementitious product forms between the xanthan gum and clay particles which causes flocculation of particles which induces reduction in the surface area [15]. According to Cho and Santamarina

[7], matric suction of soil was found to be function of surface area of soil. Increase in the surface area due to filled up of the pore spaces of the black cotton soil owing to hydrogen bonding and cation bridging caused significant reduction in the matric suction of the unsaturated expansive soil (Table 4). Addition of xanthan gum would increase the viscosity of the pore fluid; as a result, the solute concentration might imbibe in the primarily viscous and conclusively hard cementitious bond. This would harp the osmosis action within the pore spaces, resulting in extensive decline of the osmotic suction values of Bhavnagar saline black cotton soil. Similar response of the xanthan gum treatment was observed from the electrical conductivity test on black cotton soil treated at optimum percentage of xanthan gum as shown in Table 3.

From the present study it was observed that swelling pressure of xanthan gum treated soil was observed to decrease with xanthan gum treatment as compared to untreated black cotton soil as presented in Table 3. Development of hydrogen bond between xanthan gum monomers and soil particles would decrease the micro pore spacing and satisfy the unbalanced negative charge on the individual clay particle. Hence, affinity for water to create neutral equilibrium and interlayer spacing between the structural units of swelling mineral would also reduce, which would ultimately result in the lower swelling pressure of black cotton soil.

Significant improvement in the unconfined compressive strength of the xanthan gum treated black cotton soil was observed from the present study (Table 3). Growth of xanthan gum bridges within soil fabric [6] would lead to reduction in micro pore spaces of the soil. This would induce enhanced contact area and interlocking among the soil particles, which would cause increase in the strength of the soil mass macroscopically.

4.2.3 Treatment by Addition of Guar Gum

Influence of guar gum treatment was evaluated by adding varying percentage of guar gum by weight to the Bhavnagar black cotton soil. The percentage variation of guar gum was varied from 0.5, 1 and 1.5%. The total suction, matric suction and osmotic suction characteristics at different percentage of guar gum have been presented in Table 2. From the present study it was observed that sample preparation using guar gum was found to be very difficult due to the tendency of guar gum to form lumps when mixed with soil. Decrease in the suction value was observed with increase in the percentage of guar gum. It was observed that the optimum percentage of guar gum was obtained to be 1.5% (Table 4). With addition of 1.5% of guar gum, total suction and osmotic suction was reduced by 150 kPa and 287 kPa, respectively, but reduction in matric suction values was not observed. Guar gum tends to possess high specific surface area and forms hydrogel with the soil particles. It has tendency to hydrate and introduce an additional adhesive force due to creation of strong hydrogen bonds which tend to hold soil particles together reducing the micro pore spaces between the particles [29]. Formed hydrogel has acute affinity to absorb water. Matric suction would originate in an unsaturated soil due to adhesive-cohesive force between the soil particles. Guar gum treatment intensified the adhesive force within the soil particles and hydrogel. Specific surface area of the soil would also increase due to addition of guar gum. The matric suction of unsaturated soil depends upon the specific surface area of the soil particles [7]. Hence, due to intensified adhesive force and increased specific surface area of the soil, the matric suction component of the Bhavnagar black cotton soil remained unaffected by guar gum treatment. However, generation of hydrogel could restrict the osmosis process within the soil skeleton due to coagulation of the salt solutes owing to the development of hydro-gel, which could have led to decrease the osmotic suction leading to considerable decrease in the electrical conductivity of the Bhavnagar saline black cotton soil.

The swelling pressure of guar gum treated soil was observed to increase compared to untreated black cotton soil, as presented in Table 3. Development of hydrogel introduced enhanced affinity for water as compared to untreated state of unsaturated expansive soil. Enhanced adhesive force between particles due to development of enhanced forces of attraction would lead to reduced interlayer spacing between structural units. This would lead to generation of severe affinity to imbibe water among clay particles to reach equilibrium and neutralization and hence leading to higher swelling pressures in treated Bhavnagar expansive soil.

Significant improvement in the unconfined compressive strength of the guar gum treated black cotton soil was observed in the present study (Table 3). Growth of viscous hydrogel develops additional cohesive force due to formation of hydrogen bond between soil particles [29]. Improved cohesive attraction contributes to increase in the unconfined compressive strength of the guar gum treated Bhavnagar expansive soil.

4.3 Comparison of Biopolymer Treatment on Bhavnagar Saline Expansive Soil

In the current study, influence of biopolymer treatment on suction, unconfined compressive strength, swelling and electrical conductivity characteristics of Bhavnagar saline expansive soil was evaluated. Chitosan, xanthan gum and guar gum were added to the soil by weight and cured as a part of biopolymer treatment. Comparison of total suction, matric suction and osmotic suction of biopolymer treated soil by chitosan, xanthan gum and guar gum has been presented in Fig. 4a, b, c, respectively. It could be identified that soil treated by xanthan gum showed best treatment response by exhibiting the lowest total and matric suction values as compared to chitosan and guar gum. However, guar gum exhibited lower osmotic suction values as compared to chitosan and xanthan gum.

Comparison of the swelling pressure, unconfined compressive strength and electrical conductivity of the biopolymer treated expansive soil has been presented in Fig. 5a, b, c, respectively. It could be identified from Fig. 5(a) that specimen treated with chitosan exhibited lowest swelling pressure; however, not much difference (7 kPa) in the swell pressure values was obtained for specimen treated with xanthan

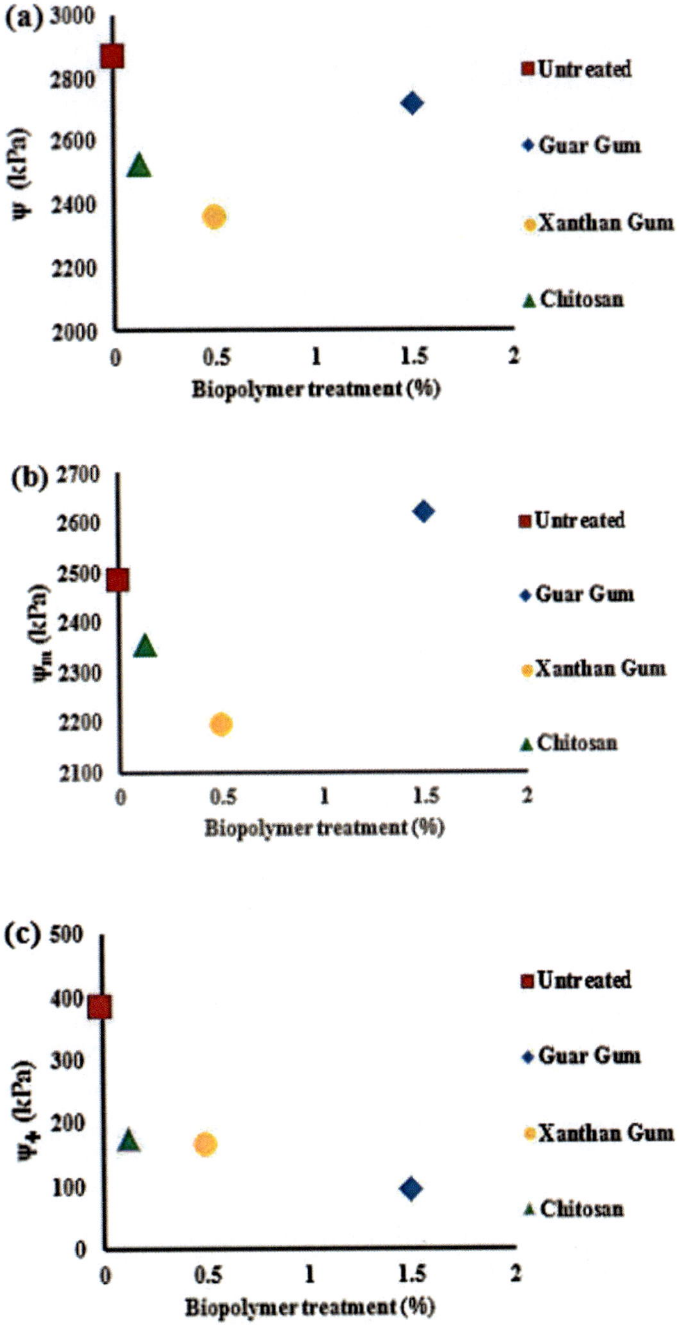

Fig. 4 Variation of suction with optimum percentage of biopolymer treatment. **a** Total suction, **b** Matric suction and **c** Osmotic suction

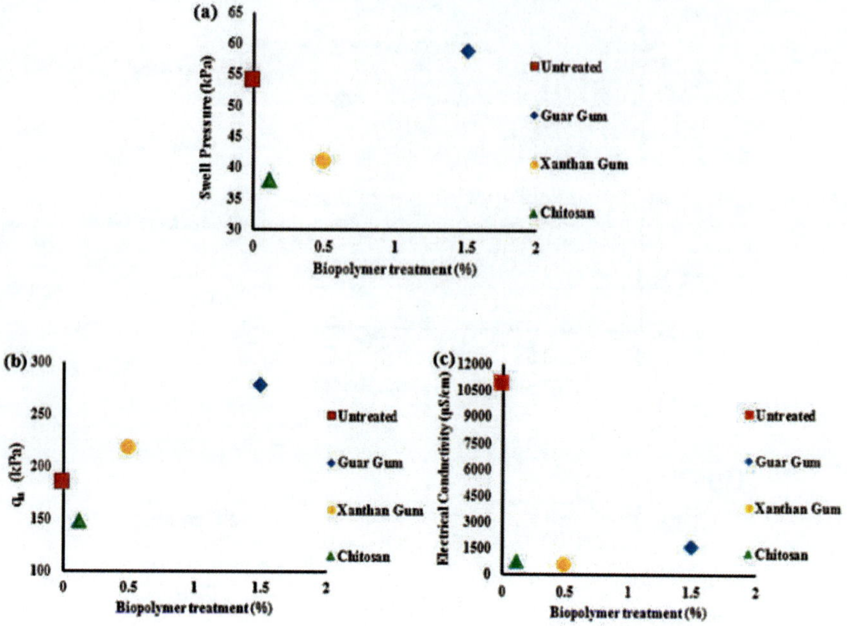

Fig. 5 Variation of swelling, strength and salinity characteristics of untreated and different biopolymer treated Bhavnagar expansive soil. **a** Swell pressure, **b** Unconfined compression strength (q_u), **c** Electrical conductivity

gum. Unconfined compression test values were found to be highest for the specimen treated with guar gum; however, substantial improvement in the unconfined compression strength was observed for specimen treated with xanthan gum (Fig. 5b). The difference between unconfined compression strength values of specimen treated with guar gum and xanthan gum was found to be 61 kPa.

Electrical conductivity test results (Fig. 5c) revealed that the specimen treated by xanthan gum exhibited least electrical conductivity as compared to specimen treated by chitosan and guar gum. Water present within the hydrogel formed by guar gum treatment would allow free diffusion of some solute molecules; as a result, electrical conductivity of guar gum was found to be considerably higher as compared to chitosan and xanthan gum treated specimens.

5 Methodology of Application of Biopolymer as Soil Stabilizer in Field

The biopolymer inclusions in the field can be made exactly in a similar manner as conventional lime treatment [18]. For the biopolymer treatment same methodology could be followed with some modifications distinctive to the biopolymer treatment employed. The available expansive soil which has to be stabilized would be pulverized by suitable equipment. In-place mixing should be performed with required amount of biopolymer up to appropriate depth. Effective pulverization and mixing should be ensured so that thorough amalgamation of biopolymer and expansive soil would be achieved. Amount of water equivalent to required water content should be sprayed biopolymer mixed soil mass and then should be left for characteristic curing process depending upon the biopolymer employed for treatment. Proper curing should be assured to enable development of proper bond between expansive soil matrix and biopolymers. Soil should be sprinkled with water from time to time or covered with subsequent impermeable material to ensure negligible evaporation of water during curing process. Light compaction could be applied if required, during curing period to keep the soil matrix and biopolymers bonded together. After the completion of curing duration final thorough mixing should be executed. Field tests should be conducted to check water content and maximum dry density of compacted soil.

6 Conclusions

Biopolymer treatment significantly reduces the total, matric and osmotic suction of Bhavnagar saline expansive soil. From the present study the following conclusions are made:

1. The optimum percentages for reducing suction (total, matric and osmotic) of xanthan gum and chitosan are 0.5% and 0.12%, respectively.
2. Biopolymer treatment by xanthan gum and chitosan led to substantial reduction in the suction (total, matric and osmotic) of Bhavnagar saline expansive soil.
3. Guar gum treatment was found to be ineffective in reducing matric suction and swell pressure of expansive soil. However, osmotic suction and electrical conductivity were found to decrease by guar gum treatment.
4. Xanthan gum and chitosan biopolymer treatment led to reduction in the electrical conductivity and swelling characteristics of the expansive soil.
5. Significant improvement in the unconfined compressive strength of biopolymer treated expansive soil was observed for xanthan gum and guar gum treatment. However, unconfined compressive strength of specimen treated with chitosan reduced due to induced brittleness owing to sol–gel development within soil matrix.
6. Xanthan gum biopolymer treatment was found to be most effective in treating Bhavnagar expansive soil as it improved salinity, suction, and swelling and strength characteristics of the soil. Easy availability of xanthan gum and efficacious treatment makes it the best biopolymer stabilization technique for Bhavnagar saline expansive soil.

References

1. ASTM D5298-10: Standard test method for measurement of soil potential (suction) using filter paper, West Conshohocken, Pennsylvania, USA (2013)
2. Bulut, R., Leong, E.C.: Indirect measurement of suction. Geotech. Geol. Eng. **26**(6), 633–644 (2008)
3. Bulut, R., Lytton, R.L., Wray, W.K.: Soil suction measurements by filter paper. In expansive clay soils and vegetative influence on shallow foundations. ASCE **115**(1), 243–261 (2001)
4. Chandler, R.J., Gutierrez, C.I.: The filter-paper method of suction measurement. Geotechnique **36**(2), 265–268 (1986)
5. Chang, I., Im, J., Cho, G.C.: Introduction of microbial biopolymers in soil treatment for future environmentally-friendly and sustainable geotechnical engineering. Sustainability **8**(3), 251–274 (2016)
6. Chang, I., Im, J., Prasidhi, A.K., Cho, G.C.: Effects of Xanthan gum biopolymer on soil strengthening. Constr. Build. Mater. **74**, 65–72 (2015)
7. Cho, G.C., Santamarina, J.C.: Unsaturated particulate materials-particle-level studies. J. Geotech. Geoenviron. Eng. **127**(1), 84–96 (2001)
8. Hataf, N., Ghadir, P., Ranjbar, N.: Investigation of soil stabilization using chitosan biopolymer. J. Clean. Prod. **170**(1), 1493–1500 (2018)
9. Houston, S.L., Dye, H.B., Zapata, C.E., Walsh, K.D., Houston, W.N.: Study of expansive soils and residential foundations on expansive soils in Arizona. J. Perform. Construct Facilities **25**(1), 31–44 (2011)
10. IS 14767: Methods of test for soils—determination of the specific electrical conductivity of soils. Bureau of Indian Standards, New Delhi (2000)
11. IS 2720-Part 10: Methods of test for soils—determination of unconfined compressive strength. Bureau of Indian Standards, New Delhi (2011)
12. IS 2720-Part 4: Methods of test for soils: Part 4 grain size analysis. Bureau of Indian Standards, New Delhi (1985)
13. IS 2720-Part XLI: Measurement of swelling pressure of soils. Bureau of Indian Standards, New Delhi (1977)
14. Joga, J.R., Varaprasad, B.J.S.: Sustainable improvement of expansive clays using xanthan gum as a biopolymer. Civil Eng. J. **5**(9), 1893–1903 (2019)
15. Latifi, N., Horpibulsuk, S., Meehan, C. L., Abd Majid, M. Z., Tahir, M. M., Mohamad, E.T.: Improvement of problematic soils with biopolymer—an environmentally friendly soil stabilizer. J. Mater. Civil Eng. **29**(2), 04016204 (1)–04016204 (11) (2017)
16. Likos, W.J., Lu, N.: Hysteresis of capillary stress in unsaturated granular soil. J. Eng. Mech. **130**(6), 646–655 (2004)
17. Mabirizi, D., Bulut, R.: A comparison of total suction measurements with thermocouple psychrometer, filter paper technique, and chilled-mirror device, GeoHunan International Conference: Challenges and Recent Advances in Pavement Technologies and Transportation Geotechnics. Geotechnical Special Publication, ASCE **189**(1), 1–6 (2009)
18. McDowell, C.: Stabilization of soils with lime, lime-fly ash and other lime reactive materials. Highw. Res. Board Bull. **231**(1), 60–66 (1959)
19. Mehta, B., Sachan, A.: Effect of mineralogical properties of expansive soil on its mechanical behavior. Geotech. Geol. Eng. **35**(6), 2923–2934 (2017)
20. Naveena, S., Reddy, G.S.: Strength characteristics of expansive soils using eco-friendly Xanthan Gum. Int. J. Sci. Res. **6**(6), 2439–2442 (2015)
21. Nelson, J., Miller, D.J.: Expansive soils: problems and practice in foundation and pavement engineering, 1st edn. Wiley, Canada (1997)
22. Pandya, S., Sachan, A.: Matric suction, swelling and collapsible characteristics of unsaturated expansive soils. J. Geotech. Transp. Eng. **4**(1), 1–9 (2018)
23. Pandya, S., Sachan, A.: Effect of matric suction and initial static loading on dynamic behaviour of unsaturated cohesive soil. Int. J. Geotech. Eng. **12**(5), 438–448 (2017)

24. Rao, S.M., Thyagaraj, T., Rao, P.R.: Crystalline and osmotic swelling of an expansive clay inundated with sodium chloride solutions. Geotech. Geol. Eng. **31**(4), 1399–1404 (2013)
25. Sawangsuriya, A., Edil, T.B., Bosscher, P.J.: Modulus-suction-moisture relationship for compacted soils. Can. Geotech. J. **45**(7), 973–983 (2008)
26. Sen, S.K., Mandal, P.: Solid matrix priming with chitosan enhances seed germination and seedling invigoration in mung bean under salinity stress. J. Central Eur. Agricul. **17**(3), 749–762 (2016)
27. Shen, M.F., Tan, X.H., Xin, Z.Y., Xie, Y., Xu, Q.: Laboratory research of soil water characteristic curve by osmotic method and filter paper method. Electron. J. Geotech. Eng. **18**(1), 5421–5434 (2013)
28. Stenke, F., Toll, D.G., Gallipoli, D.: Comparison of water retention curves for clayey soils using different measurement techniques, Unsaturated soils, Geotechnical Special Publication, ASCE **147**(1), 1451–1461 (2006)
29. Sujatha, E.R., Saisree, S.: Geotechnical behaviour of guar gum-treated soil. Soils Found. **59**(6), 2155–2166 (2019)
30. Swain, K.: Stabilization of soil using geopolymer and biopolymer. Doctoral dissertation. National Institute of Technology, Rourkela (2015)
31. Verma, S.K., Maru, S.: Behaviourial study of expansive soils and its effect on structures–A review. Int. J. Innov. Eng. Technol. **2**(2), 228–238 (2013)

A Novel Method of On-site Biochar Production from Crop Residues and Further Application to Soils for Improvement in Soil Characteristics

Mahendra Pratap Choudhary, H. D. Charan, and Biswajit Acharya

1 Introduction

Being an agricultural country, a huge amount of crop residue is generated per year in India. According to the Ministry of New and Renewable Energy (MNRE), Govt. of India, about 500 million tons (MT) of crop residue is generated per year in our country out of which about 92 MT is burned openly [1]. One of the main reasons behind it is the enhanced use agricultural machinery by the Indian framers in harvesting their crops, and it has resulted in a serious problem of managing the crop residues in the form of stubble and stalks. The crop cutters and harvesters cut only the useful upper parts of the crops, like wheat, mustard and paddy, and the remaining parts are left out in the fields and to save money and time for preparing the next crop, the farmers directly burn these residues in open fields releasing harmful greenhouse gases into the atmosphere.

The annual outburst of smog during the months of October–November in the capital city of India, New Delhi, during the last four–five consecutive years is known to occur due to open burning of crop residues by the farmers of nearby States of Punjab, Haryana, Madhya Pradesh, Uttar Pradesh and Rajasthan. In 2015, although, the National Green Tribunal (NGT) has imposed a complete ban on the open burning of crop residues, leaves and other materials which release toxic pollutants into the air, but farmers are still not complying with it in its strict sense.

We can understand the severity of the problem with an example when the city of Delhi witnessed the worst category of air quality index in winter 2017. During this period, the concentration of fine particulate matter ($PM_{2.5}$) was observed at the

M. P. Choudhary (✉) · B. Acharya
Rajasthan Technical University, Kota, Rajasthan, India

H. D. Charan
Bikaner Technical University, Bikaner, Rajasthan, India

C. N. V. Satyanarayana Reddy et al. (eds.), *Ground Improvement and Reinforced Soil Structures*, Lecture Notes in Civil Engineering 152,
https://doi.org/10.1007/978-981-16-1831-4_44

highest level ever at 640 μg/m³ [2] against the annual permissible limits of 40 μg/m³ as per the national ambient air quality standards of India [3].

At present, agriculture is known to be the third largest source of greenhouse gas emission, followed by fossil fuel burning and transportation [4]. In the near future, the entire developing world is likely to witness a rise in the growth of agro-processing industries which will require managing the waste generated from such industries in a well-organized and sustainable manner. The agricultural and associated agro-industries waste has the potential to provide feedstock for the production of biochar. By converting residual biomass into biochar, long-standing carbon confiscation and additional advantageous effects on soils can be ascertained [5].

The open burning of crop residues in agricultural fields is not an environment-friendly approach because it not only causes loss of biomass but also introduces harmful gases into the environment. Therefore, converting the agro-wastes into biochar rather than direct burning in open fields could be a more reliable and feasible way out to this problem of air pollution. Production and application of biochar to agricultural field soils can deal with several global and local environmental issues. Biochar application to soil also helps in carbon sequestration which ultimately converts bio-energy into a carbon-negative industry [6]. However, due to economic concerns, the use of biochar in agricultural fields has not been so encouraging [7, 8].

There are studies available in the literature on the factors involved in the production and use of biochar for soil amendment; but in India, not much work has been carried out yet, as it is relatively a new concept in terms of using crop residues for the production of biochar and its application. The on-site photographs taken during open burning of crop residues after harvesting are shown in Fig. 1.

Looking to severity and frequent occurrence of annual smog outbursts due to stubble burning throughout the country during winter months, it has been decided to find out an economic and feasible solution of this geo-environmental issue. A novel method of on-site conversion of crop residues into a product known as biochar has been discovered during the process which will not only solve the problem of timely clearance of agricultural fields after harvesting, but will also help in mitigating the environmental pollution, stabilizing the soil, increasing the fertility of the soil and hence an all-in-one solution to so many concerns.

Fig. 1 Mustard crop residues put on fire by farmers to prepare their fields for upcoming crops near Kota, Rajasthan, India

2 Biochar

Biochar is a carbon-affluent product formed by the pyrolysis process, in which the biomass or other organic substance is heated either in the absence of or with a limited amount of oxygen at temperatures above 250 °C [9]. The charcoal is also made with the same process but biochar is altogether taken as a different product than charcoal and other carbon products in that it is used as a soil amendment [10]. In recent times, biochar has drawn more attention due to its scope and role in alleviation of climate change and potential for soil amendment and sustainable cultivation [11].

Biochar can be made from biomasses having different physical and chemical properties. An extensive range of feedstock biomasses can be used for producing biochar, like agricultural and organic waste, forest residues, bioenergy crops, kitchen waste and even sewage sludge also [12]. The properties of each biomass feedstock, like moisture and ash content, calorific value, mass, density, particle size, fixed carbon and volatile ingredients, are different and hence the produced biochars also have unique physical and chemical properties [13].

Biochar can also be produced from waste materials including those that may otherwise produce even more harmful greenhouse gases (e.g., manure or green wastes) [10]. So, it is beneficial to make use of waste materials in a useful manner and safe disposal at the same time. Production of biochar and its presence in soils has been recommended as a way of extenuating climate change by confiscating carbon and at the same time providing energy and increased crop yields [14].

Other than crop residues, locally available weed biomass is an important source for preparing the biochar as it is not economically important as well as causes crop loss due to its presence. If biochar is prepared from the weeds, available locally, then it can reduce the weed mass in the fields on one hand, and on the other hand, it can enhance the growth of plants by improving the physical, chemical and biological characteristics of soil and hence contributing to increased crop yield [15].

The biomass can be pyrolyzed in a reactor through gasification or carbonization at different temperature and time depending on the final anticipated use of the end product [16]. Biochar obtained at high temperatures (500 °C) has been found apposite for direct use as a fuel because of high contents of carbon, retention time, pH and electrical conductivity whereas biochars obtained at low temperatures (300–350 °C) are found appropriate for land application [17].

3 Materials and Methods

3.1 A Novel Method of on-Site Biochar Production

For Indian conditions, biochar can be produced by individual farmgers on their fields in conventional kilns made by locally available material or at community kilns by using the agricultural wastes and other by-products so that the biochar produced can

Fig. 2 Field arrangement showing empty diesel drums for in situ production of biochar

be utilized again for applying in the fields for the upcoming crops. A novel method of producing biochar has been developed wherein the agricultural waste is heated in empty diesel drums of capacity 220 L, which are generally available with farmers. The arrangement of drum, feeding agricultural waste into it, igniting the residues, raw biochar produced and biochar in final finished form are shown in Figs. 2, 3 and 4.

Fig. 3 Drums are filled with crop residues, compacted and covered with top lid after lit on fire at bottom and left for some time so that pyrolysis takes place

Fig. 4 The crop residue is converted into biochar in 15–20 min, which is further pulverized (into powder form) before addition to soil for lab experiment/field application

The biochar so produced was characterized by finding out its physical and chemical characteristics in the laboratory, and the results are presented in Table 1.

3.2 Soil Sample

The soil samples for the study were collected from a field at Baran road near Kota, Rajasthan from a depth of up to 1.0 m below the ground level so as to exclude any type of organic matter. The samples were collected and sealed in plastic bags and brought to the laboratory. The samples for various tests were prepared in the soil engineering laboratory of the department of Civil Engineering, Rajasthan Technical University, Kota according to the procedures laid down by the Bureau of Indian Standards in IS: 2720 (Part 1)–1983 [18]. The samples were first of all oven dried, pulverized and then sieved through a 4.75 mm size sieve to get an idea of the primary classification. In sieve size analysis, the constituents of soil were found as 48% of sand, 45% of silt and 7% of clay. The Atterberg limits of the soil sample were found out. It was observed that liquid limit, plastic limit and shrinkage limit of the soil were 58.26%, 29.32% and 14.13%, respectively. Further, the soil sample was classified as CH (inorganic clay of high plasticity) as per the Unified Soil Classification System (USCS). This type of soil is generally expansive soil by nature which needs to be stabilized before construction of any pavement or load bearing structure on it.

3.3 Methods

The biochar produced at field level was used for stabilization of expansive soil. Many researchers had used lime, rice husk, fly ash and other conventional materials for the

Table 1 Physico-chemical characteristics of biochar

S. No.	Characteristics	Value
1	Color	Black
2	pH	10.13
3	Natural Moisture Content (%, at 105 °C)	18.64
4	Specific Gravity	1.61
5	Electrical Conductivity (mS/cm, at 25 °C)	10.54
6	Organic Carbon (%, as OC)	1.48
7	Organic Matter (%)	2.55
8	*Grain Size Distribution (%)*	
	(i) 2–4.75 mm (Coarse)	22.8
	(ii) 0.425–2 mm (Medium)	43.4
	(iii) 0.075–0.425 mm (Fine)	33.8

stabilization of soil but not much reference is available in the literature about the application of biochar for soil stabilization. Therefore, this study was carried out to find out the applicability of biochar in soil stabilization and amendment. The test methods and standards that had been followed during experimental work were the standards as prescribed by the Bureau of Indian Standards in different parts of IS: 2720 like pH, specific gravity, moisture content, grain size distribution, consistency limits, Proctor test, unconfined compressive strength, CBR, shear strength characteristics, swelling pressure and free swell index [19–31]. Table 2 presents the summary of the important physical and engineering characteristics of the soil. Figure 5 shows the grain size distribution of soil sample and biochar produced.

Table 2 Characteristics of soil sample

S. No.	Name of Characteristics	Value
1	Color	Red brown
2	pH	7.86
3	Specific Gravity	2.62
4	Electrical Conductivity (mS/cm)	4.35
5	Natural Moisture Content (%)	10.23
6	Optimum Moisture Content (%)	25.35
7	Maximum Dry Density (g/cc)	1.765
8	*Grain Size Distribution (%)*	
	(i) 2–4.75 mm (Coarse Sand)	4.4
	(ii) 0.425–2 mm (Medium Sand)	20.2
	(iii) 0.075–0.425 mm (Fine Sand)	23.4
	(iv) 0.002–0.075 mm (Silt)	45.0
	(v) < 0.002 mm (Clay)	7.0
9	*Consistency Limits (%)*	
	(i) Liquid Limit	58.26
	(ii) Plastic Limit	29.32
	(iii) Plasticity Index	28.94
	(iv) Shrinkage Limit	14.13
10	Unconfined Compressive Strength (kg/cm^2)	1.17
11	California Bearing Ratio (%)	1.55
12	Direct Shear Test	
	(i) Cohesion (C, kg/cm^2)	0.065
	(ii) Angle of internal friction (ϕ, degree)	26.24

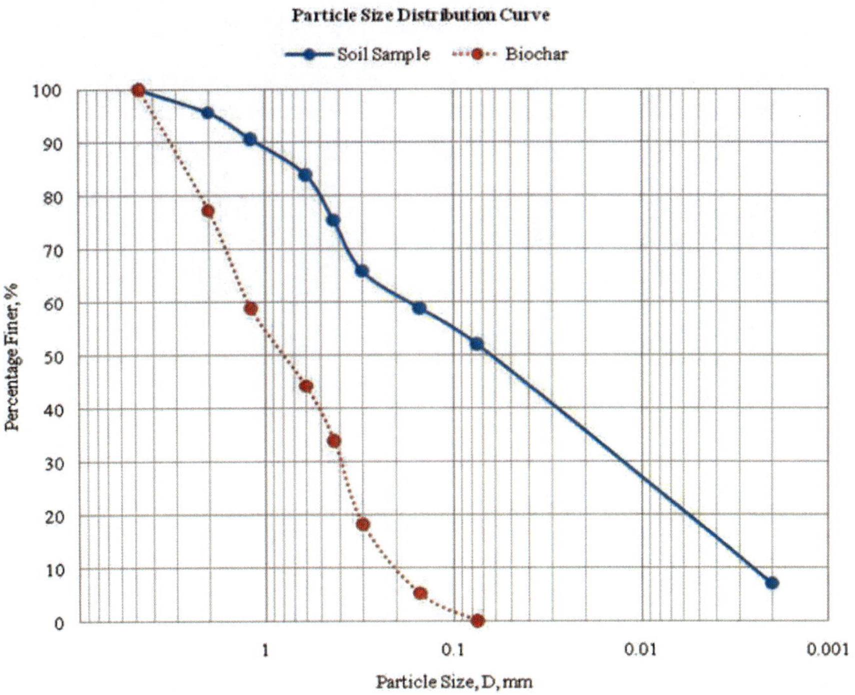

Fig. 5 Grain size distribution curve of soil sample and biochar

3.4 Biochar as Soil Amendment

To study the effect of biochar on physical and engineering properties of soil, various tests were carried out in the geotechnical engineering laboratory wherein the expansive soil was amended with different percent by weight (% w/w) of biochar, like 5%, 10%, 15%, 20% and 25%. For this purpose, the biochar was added in dry condition to the soil, and the soil-biochar mix was thoroughly mixed with each other.

4 Results and Discussion

4.1 Index Properties

The effect of biochar amendment on consistency limits of the soil is shown in Fig. 6. It was found that the liquid limit of the soil amended with biochar decreased from 58.26% to 49.32% on addition of biochar. The percentage reduction in liquid limit is approximately 4.98%, 12.22%, 13.82%, 14.98% and 15.35%, respectively. This reduction is due to replacement of soil fines by biochar because it has high affinity

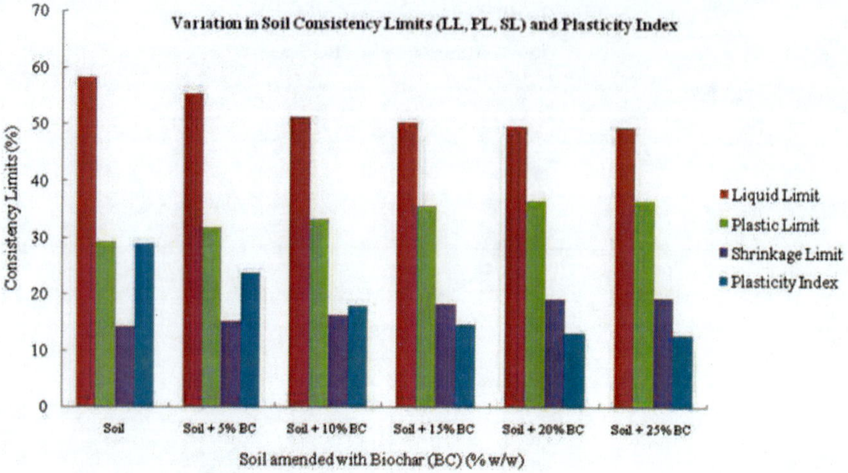

Fig. 6 Effect of biochar amendment on consistency limits of soil

for water. Biochar acts as a stabilizer and hence the liquid limit of the soil-biochar mix is reduced.

The plastic limit of the soil-biochar composite increased from 29.32% to 36.56% when biochar was mixed with soil. The percentage increase in plastic limit is observed as approximately 8.29%, 13.37%, 21.59%, 24.18% and 24.69%. Here, we can infer that the increase in plastic limit is because of the water absorbing capacity of the soil-biochar composite. The increase in shrinkage limit was observed from 14.13% to 19.25% on addition of biochar and hence the percentage increase in shrinkage limit is observed as approximately 6.86%, 14.86%, 29.16%, 35.14% and 36.23%. The plasticity index of the soil amended with biochar decreased from 28.94% to 12.76% on addition of biochar. The percentage reduction in plasticity index is approximately 18.42%, 38.15%, 49.69%, 54.66% and 55.91%, respectively. The decrease in liquid limit and plasticity index by adding biochar to the soil up to a certain percent by weight shows that the index properties of the soil have enhanced.

4.2 Compaction Characteristics

The compaction characteristics of soil and soil-biochar mix were found out using Proctor test according to the IS standards in which maximum dry density (MDD) and optimum moisture content (OMC) were found out at varying percentages of biochar amendment at a rate of 5%, 10%, 15%, 20% and 25%. The value of MDD of soil, i.e., 1.765 g/cc decreased to 1.345 g/cc at a rate of 7.99%, 12.97%, 20.23%, 23.17% and 23.80%, respectively, as we added biochar content to it as shown in Fig. 7. One of the possible reasons for reduction in maximum dry density may be low specific

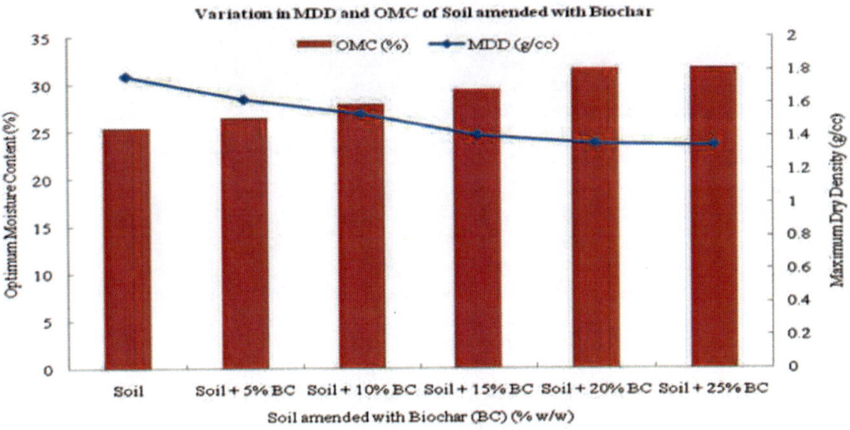

Fig. 7 Variation in MDD and OMC of soil amended with biochar

gravity of biochar in comparison to the soil. Another reason may be the coating of soil particles by biochar resulting in larger particles with increased voids and hence less density.

The value of OMC, i.e., 25.35% went on increasing up to 31.70% at a rate of 4.26%, 10.14%, 16.21%, 24.97% and 25.05%, respectively. The increase in optimum moisture content may be attributed to the absorption of moisture by the biochar. Figure 8 shows the variation in MDD with respect to OMC.

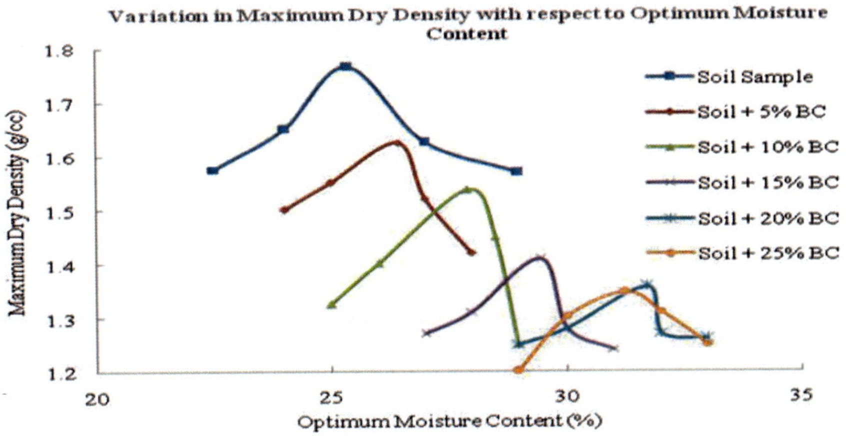

Fig. 8 Variation in MDD with respect to OMC for different percentages of biochar

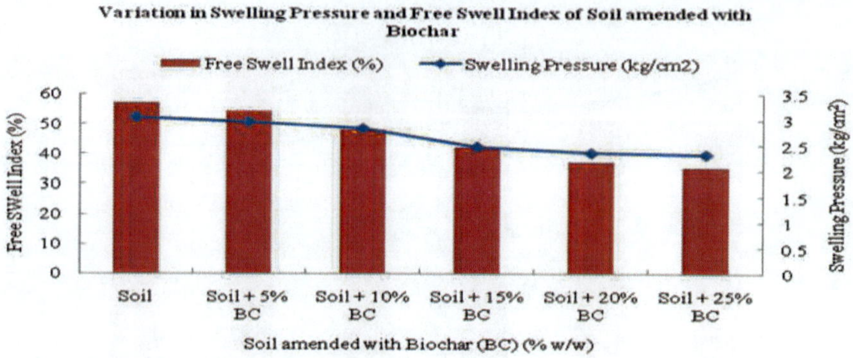

Fig. 9 Variation in swelling pressure and free swell index of soil amended with biochar

4.3 *Swelling Pressure Characteristics*

We know that expansive soils have more swelling capability as there exist clay minerals which are swelling dominant. It has been observed that both swelling pressure and free swell index (FSI) of the soil go on decreasing on addition of biochar to it. The value of swelling pressure starts decreasing from 3.05 kg/cm^2 to 2.32 kg/cm^2 at a rate of 2.95%, 7.21%, 19.02%, 22.95% and 23.93%, respectively, on addition of biochar. The variation in swelling pressure and FSI is shown in Fig. 9.

Similarly, the value of FSI of the soil goes on reducing from 57% to 35% at a rate of 5.26%, 15.79%, 26.32%, 35.09% and 38.60%, respectively. The reduction in swelling pressure implies that as more biochar is mixed, soil becomes more stable due to decrease in the swelling potential of the compacted soil.

4.4 *Unconfined Compressive Strength Characteristics*

Another important parameter of expansive soil is the unconfined compressive stress, which was found as 1.17 kg/cm^2 for the soil sample and it went on increasing on addition of biochar. The increase was observed at a rate of 7.69%, 19.66%, 32.48%, 39.32% and 41.88%, respectively. Figure 10 presents the variation in compressive stress with respect to strain for different percentages of biochar addition to soil.

4.5 *California Bearing Ratio Characteristics*

The California bearing ratio (CBR) test was performed to find out the variation in the penetration/bearing capacity of the soil on addition of biochar. It was observed that the value of CBR for the soil sample (1.55%) increased at a rate of 5.56%,

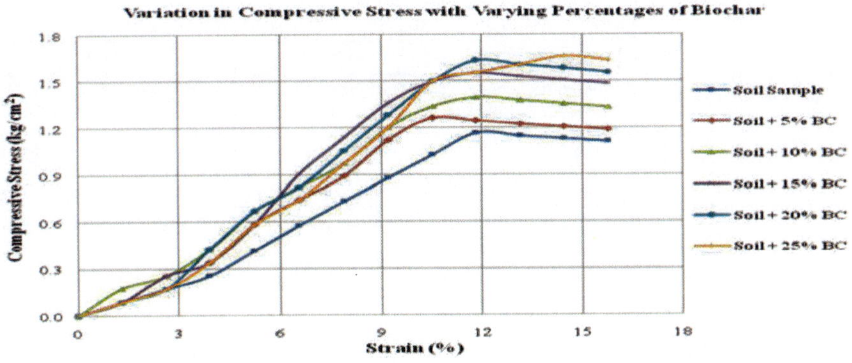

Fig. 10 Variation in compressive stress with respect to strain for soil amended with biochar

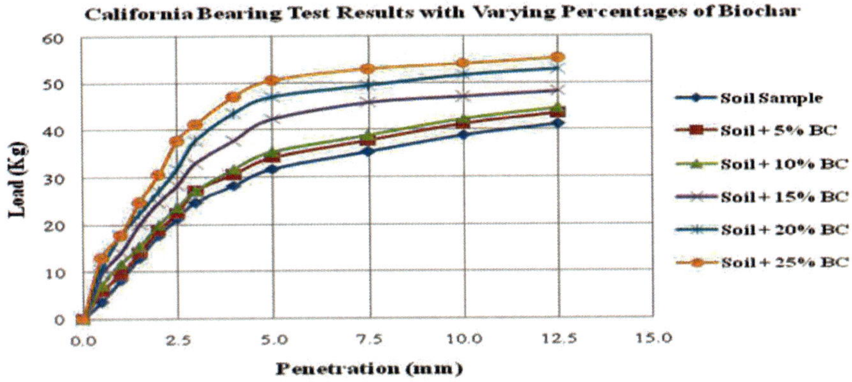

Fig. 11 Variation in load at different penetration for varying percentages of biochar

11.11%, 33.33%, 48.15% and 59.26%, respectively, on addition of biochar at 5%, 10%, 15%, 20% and 25%, respectively. Figure 11 shows the results of CBR test in which variation in load with respect to penetration is represented.

The increase in CBR value indicates that biochar addition gives positive results and proves to be an apt stabilizer for treatment and amendment of expansive soils.

4.6 Direct Shear Test Characteristics

The direct shear test was carried out to find out the effect on shear characteristics of the soil as a result of biochar application. Initially, the value of cohesion (C) and angle of internal frictions (ϕ) of the soil were observed as 0.065 kg/cm^2 and 26.24°, respectively. Upon addition of biochar, both the values of C and ϕ increased considerably as shown in Table 3. The increase in value of C was observed up to a

Table 3 Variation in values of C and ϕ

S. No.	Soil Sample amended with Biochar	ϕ (in degrees)	Value of C (kg/cm^2)
1	Soil	26.24	0.065
2	Soil + 5% BC	27.02	0.076
3	Soil + 10% BC	27.92	0.090
4	Soil + 15% BC	27.47	0.130
5	Soil + 20% BC	28.37	0.156
6	Soil + 25% BC	28.81	0.163

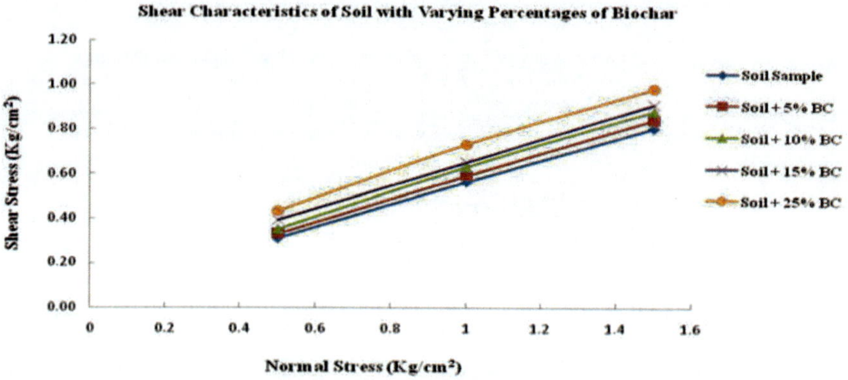

Fig. 12 Variation in shear stress with respect to normal stress for soil amended with biochar

level of 150.77%, which is quite significant and verifies the hypothesis that biochar is a good stabilizer for expansive soils. Figure 12 presents the variation in shear stress with regard to normal stress.

5 Conclusions

Biochar production from crop residues and agricultural wastes and its application in soils puts forward numerous environmental and economical benefits. Applying biochar in soils for amendment of soil properties has shown positive results. The major engineering characteristics of expansive soils including consistency limits, maximum dry density, optimum moisture content, swelling pressure and free swell index, unconfined compressive strength, bearing capacity and shear strength have improved by addition of biochar at the rate of 20–25% (w/w). Biochar application has a very promising prospective for development of sustainable agricultural systems in our country as well as the problem of air pollution in northern India, especially New Delhi through direct burning of agricultural waste in the nearby states of Haryana,

Punjab, Uttar Pradesh and Rajasthan can be minimized. We can call it as a multi-benefit concept in which agricultural waste is utilized, soil amendment is achieved and air pollution is controlled. Finally, we can conclude in nutshell that in recent years, the research activities on the use of biochar in soils have increased manifold and the trend is expected to persist over the years due to the numerous potential benefits associated with the biochar application.

References

1. NPMCR (National Policy for Management of Crop Residues): Available online: http://agricoop.nic.in/sites/default/files/NPMCR_1.pdf. Last accessed 7 Jan 2020 (2014)
2. ETB (Economics Times Bureau): Odd-even plan for five days in Delhi from Monday. https://economictimes.indiatimes.com/news/politics-and-nation/odd-even-plan-for-five-days-in-delhi-from-monday/articleshow/61575418.cms. Last accessed 7 Jan 2020 (2017)
3. NAAQS (National Ambient Air Quality Standards): Central Pollution Control Bureau, India. http://www.moef.nic.in/sites/default/files/notification/Recved%20national.pdf. Last accessed 7 Jan 2020 (2009)
4. Laura, R.: Agriculture and Livestock remain major sources of greenhouse gas emissions. http://www.worldwatch.org/agriculture-and-livestock-remain-major-sources-greenhouse-gas-emissions-0. Last accessed 7 Jan 2020 (2013)
5. Aditya, P., et al.: Biochar production from agro-food industry residues: a sustainable approach for soil and environmental management. Curr. Sci. **107**(10), 1673–1682 (2014)
6. Johannes, L.: A handful of carbon. Nature **447**, 143–144 (2007)
7. Lehmann, J., et al.: Bio-char sequestration in terrestrial ecosystems–a review. Mitigat. Adapt. Strategies Glob. Change **11**, 403–427 (2006)
8. Ranjan, S. T. et al.: Use of biochar for greenhouse gas mitigation. Rashtriya Krishi **11**(2), 65–67 (2016)
9. Lehmann, J.: Bio-energy in the black. Frontiers Ecol. Env. **5**, 381–387 (2007)
10. Lehmann, J., Joseph, S.: Biochar for environmental management: Scientific Technology Implementation. 2nd edn. Routledge, London (2015)
11. Lehmann, J., et al.: Biochar effects on soil biota–A review. Soil Biol. Biochem. **43**(9), 1812–1836 (2011)
12. Obemah, N. D., Baowei, Z.: Biochar preparation, characterization, and adsorptive capacity and its effect on bioavailability of contaminants: an overview. Adv. Mater. Sci. Eng. http://dx.doi.org/10.1155/2014/715398 (2016)
13. Shareef, T.M.E., Zhao, B.W.: The fundamentals of biochar as a soil amendment tool and management in agriculture scope: an overview for farmers and gardeners. J. Agric. Chem. Env. **6**, 38–61 (2017)
14. Dominic, Woolf., et al.: Sustainable biochar to mitigate global climate change. Nat. Commun. **1**, 56 (2010)
15. Das, S.K., et al.: Carbon-negative biochar from weed biomass for agricultural research in India. Curr. Sci. **110**(11), 2045–2046 (2016)
16. Uchenna, O., Kirk, S. T.: Impact of biochar on organic contaminants in soil: a tool for mitigating risk. Agronomy **3**, 349–375 (2013
17. Anita, Singh., et al.: Effect of pyrolysis temperature and retention time on mustard straw-derived biochar for soil amendment. J. Basic Appl. Sci. Res. **5**(9), 31–37 (2015)
18. IS 2720-Part 1: Methods of test for soils—Preparation of dry soil samples for various tests (Second Revision) (Reaffirmed May, 2020), Bureau of Indian Standards, New Delhi (1983)
19. IS 2720-Part 26: Methods of test for soils – Determination of pH value (Second Revision) (Reaffirmed Dec., 2016), Bureau of Indian Standards, New Delhi (1987)

20. IS 2720-Part 3: Methods of test for soils – Determination of specific gravity, Section 2 – Fine, medium and coarse grained soils (First Revision) (Reaffirmed Dec. 2016), Bureau of Indian Standards, New Delhi (1980)
21. IS 14767: Determination of the specific electrical conductivity of soils—Method of test, Bureau of Indian Standards, New Delhi (2000)
22. IS 2720-Part 2: Methods of test for soils—Determination of water content (Second Revision) (Reaffirmed May, 2020), Bureau of Indian Standards, New Delhi (1973)
23. IS 2720-Part 4: Methods of test for soils—Grain size analysis (Second Revision) (Reaffirmed May, 2020), Bureau of Indian Standards, New Delhi (1985)
24. IS 2720-Part 5: Methods of test for soils—Determination of liquid and plastic limit water content (Second Revision) (Reaffirmed May, 2020), Bureau of Indian Standards, New Delhi (1985)
25. IS 2720-Part 6: Methods of test for soils—Determination of shrinkage factors (First Revision) (Reaffirmed Dec., 2016), Bureau of Indian Standards, New Delhi (1972)
26. IS 2720-Part 8: Methods of test for soils—Determination of water content – Dry density relation using heavy compaction (Second Revision) (Reaffirmed May, 2020), Bureau of Indian Standards, New Delhi (1983)
27. IS 2720-Part 10: Methods of test for soils—Determination of unconfined compressive strength (Second Revision) (Reaffirmed May, 2020), Bureau of Indian Standards, New Delhi (1991)
28. IS 2720-Part 16: Methods of test for soils—Laboratory determination of CBR (Second Revision) (Reaffirmed Dec., 2016), Bureau of Indian Standards, New Delhi (1987)
29. IS 2720-Part 13: Methods of test for soils—Direct shear test (Second Revision) (Reaffirmed Dec., 2016), Bureau of Indian Standards, New Delhi (1986)
30. IS 2720-Part 41: Methods of test for soils—Measurement of swelling pressure of soils (Reaffirmed Dec., 2016), Bureau of Indian Standards, New Delhi (1977)
31. IS 2720-Part 40: Methods of test for soils—Determination of free swell index of soils (Reaffirmed Dec., 2016), Bureau of Indian Standards, New Delhi (1977)

Analytical Studies on the Use of Stiffer Drains for Soft Soil Improvement

B. V. S. Viswanadham and Vadlamani Manaswini

1 Introduction

Stone columns have gained widespread importance as a ground improvement technique in soft clay to increase bearing capacity and reduce settlement. Several researchers have analysed this through experimental and analytical means. Stone columns also accelerate the process of consolidation by providing an additional drainage path. This was concluded from field studies [7], analytical studies [8] and numerical solutions [6]. Load transfer from soil to stone column also caused dissipation of excess pore water pressure. Analytical studies by Han and Ye [8] developed simplified closed form solutions to estimate the rate of consolidation by stone columns. Practically, stone columns lack sufficient lateral confinement in very soft soils (c_u< 15 kPa) which hinders the gain of stiffness. They would undergo a bulging failure thereby preventing further load transfer from the soil. So the objective of this paper is to analyse the rate of consolidation due to stiffer drains analytically. Numerical solution by Balaam and Booker [1] also showed that increase in modular ratio also increases the rate of consolidation. The objective of this study is to estimate the increased rate of consolidation in stiffer drains, i.e. drains with increased modular ratio. Han and Ye [8] also analysed the dissipation of pore water pressure as a result of drainage and vertical stress reduction. This study also aims at analysing the contribution of drainage and vertical stress reduction in pore water pressure dissipation due to stiffer drains. Equal strain rate of consolidation is adopted in this study. Effect of smear and clogging is not considered.

B. V. S. Viswanadham · V. Manaswini (✉)
Department of Civil Engineering, IIT Bombay, Powai, Mumbai 400076, India

B. V. S. Viswanadham
e-mail: viswam@civil.iitb.ac.in

C. N. V. Satyanarayana Reddy et al. (eds.), *Ground Improvement and Reinforced Soil Structures*, Lecture Notes in Civil Engineering 152,
https://doi.org/10.1007/978-981-16-1831-4_45

2 Review of Analytical Solutions

Barron [3] solution considered consolidation of fine grained soils by drained wells. The modular ratio (ratio of elastic moduli of drain to soft soil) typically varies from 10 to 20 as given by Barksdale and Bachus [2]. This difference in stiffness was ignored in Barron's solution. Assumptions made by Barron are: (i) Water is incompressible in saturated soil and initially all the loads are taken by excess pore water pressure; (ii) Only vertical deformation of soil is allowed; (iii) A circular influence zone of drain well is considered and (iv) Uniform distribution of load over the compressible soil zone. Barron's theory yields a partial differential equation for axisymmetric flow considering that reduction in soil volume and discharge of water from soil are equal.

$$c_r\left(\frac{1}{r}\frac{\partial u}{\partial r}+\frac{\partial^2 u}{\partial r^2}\right)+c_v\frac{\partial^2 u}{\partial z^2}=\frac{\partial \overline{u}}{\partial t} \tag{1}$$

where c_r = coefficient of radial consolidation, c_v = coefficient of vertical consolidation, u = excess pore water pressure anywhere in the soil, $\overline{u}$ = average excess pore water pressure.

The flow in (1) is decomposed into vertical and radial flows.

$$\frac{\partial u_z}{\partial t}=c_v\frac{\partial^2 u_z}{\partial z^2} \tag{2}$$

$$\frac{\partial u_r}{\partial t}=c_r\left(\frac{1}{r}\frac{\partial u_r}{\partial r}+\frac{\partial^2 u_r}{\partial r^2}\right) \tag{3}$$

where u_z and u_r are excess pore water pressures due to vertical flow alone and radial flow alone, respectively.

Solution of (2) is the solution of Terzaghi 1D consolidation. Solution of (3) can be written as

$$u_r=\frac{4\overline{u}}{d_e^2F(N)}\left[r_e^2\ln\left(\frac{r}{r_c}\right)-\frac{r^2-r_c^2}{2}\right] \tag{4}$$

where $\overline{u}=u_0e^{\lambda}$, $\lambda=-\frac{8T_r}{F(N)}$, u_0 = excess pore water pressure initially, $F(N)=\left[\frac{N^2}{N^2-1}\right]\ln(N)-\frac{3N^2-1}{4N^2}$, N = diameter ratio ($d_e/d_c$), $T_r=c_rt/d_e^2$. Average degree of consolidation in radial direction is given by

$$U_r=1-exp^{-\left[\frac{8}{F(N)}\right]T_r} \tag{5}$$

Han and Ye [8] modified Barron's solution by taking stiffness of stone columns into consideration. They assumed equal strain rate of consolidation at any depth and considered effective drained modulus for modular ratio. Water was considered to be incompressible and smear and clogging effects were neglected. A modified equation for axisymmetric flow was proposed.

$$\frac{k_r}{\gamma_w}\left(\frac{1}{r}\frac{\partial u}{\partial r}+\frac{\partial^2 u}{\partial r^2}\right)+\frac{k_v}{\gamma_w}\frac{\partial^2 u}{\partial z^2}=\frac{m_{v,s}m_{v,c}(1-a_s)}{m_{v,c}(1-a_s)+m_{v,s}a_s}\frac{\partial \overline{u}}{\partial t} \tag{6}$$

where $m_{v,s}$ and $m_{v,c}$ are coefficients of volume compressibility of soil and stone column, respectively, k_r and k_v are coefficients of radial and vertical permeability respectively, a_s is the area replacement ratio, i.e. ratio of area of stone column to the total influence area. Equation (6) is simplified and written as

$$c_r^{'}\left(\frac{1}{r}\frac{\partial u}{\partial r}+\frac{\partial^2 u}{\partial r^2}\right)+c_v^{'}\frac{\partial^2 u}{\partial z^2}=\frac{\partial \overline{u}}{\partial t} \tag{7}$$

where $c_r^{'}$ and $c_v^{'}$ are modified coefficients of consolidation in radial and vertical directions, respectively, and can be written as

$$c_r^{'}=\left(\frac{k_r}{\gamma_w}\right)\frac{m_{v,c}(1-a_s)+m_{v,s}a_s}{m_{v,s}m_{v,c}(1-a_s)} \tag{8}$$

$$c_v^{'}=\left(\frac{k_v}{\gamma_w}\right)\frac{m_{v,c}(1-a_s)+m_{v,s}a_s}{m_{v,s}m_{v,c}(1-a_s)} \tag{9}$$

For an elastic body, coefficient of compressibility can be expressed as

$$m_v=\frac{(1+\nu)(1-2\nu)}{E(1-\nu)} \tag{10}$$

where ν is Poisson's ratio and E is elastic modulus. Thus,

$$\frac{m_{v,s}}{m_{v,c}}=\frac{E_c}{E_s}\frac{(1+\nu_s)(1-2\nu_s)(1-\nu_c)}{(1+\nu_c)(1-2\nu_c)(1-\nu_s)}=\xi\frac{E_c}{E_s} \tag{11}$$

where E_c and E_s are drained elastic moduli of column and soil, respectively, ν_c and ν_s are Poisson's ratios of column and soil, respectively, ξ is Poisson ratio factor.

The steady vertical stresses after consolidation is complete, σ_{cs} and σ_{ss} in the column and soil, respectively, can be written as

$$n_s=\frac{\sigma_{cs}}{\sigma_{ss}}=\frac{m_{v,s}}{m_{v,c}}=\xi\frac{E_c}{E_s} \tag{12}$$

where n_s is the steady-state concentration ratio. Therefore, the modified coefficients of consolidation after simplification are given by

$$c_r' = c_r\left(1 + n_s\frac{1}{N^2 - 1}\right) \quad (13)$$

$$c_v' = c_v\left(1 + n_s\frac{1}{N^2 - 1}\right) \quad (14)$$

where N is diameter ratio.

The overall rate of consolidation including vertical and radial effects given by Carillo [4] is expressed as

$$U_{rv} = 1 - (1 - U_r)(1 - U_v) \quad (15)$$

Han and Ye [9] further developed solutions to account for smear effect and well resistance by considering size and permeability of smeared zone which depends on the method of installation of stone columns. However, these are not considered in the present study. Castro and Sagaseta [5] analysed the problem by accounting for radial deformation of stone column and considered both elastic and plastic deformations. As stiffer drains are considered in the present study which do not deform laterally and whose stiffness does not depend on the lateral confinement provided by the surrounded soil, radial deformations need not be considered.

3 Analysis and Discussions

The current analysis makes the following assumptions:

(i) Water is assumed to be incompressible and drained elastic modulus is considered.

(ii) Equal rate of strain approach at any depth.

(iii) Smear and clogging effects are neglected.

A modular ratio (E_c/ E_s) = 30 is considered to simulate the behaviour of stiffer drains. $\nu_c = 0.15$, $\nu_s = 0.45$, N (Diameter ratio) = 3 are taken from Han and Ye [8]. From Eq. (12) we get $n_s = 8.34$ for the above values. The results obtained from n_s = 8.34 are compared to that of $n_s = 5$ as the latter was considered by the study in Han and Ye [8] (Fig. 1).

3.1 Stress Transfer

Increase in effective stress in the surrounding soil is calculated as

$$\Delta\sigma_s' = \Delta\overline{u}\frac{1 - a_s}{1 + a_s(n_s - 1)} = u_0 U_{rv}\frac{1 - a_s}{1 + a_s(n_s - 1)} \quad (16)$$

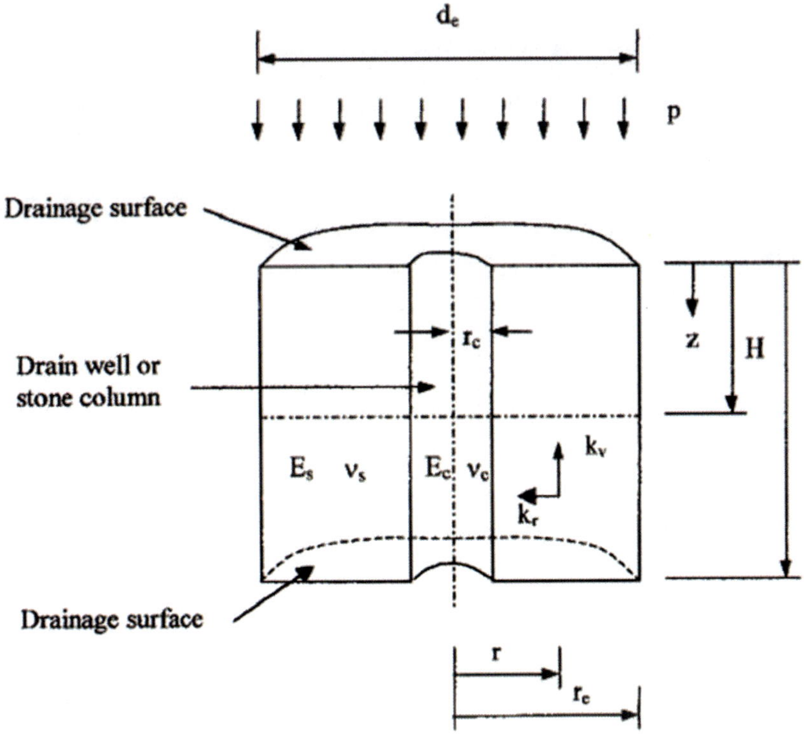

Fig. 1 Terms used in model (adopted after Han and Ye [8])

The transfer of stress from soil to columns is called stress concentration. With time, the vertical stress on the columns increases and that on soil decreases. The stress concentration ratio increases with time and reaches steady stress concentration ratio. From Fig. 2, we can see that with an increase in modular ratio (or steady-state stress concentration ratio), there is an increase in vertical stress on the columns and the vertical stress on the soil decreases. The increase in vertical stress on the columns is higher as compared to the decrease in vertical stress on the soil. Modified coefficients of consolidation are considered to get a modified time factor in the stress calculation and the results are presented against vertical stresses in Fig. 3.

3.2 *Stress Concentration Ratio*

The stress transfer process is shown in terms of stress concentration ratio in Figs. 4 and 5 which approach a steady-state concentration ratio and the consolidation ceases. Modified time factor is calculated corresponding to the conventional time factor and

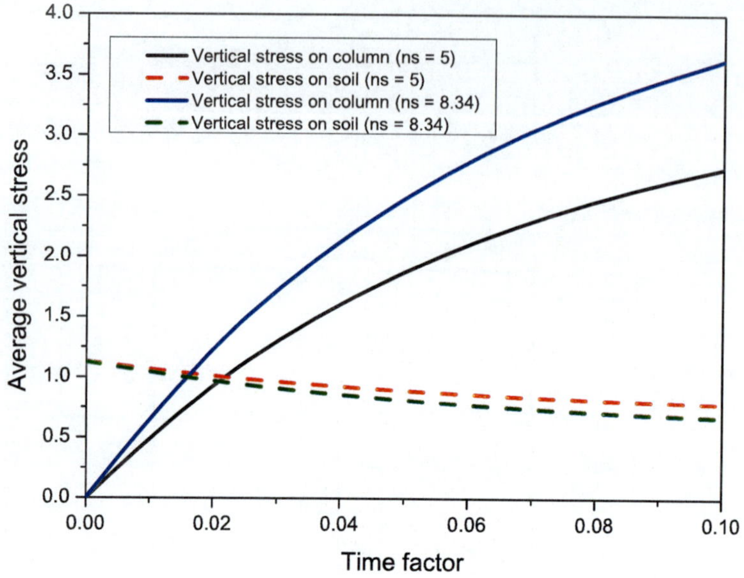

Fig. 2 Variation of normalized average vertical stress with time factor

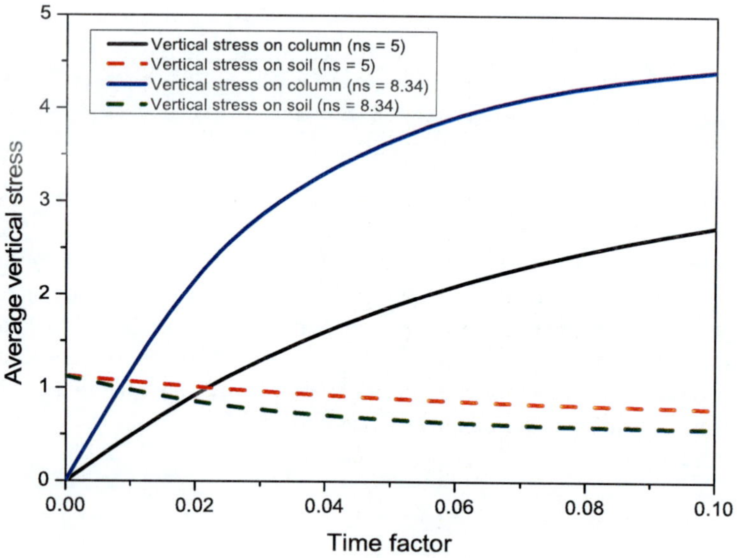

Fig. 3 Normalized average vertical stress versus time factor introducing modified time factor in stress calculation

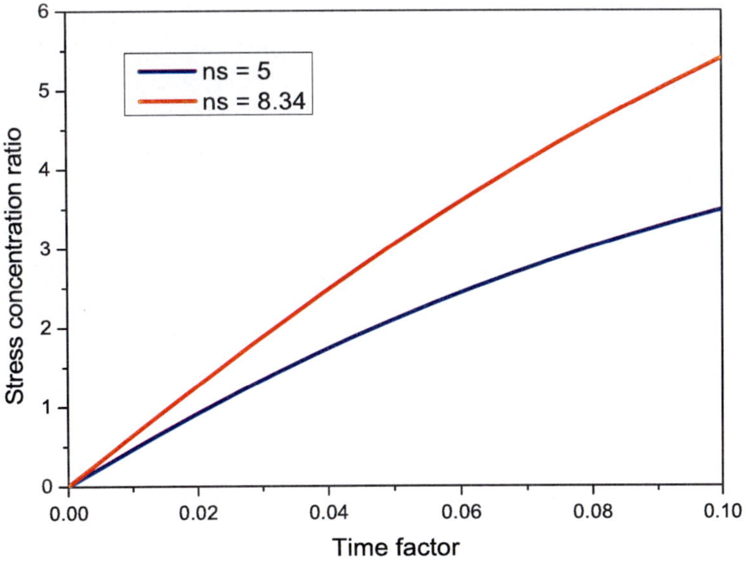

Fig. 4 Variation of stress concentration ratio versus time factor

used in the equations for calculating vertical stress and pore water pressure dissipation and plotted against conventional time factor for comparison.

3.3 *Excess Pore Water Pressure*

Excess pore water pressure variation in the soil at any time can be written as

$$u_t = u_0 - \Delta u_{\sigma_v} - \Delta u_d \tag{17}$$

where u_t and u_0 are excess pore water pressures at t > 0 and t = 0, respectively, Δu_{σ_v} is reduction in excess pore water pressure due to reduction in vertical stress on the soil or due to stress transfer and Δu_d excess pore water pressure reduced due to drainagze.

In Fig. 6, excess pore water pressure dissipated through stress reduction is the difference between the total excess pore water pressure dissipated and the excess pore water pressure dissipated through drainage. This contribution of stress reduction towards the dissipation of pore water pressure is dependent on stress concentration ratio. Larger the stress concentration ratio, greater the dissipation. But we can see that in Fig. 6, the total pore water pressure dissipated is not changed when the steady-state stress concentration ratio increases but the pore pressure dissipated through stress reduction increases. This is because we have used the same degree of consolidation for both the steady-state stress concentration ratios. Therefore modified coefficients of consolidation given by Han and Ye [8] should be used to bring the effect of stiffness of drains into picture.

From Fig. 7, we can see that though the dissipation of excess pore water pressure due to drainage accelerates in the current study in the beginning, but the final pore pressure dissipation due to drainage in the present study and by Han and Ye [8] is identical, which attributes to the fact that the drainage conditions are the same and the stiffer drains are equally permeable as that of drains assumed in the study of Han and Ye [8]. We can also observe that $\Delta u/p$ is greater than one. The presumption of all loads being carried by water in the soil at the beginning explains this. That is why stress on column in the beginning is zero.

$$\frac{\sigma_s}{p}(1-a_s)+\frac{\sigma_c}{p}a_s=1 \tag{18}$$

Substituting $\sigma_c = 0$ in the above equation and a_s (area replacement ratio) = 1/9 for N = 3, we get $\sigma_s = 1.125p$. Since initially all loads are carried by water in the soil, the initial pore water pressure u_o becomes 1.125p which dissipates gradually and gives the ratio $\Delta u/p>1$ eventually.

4 Conclusions

The rate of consolidation in the soil by using stiffer drains is studied. Stress transfer is higher if stress concentration ratio is higher, thus stiffer drains relieve more stress on the soil. The contribution of stress transfer in dissipation of excess pore water pressure increases with the increase in stress concentration ratio. These effects can be modelled accurately by using the modified coefficients of consolidation. Experimental investigations using stiffer drains need to be carried out to investigate the consolidation characteristics and economic feasibility for this to become an established field practice. Further analyses considering the smear and well resistance effects and clogging are warranted.

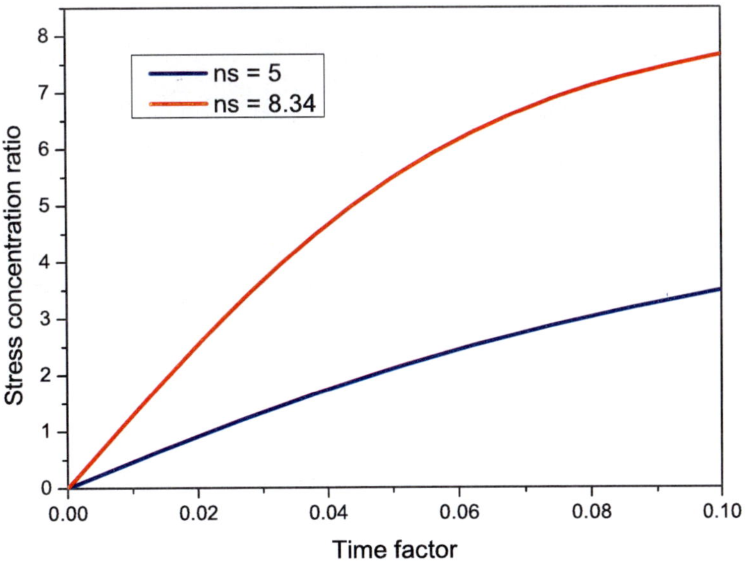

Fig. 5 Stress concentration ratio versus time factor introducing modified time factor in stress calculation

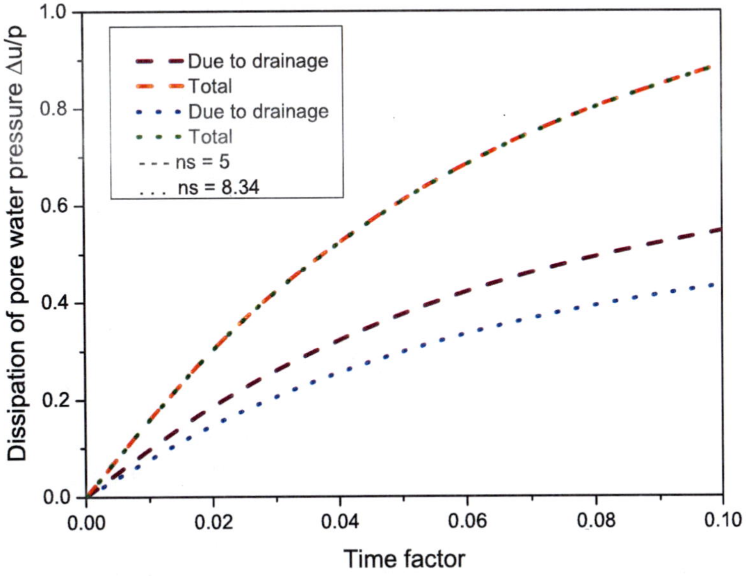

Fig. 6 Variation of dissipation of pore water pressure with time factor

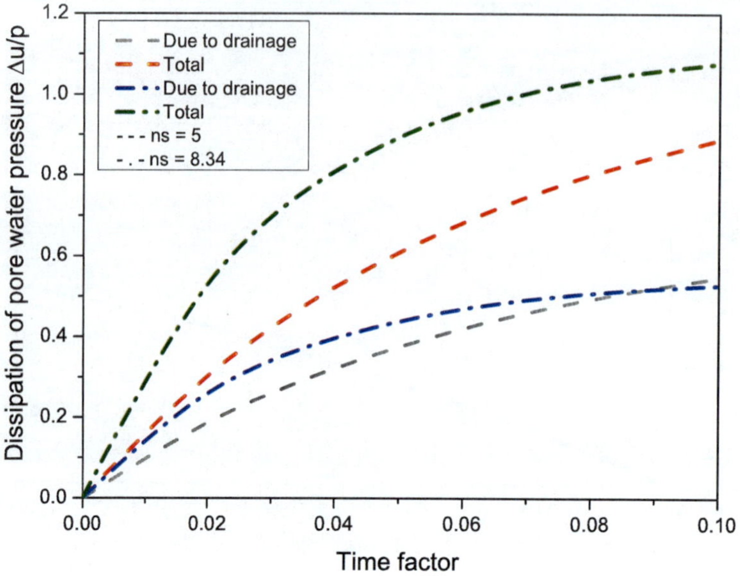

Fig. 7 Dissipation of excess pore water pressure vs time factor introducing modified time factor in stress calculation

References

1. Balaam, N.P., Booker, J.R.: Analysis of rigid rafts supported by granular piles. Int. J. Numer. Analy. Methods Geomech. **5**, 379–403 (1981)
2. Barksdale, R. D., Bachus, R. C.: 'Design and construction of stone columns. Federal Highway Administration, RD-83/026 (1983)
3. Barron, R. A.: Consolidation of fine-grained soils by drain wells. Proceedings on ASCE **73**(6), 811–835 (1947)
4. Carillo, N.: Simple two and three dimensional cases in the theory of consolidation of soils. J. Math. Phys. **21**(1), 1–5 (1942)
5. Castro, J., Sagaseta, C.: Consolidation around stone columns. Influence of column deformation. Int. J. Numer. Analy. Methods Geomech **33**(7), 851-877 (2009)
6. Guetif, Z., Bouassida, M., Debats, J.M.: Improved soft clay characteristics due to stone column installation. Comput. Geotech. **34**(2), 104–111 (2007)
7. Han, J., Ye, S. L.: Settlement analysis of buildings on the soft clays stabilized by stone columns. In: Proceedings on International Conference on Soil Improvement and Pile Foundation, 446–451 (1992)
8. Han, J., Ye, S.L.: Simplified method for consolidation rate of stone column reinforced foundations. J. Geotech. Geoenviron. Eng. ASCE **127**(7), 597–603 (2001)
9. Han, J., Ye, S.L.: A theoretical solution for consolidation rates of stone column-reinforced foundations accounting for smear and well resistance effects. Int. J. Geomech. ASCE **2**(2), 135–151 (2002)

Advances in Bioremediation of Extremely Alkaline Bauxite Residue: A Review

Manas Chandan Mishra, Bendadi Hanumantha Rao, and Swagatika Senapati

1 Introduction

Insoluble residue of Bayer's process of alumina extraction from bauxite is known as red mud or bauxite residue (BR). It is reported to be rich in iron oxide with a plethora of other elements (such as aluminum, sodium, silicon, calcium, titanium, etc.) available in different forms [1]. BR has been observed to demonstrate extortionate alkalinity (pH > 11.5) with high salinity and sodicity [2]. It is also reported that BR exhibits several geotechnical issues such as unreliable compaction characteristics, high collapse potential and small shear strength [3]. Moreover, this industrial waste has been associated with several geoenvironmental impediments counting the abundance of metal oxides, erratic and high sedimentation properties, and leaching characteristics in addition to dusting and dispersion features [1, 4].

Moreover, construction and maintenance of impoundment for BR needs skilled manpower along with machineries incurring extravagant expenses. Needless to say, improper impoundment facility creates the risk of caustic exposure to ecosystem, due to leakage of alkaline leachate into groundwater, overflow of materials during inundations and storms, and dusting (mostly Na_2CO_3) from pulverized dry surfaces [2]. Such catastrophes have been witnessed in the past during dike failure in Hungary (2010), China (2016) and India (2019), which led to massive loss of human and animal

M. C. Mishra · B. H. Rao (✉)
School of Infrastructure, IIT Bhubaneswar, Khordha, Odisha 752050, India
e-mail: bhrao@iitbbs.ac.in

M. C. Mishra
e-mail: mcm10@iitbbs.ac.in

S. Senapati
Department of Civil Engineering, IIT Madras, Chennai, Tamil Nadu 600036, India

C. N. V. Satyanarayana Reddy et al. (eds.), *Ground Improvement and Reinforced Soil Structures*, Lecture Notes in Civil Engineering 152,
https://doi.org/10.1007/978-981-16-1831-4_46

life, vegetation along with disruption of ecological balance for substantial time post-incident. Thus, it is safe to deduce that aggressive and beneficial utilization minimizes the environmental risk and pollution footprint on the eco-system. However, the associated complications of BR restrict its utilization to a bare 2–3% [3].

The available literature demonstrates various conventional and novel amelioration techniques being employed on BR, such as use of carbon dioxide, sea water, organic and mineral acids, industrial and other wastes (fly ash, gypsum, blast furnace slag, copper slag), lime, cement, chemical reagents, physical and mechanical treatments (heating, compaction, consolidation) [5–8]. However, a rebound of pH post-treatment has been observed for these treatments, which poses difficulty in utilizing BR in practical applications [7].

Soil microbial bio-diversity is known to alleviate salinity and pH associated problems in extremely alkaline soils when provided with a supportive environment [9]. Such microbial communities further proliferate by synthesizing organic and inorganic acids, carbon dioxide, extracellular polymeric substances (EPS) and modifying ionic balance of the extreme prevailing conditions. However, the exact mechanism by which microorganisms induce pH reduction in alkaline soils and wastes like BR has not been fully understood.

Furthermore, it is observed that organic acids such as citric acid (CA), lactic acid (LA), acetic acid (AA) and propionic acid (PA) are produced after metabolic fermentation of plant residues [10]. It is a well-known fact that microorganisms adapt to their extreme alkaline environment by synthesizing various forms of acidic secretions such as polysaccharides, proteins, nucleic acids and lipids [11]. The ability of CA and oxalic acid (OA) to reduce BR pH has been explored in depth [7]. In that study, the rebound characteristics of pH have been quantified, in terms of rebound rate of pH (RRP) and rebound termination period (t_{rt}), to facilitate analytical comparison between the efficiency of remediation techniques to reduce and maintain BR pH. Further, these studies give an insight into the application of microorganisms, which can synthesize organic acids for BR remediation. It is interesting to note that several studies have attempted bioremediation techniques on BR, which include microbial treatments, vegetation, organic amendments and combinations thereof.

In this study, advancements and recent developments in bioremediation techniques are discussed in order to get an overview of the situation status quo. It also encompasses brief comparisons of factors considered for these studies. It is interesting to note that biochemical mechanisms as well as the prevailing physico-chemico-mineralogical conditions are studied to incorporate better understanding of studied techniques. However, the primary contribution of this study is the experimental findings observed while investigating the efficiency of CA in mitigating BR alkalinity.

2 Bioremediation Techniques

Several bio-based remediation techniques including plantation, use of biopolymers and microbial remediation have been discussed in this section. Plantation, mostly growing grass, is suggested on BR impoundments in order to abate its associated hazards such as dusting, erosion and leaching [12]. The dusting behavior of BR has also been addressed by utilizing biopolymers such as guar gum or xanthan gum, along with the improvements in its engineering properties [13]. It is, nonetheless, important to note here that exorbitant alkalinity of BR hinders the efficiency of plantation techniques.

On the contrary, the microbial treatment involves identification of strains of microorganisms with metabolic ability to (1) produce EPS inducing aggregation of particles, (2) synthesize acidic secretions to reduce local pH, (3) produce CO_2 which could further reduce the pH of BR and (4) mutate the ionic balance by solubilizing minerals and selective extraction of ions from pore water.

Advancements in these bioremediation techniques of BR are discussed in the following sections. The observations and results of these studies have been analyzed to understand their efficacy and compared to the basics of soil chemistry and remediation, followed by summary of limitations and research prospects in this field.

2.1 *Producing EPS to Induce Aggregation of Particles*

Studies have reported production of EPS by various strains of bacteria and fungi, which led to precipitation, bridging and cementation around particles, resulting in improvement of engineering properties at microstructure level [14]. EPS is produced by microbes as a defense mechanism to adapt and survive in extreme conditions along with providing mechanical stability by increasing inter-particle cohesion. Microbial EPS generally consists of lipids, genetically modified nucleic acids, oligosaccharides and polysaccharides, and other forms of proteins [14]. It was also reported in the study that the particular structure of EPS produced varies for each strain bearing genetical mark of the microbial strain or symbiotic community. However, it has been observed in various studies that the composition and structure of EPS is considerably affected by the prevailing environment of microbial communities [15, 16]. It has been identified that nutrient availability, pH and temperature are the most important factors for bacterial communities [15]. However, the most crucial factor affecting EPS for fungal strains is water potential, in addition to prevailing pH and nutrient concentration [16].

Remediation using EPS produced by microbes is an important alternative for improving the engineering behavior of BR. The efficiency of this technique can be further enhanced by physico-chemically modifying the conditions for these microbial strains. It can be hypothesized based on the available literature that the local conditions for these microbes can be rendered more amicable by identifying microbial strains which can alter the pH and chemical composition of BR by any of the aforementioned mechanisms such as synthesizing acidic secretions, producing CO_2, metabolically fermenting vegetation residues or mutating ionic balance. However, the interaction of these microbial communities in conditions prevailing in BR and their combined efficiency should be explored in detail before recommending the remediation technique for industrial applications.

2.2 *Synthesizing Acidic Secretions to Reduce Local pH*

The use of acid synthesizing microorganisms in reducing pH poses as a practicable and an alluring alternative for BR remediation. Previous studies have identified various species and strains of microbes which produce acidic secretions such as enzymes, polysaccharides, proteins, nucleic acids, mineral acids and lipids [10, 17–21]. It is reported that microorganisms produce acidic secretions by fermenting organic residues of plants and insects available in the BR ponds [17]. These acidic secretions can be effective in reducing BR pH by donating available H^+ to react with alkalinity present in both liquid and solid phases. In addition to that, it has also been observed that various microorganisms undergo autotrophic or heterotrophic metabolisms in the presence of certain chemicals and produce inorganic acids such as sulfuric acid [20, 21]. These microbes present a practicable alternative which can be studied further in order to provide a viable remediation technique of pH mitigation for BR.

2.3 *Producing CO_2 Further Abating BR pH*

Carbon dioxide, whether added mechanically or sequestered from atmosphere or produced by microorganisms, has proved to reduce pH of BR as it reacts with available pore water producing carbonic acid and metal carbonates in some cases. CO_2 is generated as an exhaled gas of aerobic and a fermentation byproduct of anaerobic microbes. It is understood that the aerobic CO_2 is not efficiently trapped in the pores before releasing into the atmosphere, but fermentation in anaerobic reactions leads to more formation of carbonic acids, thus leading to higher reaction rates with BR alkalinity [9].

2.4 *Mutation of Local Ionic Balance*

It is well established that BR is rich in sodium ions, most of which are in reactive or exchangeable form [4]. These exchangeable sodium ions can alter the cation exchange capacity of BR, which has been reported as a cause of pH rebound. In order to reduce the pH and sustain it below the prescribed level, these exchangeable sodium ions have to be extracted or converted to forms that are more inert. Another prospect that can be explored is substituting sodium ions with other metal ions (such as calcium) at these cation exchange sites, which can restrict the adsorbed free sodium from altering BR alkalinity. This has been observed in the case of reduction of BR pH with gypsum. Gypsum solubilizes Ca^{2+} replacing exchangeable sodium by simultaneously producing sulfuric acid, leading to reduction of pH in BR. Thus, identifying suitable microorganisms which can solubilize other metal ions or extract Na^{+} can provide a viable alternative to reduce BR alkalinity. Several strains have been identified [22], which can solubilize Ca^{2+} with change in pH of alkaline media, and surface etching of calcium-based minerals. Further, it has been reported that few strains of fungi can selectively extract ions from sodic soils [23]. Identification of such microbial communities, which can alter the ionic balance of BR by selectively extracting alkalinity inducing ions like sodium, can lead to further detailed investigations toward finding a strong bioremediation technique.

3 Useful Studies on Microbial Bioremediation

Alkaliphiles are microorganisms which find alkaline medium (pH > 9) favorable for their proliferation [24]. They utilize the available organic matter as nutrition and synthesize organic acids. Four strains of alkaliphile bacilli were studied by Paavilainen et al. [24] in sodium chloride media. It was reported that various organic acids (carboxylic acids, amino acids, aliphatic and aromatic acids) were produced by these strains depending on the nutrition provided.

On the other hand, haloalkaliphiles require extremely high pH as well as abundance of sodium ions for their proliferation. Alkaliphiles are also found in neutralized or acidic soil and deep sea, but haloalkaliphiles are mostly reported in highly alkaline and saline soils. Few strains of aerobic microbes (among genus *Bacillus*, *Micrococcus*, *Pseudomonas* and *Streptomyces*) [25] and anaerobic microbes (genus *Clostridium*, *Amphibacillus*, *Thermococcus*) [26–28] are identified from alkaline soils and wastes, which are known to reduce the local pH for their faster proliferation.

Few species of alkaliphiles and haloalkalihiles (such as *Vagococcus*, *Marinobacter*, *Paenibacillus*, *Roseinatrobacter*, *Rhodobaca*) have been isolated from various soda lakes of India, which prefer a highly alkaline environment of pH 10 to 10.5 and high sodium concentration [29–32]. *Vagococcus* generate exopolysaccharide (a sugar-based EPS), whereas Marinobacter is known to be highly adaptive to its local environment, and used as an alternative to treat contaminated soils [33]. It is interesting to note that *Paenibacillus polymyxa* is resistant to chemicals while presenting favorable response to mutations and genome engineering [34].

4 Microbial Bioremediation of BR

Bioremediation of BR by certain microbes has proved to be sustainable and ecofriendly [35, 36]. Low levels of injured bacterial cells grew (from <10 to more than 109 cells/g) with addition of required nutrients leading to formation of organic acids in the BR, which might have lowered its pH from 13 to 7 [37].

The microbial strain identified in the aforementioned studies belongs to a wide variety of species: *Bacillus, Lactobacillus, Leuconostoc, Micrococcus, Staphylococcus, Pseudomonas, Flavobacterium* and *Enterobacter*. It is interesting to note that vegetation growth was observed to be supported in the BR post-treatment. *Cyanobacteria*, also known as blue green algae, is considered efficient for remediation of problematic soils [36] and extraction of various elements from them as well as resistant to different forms of various metals [38]. With this in mind, four different strains of *Cyanobacteria* (i.e. *Oscillatoria* sp., *Lyngbya* sp., *Phormidium* sp., and *Microcystis* sp.) were investigated upon in order to identify their efficacy in bioremediation of BR [36]. It was reported that *Oscillatoria* and *Phormidium* when combined with *Aspergillus tubingensis* could act as an effective remediation technique of BR. Microbial technique for pH reduction was also explored by [39], which led to the isolation of EEEL02 strain of *Bacillus thuringiensis* from the BR impoundments. It was reported that PA, AA and CO_2 (leading to the formation of carbonic acid) were generated by this strain with glucose and peptose as nutrition and at optimal conditions of 25 °C and pH 10, which led to substantial decrease in BR pH within a short span of time. However, no study on the efficacy of these microbes in abating the post-treatment pH rebound of BR has been conducted. Table 1 summarizes the key findings of pertinent studies.

5 Experimental Findings

5.1 Materials Used

BR samples used in the present study were collected in disturbed and wet state from the impoundment site of Vedanta Aluminium Limited at Kalahandi district in the state of Odisha, India. Samples were subjected to oven drying at 110 ± 0.5 °C for at least 24 h to remove entrapped moisture before pulverizing within the required size. Atterberg's limits, specific gravity, particle size distribution and pH of different samples were determined with the help of standard equipment in compliance with relevant ASTM standards. Table 2 presents the results obtained on basic characterization of BR. This sample of BR can be classified as inorganic silt of low plasticity (ML) according to the Unified Soil Classification System (USCS).

In the present study, to examine the effect of organic acid on pH of BR, CA procured from CDH (Central Drug House (P) Ltd, New Delhi, India) was used as

Table 1 Summary of available literature

Authors	Key findings
Bioremediation Techniques	
Fuller et al. [12]	BR impoundments abate the hazards associated with it: such as dusting, erosion, and leaching
Producing EPS to Induce Aggregation of Particles	
Flemming and Wingender [14]	EPS production by various strains of bacteria and fungi results in improvement of engineering properties
Gorret et al. [15] and Papinutti [16]	The composition and structure of EPS is affected by the prevailing environmental conditions (i.e. nutrient availability and concentration, pH, temperature and water potential) of microbial communities
Synthesizing Acidic Secretions to Reduce Local pH	
Sugaya et al. [10], Mussel et al. [17], Kubicek et al. [18], Wu et al. [19], Schippers and Sand [20], Friedrich et al. [21]	Microbes produce acidic secretions (such as enzymes, polysaccharides, proteins, nucleic acids, mineral acids and lipids) by fermenting organic residues of plants and insects available in the BR ponds which helps in reducing BR pH
Schippers and Sand [20], Friedrich et al. [21]	Various microorganisms undergo autotrophic or heterotrophic metabolisms to produce sulfuric acid which regulates pH mitigation
Producing CO_2 *Further Abating BR pH*	
Santini et al. [9]	Fermentation in anaerobic reactions leads to more formation of carbonic acids leading to reduction in pH of BR
Mutation of Local Ionic Balance	
Kolo et al. [22]	Suitable microorganisms solubilize Ca^{2+} with change in pH of alkaline media which reduces BR alkalinity
Wu et al. [23]	Extraction of ions from sodic soils by few strains of fungi provides strong bioremediation technique
Useful Studies on Microbial Bioremediation	
Paavilainen et al. [24]	Alkaliphiles utilize available organic matter as nutrition and proliferate in alkaline medium. It produces organic acids (i.e. carboxylic acids, amino acids, aliphatic and aromatic acids)
Duckworth et al. [25], Podkovyrov and Zeikus [26], Kodama and Koyama [27], Keller et al. [28],	Certain aerobic microbes and anaerobic microbes reduce the local pH for their faster proliferation

(continued)

Table 1 (continued)

Authors	Key findings
Kulkarni et al. [29], Nandy and Deo [30], Malu et al. [31], Upasani and Desai [32]	Alkaliphiles and haloalkalihiles (such as *Vagococcus, Marinobacter, Paenibacillus, Roseinatrobacter, Rhodobaca*) prefer a highly alkaline environment of pH 10 to 10.5 and high sodium concentration
Microbial Bioremediation of BR	
Prasanna et al. [35], Dubey and Dubey [36]	Microbial bioremediation of BR by certain microbes (*Bacillus, Lactobacillus, Leuconostoc, Micrococcus, Staphylococcus, Pseudomonas, Flavobacterium* and *Enterobacter*) is sustainable and ecofriendly
Hamdy and Williams [37]	Growth of bacterial cells leads to formation of organic acids in the BR by lowering its pH from 13 to 7

Table 2 Characteristics of bauxite residue

Property	Value
Specific gravity, G_s	3.06
Consistency limits (%)	
w_L	42
w_P	35
w_{PI}	7
Percent fraction	
Sand (%)	04
Silt (%)	71
Clay (%)	25
pH value	11.4
USCS	ML

an additive at different concentrations varying from 0.1 to 6 M. CA is a naturally occurring organic acid in plants that can also be synthesized by *Aspergillus niger.*

5.2 *Sample Preparation*

Acid solution, instead of water, of pre-decided concentrations was added to BR samples of quantity calculated based on optimum water content. These CA admixed BR samples were then used to prepare compacted cylindrical samples using the moist tamping method to attain compaction state of maximum dry unit weight at optimum water content. The samples were sealed in zip-lock bags and preserved inside a humidity chamber in order to maintain the moisture equilibrium.

5.3 *Testing Methodology*

Morphological characteristics of BR samples are examined using MERLIN compact field emission scanning electron microscope (FE-SEM) (make, ZEISS, Germany). Representative samples in briquette form are prepared from the cured compacted cylindrical samples stored in humidity chamber. Each briquette is sputtered with gold coating using Q150R ES Sputter Coater (make, Quorum, UK) before conducting SEM analysis in order to obtain clear morphological characteristics.

5.4 *Results and Discussions*

Figure 1 presents SEM images of BR samples preserved under maintained moisture and temperature conditions after admixing with CA solutions. The overall morphological features of untreated BR can be observed in Fig. 1a. It can be observed from the figure that the untreated BR particles are mostly angular with few fluffy aggregations, which might be due to sodalite and calcite found in it [40]. Figure 1b shows the CA treated BR compacted sample. Here prominent aggregations can be observed over almost every BR particle. This might have occurred because of changes in mineralogical phases of BR due to its reaction with CA.

Figure 1c and d presents the microbial growth observed in CA treated and compacted BR samples. It is interesting to note here that microbial colonies have proliferated, connecting and bridging between BR particles. This observation is encouraging as it is difficult for microorganisms to adapt and proliferate in the extreme conditions of high pH and complex chemico-mineralogical properties. However, further detailed investigation is underway to characterize the microbial communities observed in the present study.

6 Conclusions

Bioremediation technique that primarily implements microorganisms seems to provide promising avenues for pH reduction of BR with the scope for enhancement of geotechnical behavior. These techniques are considered significantly lesser energy-intensive as well. This review study has discussed various possible approaches for utilizing microorganisms-based bioremediation techniques. However, the climatic, geoenvironmental and biochemical variations of BR generated and disposed by alumina industry worldwide need to be taken into consideration while choosing

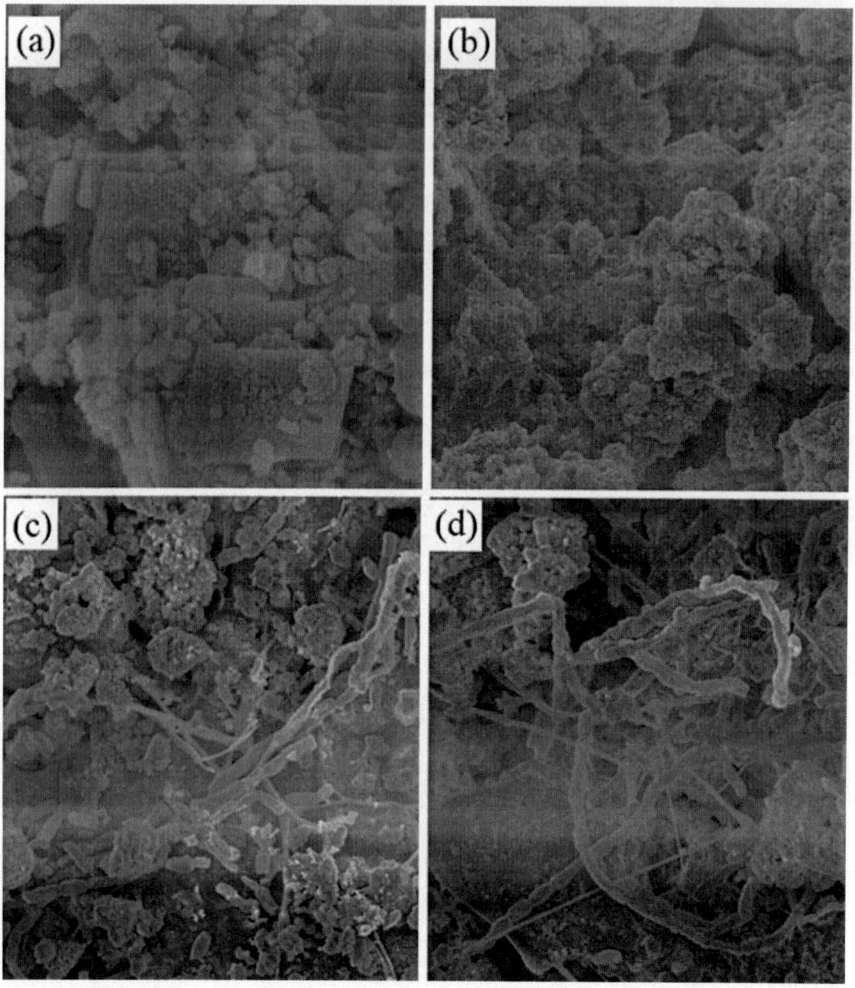

Fig. 1 SEM images of **a** untreated BR, **b**, **c** and **d** CA treated BR samples

any of these techniques for further detailed study. BR impoundments might also be hosting a few species of microorganisms, which can be studied upon. Furthermore, pH abatement of BR through acid secretion, CO_2 generation, selective ions extraction/solubilization as well as aggregation of BR particles through EPS synthesis seems to provide practicable mechanisms for bioremediation of the alkaline industrial waste. The efficacy of such native microorganisms along with externally procured microbial communities in remediating the extremely alkaline BR needs to be investigated in order to provide a more ecofriendly and cost-effective alternative, which can render BR more environmentally benign.

Future endeavors in this regard should be dovetailed to address the bioremediation techniques from three aspects: (a) isolating suitable microorganisms for bioremediation for pH reduction and/or EPS generation, (b) identifying and characterizing biochemical additives that could mutually benefit with the identified bioremediation techniques and (c) conducting a comparative study on the identified techniques to recommend the most ecofriendly and cost-effective approach for industrial application.

Appreciable contribution can be done to the vast ocean of knowledge in the field of geoenvironmental engineering in general and waste utilization, in particular, while attending to the technical and economic aspects of research avenues discussed in this review. The outcome of identified research approaches might assist in providing a green and economical alternative towards utilizing BR.

References

1. Power, G., Gräfe, M., Klauber, C.: Bauxite residue issues: I. Current management, disposal and storage practices. Hydrometallurgy **108** (1–2), 33–45 (2011)
2. Alam, S., Das, B.K., Das, S.K.: Dispersion and sedimentation characteristics of red mud. J. Hazard. Toxic Radioactive Waste **22**(4), 04018025 (2018)
3. Reddy, N.G., Rao, B.H.: Evaluation of the compaction characteristics of untreated and treated red mud. GSP, ASCE **272**, 23–32 (2016)
4. Mishra, M.C., Reddy, N.G., Rao, B.H.: Potential of citric acid for treatment of extremely alkaline bauxite residue: effect on geotechnical and geoenvironmental properties. J. Hazard. Toxic Radioactive Waste **24**(4), 04020047 (2020)
5. Hanahan, C., McConchie, D., Pohl, J., Creelman, R., Clark, M., Stocksiek, C.: Chemistry of seawater neutralization of bauxite refinery residues (red mud). Environ. Eng. Sci. **21**(2), 125–138 (2004)
6. Khaitan, S., Dzombak, D.A., Swallow, P., Schmidt, K., Fu, J., Lowry, G.V.: Field evaluation of bauxite residue neutralization by carbon dioxide, vegetation, and organic amendments. J. Environ. Eng. **136**(10), 1045–1053 (2010)
7. Mishra, M.C., Rao, B.H.: Neutralization of red mud with organic acids and assessment of their usefulness in abating pH rebound. J. Hazard. Toxic Radioactive Waste **24**(1), 04019026 (2019)
8. Sutar, H., Mishra, S.C., Sahoo, S.K., Chakraverty, A.P.: Progress of red mud utilization: an overview. Am. Chem. Sci. J. **4**(3), 255–279 (2014)
9. Santini, T.C., Kerr, J.L., Warren, L.A.: Microbially-driven strategies for bioremediation of bauxite residue. J. Hazard. Mater. **293**, 131–157 (2015)
10. Sugaya, K., Tuse, D., Jones, J.L.: Production of acetic acid by Clostridium thermoaceticum in batch and continuous fermentations. Biotechnol. Bioeng. **28**(5), 678–683 (1986)
11. Flemming, H.C., Wingender, J.: The biofilm matrix. Nat. Rev. Microbiol. **8**(9), 623–633 (2010)
12. Fuller, R.D., Nelson, E.D., Richardson, C.J.: Reclamation of red mud (bauxite residues) using alkaline-tolerant grasses with organic amendments. J. Environ. Qual. **11**(3), 533–539 (1982)
13. Reddy, N.G., Rao, B.H., Reddy, K.R.: Biopolymer amendment for mitigating dispersive characteristics of red mud waste. Géotech. Lett. **8**(3), 201–207 (2018)
14. Flemming, H.C., Wingender, J.: The biofilm matrix. Nat. Rev. Microbiol. **8**, 623–633 (2010)
15. Gorret, N., Maubois, J.L., Engasser, J.M., Ghoul, M.: Study of the effects of temperature, pH and yeast extract on growth and exopolysaccharides production by Propionibacterium acidipropionici on milk microfiltrate using a response surface methodology. J. Appl. Microbiol. **90**, 788–796 (2001)

16. Papinutti, L.: Effects of nutrients, pH and water potential on exopolysaccharides production by a fungal strain belonging to Ganoderma lucidum complex. Bioresour. Technol. **101**, 1941–1946 (2010)
17. Mussel, G., Sparling, G., Summers, J.: Bioremediation of Bauxite Residue in Western Australia: An Initial Feasibility Study. Alcoa of Australia Limited, Kwinana Australia (1993)
18. Kubicek, C.P., Röhr, M., Rehm, H.J.: Citric acid fermentation. Crit. Rev. Biotechnol. **3**, 331–373 (1985)
19. Wu, C. Y., Zhuang, L., Zhou, S. G., Li, F.B., He, J.: Corynebacterium humireducens sp. nov. an alkaliphilic, humic acid-reducing bacterium isolated from a microbial fuel cell. Int. J. Syst. Evol. Microbiol. **61**, 882–887 (2011)
20. Schippers, A., Sand, W.: Bacterial leaching of metal sulfides proceeds by two indirect mechanisms via thiosulfate or via polysulfides and sulfur. Appl. Environ. Microbiol. **65**, 319–321 (2001)
21. Friedrich, C.G., Rother, D., Bardischewsky, F., Quentmeier, A., Fischer, J.: Oxidation of reduced inorganic sulfur compounds by bacteria: emergence of a common mechanism? Appl. Environ. Microbiol. **67**, 2873–2882 (2001)
22. Kolo, K., Keppens, E., Préat, A., Claeys, P.: Experimental observations on fungal diagenesis of carbonate substrates, J. Geophys. Res. Biogeosci. **112** (2007)
23. Wu, Q.S., Zou, Y.N., He, X.H.: Contributions of arbuscular mycorrhizal fungi to growth photosynthesis, root morphology and ionic balance of citrus seedlings under salt stress. Acta Physiol. Plant **32**, 297–304 (2010)
24. Paavilainen, S., Helistö, P., Korpela, T.: Conversion of carbohy-drates to organic acids by alkaliphilic bacilli. J. Ferment. Bio-eng. **78**(3), 217–222 (1994)
25. Duckworth, A.W., Grant, W.D., Jones, B.E., Van Steenbergen, R.: Phylogenetic diversity of soda lake alkaliphiles. FEMS Microbiol. Ecol. **19**(3), 181–191 (1996)
26. Podkovyrov, S.M., Zeikus, J.G.: Structure of the gene encoding cyclomaltodextrinase from Clostridium thermohydrosulfuricum 39E and characterization of the enzyme purified from Escherichia coli. J. Bacteriol. **174**(16), 5400–5405 (1992)
27. Kodama, H., Koyama, N.: Unique characteristics of anaerobic alkalophiles belonging to Amphibacillus xylanus. Microbios **89**(358), 7–14 (1997)
28. Keller, M., Braun, F.J., Dirmeier, R., Hafenbradl, D., Burggraf, S., Rachel, R., Stetter, K.O.: Thermococcus alcaliphilus sp. nov., a new hyperthermophilic archaeum growing on polysulfide at alkaline pH. Arch. Microbiol. **164**(6), 390–395 (1995)
29. Kulkarni, S., Dhakar, K., Joshi, A.: Alkaliphiles: diversity and bioprospection. In: Microbial Diversity in the Genomic Era. Academic Press, pp. 239–263 (2019)
30. Nandy, N.C., Deo, V.B.: Origin of the Lonar lake and its alkalinity. TISCO **8**(3), 1–12 (1961)
31. Malu, R.A., Dhabhade, D.S., Kodarkar, M.S.: Diversity in Lonar lake. J. Aquat. Bio. **15**, 16–18 (2000)
32. Upasani, V., Desai, S.: Sambhar salt lake. Archiv. Microbiol. **154**(6), 589–593 (1990)
33. Handley, K.M., Lloyd, J.R.: Biogeochemical implications of the ubiquitous colonization of marine habitats and redox gradients by Marinobacter species. Front. Microbiol. **4**, 136 (2013)
34. Jeong, H., Choi, S.K., Ryu, C.M., Park, S.H.: Chronicle of a soil bacterium: Paenibacillus polymyxa E681 as a tiny guardian of plant and human health. Front. Microbiol. **10**, 467 (2019)
35. Prasanna, R., Jaiswal, P., Kaushik, B.D.: Cyanobacteria as potential options for environmental sustainability—promises and challenges. Indian J. Microbiol. **48**(1), 89 (2008)
36. Dubey, K., Dubey, K.P.: A study of the effect of red mud amendments on the growth of cyanobacterial species. Bioremediat. J. **15**(3), 133–139 (2011)
37. Hamdy, M.K., Williams, F.S.: Bacterial amelioration of bauxite residue waste of industrial alumina plants. J. Ind. Microbiol. Biotechnol. **27**(4), 228–233 (2001)
38. Wilde, E.W., Benemann, J.R.: Bioremoval of heavy metals by the use of microalgae. Biotechnol. Adv. **11**(4), 781–812 (1993)

39. Wu, H., Liao, J. X., Zhu, F., Millar, G., Courtney, R., Xue, S.G.: Isolation of an acid producing Bacillus sp. EEEL02: potential for bauxite residue neutralization. J. Central South Univ. **26**(2), 343–352 (2019)
40. Mishra, M.C., Babu, S.K., Reddy, N.G., Dey, P.P., Rao, B.H.: Performance of lime stabilization on extremely alkaline red mud waste under acidic environment. J. Hazard. Toxic Radioactive Waste **23**(4), 04019012 (2019)

Soil Nail Wall Design Using Simplified Charts

Vikas Pratap Singh

1 Introduction

Soil nailing is one of the most widely used in situ earth retaining techniques. Typical geotechnical engineering applications of soil nailing technique include stabilization of road/rail side slopes, landslides, bridge abutments, and vertical cuts for basement excavations and approach roads for subways [1–6].

Over the years, with the extensive use of soil nailing technique, various design methodologies evolved and are being used extensively in practice. In literature, a number of prominent design methodologies [7–14] for the analysis and design of soil nail walls are readily available. However, for soil nail wall design in practice, very limited information in the form of charts and tables is readily available. In this study, using allowable stress design approach [12], simplified design charts are proposed that can facilitate easy design of soil nail walls and also assist in internal stability analysis. The proposed design charts are prepared by considering a variety of geometric and in situ soil conditions.

2 Failure Modes of Soil Nail Walls

Various failure modes [12] for a typical soil nail wall can be broadly classified into three distinct groups as: external failure modes, internal failure modes, and facing failure modes. The two prominent external failure modes of soil nail walls are global stability and sliding stability. Global stability refers to the overall stability of the reinforced soil nail wall mass. On the other hand, sliding stability indicates the sliding

V. P. Singh (✉)
National Institute of Technology Uttarakhand, Srinagar Garhwal, Uttarakhand 246174, India
e-mail: vpsingh@nituk.ac.in

C. N. V. Satyanarayana Reddy et al. (eds.), *Ground Improvement and Reinforced Soil Structures*, Lecture Notes in Civil Engineering 152,
https://doi.org/10.1007/978-981-16-1831-4_47

Table 1 Minimum recommended factors of safety for soil nailing [12]

Mode type	Failure type	Symbol	Factors of safety
External	Global stability	FS_G	1.35–1.50
	Sliding stability	FS_{SL}	1.30–1.50
Internal	Nail pullout failure	FS_P	2.00
	Nail tensile failure	FS_T	1.80

resistance along the base of the soil nail wall in response to the additional lateral earth pressure mobilized due to the excavation. Further, the two most prominent internal failure modes of soil nail walls are nail-soil interface pullout failure (simply, pullout failure) and tensile failure of nail tendon (i.e., reinforcing element). Pullout failure occurs due to insufficient intrinsic bond strength and/or insufficient nail length in the passive zone. Tensile failure of a soil nail takes place when the maximum tensile axial force in the soil nail is greater than the nail tensile capacity. Facing failure modes are not considered in the present study being often attributed to the poor construction practices more than the in situ soil, boundary and loading conditions. A detailed discussion on various failure modes is readily available in the literature [12, 15]. Table 1 shows minimum recommended factors of safety for various external and internal failure modes.

In the present study, design charts are proposed to address the following four prominent failure modes: (a) global stability, (b) sliding stability, (c) pullout failure, and (d) nail tensile failure. Allowable stress design (ASD) methodology is adopted for obtaining expressions of factor of safety for each of the above failure modes of soil nail walls. As shown in Fig. 1, a simplified single wedge failure mechanism is assumed to be inclined at an angle $\psi = 45 + (\phi/2)$ with respect to horizontal, where ϕ is the angle of internal friction of the in situ soil [12, 16].

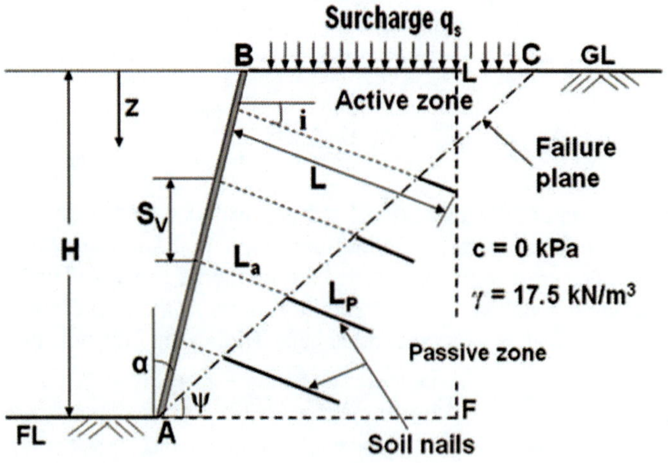

Fig. 1 Schematic reference diagram for design charts

2.1 Global Stability

Referring to Fig. 1, the general expression of factor of safety for global stability FS_G is given by Eq. (1).

$$FS_G = \frac{\sum R}{\sum D} = \frac{cL_F + T_{eq}\cos(\psi - i) + \left[(W + Q_T)\cos\psi + T_{eq}\sin(\psi - i)\right]\tan\phi}{(W + Q_T)\sin\psi} \tag{1}$$

where ΣR and ΣD = summation of resisting and driving forces along the potential failure plane AC having length L_F, respectively; c = in situ cohesion (kPa); $i(degrees)$ = nail inclination with respect to horizontal; W (kN/m) = weight of failure wedge ABC; Q_T (kN/m) = total surcharge load; and T_{eq}(kN/m) $= \frac{1}{S_h}\sum_{j=1}^{n}(T_{all})_j$ = equivalent nail force, where T_{all} is the *minimum* of pullout capacity R_P and tensile capacity R_T given by Eqs. (2) and (3), respectively, of the nail j embedded at depth z out of the total n nails embedded at different depths in a section.

$$(R_P)_z\,(kN) = (\pi D\, L_P\, q_u)/1000 \tag{2}$$

$$(R_T)_z\,(kN) = \left(0.25\pi d^2 f_y\right)/1000 \tag{3}$$

where q_u = ultimate bond strength [kPa]; f_y = yield strength of steel [MPa]; D (m) = drill hole diameter D_{DH} (in case of grouted nails) or tendon diameter d (in case of driven nails); and for nail length L (m), effective bond length L_P (m) is given by Eq. (4).

$$(L_P)_z = L - \left[\frac{(H - z)\cos\psi}{\sin(\psi + i)}\right] \tag{4}$$

2.2 Sliding Stability

Referring to Fig. 1, expression of factor of safety for sliding stability FS_{SL} is given by Eq. (5).

$$FS_{SL} = \frac{\sum R}{\sum D} = \frac{c_b(AF) + \left(W' + Q_T'\right)\tan\phi_b}{P} \tag{5}$$

where ΣR and ΣD = summation of resisting and driving forces along the potential failure plane AF, respectively; for the rigid sliding block $ABLF$: c_b and ϕ_b = cohesion and friction angle along the base (AF), respectively; W' (kN/m) = its weight; Q'_T (kN/m) = q_s x BL = total surcharge load acting on it; and P (kN/m) = total lateral active thrust acting behind it.

$$P = \frac{\gamma H^2}{2} K_a \left[1 + \frac{2q_S}{\gamma H} \right] \tag{6}$$

where K_a = coefficient of Rankine's lateral active earth pressure; q_s (kPa) = distributed surcharge loading, H (m) = vertical height of the soil nail wall; and γ (kN/m^3) = in-situ soil unit weight.

2.3 Pullout Failure Mode

The expression for factor of safety against pullout failure FS_P is given by Eq. (7).

$$(FS_P)_z = \frac{(R_P)_Z}{(T)_Z} \tag{7}$$

where $(R_P)_z$ is as determined by Eq. (2); and the maximum axial force T at depth z can be obtained as:

$$(T)_z(kN) = K_a(q_s + \gamma z)S_h S_v \tag{8}$$

2.4 Nail Tensile Failure Mode

The expression for factor of safety against nail (tendon) tensile failure FS_T is given by Eq. (9).

$$(FS_T)_z = \frac{(R_T)_Z}{(T)_Z} \tag{9}$$

where $(R_T)_z$ and $(T)_z$ are as determined by Eqs. (3) and (8), respectively.

3 General Design Procedure

In general, the design procedure for a soil nail wall includes the following steps:

1. For the specified structure geometry (depth and cut slope inclination), ground profile, and boundary (surcharge) loadings, working nail forces are estimated and the location of the potential failure surface are determined.
2. Selection of the reinforcement type (type, cross-sectional area, length, inclination, and spacing) is done and verification of local stability at each reinforcement level is assured. Further, global stability of the nailed-soil structure is also done.
3. Estimation of the system of forces acting on the facing (i.e., lateral earth pre sure and nail forces at the connection) and hence, design of the facing for specified architectural and durability criteria is to be carried out.
4. For permanent structures, corrosion protection relevant to site conditions is selected. Also, the suitable drainage system for groundwater piezometric levels is adopted.
5. A few of the usual guidelines for preliminary design of soil nail wall are presented in Table 2.

Table 2 Recommended guidelines for soil nail wall preliminary design [12]

Item	Recommended guideline
Nail installation process	Drilled and Grouted/Driven
Nail spacing	Grouted nails: 1.25–2 m; Driven nails: 0.5–1.20 m. Influence area: (Horizontal, S_h x vertical, S_v) ≤ 4 m^2
Nail diameter	Grouted nails: 100–200 mm drill hole diameter for grouted nails with minimum 20 mm reinforcement bar Driven nails: 20–36 mm reinforcement bar
Nail length	Usually 0.6–0.8 times the vertical wall height, H
Nail inclination (w.r.t. horizontal)	10–20° (usually 15°)
Nail pattern at wall face	Square or staggered
Yield strength of nail tendon	≥415 MPa
Unconfined compressive strength of grout / shotcrete	≥20 MPa
Minimum cover to reinforcement for corrosion protection	25 mm (minimum)
Temporary facing thickness	75–100 mm (shotcreted welded wire mesh)
Permanent facing thickness	150–200 mm (cast in situ RCC or precast concrete facing)
Wall face batter (w.r.t. vertical)	0–10°

4 Design Charts for Soil Nail Walls

A series of charts have been prepared as a design aid to provide most frequently required parameters in the design of soil nail walls. It is worth stating that these charts are only applicable for the conditions they are developed for. Details of the range of variables considered for the development of the design charts are as shown in Table 3. Design charts are developed with reference to Fig. 1 and are based on the following assumptions:

a. Single-wedge failure mechanism (see Fig. 1) with failure plane inclined at an angle $\psi = 45 + (\phi/2)$ (in degrees) w.r.t. the horizontal is considered.
b. In situ soil is considered to be dry, cohesionless, and homogenous. It is reported in the literature [15, 17] that the unit of soil has negligible influence on the

Table 3 Details of various design charts and their application

Design chart shown in	Variable(s)	Unit	Variable range (or values)	Output parameter
Figure 2	(a) Vertical height of the wall, *H*	m	0–20	Max. axial force per unit nail influence area t_{max}
	(b) Angle of internal friction, ϕ	degrees	20, 24, 28, 32, 36	
Figure 3a–b	(a) Face batter, α	degrees	0, 10	Nail length in the active zone L_a embedded at depth *z*
	(b) Angle of internal friction,ϕ	degrees	20, 24, 28, 32, 36	
Figure 4a–e	(a) Normalized bond strength, μ	–	0.10, 0.15, 0.25, 0.35, 0.45	Minimum pullout length $L_{P,\,min}$ (i.e., length of nail in the passive zone) required at any embedment depth *z*
	(b) Angle of internal friction,ϕ	degrees	20, 24, 28, 32, 36	
Figure 5a–f	(a) *L*/*H* ratio	–	0.6, 0.7, 0.8	Available pullout length $L_{P,\,avail}$ at any embedment depth *z*
	(b) Face batter,α	degrees	0, 10	
	(c) Angle of internal friction,ϕ	degrees	20, 24, 28, 32, 36	
Figure 6	(a) Characteristic strength of steel, f_y	MPa	250, 415, 500	Nail tensile capacity R_T
	(b) *Nail or reinforcement bar diameter, d*	mm	6–36	

Note Normalized bond strength, $\mu = \frac{q_u D}{FS_P \gamma S_h S_v}$; Maximum axial force per unit nail influence area, $t_{max} = \left(\frac{1-\sin\phi}{1+\sin\phi}\right)\gamma H$; Maximum design axial force, $T_{max} = t_{max} S_h S_v$; Equivalent height for surcharge, $h_{eq} = q_s/\gamma$

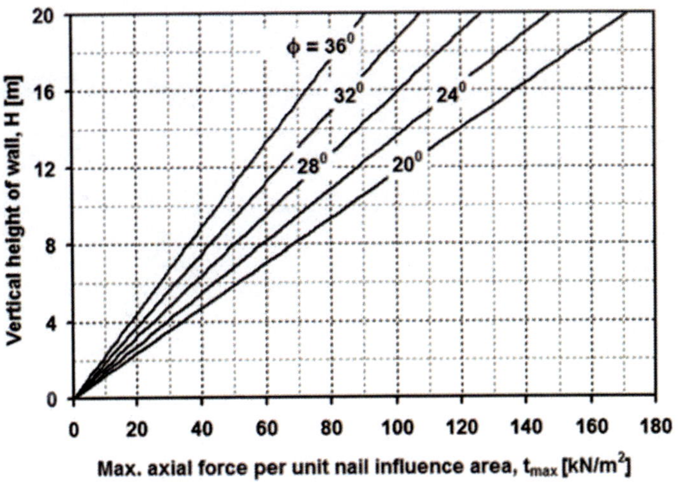

Fig. 2 Determination of maximum axial tensile force per unit nail influence area (t_{max}) with vertical height of the soil nail wall (H)

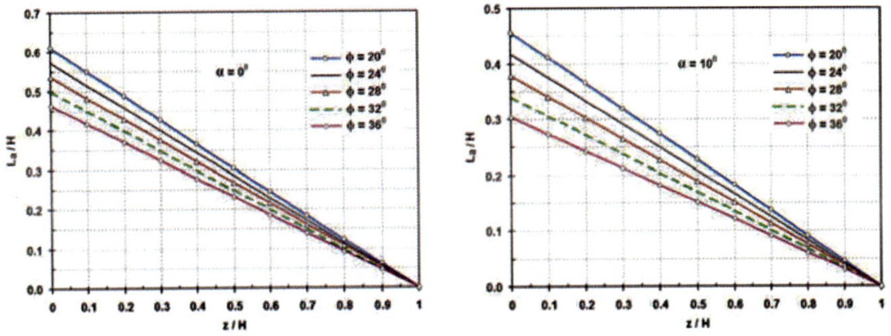

(a) for $\alpha = 0^0$

(b) for $\alpha = 10^0$

Fig. 3 Determination of nail length in active zone (L_a) with depth of embedment of nail (z)

significant failure modes of soil nail walls. Therefore, all charts have been prepared for the fixed value of unit weight of in situ soil $\gamma = 17.5\,\text{kN/m}^3$.

c. A nominal inclination of about 10–15° is usually provided to the grouted soil nails to facilitate grout flow through gravity. Inclination of nails mobilizes the bending and shearing resistance of the nails; however, this has negligible influence on the overall stability [12]. Therefore, charts are prepared assuming a fixed nail inclination $i = 15°$ with respect to the horizontal.

d. Soil nail wall is subjected only to the dead load due to its self-weight, and backslope is considered to be horizontal, i.e., $\beta = 0$.

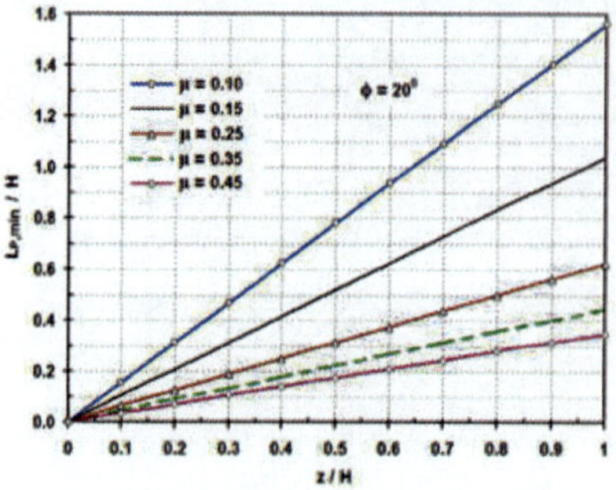

(a) for $\phi = 20^0$

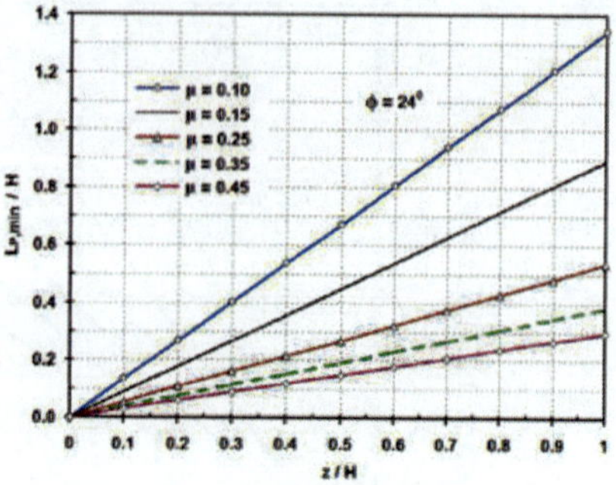

(b) for $\phi = 24^0$

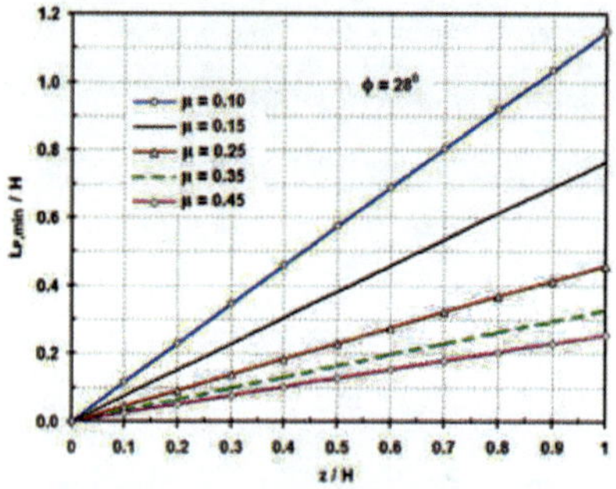

(c) for $\phi = 28^0$

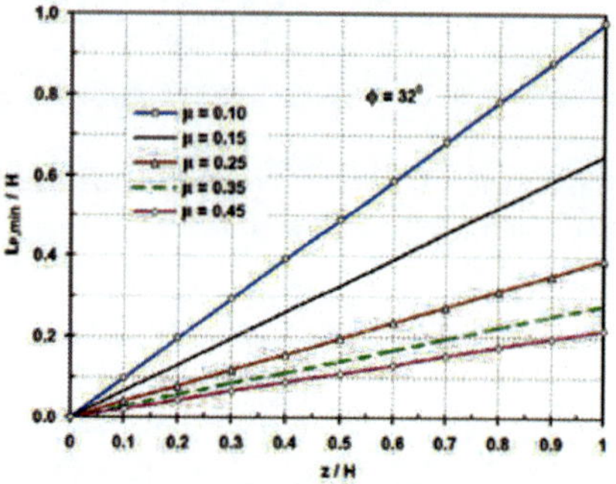

(d) for $\phi = 32^0$

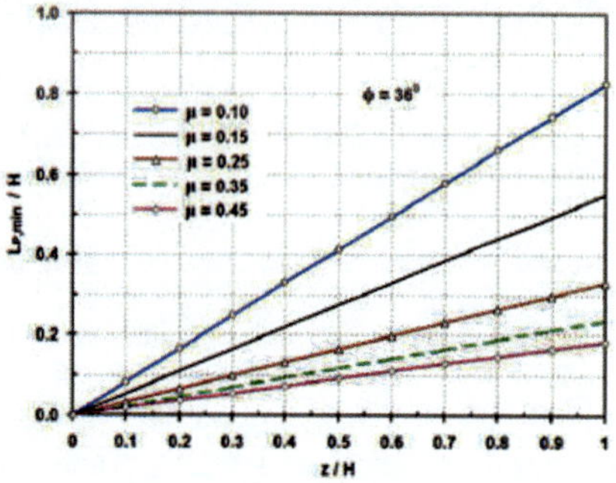

(e) for $\phi = 36^0$

Fig. 4 Determination of minimum pullout length ($L_{P,\,\min}$), i.e., length of nail in passive zone with depth of embedment of nail (z)

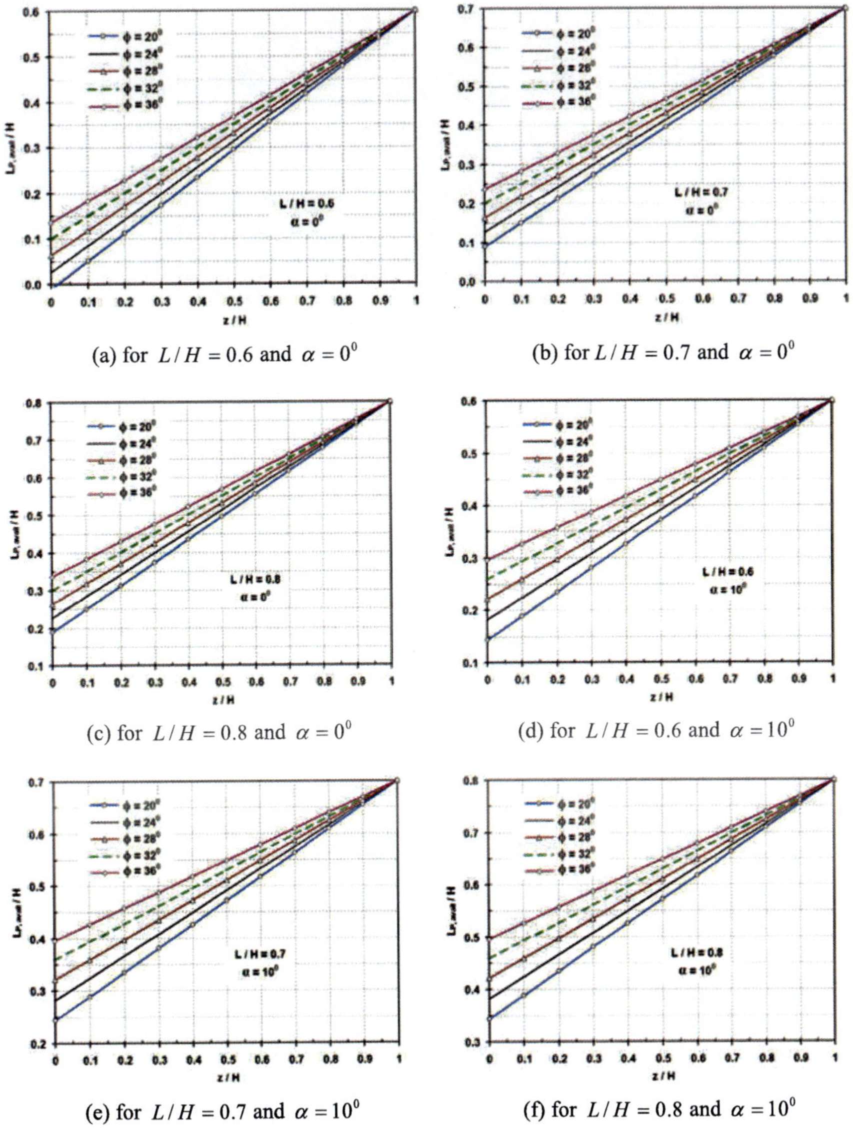

(a) for $L/H = 0.6$ and $\alpha = 0^0$

(b) for $L/H = 0.7$ and $\alpha = 0^0$

(c) for $L/H = 0.8$ and $\alpha = 0^0$

(d) for $L/H = 0.6$ and $\alpha = 10^0$

(e) for $L/H = 0.7$ and $\alpha = 10^0$

(f) for $L/H = 0.8$ and $\alpha = 10^0$

Fig. 5 Determination of available pullout length ($L_{P,\ avail}$), i.e., length of nail in passive zone with depth of embedment of nail (z)

5 Applications of the Proposed Design Charts

The proposed design charts can be suitably used for the following purposes:

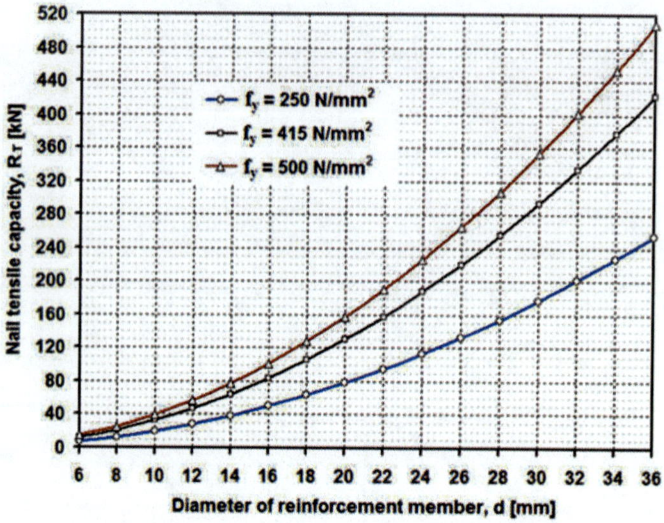

Fig. 6 Determination of nail tensile capacity (R_T) with diameter of tendon in soil nail (d)

a. Preliminary design to obtain preliminary nail length L and maximum design tensile force $T_{\max}$ using charts shown in Figs. 2, 3a–b, and 4a–e, accordingly. Minimum length of soil nail L is adopted as the maximum of L_1 and L_2: $L_1 = L_a + L_{P,\min}$ and $L_2 = 0.6H$. It is to be noted that $L_{P,\min}$ shall be determined with respect to the topmost nail.

b. To obtain available pullout length L_P, avail using charts shown in Figs. 4a–e and 5a–f, which is required for: (i) determination of equivalent nail force T_{eq} for obtaining global stability and (ii) check for pullout failure mode at each nail level.

c. To obtain nail tensile capacity R_T required for checking nail tensile failure mode using charts shown in Figs. 2 and 6.

6 Illustration of Typical Soil Nail Wall Design Using Charts

Using the charts as developed in the earlier sections, a design example for a typical soil nail wall is presented. For the known soil nail wall parameters, including both geometric and soil conditions, the required nail (rebar) diameter and length are obtained. Also, internal failure modes, namely pullout failure and nail tensile failure modes, are evaluated at each nail level.

Known Soil Nail Wall Parameters.

(a) Vertical height of wall: $H = 8$ m.
(b) Face batter: $\alpha = 0.0°$; Backslope angle: $\beta = 0.0°$

(c) Soil nail spacing: $S_h = S_v = 1.5$ m (Note: vertical spacing of first nail $S_{vI} =$ 0.75 m).
(d) Soil nail inclination: $i = 15°$
(e) Drill hole diameter: $D_{DH} = 130$ mm.
(f) Soil nail material: Grade Fe415; $f_y = 415$ MPa.
(g) Representative soil properties from soil investigation report: Soil type: dense to very dense silty sands; Cohesion: $c = 0$ kPa; Friction angle: $\phi = 32°$; Unit weight: $\gamma = 17.5$ kN/m^3; ultimate bond strength: $q_u = 100$ kPa.

Nail (Rebar or Tendon) Diameter d.

From Fig. 2, for $H = 8$ m, $\phi = 32°$, $t_{\max} = 43$ kN/m^2.

Therefore, maximum design axial force, $T_{\max} = t_{\max} S_h S_v = 43 \times 1.5 \times 1.5 = 9$ 6.75 kN.

For a minimum factor of safety of against nail tensile failure $FS_T = 1.80$, the required cross-sectional area A_t of the nail bar can be determined as:

$$A_t[\text{mm}^2] = \frac{T_{\max} FS_T}{f_y} = \frac{96.75 \times 1000 \times 1.80}{415} = 419.63$$

Select reinforcement bar of diameter $d = 25$ mm providing cross-sectional area $A_t = 490$ mm^2 (>419.63 mm^2).

Nail Length L.

Minimum length of soil nail L is adopted as the maximum of L_I and L_2: $L_1 = L_a + L_{P,\min}$ and $L_2 = 0.6\ H$.

Here: $z = S_{vI} = 0.75$ m (i.e., topmost nail).

For $z/H = 0.75/\ 8 = 0.09$, $\alpha = 0.0$; using Fig. 3a, $L_a = 0.45\ H$.

Further, for $FS_P = 2.0$, $D = D_{DH} = 0.13$ m;

$$\mu = \frac{q_u D}{FS_P \gamma S_H S_V} = \frac{100 \times 0.13}{2 \times 17.5 \times 1.5 \times 1.5} = 0.16$$

for $\mu = 0.16$, $z/H = 0.09$; using Fig. 4d, $L_{P,\min} = 0.06\ H$.

Therefore, $L_I = 0.45\ H + 0.06\ H = 0.51\ H$ and $L_2 = 0.6\ H$.

Hence, adopt nail length: $L = 0.6\ H = 0.6 \times 8 = 4.20$ m.

Evaluation of Internal Stability. To evaluate the internal stability, i.e., checking for required factors of safety against nail pullout and tensile failures at each nail level, the charts given in Figs. 5 and 6 can be used to determine pullout capacity and tensile capacity of nails installed at different levels. For the current illustrative example, from the chart given in Fig. 5a, the available length in passive zone, i.e., $L_{P,\text{ avail}}$ is obtained at different levels. Similarly, from the chart given in Fig. 6, nail tensile capacity can be obtained for the known nail diameter and yield strength. Table 4 summarizes the internal stability computations for the illustrative example.

Table 4 Evaluation of the internal stability of illustrative soil nail wall using charts

Nail no	z [m]	z/H ratio	$L_{P,\ avail}/H$ (from Fig. 5a)	$L_{P,\ avail}$ [m]	R_P [kN] (using Eq. 2)	T [kN] (using Eq. 8)	FS_P (using Eq. 7)	R_T [kN] (from Fig. 6)	FS_T (using Eq. 9)
1	0.75	0.1	0.15	1.20	49.01	9.07	5.40	200	22.05
2	2.25	0.3	0.25	2.00	81.68	27.22	3.00	200	7.34
3	3.75	0.5	0.35	2.80	114.35	45.37	2.52	200	4.41
4	5.25	0.7	0.45	3.60	147.03	63.52	2.31	200	3.15
5	6.75	0.8	0.50	4.00	163.36	81.66	2.00	200	2.45

7 Conclusions

Considering the widespread use of soil nailing technique, it is desirable that design aids in the form of charts and tables be readily available for the practicing engineers. As a contribution to this important aspect, this study proposed a set of charts for variety of geometric and in situ soil conditions based on one of the most popular design methodology in practice. From the illustrative design example, it is evident that the simplified charts similar to those proposed in the present study can be easily used for the preliminary design of soil nail walls, and also assist in the evaluation of the prominent internal failure modes, namely pullout failure and tensile failure.

References

1. Briaud, J.L., Lim, Y.: Soil nailed wall under piled bridge abutment: simulation and guidelines. J. Geotech. Geo-Environ. Eng. **123**(11), 1043–1050 (1997)
2. Wong, I.H., Low, B.K., Pang, P.Y., Raju, G.V.R.: Field performance of nailed soil wall in residual soil. J. Perform. Construct. Facilit. **11**(3), 105–112 (1997)
3. Murthy, B.R.S., Babu, G.L.S., Srinivas, A.: Analysis of prototype soil nailed retaining wall. Ground Improv. **6**(3), 129–136 (2002)
4. Turner, J.P., Jensen, W.G.: Landslide stabilization using soil nail and mechanically stabilized earth walls: case study. J. Geotech. Geo-Environ. Eng. **131**(2), 141–150 (2005)
5. Yang, Y.: Remediating a soil-nailed excavation in Wuhan, China. Proc. Inst. Civil Eng. Geotech. Eng. **160**(4), 209–214 (2007)
6. Babu, G.L.S., Rao, R.S., Dasaka, S.M.: Stabilisation of vertical cut supporting a retaining wall using soil nailing: a case study. Ground Improv. **11**(3), 157–162 (2007)
7. FHWA-SA-93–026: French national research project Clouterre—Recommendations Clouterre, English Translation by Federal Highway Administration (FHWA), Washington D.C. (1991)
8. HA68/94: Design methods for the reinforcement of highway slopes by reinforced soil and soil nailing techniques, Highway Agency, United Kingdom (1994)
9. BS8006: Code of practice for strengthened/reinforced soils and other fills, British Standards Institution (BSI), London (1995)
10. pr EN 14490: Execution of special geotechnical works—Soil nailing, European Standard. European Committee for Standardization (CEN), Brussels (2002)
11. FHWA-SA-96–069R: Manual for design and construction monitoring of soil nail walls, Federal Highway Administration (FHWA), Washington D.C. (1998)
12. FHWA0-IF-03–017: Geotechnical engineering circular no. 7 soil nail walls, Federal Highway Administration (FHWA), Washington D.C. (2003)

13. FHWA-NHI-14–007: Geotechnical engineering circular No. 7: Soil nail wall reference manual, Federal Highway Administration (FHWA), Washington D.C. (2015)
14. Geoguide 7: Guide to soil nail design and construction, Civil Engineering and Development Department, The Government of the Hong Kong Special Administrative Region (GHKSAR), Hong Kong (2008)
15. Babu, G.L.S., Singh, V.P.: Reliability analysis of soil nail walls. Georisk: Assess. Manag. Risk Eng. Syst. Geohazards **3**(1), 44–54 (2009)
16. Sheahan, T.C., Ho, C.L.: Simplified trial wedge method for soil nailed walls analysis. J. Geotech. Geo-Environ. Eng. **129**(2), 117–124 (2003)
17. Babu, G.L.S., Singh, V.P.: Reliability based load and resistance factors for soil nailing. Can. Geotech. J. **48**(6), 915–930 (2011)

Bearing Capacity of Thin Ring Footing on Reinforced Foundation Bed Over Soft Ground

K. Rajyalakshmi and M. R. Madhav

1 Introduction

Ring footings supporting chimneys, silos, storage tanks and bridge piers, etc., on soft grounds having high water content, low undrained shear strength and high compressibility pose very interesting stability and deformation problems, for geotechnical engineers. Boushehrian and Hataf [1] performed a study to investigate the bearing capacity of ring footings on reinforced sand by conducting laboratory model tests along with numerical analysis. Sawwaf and Nazir [7] presented an experimental study of the behavior of an eccentrically loaded model ring footing resting on a compacted dense layer of sand that overlies an extended layer of loose sand with emphasis on the potential benefits of reinforcing the replaced sand layer with geogrid reinforcement. Shalaby [8] investigated the use of a combination of stone piles and sand trench instead of using complete sand replacement cushion above weak soft clay soil, experimentally using a model ring footing.

2 Problem Definition

A ring footing of outer diameter d_o and inner diameter d_i, embedded at a shallow depth D below the ground surface in a relatively thin, dense granular fill of thickness H, unit weight γ and angle of shearing resistance ϕ over thick, homogeneous, saturated soft ground with undrained shear strength s_u is shown in Fig. 1b. The plan view of a

K. Rajyalakshmi (✉)
Department of Technical Education, A.P Bheemunipatnam, India
e-mail: dhanista123@gmail.com

M. R. Madhav
Department of Civil Engineering, JNT University & IIT, Hyderabad, India

C. N. V. Satyanarayana Reddy et al. (eds.), *Ground Improvement and Reinforced Soil Structures*, Lecture Notes in Civil Engineering 152,
https://doi.org/10.1007/978-981-16-1831-4_48

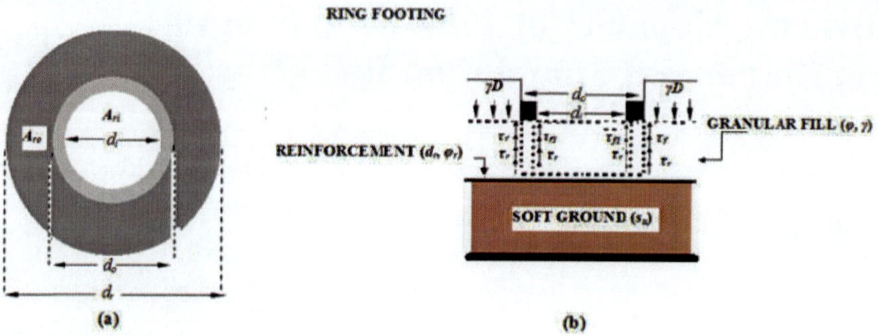

Fig. 1 **a** Plan view of ring footing over circular geosynthetic sheet reinforcement **b** Schematic of ring footing

circular geosynthetic sheet reinforcement of diameter d_r placed below a ring footing of outer diameter d_o and inner diameter $d_i$is shown in Fig. 1a. The reinforcement–fill interface angle of shearing resistance is ϕ_r. The ultimate bearing capacity of the ring footing in reinforced granular fill over soft ground is q_{ur}*. The additional shear stresses τ_f, τ_{fl} and τ_r are those that develop along a vertical cylindrical surface passing through the outer and inner edges of the footing due to the shear layer effect of granular fill [9] and axial tension [5, 6], respectively.

Meyerhof [3] proposed a punching mode of failure for a circular footing ($d_i = 0$) of diameter $d_o = B$ embedded at depth D in a relatively thin, dense sand layer of thickness H with angle of shearing resistance ϕ and unit weight γ, overlying thick soft clay with undrained cohesion s_u, by considering the failure as an inverted uplift problem.

2.1 *Formulation*

According to Meyerhof [3], the ultimate bearing capacity q_u of a circular footing in a thin, dense sand layer over soft clay is

$$q_u = 1.2cN_c + \frac{2\gamma H^2}{B}\left[1 + \frac{2D}{H}\right]sK_s tan\varphi + \gamma D \tag{1}$$

which is limited by the ultimate bearing capacity q_t of the footing in a thick deposit of sand ($c = 0$) as

$$q_t = \gamma D N_q + 0.3\gamma B N_\gamma \tag{2}$$

where s is a shape factor governing the passive earth pressure on a cylindrical wall (taken as unity for relatively small H/B ratios); K_s is a coefficient of punching shear resistance; $N_c(= 2 + \pi)$, N_q and N_γ are Meyerhof's bearing capacity factors.

Substituting $B = (d_o - d_i)$ for a ring/annular footing and normalizing by s_u, Eq. 2 becomes

$$N_{limiting} = \left[\frac{\gamma d_0}{s_u}\right]\left[\frac{D}{d_0}\right] N_q + 0.3\left[\frac{\gamma d_0}{s_u}\right]\left[1 - \frac{d_i}{d_0}\right] N_\gamma \tag{3}$$

Keshavarz and Kumar [2] used the method of characteristics and the finite difference technique to estimate the ultimate bearing capacity of smooth and rough ring footings in c–ϕ soil considering stress singularities at the inner as well as outer edges of the footing as

$$q_u = cN_c + qN_q + 0.5\gamma(d_0 - d_i)\left[1 - 0.5\frac{d_i}{d_0}\right] N_\gamma \tag{4}$$

where $q = \gamma D$, the overburden pressure at the base of the footing, N_c, N_q and N_γ are bearing capacity factors, which depend on the angle of shearing resistance ϕ of soil and the ratio r_i/r_o of the ring footing, where r_i and r_o are the internal and external radii of the footing, respectively.

3 Bearing Capacity of Ring Footing in an Unreinforced Granular Fill Over Soft Ground

3.1 Ring Footing in an Unreinforced Granular Fill Over Soft Ground Considering Only Shear Stresses Along the Outer Periphery

The ultimate bearing capacity q_u of a ring footing in an unreinforced granular fill over soft ground is obtained by extending Meyerhof's two-layered theory for a solid circular footing given in Eq. (1), by substituting $B = d_o$ for the shear stresses acting along only the outer periphery, of the ring footing, as

$$q_{u/outer} = s_u N_c + \left\{\frac{2\gamma H^2}{(do)}\left[1 + \frac{2D}{H}\right] sK_s tan\varphi\right\} + \gamma D \tag{5}$$

where N_c is the bearing capacity factor obtained by Keshavarz and Kumar [2]. No shape factor is required for the first term because the bearing capacity factor N_c corresponds directly for a ring footing itself and not for a footing of any other shape. K_s is the coefficient of punching shear resistance, which is a function of the angle of shearing resistance ϕ of granular fill and the ratio q_2/q_1[4] where q_1 and q_2 are the ultimate bearing capacities of a ring footing on the surfaces ($D = 0$) of homogeneous thick beds of granular fill ($c = 0$) and soft ground, respectively, defined as

$$\frac{q_2}{q_1} = \frac{s_u N_c}{0.5\gamma(d_0 - d_i)\left[1 - 0.5\frac{d_i}{d_0}\right]N_\gamma} \quad (6)$$

where N_γ can be obtained from values given by Keshavarz and Kumar [2] for different radii ratios, r_i/r_o. Since the shear strength of granular fill is greater than that of soft ground, the ratio q_2/q_1 ranges between 0 and 1.

Meyerhof [3] represented the thickness of the granular bed below the footing as H. However, since the entire thickness of the granular fill is considered as H (Fig. 1), the thickness of the granular bed below the footing is now equal to $H - D$ and Eq. (5) gets reduced as below

$$q_{u/outer} = s_u N_c + \left\{\frac{2\gamma(H-D)^2}{(do)}\left[1 + \frac{2D}{(H-D)}\right]sK_s tan\varphi\right\} + \gamma D \quad (7)$$

$$q_{u/outer} = s_u N_c + \left\{2\gamma(H-D)^2\frac{1}{(do)}K_s tan\varphi\right\} + \gamma D \quad (8)$$

Normalizing Eq. (8) with the undrained shear strength s_u of soft ground, the normalized ultimate bearing capacity N_u of a ring footing in an unreinforced two-layered system of granular fill over soft ground, considering the shearing resistances acting along vertical cylindrical surfaces passing through the outer edge of the footing is

$$N_{u,o} = N_c + \left\{2\left(\frac{\gamma d_0}{s_u}\right)\left[\left(\frac{H}{d_0}\right)^2 - \left(\frac{D}{d_0}\right)^2\right]K_s tan\varnothing\right\} + \left\{\left(\frac{\gamma d_0}{s_u}\right)\left(\frac{D}{d_0}\right)\right\} \quad (9)$$

3.2 *Ring Footing in/an Unreinforced Granular Fill Over Soft Ground Considering Shear Stresses Along Both Outer and Inner Peripheries*

The ultimate bearing capacity q_u of a ring footing, on unreinforced granular fill over soft ground is obtained by extending Meyerhof's two-layered theory for a solid circular footing given in Eq. (1), by substituting $B = d_o$ and $B = d_i$ for the shear stresses acting along both outer and inner peripheries, respectively, of the ring footing, as

$$q_{u/outer+inner} = s_u N_c + \left\{ \frac{2\gamma H^2}{(do)} \left[1 + \frac{2D}{H} \right] sK_s tan\varphi \right\} + \left\{ \frac{2\gamma H^2}{(d_i)} \left[1 + \frac{2D}{H} \right] sK_s tan\varphi \right\} + \gamma D \tag{10}$$

$$q_{u/outer+inner} = s_u N_c + \left\{ 2\gamma H^2 \left[1 + \frac{2D}{H} \right] sK_s tan\varphi \left(\frac{1}{d_0} \right) \left[1 + \left(\frac{d_i}{d_o} \right)^{-1} \right] \right\} + \gamma D \tag{11}$$

Meyerhof [3] represented the thickness of the granular bed below the footing as H. However, since the entire thickness of the granular fill is considered as H (Fig. 1), the thickness of the granular bed below the footing is now equal to $H - D$ and Eq. (11) gets reduced as below

$$q_{u/outer+inner} = s_u N_c + \left\{ 2\gamma (H - D)^2 \left[1 + \frac{2D}{(H - D)} \right] K_s tan\varphi \left(\frac{1}{d_0} \right) \left[1 + \left(\frac{d_i}{d_o} \right)^{-1} \right] \right\} + \gamma D \tag{12}$$

or

$$q_u/_{outer+inner} = s_u N_c + \left\{ 2\gamma (H^2 - D^2) \left(\frac{1}{d_0} \right) \left[1 + \left(\frac{d_i}{d_o} \right)^{-1} \right] K_s tan\varphi \right\} + \gamma D \tag{13}$$

Normalizing Eq. (13) with the undrained shear strength s_u of soft ground, the normalized ultimate bearing capacity N_u of a ring footing in an unreinforced two layered system of granular fill over soft ground, considering the shearing resistances acting along vertical cylindrical surfaces passing through the outer and inner edges of the footing is

$$N_{u,oi} = N_c + \left\{ 2 \left(\frac{\gamma d_0}{s_u} \right) \left[1 + \left(\frac{d_i}{d_o} \right)^{-1} \right] \left[\left(\frac{H}{d_0} \right)^2 - \left(\frac{D}{d_0} \right)^2 \right] K_s tan\varnothing \right\} + \left\{ \left(\frac{\gamma d_0}{s_u} \right) \left(\frac{D}{d_0} \right) \right\} \tag{14}$$

4 Bond Resistance of Geosynthetic Reinforcement

As the ring footing moves down into the granular fill, it tries to displace the fill particles radially over the effective outer area A_{ro} (dark shaded area) and A_{ri} (inner area of the ring footing) in Fig. 1a of the reinforcement. However, the radial motion

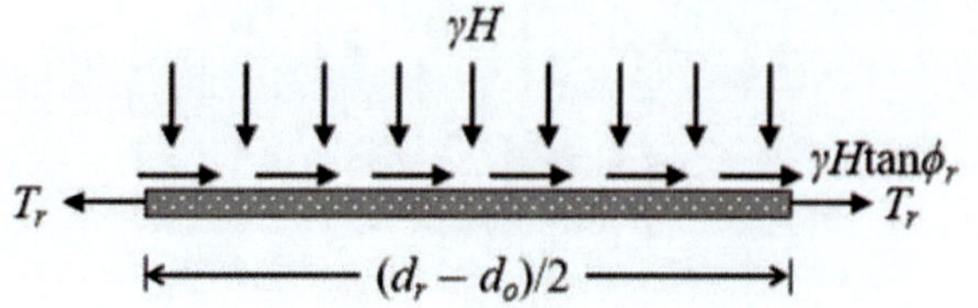

Fig. 2 Free body diagram of reinforcement segment beyond outer edge of ring footing

of the fill particles is restrained by frictional or shear stresses $\gamma H\tan\phi_r$ mobilized along the interface between the reinforcement and the fill (Fig. 1b).

Subsequently, axial tensile stresses and strains are induced in the reinforcement (Fig. 2). Thus, the reinforcement provides additional lateral confinement to the fill, which leads to additional shearing resistance along the vertical cylindrical surface passing through the outer and inner edges of the ring footing—this in turn enhances the ultimate bearing capacity of the footing.

The axial tension T_r that develops in the reinforcement due to radial shear stresses mobilized along the interface between the reinforcement and the fill for a circular footing is

$$T_r = \int_0^{2\pi}\int_{B/2}^{L_r/2} 2\sigma_z tan\varphi_r r dr d\theta$$

$$= \int_{B/2}^{L_r/2} 2\pi 2\sigma_z tan\varphi_r r dr$$

$$= 2\pi 2\sigma_z tan\varphi_r \frac{\left(\frac{L_r}{2}\right)^2 - \left(\frac{B}{2}\right)^2}{2} \& = 2\pi\sigma_z tan\varphi_r\left[\frac{L_r{}^2 - B^2}{4}\right]$$

$$= \frac{1}{4}\pi B^2\left(2\sigma_z tan\varphi_r\left[\left(\frac{L_r}{B}\right)^2 - 1\right]\right) \tag{15}$$

$$T_r = \frac{1}{4}\pi B^2\left(2\gamma H tan\varphi_r\left[\left(\frac{L_r}{B}\right)^2 - 1\right]\right) \tag{16}$$

4.1 *Bond Resistance of Geosynthetic Reinforcement Considering the Outer Area*

For a ring footing of outer diameter d_o, inner diameter d_i and reinforcement diameter d_r, the axial resistance mobilized in the reinforcement, considering the outer area of the geosynthetic, is obtained by substituting $L_r = d_r$ and $B = d_o$, in Eq. (16), as

$$T_{r,outer} = \frac{1}{4}\pi d_o^2\left(2\gamma H tan\varphi_r\left[\left(\frac{d_r}{d_o}\right)^2 - 1\right]\right) \tag{17}$$

where σ'_v (at $z = H$) is the vertical effective stress/normal stress acting on the reinforcement, $\mu = \tan\phi_r$ is the coefficient of friction between the reinforcement and the fill, $\phi_r$is the reinforcement–fill interface angle of shearing resistance.

The contribution from the interface/bond resistance q_r of the reinforcement toward the ultimate bearing capacity of the footing considering the outer area, A_{eo} of the reinforcement is

$$q_{r,ax/outer} = \frac{T_{r,outer}}{\frac{\pi}{4}\left(d_0^2 - d_i^2\right)} = \frac{\left[\left(\frac{1}{4}\pi d_o^2\left(2\gamma H tan\varphi_r\left[\left(\frac{d_r}{d_o}\right)^2 - 1\right]\right)\right)\right]}{\frac{\pi}{4}\left(d_0^2 - d_i^2\right)} \tag{18}$$

$$q_{r,ax/outer} = 2\gamma H tan\varphi_r\left[\frac{\left(\frac{d_r}{d_0}\right)^2 - 1}{1 - \left(\frac{d_i}{d_0}\right)^2}\right] \tag{19}$$

Normalizing $q_{r,ax/outer}$ with the undrained shear strength s_u of soft ground, $N_{r,ax/outer}$ is obtained as

$$N_{r,ax/outer} = 2\left[\frac{\gamma d_0}{s_u}\right]\left[\frac{H}{d_0}\right]tan\varphi_r\left[\frac{\left(\frac{d_r}{d_0}\right)^2 - 1}{1 - \left(\frac{d_i}{d_0}\right)^2}\right] \tag{20}$$

4.2 Bond Resistance of Geosynthetic Reinforcement Considering Both the Outer and Inner Areas

The resistance mobilized in the reinforcement, considering the effective inner circular area of diameter d_i of the geosynthetic, is obtained as

$$T_{r,inner} = \frac{1}{4}\pi d_i^2(2\gamma H tan\varphi_r) \tag{21}$$

The total axial resistance mobilized in the reinforcement, considering the outer and inner areas of the reinforcement is obtained by adding Eqs. 17 and 21 as

$$\begin{aligned} T_r &= T_{r,outer} + T_{r,inner} \\ &= \left(\frac{1}{4}\pi d_o^2\left(2\gamma H tan\varphi_r\left[\left(\frac{d_r}{d_o}\right)^2 - 1\right]\right)\right) + \left(\frac{1}{4}\pi d_i^2(2\gamma H tan\varphi_r)\right) \end{aligned} \tag{22}$$

The contribution from the interface/bond resistance q_r of the reinforcement toward the ultimate bearing capacity of the footing is

$$q_{r,ax(outer+inner)} = \frac{T_r}{\frac{\pi}{4}\left(d_0^2 - d_i^2\right)}$$
$$= \frac{\left[\left(\frac{1}{4}\pi d_o^2\left(2\gamma H tan\varphi_r\left[\left(\frac{d_r}{d_o}\right)^2 - 1\right]\right)\right) + \left(\frac{1}{4}\pi d_i^2(2\gamma H tan\varphi_r)\right)\right]}{\frac{\pi}{4}\left(d_0^2 - d_i^2\right)} \tag{23}$$

$$q_{r,ax(outer+inner)} = 2\gamma H tan\varphi_r\left[\frac{\left(\frac{d_r}{d_0}\right)^2 - 1 + \left(\frac{d_i}{d_o}\right)^2}{1 - \left(\frac{d_i}{d_0}\right)^2}\right] \tag{24}$$

Normalizing $q_{r,ax(outer+inner)}$ with the undrained shear strength s_u of soft ground, $N_{r,ax(outer+inner)}$ is obtained as

$$N_{r,ax(outer+inner)} = 2\left[\frac{\gamma d_0}{s_u}\right]\left[\frac{H}{d_0}\right]tan\varphi_r\left[\frac{\left(\frac{d_r}{d_0}\right)^2 - 1 + \left(\frac{d_i}{d_o}\right)^2}{1 - \left(\frac{d_i}{d_0}\right)^2}\right] \tag{25}$$

5 Bearing Capacity of Ring Footing in Reinforced Granular Fill Over Soft Ground

5.1 *Ring Footing in Reinforced Granular Fill Over Soft Ground Considering Response of Reinforcement Over Outer Area*

The ultimate bearing capacity $q_{ur(outer)}{}^*$ of a ring footing in reinforced granular fill over soft ground (Fig. 1) is obtained by adding the contribution of the bond resistance $q_{r,ax/outer}$ of the reinforcement (considering axial response of the outer area of reinforcement to pullout) in Eq. 19 to the ultimate bearing capacity $q_{u/outer}$ of a ring footing, on unreinforced granular fill over soft ground (considering the shear stresses on the outer periphery of the footing) in Eq. 8 as.

$$q_{ur(outer)}{}^* = q_{u/outer} + q_{r,ax/outer}$$
$$q^*_{ur(outer)} = s_u N_c + \left\{2\gamma\left(H^2 - D^2\right)\left(\frac{1}{d_0}\right)K_s tan\varphi\right\}$$

$$+\gamma D+\left\{2\gamma Htan\varphi_r\left[\frac{\left(\left(\frac{d_r}{d_0}\right)^2-1\right)}{\left(1-\left(\frac{d_i}{d_0}\right)^2\right)}\right]\right\} \tag{26}$$

The normalized ultimate bearing capacity of a ring footing in reinforced granular fill over soft ground considering axial resistance of reinforcement in the outer area to pullout is obtained by adding Eqs. 9 and 20 as.

$$N_{ur,o}{}^* = N_{u/outer} + N_{r,ax/outer}$$

$$N_{ur,o}{}^* = N_c + \left\{2\left(\frac{\gamma d_0}{s_u}\right)\left[\left(\frac{H}{d_0}\right)^2 - \left(\frac{D}{d_0}\right)^2\right]K_s tan\varnothing\right\} + \left\{\left(\frac{\gamma d_0}{s_u}\right)\left(\frac{D}{d_0}\right)\right\}$$

$$+\left\{2\left[\frac{\gamma d_0}{s_u}\right]\left[\frac{H}{d_0}\right]tan\varphi_r\left[\frac{\left(\frac{d_r}{d_0}\right)^2-1}{1-\left(\frac{d_i}{d_0}\right)^2}\right]\right\} \tag{27}$$

5.2 *Ring Footing in Reinforced Granular Fill Over Soft Ground Considering Response of Reinforcement Over Both Outer and Inner Areas*

The ultimate bearing capacity $q_{ur(outer+inner)}{}^*$ of a ring footing in reinforced granular fill over soft ground (Fig. 1) is obtained by adding the contribution of the bond resistance $q_{r,ax(outer+inner)}$ of the reinforcement (considering axial response of reinforcement to pullout along the outer and inner areas of reinforcement to pullout) to the ultimate bearing capacity $q_{u(outer+inner)}$ of a ring footing, on unreinforced granular fill over soft ground by adding Eqs. 13 and 24, as.

$$q_{ur(outer+inner)}{}^* = q_{u(outer+inner)} + q_{r,ax(outer+inner)}$$

$$q^*_{ur(outer+inner)} = s_u N_c + \left\{2\gamma\left(H^2 - D^2\right)\left(\frac{1}{d_0}\right)\left[1+\left(\frac{d_i}{d_o}\right)^{-1}\right]K_s tan\varphi\right\}$$

$$+\gamma D+\left\{2\gamma Htan\varphi_r\left[\frac{\left(\frac{d_r}{d_0}\right)^2-1+\left(\frac{d_i}{d_o}\right)^2}{1-\left(\frac{d_i}{d_0}\right)^2}\right]\right\} \tag{28}$$

The normalized ultimate bearing capacity of a ring footing in reinforced granular fill over soft ground considering axial resistance of reinforcement to pullout is obtained by adding Eqs. 14 and 25 as.

$$N_{ur,oi}^{*} = N_{u(outer+inner)} + N_{r,ax(outer+inner)}$$

$$N_{ur,oi}^{*} = N_c + \left\{2\left(\frac{\gamma d_0}{s_u}\right)\left[1+\left(\frac{d_i}{d_o}\right)\right]^{-1}\left[\left(\frac{H}{d_0}\right)^2-\left(\frac{D}{d_0}\right)^2\right]K_s tan\varnothing\right\} + \left\{\left(\frac{\gamma d_0}{s_u}\right)\left(\frac{D}{d_0}\right)\right\} + \left\{2\left[\frac{\gamma d_0}{s_u}\right]\left[\frac{H}{d_0}\right]tan\varphi_r\left[\frac{\left(\frac{d_r}{d_0}\right)^2-1+\left(\frac{d_i}{d_o}\right)^2}{1-\left(\frac{d_i}{d_0}\right)^2}\right]\right\} \tag{29}$$

Bearing capacity ratios (BCRs) are defined to quantify the degrees of improvement of the bearing capacity of ring footing as.

$(BCR)_{f,o} = N_{u,o}/N_c$; $(BCR)_{f,oi} = N_{u,oi}/N_c$; $(BCR)_{fr,o}^{*} = N_{ur,o}^{*}/N_c$; $(BCR)_{fr,oi}^{*} = N_{ur,oi}^{*}/N_c$; The ratios $(BCR)_{f,o}$ and $(BCR)_{fr,o}^{*}$ quantify the contribution of granular fill and that of the reinforced granular fill (considering the shear resistance along the outer periphery of the ring footing and the outer area of reinforcement), while $(BCR)_{f,oi}(BCR)_{fr,oi}^{*}$ quantify the contribution of granular fill and that of the reinforced granular fill (considering shear resistances on both the outer and inner peripheries of the ring footing and the response of reinforcement on the outer and inner areas of reinforcement).

6 Results and Discussion

Figure 3 presents the variations of the normalized ultimate bearing capacity of a ring footing in an unreinforced two-layered system of granular fill over soft ground, $N_{u,o}$ (considering the shearing resistances acting along vertical cylindrical surfaces passing through the outer edge of the ring footing), and $N_{u,oi}$ (considering the shearing resistances acting along vertical cylindrical surfaces passing through the outer and inner edges of the footing) while Fig. 4 presents the variations of the normalized ultimate bearing capacity of a ring footing in a reinforced two-layered system of granular fill over soft ground, $N_{ur,o}^{*}$ (considering the shearing resistances acting along vertical cylindrical surfaces passing through the outer edge of the footing), and $N_{ur,oi}^{*}$ (considering the shearing resistances acting along vertical cylindrical surfaces passing through the outer and inner edges of the footing) with the footing diameter ratio, d_i/d_0, for $\gamma B/s_u$ equal to 5, 15 and 25, for φ of 30°, H/d_0 of 0.15, D/d_0 of 0.1, φ_r/φ of 0.75 and d_r/d_0 of 3. The values of $N_{u,o}$, $N_{u,oi}$, $N_{ur,o}$ and $N_{ur,oi}$ of thin ring footings (for d_i/d_o varying from 0.5 to 0.9) for different values of $\gamma d_0/s_u$ are tabulated in Table 1.

Improved normalized bearing capacity values are obtained by considering shear resistance along both the outer and inner peripheries of ring footings in a two-layered system of unreinforced and reinforced granular fill over soft ground, when compared to those obtained by considering the shear resistance along the outer periphery alone, due to the increased length of vertical cylindrical surface considered and thereby

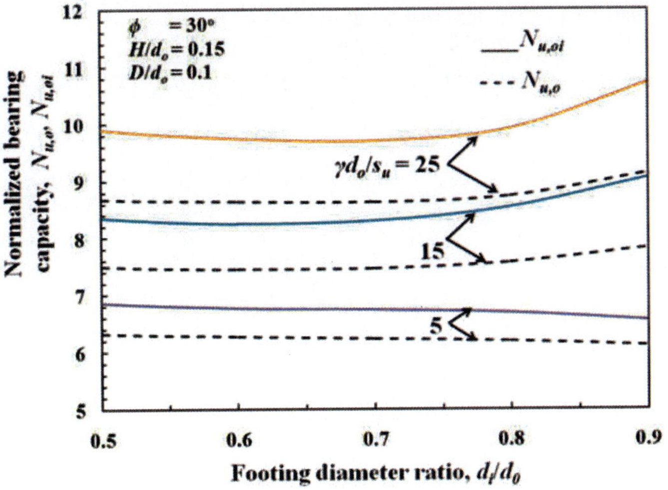

Fig. 3 Variation of normalized bearing capacities $N_{u,o}$ and $N_{u,oi}$ with footing diameter ratio (d_i/d_o): Effect of $\gamma d_0/s_u$

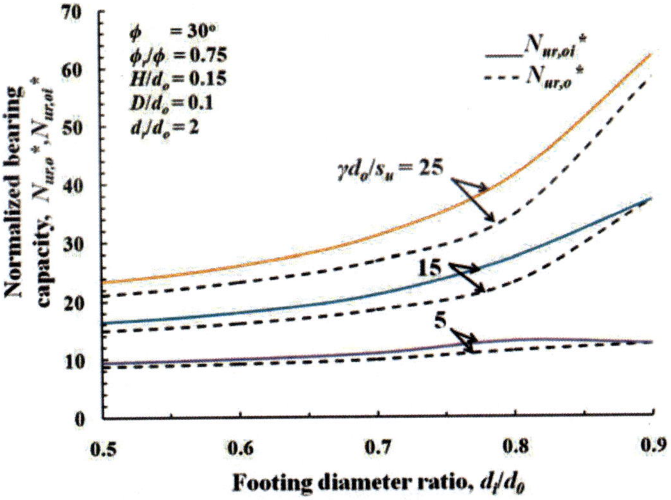

Fig. 4 Variation of normalized bearing capacities $N_{ur,o}$* and $N_{ur,oi}$* with footing diameter ratio (d_i/d_o)—Effect of $\gamma d_0/s_u$

increased contribution of shear resistance. Enhanced normalized bearing capacity values are obtained for a ring footing in a two-layered system of reinforced granular fill over soft ground, when compared to those in an unreinforced two-layered system. Thinner ring footings in/on a two-layered system of reinforced granular fill over soft ground show improved normalized bearing capacity values, due to the increased

Table 1 Variation of normalized bearing capacities, $N_{u,o}$, $N_{u,oi}$, $N_{ur,o}*$ and $N_{ur,oi}*$ with footing diameter ratio (d_i/d_o), for varying $\gamma d_0/s_u$

$\frac{d_i}{d_0}$	0.5	0.6	0.7	0.8	0.9	0.5	0.6	0.7	0.8	0.9
$\frac{\gamma d_0}{s_u}$	$N_{u,o}$					$N_{u,oi}$				
5	6.33	6.27	6.24	6.20	6.12	6.86	6.77	6.74	6.69	6.56
15	7.50	7.46	7.47	7.58	7.83	8.36	8.26	8.31	8.55	9.06
25	8.68	8.64	8.63	8.74	9.14	9.90	9.74	9.70	9.91	10.72
	$N_{ur,o}*$					$N_{ur,oi}*$				
5	8.82	9.19	9.89	11.37	12.39	9.55	10.03	11.00	12.98	12.39
15	14.96	16.20	18.44	23.11	37.17	16.44	18.05	21.06	27.40	37.17
25	21.11	23.20	26.91	34.63	58.19	23.37	26.05	30.96	41.32	61.94

contribution from the granular fill and reinforcement. Softer clays or relatively wider footings with higher values of $\gamma d_0/s_u$ show improved bearing responses, due to inclusion of granular fill and reinforcement.

Figure 5 presents the variations of the bearing capacity ratios of a ring footing in an unreinforced two-layered system of granular fill over soft ground, $(BCR)_{f,o}$ (considering the shearing resistances acting along vertical cylindrical surfaces passing through the outer edge of the footing), and $(BCR)_{f,oi}$ (considering the shearing resistances acting along vertical cylindrical surfaces passing through the outer and inner edges of the footing) with the footing diameter ratio, d_i/d_0,, for $\gamma B/s_u$equal to 5, 15 and 25, for φ of 30°, H/d_0 of 0.15 and D/d_0 of 0.1.

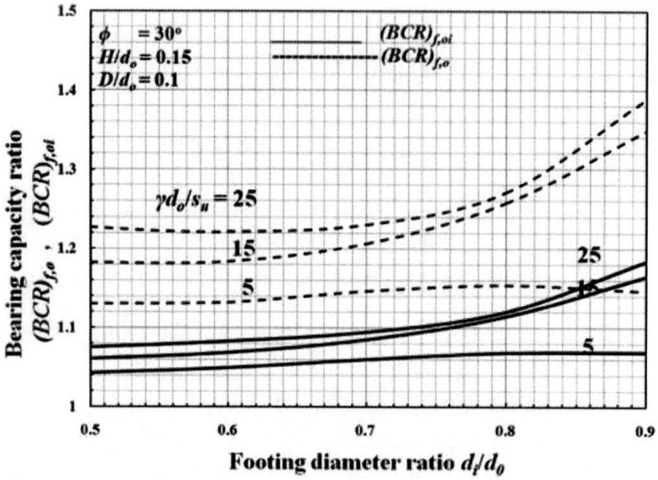

Fig. 5 Variation of bearing capacity ratios $(BCR)_{f,o}$ and $(BCR)_{f,oi}$ with footing diameter ratio (d_i/d_o): Effect of $\gamma d_0/s_u$

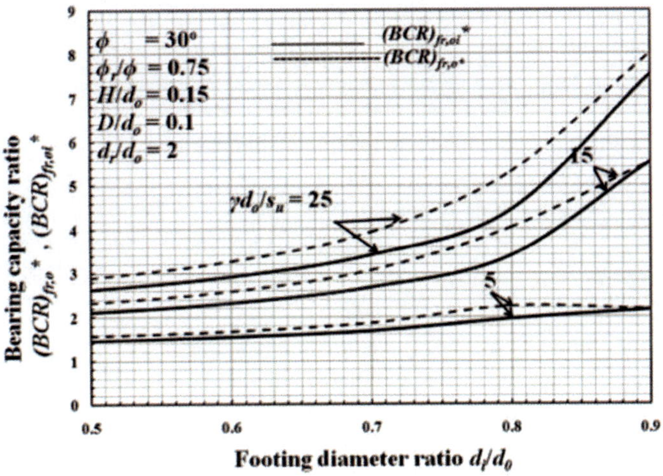

Fig. 6 Variation of normalized bearing capacities $(BCR)_{fr,o}$* and $(BCR)_{fr,oi}$*with footing diameter ratio (d_i/d_o): Effect of $\gamma d_0/s_u$

Figure 6 presents the variations of the bearing capacity ratios of a ring footing in a reinforced two-layered system of granular fill over soft ground, $(BCR)_{fr,o}$* (considering shear resistance on the outer periphery of the ring footing and the response of reinforcement on the outer area of reinforcement), and $(BCR)_{fr,oi}$* (considering shear resistances on both the outer and inner peripheries of the ring footing and the response of reinforcement on the outer and inner areas of reinforcement) with the footing diameter ratio, d_i/d_0, for $\gamma B/s_u$equal to 5, 15 and 25, for φ of 30°, H/d_0 of 0.15, D/d_0 of 0.1, φ_r/φ of 0.75 and d_r/d_0 of 3.

Improved normalized bearing capacity ratio values are obtained by considering shear resistance along both the outer and inner peripheries of ring footings in a two-layered system of unreinforced and reinforced granular fill over soft ground, when compared to those obtained by considering the shear resistance along the outer periphery alone, due to the increased contribution of shear resistance. Enhanced normalized bearing capacity ratio values are obtained for a ring footing in/on a two-layered system of reinforced granular fill over soft ground, when compared to those in an unreinforced two-layered system. Thinner ring footings in a two-layered system of reinforced granular fill over soft ground show improved normalized bearing capacity ratio values, due to the increased contribution from the granular fill and reinforcement. Softer clays or relatively wider footings with higher values of $\gamma d_0/s_u$show improved bearing responses. The bearing capacity ratios $(BCR)_{f,o}$, $(BCR)_{f,oi}$, $(BCR)_{fr,o}$* and $(BCR)_{fr,oi}$* of thin ring footings (d_i/d_o varying from 0.5 to 0.9) for different values of $\gamma d_0/s_u$and $d_i/d_o$are tabulated in Table 2.

Figure 7 shows the analytical results from the present study, obtained for ultimate bearing capacity of a ring footing on the surface ($z = 0$) of an unreinforced and reinforced (considering axial response of reinforcement to pull out) granular fill over

Table 2 Variation of normalized bearing capacity ratios $(BCR)_{f,o}$, $(BCR)_{f,oi}$, $(BCR)_{fr,o}$* and $(BCR)_{fr,oi}$* with footing diameter ratio (d_i/d_o), for varying $\gamma d_0/s_u$

$\frac{d_i}{d_0}$	0.5	0.6	0.7	0.8	0.9	0.5	0.6	0.7	0.8	0.9
$\frac{\gamma d_0}{s_u}$	$(BCR)_{f,o}$					$(BCR)_{f,oi}$				
5	1.04	1.05	1.06	1.07	1.07	1.13	1.13	1.15	1.15	1.15
15	1.06	1.07	1.09	1.11	1.17	1.18	1.18	1.21	1.26	1.35
25	1.08	1.08	1.09	1.12	1.18	1.23	1.22	1.23	1.27	1.39
	$(BCR)_{fr,o}$*					$(BCR)_{fr,oi}$*				
5	1.45	1.54	1.68	1.96	2.17	1.57	1.68	1.87	2.24	2.17
15	2.12	2.32	2.68	3.40	5.53	2.33	2.59	3.06	4.03	5.53
25	2.62	2.91	3.41	4.44	7.54	2.90	3.27	3.93	5.30	8.02

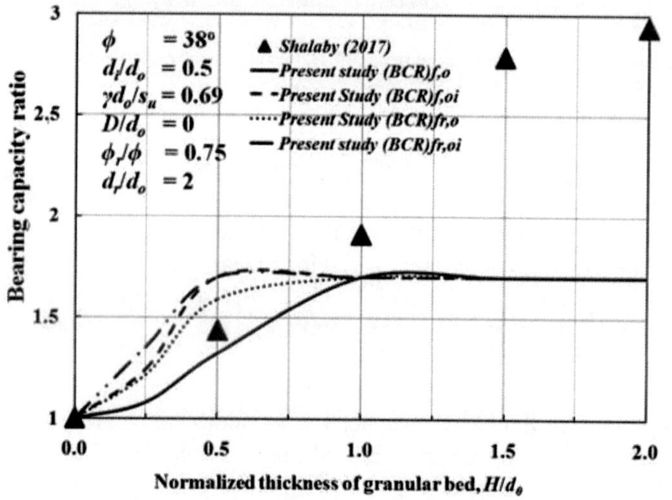

Fig. 7 Variation of bearing capacity ratio (BCR) with normalized thickness of granular bed (H/d_0)

soft ground, $(BCR)_{f,o}$, $(BCR)_{f,oi}$, $(BCR)_{fr,o}$* and $(BCR)_{fr,oi}$*, in comparison with the experimental results of Shalaby [8] for soft clay partly replaced with sand and without stone piles for $(B/b) = 2$, and the results are tabulated in Table 3. Figure 7 shows that the results from the present study compare well for relatively smaller thicknesses ($H/d_0 = 0$ to 1.0) of granular layer, but for relatively thick granular fill (with $H/d_0 >$ 1.0) "conservative" values of *BCR* are obtained, when compared to those obtained by Shalaby [8]. Possible reasons for the conservative values obtained for relatively thicker granular fills could be the surcharge effect which has been neglected in this study and/or the application of Meyerhof's limiting criterion.

Table 3 Variation of (BCR) with (H/d_o)

$\frac{H}{d_0}$	Reinforced/unreinforced	0.0	0.5	1.0	1.5	2.0
		(BCR) values from				
Shalaby [8] $(BCR)_f$	Unreinforced	1.0	1.43	1.91	2.79	2.94
Present study $(BCR)_{f,o}$	Unreinforced	1.0	1.32	1.70	1.70	1.70
Present study $(BCR)_{f,oi}$	Unreinforced	1.0	1.70	1.70	1.70	1.70
Present study$(BCR)_{fr,o}$*	Reinforced	1.0	1.59	1.70	1.70	1.70
Present study$(BCR)_{fr,oi}$*	Reinforced	1.0	1.70	1.70	1.70	1.70

7 Conclusions

The proposed model considers the effect of axial resistance mobilized by a single circular sheet of geosynthetic reinforcement due to interfacial shear stresses developed over its top and bottom surfaces, shear resistances of the granular fill along the outer and inner edges of the ring and that of the soft ground to arrive at the total ultimate bearing capacity of the reinforced two-layer system. The values provided by Keshavarz and Kumar [2] for the bearing capacity factor N_c for rough ring footings on a cohesive–frictional soil are incorporated in the formulation. Predictions compare well with the experimental results of Shalaby [8] in literature for relatively smaller thicknesses of granular layer.

References

1. Boushehrian, J.H., Hataf, N.: Experimental and numerical investigation of the bearing capacity of model circular and ring footings on reinforced sand. Geotext. Geomembr. **21**(2003), 241–256 (2003)
2. Keshavarz, A., Kumar, J.: Bearing capacity computation for a ring foundation using the stress characteristics method. Comput. Geotech. **89**, 33–42 (2017)
3. Meyerhof, G.G.: Ultimate bearing capacity of footings on sand layer overlying clay. Can. Geotechnical J **11**(2), 223–229 (1974)
4. Meyerhof, G.G., Hanna, A.M.: Ultimate bearing capacity of foundations on layered soils under inclined load. Can. Geotech. J. **15**(4), 565–572 (1978)
5. Rajyalakshmi, K., Madhav, M.R., Ramu, K.: Bearing capacity of foundations on geosynthetic reinforced foundation beds on soft non-homogeneous ground. Indian J. Geosynth. Gr. Improv. **1**(1):11–19 (2012)
6. Rethaliya, R.P., Verma, A.K.: Strip footing on sand overlying soft clay with geotextile interface. Indian Geotech. J. **39**(3), 271–287 (2009)
7. Sawwaf, M.E.I., Nazir, A.: Behavior of eccentrically loaded small-scale ring footings resting on reinforced layered soil. J. Geotech. Geoenviron. Eng. **138**(3), 376–384 (2012)
8. Shalaby, S.I.: Bearing capacity of ring footing on stabilized clay with sand trench-stone pile combination. Int. J. Scient. Eng. Appl. Sci. **3**(5), 218–226 (2017)
9. Shivshankar, R., Madhav, M.R., Miura, N.: Reinforced granular beds overlying soft clay. In: 11th SEAGC, Singapore, pp. 409–414 (1993)

Slope–Reinforcement Interactions: Effect of Strength Parameters

Akshay Kumar Jha, Madhav Madhira, and G. V. N. Reddy

1 Introduction

Analysis of earth slopes is one of the oldest geotechnical engineering problems that engineers have been dealing with using various techniques. The methods can be classified as limit equilibrium methods (LEM), finite element method (FEM) based on c and ϕ reduction, finite difference method (FDM), combination of FEM and LEM, limit analysis (LA), etc. Geosynthetic reinforcement of earth slope results in reduction in the land requirement and preservation of natural resources (land and backfill requirements) apart from time and cost. Designing geosynthetic reinforced slope with minimum length of geosynthetics leads to further economy. Jewell et al. [7], Bonparte et al. [3], and Verduin and Holtz [16] present design methods for earth slopes reinforced with geotextiles or/and geogrids using LEM assuming different types of failure surfaces such as circular or/and bilinear wedges. Jewell et al. [7] used limit equilibrium analysis and local stress calculation for design of reinforced slope. Jewell [8] presented revised design charts for steep slopes valid for all polymer reinforcement materials. These revised charts lead to savings of the order of 20–30% in reinforcement quantity. Leshchinsky [9] and Leshchinsky et al. [12] used log spiral failure mechanism to determine the required reinforcement long-term strength. Zhao [17] and Michalowski [14] present kinematic limit analysis solutions for the stability

A. K. Jha (✉)
South Central Railway, Hyderabad, India
e-mail: akshayghungru@gmail.com

M. Madhira · G. V. N. Reddy
JNTUH Hyderabad, Hyderabad, India
e-mail: madhavmr@gmail.com

G. V. N. Reddy
e-mail: gvnreddy@jntu.ac.in

C. N. V. Satyanarayana Reddy et al. (eds.), *Ground Improvement and Reinforced Soil Structures*, Lecture Notes in Civil Engineering 152,
https://doi.org/10.1007/978-981-16-1831-4_49

of reinforced soil slopes. Shiwakoti et al. [15] conducted parametric studies to investigate the effect of geosynthetic strength, soil–geosynthetic interaction coefficients, vertical spacing of geosynthetics for soil slope/wall on competent foundation. Baker and Klein [1, 2] modified the top-down approach of Leshchinsky [9] to obtain the reinforcement force needed for a prescribed factor of safety everywhere within the reinforced mass. Han and Leshchinsky [5] present a general analytical framework for design of flexible reinforced earth structures, i.e., walls and slopes. Leshchinsky et al. [13] presented a limit equilibrium methodology to determine the unfactored global geosynthetic strength required to ensure sufficient internal stability in reinforced earth structures. Leshchinsky et al. [11] introduced a limit state design framework for geosynthetic reinforced slopes and walls. Leshchinsky and Ambauen [10] presented use of upper bound limit analysis (LA) in conjunction with discretization procedure known as discontinuity layout optimization (DLO) for comparison with rigorous LE methods. DLO-LA is an effective tool for establishing a critical failure mechanism and its stability without the constraint or assumptions required in LE analysis. Gao et al. [4] in their study considered three-dimensional effects on reinforced earth structure stability and to determine required strength and length of reinforcement using limit analysis approach. The three-dimensional effects are more significant for the minimum required length of reinforcement than for the minimum required tensile strength.

1.1 Problem Definition

An embankment of height, H, of 6.0 m with side slopes of 1.5 horizontal to 1 vertical (Fig. 1) is considered. H is height of embankment. L_r is the total length of geosynthetic reinforcement, L_f is the length of reinforcement in unstable zone and L_e is the length of reinforcement in stable zone and is contributing directly to slope stability. The embankment and the foundation soil have the same soil parameters. Cohesion (c) of 5 kPa and angle of shearing resistance of soil of 23° have been considered. The unreinforced slope has minimum factor of safety (FS_{min}) of 1.22, which is less than the desired minimum factor of safety (FS) of 1.5 and hence considered to be unsafe. The slope is reinforced with a single layer of reinforcement to ensure the required minimum FS. Ultimate tensile strength (T_{ult}) of geosynthetic reinforcement is 200 kN/m and allowable tensile strength (T_{all}) is 100 kN/m. The adhesion between soil and reinforcement (c_a) of 3 kPa and interface friction angle or bond resistance between soil and reinforcement (δ) of 17° have been considered for analysing the reinforced slope. The embankment with reinforcement placed at $Z_0 = 3.0$, 4.0 or 5.0 m from the top of embankment has been analysed for the above soil and soil–reinforcement interaction parameters. The initial lengths of reinforcement considered are 8.0 m, 7.8 m and 8.2 m for reinforcement at $Z_0 = 3.0$ m, 4.0 m and 5.0 m, respectively. Unreinforced and reinforced slopes are analysed using Morgenstern-Price method to obtain the critical factors of safety. Geostudio 2004 SLOPEW has been used for the analysis.

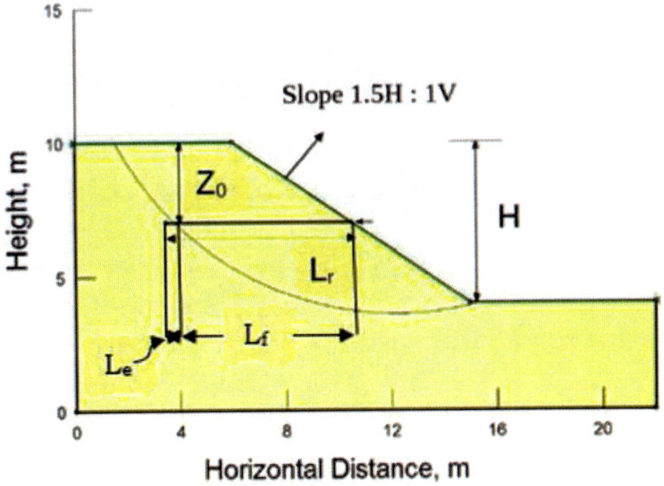

Fig. 1 Definition sketch, H: Height of embankment, L_T: Length of geosynthetic reinforcement, L_f: Length of geosynthetic reinforcement in unstable zone, L_e: Effective length of reinforcement in stable zone

1.2 Analysis

The slope with reinforcement at 3.0 m from top of embankment analysed and reinforcement length optimized from non-face end is detailed in Akshay and Madhav [6]. The length of reinforcement after optimization form non-face end was 7.27 m (Fig. 2).

The slope in Fig. 2 is reanalysed by curtailing the reinforcement from the face end by moving point P inside the slope but still maintaining minimum factor of safety of 1.5. The face end optimization of length resulted in reduction in length of reinforcement to 5.08 m (Fig. 3).

The face end optimization has led to further reduction in length and saving is length of the order of 30%. Similar analysis was carried for reinforcement at 4.0 m and 5.0 m depths from the top of the embankment. The analysis was repeated with two more sets of soil parameters, c = 6 kPa and $\phi = 20°$ and c = 4 kPa and $\phi = 26°$. The summary of soil parameters used for analysis is being detailed in Table 1.

2 Effect of Soil Parameters on Factor of Safety

The effect of soil parameters on slope stability requires analysis of same slope with different sets of soil parameters.

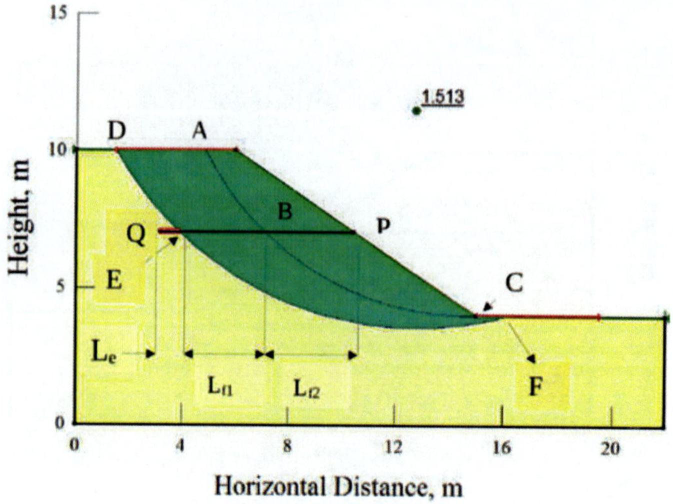

Fig. 2 Critical slip circle for reinforced slope for $Z_0 = 3.0$ m, $FS_{min} = 1.51$, $L_T = 7.27$ m

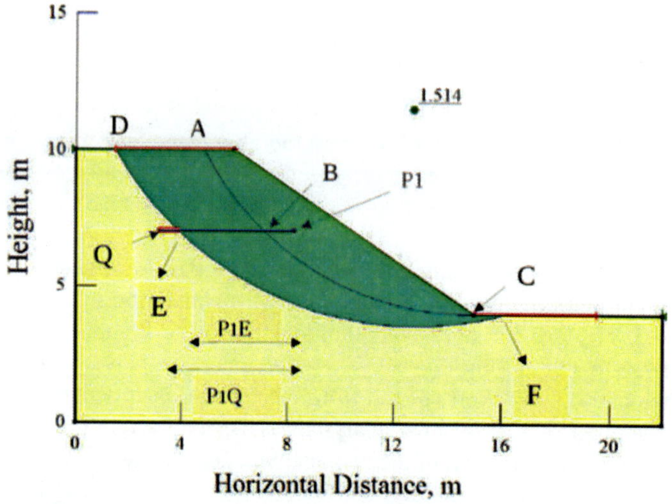

Fig. 3 Critical slip circle for slope with reinforcement length optimized from face end with $Z_0 = 3.0$ m, $L_T = 5.08$ m and $FS_{min} = 1.51$

2.1 *Variation of Reinforcement Length with Angle of Shearing Resistance and Cohesion*

As mentioned in Sect. 1.2, the embankment slope was analysed with soil and soil–reinforcement parameters as detailed at S. no. 1, 2 and 3 of Table 1 with reinforcement at 3, 4 and 5 m from the top of the embankment. The results of analysis for all the

Table 1 Parameters of soil and reinforcement for stability analysis

S. no	Unit weight (γ), kN/m^3	Cohesion (c), kPa	Angle of shearing resistance ($\phi°$)	Interface friction angle between soil and reinforcement, $\delta°$	Adhesion between geosynthetic and backfill, c_a, kN/m	Ultimate tensile strength of geosynthetics (T_{ult}), kN/m
1	18	5	23	17	3	200
2	18	6	20	16	4.5	200
3	18	4	26	20	3	200

Table 2 Summary of results for three sets of soil parameters

Location of reinforcement from top, m	Soil cohesion (C), kPa	Angle of shearing resistance, $\phi°$	Non-face optimized length of reinforcement, L_T, m	Optimal length of reinforcement (L_{Topt}), m	Saving in length of reinforcement (L_T-L_{opt}), m	Effective length of reinforcement (L_e), m
3	4	26	6.11	3.88	2.23	0.62
3	5	23	7.27	5.08	2.19	0.76
3	6	20	8.12	5.87	2.25	1.68
4	4	26	6.15	4.15	2.00	0.22
4	5	23	7.33	5.26	2.07	0.25
4	6	20	8.56	6.50	2.06	0.29
5	4	26	6.45	4.94	1.51	0.15
5	5	23	7.64	6.04	1.60	0.15
5	6	20	9.02	7.47	1.55	0.22

three cases of single-layer reinforced slope as analysed in detail are summarized in Table 2.

2.2 *Discussion of Results*

The saving in length of reinforcement ($L_T - L_{opt}$) is maximum for the reinforcement at 3 m depth from the top of the embankment for the three sets of soil parameters. For a given location of reinforcement, saving in length of reinforcement increases with decreasing angle of shearing resistance and increasing cohesion. Effective length of reinforcement (Le) increases with decreasing angle of shearing resistance (ϕ) as is expected since mobilised force in reinforcement is highly dependent on ϕ. The savings in reinforcement length is maximum for $Z_0 = 3$ m case for all three sets of soil parameters and least for $Z_0 = 5$ m case.

2.3 Variation of Reinforcement Length with Angle of Shearing Resistance and Cohesion

Figure 4 shows variation of length of reinforcement with angle of shearing resistance of soil.

Length of reinforcement, L_T, decreases (Fig. 4) with increase of angle of shearing strength of soil. This is expected as increasing angle of shearing resistance of soil contributes more towards slope stability by way of increasing stabilizing force and moment. The variation of L_T with angle of shearing resistance is nearly the same for 4 m and 5 m cases.

The savings in length of reinforcement (difference in length of reinforcement for non-face optimized length and face optimized length) ($L_T - L_{Topt}$) are nearly independent of angle of shearing resistance of soil and cohesion, as shown in Figs. 5 and 6. The variation of savings in length with angle of shearing resistance of soil (Fig. 5) is practically straight line for given depth of reinforcement from the top of the embankment. The possible explanation for this behaviour could be the fact that reduction in length from face end is possible only for shallow slip circle that do not become critical. As depth of reinforcement from top of embankment increases the savings in length of reinforcement decreases (Fig. 5). As depth of reinforcement from the top of the embankment increases with curtailment of length of reinforcement from face end, the shallow slip circle becomes critical. Hence further curtailment of reinforcement from face gives factor of safety less than the desired value of 1.5. The variation of savings in length of reinforcement with cohesion varies in a similar fashion as with angle of shearing resistance (Fig. 6).

The length of reinforcement in the stable zone (L_e) (Fig. 1) has been found to be constant for 4.0 and 5.0 m depths of reinforcement but decreases with angle of

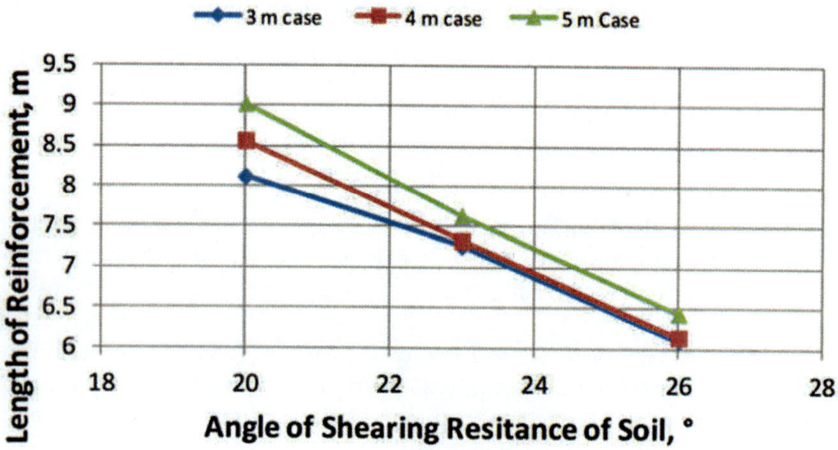

Fig. 4 Variation of L_T with angle of shearing resistance of soil

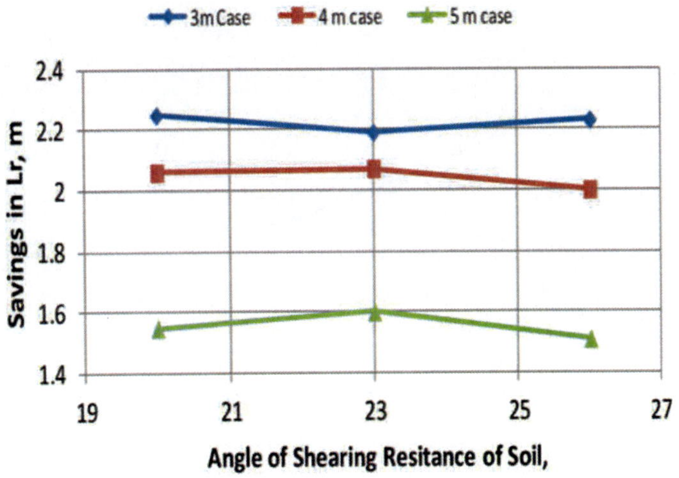

Fig. 5 Saving in length of reinforcement versus angle of shearing resistance, ϕ

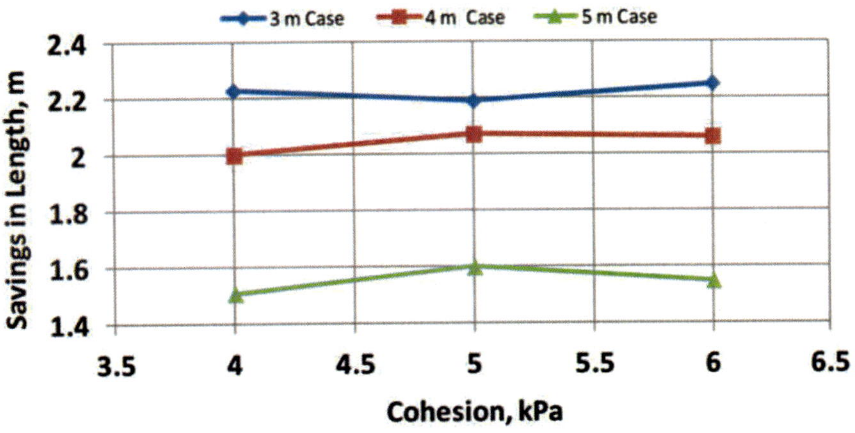

Fig. 6 Saving in length of reinforcement versus cohesion

shearing resistance for 3.0 m depth case (Fig. 7). Figure 7 indicates the effective length of reinforcement, L_e, in the stable zone is independent of angle of shearing resistance of soil for Z_0 equal to 4.0 and 5.0 m.

2.4 *Effect of Adhesion on Factor of Safety*

To study the effect of interface adhesion between soil and reinforcement on slope stability, 6 m high slope 1.5H:1 V with the soil parameters as detailed in Sect. 1.2

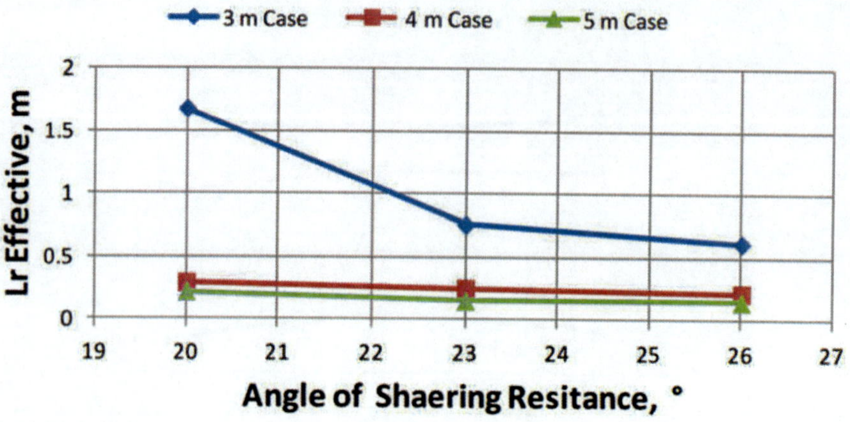

Fig. 7 Variation of effective length of reinforcement versus ϕ

with reinforcement at 3, 4 and 5 m from the top of the embankment but with zero adhesion has been re-analysed, and FS_{min} and reinforcement loads, F_r obtained have been compared with the values obtained with $c_a = 3.0$ kPa.

The results of analysis are summarized in Table 3. Critical study of data in the table indicates that (i) with increasing depth of reinforcement from the top of the slope (i.e. with increasing Z_0) the effect of neglecting adhesion between soil and reinforcement

Table 3 FS_{min} and F_r with and without considering adhesion between reinforcement and soil

Fixed parameters of study: Slope 1.5H:1 V, c = 5 kPa, $\phi = 23°$, Unit Weight = 18 kN/m^3, Tall = 100 kN/m

Study of effect of adhesion on FS_{min} and reinforcement load (F_r)

S. no	Z_0, m	c_a (kPa)	FS_{min}	F_r, kN	FS_{min}	F_r, kN
1	3	3	$L_r = 7.27$ m		$L_{ropt} = 5.08$ m	
			1.513	19.59	1.513	19.59
2		0	1.499	16.74	1.499	16.74
% Change in parameters			0.93	14.55	0.93	14.55
3	4	3	$L_r = 7.33$ m		$L_{ropt} = 5.26$ m	
			1.514	8.38	1.514	8.38
4		0	1.51	7.43	1.51	7.43
% Change in parameters			0.26	11.34	0.26	11.34
5	5	3	$L_r = 7.64$ m		$L_{ropt} = 6.06$ m	
			1.508	6.23	1.508	6.23
6		0	1.505	5.64	1.505	5.64
% Change in parameters			0.20	9.47	0.20	9.47

% Change in FSmin $= \frac{(\text{FSminwithAdhesion} - \text{FSminwithoutadhesion}) * 100}{\text{FSminwithAdhesion}}$

% Change in F_r is also defined in similar way

(c_a) reduces marginally on factor of safety (FS_{min}) and significantly on mobilised force in reinforcement, F_r. FS_{min} reduces by 0.93–0.20% as reinforcement position from the top of embankment increases from 3.0 to 5.0 m. Thus the contribution of adhesion is practically negligible so far slope stability is concerned. However, % change of F_r reduces from 16.8 to 9.8 for the same positions of the reinforcement. Absolute change in Fr (i.e. difference of F_r with and without considering c_a) reduces from 1.85 to 0.59 kN/m as reinforcement position changes from 3.0 to 5.0 m probably due to increase in normal stress as reinforcement load increases with increasing depth (increase in normal stress).

3 Conclusions

The analysis of 6 m high slope with 1.5H:1 V side slope with different sets of soil parameters show that length of reinforcement (L_T) decreases with increasing angle of shearing resistance (ϕ) as is expected (Fig. 4). With varying depth of reinforcement from top of embankment for same value of ϕ, L_T is lowest for 3 m depth followed by 4 and 5 m depth cases. Savings in length of reinforcement were found to be practically independent of c and ϕ but dependent on depth of reinforcement. Adhesion has negligible effect on slope stability.

References

1. Baker, R., Klein, Y.: An integrated limiting equilibrium approach for design of reinforced soil retaining structures: Part I—formulation. Geotextile & Geomembrane **22**(3), 119–150 (2004)
2. Baker, R., Klein, Y.: An integrated limiting equilibrium approach for design of reinforced soil retaining structures: Part II—design examples. Geotextile & Geomembrane **22**(3), 151–177 (2004)
3. Bonaparte, R., Holtz, R.D., Giroud, J.P.: Soil reinforcement design using geotextile and geogrids. In: Fluet, J.E. (ed.) Geotextile Testing and Design Engineer, ASTM STP 952, pp. 69–116. American Society for Testing Materials, Philadelphia (1987)
4. Gao, Y., Yang, S., Zhang, F., Leshchinsky, B.: Three dimensional reinforced slopes: evaluation of required reinforcement strength and embedment length using limit analysis. Geotext. Geomembr. **44**, 133–142 (2016)
5. Han, J., Leshchinsky, D.: General analytical framework for design of flexible reinforced earth structures. J Geotech Geoenviron Eng ASCE **132**, 1427–1435 (2006)
6. Jha, A.K., Madhav, M.R.: Analysis of Effect of Reinforcement on Stability of Slopes. In: Proceedings 50th Indian Geotechnical Conference, Pune, India (2015)
7. Jewell, R.A., Paine, N., Woods, R.I.: Design methods for steep reinforced embankments. Polymer grid reinforcement, Thomas Telford Limited, London, pp. 70–81 (1985)
8. Jewell, R.A.: Application of the revised design charts for steep slopes. Geotext. Geomembr. **10**, 203–233 (1991)
9. Leshchinsky, D.: Issues in geosynthetic reinforced soil. Keynote Paper Proceedings of International Symposium on Earth Reinforcement practice Kyushu Japan Balkema Rotterdam, The Netherlands, vol. 2, pp. 871–897 (1992)

10. Leshchinsky, B., Ambauen, S.: Limit equilibrium and limit analysis: comparison of benchmark slope stability problems. J. Geotech. Geoenviron. Eng. ASCE **04015043**, 1–8 (2015)
11. Leshchinsky, D., Kang, B., Han, J.: Framework for limit state design of geosynthetic reinforced walls and slopes. Transport Infrastruct. Geotech. **1**, 129–164 (2014)
12. Leshchinsky, D., Ling, H., Hanks, G.: Unified design approach to geosynthetic reinforced slope and segmental walls. Geosynth. Int. **2**(5), 845–881 (1995)
13. Leshchinsky, D., Zhu, F., Meehan, C.L.: Required unfactored strength of geosynthetic in reinforced earth structures. J. Geotech. Geoenviron. Eng. **136**(2), 281–289 (2010)
14. Michalowski, R.L.: Stability of uniformly reinforced slopes. J. Geotech. Geoenviron. Eng. ASCE **23**(6), 546–556 (1997)
15. Shiwakoti, D.R., Pradhan, T.B.S., Leshchinsky, D.: Performance of geosynthetic reinforced soil structures at limit equilibrium state. Geosynth. Int. **5**(6), 555–587 (1998)
16. Verduin, J.R., Holtz, R.D.: Geosynthetically reinforced slopes: a new procedure. In: Proceedings Geosynthetics, San Deigo, IFAI (1989)
17. Zhao, A.: Limit analysis of geosynthetic reinforced slopes. Geosynth. Int. **3**(6), 721–740 (1996)

Influence of Excavation Phase on the Performance of Soil Nail System

H. R. Krupa and S. K. Prasad

1 Introduction

Soil nails are reinforcing, passive elements that are drilled and grouted sub horizontally in the ground to support excavations in soil, or in soft or weathered rock. It contributes to the stability of the earth-resisting systems mainly through tension as a result of the deformation of the retained soil or weathered rock mass. They transfer the tensile forces to the surrounding soil through bond stresses (shear stresses) along the grout–ground interface and ensure long-term performance of the system. Soil nail walls can be used in roadway applications involving roadway cuts, road widening under existing bridge abutments, tunnel portals, repair and reconstruction of existing retaining structures [1]. The behavior and performance of the soil nail system are influenced by many factors [2]. Construction sequence being one such parameter, it is important to decide the depth of excavation at each stage of installation, knowing the number of nails installed at each stage and for overall system. Some researchers have done a detailed study on the reinforced earth structures, and the depth of excavation to be carried out at each stage is still uncertain. Hence, efforts have been made in this paper to study the effect of depth of excavation at each stage and thus analyzing the pattern of behavior of the nail forces and deformation of the system. Elias and Juran [3] have found that shear and bending nail strengths contribute less than 10% to the overall stability of the wall. Due to this relatively modest contribution, the shear and bending strengths of the soil nails are conservatively disregarded in the conventional design procedure. In this paper, the variation of flexural forces and its effect on the soil nail system is also studied.

H. R. Krupa
JSS Science and Technology University, Mysuru 570006, India

S. K. Prasad (✉)
Vidyavardhaka College of Engineering, Mysuru 570002, India

C. N. V. Satyanarayana Reddy et al. (eds.), *Ground Improvement and Reinforced Soil Structures*, Lecture Notes in Civil Engineering 152,
https://doi.org/10.1007/978-981-16-1831-4_50

2 Methodology

In order to study the behavior and stability analysis, numerical simulations were done using PLAXIS-2D. For preliminary design values to be used in the software, FHWA [4] soil reference manual was considered. A 2D, 15-noded, plane strain, finite element model was modeled. The model parameters are mentioned in Table 1.

2.1 Conventional Design Procedure

The manual design procedure is based on the FHWA [4] soil reference manual, which consists of two parts in the design: a preliminary design and the final design. The

Table 1 Model/material parameters

Parameter	Value
Vertical height of wall, H	10 m
Face batter, α (wrt vertical)	0°
Slope of backfill, β (wrt horizontal)	0°
Soil type	Dense silty sand
Cohesion, c	5 kPa
Friction angle, φ	31.5°
Unit weight, γ	17 kN/m^3
Modulus of elasticity of soil, Es	20 MPa
Poisson's ratio, ν	0.3
Nail installation method	Rotary drilled and grouted
Grade of steel	Fe415
Modulus of elasticity of nail, En	200 GPa
Nail spacing, $S_V \times S_H$	1 × 1 m × m
Nail inclination (wrt horizontal), i	15°
Drill hole diameter, D_{DH}	100 mm
Nail diameter, D_{DH}	20 mm
Compressive strength of grout, f_{ck}	20 MPa
Ultimate bond strength, qu	100 kPa
Modulus of elasticity of grout, E_g	22 GPa
Target Factors of Safety	
FS for global stability, FS_G	1.35
FS for pull-out, FS_P	2
FS for tensile strength, FS_T	1.8
FS for flexure failure, FS_{FF}	1.35
FS for punching shear, FS_{FP}	1.35

final design includes analysis of external failure modes (global stability and sliding stability), analysis of internal failure modes (nail pullout failure and nail tensile strength failure), design of permanent facing and verification of important facing failure modes (facing flexure failure and facing punching shear failure), and influence of other site-specific considerations, such as seismic loading. In the present study only the important internal failure modes and facing failure modes are considered to assess and compare the performance of the (conventionally designed) soil nail wall. Table 2 comprises conventional design details. It is observed that the factors of safety are greater than/equal to the target assumed, thus making the design procedure dependable.

2.2 *Numerical Simulations*

Using the above values from the preliminary design, a model was created using PLAXIS-2D. As mentioned earlier, soil nail wall is modeled as a plane strain problem and 15-noded triangular elements are used for generating finite element mesh of appropriate density. The model considered was of total width 30 m, with width of excavation kept as 5 m to the left. The depth of excavation was taken as 10 m and total depth of model was taken as 25 m. Medium mesh density was adopted globally, which was refined to fine density in the vicinity of the soil nail wall (Fig. 1). Mesh boundaries were placed far enough so as to minimize the influence of mesh boundaries on the results of the numerical simulation. The validation is done with an already published paper and the results for axial force developed in the nails were comparable with the published information [5]. The soil nail system for three cases of varying depth of excavation at each stage for each case, considered as 1 m, 2 m and 5 m, respectively, are considered. The effect of other parameters is ignored in this paper. The results from the numerical simulation are summarized in Table 3.

3 Results and Discussion

3.1 *Axial Forces Along Nail Length*

Figures 2, 3 and 4 show the axial forces developed in the nails of the soil nail system for cases 1, 2 and 3, respectively. It is observed that Case 1 with lower depth of excavation for each stage shows lower axial force compared to Case 2 and Case 3. Also, we observe that the maximum tensile force of the system is developed for the bottom-most nail of the system in Case 2 and Case 3. But for Case 1, it is maximum for the 9th nail (2nd nail from the bottom) and decreases for the bottom-most nail. Figure 5 shows the axial force for bottom-most nail for all three cases along the length of the nail. FHWA [4] says that the value of axial (tensile forces)

Table 2 Conventional design summary

Description	Symbol (unit)	Formulae	Value
Normalized bond resistance	μ	$(quD_{DH})/(FS_{PO}\gamma_s S_H S_V)$	0.29
Normalized SN length	L/H	from the charts of FHWA [4]	0.65
Normalized T_{max}	t_{max}		0.29
Correction for drill hole dia	C1	$C1 = 1.50–0.15D_{DH} + 0.0065D_{DH}^2$	1.00
Correction for cohesion	C2	C2 = -4.0c* + 1.09	0.97
Correction for F_{OS}	C3	$C3 = 0.52FS_{OS} + 0.30$	1.00
Correction for drill hole dia	T1	$T1 = -0.3 + 0.4D_{DH} - 0.017D_{DH}^2$	1.00
Correction for cohesion	T2	T2 = −4.0c* + 1.09	0.97
Corrected length factor	L/H *	C1*C2*C3*L/H	0.63
Corrected T_{max} factor	t_{max*}	$T1*T2*T_{max}$	0.28
Length of the Nail	L (m)	Rounded up	**7.00**
Maximum tension in the Nail	T_{max} (kN)	$\gamma_s S_H S_V t_{max*}$	**47.60**
Tension at face	T_o (kN)	$T_{max*}(0.6 + 0.2(S_{max}-1)$	**28.56**
Check for factor of safety for nail			
Nominal pullout resistance/m	r_{po}(kN/m)	$\prod q_u D_{DH}$	31.42
Pullout length	Lp (m)	0.35L	2.45
Nominal pullout resistance	R_{po} (kN)	$r_{PO}*L_p$	76.98
FS for pullout+	FS_{PO}*	R_{po}/T_{max}	**1.62**
Cross sectional area	A_t (mm^2)	$(FS_T*T_{max})/f_y$	206.46
Nominal tensile resistance	R_T (kN)	A_t*f_y	85.68
FS for tensile strength	FS_T*	R_T/T_{max}	**1.80**
Check for factor of safety for facing			
Type	Initial + Final Facing		
Thickness, h, mm	100 each		
Reinforcement*	WWM – 102 × 102 (reinforcement ratio 1%), 2 Waler bars of Φ 10 mm		
Bearing plate grade	Fe 250		
Bearing plate dimensions	225 mm × 225 mm × 25 mm		
Flexure capacity	R_{FF} (kN)	Table 6.6 of FHWA manual	80.07
Punching shear capacity	R_{FP} (kN)	Table 6.7a of FHWA manual	155.69
FS against flexure failure	FS_{FF}	R_{FF}/T_o	**2.80**
FS against punching shear	FS_{FP}	R_{FP}/T_o	**5.45**

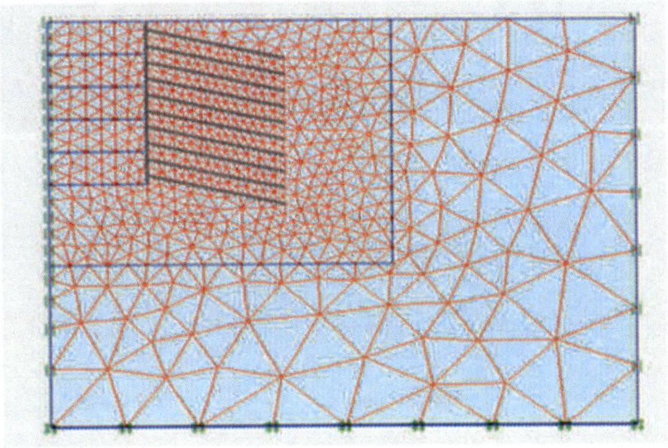

Fig. 1 Typical finite element model of soil nail system (H = 10 m)

Table 3 Results from numerical simulation

Case		1	2	3
Depth of excavation at each stage		1 m	2 m	5 m
Parameter	Symbol (unit)	Value		
Horizontal displacement of the system	H (mm)	21.52	20.88	18.15
Maximum tension in the nail	T_{max} (kN)	56.21	79.10	83.32
Error from theoretical value	%	+15.32	+39.82	42.87
Tension at face	T_o (kN)	51.41	70.13	74.32
Maximum shear force	Q (kN)	19.63	17.76	16.07
Maximum bending moment	M (kN-m)	2.74	2.92	3.68
Pullout length	Lp (m)	6.53	6.53	6.53
FS for pullout of nail	FS_{PO}	**3.65**	**2.60**	**2.46**
FS for tensile strength of nail	FS_T	**2.31**	**1.86**	**1.75**
FS against flexure failure of face (R_{FF} = 157.91 kN/m^2)	FS_{FF}	**3.07**	**2.25**	**2.12**
FS against punching shear of face (R_{FP} = 204.62 kN/m^2)	FS_{FP}	**3.98**	**2.92**	**2.75**

range between 0.5 and 1.1 times of $K_a \Upsilon H S_H S_V$, i.e., between 40.43 and 89.37 kN for the model considered in this paper. The maximum axial forces obtained from numerical simulations in all three cases are well within these limits. But Case 1 giving out the lower force value suits good by yielding higher factors of safety for internal and external stabilities. Also, theoretically, it is said that the axial force at nail head (at wall face) varies about 60–100% of the maximum axial force and from the simulations, it is about 90% of the maximum tensile force of the nail for all three cases.

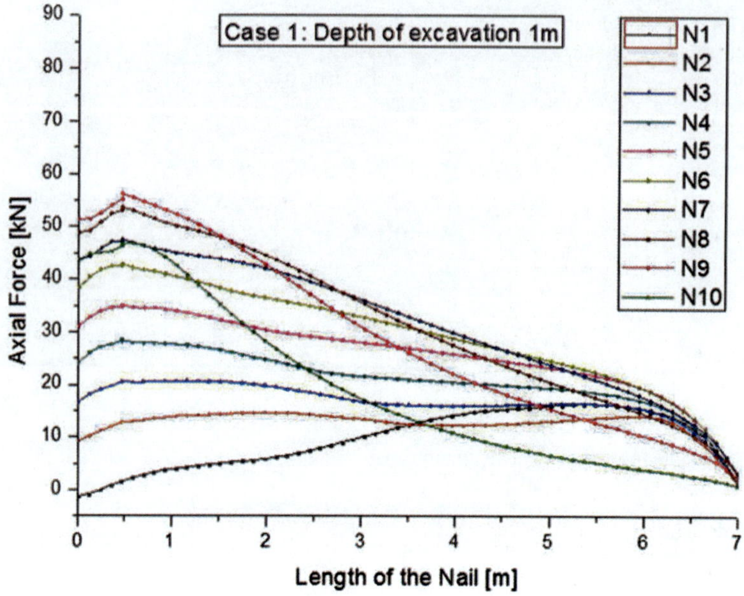

Fig. 2 Axial force along nail length for Case 1

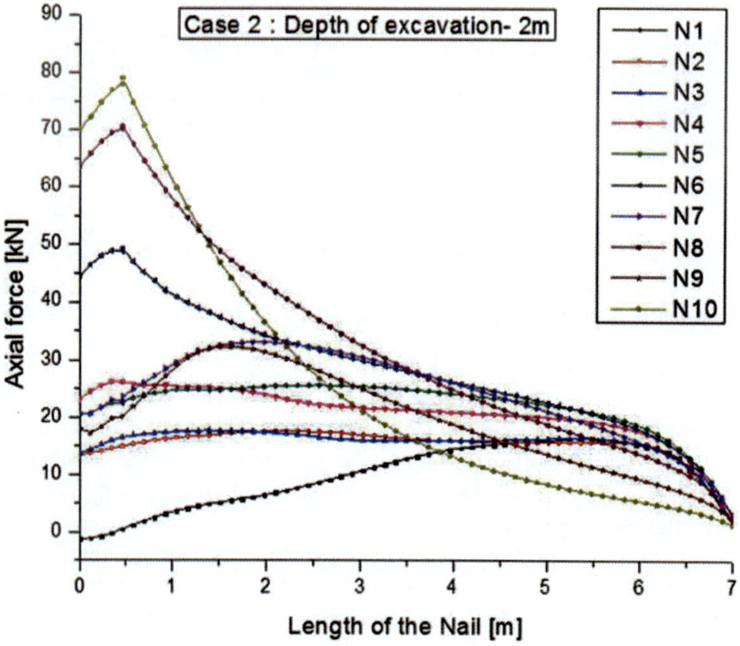

Fig. 3 Axial force along nail length for Case 2

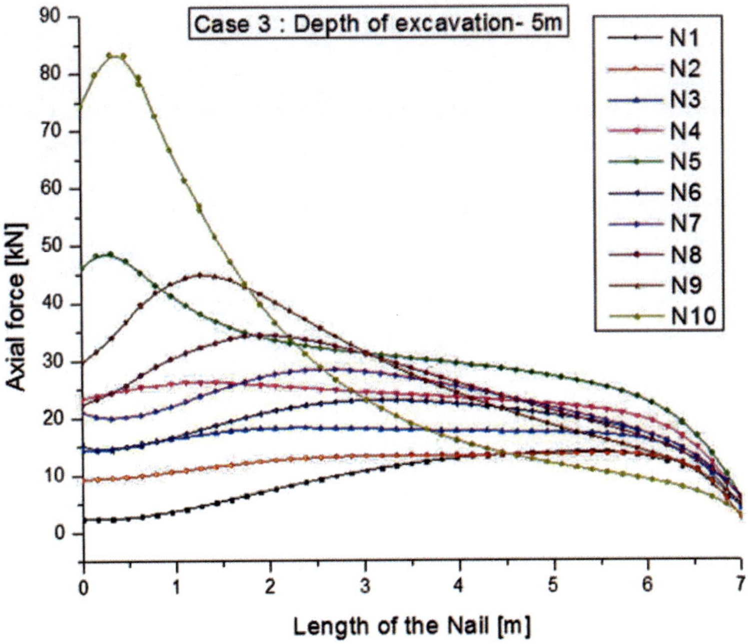

Fig. 4 Axial force along nail length for Case 3

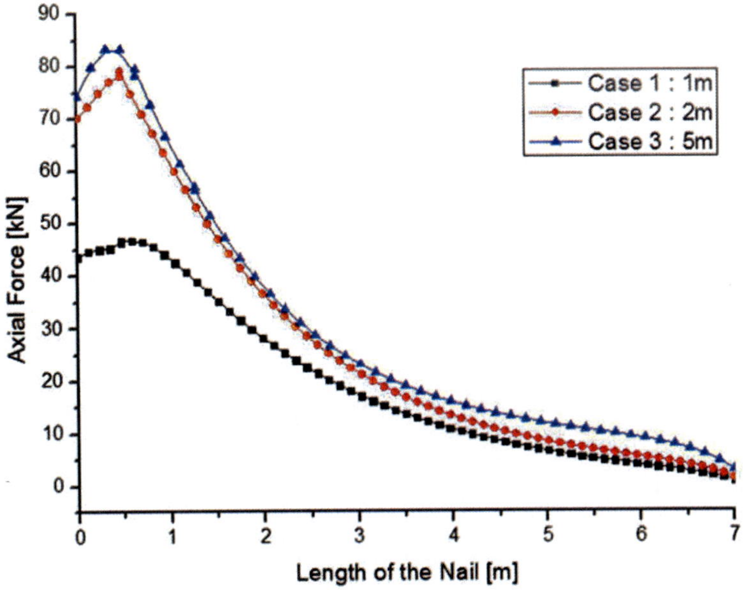

Fig. 5 Axial force variation in bottom-most (10th) nail along the length for all three cases

3.2 *Variation of Forces Along the Depth of Excavation*

Maximum Axial Force and Nail Head Axial Force: Figure 6 shows the variation of maximum axial forces of nails from nail to nail (also along the depth of excavation). It is observed that for Case 1, this variation is linear. We see a constant increase along the depth of cut and then decreases for the bottom-most nail (at the final stage of excavation). For Case 2, we see that the axial force in the first nail is lesser than that of the second nail installed at that stage of excavation. Also the axial force 1st nail of the next excavation process is lesser than the axial force of the last nail of the previous stage, giving an irregular pattern as shown in Fig. 6. Similarly for Case 3, this process is repeated for every stage of excavation. The first nail of the construction process has lesser axial force and increases along the depth. But for the next stage, the 1st nail shows a decreased axial force than the last nail of the previous excavation stage as observed in Fig. 6. Also, the forces at the nail head follow a similar pattern (Fig. 7).

Shear Force and Bending Moment: Figures 8 and 9 show the variation of shear force and bending moment along the length of bottom-most nail of the system for all three cases. It is observed that the flexural forces show less variation in all three cases. It is also observed that these forces exist only near the face and nail junction

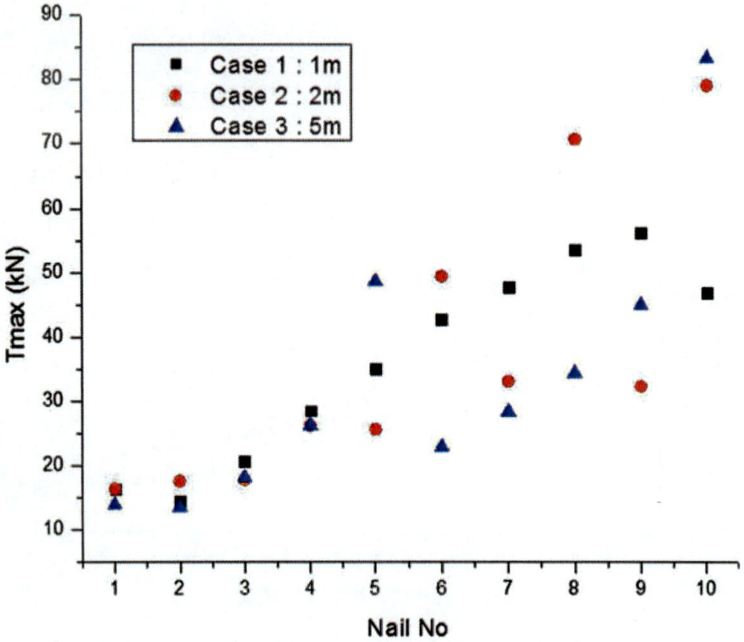

Fig. 6 Variation of maximum axial force for each nail along the depth for all three cases

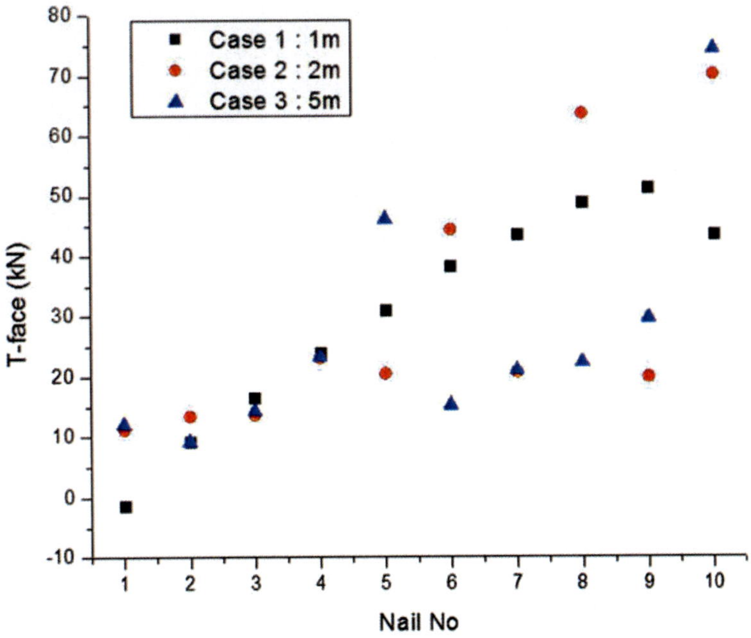

Fig. 7 Variation of axial force at nail head of each nail along the depth for all three cases

and not all along the nail. This shows good interaction between the facing element and the nail head. Also, the values of the bending moment are negligible.

In Figs. 10 and 11, we can see the maximum shear force and bending moment for each nail plotted against the nail no along the depth of excavation. We see that these forces for Case 1 follow a similar pattern of the axial forces in Case 1 and thus linearly varying. But for Case 2 and Case 3 as discussed before, it is dependent on the depth of cut at each stage. This also suggests that there is good interaction between the soil, nail and facing element in Case 1 than in Cases 2 and 3. Also as mentioned earlier, shear and bending forces contribute less than 10% in the performance of the soil nail system. It can be only considered for designing the facing element.

3.3 Horizontal Displacements

In general, soil undergoes deformation in two directions, horizontal and vertical. Vertical (upward) displacements are nothing but the basal heaving property of the soil in case of soil nail system. In this paper, only horizontal displacement of the soil at the face of the wall is considered and studied. Figure 12 gives the horizontal displacement of the soil nail wall. We observe that the displacement is slightly higher in Case 1 compared to Cases 2 and 3. Juran [6] states that the displacement of the

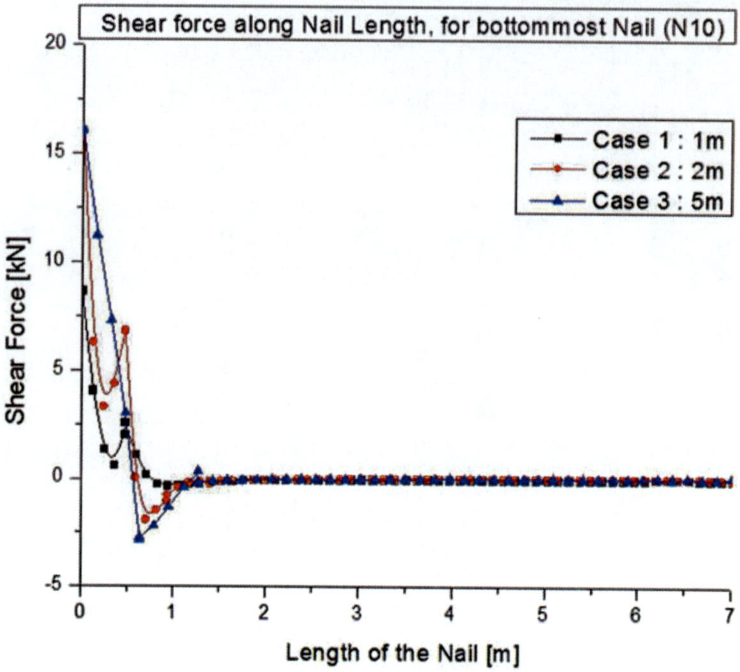

Fig. 8 Shear force along the nail length for bottom-most nail (N10) for all three cases

vertical cut has to be approximately 0.2% of the height of the wall, which accounts to 20 mm for 10 m height wall considered in this paper. Even though the displacement of the system in Case 1 is slightly higher, i.e., 21.52 mm the system is stable compared to the other two cases when the factors of safety considered.

4 Conclusions

Axial forces developed in nails are due to the frictional resistance or bond shear resistance between the soil and the nail. Lower axial force means the higher bond shear resistance developed between the soil and the nail interface, thus keeping the soil nail system intact. From the observations made above, the axial forces in Case 1 are much lesser than Case 2 and Case 3. The axial force obtained in Case 1 also has a less error compared to theoretical than the other two cases. Further, the variation of axial forces along the depth of excavation is constant or it gradually increases down the excavation depth whereas in Cases 2 and 3 we observe the variation alternatively and depth-dependent, respectively. We also observed that the nail head axial forces also follow the same pattern as that of the maximum tensile forces of the system and are about 90% of the maximum forces.

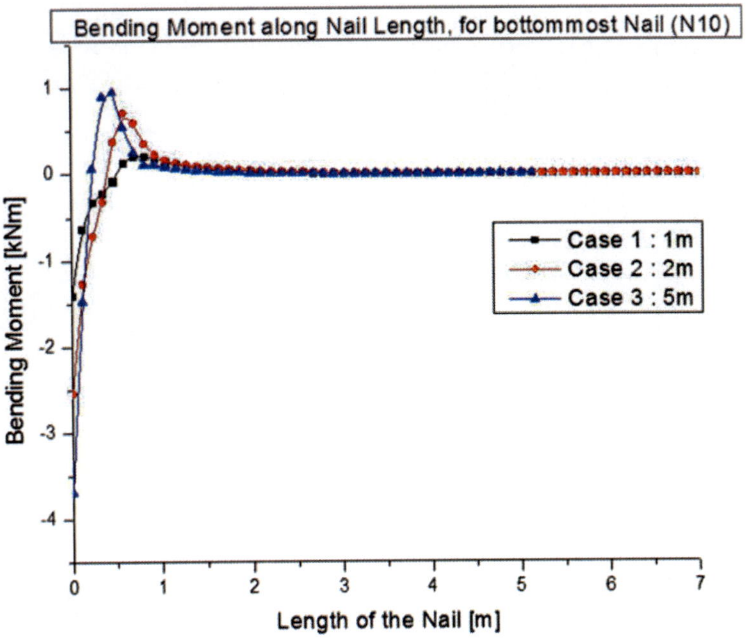

Fig. 9 Bending moment along the nail length for bottom-most nail (N10) for all three cases

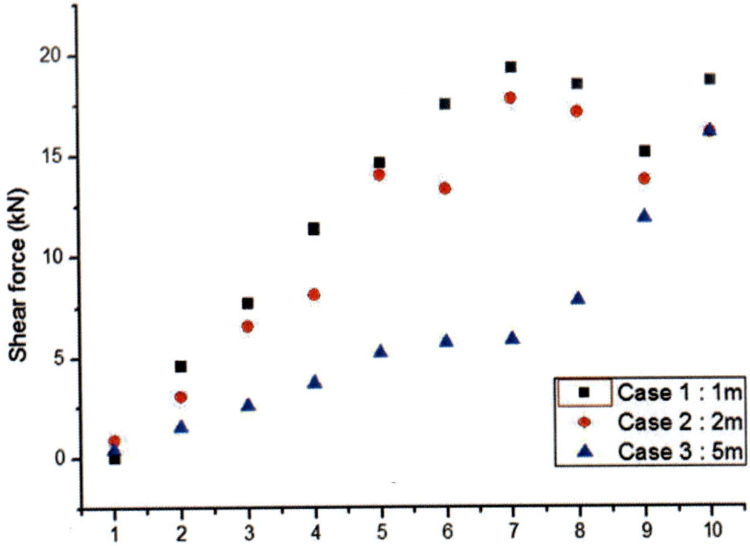

Fig. 10 Variation of maximum shear force of nails along the depth for all three cases

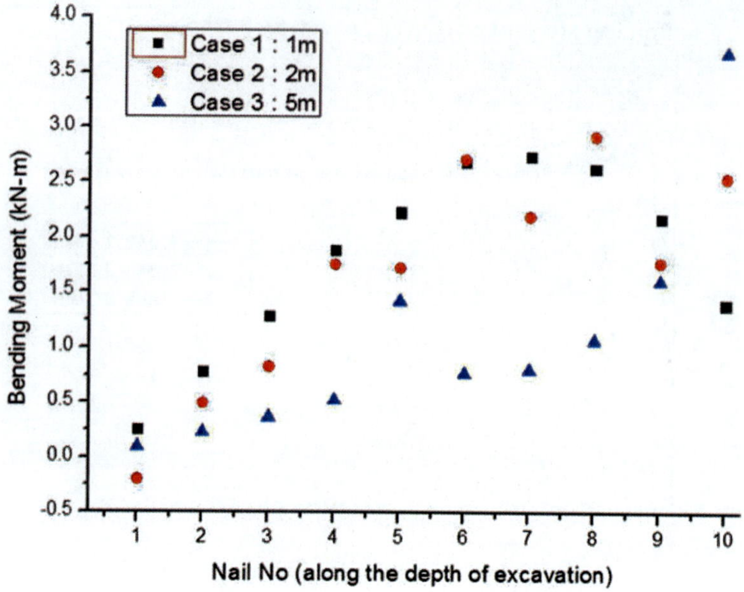

Fig. 11 Variation of maximum bending moment of nails along the depth for all three cases

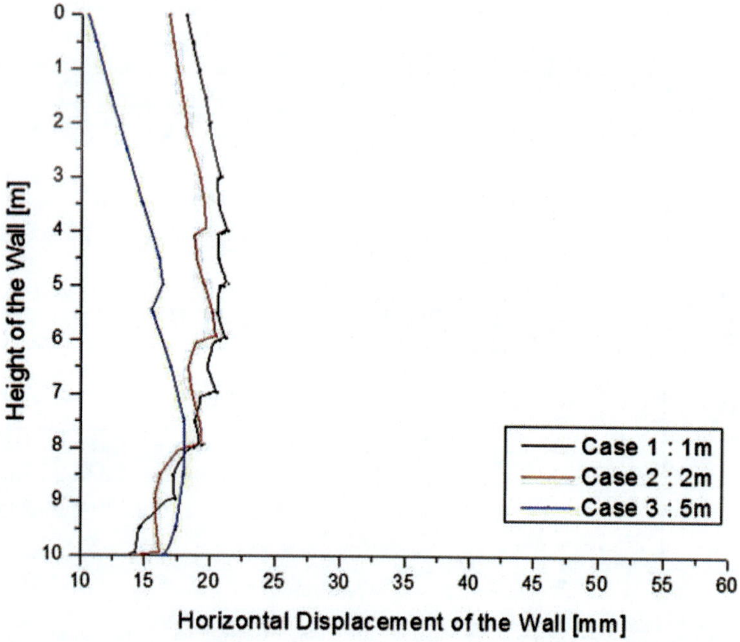

Fig. 12 Horizontal displacement of the wall in all three cases (displacement-away from wall)

The factors of safety for Case 1 is higher than those of Case 2 and Case 3. The factors of safety for internal stability (against nail pullout strength and nail tensile strength) are higher in Case 1, proving that lower axial forces always yield higher factors of safety. Also, the face flexural and shear resistance is also higher in Case 1 than in Case 2 and Case 3 for each stage. The shear force variation of the nails shows a good interaction between the soil nail and facing element.

References

1. Briaud, J.L., Lim, Y.: Soil-nailed wall under piled bridge abutment: simulation and guidelines. J. Geotech. Geo-environ. Eng. ASCE **123**(11), 1043–1050 (1997)
2. Shivakumar Babu, G.L., Murthy, B.R.S., Srinivas, A.: Analysis of construction factors influencing the behaviour of soil-nailed earth retaining walls. Ground Improv. **6**(3), 137–143 (2002)
3. Elias, V., Juran, I.: Soil nailing for stabilization of highway slopes and excavations. Publication FHWA-RD-89–198, Federal Highway Administration, Washington D.C (1991)
4. FHWA Soil Nail Walls, Geotechnical Engineering Circular No. 7, Report No. FHWA0-IF-03–017, Federal Highway Administration (2003)
5. Shivakumara Babu, G.L., Singh, V.P.: Appraisal for soil nailing. Indian Geotech. J. **39**(1), 81–95 (2009)
6. Juran, I.: Reinforced soil systems—application in retaining structures. Geotech. Eng. **16**, 39–81 (1985)

Dynamic Response of Dry Rubber Tire Chips and Sand Mixture

Adyasha Swayamsiddha Amanta and Satyanarayana Murty Dasaka

1 Introduction

The accumulation of the discarded tires in large volume increases loads on the landfills and hence raising a concern to find an alternative beneficial use of these used tires. Again improper disposal of these scrap tires leads to health risks. These stockpiled tires turn into ideal breeding places for mosquitoes, vermin, etc., which may lead to the spread of many diseases. Apart from these, tire heaps are also vulnerable to fires.

ASTM D6270 [4] specifies three different categories of the scrap tire types used for the application of civil engineering works. They are tire crumbs (length less than 10 mm), tire chips (length in between 10 and 50 mm) and tire shreds (the length is greater than 50 mm). Use of scrap tire derivatives in various civil engineering applications has been reported, such as lightweight backfill materials [1, 10, 11, 14, 15], highway embankments [5], soil reinforcement [7, 9], drainage material [6, 13], vibration reduction [8, 9, 16], ground improvement [12] to name a few.

This study investigates the effect of dynamic loads on a mixture of rubber tire chips, a scrap tire derivative and sand in its dry state, for the possible use of tire chips in various geotechnical applications.

A. S. Amanta (✉) · S. M. Dasaka
Department of Civil Engineering, IIT Bombay, Mumbai 400076, India

S. M. Dasaka
e-mail: dasaka@civil.iitb.ac.in

C. N. V. Satyanarayana Reddy et al. (eds.), *Ground Improvement and Reinforced Soil Structures*, Lecture Notes in Civil Engineering 152,
https://doi.org/10.1007/978-981-16-1831-4_51

2 Materials and Methods

The scrap tire chips used in the study are cut from the available scrap tires. The properties of the tire chips and sand are presented in Table 1. The ratio of the mixture of sand to tire chips was taken as 80:20 by weight.

The experimental program consisted of 63 cyclic triaxial tests on dry scrap tire chips. The tests were performed using a computer-controlled cyclic triaxial apparatus. All the tests were performed in strain-controlled conditions at different strain amplitudes, confining pressures and frequencies. The different confining pressures (CP) considered were 50, 75 and 100 kPa. The strain amplitudes considered in the study were 0.075, 0.1125, 0.15, 0.1875, 0.225, 0.2625 and 0.3%. The tests were carried out at frequencies of 0.5, 1 and 3 Hz.

The tests were carried out on the samples of 100 mm diameter and 200 mm height. The sample was prepared at a relative density of 90%. The cyclic triaxial tests were carried out as per the ASTM D5311 [3] and ASTM D3999 [2]. There was no saturation stage in the present study as the samples were to be tested dry. Dry pluviation method was employed for the sample preparation and the vacuum method was used to properly mount the sample on the pedestal. The preparation of the sample confining pressure was applied to the sample, which was then followed by the application of the desired strain amplitude in the form of sinusoidal cyclic loading. A data acquisition system was equipped to continuously monitor and record the data during the tests. The sampling frequency was 500 Hz.

Table 1 Properties of the tire chips used in the study

Description	Tire chips	Sand (Gade and Dasaka 2016)
Specific gravity	1.136	2.62
Maximum dry density (kN/m^3)	5.32	14.18
Minimum dry density (kN/m^3)	5.17	17.04
Cross sectional dimension (mm x mm)	12 × 12 (average)	–
Length (mm)	20 (average)	–
Coefficient of uniformity (C_u)	–	1.36
Coefficient of curvature (C_c)	–	0.95
Soil classification as per USCS	–	SP

3 Results and Discussions

The basic properties of the material used in the study are presented in Table 1. The study investigates the dynamic behavior of the sand and scraps tire chips mixture when subjected to cyclic loading. The typical behaviors of the mixture are shown in Figs. 1, 2, 3. The figure shows the results obtained from the sample tested at 50 kPa confining pressure, 1 Hz loading frequency and 0.3% strain amplitude. Figure 1 shows the response of the deviatoric stress with the strain amplitude at different loading cycles, Fig. 2 shows the variation of the deviatoric stress with a number of loading cycles and Fig. 3 shows the strain amplitude with time. From Fig. 1, it can be said that the stiffness of the material did not degrade much with the application of the loading cycles. Figure 2 shows that the magnitude of maximum deviatoric stress was 123 kPa for the initial cycle, which decreased to around 115 kpa in 30 cycles and then decreased very gradually afterward, i.e. to 109 kPa at the end of 500 cycles.

The dynamic properties, i.e. the shear modulus and the damping ratio, of the mixture were calculated from deviatoric stress versus strain graph. The values were calculated for each loading cycle. The values obtained at the third cycle of the loop were considered as the dynamic properties of the tested sample.

The variation of shear modulus and damping ratio with strain are presented in Figs. 4 and 5, respectively, for different conditions considered in the study. The effect of confining pressure and strain amplitude at different frequencies is also presented.

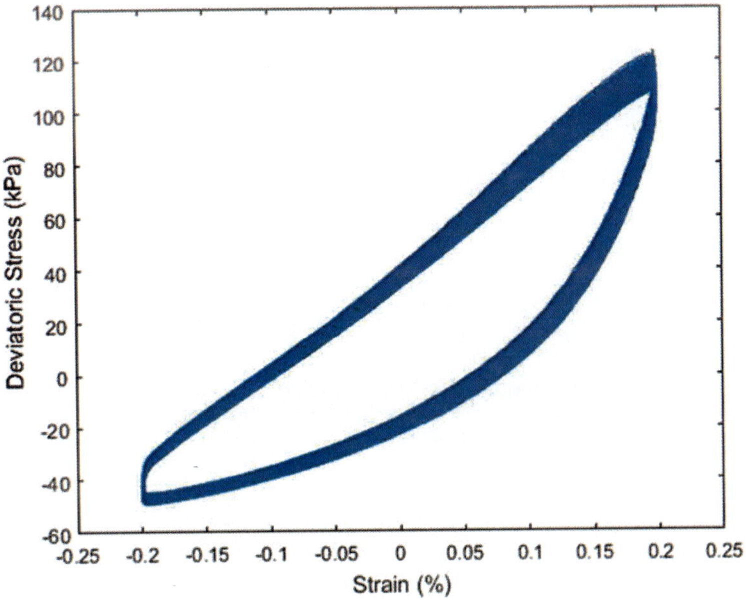

Fig. 1 Variation of deviatoric stress with axial strain amplitude at 0.3% strain amplitude under 50 kPa confining pressure at 1 Hz frequency

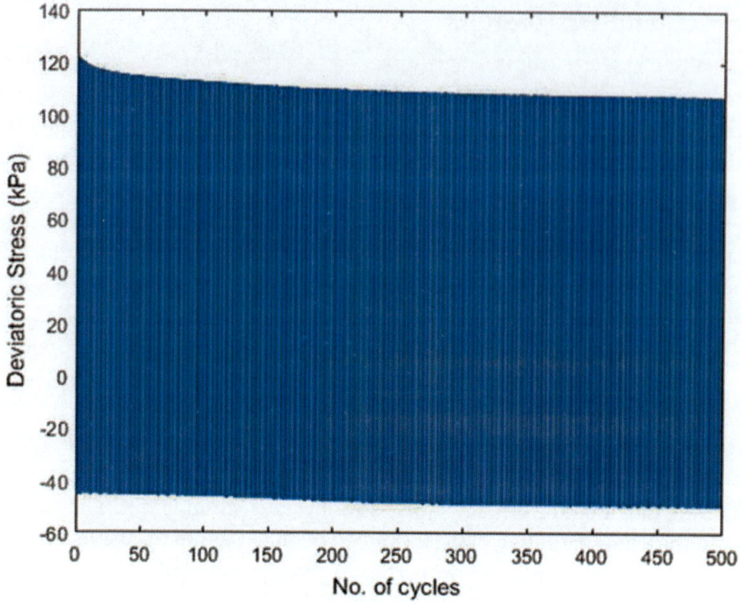

Fig. 2 Variation of deviatoric stress with a number of loading cycles at 0.3% strain amplitude under 50 kPa confining pressure at 1 Hz frequency

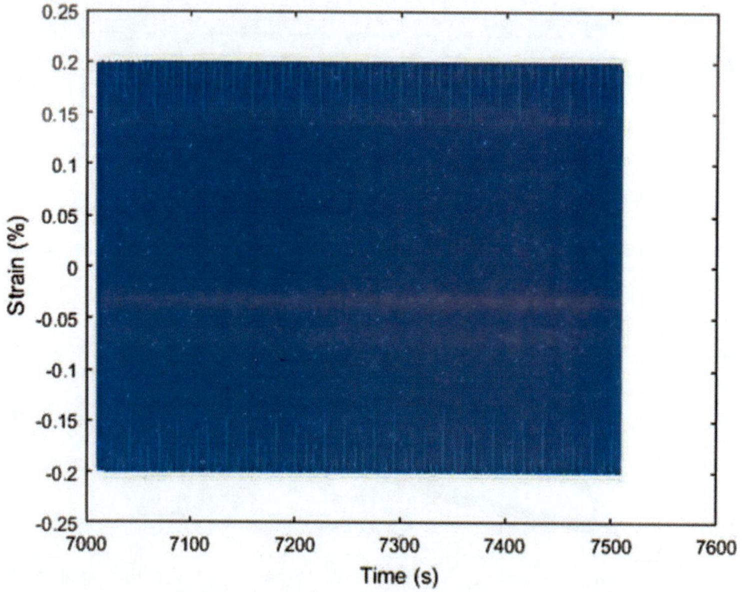

Fig. 3 Variation of axial strain amplitude with time under 50 kPa confining pressure at 1 Hz frequency

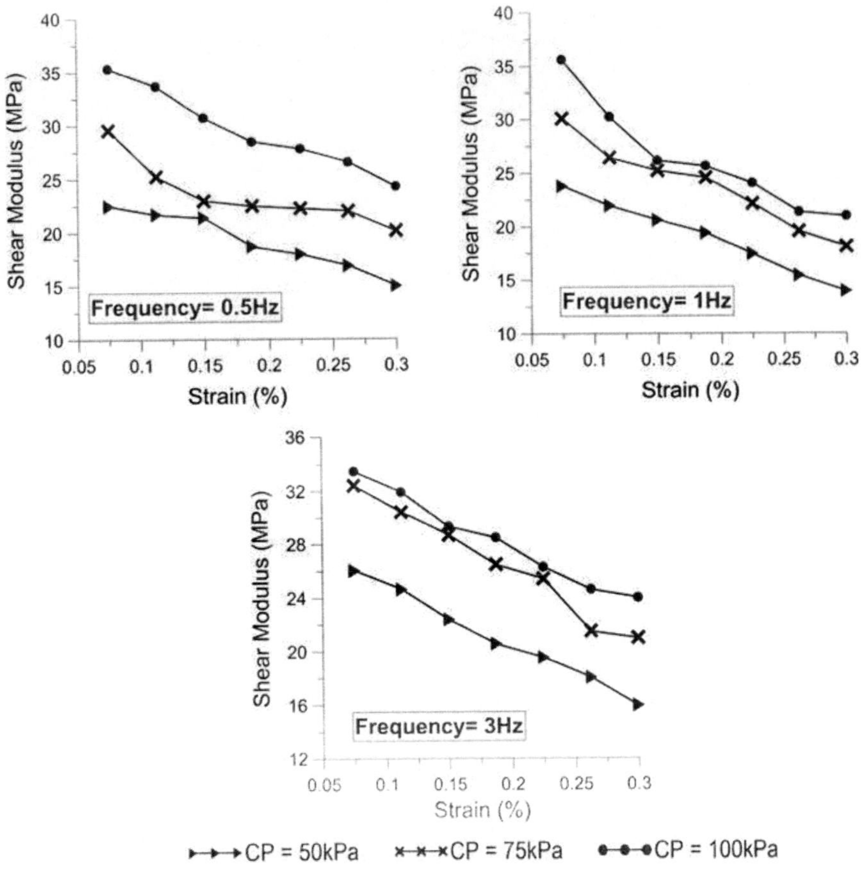

Fig. 4 Variation of shear modulus with strain

From the figures, it can be said that the shear modulus, representing the stiffness of the mixture, and the damping ratio vary with the confining pressure and the strain amplitude. The stiffness of the mixture increases with an increase in the confining pressure and decreases with an increase in the strain amplitude, whereas vice versa is true for the damping ratio. Similar behavior is seen for all the three frequencies considered.

The effect of frequency on the shear modulus and damping ratio is not very clear. For the frequencies of 0.5 Hz, 1 Hz and 3 Hz, the shear modulus values varied between 15 and 35 MPa, 14 MPa and 35 MPa and 16 MPa and 34 MPa, respectively, for different confining pressures and strain amplitudes considered. The damping ratios varied from 0.09 to 0.13 for 0.5 Hz frequency, 0.1–0.13 for 1 Hz frequency and 0.097–0.135 for 3 Hz frequency. For all the three frequencies considered, the values remained similar for different conditions considered.

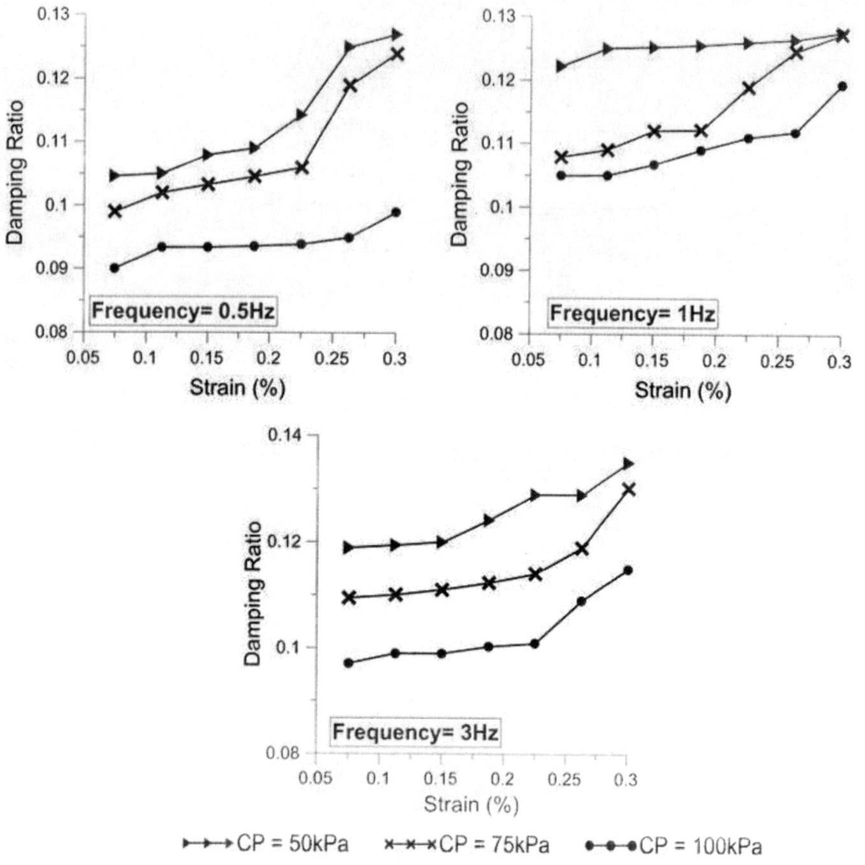

Fig. 5 Variation of damping ratio with strain

4 Conclusions

This study investigated the dynamic properties of a mixture of rubber tire chips and sand using cyclic triaxial apparatus. Sixty-three strain controlled cyclic triaxial tests were carried out on the dry samples. For a better understanding of the property of the mixture, the loading frequency, confining pressures and strain amplitudes were kept as variables.

- The stiffness and damping ratio of the material get highly affected by the confining pressure. The value of the shear modulus increases, whereas the damping ratio decreases with an increase in confining pressure.
- The strain amplitudes also remarkably affected the dynamic properties of the mixture. The shear modulus value decreases and damping ratio value increases with an increase in the strain amplitude.

- The frequency of loading did not show much effect on the dynamic properties. For all the three frequencies considered, the material responded a similar way.

References

1. Abiuchou, T., Tawfiq, K., Edil, T.B., Benson, C.H.: Behavior of a soil-tire shreds backfill for modular block-wall. Recycled Materials in Geotechnics, ASCE 162–172 (2004)
2. ASTM D3999–91: Standard Test Methods for the Determination of the Modulus and Damping Properties of Soils Using the Cyclic Triaxial Apparatus. Reproduced (1996)
3. ASTM D5311–92: Standard Test Method for Load Controlled Cyclic Triaxial Strength of Soil.(1992)
4. ASTMD6270–17: Standard practice for use of scrap tires in civil engineering applications (2017)
5. Bosscher, P.J., Edil, T.B., Kuraoka, S.: Design of highway embankments using tire chips. Geotech. Geo-environ. Eng. ASCE **123**(4), 295–304 (1997)
6. Edil, T.B., Park, J.K., Kim, J.Y.: Effectiveness of scrap tire chips as sorptive drainage material. J. Environ. Eng. ASCE **130**(7), 824–831 (2004)
7. Foose, G.J., Benson, C.H., Bosscher, P.J.: Sand reinforced with shredded waste tires. J. Geotech. Eng. ASCE **122**(9), 760–767 (1996)
8. Hazarika, H., Kohama, E., Sugano, T.: Underwater shake table tests on waterfront structures protected with tire chips cushion. J. Geotech. Geoenviron. Eng. ASCE **134**(12), 1706–1719 (2008)
9. Hazarika, H., Yasuhara, K., Kikuchi, Y., Karmokar, A.K., Mitarai, Y.: Multifaceted potentials of tire-derived three dimensional geosynthetics in geotechnical applications and their evaluation. Geotext. Geomembr. **28**(3), 303–315 (2010)
10. Lee, J.H., Salgado, R., Bernal, A., Lovell, C.W.: Shredded tires and rubber-sand as lightweight backfill. Geotech. Geo-environ. Eng. ASCE **125**(2), 132–141 (1999)
11. Masad, E., Taha, R., Ho, C., Papagiannakis, T.: Engineering properties of tire/soil mixtures as a lightweight fill material. Geotech. Test. J. ASTM **19**(3), 297–304 (1996)
12. Mashiri, M.S., Vinod, J.S., Sheikh, M.N.: Liquefaction potential and dynamic properties of sand-tyre chip (STCh) mixtures. Geotech. Test. J. **39**(1), 69–79 (2016)
13. Reddy, K.R., Stark, T.D., Marella, A.: Beneficial use of shredded tires as drainage material in cover systems for abandoned landfills. Pract. Period. Hazard. Toxic Radioact. Waste Manag. ASCE **14**(1), 47–60 (2010)
14. Shrestha, S., Ravichandran, N., Raveendra, M., Attenhofer, J.A.: Design and analysis of retaining wall backfilled with shredded tire and subjected to earthquake shaking. Soil Dyn. Earthquake Eng. **90**, 227–239 (2016)
15. Tweedie, J.J., Humphrey, D.N., Sandford, T.C.: Tire shreds as lightweight retaining wall backfill: active conditions. Geotech. Geo-environ. Eng. ASCE **124**(11), 1061–1070 (1998)
16. Wolfe, S.L., Humphrey, D.N., Wetzel, E.A.: Development of tire shred underlayment to reduce ground borne vibration from LRT track. Geotechnical engineering for transportation projects, ASCE, pp. 750–759 (2004)

Comparative Analysis of Single and Two-Tiered Geo-Synthetic Reinforced Soil Walls Subjected to Dynamic Excitation

Anindita Gogoi and Arup Bhattacharjee

1 Introduction

Geo-synthetic reinforced soil (GRS) retaining wall is one of the most common applications of reinforced soil where polymeric material geo-synthetic is used as reinforcement. It has gained widespread acceptance in the engineering field as an economic and innovative alternative to earth retaining wall. This technique has been chosen more and more often due to its aesthetics, stability, cost-effectiveness and sound performance during earthquakes. In seismically active areas, GRS walls are constructed in the tiered configuration as it helps to reduce maximum lateral deformation of the wall caused by earthquake loading. Researchers suggested that tiered configuration in geo-synthetic reinforced soil (GRS) retaining wall is needed when considering tall walls since both internal and external stability of the retaining wall are affected by increasing wall height. The use of multi-tiered wall is applicable when the height is greater than 6 m [4]. In FHWA [1] guideline, the tiered wall is termed as superimposed wall and suggests that GRS walls in tiered configuration with smaller wall height reduce vertical stress on the facing element as well as the lateral stress in the whole wall system. For multi-tiered walls, Liu et al. [5] showed that depending upon the offset distance, tiered configurations could considerably reduce residual lateral facing displacement and average reinforcement load.

The objective of the present work is to study the dynamic behavior of single-tiered and two-tiered Geo-synthetic reinforced soil (GRS) retaining wall through numerical simulation. The study is executed by comparing the dynamic behavior of the single-tiered and two-tiered walls with different offset distances under dynamic loading conditions. In the present work, a set of numerical models have been developed using finite element program PLAXIS 2D, which can describe the seismic

A. Gogoi (✉) · A. Bhattacharjee
Department of Civil Engineering, Jorhat Engineering College, Jorhat 785007, India

C. N. V. Satyanarayana Reddy et al. (eds.), *Ground Improvement and Reinforced Soil Structures*, Lecture Notes in Civil Engineering 152,
https://doi.org/10.1007/978-981-16-1831-4_52

behavior of tiered GRS retaining wall under dynamic conditions. To assess the accuracy of the numerical procedure employed for this research work, a validation analysis is performed. After that, numerical analyses of multi-tiered GRS retaining walls are performed for different offset distances. Reinforcement and offset distance are considered as per FHWA [1] guideline.

2 Modeling of Geo-Synthetic Reinforced Soil (GRS) Retaining Wall

In this study, finite element software PLAXIS 2D is used for numerical modeling. Plane strain model of 15 nodded triangular elements is used to discretize the soil medium and other material clusters. The 15 nodded triangular elements are used as it gives high-quality stress results. The geo-synthetic reinforcements are simulated using the five-node geo-grid element and soil-structure interactions are simulated using the five-node thin layer interface element. To simulate the effect of the real construction process of GRS retaining wall, stage construction procedure is implemented. In PLAXIS 2D, this procedure allows for a realistic determination of stresses and displacements. An experimental modular block reinforced soil wall reported by Ling et al. [2] is selected for validation analysis. The experimental model is developed using finite element program PLAXIS 2D and results are compared with experimental results. The comparisons are carried out in terms of parameters such as horizontal displacement, lateral earth pressure and acceleration amplification factor.

2.1 Validation

Ling et al. [2] conducted an experimental study by using large-scale shake table test to observe the seismic performance of modular block reinforced soil retaining walls. The large-scale 2.8 m high modular block geo-synthetic-reinforced soil wall was subjected to significant shaking using the 0.8 g Kobe earthquake. The wall was 2.8 m high, 5 m long and 2 m wide constructed on 20 cm thick foundation. The facing blocks were 24 cm high, 30 cm deep and 45 cm wide by creating an angle of 78 degree with foundation. The wall was backfilled with fine Tokachi sand with a relative density 55% and polyester geogrid (PET) of length 205 cm were placed at a vertical distance of 60 cm. The foundation was constructed with the same type of sand as backfilled soil. To prevent the waves reflecting from the steel walls during shaking, 10 cm thick expanded polystyrene (EPS) boards were placed at the front and back ends of the steel container (Fig. 1).

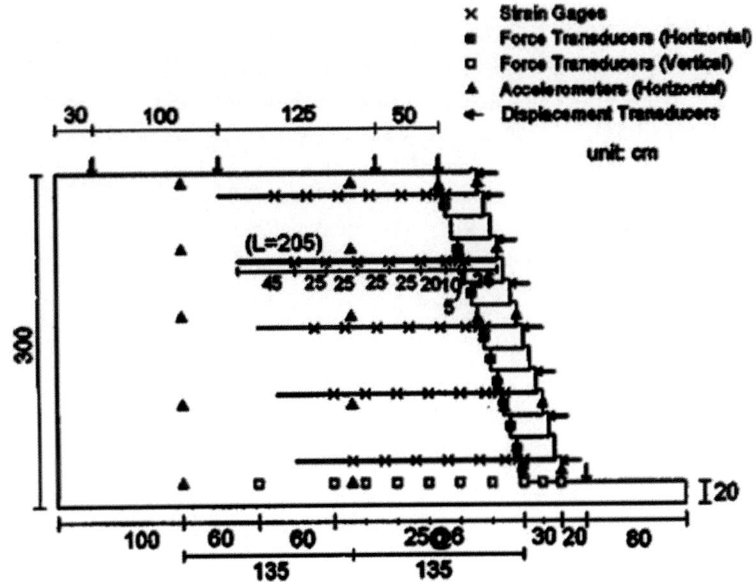

Fig. 1 The geometry of the model used for validation (after Ling et al. [2])

Backfill soil

The linear elastic perfectly plastic Mohr–Coulomb model is considered to represent the sand used in foundation and backfill in the physical model. Basic input parameters for Mohr–Coulomb models are elastic modulus (E), cohesion (c), frictional angle (ϕ), Poisson's ratio (ν) and dilatancy angle (ψ).

Facing Block, Geo-grids and EPS board

The geo-grids are modeled using geo-grid element with modulus of axial stiffness (EA). The facing blocks are modeled as linear elastic material using plate element, which consists of input parameters including elastic modulus (E), Poisson's ratio (ν) and unit weight (γ). The EPS boards are modeled as linear elastic and the input parameters are modulus of elasticity (E), Poisson's ratio (ν) and density (γ).

Interface element

In order to properly simulate the soil-structure interaction, interface elements are used between two different materials. The interface elements are modeled as linear elastic material using interface elements. The roughness of the interaction is modeled by choosing a suitable value for strength reduction factor R_{inter}, which is interlinked to the strength properties of the soil layer. The R_{inter} is chosen as 0.7, 0.65 and 0.5 for the interface between soil–geogrid, soil–concrete and concrete–geogrid.

Dynamic boundary condition

In the numerical model, the side boundary nodes are fixed in the horizontal direction and bottom boundary nodes are fixed in both horizontal and vertical directions. In dynamic analysis, to reduce reflections of seismic waves reaching the model boundaries, special dynamic boundaries are provided. Earthquake load is defined by prescribe displacement and applied at the bottom boundary with a maximum horizontal acceleration of 0.8 g Kobe Earthquake (1995). To simulate the damping of soil, damping ratio of 5% is taken for soil (Fig. 2 and Table 1).

Horizontal Displacement of facing

Figure 3 shows the maximum displacements measured on the experimental wall and the maximum displacements are calculated using the finite element (FE) model. The maximum displacement of the FE model is found to be 75 mm at the top, which is very close to the measured value 72 mm as discussed by Ling et al. [2]. Therefore, good agreement is shown between the FE model and the experimental model.

Lateral soil stress

Figure 4 illustrated the lateral stress obtained at facing during dynamic excitation from the experimental model and FE analysis. Both the walls have maximum lateral stress at the bottom and minimum at the top of the wall. The maximum lateral stresses are found to be 31 kPa and 28 kPa for PLAXIS 2D model and experimental model respectively.

Horizontal acceleration amplification

Horizontal acceleration amplification along the height of the wall obtained from numerical analysis and experimental analysis at the end of dynamic excitation is presented in Fig. 5. The amplification factor is given for the ratio of maximum acceleration in the backfill, typically at the top of the backfill, to the acceleration at the foundation level [2]. It is observed from the figure that the measured and

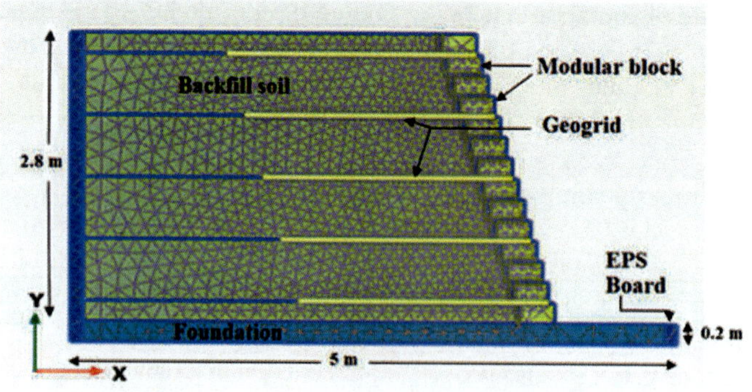

Fig. 2 Numerical mesh (FE model)

Table 1 Material properties of Finite element (FE) model

Backfill soil properties	
Elastic Modulus (kN/m^2)	159 × 10^3
Cohesion (kPa)	1
Mass density (kN/m^3)	14.30
Poisson's ratio	0.33
Angle of friction (°)	38
Dilatancy angle (°)	8
Facing wall	
Elastic Modulus (kN/m^2)	2 × 10^6
Mass density (kN/m^3)	23
Poisson's ratio	0.2
EPS Board	
Elastic Modulus (kN/m^2)	2 × 10^6
Mass density (kN/m^3)	1
Poisson's ratio	0.2
Geogrid	
Axial stiffness (kN/m)	680

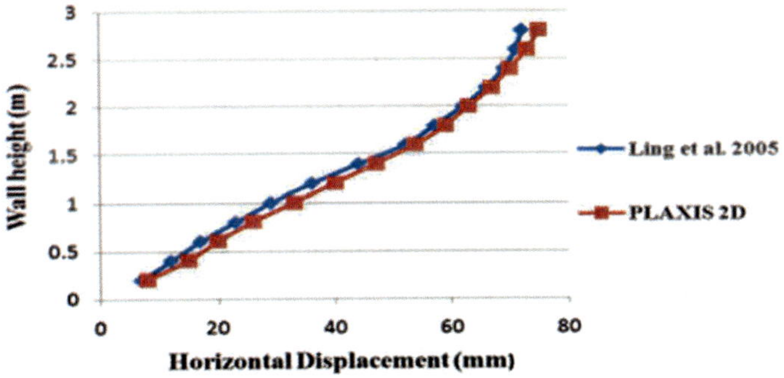

Fig. 3 Horizontal displacement measured by Ling et al. [2] and FE Model

predicted acceleration amplification is in reasonable agreement. The peak acceleration computed by the FE model is 1.2 while the measured value suggested by Ling et al. [2] is 1.1.

The validation procedure adopted for the present work shows that numerical results are in good agreement with experimental results. Therefore, it can be suggested that that the finite element method is capable of simulating the construction behavior of geo-synthetic reinforced soil retaining wall.

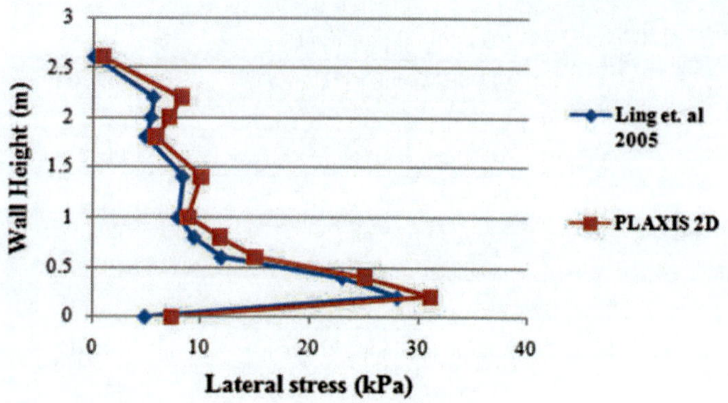

Fig. 4 Lateral stress measured by Ling et al. [2] and FE model

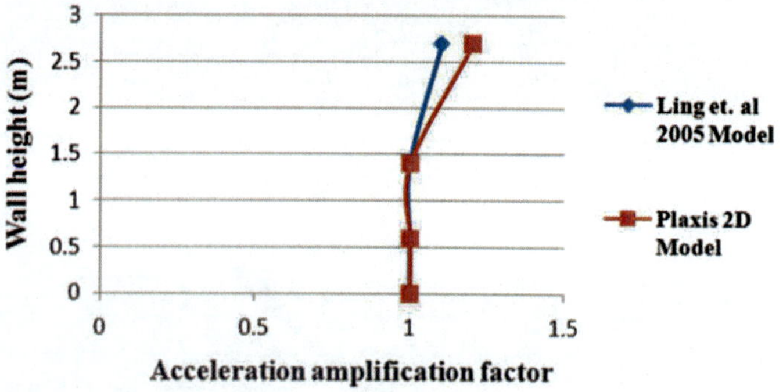

Fig. 5 Horizontal acceleration amplification factor measured by Ling et al. [2] and FE model

3 Numerical Modeling of Multi-Tiered GRS Wall with Offset Distances

In the present work, with the aim of investigating the seismic behavior of tiered geo-synthetic reinforced soil retaining wall under dynamic load, a 9 m height wall is selected for numerical analysis. Using the same model parameters from validate model, single-tiered and two-tiered walls are developed. First, single-tiered wall with 12 geo-grid reinforcement of length 0.7 times of wall is considered for simulation. Furthermore, two-tiered wall of each 4.5 m tier height with different offset distances of 0.5 m, 1.5 m, 2 m, 2.5 m and 3 m is considered for numerical simulations. The offset length and reinforcement length are calculated as per FHWA [1]. A record of the 1989 Loma Prieta with a peak acceleration of 0.25 g is used as input and applied at the base of the walls. The dynamic behavior of walls is studied in terms of parameters such as horizontal displacement, lateral earth pressure, reinforcement load and acceleration amplification factor (Figs. 6, 7, 8).

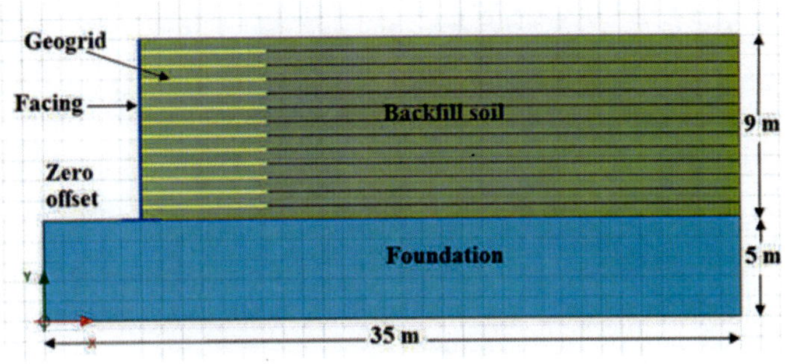

Fig. 6 FE model of Single-tiered GRS wall (Zero offset)

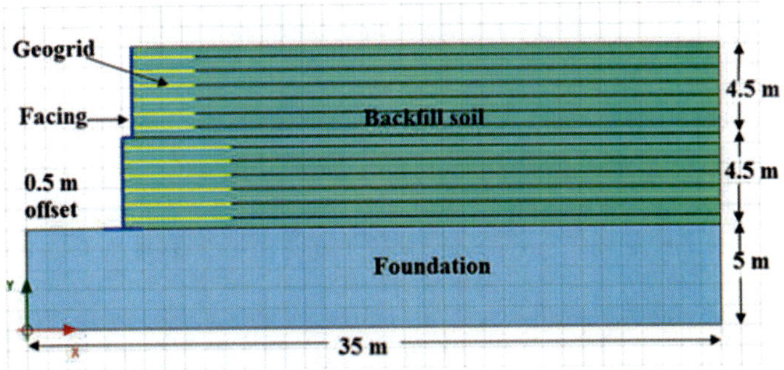

Fig. 7 FE model of Two-tiered GRS wall (0.5 m offset)

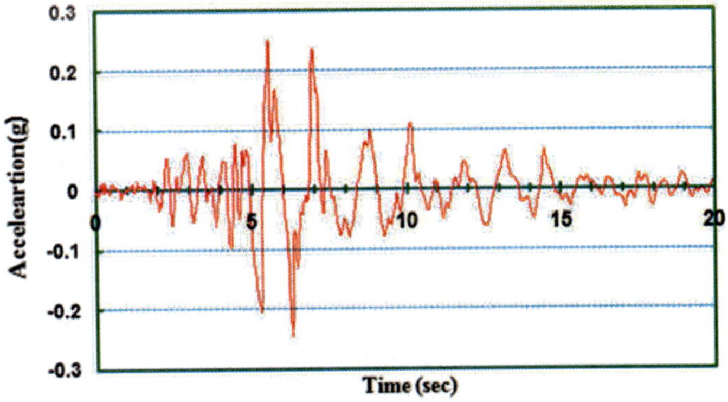

Fig. 8 Acceleration time history for Loma Preita Earthqauke (1989)

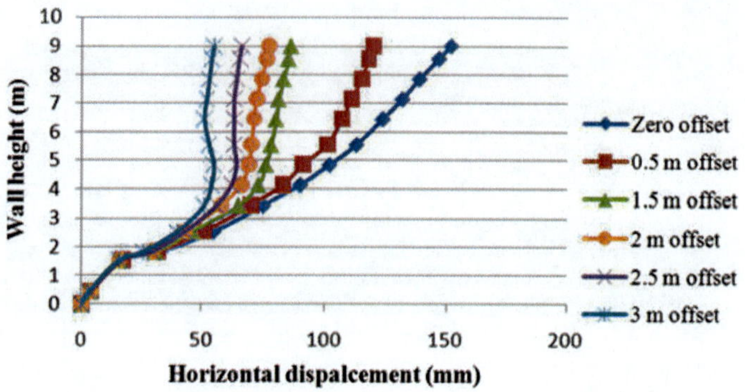

Fig. 9 Wall deformation for single-tiered and two-tiered walls with different offset distances

3.1 Horizontal Displacement of Facing

Figure 9 shows the comparison of lateral displacement of the single-tiered wall and two-tiered wall with different offset lengths. The maximum displacement at the top of the walls is found to be 152 mm, 120 mm, 86 mm, 72 mm, 66 mm and 52 mm for zero offset, 0.5 m offset, 1.5 m offset, 2 m offset, 2.5 m offset and 3 m offset respectively. The results show that maximum displacement decreases with an increase in tier offset length. During an earthquake, retaining wall experienced an additional thrust behind it, which is known as dynamic earth pressure. Due to the effect of dynamic earth pressure, wall exhibits excessive horizontal deformation, but in retaining wall with tiered configuration shows smaller displacement, unlike vertical wall. The smaller displacement is the result of the smaller inertial force comes from the smaller mass in the upper tier. Hence, it can be suggested that by providing adequate offset distances in vertical wall, notable displacement can be reduced.

3.2 Reinforcement Load

The variation of reinforcement load distribution along the height of the wall of all cases is shown in Fig. 10. As can be seen in the figure, the occurrence of the maximum reinforcement load decreases significantly with an increase in tier offset. The maximum reinforcement loads are found to be 14.3 kN/m, 12.89 kN/m, 10.9 kN/m, 9.6 kN/m, 8.4 kN/m and 7.2 kN/m for offset distances of zero offset, 0.5 m offset, 1.5 m offset, 2 m offset, 2.5 m offset and 3 m offset respectively. In reinforced soil retaining wall, internal stability depends on the reinforcement layer, which is basically known as tension resisting component. The role of tensile reinforcement is to resist induced shear deformation due to dynamic force. Therefore, from the results presented in Fig. 10, it can be concluded that by providing offset distance in single-tiered wall, reinforcement load can be significantly reduced.

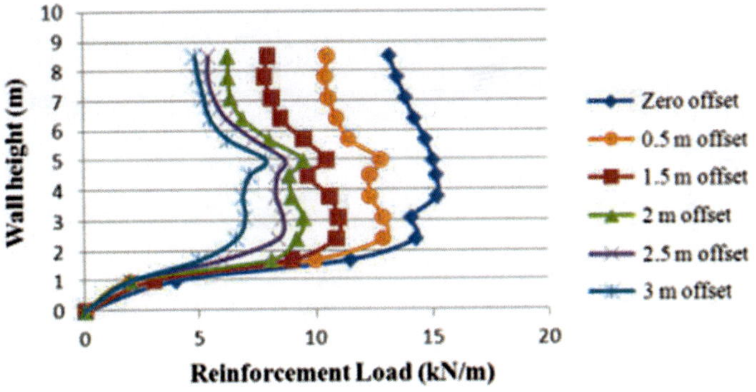

Fig. 10 Reinforcement laod for single-tiered and two-tiered wall with different offset

3.3 *Lateral Soil Pressure at the Face of the Wall*

Figure 11 compares the lateral soil pressure of single-tiered and two-tiered walls with offset distances. As can be seen from the figure, the maximum soil pressure occurs near the base of the wall and decreases linearly toward the top, attained a very small value at the top of the wall for all cases. The maximum lateral soil stresses are found to be 189 kPa, 171 kPa, 160 kPa, 151 kPa, 148 kPa and 135 kPa for zero offset, 0.5 m offset, 1.5 m offset, 2 m offset, 2.5 m offset and 3 m offset, respectively, near the bottom of the walls. The lateral soil pressure at the mid-height is found to be 30.3 kPa, 28.12 kPa, 26 kPa, 24.7 kPa and 24 kPa for 0.5 m offset, 1.5 m, 2 m, 2.5 m offset and 3 m offset respectively. Thus, it can be observed that maximum lateral stress decreases with an increase in tier offset distance.

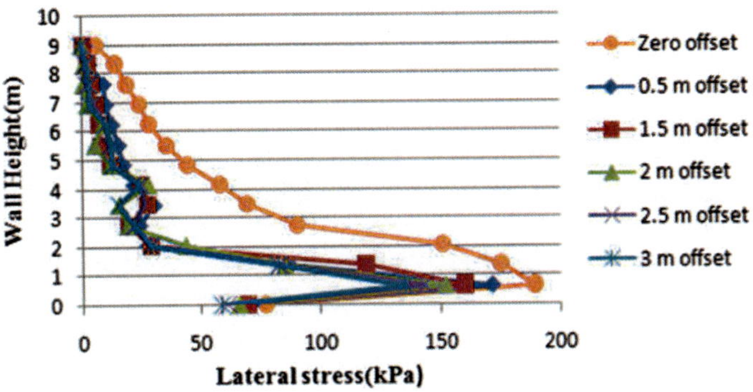

Fig. 11 Lateral soil pressure on the face of the wall for single-tiered and two-tiered walls with different offset distances

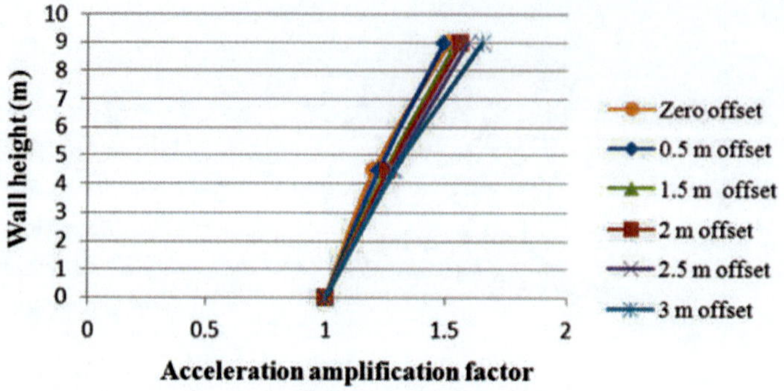

Fig. 12 Horizontal acceleration amplification factor recorded at backfill soil for different tier offset lengths

3.4 Acceleration Amplification Factor

Figure 12 shows the horizontal acceleration amplification recorded at backfill soil for different tier offset lengths. The acceleration amplification factor is defined as the ratio of the maximum acceleration at that point to the acceleration applied at the foundation. The measuring points are selected at a distance of 15 m away from the toe. The acceleration amplification is minimum at the base of the wall and gradually increases along the height of the wall. The peak acceleration amplifications are observed at the top surface of the walls and are found to be 1.51, 1.49, 1.542, 1.56, 1.59 and 1.65 for offset lengths of 0 m, 0.5 m, 1.5 m, 2 m, 2.5 m and 3 m respectively. As can be seen in Fig. 12, the acceleration amplifications are slightly higher in the multi-tiered walls due to the effect of wall facing.

4 Conclusions

In the present work, verified finite element models are used to conduct numerical analysis on the seismic performance of tiered GRS retaining walls. Finite element program PLAXIS2D is used to simulate the construction procedure of single-tiered and two-tiered walls. A record of the 1989 Loma Prieta is used as input earthquake for dynamic excitation. The seismic responses are studied in terms of horizontal displacement of facing, maximum reinforcement load, lateral soil pressure and acceleration amplification.

The following observations are made from the present study:

1. A two-tiered wall can significantly reduce the horizontal displacement and maximum reinforcement load in comparison to the wall with single-tiered configuration. The magnitude of smaller displacement and smaller reinforcement load is due to the effect of smaller soil mass in upper tiers.

2. A two-tiered wall also reduces the maximum lateral stress considerably compared to the single-tiered wall.
3. The acceleration amplifications are high in multi-tiered walls due to the influence of wall facing.

Therefore, it can be seen from the present study that multi-tiered configuration significantly improves the seismic performance of the GRS wall.

Reference

1. FHWA Publication: Mechanically Stabilized Earth Walls and Reinforced Soil Slopes De sign and Construction Guidelines, Vol. I & II. Publication No. FHWA-NHI-10–024, US Department of Federal Highway Administration (FHWA), US (2010)
2. Ling, H.I., Mohri, Y., Leshchinsky, D., Burke, C., Matsushima, K., Liu, H.: Large-scale shaking table tests on modular-block reinforced soil retaining walls. J. Geotech. Geoenviron. Eng. **131**, No. 4©ASCE, ISSN 1090–0241/2005/4 pp.465–476 (2005)
3. Ling, H.I., Yang, S., Leshchinsky, D., Liu, H., Burke, C.: Finite-element simulations of full scale modular-block reinforced soil retaining walls under earthquake loading. J. Eng. Mech. ASCE **135**(5), 653–661 (2010)
4. Liu, H.: Comparing the seismic responses of single and multi-tiered geo-synthetic reinforced soil walls: Geo-Frontiers, ASCE (2011)
5. Liu, H., Yang, G., Ling, H.I.: Seismic response of multi-tiered reinforced soil retaining walls. Soil Dyn. Earthquake Eng. **61**, 1–12 (2014)
6. Reference Manual: PLAXIS 2D, Brinkgreve, R.B.J. (eds): Delft University of Technology & PLAXIS, The Netherlands (2016)
7. Tutorial Manual: PLAXIS 2D, Brinkfreve, R.B.J. (eds): Deft University of Technology & PLAXIS, Netherlands (2016)

Effect of Submergence on Settlement and Bearing Capacity of Sand Reinforced with Pet Bottle Geocell

A. Vismaya, Monica Simon, and P. K. Jayasree

1 Introduction

Natural catastrophes, for example, floods or heavy rain, can lift the groundwater table up to or past the footing level and cause further settlements of shallow foundations. Under submerged conditions, the soil loses its stiffness and settles more. Unlike in cohesive soils, settlement is quick in cohesion-less soil.

Polyethylene terephthalate (PET or PETE) is the material that is used for making most of the single-use plastic bottles, comprising those for water, juices and soft drinks. A large number of plastic bottles are disposed of every year worldwide. Landfills or incinerators are the ultimate destinations of most of the collected plastic wastes and the remaining is reused. To avoid the disposal issues, reuse of plastic is gainful rather than recycling; it is likewise efficient.

Geocells have a very significant role in controlling erosion and for the stabilization of soil on level ground and steep slopes. Also, this cellular system provides channel protection and structural reinforcement for load support and soil retention. A geocell when in-filled with compacted soil makes another composite element that has improved mechanical and geotechnical properties. Lal and Mandal [7] carried out laboratory model load tests using strip footing on fly ash reinforced with cellular reinforcements fabricated using waste PET bottles. They found out that with an increase in height and coverage ratio failure, surcharge pressure is getting improved. Dutta and Mandal [3] conducted laboratory model load tests on fly ash bed overlying soft clay strengthened with cellular reinforcement made using waste PET bottles. A set of model tests were carried out to study the adequacy of cellular mattress

A. Vismaya (✉) · M. Simon · P. K. Jayasree
College of Engineering Trivandrum, Trivandrum 695016, Kerala, India

P. K. Jayasree
e-mail: jayasreepk@cet.ac.in

C. N. V. Satyanarayana Reddy et al. (eds.), *Ground Improvement and Reinforced Soil Structures*, Lecture Notes in Civil Engineering 152,
https://doi.org/10.1007/978-981-16-1831-4_53

Table 1 Results of wide width tensile strength test

Diameter of PET bottle (mm)	Thickness (μm)	Tensile strength (kN/m)	Tensile stiffness (kN/m)
75 (Type A)	24	25	81
65 (Type B)	18	15	62

in improving the bearing capacity of soft clay. Dutta et al. [4] carried out laboratory strain-controlled compression tests on unit cells formed using waste PET bottles and in filled with compacted fly ash or stone aggregates. The quantitative results showed the effectiveness of geocell fabricated using waste PET bottles in load-carrying capacity. Choudhary et al. [2] conducted model load tests by varying different parameters of geocell like the depth of installation of geocell mattress (u—from footing base to the top of geocell mattress) and width of the cellular confinement system (b). Nadaf and Mandal [6] studied the performance of reinforced fly ash slopes with cellular mattresses and strips under strip loading. With the change in edge distances and length of the geocell mattress, significant improvement was observed on the settlement of the backfill. Kazi et al. [5] studied the submergence effect on settlement and load-carrying capacity of surface strip footing on sand bed reinforced with geotextiles with and without wrap-around ends. The results indicate that the rising of water table level is a reason for significant settlement of the footing for both unreinforced and reinforced cases.

The present study states the possibility of using waste PET bottles for ground improvement purposes by installing geocell mattress. The model load tests were conducted under dry and submerged conditions with and without geocell mattress, and the effect on settlement and bearing capacity of sand was studied.

2 Materials

2.1 *Plastic Bottles*

PET bottles were collected locally based on the diameter of PET bottles, they were classified as Type A and Type B. Thickness and tensile stiffness [1] of the plastic bottles were found and are given in Table 1.

2.2 *Soil*

Soil sample was also collected locally. Dry sieve analysis was conducted and soil is categorized as SP (poorly graded sand) as per IS 2720. The properties of soil are summarized in Table 2.

Table 2 Properties of soil

Properties	Values
Specific gravity	2.60
Gravel (%)	0.1
Sand (%)	95
Silt and clay (%)	4.9
Uniformity coefficient, C_u	4
Coefficient of curvature, C_C	1
Classification	SP
Max. dry density (kN/m3)	20.10
Min. dry density (kN/m3)	14.49
Cohesion	Nil

Table 3 Test variables for Type A and Type B PET bottles

Width of geocell (b/B)	1.21	1.79	2.52	3.12
Height of geocell, (h/B)	0.25	0.42	0.58	–

3 Laboratory Model Tests

3.1 Test Variables

After classifying the soil and finding the properties of plastic bottles used, pressure settlement behavior of sand reinforced with PET bottle geocell was studied under submerged and dry conditions. Effect of footing pressure during footing settlement on sand was found out by varying height and width of geocell mattress. Test variables are shown in Table 3.

3.2 Test Setup and Procedure

Laboratory model load tests were conducted on sand with and without reinforcement under submerged and dry conditions in a circular tank of inner dimensions as 600 mm diameter and 400 mm height. A 20-mm-thick rigid steel plate with a diameter of 120 mm was used as a footing material. Vertical loads were exerted to the footing plate using a hand-operated mechanical jack of capacity 78.5 kN. The applied load was measured using a proving ring of capacity 30 kN. The settlements were determined using two dial gauges kept diametrically opposite to each other.

For testing under dry condition, the tank was filled in four layers constituting 350-mm-thick sand bed. Each layer was compacted uniformly using a rammer to attain a relative density of 45%. For testing the sand under submerged condition, a certain amount of water was filled in the tank, and then the first layer of soil was filled such that level of water should be above the sand layer. Then again water was filled up to certain level and next layer of sand was added. The process was repeated for the next two layers also. At this submerged condition, also each layer was compacted uniformly to attain a relative density of 45%.

Figure 1 shows a diagrammatic representation and image of the test setup under dry and submerged conditions.

Figure 2 shows photographs of geocell-reinforced sand at its preparation stage. The plastic bottles were cut manually using scissors and knife and were connected with plastic cable tie (nylon 66 cables) to make the geocell mattress.

Fig. 1 Diagrammatic representation and image of the test setup under dry and submerged conditions

Fig. 2 Preparation of geocell-reinforced sand bed

4 Results and Discussion

For analyzing the results, the bearing pressure (p) and settlement of the footing (s) are represented in non-dimensional forms, so that it can be scaled up to obtain the results in practical cases. Footing pressure (p) is given in terms of bulk unit weight of sand (γ) and footing width (B) as p/γB. The settlement of the footing is given in terms of the footing width (B) as s/B (%).

4.1 *Effect of Mattress Width on Pressure Settlement Response*

The width of geocell mattress is a key element that affects the performance of reinforced sand. The mattress width was differed, such that the height of mattress kept constant at h/B = 0.58 and the mattresses was placed at a constant depth (u/B) = 0.1. Effect of width of mattress was studied under dry and submerged conditions. Figure 3 shows the comparison of results of pressure settlement behavior of sand when the width of geocell mattress is varied under dry and submerged conditions at s/B = 10%.

The results show that the bearing capacity of sand becomes better when reinforced with a mattress having a wider area. From this graph, it can be understood that under submerged conditions, when the sand is reinforced with a geocell mattress of width (b/B) = 1.21, the BC is 1.56 times the BC of unreinforced sand. When b/B is increased 2.57 times (b/B = 3.12), the BC obtained is 2.77 times the BC of unreinforced sand. It implies that the BC can be increased 1.78 times if b/B is increased from 1.21 to 3.12.

On comparing the load-bearing capacity of sand under dry and submerged conditions, it can be seen that the strength of unreinforced submerged sand is 0.52 times the strength of unreinforced dry sand. When the submerged sand is reinforced with

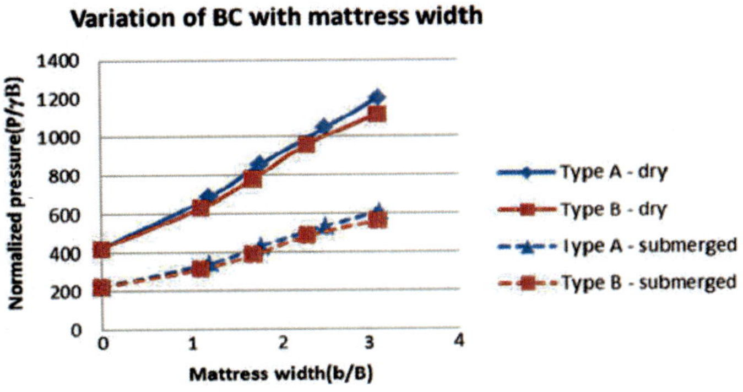

Fig. 3 Variation of bearing capacity with mattress width under dry and submerged condition

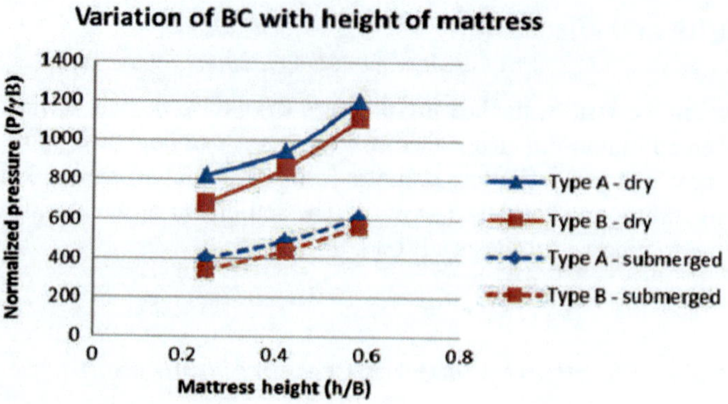

Fig. 4 Variation of bearing capacity with mattress height under dry and submerged condition

the mattress of size b/B = 1.21, the strength is 0.81 times the strength of unreinforced dry sand. When the submerged sand is reinforced with the mattress of greater width (b/B = 3.12), the bearing capacity becomes 1.44 times that of unreinforced dry sand, which means that the bearing capacity of submerged sand can be improved effectively.

4.2 Effect of Mattress Height on Pressure Settlement Response

Height of cellular mattress is another important element and its effect on the pressure settlement response of footing was found out by varying the height for a constant mattress width (b/B) = 3.12 and a constant depth of placement (u/B) = 0.1. Figure 4 shows the comparison of results of pressure settlement behavior of sand when the height of geocell mattress is varied under dry and submerged conditions at s/B = 10%.

From this graph, it can be inferred that the more the height of mattress, the more will be the improvement in pressure settlement response of the sand. Also under submerged conditions, when the sand is reinforced with a geocell mattress of height (h/B) = 0.25, the BC is 1.83 times the BC of unreinforced sand. When the height of mattress is increased 2.33 times (h/B = 0.58), the BC obtained is 2.77 times the BC of unreinforced sand. It implies that the BC can be increased 1.58 times if h/B is increased from 0.25 to 0.58.

When the load-bearing capacity of sand under dry and submerged conditions is compared and it can be seen that the strength of unreinforced submerged sand is 0.52 times the strength of unreinforced dry sand. When the submerged sand is reinforced with the mattress of height, h/B = 0.25, the strength is 0.94 times the strength of unreinforced dry sand. When the submerged sand is reinforced with the mattress of greater height (h/B = 0.58), the bearing capacity becomes 1.44 times that of unreinforced dry sand, which means that load-carrying capacity of submerged sand can be improved significantly.

5 Conclusions

1. Increment in width of the mattress improved the load-carrying capacity of sand significantly. The submerged sand reinforced with the mattress of width b/B = 3.12 improves the bearing capacity by 1.78 times that of submerged sand reinforced with mattress of width b/B = 1.21.
2. Height of the mattress also has a significant role in improving the load-carrying capacity of sand. The bearing capacity of submerged sand can be improved 1.58 times if h/B is increased from 0.25 to 0.58.
3. As the type A bottle has a diameter 15% higher than that of type B and stiffness 31% higher than that of type B, it carried 10% higher footing pressure compared with type B.
4. Comparison of pressure settlement behavior at s/B = 10% shows that
   i. Strength of unreinforced submerged sand can be increased 2.77 times, when the best configuration of reinforcement is chosen
   ii. At this state, it is 1.44 times stronger than unreinforced dry sand.

The tests were carried out on small scale. To validate the outcomes, the tests should be conducted on large scale. Based on the availability, the maximum height of the PET bottle used for making the mattress in the present study was limited to h = 70 mm (or h/B = 0.58). It will be more effective if PET bottles of greater height are used. The type of plastic used is PET (polyethylene terephthalate), if other sorts of plastics such as PP (polypropylene), HDPE or LDPE are used, then the improvement in bearing capacity of sand would be different. Only two types of PET bottles were used in the present study. If bottles with higher stiffness are used, the load-carrying capacity of the soil can be improved further.

References

1. ASTM part 11: Standard test method for breaking force and elongation of textile fabrics (strip method). D5035, West Conshohocken, PA (2011)
2. Choudhary, A.K., Jha, J.N., Sujata Fulambarkar.: Strength and deformation characteristics of bottom-ash reinforced with single geo-cell mattress made of waste pet bottles. Geo-Congress 2019, ACSE, GSP 312, 263–272 (2019)
3. Dutta, S., Mandal, J.N.: Model studies on geocell-reinforced fly ash bed overlying soft clay. J. Mater. Civ. Eng. ASCE **04015091**, 1–13 (2015)

4. Dutta, S., Nadaf, M.B., Mandal, J.N.: An overview on the use of waste plastic bottles and fly ash in civil engineering applications. Proc. Environ. Sci. **35**(2016), 681–691 (2016)
5. Kazi, M., Shukla, S. K., Habibi, D.: Effect of submergence on settlement and bearing capacity of surface strip footing on geo-textile reinforced sand bed. Int. J. Geo-synt. Ground Eng. **1**:4, 1–11 (2015)
6. Nadaf, M.B., Mandal, J.N.: Behavior of reinforced fly ash slopes with cellular mattress and strops under strip loading. J. Hazard. Toxic Radioactive Waste Manage ASCE **21**(4) (2017)
7. Ram RathanLal, B., Mandal, J.N.: Behavior of cellular-reinforced fly-ash walls under strip loading. J. Hazard. Toxic Radioactive Waste Manage ASCE **18**(1), 45–55 (2014)

Effect of Surface Modification on the Performance of Natural Fibres—A Review

R. B. Jiniraj, P. K. Jayasree, and S. P. Anusha

1 Introduction

Ecological sustainability and environmental protection are the biggest concerns of current development. Current developments focus on the utilization of natural geotextiles over their synthetic counterpart because of their environmentally friendly and biodegradable nature [7, 9, 20]. Natural geotextiles produced from natural fibres such as coir, jute, sisal, hemp etc. are non-toxic materials with low cost and renewability [5, 6]. They are energy efficient and hence have less carbon footprint and are also easy to process. They help in the reduction of CO_2 emission and consumption of fuel. There are different sources such as plants, animals and minerals from which natural fibres can be extracted. Fibres originate from plants that include coir, cotton, kapok, sisal, flax jute, kenaf and hemp. Wool and silk are natural fibres from animal sources. Natural fibres possess unique properties because of their growth in natural environment with the assistance of air, soil, sunlight and water. Mechanical and thermal properties of fibres vary with their geographical locations [16]. Natural fibres show fluctuating length, colour, texture, drapability, flexibility, tensile strength and degradation resistance depending on their source of origin, age, environmental conditions and mode of extraction [28].

Nowadays, natural fibres and geotextiles are used in solving various engineering problems. Natural fibre-based geotextiles play a vital role in addressing different engineering problems safely, efficiently and economically. Geotextiles from natural fibres are widely used in slope protection and erosion control [17, 45, 46]. The use of natural fibres such as jute, coir and bamboo as a reinforcing material in soil is an

R. B. Jiniraj (✉) · P. K. Jayasree · S. P. Anusha
College of Engineering Trivandrum, Trivandrum 695016, Kerala, India

P. K. Jayasree
e-mail: jayasreepk@cet.ac.in

C. N. V. Satyanarayana Reddy et al. (eds.), *Ground Improvement and Reinforced Soil Structures*, Lecture Notes in Civil Engineering 152,
https://doi.org/10.1007/978-981-16-1831-4_54

accepted engineering practice. They were used as a reinforcing material in retaining walls, in embankments [31, 45], in pavements [15, 40, 43] and under foundations [12].

Natural fibres lose their strength over time and this loss of strength varies among fibres. The long-term application of natural fibre-based geotextiles is limited due to their higher rate of biodegradation [30]. The fundamental mechanisms involved in the degradation of natural fibre-based geotextiles are hydrolysis and biological action. The numerous pores on the surface of natural fibres have a tendency to absorb moisture and this weakens the cell wall and provides access to microbial attack under favourable conditions of nutrients, pH and temperature. Hence, there is a need for improving the durability characteristics of natural geotextile. The durability of natural geotextiles can be enhanced by surface modification, which cleans and alters the surface properties of fibres. This paper provides an overview of already existing surface modification methods and discusses the effect of surface modification on the performance of natural fibres.

2 Degradation Behaviour of Natural Fibres

Degradation of natural fibres is mainly due to the change in moisture content, microbial activity, temperature and ultraviolet radiation [38, 22, 32]. The chemical composition of natural fibres also plays a vital role in their degradation behaviour. Natural fibres are mainly composed of cellulose, hemicellulose and lignin [6, 25, 29]. Hemicellulose in natural fibres is responsible for moisture absorption, which leads to the swelling of the cell wall of fibre [32]. Microorganisms hydrolyze the cellulose in the cell wall and thereby enhance the degradation [32]. By thermal degradation, hemicellulose, cellulose and lignin in the natural fibres get decomposed [1]. The degradation resistance of natural fibres primarily depends on the lignin content. Lignin in natural fibres reduces the water uptake and microbial activity [2, 25].

The durability of natural fibres exposed to natural soil is influenced by the soil type, organic content of soil, moisture content in soil, type of vegetation and climatic conditions [4, 17]. Environmental exposure of natural fibres leads to the oxidation of lignin and the formation of low mass lignin chains. The newly formed chains are chemically unstable and susceptible to thermal degradation [21]. The degradation behaviour of an environmentally exposed fibre is a function of its depth of burial. Natural fibre, which is placed in the top layer of soil, is the most degradable [13]. Due to the degradation of fibres, major strength loss of fibre-based geotextiles takes place within 1 year of installation [6, 17, 21, 45].

3 Surface Modification of Natural Fibres

Natural fibres possess certain drawbacks such as degradability and higher moisture absorption, which limits their application in long-term engineering problems. Hence, there is a requirement for modification of natural fibres for their long-lasting performance. Surface modification is an effective way of enhancing the durability of natural fibres [26, 34, 37, 42]. There are two methods of surface modification—physical and chemical methods. Physical methods of surface modification are cold plasma, UV bombardment, corona discharge and γ ray. Mercerization, silanization, acrylation and grafting are the commonly used chemical surface modification methods [24, 36]. Alkali treatment [14, 26, 33], enzyme treatment [11], transesterification with vegetable oils [8, 35], UV grafting with monomers [27], surface coating [34, 37, 41, 42, 44] and treatment with specific chemicals [23, 47] are the usually adopted surface modification techniques. Enzyme treatment is an effective method for improving the thermal properties, by removing cellulose and pectin from the fibre surface [11]. Vegetable oils help to remove the water-soluble components and lignin in natural fibres, which leads to the reduction of swelling behaviour [34]. In UV grafting method, monomers form a mechanically stable coating around the fibres by penetrating into the pores on the surface of the fibres. This is the reason for the improved strength of fibres after surface modification by monomers [27]. Surface coating of natural fibres reduces moisture absorption and improves the tensile strength and degradation period [34, 37]. The surface coating is effective in controlling the penetration of aqueous chemicals through the micropores and thereby inhibits the development of fungal growth on fibre surface by 95% [41, 42].

4 Alkaline Treatment

Alkaline treatment using sodium hydroxide is one of the commonly used surface modification methods [14]. Modifications of the surface of coir fibres with sodium hydroxide lead to an increase in wettability and prevent the flotation and segregation of coir fibres. Alkali treatment on coir fibre improves the tensile strength, modulus and debonding strength from polyester [26]. Sodium hydroxide-treated kenaf fibre-based geotextile shows an improvement in tensile strength about 45.5 and 51% compared to untreated geotextiles for wet and dry conditions respectively [39]. Sodium hydroxide treatment on natural fibres forms an amorphous region in which cellulose molecules are separated by large spaces. The sensible OH groups in fibre react with the water molecules occupied in the large paces in the amorphous region and cause a reduction in hydrophilic groups [14]. The alkaline treatment removes the hemicelluloses, pectin, lignin, oil and wax from the surface of fibre and makes the fibre surface becomes more rough uniform [14, 33]. Surface modification by alkali treatment makes the fibre more stiff and strong by changing the structure of cellulose I to cellulose II [19]. It also causes a reduction in fibre diameter and weight of fibres by the removal of surface impurities [19, 33].

5 Chemical Treatments

Alkali, acetylation, silane, acrylation, benzoylation, maleated coupling agents, isocyanates and permanganate treatments are commonly used chemical treatments on natural fibres. Chemical treatments are found to be effective in enhancing the strength of fibre, fibre rigidity and fibre matrix adhesion [18]. Chemical modification on natural fibres such as jute, sisal, hemp and kapok fibres was effective in changing the surface characteristics of the fibre. Chemical modification on natural fibres results in the removal of surface impurities and thereby improved the matrix-fibre adhesion, mechanical interlocking and bonding with chemicals [23]. Structural properties of coir geotextile depend on the cellulose retention capacity and lignin degradation. Coir geotextile treated with lime was effective in conserving the cellulose content during the early stages of degradation [21]. Anggraini et al. [3] investigated the effect of surface modification of coir fibre with nanoparticle on the mechanical properties of lime-treated marine clay. Nanoparticles of Ca(OH)2 and Mg(OH)2 occupy both internal spaces and external surface of the fibre. Nanoparticle in the internal spaces causes an improvement in mechanical properties whereas particle in external surface causes an increased interaction between soil and fibre. The load-carrying capacity and peak strength of lime-treated marine clay increase within cooperation of nanoparticle modified coir fibre. Sodium bicarbonate treatment on sisal fibres removes a large amount of hemicelluloses, which leads to the close packing of fibres and the formation of hydrogen bonds between chains of cellulose [10]. Chemical treatment using chemicals such as sodium periodate, p aminophenol and sodium hydroxide enhances the tensile strength of both non-woven and woven coir geotextiles [47].

6 Summary and Conclusions

1. Surface modification of natural fibres significantly improves the performance of natural fibres.
2. The degradation behaviour of natural fibres mainly depends on their chemical composition. Natural fibres with high cellulose content are susceptible to more degradation. Lignin in fibres is responsible for degradation resistance and reduces microbial activity.
3. Alkaline treatment of natural fibres removes the surface impurities and makes the surface more rough and strong. Sodium hydroxide treatment improves the hydrophobicity and tensile strength of fibres.
4. Chemical treatment on natural fibres enhances the surface characteristics of fibres. Such fibres provide better tensile strength, bonding reaction with chemicals and fibre matrix adhesion.

References

1. Adeniyi, A.G., Onifade, D.V., Ighalo, J.O., Adeoye, A.S.: A review of coir fiber reinforced polymer composites. Compos. Part B Eng. **107305**, 1–10 (2019)
2. Akin, D.E., Eder, M., Burgert, I., Mussig, J.: What are natural fibres? In: Müssig, J. (ed.) Indutrial Applications of Natural Fibres: Structure, Properties and Technical Applications. Wiley, pp. 10–48 (2010)
3. Anggraini, V., Asadi, A., Farzadnia, N., Jahangirian, H., Huat, B.B.K.: Effects of coir fibres modified with Ca(OH)2and Mg(OH)2nanoparticles on mechanical properties of lime-treated marine clay. Geosyn. Int. **23**(3), 206–218 (2016)
4. Balan, K.: Studies on engineering behaviour and uses of geotextiles with natural fibres. Ph.D. thesis, Indian Institute of Technology Delhi, India (1995)
5. Baltazar-Y-Jimenez, A., Bismarck, A.: Surface modification of lignocellulosic fibres in atmospheric air pressure plasma. Green Chem. **9**(10), 1057 (2007)
6. Carvalho, R., Fangueiro, R., Neves, J.: Durability of natural fibers for geotechnical engineering. Key Eng. Mater. **634**, 447–454 (2014)
7. Chattopadhyay, S.N., Pan, N.C., Roy, A.K., Khan, A.: Finishing of jute fabric for value-added products. J. Nat. Fibers **7**(3), 155–164 (2010)
8. Dankovich, T.A., Hsieh, Y.-L.: Surface modification of cellulose with plant triglycerides for hydrophobicity. Cellulose **14**(5), 469–480 (2007)
9. Desai, A. N., Kant, R.: Geotextiles made from natural fibres. In: Geotextiles. Woodhead publication, Cambridge, 61–87 (2016)
10. Fiore, V., Scalici, T., Nicoletti, F., Vitale, G., Prestipino, M., Valenza, A.: A new eco-friendly chemical treatment of natural fibres: Effect of sodium bicarbonate on properties of sisal fibre and its epoxy composites. Compos. Part B Eng. **85**, 150–160 (2016)
11. George, M., Mussone, P.G., Bressler, D.C.: Surface and thermal characterization of natural fibres treated with enzymes. Ind. Crops Prod. **53**, 365–373 (2014)
12. Ghosh, A., Ghosh, A., Bera, A.: Bearing capacity of square footing on pond ash reinforced with jute-geotextile. Geotext. Geomembr. **23**(2), 144–173 (2005)
13. Joy, S., Balan, K., Jayasree, P.K.: Biodegradation of coir geotextile in tropical climatic conditions. In: Proceedings of Indian Geotechnical Conference, December 15–17, pp. 604–606, Kochi, India (2011)
14. Kabir, M.M., Wang, H., Lau, K.T., Cardona, F.: Chemical treatments on plant-based natural fibre reinforced polymer composites: an overview. Compos. Part B Eng. **43**(7), 2883–2892 (2012)
15. Khan, A.J., Huq, F., Hossain, S.Z.: Application of jute geotextiles for rural road pavement construction. Ground Improv. Geosyn. ASCE 370–379 (2014)
16. Komuraiah, A., Kumar, N.S.,Prasad, B.D.: Chemical composition of natural fibers and its influence on their mechanical properties. Mech. Compos. Mater. **50**(3), 359–376 (2014)
17. Lekha, K.: Field instrumentation and monitoring of soil erosion in coir geotextile stabilised slopes—A case study. Geotext. Geomembr. **22**(5), 399–413 (2004)
18. Li, X., Tabil, L.G., Panigrahi, S.: Chemical treatments of natural fiber for use in natural fiber-reinforced composites: a review. J. Polym. Environ. **15**(1), 25–33 (2007)
19. Liu, L., Yu, J., Cheng, L., Qu, W.: Mechanical properties of poly(butylene succinate) (PBS) biocomposites reinforced with surface modified jute fibre. Compos. Part A Appl. Sci. Manufac **40**(5), 669–674 (2009)
20. Maity, S.: Jute needle punched nonwovens: manufacturing, properties, and applications. J. Nat. Fibers **13**(4), 383–396 (2016)
21. Marques, A. R., Santiago de Oliveira Patrício, P., Soares dos Santos, F., Monteiro, M. L., de Carvalho Urashima, D., de Souza Rodrigues, C.: Effects of the climatic conditions of the southeastern Brazil on degradation the fibers of coir-geotextile: evaluation of mechanical and structural properties. Geotext. Geomembr. **42**(1), 76–82 (2014)
22. Miller, D. E., Hoitsma, T. R., White, D. J.: Degradation rates of woven coir fabric under field conditions. Eng. Approaches Ecosyst. Restor. (1998)

23. Mwaikambo, L. Y., Ansell, M. P.: Chemical modification of hemp, sisal, jute, and kapok fibers by alkalization. J. Appl. Polym. Sci. **84**(12), 2222–2234 (2002)
24. Obi Reddy, K., Uma Maheswari, C., Shukla, M., Song, J.I., Varada Rajulu, A.: Tensile and structural characterization of alkali treated Borassus fruit fine fibers. Compos. B Eng. **44**(1), 433–438 (2013)
25. Prambauer, M., Wendeler, C., Weitzenböck, J., Burgstaller, C.: Biodegradable geotextiles – An overview of existing and potential materials. Geotext. Geomembr. **47**(1), 48–59 (2019)
26. Prasad, S.V., Pavithran, C., Rohatgi, P.K.: Alkali treatment of coir fibres for coir-polyester composites. J. Mater. Sci. **18**(5), 1443–1454 (1983)
27. Rahman, M. M., Khan, M. A.: Surface treatment of coir (Cocos nucifera) fibers and its influence on the fibers' physico-mechanical properties. Compos. Sci. Technol. **67**(11–12), 2369–2376 (2007)
28. Rajan, A., Senan, R.C., Pavithran, C., Abraham, T.E.: Biosoftening of coir fiber using selected microorganisms. Bioprocess Biosyst. Eng. **28**(3), 165–17 (2005)
29. Rajan, A., Abraham, T.E.: Coir fiber-process and opportunities. J. Nat. Fibers **3**(4), 29–41 (2007)
30. Ramesh, H. N., Krishna, M. K. V, Meena: Performance of coated coir fibers on the compressive strength behaviour of reinforced soil. Int. J. Earth Sci. Eng. **4**(6), pp. 26–29 (2011)
31. Ranganathan, S.R.: Development and potential of jute geotextiles. Geotext. Geomembr. **13**(6–7), 421–433 (1994)
32. Rowell, R.M.: A new generation of composite materials from agro based fiber. In: Proceedings of the Third International Conference on Frontier of Polymers and Advanced Materials, pp. 659–665, Kuala Lumpur, Malaysia (1995)
33. Saha, P., Manna, S., Chowdhury, S.R., Sen, R., Roy, D., Adhikari, B.: Enhancement of tensile strength of lignocellulosic jute fibers by alkali-steam treatment. Biores. Technol. **101**(9), 3182–3187 (2010)
34. Saha, P., Roy, D., Manna, S., Adhikari, B., Sen, R., Roy, S.: Durability of transesterified jute geotextiles. Geotext. Geomembr. **35**, 69–75 (2012)
35. Saha, P., Manna, S., Sen, R., Roy, D., Adhikari, B.: Durability of lignocellulosic fibers treated with vegetable oil–phenolic resin. Carbohydr. Polym. **87**(2), 1628–1636 (2012b)
36. dos Santos, J.C., de Oliveira, L.Á., Gomes Vieira, L.M., Mano, V., Freire, R.T.S., Panzera, T.H.: Eco-friendly sodium bicarbonate treatment and its effect on epoxy and polyester coir fibre composites. Constr. Build. Mater. **211**, 427–436 (2019)
37. Sanyal, T., Chakraborty, K.: Application of bitumen coated jute geotextile in bank protection works in Hoogly estuary. Geotext. Geomembr. **13**(2), 127–132 (1994)
38. Sarsby, R. W.: Use of limited life geotextiles (LLGs) for basal reinforcement of embankments built on soft clay. Geotext. Geomembr. **25**(4–5), 302–310 (2007)
39. Shirazi, M. G., A Rashid, A. S., Nazir, R. B., Abdul Rashid, A. H., Kassim, A., Horpibulsuk, S.: Investigation of tensile strength on alkaline treated and untreated kenaf geotextile under dry and wet conditions. Geotext. Geomembr. **47**, 522–527 (2019)
40. Subaida, E. A., Chandrakaran, S., Sankar, N.: Laboratory performance of unpaved roads reinforced with woven coir geotextiles. Geotext. Geomembr. **27**, 204–210 (2009)
41. Sumi, S., Unnikrishnan, N., Mathew, L.: Experimental investigations on biological resistance of surface modified coir geotextiles. Int. J. Geosyn. Ground Eng. **2**(4) (2016)
42. Sumi, S., Unnikrishnan, N., Mathew, L.: Durability studies surface—modified coir geotextiles. Geotext. Geomembr. **46**, 699–706 (2018)
43. Sudarsanan, N., Mohapatra, S. R., Karpurapu, R., Amirthalingam, V.: Use of natural geotextiles to retard reflection cracking in highway pavements. J. Mater. Civil Eng. **30**(4), 04018036 (2018)
44. Tiwari, N., Satyam, N.: An experimental study on the behavior of lime and silica fume treated coir geotextile reinforced expansive soil subgrade. Eng. Sci. Technol. Int. J. 1–9 (2020)
45. Vishnudas, S., Savenije, H.H.G., Van der Zaag, P., Anil, K.R., Balan, K.: The protective and attractive covering of a vegetated embankment using coir geotextiles. Hydrol. Earth Syst. Sci. **10**(4), 565–574 (2006)

46. Vishnudas, S., Savenije, H.H.G., Van der Zaag, P., Anil, K.R.: Coir geotextile for slope stabilization and cultivation—a case study in a highland region of Kerala, South India. Phys. Chem. Earth Parts A/B/C **47–48**, 135–138 (2012)
47. Vivek, R.K., Parti, R.: Effect of chemical treatment on the tensile strength behavior of coir geotextiles. J. Nat. Fibers 1–15 (2018)

Comparison of Factor of Safety Between LEM and FEM for Geotextile Reinforced Embankment on Difficult Foundation

Jigisha Vashi, Atul Desai, Chandresh Solanki, and Babu V. Sundararaman

1 Introduction

Embankment slope stability analysis using computers is an easy task when the slope configuration and the soil parameters are known. However, the selection of slope stability analysis method is not an easy task and effort should be made to accumulate the field conditions and the failure observations in order to understand the failure mechanism, which determines the slope stability method that should be used in the analysis. Deformation and stability are the two main concerns for reinforced slopes. An accurate assessment of deformation in a reinforced slope can only be achieved through stress deformation analysis, such as finite element analysis. The stability of reinforced slope, on the other hand, can be evaluated using either a limit equilibrium method or a stress deformation analysis.

Limit equilibrium methods (LEMs) are still the most common analytical approaches in recent design practices for reinforced soil slopes. They have been used extensively for many years for the analysis of natural and manmade slopes. LEM uses the Mohr–Coulomb failure criterion to determine the shear strength along the slip surface. They are based on force and moment equilibrium. At failure, the shear strength is fully mobilized along the critical slip surface. The FS against slope failure is calculated as the ratio of the available shear strength to the mobilized shear strength. In LEM analysis, the sliding mass is divided into slices, determination of the shear

J. Vashi (✉)
SVKM's NMIMS Mukesh Patel School of Technology Management & Engineering, Mumbai 400056, India

A. Desai · C. Solanki
Civil Engineering Department, NIT Surat, Gujarat 395007, India

B. V. Sundararaman
Civil Engineering Division, NFC, Hyderabad 500062, India

C. N. V. Satyanarayana Reddy et al. (eds.), *Ground Improvement and Reinforced Soil Structures*, Lecture Notes in Civil Engineering 152,
https://doi.org/10.1007/978-981-16-1831-4_55

and normal inter-slice forces is made, and appropriate force and/or moment equilibrium equations are satisfied for static equilibrium conditions. The LEM chosen for this study is Bishop's simplified method (BSM) and is mainly applied to determine the Global Stability of GRE.

The finite element method (FEM) has been used to analyze different types of geotechnical structures such as earth dams, embankments, shallow and deep foundations, slopes and retaining walls. The application of the finite element method to reinforced soil structures is relatively recent. Reinforced soil is a complex system that involves interactions between different structural components and soil. Since the procedure itself is very sophisticated, the application to design is rare. However, FE analysis renders additional information compared to traditional LE analysis. FEM uses the soil stress–strain behavior for slope stability modeling. FEM preserves global equilibrium until failure is reached and monitors progressive failure up to and including overall shear failure. FEM approach divides the model into a number of pieces or elements of a mesh. Stresses and strains are calculated using the constitutive laws for materials comprising of the slope stability model. Failure occurs naturally through the zones in which the soil shear strength is unable to sustain the applied shear stresses. The shear strength reduction technique enables the FEM to calculate equivalent FS. FEM has been mainly applied to predict the reinforcing stress, strains and the deformation of the GRE (internal stability).

Evaluating the stability of slopes is through the determination of FS against failure under a given set of conditions. The FS is commonly defined as the ratio of the resisting forces to driving forces along a potential failure surface. FS of 1.0 means the driving and resisting forces are at equilibrium. Greater FS indicates increased stability, whereas a lower FS suggests that the slope is unstable. In the present study, an analytical GRE slope stability analysis was carried out by software using LEM and FEM. The purpose of the study was to compare the LEM and FEM analyses, compare the similarities and differences in the FS results computed using these different methods. This study attempts to guide the development of more accurate LEM analysis or arrive at a suitable LEM procedure to be adopted in given boundary conditions based on the results obtained from FEM analysis or vice versa.

2 Slope Geometry and Modeling

A typical analytical model of 8 m high embankment, crest width 20 m and having slope angles of 58° at base with a berm of 1 m width at 4 m height and slope angle 64° above the berm was considered. Providing a berm in the embankment can increase the factor of safety of a reinforced structure [1–3] (Table 1). The embankment was assumed to be placed over a 2 m thick embankment foundation overlying a relatively soft layer of 5 m thickness. A nominal height of 8 m was assumed based on commonly adopted industry practice of vertical clearance required for flyover openings, which is 6 m as per [4]. The horizontal crest width of 20 m was used considering a four-lane highway each of 7.5 m wide carriageway with 2 m wide median at center

Table 1 Geometry ranges

Parameter (Fig. 1)	Range of values examined
D_1	2 m
D_2	5 m
B	20 m
β_1	58°
β_2	64°
H	8 m
X	5–24 m

and walkways of 1.5 m on either side of the highway. The range of slopes 58°–64° was selected by running pilot model of GRE with PET geotextile. The present study covers all practical ranges of slope angle feasible for site application from this configuration because, in practice, we encounter slopes that are steeper than 26.57° (2H: 1 V) or 45° (IH: 1 V). The modeled embankment was reinforced by layers of geotextile, covering the whole width of embankment. The vertical spacing of geotextile is varied from 0.5 m and 0.4 m. As suggested by Shukla [1], the placement of the reinforcement layer, with respect to the foundation soil surface, was limited to 0.3–0.5 m. A nominal surcharge of 50 kPa has been used for modeling the traffic load as commonly adopted in practice [5]. Figure 1 shows the geometry of reinforced earth embankment modeled in software for both LEM and FEM analyses. The soil properties in embankment fill and foundation to have a realistic model are mentioned in Table 2.

For FEM analysis, the above model was run for mesh generation in GEO5-FEM software. The finite element mesh used in these analyses involved 2037 elements with six nodes. Figure 1 shows the assumed boundary conditions and distinguished

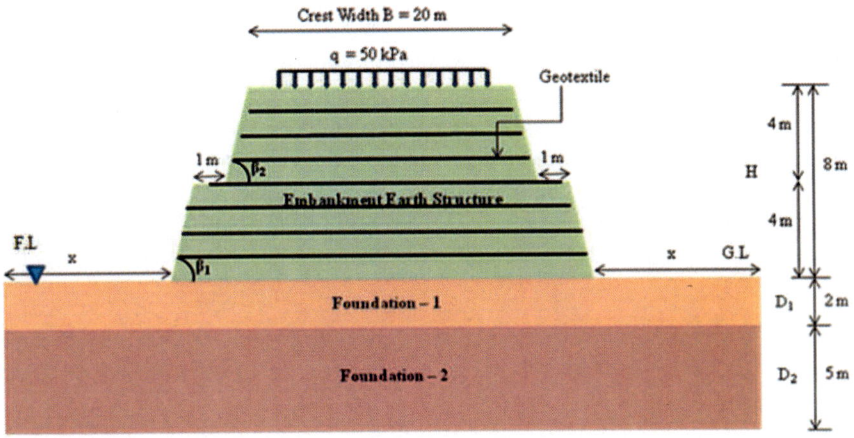

Fig. 1 Geometry of models (reinforced earth embankment)

Table 2 Properties of soil material of foundation and earth structure

Properties	Type		
	Earth structure	Foundation 1	Foundation 2
Unit weight, γ (kN/m^3)	14.12	14.12	20.5
Saturated unit weight, γ_{sat} (kN/m^3)	19.06	19.06	25.0
Cohesion, c_{ef} (kPa)	15	15	5
Angle of internal friction, φ_{ef} (deg)	30	30	15
Poisson's ratio, ν	0.30	0.30	0.42
Elastic modulus, E (MPa)	0.8–16.66[a]	0.8–16.66[a]	3.0
Dilation angle, Ψ (deg)	0.0	0.0	0.0
Biot parameter, α	1.0	1.0	1.0
Material model	Mohr–Coulomb		

[a]Same as the embankment fill material

layers according to the representative materials. The base of the foundation has been fixed at the boundary condition. Boundary conditions for the right and left sides of the embankment foundation are considered to be rollers allowing only vertical deformation and were typically located at $x = 5$, 12 and 24 m from the toe of the embankment (Table 1).

The geotextile in the reinforced earth embankment is considered as a bar element to allow both vertical and horizontal deformations but not to allow rotation. The analyses were performed for embankments reinforced with geotextiles ranging in "moduli" from 50 kN/m to 2000 kN/m (axial stiffness of a geotextile is expressed as the force per unit width per unit strain (kN/m) commonly referred to as the "secant modulus" of the geotextile). Also it was assumed that for a given model, each layer of geotextile has same tensile strength or stiffness and is placed horizontally. Soil parameters of the backfill are based on lab test results [6]. Parameters of foundation 2 are determined by analysis based on the measured data from the literature [7, 8]. Thus in this study; the analytical modeling of earth embankment with geotextile-reinforcement was performed using the GEO5-slope stability software and GEO5-FEM software.

GEO5-slope stability analysis was carried out for both normal and flooded conditions. FEM analysis was carried out considering the worst condition, i.e. the model was analyzed for flooded condition only. Flooded or saturated condition means environmental condition considering F.L at G.L in heavy rainfall area and normal condition means non-flooding condition with W.T at 8–10 m below the Foundation 2.

3 Results and Discussion

Geotechnical software is used to analyze the stability of embankment slopes. Most of these software are designed to calculate the weakest points of the prepared model and draw a slip surface. The behavior of the model is strongly controlled by the geological built-up and water content especially when foundation subsoil is soft desiccated clay-type deposits. In the present study, GEO5 slope stability and GEO5-FEM geotechnical software packages were used to perform the analysis.

GEO5-slope stability package is based on LEM and GEO5-FEM package is based on FEM, specifically intended for the two-dimensional analysis of deformation and stability in geotechnical engineering projects, to obtain the estimates of factor of safety by critical slip surfaces and is built on the same original user-friendly interface platform as the GEO5-geotechnical software. For a better understanding of the differences and method of analysis between GEO5-LEM and GEO5-FEM, FS results obtained from both GEO5-LEM and GEO5-FEM were compared.

The weakest failure plane of the reinforced embankment model was determined, where the factor of safety had a minimum value. GEO5-slope stability software calculations were applied (Fig. 2a; Table 3). The model was considered safe at minimum calculated factor of safety of 1.5. For comparison and to understand the effect of critical slip circle, the same model was analyzed with GEO5-FEM software. The analyses with the GEO5-FEM underestimated the nature of slip surface (Fig. 2b). The nature of slip surface analyzed with GEO5-FEM was different from that analyzed with GEO5-slope stability (Fig. 2a).

GEO5-FEM works with combined factor of safety. Consequently, a factor of safety 1.0 calculated with GEO5-FEM is equal to a factor of safety of 1.5 calculated with the conventional method of GEO5-slope stability software. [7] also observed the differences between two methodologies. Thus, the FEM model was stated or considered safe with minimum calculated factor of safety of 1.0 as shown in Table 3.

Global FS of an embankment slope cannot be analyzed with GEO5-FEM software. Therefore, the global stability factor was also determined using GEO5-slope stability software. This can be related to the fact that in finite element modeling, the location of the slip surface cannot be manually adjusted or determined [9].

In order to analyze the effects of boundary distance from the embankment toe on the FS, the proposed model of unreinforced and reinforced embankment adopting a berm of 1 m at 4 m height was analyzed by GEO5-slope stability (Bishop's limit state stability analysis) with boundaries on either side of the of the foundation typically located at 5 m, 12 m and 24 m from the toe of the embankment. No noteworthy influence on FS was observed for variation on boundary distance beyond 12 m on either side of the toe of embankment.

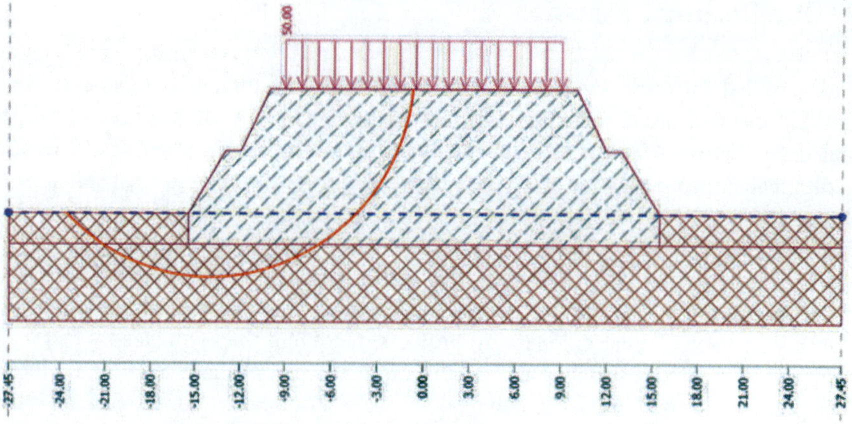

(a) Model analyzed in *GEO5*-slope stability (*LEM* based)

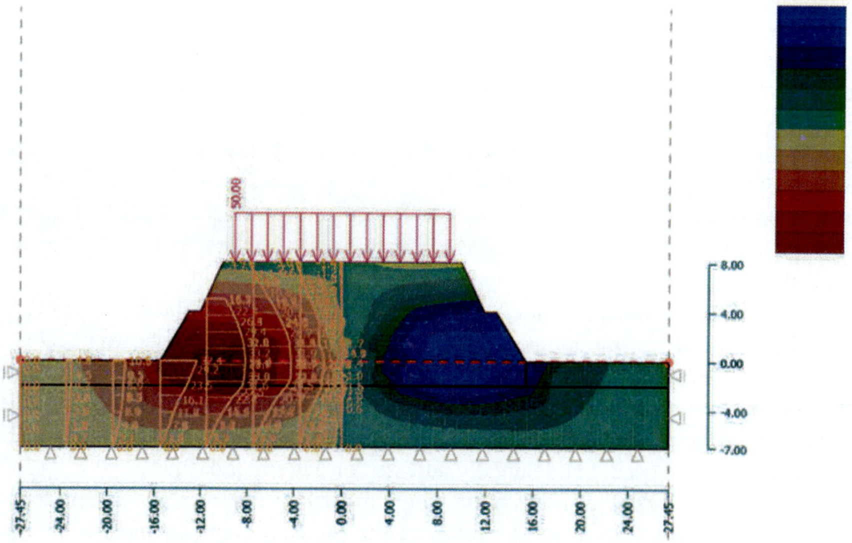

(b) Model analyzed in *GEO5-FEM* version

Fig. 2 Nature of failure slip circle (indicating range of movement as per color index)

4 Conclusions

The overall stability of GRE on difficult soils was analyzed with software having two different methodologies and fundamental approaches:

i. GEO5-FEM, a model based on the finite element method and

Table 3 Analysis of *FS* for unreinforced and reinforced earth embankment by GEO5-slope stability and GEO5-FEM

Tensile strength/stiffness (kN/m)		10/50		10/100		10/200		20/500		40/1000		80/2000	
Unreinforced	Reinforcement	With FL	Without FL	With FL	Without FL	With FL	Without FL	With FL	Without FL	With FL	Without FL	With FL	Without FL
F.S (Bishops) $FS_{without\ FL} =$ 1.40 $FS_{with\ FL} =$ 1.20	Model with berm of 1 m and 5.0 m extra margin (GEO5-slope stability)[a]												
	$S_V = 0.5$ m	1.45	1.55	1.46	1.56	1.47	1.57	1.56	1.65	1.72	1.82	2.07	2.16
	$S_V = 0.4$ m	1.47	1.57	1.48	1.58	1.49	1.59	1.60	1.70	1.81	1.91	2.14	2.27
Model with berm of 1 m and 12 m extra margin (GEO5-slope stability)[a]													
F.S (Bishops) $FS_{without\ FL} =$ 1.25 $FS_{with\ FL} =$ 1.24	$S_V = 0.5$ m	1.40	1.52	1.41	1.53	1.51	1.54	1.51	1.63	1.64	1.78	1.87	2.03
	$S_V = 0.4$ m	1.43	1.55	1.44	1.56	1.45	1.57	1.55	1.67	1.71	1.86	2.01	2.15
Model with berm of 1 m and 24 m extra margin (GEO5-slope stability)[a]													
F.S (Bishops) $FS_{without\ FL} =$ 1.20 $FS_{with\ FL} =$ 1.30	$S_V = 0.5$ m	1.41	1.53	1.42	1.54	1.43	1.55	1.51	1.63	1.64	1.78	1.88	2.03
	$S_V = 0.4$ m	1.44	1.55	1.45	1.56	1.46	1.57	1.55	1.67	1.71	1.86	2.01	2.16
Model with berm of 1 m and 5.0 m extra margin (GEO5-FEM)[b]													
F.S (FEM) $FS_{without\ FL} =$ 0.9 $FS_{with\ FL} =$ 1.01	$S_V = 0.5$ m	0.90	0.90	0.90	0.90	0.90	0.90	0.90	0.90	0.90	0.90	0.90	0.90
	$S_V = 0.4$ m	0.90	1.11	0.90	1.11	0.90	1.11	0.90	1.11	0.90	1.11	0.90	1.11

(continued)

Table 3 (continued)

Tensile strength/stiffness (kN/m)		10/50		10/100		10/200		20/500		40/1000		80/2000	
Unreinforced	Reinforcement	With FL	Without FL	With FL	Without FL	With FL	Without FL	With FL	Without FL	With FL	Without FL	With FL	Without FL
Model with berm of 1 m and 12 m extra margin (GEO5-FEM)[b]													
F.S (FEM)	$S_V = 0.5$ m	0.73	0.90	0.73	0.90	0.73	0.90	0.73	0.90	0.73	0.90	0.73	0.90
$FS_{without\ FL} =$ 0.81 $FS_{with\ FL} = 0.90$	$S_V = 0.4$ m	0.66	1.03	0.66	1.03	0.66	1.03	0.66	1.03	0.66	1.03	0.66	1.03
Model with berm of 1 m and 24 m extra margin (GEO5-FEM)[b]													
F.S (FEM)	$S_V = 0.5$ m	0.81	0.81	0.81	0.81	0.81	0.81	0.81	0.81	0.81	0.81	0.81	0.81
$FS_{without\ FL} =$ 0.90 $FS_{with\ FL} = 0.81$	$S_V = 0.4$ m	0.73	0.59	0.73	0.59	0.73	0.59	0.73	0.59	0.73	0.59	0.73	0.59

Note (a): For same Sv with margin 5, 12, 24 m shows for limit state (Bishop's) analysis that FS is constant for extra margin exceeding 12 m. FS is maximum within the margin of 5–12 m. (b): Table shows increased stiffness for same tensile strength, increases FS of slip circle. The increase is apparent for higher tensile strength /stiffness in 20/500 to 80/2000 range.

ii. GEO5-slope stability model based on limit equilibrium method.

LEM was simpler and faster whereas FEM required tedious input parameters for the same model to be analyzed. It is seen that, for analyzing FS, the critical slip circle behavior of GRE by LEM is best suited compared to FEM analysis, which is unable to give a clear-cut idea about the FS and nature of slip circle analysis. Therefore, LEM checked the overall safety of model whereas approximate durability and displacement of model were attempted by FEM analysis.

The LEM analysis does not evaluate the inter-slice forces, which is dependent on a number of factors including stress–strain deformation characteristics of the materials in GRE. The FEM is an appropriate tool for the investigations of deformation and the behavior of the GRE. The greater rigidity of the embankment and foundation fill material, greater the deformation of the embankment. But the deformation of the embankment face cannot be exactly estimated by integrating the strains of the reinforcement since these strains do not include the external factors (e.g., foundation settlement or global embankment rotation). Therefore, FEM checked the local stability of the model.

The LEM and FEM methods used in this study provide fairly consistent FS. When a relatively same critical failure surface is analyzed, smaller values of FS are related to some stress redistribution that occurs inside of the soil mass simulated by the FEM but not considered in the simplified hypotheses of the LEM. Also, FEM shows a high concentration of strain near the toe of the slope.

With the understandings of different analysis approaches between LEM and FEM, as well as knowing the advantages and limitations, it is suggested to use both methods for GRE stability analysis. Both methods have their own benefits and limitations and both methods can be used to provide an estimate of safety factors and slip surfaces. GEO5-slope stability for nature of failure of circular slip surface and overall/global stability, while GEO5-FEM for deformation, strain behavior and local stability thus providing the best economical and safe design.

References

1. Shukla, S.K.: Geosynthetics and their applications. Embankments, Thomas Telfords Ltd (2002)
2. Silva, A.R.L., Palmeria, E.M.: Stability of geosynthetic reinforced embankment on soft soil. In: Proceeding of 12th Brazilian Conference on Geotechnical Engineering, pp. 1213–122. Brasilia, Brazil (1998)
3. Jakobson, B.: The design of embankment on soft soil. Geotechnique **1**, 80–90 (1948)
4. IRC 6: Standard specification and code of practice for road bridges, Section–II (Load & Stresses), Fourth Revision (2000)
5. Wulandari, P.S., Tjandra, D.: Determination of optimum tensile strength of geogrid reinforced embankment. In: International Civil Engineering Conference, Towards Sustainable Civil Engineering Practice, pp. 187–194 (2006)
6. Vashi, J.M., Desai, A.K., Solanki, C.H.: Behavior of geotextile reinforced flyash + clay-mix by laboratory evaluation. Geomech. Eng. **5**(4), 331–342 (2013)
7. Borges, J.L., Cardoso, A.S.: Overall stability of geosynthetic—reinforced embankments on soft soils. Geotext. Geomembr. (20), 395–421 (2002)

8. Desai, M.D.: Ground property characterization from in-situ testing. Published by IGS Surat Chapter (2005)
9. Gorog, P., Torok, A.: Slope stability assessment of weathered clay by using field data and computer modelling: a case study from Budapest. Nat. Hazards Earth Syst. Sci 417–422 (2007)

Experimental Investigation on Bamboo-Made Cellular Mattress Reinforced Fly Ash Beds Overlying Soft Clay

Sushovan Dutta, Ankita Kumar, and J. N. Mandal

1 Introduction

Planar 2D reinforcements like geotextiles, geogrids can only provide tensile membrane support. Three-dimensional cellular reinforcement also called as geocell mattress, basically made of interconnected multiple cells, can provide the anchorage support to the loaded area through tensile membrane effect as well as provides the lateral confinement and interface wall shear resistance to the infill material, delivering improved results as compared to the planar form of reinforcements. The footing load applied over geocell mattress reinforced composite bed gets distributed over a wider area, while resulted in the reduced settlement as compared to other planar and randomly distributed mesh elements [1]. Different Geocell systems were successfully used in roads, railways and embankments [2–5]. Laboratory model studies were performed by several researchers to optimize the performance of cellular mattress by varying different influencing parameters, like shape and size of cells, tensile strength of cell material, height and width of geocell mattress and infill material [6–10, 11].

S. Dutta
Civil Engineering Department, Datta Meghe College of Engineering, Airoli, Navi Mumbai 400708, India

A. Kumar (✉) · J. N. Mandal
Civil Engineering Department, Indian Institute of Technology Bombay, Powai, Navi Mumbai 400076, Maharashtra, India
e-mail: ankitakumar@iitb.ac.in

J. N. Mandal
e-mail: cejnm@civil.iitb.ac.in

C. N. V. Satyanarayana Reddy et al. (eds.), *Ground Improvement and Reinforced Soil Structures*, Lecture Notes in Civil Engineering 152,
https://doi.org/10.1007/978-981-16-1831-4_56

2 Materials

The model study involves five different materials, bamboo geogrid, bamboo-made cellular mattress, jute geotextile, fly ash and marine clay. The bamboo geogrids, used as a basal reinforcement underneath the geocell mattress, were prepared in laboratory. Width of the fabricated well-finished individual bamboo stick was 10 mm with a thickness of 0.9 mm. The sticks were interwoven orthogonally by maintaining 10 mm × 10 mm square openings to form the geogrids. Mass per unit area of the geogrid evaluated as per ASTM D5261-2010 was 0.5 kg/m^2. Peak tensile strength of the bamboo geogrid determined as per ASTM D4595-2011 [12] was 110 kN/m with tensile stiffness as 2444 kN/m at 4.5% tensile strain. The tensile strength–strain curve for bamboo geogrid is shown in Fig. 1. Cylindrical perforated bamboo cells of 100 mm internal diameter, 150 mm height maintaining average 10 mm square openings were prepared by using the same bamboo sticks. Square cellular mattresses of varying widths were obtained by combining a required number of individual cells with the aid of tie wires.

A woven jute geotextile was used as separator between the clay bed and fly ash bed. The thickness of the jute geotextile under 2 kPa normal pressure [13] was 1.3 mm having mass per unit area of 0.7 kg/m^2 [14]. The apparent opening size of the geotextile was 135μ (micron) determined as per ASTM D4751-2012 [15]. The peak tensile strength of the jute geotextile was 10 kN/m with tensile stiffness as 139 kN/m at 7.2% tensile strain. The tensile strength–strain curve for jute geotextile is shown in Fig. 2.

Fly ash is a by-product waste material produced in enormous quantity from coal-based thermal power plants. Fly ash, having lower dry unit weight as compared to soil, may produce similar strength and compressibility [16]. X-ray fluorescence (XRF) test was conducted on fly ash samples to identify its basic chemical composition. It mainly consists of silicon dioxide (SiO_2) 63.52%, aluminum oxide (Al_2O_3) 26.89%, iron oxide (Fe_2O_3) 5% and calcium oxide (CaO) 1.23%. Sulfur trioxide (SO_3) is

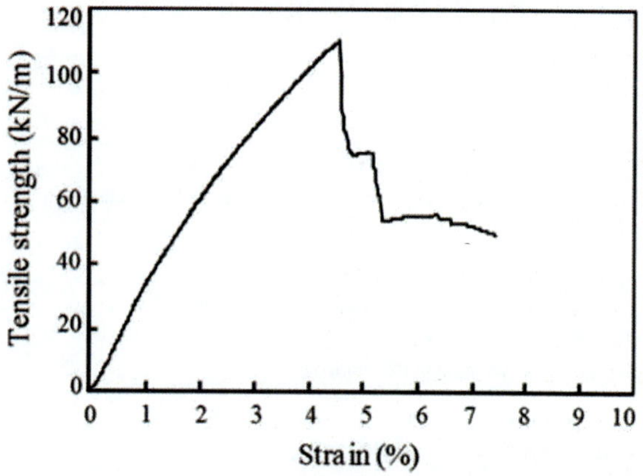

Fig. 1 Tensile strength–strain response of bamboo geogrid

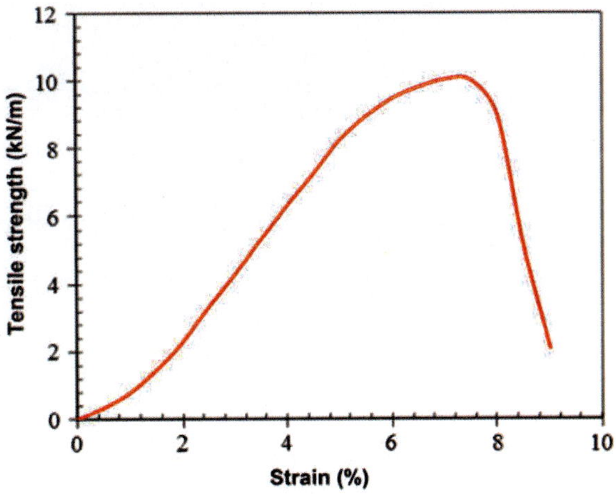

Fig. 2 Tensile strength–strain curve of jute geotextile

0.072% and total loss on ignition is 1.49%. As per ASTM C618-12 [17], it is Class F fly ash. The specific gravity of fly ash is 2.15. It contains silt-sized particles (78%), sand-sized particles (15%) and clay-sized particles (7%). It is non-plastic with a liquid limit 29.54%. The maximum dry unit weight and optimum moisture content (OMC) as obtained from standard Proctor tests according to ASTM D698-2012 are 12.1 kN/m^3 and 24% respectively. Although at dry state, fly ash has no cohesion value, it develops apparent cohesion while compacted in the presence of water. The cohesion and angle of internal friction of the fly ash at OMC are 26 kPa and 28°, respectively, as determined from consolidated drained (CD) triaxial tests as per ASTM D7181-2011 [18]. The deviator stress–axial strain curves at low confining pressures of 5, 10, 15 and 25 kPa are shown in Fig. 3. The confining pressures were kept low so as to simulate the laboratory test conditions.

Marine clay used in the present study was collected from Dronagiri area near Navi Mumbai, India. The specific gravity of the marine clay is 2.6. It contains clay particles 51%, silt particles 47.55% and sand particles 1.45%. Liquid limit, plastic limit and plasticity index are 82%, 35% and 47% respectively. As per ASTM D2487-11 [19] (Unified Soil Classification System), the soil is classified as inorganic clay of high plasticity (CH). The maximum dry unit weight and optimum moisture content (OMC) obtained from standard Proctor test are 14.2 kN/m^3 and 28% respectively.

3 Test Setup and Procedure

The cellular mattresses with infill compacted fly ash along with a jute geotextile separator without and with a basal bamboo-grid were placed over a 400-mm-thick soft clay bed, formed inside a rectangular steel tank of 850 mm × 750 mm × 620 mm inner dimension. Photograph and schematic of the test setup are shown in Figs. 4 and 5, respectively. The test setup strain-controlled loading at a constant

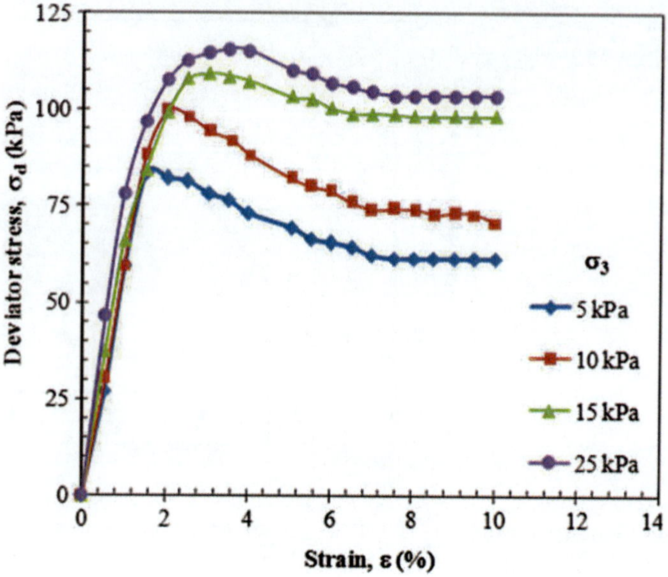

Fig. 3 Deviator stress–axial strain curves of fly ash from consolidated drained (CD) triaxial tests

Fig. 4 Photograph of the test setup

rate of 2 mm/min was applied over the center of prepared different testbeds using a rigid steel plate of 10 mm thickness having contact area as 170 mm × 170 mm. A spherical recess was made on the loading plate at its center to accommodate a steel ball bearing, over which the groove of a 50 kN load cell seated perfectly, while the top of the load cell was attached to a strain-controlled threaded jack supported against a 100 kN reaction frame. Linear variable differential transformers (LVDTs) were used to measure the footing settlement as well as the surface deformation on the fly ash bed. Rear ends of all the LVDTs were fixed to a slotted angle as fixed rigid

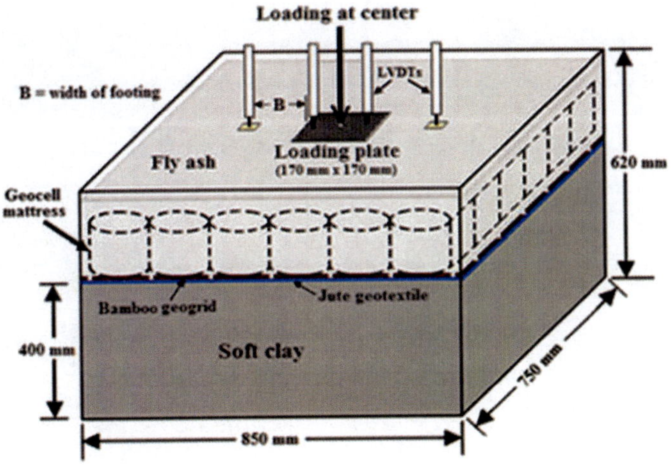

Fig. 5 Schematic of the test setup

datum. Two LVDTs (front movable spindle) were placed at the diagonally opposite side of the centerline of the loading plate to measure its vertical settlement during loading. Deformation (heave/settlement) on the fly ash surface was measured on either side of the loading plate along the center line in the direction of tank length using two LVDTs, while the LVDT spindles were placed on 20 mm length × 20 mm width × 4 mm thick Perspex plates (almost weightless), placed over the fly ash bed at 1.5B distance from the centerline of the loading plate. A small recess was made at the center of the top surface of the Perspex plates so as to accommodate the LVDT spindle avoiding any slippage during model test. As the surface deforms, it will move the Perspex plates upward during heaving or downward during settlement and causes the consequent movement of the LVDT spindles creating data logged as the surface deformation. The loading was continued up to a maximum 60 mm footing settlement considering the maximum capacity of the LVDTs.

4 Preparation of Clay Bed and Fly Ash Bed

The same compaction procedure as reported by Dutta and Mandal [6] was adopted to prepare a uniform clay bed inside the model tank. Moisture content of the soft clay beds was maintained around 70%, with undrained shear strength 8–10 kPa and bulk unit weight around 20 kN/m^3.

The unreinforced and bamboo cellular mattress reinforced fly ash beds were prepared by compaction following the same procedure as adopted by Dutta and Mandal [6] to prepare unreinforced and geocell mattress reinforced fly ash beds. Unreinforced fly ash bed refers to the fly ash beds without geocell mattress, but it may have basal bamboo grid and jute geotextile separator. By compacting fly ash

Fig. 6 Photograph of the preparation of bamboo cellular mattress reinforced fly ash bed

beds over soft clay bed, achieved dry unit weights for unreinforced and mattress reinforced beds were 87 and 84% of the maximum dry unit weight, respectively. Photograph of the preparation of bamboo cellular mattress reinforced fly ash bed is shown in Fig. 6.

A 15 mm fly ash head, almost equal to 10% of the width of the loading plate, was always kept over the geocell mattress to avoid any uneven placement of the loading plate, which may occur if it was placed directly over the geocell mattress. Dash et al. [20, 21] as well as Moghaddas Tafreshi and Dawson [22] recommended optimum geocell performance using a top fill cover as 10% of diameter or width of loading plate.

5 Test Series

Test results reported in this paper have been obtained from the test series reported in Table 1.

6 Results and Discussions

The pressure-settlement responses of the footing on different test beds are plotted, while the footing pressure (p) as abscissa and settlement (s) as ordinate is expressed in non-dimensional forms as $p/\gamma B$ and $s/B(\%)$, respectively. B is the footing width

Table 1 Test series for the model study

Test series	Fly ash bed	Influencing parameters
A	Absent, i.e. test over clay bed	
B	Unreinforced (F): a) Fly ash bed directly over clay b) J + Fly ash bed	Constant height of fly ash bed: H = 150 mm
C	Reinforced with bamboo cellular mattress: (D = 100 mm; E = 2444 kN/m) J + M + infill fly ash J + BG + M + infill fly ash	Variation in mattress width: b/B = 1.24, 1.85, 2.47, 3.09, 3.71 Constant height: h = 150 mm

F = Unreinforced fly ash bed; J = Jute geotextile; BG = Bamboo geogrid; M = Bamboo geocell mattress; H = Height of unreinforced fly ash bed; h = Height of bamboo cellular mattress; b = Width of mattress; B = Width of footing; D = Diameter of mattress pockets; E = Tensile stiffness of bamboo cell corresponding to the peak tensile strength

and γ is the bulk unit weight of clay. During the settlement of the footing, surface deformation (settlement/heave) was recorded over the footing adjacent area. The surface deformation (δ) is expressed in non-dimensional form as $\delta/B(\%)$ and is plotted against $s/B(\%)$.

6.1 *Effect of Jute Geotextile*

The variation of pressure-settlement responses of footing over only clay bed as well as over the unreinforced fly ash beds without and with jute geotextile separator are depicted in Fig. 7. It can be observed that punching failure occurred in the only clay bed. Placement of fly ash over the clay bed increased the footing capacity as fly ash has more modulus of elasticity as compared to clay. In the presence of jute geotextile separator, the footing capacity further increased owing to the tensile membrane influence from the jut geotextile.

6.2 *Effect of Basal Bamboo Grid*

The influence of basal bamboo grid underneath the geocell mattress is illustrated in Fig. 8 for a cellular mattress of height 150 mm and width b/B = 3.09. The placement of a basal bamboo grid along with jute separator underneath the bamboo cellular

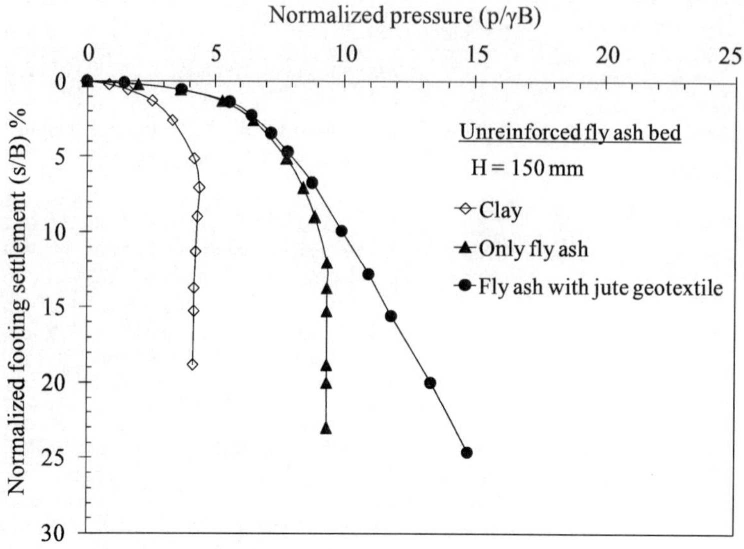

Fig. 7 Variation in footing pressure over only clay and unreinforced fly ash beds

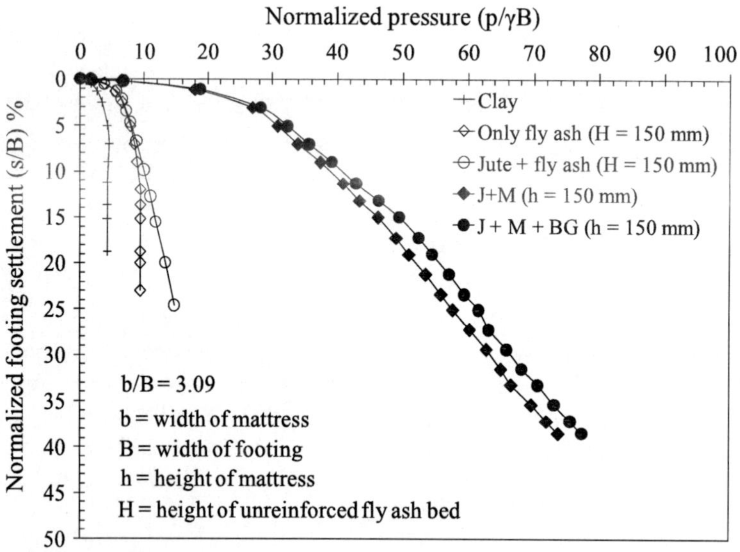

Fig. 8 Variation in footing pressure over geocell mattress without and with basal bamboo grid

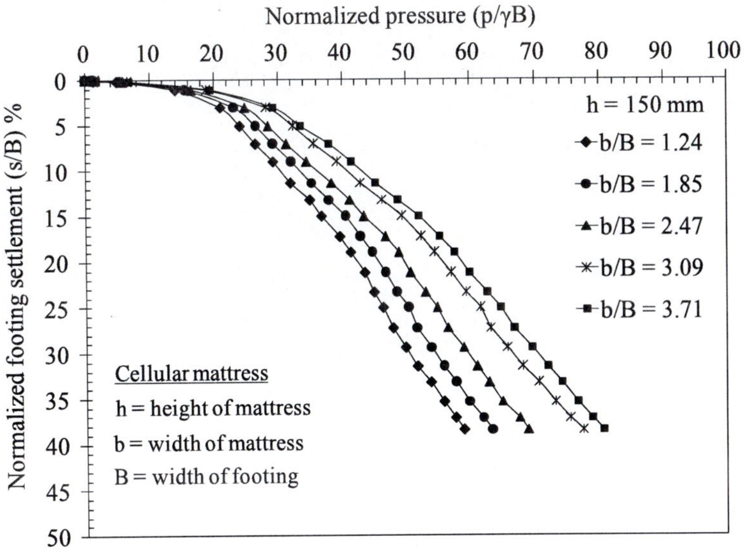

Fig. 9 Variation in footing pressure over geocell mattress of varying widths along with basal bamboo grid

mattress improved the footing capacity as compared to the footing capacity obtained from placing the mattress along with only jute separator. It may be attributed to the added tensile resistance from the bamboo grid.

6.3 *Effect of Width of Cellular Mattress*

With an increase in the width of cellular mattress, the footing pressure increased as can be observed from Fig. 9 that represents the variation of footing pressure with settlement for different mattress widths (h = 150 mm) along with basal bamboo grid. Geocell mattress supported the footing pressure through mobilization of shear resistance with fly ash at its outer periphery throughout the height of the geocell wall. The mattress with higher plan area intercepted the rupture plane more as well as produced increased anchorage support to the footing settlement and consequently, redistributed the footing pressure over a wider area. These may be the reasons for the improved footing performance with an increase in the width of mattress. At b/B = 3.09, the footing capacity improved around 5.5 times as compared to only fly ash bed of same height over the clay bed. The footing adjacent surface heaving decreased with an increase in the width of mattress as can be observed from Fig. 10. As the width of mattress increases, resistance to the upward heaving of the footing adjacent cells increased from the cells attached to those adjacent cells.

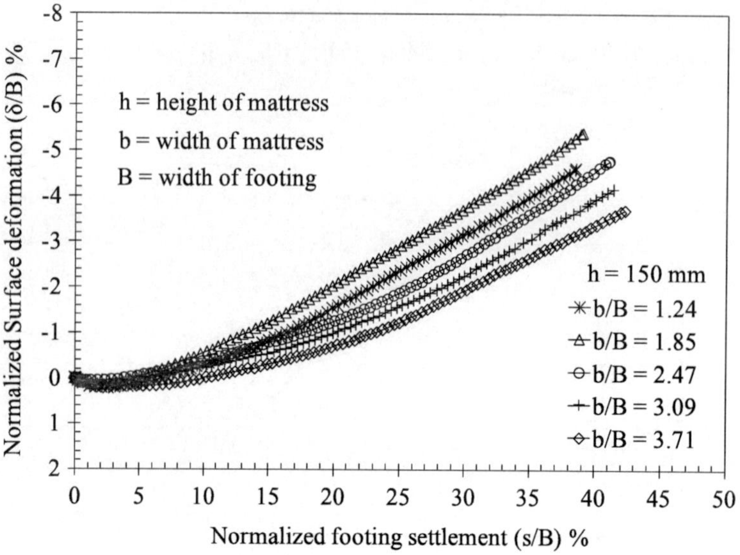

Fig. 10 Surface deformation surrounding the loaded area with varying mattress widths

7 Conclusion

The placement of fly ash bed over only clay increased the footing capacity, which further increased in the presence of a jute geotextile separator. Optimum performance of the cellular mattress reinforced fly ash bed was obtained along with jute separator and basal bamboo grid, while the footing capacity improved around 5.5 times. An increase in the width of cellular mattress maintaining a particular cell height increased the footing capacity, while from the present study, the optimum width of cellular mattress was obtained as 3.09 times the width of square loading plate.

References

1. de Garidel, R., Morel, G.: New soil strengthening techniques by textile elements for low volume roads, road and railway applications. In: Proceedings of 3rd International Conference on Geotextiles, Vienna, Austria (1986)
2. Bush, D.I., Jenner, C.G., Basset, R.H.: The design and construction of geocell foundation mattress supporting embankments over soft ground. Geotext. Geomembr. **9**(1), 83–98 (1990)
3. Dong, P.: The application of geocell mattress in Qin-Yin highway. Transp. World **14**(10), 96–98 (2007)
4. Li, M.T.: An analysis of the stability of roads and bridges in the mined-up region of the Taigu highway and the treating scheme. J. Chongqing Jiaotong Inst. **19**(3), 85–87 (2000)
5. Xie, Y.L., Yu, Y.H., Yang, X.H.: Application study of treating differential settlement of subgrade with geocell. China J. Highway Transp. **17**(4), 7–10 (2004)

6. Dutta, S., Mandal, J.N.: Model studies on geocell reinforced fly ash bed overlying soft clay. J. Mater. Civil Eng. (SCI and Scopus Indexed) ASCE **28**(2) (2015)
7. Dutta, S., Mandal, J.N.: Model studies on encased fly ash column-geocell composite systems in soft clay. J. Hazard. Toxic Radioactive Waste ASCE **21**(3) (2017)
8. Moghaddas Tafreshi, S.N., Dawson, A.R.: A comparison of static and cyclic loading responses of foundations on geocell-reinforced sand. Geotext. Geomembr. **32**, 55–68 (2012)
9. Ravindran, S., Kumar, A., Dutta, S., Nadaf, M.B., Mandal, J.N., Shin, E.C.: Unpaved road stabilization using bamboo grid and bitumen coated bamboo cells. Soil Mech. Found. Eng. (Indexed in SCI Expanded and Scopus) **56**(5), 346–351 (2019). https://doi.org/10.1007/s11204-019-09613-7
10. Sitharam, T.G., Sireesh, S., Dash, S.K.: Performance of surface footing on geocell-reinforced soft clay beds. Geotech. Geologic. Eng. **25**(5), 509–524 (2007)
11. Mahiskar, S.Y., Mandal, J.N.: Investigation on soft clay subgrade strengthening using geocells. Construct. Build. Mater. **10**(4), 281–286 (1996)
12. ASTM D4595: Standard test method for tensile properties of geotextiles by the wide-width strip method. Annual Book of ASTM Standards, ASTM International, 04.13, West Conshohocken, PA, USA (2011)
13. ASTM D5199: Standard test method for measuring nominal thickness of geosynthetics. Annual Book of ASTM Standards, ASTM International, 04.13, West Conshohocken, PA, USA (2012)
14. ASTM D5261: Standard test method for measuring mass per unit area of geotextiles. Annual Book of ASTM Standards, ASTM International, 04.13, West Conshohocken, PA, USA (2010)
15. ASTM D4751: Standard test method for determining apparent opening size of a geotextile. Annual Book of ASTM Standards, ASTM International, 04.13, West Conshohocken, PA, USA (2012)
16. Kim, B., Prezzi, M., Salgado, R.: Geotechnical properties of fly and bottom ash mixtures for use in highway embankments. J. Geotech. Geoenviron. Eng. 0.1061/(ASCE)1090-0241(2005)**131**:7(914), 914–924 (2005)
17. ASTM C618: Standard specification for coal fly ash and raw or calcined natural pozzolan for use in concrete. Annual Book of ASTM Standards, ASTM International, 04.02, West Conshohocken, PA, USA (2012)
18. ASTM D7181: Standard test method for consolidated drained triaxial compression test for soils. Annual Book of ASTM Standards, ASTM International, 04.13, West Conshohocken, PA, USA (2011)
19. ASTM D2487: Standard practice for classification of soil for engineering purpose (Unified Soil Classification System). Annual Book of ASTM Standards, ASTM International, 04.08, West Conshohocken, PA, USA (2011)
20. Dash, S.K., Krishnaswamy, N.R., Rajagopal, K.: Bearing capacity of strip footings supported on geocell-reinforced sand. Geotext. Geomembr. **19**, 235–256 (2001)
21. Dash, S.K., Sireesh, S., Sitharam, T.G.: Model studies on circular footing supported on geocell reinforced sand underlain by soft clay. Geotext. Geomembr. **21**(4), 197–219 (2003)
22. Moghaddas Tafreshi, S.N., Dawson, A.R.: Comparison of bearing capacity of a strip footing on sand with geocell and with planar forms of geotextile reinforcement. Geotext. Geomembr. **28**(1), 72–84 (2010)
23. Zhou, H., Wen, X.: Model studies on geogrid or geocell-reinforced sand cushion on soft soil. Geotext. Geomembr. **26**, 231–238 (2008)

Experimental and Numerical Studies of Three-Layered Reinforced Soil Slope Under Dynamic Loading Condition

Kingshuk Jana, Suman Hazari, and Sima Ghosh

1 Introduction

Slopes are the exposed ground surface that stands at an angle with the horizontal. The stability of slope is one of the major concerns and trending research topics in geotechnical engineering for designing highway embankments, earth dams etc. Slopes may be in stable condition under static loading but under seismic loading condition, it collapses due to generation of inertia force, which causes the reduction of the strength of soil mass. In such a distress situation, soil should have enough strength to negotiate seismic force. Inclusion of reinforcement may provide required stability by absorbing seismic energy and it can transmit a small amount of seismic wave to the overlying structure to mitigate earthquake-related hazards. Therefore, soil should provide enough strength and desired factor of safety during and after earthquake shock. The study of slope under static loading conditions is done by several researchers [1–4]. But to understand the response under seismic loading, Clough and Pirtz [5] first performed a shaking table test on model rockfill dams and concluded that earth-fill dams are very much resistant to earthquake due to their flexible nature. Seed and Clough [6] studied the earthquake-resistant dams and concluded that catastrophic failure is not caused by earthquake; the major effect is there is a settlement in the upper section of slope and sliding in the base of the slope. Hazari et al. [7] performed a small-scale shaking table test with the variation of water content and the obtained results are verified by the numerical analysis.

K. Jana (✉) · S. Hazari · S. Ghosh
NIT Agartala, Tripura 700046, India

C. N. V. Satyanarayana Reddy et al. (eds.), *Ground Improvement and Reinforced Soil Structures*, Lecture Notes in Civil Engineering 152,
https://doi.org/10.1007/978-981-16-1831-4_57

2 Objective

This study is preferred to show the effect of non-homogeneity of soil mass by introducing three-layered soil consisting of two c-ϕ soils and one sandy soil. The stability of layered soil slopes is improved by introduction of geogrid and geotextile reinforcement in single, two and three layers separately. To measure the stability of slopes under seismic loading, model tests are conducted on shake table by varying base frequency, base shaking acceleration, quantity and type of reinforcement. The numerical analysis has been performed by PLAXIS 2D [8] using parameters obtained from laboratory investigations so that the output results can be compared with those obtained from the experimental investigations.

3 Equipment and Materials

3.1 Shake Table

A computer-controlled uniaxial shaking table has been used to simulate seismic conditions or any other vibrating conditions during tests. The shaking table has a 1 m × 1 m square loading platform of capacity 100 kg fitted with symmetrically placed four (4) number ball bearings below the platform in such a way that it can run through the base channels, which are firmly fixed with the foundation through nuts and bolts. One end of the crankshaft is fitted with the shake table and the other end is fitted with 7.5 H.P. reciprocating motor whose design speed is about 1400 rpm. The shake table can be operated within the acceleration range of 0.05–0.5 g and frequency range of 1–7 Hz with the amplitude of ±100 mm. Model slopes are constructed in a 12 mm thick glass perspex box having inside dimension 60 cm × 40 cm × 40 cm (l × b × h) for conducting shake table test. The setup of the shake table instrument used in the present study is shown in Fig. 1.

3.2 Soil

Three types of locally available soils are used for model slopes. Soil 1 and soil 2 are typical c-ϕ soils, which are classified as silty sand and sandy clay according to IS classification. Sand is another one, which is classified as poorly graded (SP) as per IS classification. The properties of each soil are tabulated in Table 1.

Fig. 1 Arrangement of the shake table instrument

Table 1 Properties of soil used in the experiment

Parameters	Soil 1	Sand	Soil 2
Zone	Nil	IV	Nil
Specific gravity	2.59	2.66	2.61
D60, D30, D10 (mm)	00.125,0.25 and 0.32	0.35, 0.2 and 0.28	0.25,0.30 and 0.40
Cu and Cc	1.56,2.56	1.12,1.75	1.68,2.62
Maximum dry density (kN/m3)	20.6	Nil	19.3
Classification	Silty sand	Poorly graded sand	Sandy clay

3.3 *Reinforcement*

A bi-axial geogrid having an aperture size 10 mm x 10 mm and a geotextile are used to reinforce the model slopes. Ultimate strength of geogrid is determined as per ASTM D6637 [9]. Ultimate strength of geotextile is determined as per ASTM D4595 [10]. Properties of reinforcements are given in Table 2.

3.4 *Model Construction and Methodology*

A rainfall falling system is used for uniform filling of soils in perspex box. Each model is constructed by using three types of soil (soil 1 at top, sand at middle and soil 2 at bottom of equal height). Approximately, 45 kg of soil is used for model slope consisting of 15 kg each. Uniform weight of soil in each layer in each case is achieved by providing a fixed number of blows with a hammer of 4.9 kg weight. Base accelerations are varied at 0.1, 0.2 and 0.3 with frequencies 1, 2 and 3 Hz for each base acceleration for duration of 10 s. In this study, a total of 35 different shaking table tests are performed on un-reinforced and reinforced slope model. Schematic

Table 2 Properties of reinforcements

Parameters	Geogrid	Geotextile
Material type	Polypropylene	Polyethylene
Ultimate tensile strength (kN/m)	25	19
Elongation at specified tensile strength	16.60	19.57
Aperture shape	Square	–
Thickness (mm)	3	1
Secant modulus at 2% strain (kN/m^2)	219	162
Secant modulus at 5% strain (kN/m^2)	169	155.8
Mass per unit area (kg/m^2)	0.22	0.21

diagram of a typical three-layer slope with instrumentation is shown in Fig. 2. Law of similitude is applied in order to simulate the prototype slope and model slope with a scaling factor of 10 (λ). Accordingly, the scaling parameters between model and prototype are derived in Table 3.

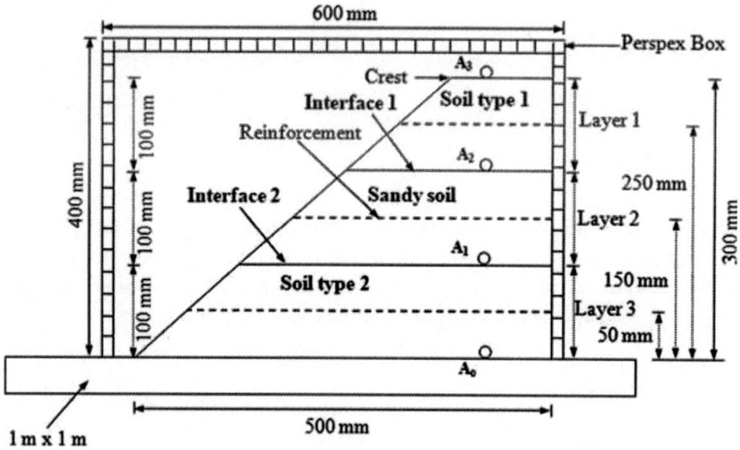

Fig. 2 Schematic diagram of three-layered soil slope

Table 3 Relationship between prototype and model

Parameters	Value of model parameters	Equation for scaling factor	Scaling factor	Prototype parameters
Unit weight of soil (kN/m3)	14.5, 16, 14	1	1	14.5, 16, 14
Dimension (L × B × H)	0.50 × 0.40 × 0.3	λ	10	5 × 4 × 3
Acceleration (g)	0.1, 0.2, 0.3	1	1	0.1, 0.2, 0.3
Frequency (Hz)	1, 2, 3	$1/(\lambda^{3/4})$	0.17	0.17, .34, .51
Time (s)	t	$\lambda^{3/4}$	5.623	5.623 × t
Displacement (m)	u	λ	10	10 × u

4 Results and Discussions

4.1 *Effect of Different Frequencies*

It is observed from Fig. 3a–c that horizontal deformation decreases with an increase in base frequency but RMSA amplification and crest deformation go on increasing with an increase in frequency level. Inclusion of reinforcement in the slope has no significant effect on reducing horizontal and crest deformation but it has substantial effect in reducing RMSA amplification factor. Figure 3b reflects that at low frequency of 1 Hz, reinforced slopes are deamplified up to 0.33 normalized heights and further amplified up to a maximum value of 1. But unreinforced slopes are amplified throughout the full height of the slope.

4.2 *Effect of Different Base Shaking Accelerations*

Figure 4a–c shows the variation of horizontal deformation, RMSA amplification factor and crest deformation along with normalized height at different base shaking accelerations of 0.1, 0.2 and 0.3 g for a constant frequency of 3 Hz. It is seen from the plots that all the above parameters are increased with an increase in base shaking accelerations. The inclusion of reinforcements reduces RMSA amplification factor drastically than other two parameters. Slopes with three-layer geogrid reinforcement are found to be better performing than slopes with three-layer geotextile reinforcement.

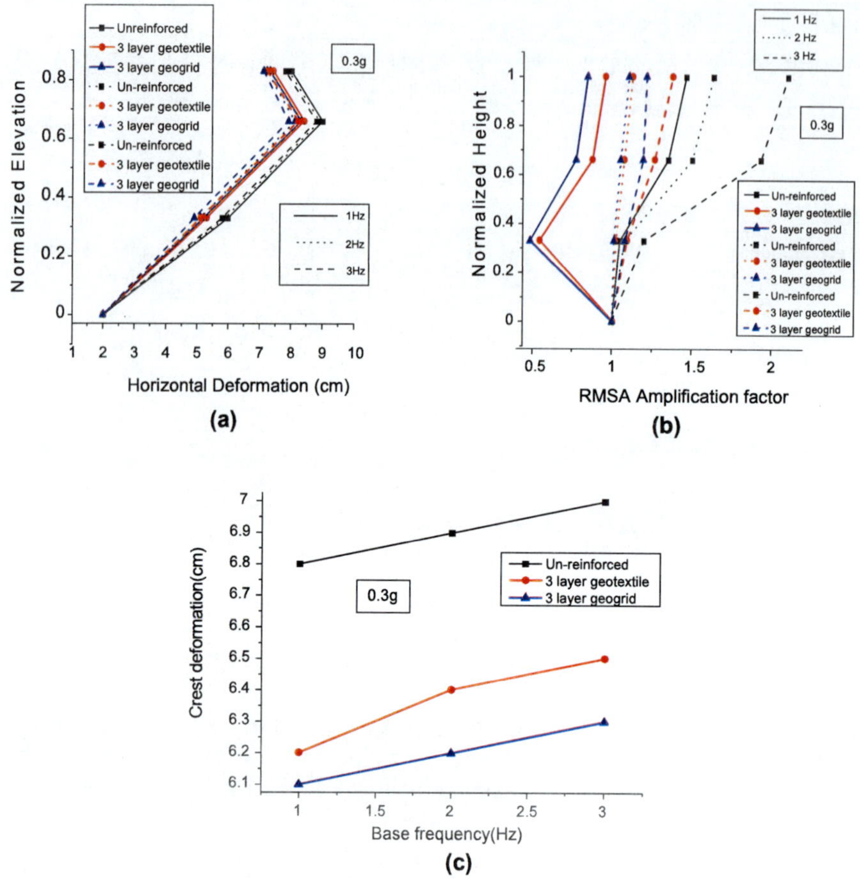

Fig. 3 Variation with frequency **a** Horizontal deformation, **b** RMSA amplification factor and **c** Crest deformation

4.3 Effect of Type and Quantity of Reinforcement

Figure 5a shows the response of horizontal deformation with respect to the normalized height of the slope reinforced with single-, two-layer and three-layer geogrid/geotextile subjected to 0.3 g base shaking acceleration and 3 Hz frequency. Maximum horizontal deformation decreases with an increase in reinforcement quantity. Horizontal deformations are reduced to 2.2, 4.4 and 6.6% with single-, two- and three-layer geotextile reinforcements, respectively, as compared to unreinforced slope. The corresponding percentage of reduction is 4.4, 5.5 and 10% for slopes with geogrid reinforcement.

Figure 5b presents the response of RMSA amplification factor with respect to the normalized height of the slopes. Recorded response of acceleration amplification factors reflects the reduction up to 4.7%, 24.8% and 34.7% for single-, two- and

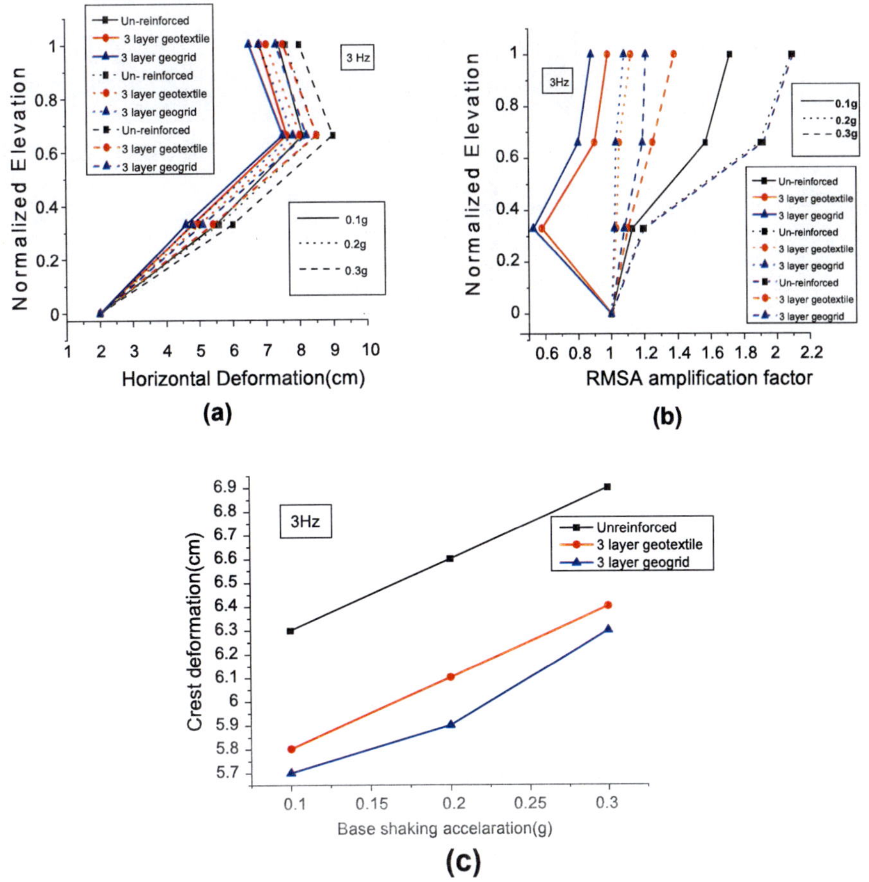

Fig. 4 Variation with base acceleration **a** Horizontal deformation **b** RMSA amplification factor **c** Crest deformation

three-layer geotextile reinforced slopes respectively. The corresponding percentage of reduction is 6.2, 28.5 and 42.3% for slopes with geogrid reinforcement. Both geogrid and geotextile are effective in reduction of RMSA amplification factor but the reduction is significant for geogrid reinforced slope.

4.4 *Comparison Between Homogeneous (Single Layered of Soil 1) and Layered Soil Slope*

Figure 6a, b shows the comparison of horizontal deformation between homogeneous and three-layered slopes. It is observed that in the case of three-layered soil slope, the

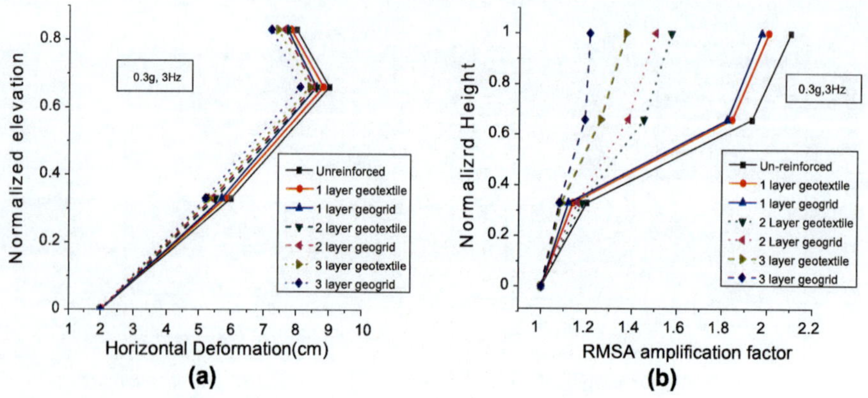

Fig. 5 Effect of type and quantity of reinforcement **a** Horizontal deformation **b** RMSA amplification factor

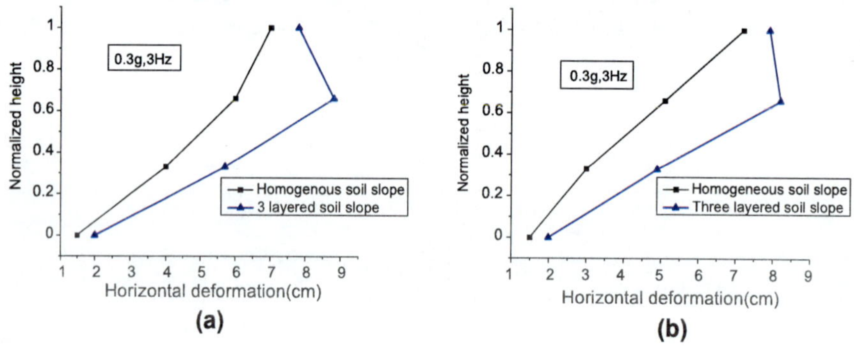

Fig. 6 Comparison of horizontal deformation between homogeneous soil slope and three-layered soil slopes **a** unreinforced slope **b** reinforced slope

horizontal deformation is maximum at normalized height 0.66, after that the horizontal deformation decreases. But in the case of homogeneous soil slope, horizontal deformation is increased with elevation and gets maximum at top of the slope. On the other hand, Fig. 7a, b shows the comparison of the RMSA amplification factor between homogeneous and three-layered slopes. It is seen that in the case of three-layered soil slope, the rate of increase in RMSA amplification factor is different at different soil layers and it is maximum in sand layer. But in the case of homogeneous soil slope, the RMSA amplification factor is not changed significantly throughout the height of the slope. The variations of the horizontal deformation and RMSA amplification factor in case of layered slope are due to the influence of the interfaces between the soils.

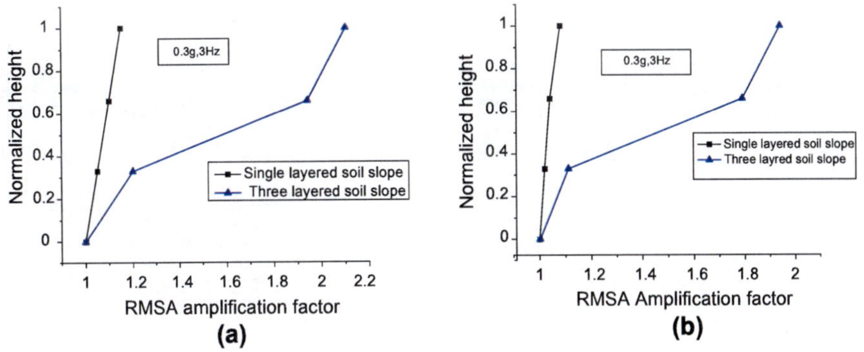

Fig. 7 Comparison of RMSA amplification factor between single-layered soil slope and three-layered soil slopes **a** unreinforced slope **b** reinforced slope

5 Numerical Modeling

In the present study, a two-dimensional finite element model of the experimental slope model during dynamic loading condition is developed and analyzed using PLAXIS 2D. To verify the model, the results as obtained from the experimental observation are compared with those obtained from the numerical analysis.

5.1 Geometry and Boundary Condition

Figure 8 presents the geometry of the numerical model with a slope angle of 35°. The boundary of the model slope is considered at the base at a depth 300 mm below the top level of slope and roller at the two vertical sides; one vertical side passes

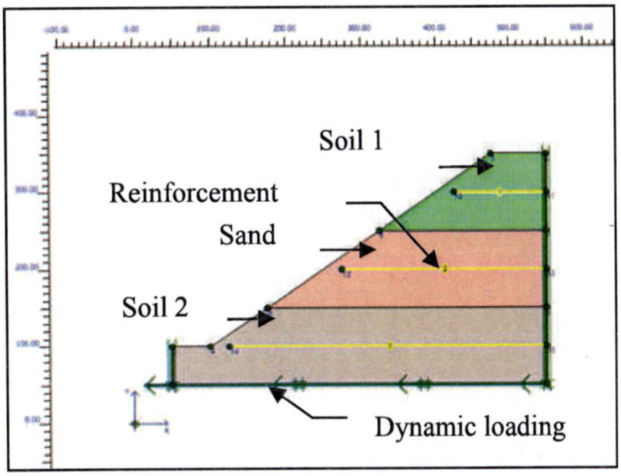

Fig. 8 Numerical model of slope at PLAXIS 2D

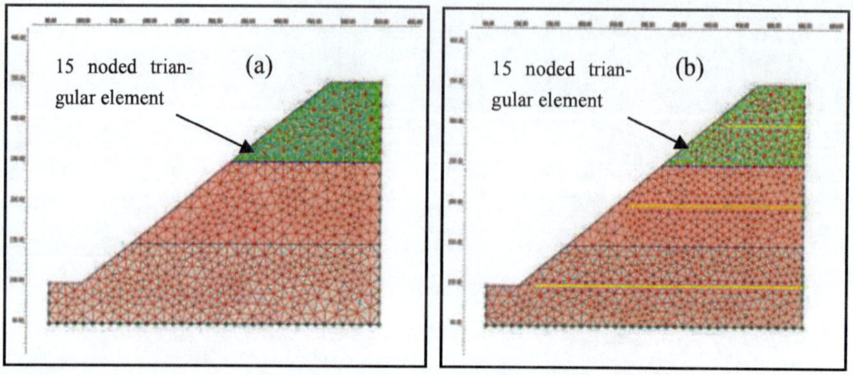

Fig. 9 Mesh structure of numerical model slope **a** Un-reinforced **b** Reinforced

through the center line of slope and another at a distance of 550 mm away from the toe of the slope. A full fixed boundary condition is considered at the outer boundary of the slope. The boundaries are extended from the model to avoid the disturbances due to the possible reflections. Therefore, absorbent boundaries are adopted to avoid the reflections of the waves at left, right and bottom of the slope.

5.2 Mesh Generation

The slope model is divided into a number of 15-noded triangular elements having three degrees of freedom in each node i.e. vertical displacement, horizontal displacement and rotation. Figure 9a, b shows the mesh structure of the numerical model for unreinforced and reinforced slopes respectively.

5.3 Numerical Results and Discussions

Figure 10a, b shows the deformed mesh of unreinforced and reinforced slopes due to the application of the dynamic loading. Top layer of the slope goes down significantly due to the presence of weak sand layer in the middle. However, the subsidence of third layer is negligible. In case of reinforced slope, the top reinforcement deforms significantly. The middle and bottom reinforcements also distort after the application of dynamic loading but it is very nominal. It is also observed that meshes are deformed in the horizontal direction at the middle part of the slope for both un-reinforced and reinforced slopes but the application of the reinforcement reduces deformation.

Figure 11a, b shows the shaded horizontal deformation of un-reinforced and reinforced slopes. It is observed that horizontal deformations are appeared at the inclined face of slopes indicating rotational failures. But the middle layer of the slope i.e. sand

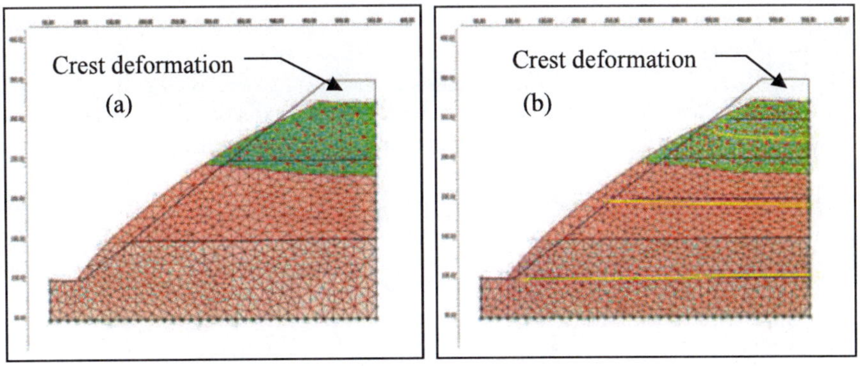

Fig. 10 Deformed mesh of model slope **a** Un-reinforced **b** Reinforced

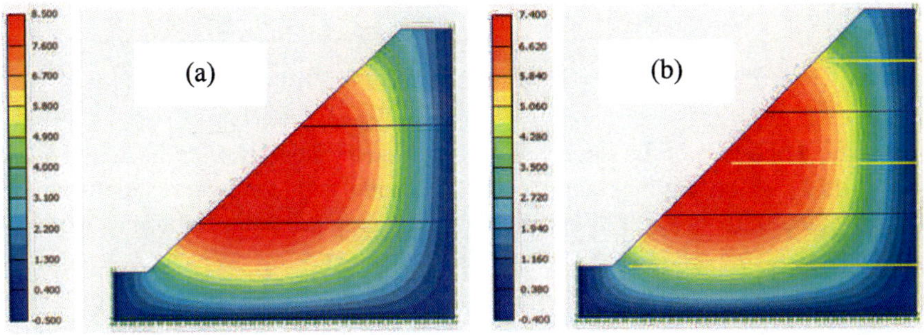

Fig. 11 Horizontal shaded deformation **a** Un-reinforced **b** Reinforced

deforms more due to its cohesion less property. Figure 11b reveals that the provision of reinforcement reduces horizontal deformation.

Figure 12a, b shows the shaded vertical deformation of un-reinforced and reinforced slopes. From the plots, it is seen that the vertical deformation is very less or

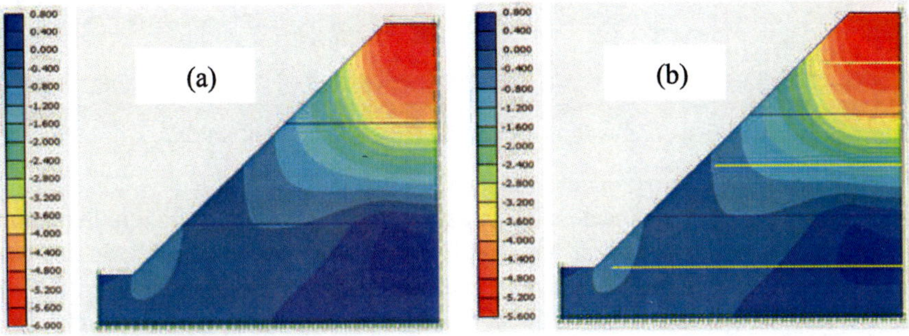

Fig. 12 Vertical shaded deformation **a** Un-reinforced **b** Reinforced

Table 4 Comparison of deformation values obtained from experimental and numerical studies at 0.3 g base shaking

		Experimental study		Numerical study	
Reinforcement type	Frequency (Hz)	Maximum horizontal deformation (cm)	Crest deformation (cm)	Maximum horizontal deformation (cm)	Crest deformation (cm)
Unreinforced	1	9.0	6.7	8.5	6.0
	2	8.9	6.9	8.0	6.4
	3	8.8	7.0	7.5	6.5
Three-layer geotextile	1	8.4	6.2	7.4	5.8
	2	8.3	6.4	7.3	6.0
	3	8.2	6.5	7.2	6.2
Three-layer geogrid	1	8.3	6.0	7.3	5.7
	2	8.2	6.3	7.2	5.9
	3	8.0	6.4	7.1	6.0

negligible at the base of the slope which is increased gradually with the height of the slope and at the top it is maximum. Figure 12b reflects that the vertical deformation of reinforced slope is lesser in magnitudes in comparison to un-reinforced slope.

6 Comparison of Experimental and Numerical Results

The results obtained from the experimental study performed by shake table tests are compared with the results obtained from two-dimensional numerical analysis. Table 4 compares soil stability parameters on horizontal and crest deformation. From comparisons, it is concluded that the results of experimental and numerical studies have good agreement with each other.

7 Conclusions

The following major conclusions are drawn from the present study.

1. Inclusion of reinforcement in layered soil slope is very effective for reducing the soil stability parameters like maximum horizontal deformation, maximum RMSA amplification and crest deformation for all base accelerations and frequencies but more effective in reducing RMSA amplification.
2. It is observed that at low frequency of 1 Hz, acceleration response deamplified up to 0.33 normalized height and amplified further up to 1.0 for reinforced slopes. But un-reinforced slopes do not show the same response.

3. Soil interfaces influence the accelerations and deformations of the slope.
4. The rate of increase of RMSA amplification factor in the middle layered soil (sand) is more than other two soil layer (c-ϕ) indicating that sand layer is the weakest layer than other two c-ϕ soil layers in mitigating stability of slopes.
5. Layered soil slope is less stable in mitigating stability parameters than slopes made up of single-layered soil (homogeneous soil).
6. Experimental and numerical studies have good agreement with each other.

References

1. Fellenius, W.: Calculation of the stability of earth dams. In: Proceedings of the Second Congress of Large Dams, Washington, DC, v. 4, pp. 445–463 (1936)
2. Bishop, A.W.: The use of slip circle in the stability analysis of earth slopes. Geotechnique **5**(1), 7–17 (1955)
3. Janbu, N.: Applications of composite slip surfaces for stability analysis. In: Proceeding of European Conference on the Stability of Earth Slopes, Stockholm, Sweden, v. 3, pp. 43–49 (1954)
4. Morgenstern, N.R, Price, V.E.: The analyses of the stability of general slip surfaces. Geotechnique **15**(1), 79–93 (1965)
5. Clough, R.W., Pirtz, D.: Earthquake resistance of rockfill dams. Soil Mech. Found. Div **82**(2), 1–26 (1956)
6. Seed, H.B., Clough, R.W.: Earthquake resistance of sloping core dams. J. Soil Mech. Found. Div. ASCE **89**(1), 209–242 (1963)
7. Hazari, S., Ghosh, S., Sharma, R.P.: Experimental and numerical study of soil slopes at varying water content under dynamic loading condition. Int. J. Civil Eng. **18**(2), 215–22 (2019)
8. PLAXIS 2D 8.2 [Computer software]. Delft, Netherlands, Plaxis
9. ASTM D6637-01. Standard Test Method for Determining Ten sile Properties of Geogrid by the Single or Multi-Rib Tensile Method. ASTM International, West Conshohocken, PA. www.astm.org (2001)
10. ASTM D4595-11, Standard Test Method for Tensile Properties of Geotextiles by the Wide-Width Strip Method. ASTM International, West Conshohocken, PA. www.astm.org (2011)

Effect of a Coarse Material Sandwich Technique on the Behavior of Geotextile Reinforced Clay

E. Dayana Aravind and A. K. Vasudevan

1 Introduction

A large part of our world consists of weak cohesive soil deposits. These soils are unsuitable for construction purposes, both as a foundation material and as a construction material. However, this type of soil can be strengthened by including various types of synthetic and natural materials and making it suitable for construction purposes. Ground improvement using geosynthetic materials are in which geosynthetic reinforcements are embedded in a soil mass, have several distinct advantages because of their ductility, high tolerance to the differential settlement without structural distress, rapid method for construction and cost effectiveness. Since Natural materials are biodegradable, eco-friendly, locally available and cost effective, these materials are widely using in various soils improvement projects. One of the most common geosynthetic materials used to reinforced soil is geotextiles. These elements' main function is to redistribute stresses within the soil mass to enhance the internal stability of reinforced soil structures [1, 2]. Thin sand-layer inclusions could increase the interface interaction between the clay and reinforcement, improving the reinforced clay's overall shear strength [3, 4]. The sand also acted as a lateral drainage layer to dissipate excess pore water pressure during shearing. In addition to its mechanical function, the sandwich technique has been demonstrated to increase a soil–geotextile or soil–geocomposite system's lateral drainage capacity, accelerate pore water pressure dissipation within reinforced soil.

E. Dayana Aravind · A. K. Vasudevan (✉)
Thejus Engineering College, Vellarkkad, Thrissur, Kerala, India

C. N. V. Satyanarayana Reddy et al. (eds.), *Ground Improvement and Reinforced Soil Structures*, Lecture Notes in Civil Engineering 152,
https://doi.org/10.1007/978-981-16-1831-4_58

Table 1 Properties of clay

Properties	Value
Specific gravity	2.6
Liquid limit (%)	67
Plastic limit (%)	38
Shrinkage limit (%)	21
Plasticity Index (%)	29
IS classification	OH
Optimum moisture content (%)	27
Maximum dry density (kN/m^3)	14

2 Experimental Program

A series of unconsolidated undrained triaxial compression tests were conducted for investigating the shear strength and stiffness behaviour of unreinforced clay, jute geotextile reinforced clay. The effects of jute geotextile sandwiched in sand layer of varied thickness is assessed in terms shear strength and stiffness of reinforced clay. The tests were performed under different confining pressures (50, 100 and 150 kPa) and various numbers of geotextiles (one, two and three layers). For sandwich technique, the thickness of sand layers is 2, 4, 6, and 8 mm.

3 Materials

3.1 Clay

The clayey soil sample was collected from Kannapuram, Thaliparamba Taluk, Kannur District, Kerala, India. Table 1 presents the basic properties of clayey soil.

3.2 Sand

Sand is used as the coarse material for the sandwich technique. River sand is used for the study. The sand was collected from Mambaram, Thalassery Taluk, Kannur District, Kerala, India. Table 2 presents the basic properties of sand.

Table 2 Properties of sand

Properties	Value
Coefficient of uniformity (C_u)	2.875
Coefficient of curvature (C_c)	0.96
Specific gravity	2.63
IS classification	SP
Maximum dry density (kN/m^3)	17

3.3 Geotextile

The use of waste material and natural geotextile for improving soil property is advantageous. Commercially available woven jute geotextile is used for this study. It was collected from Weldone Industries, Kannur district, Kerala, India. Properties of woven jute geotextile obtained from the National jute board are shown in Table 3.

4 Sample Preparation

In the form of wet condition, the collected clayey soil was placed in an oven for 24 h and then crushed into dry powder form. The soil sample was allowed to sieve through a 425 micron IS sieve. Figure 1 represents the diagram showing the location of geotextile layers and the sandwich specimen. The cylindrical soil specimens with 38 mm diameter and 72 mm height were prepared. The diameter of the geotextile layer and sand layer was taken as 32 mm. For unreinforced soil, the sample was filled in layers and compacted by using a standard compaction approach so as to attain the maximum dry density obtained from the compaction test. For reinforced soil, the samples were filled in several layers keeping the same density of unreinforced specimens. For the sandwich specimens, clay was placed into a mould. After the clay was compacted and leveled to the desired height, half of the predetermined quantity of dry sand was placed and compacted with a small tamper to achieve the required thickness and density. Afterward, the reinforcement layer was introduced above the lower part of the sand. Then the remaining portion of sand and clay was placed. Figure 2 shows the preparation of reinforced soil specimen with one layer of geotex-

Table 3 Properties of jute geotextile (National jute board)

Properties	Value
Thickness (mm)	2
Weight (gm/m^2)	760
Permittivity (m/s)	0.5
Strength (warp x weft) (kN/m)	20 × 20

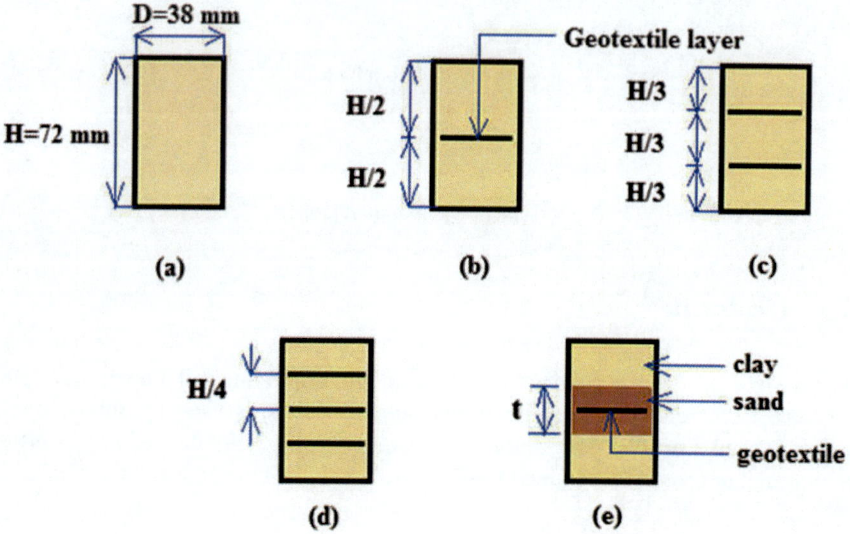

Fig. 1 Arrangement of geotextile layers in reinforced soil **a** Plane soil **b** Soil specimen with one-layer geotextile **c** Soil specimen with two-layer geotextile **d** Soil specimen with three-layer geotextile **e** Sandwich specimen

Fig. 2 Specimen preparation with one layer of geotextile

tile. Where geotextile layer was placed at the center of specimen and compacted to same density obtained for unreinforced specimens.

5 Testing Program

Triaxial tests were performed on the unreinforced, reinforced clay and the sandwich specimen under different confining pressures (50,100 and 150 kPa). The axial loading was increased gradually at the rate of 1.2 mm per minute. The loading was continued until the strain level of the reinforced specimen reached 20%. This strain is known as strain at failure.

6 Results and Discussion

6.1 Triaxial Test Results of Clayey Soil Reinforced with Geotextile Layers

Type of failure. Figure 3 shows photos of deformed failure specimens. In unreinforced samples, failure was observed as a bulging failure at the center of the specimen. In reinforced samples, bulging occurred between two adjacent geotextile layers. As the number of geotextile layers increased, the bulging gets reduced. Thereby deformation becomes comparatively less.

Stress–Strain behavior. Figure 4 shows a comparison of Stress–Strain behavior

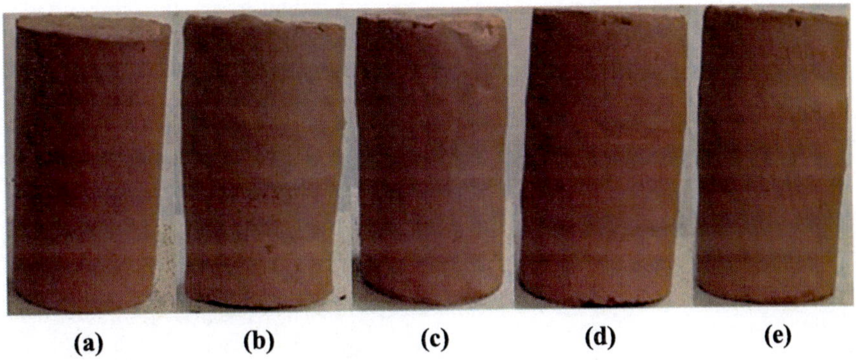

Fig. 3 Soil specimen before and after loading with and without geotextile **a** Plane soil specimen before loading **b** Plane soil specimen after loading **c** Soil specimen with one-layer geotextile after loading **d** Soil specimen with two-layer geotextile after loading **e** Soil specimen with three-layer geotextile after loading

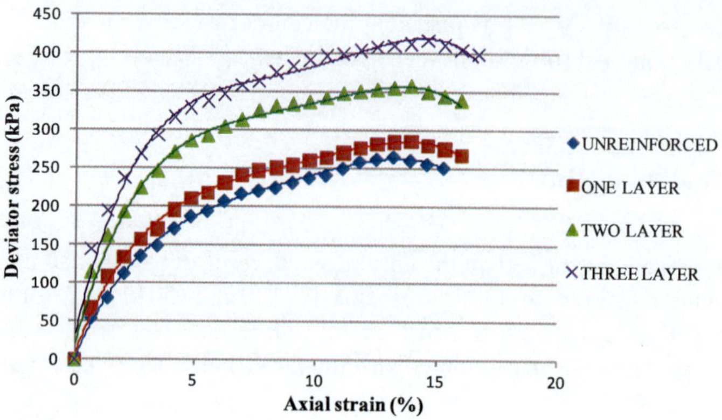

Fig. 4 Deviator stress versus Axial strain on confining pressure 150 kPa

Table 4 Deviator stress at the failure of reinforced clay specimens

Configuration	Deviator stress at failure (kPa)		
	Confining pressure (kPa)		
	50	100	150
Unreinforced clay	173.5	211	265.4
One layer	187.2	227.8	284.2
Two layer	236.8	278.2	357.2
Three layer	275	328.5	416.3

of unreinforced and reinforced clayey soil on confining pressure of 150 kPa. The deviator stress and axial strain at failure were increased as the number of geotextile layers were increased.

Table 4 presents the values of deviator stress at failure in unreinforced and reinforced clayey soil specimens. The deviator stress at failure was increased with a number of geotextiles due to good soil–geotextile interaction.

Improvement in deviator stress. The strength improvement was expressed as an improvement factor. The improvement factor is defined as the ratio of deviator stress of reinforced soil to that of unreinforced soil. Table 5 presents the improvement in

Table 5 Strength improvement of reinforced clayey soil

Configuration	Improvement factor		
	Confining pressure (kPa)		
	50	100	150
One layer	1.07	1.08	1.07
Two layer	1.36	1.31	1.34
Three layer	1.58	1.55	1.56

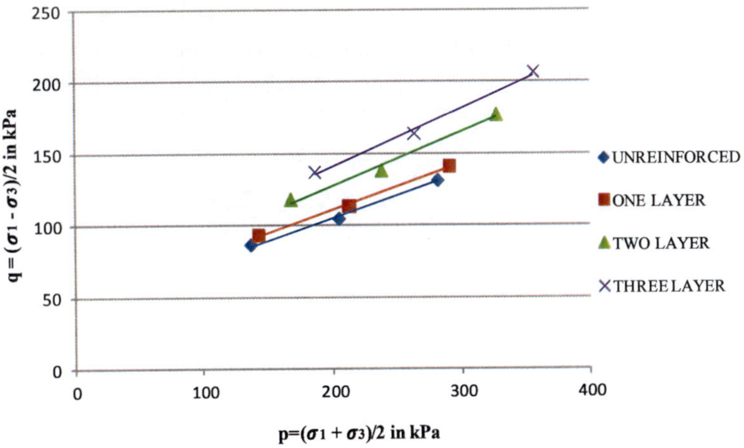

Fig. 5 Comparison of modified failure envelope of clayey soil

deviator stress at the failure of reinforced clayey soil on three confining pressures. The improvement factor was increased with a number of geotextile layers.

Modified failure envelope. The shear strength characteristics such as undrained cohesion (C_u) and angle of internal friction (ϕ) of soil samples can be determined from modified failure envelopes. Figure 5 presents a comparison of modified failure envelopes of unreinforced and reinforced clayey soil specimens.

As the number of geotextile layers was increased, the failure envelope of reinforced specimens shifted upward. The undrained cohesion was increased because the development of pseudo cohesion on confining soil layers from geotextile layers and angle of internal friction also increased due to increased passive resistance as the confining pressure increased. The increase in shear strength of reinforced clay is mainly due to the increase in undrained cohesion of the clay.

Modulus of Elasticity, E. Modulus of elasticity is a measure of soil stiffness. It is calculated by taking the slope of the initial tangent drawn to the stress–strain curve. Table 6 presents the Modulus of elasticity obtained for unreinforced and reinforced

Table 6 Modulus of elasticity (kPa) of reinforced clayey soil

Configuration	Modulus of Elasticity (kPa)		
	Confining pressure (kPa)		
	50	100	150
Unreinforced clay	3125	4375	4667
One	4000	4600	6500
Two	7500	8000	11111
Three	9285	11250	13333

clay specimens on three confining pressures. The Modulus of elasticity was increased as the number of geotextile layers and confining pressures were increased.

6.2 *Triaxial Test Results of Clay Reinforced with a Geotextile Encapsulated in Thin Layer of Sand (Sandwich Specimen)*

Type of failure. Figure 6 presents a typical image of the deformed sandwich specimen. The failure was caused by a bulging of the clay and a discontinuous deformation at the sand–clay interface. The bulging and deformation were attributed to a lateral expansion of clay restrained by the sand–clay interface. Because the sand particles were prone to penetrate the clay, the sand–clay interface shear strength could be stronger than the shear strength of the clay itself. Thus, the zone of maximum lateral deformation moved away from the sand geotextile interface to the clay.

Stress–Strain behavior. Figure 7 shows comparison of Stress–Strain behavior of unreinforced clay, clay reinforced with one-layer geotextile and sandwich specimen of clayey soil on confining pressure of 150 kPa.

The sandwich soil specimens reached peak deviator stress at specified confining pressure than that of the clay–geotextile specimen and unreinforced soil specimens. The deviator stress and axial strain at failure were increased as the thickness of the sand layer were increased due to good soil–geotextile interaction.

Table 7 presents the values of deviator stress at the failure of unreinforced clay, clay–geotextile specimen and sandwich specimen of clay on three confining pressures. The deviator stress and axial strain at failure increased with sand layers' thickness and confining pressures. The maximum value of deviator stress was found to be 569.4 kPa with a sand thickness of 8 mm on confining pressure of 150 kPa.

Improvement in deviator stress. The strength improvement was expressed as an improvement factor. The improvement factor is defined as the ratio of deviator stress of reinforced soil to that of unreinforced soil. Table 8 presents the improvement in deviator stress at the failure of sandwich specimens under three confining pressures.

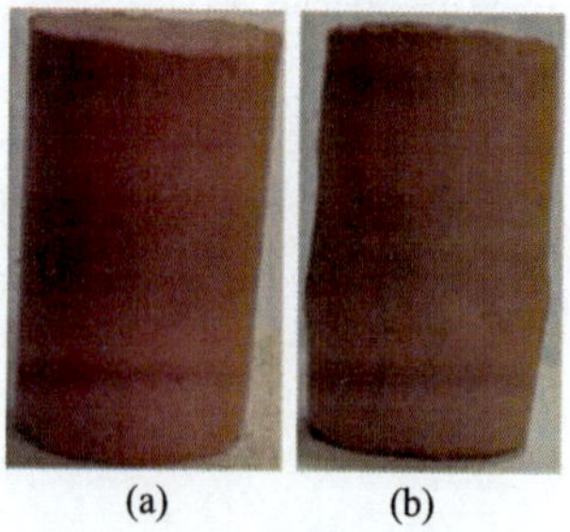

Fig. 6 **a** Sandwich specimen before loading **b** Sandwich specimen after loading

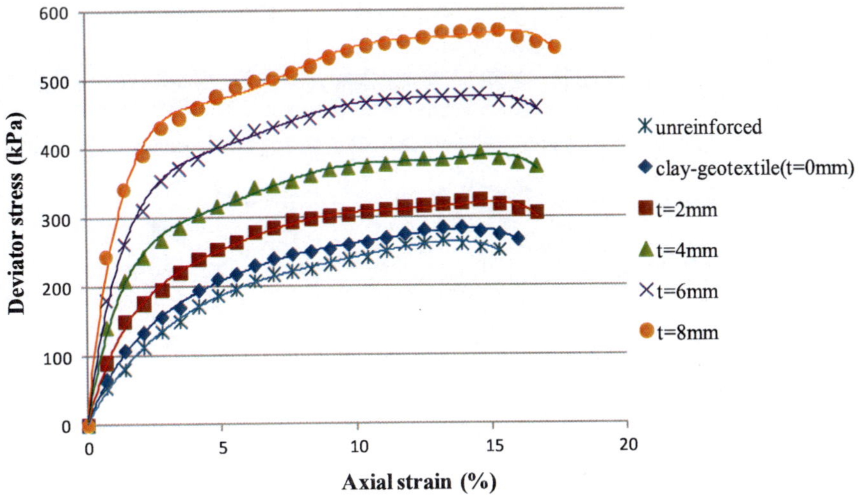

Fig. 7 Deviator stress versus Axial strain on confining pressure 150 kPa

Table 7 Deviator stress at failure of sandwich specimen

Configuration	Deviator stress at failure (kPa)		
	Confining pressure (kPa)		
	50	100	150
Unreinforced clay	173.5	211	265.4
Clay–geotextile (t = 0 mm)	187.2	227.8	284.2
t = 2 mm	204.5	257	323.2
t = 4 mm	242.1	310.1	393
t = 6 mm	291	372.8	478.4
t = 8 mm	341.7	445.7	569.4

Table 8 Strength improvement of sandwich specimens

Configuration	Improvement factor		
	Confining pressure (kPa)		
	50	100	150
Clay–geotextile (t = 0 mm)	1.07	1.08	1.07
t = 2 mm	1.17	1.21	1.22
t = 4 mm	1.39	1.46	1.48
t = 6 mm	1.67	1.76	1.8
t = 8 mm	1.96	2.11	2.14

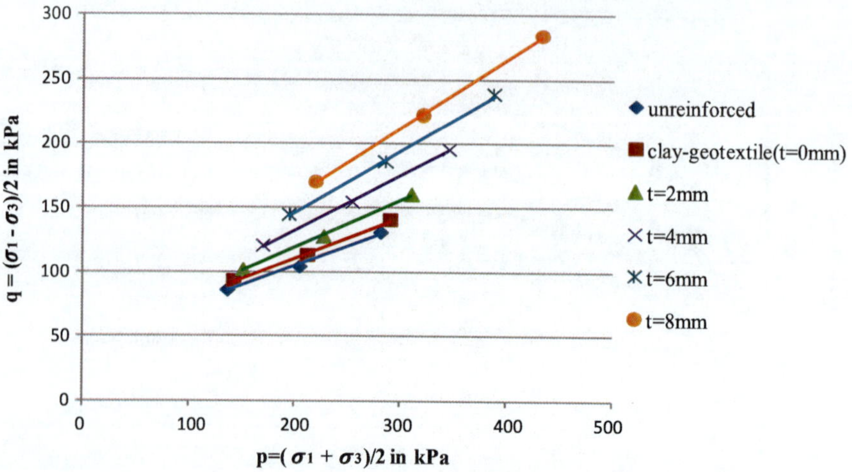

Fig. 8 Comparison of modified failure envelope of sandwich specimen

The improvement factor was increased with an increase in the thickness of sand layers. The maximum value of the improvement factor was found to be 2.14 when the sand thickness is 8 mm on confining pressure of 150 kPa. The improvement factor for the sandwich specimen is more when compared to the improvement factor of clay reinforced with one-layer geotextile.

Modified failure envelope. The shear strength characteristics such as undrained cohesion (C_u) and angle of internal friction (ϕ) of soil samples can be determined from modified failure envelopes. Figure 8 presents the comparison of modified failure envelopes of the unreinforced, clay–geotextile specimen and sandwich specimens. The slopes of the failure envelopes steepened as the thicknesses of the sand layers were increased, resulting in increases in the friction angles and slight decreases in the cohesion. The inclusion of the sand layer increases the interface interaction between clay and geotextile. Regarding the sandwich technique, the increase in shear strength of clay is mainly due to the increase in friction angle of the clay.

Modulus of Elasticity, E. Modulus of elasticity is calculated by taking the slope of the initial tangent drawn to the stress versus strain curve. Table 9 presents the Modulus of elasticity obtained for unreinforced clay, clay–geotextile specimen and sandwich specimens on three confining pressures. The Modulus of elasticity was increased as the thickness of the sand layer and confining pressure was increased.

Table 9 Modulus of elasticity (kPa) of sandwich specimen

Configuration	Modulus of elasticity, E		
	Confining pressure, kPa		
	50	100	150
Unreinforced clay	3125	4375	4667
t = 0 mm	4000	4500	6500
t = 2 mm	6667	8333	10666
t = 4 mm	10000	11000	16000
t = 6 mm	12500	14000	26000
t = 8 mm	17000	20000	35000

7 Conclusions

The conclusions of this study can be summarized as follows:

1. The reinforced clay specimens failed when bulging occurred between two adjacent reinforcement layers. The sandwich specimens failed because of the bulging of the clay and discontinuous deformation at the sand–clay interface.
2. Both the reinforced clay and sandwich specimens enhanced the peak shear strength of the clay. The deviator stress was increased as the number of geotextile layers, confining pressures, and the sand-layer thickness was increased.
3. The Axial strain at failure also increased as the number of geotextile layers, confining pressures, and sand layers were increased. The maximum deviator stress at failure occurred in sandwich specimens of clayey soil compared to reinforced clay with geotextile layers.
4. The improvement factor was increased with the number of geotextile layers and thickness of sand layers on different confining pressures. The improvement factor of the sandwich specimen was found to be 2.14 on confining pressure of 150 kPa, which is more compared to reinforced clay.
5. The undrained cohesion and angle of internal friction were increased as the number of geotextile layers and thickness of the sand layer was increased. In sandwich specimen of clay, a slight decrease in cohesion and an increase in friction angle were found when compared with reinforced clay.
6. The modulus of elasticity was increased with the number of geotextile layers, confining pressures and thickness of sand layers. The modulus of elasticity of sandwich specimen was found to be more compared to reinforced clay.

This study clarifies that small reinforcement spacing and thick sand layers can effectively strengthen the soil–geotextile interaction. Therefore, as a practical application, low permeability and fine-grained soils can be strengthened by using permeable geotextile layers and coarse-grained soil layers around reinforcement.

References

1. Sridharan, A., Srinivasa Murthy, B.R., Bindumadhava, Revanasiddappa, K.: Technique for using fine-grained soil in reinforced earth. J. Geotech. **117**(8) (1997)
2. Ingold, K., Miller, S.: The performance of impermeable and permeable reinforcement in clay subject to undrained loading. J. Eng.Geol. London **15** (1982)
3. Yang, K.-H., Yalew, W.M., Nguyen, M.D.: Behavior of geotextile-reinforced clay with a coarse material sandwich technique under unconsolidated-undrained triaxial compression. Int. J. Geomech. 04015083-1 (2016)
4. Abdi, M.R., Sadrnejad, A., Arjomand, M.A.: Strength enhancement of clay by encapsulating geogrids in thin layers of sand. Geotext. Geomembr. 447–455 (2009)

Finite Element Analysis of Soil Reinforced Canal Tunnel

Archana Singh Bagri and A. K. Singh

1 Introduction

In developing countries, scarcity of land occurs due to rapid increasing industrialization, development and civilization. Industrialization and urbanization cannot be stopped instead, it may be limited up to an extent. One of the best alternatives is to use underground space, that is, basement, cavern, tunnel, etc. The need for tunnel and underground structures is progressively growing for various facilities that need to be provided, that is, water conductor system for power generation, irrigation, road traffic, and underground metro railway. Also, sewerages, electrical cables in metropolitan cities, underground oil storage and for extraction of valuable ore require the construction of tunnel.

In hilly areas, so much undulation is encountered that it makes construction difficult, costlier and hectic. It is necessary to make an underground or elevated structure. If a hill is encountered in the canal then, it becomes a compulsion to pass the canal through a tunnel, as it will be costly and tedious to cut the hill completely.

In the process of designing a tunnel, factors taken into consideration are geological strata, type of soil, seepage pressure, type of surcharge, etc. Loads considered are pressure exerted by ground, the pressure exerted by water, self-weight, surcharge, subgrade reaction, etc. Subsequent loading conditions may also occur, that is, loads from the inner side of the tunnel, loads throughout the construction stage, earthquake loads, effect of the adjoining tunnel, settlement effects and other loads.

A. S. Bagri (✉)
National Institute of Technology, Jamshedpur 831014, Jharkhand, India
e-mail: archanasinghbagri1994@gmail.com

A. K. Singh
Civil Engineering Department, National Institute of Technology, Jamshedpur 831014, Jharkhand, India
e-mail: aksingh.civil@nitjsr.ac.in

C. N. V. Satyanarayana Reddy et al. (eds.), *Ground Improvement and Reinforced Soil Structures*, Lecture Notes in Civil Engineering 152,
https://doi.org/10.1007/978-981-16-1831-4_59

Many Finite Element method and Finite Difference method software are available, such as FLAC-3D, PLAXIS, ANSYS, GTS-NX, ABAQUS, UDEC, etc., to analyse tunnels and other structures. Results generated from the software are precise thus reliable and is able to compare with analytical solution.

Geogrid is a geosynthetic material used for reinforcing the soil. This material has high tensile strength and low elongation (high stiffness). The shearing resistance is high along the soil reinforcement interface.

Karparapu and Bathurust [1] have done finite element modelling (discrete type) for soil walls reinforced by geosynthetic in addition to the material models employed to replicate the behaviour of a variety of components in these types of structures. Tsukamoto et al. [2] have studied the effect of geogrid on earth pressure against model retaining wall. It has been demonstrated that the active earth pressure coefficient was effectively reduced in the case of unattached geogrid to the face of the wall; consequently, the earth pressure on the wall was found to be reduced. Zeigler [3] has investigated the confining effect of geogrid with plane strain model wall tests. With the help of series of triaxial tests, he showed the confining effect of geogrid. He concluded that the horizontal earth pressure was decreased due to geogrids and also bearing capacity of the triaxial specimen increased. Ahmed et al. [4] have done the experimental and numerical analysis to find out the role of geogrid reinforcement in reducing earth pressure on buried pipes. He concluded that the radial pressure was reduced in the reinforced case as compared to the unreinforced case due to the involvement of geogrid. Banerjee et al. [5] have done experimental and 3D finite element analysis o geocell reinforced embankments using GTS NX software. They concluded that the reinforced embankment's ultimate capacity was four times the capacity of the unreinforced embankment. Mandal and Singh [6] have done the finite element analysis of tunnel lining. They verified their results with a USBR monogram.

The finite element analysis of tunnel lining with unreinforced and reinforced backfill has been done in this study. Geogrid is used as a reinforcing material. Analysis has been done with varying overburden pressures (4, 8 and 12 m). Thrust, Shear force, and bending moment have been found for the unreinforced and reinforced case and then compared for both cases.

2 Finite Element Analysis

GTS-NX software has been used for finite element analysis in this work.

2.1 Problem Description

While doing the structure design of lining, earth pressure plays an important role. Two cases are analysed and then compared. The first case consists of the finite

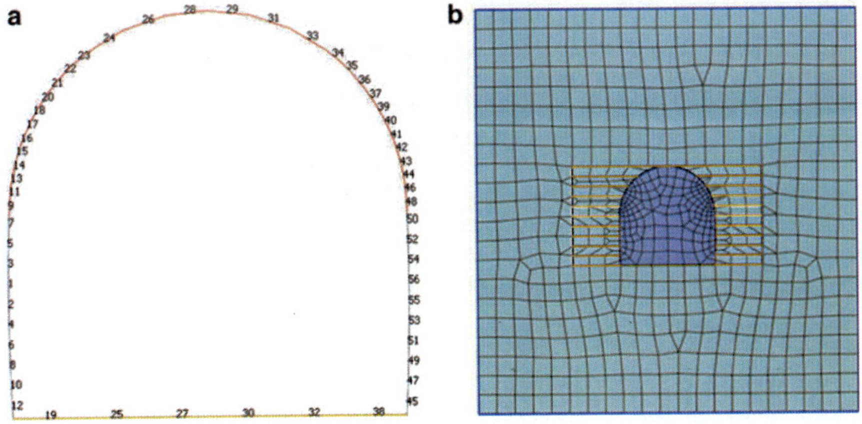

Fig. 1 **a** Tunnel lining model (showing element number). **b** Ground and tunnel lining attached with geogrid

element analysis of tunnel lining with unreinforced backfill and then the results were compared with USBR monogram [6]. The second case consists of finite element analysis of tunnel lining with reinforced backfill. Then both the cases have been compared and accordingly thickness of lining has been designed. The concept behind placing the geogrid in the backfill is that the geogrid offers high frictional resistance that counterbalances a part of earth pressure induced in the backfill. As a result, the resultant earth pressure gets reduced; consequently, the thickness of lining gets reduced. In the case of reinforced backfill, the geogrid was placed in the backfill and attached to the walls of the tunnel in a perpendicular direction, as shown in Fig. 1. Analysis has been done by varying the overburden pressure (4, 8 and 12 m).

2.2 Geometry and Model Creation

The model is created using finite element analysis software GTS-NX. A tunnel lining having 5 m height is modelled. After that, meshing has been done and 56 elements and 57 nodes has been created. After that ground meshing has been done having dimensions according to the amount of overburden to be placed above tunnel lining. Then, modelling of geogrid is done. Geogrid having 1 m length is placed at a distance of 0.5 m throughout the wall of tunnel lining (perpendicular to lining). After that boundary conditions are placed. Fixed supports are provided to the ground and geogrids are restrained against moment as it cannot bend. The forces acting on the tunnel lining are (1) Self-weight (2) Horizontal earth pressure (3) Vertical overburden pressure. Uplift force is neglected as it is on the advantageous side because it counteracts the vertical overburden pressure. Also, the empty condition of the canal is taken into consideration because hydrostatic pressure exerted by water flowing through the canal will reduce the horizontal earth pressure. So, analysis is done for critical conditions: empty condition of canal and no uplift pressure.

Model validation. The model of this current study is based on a live project constructed at Sundernagar (Jharkhand) for a canal starting from Ganjia Barrage (Kharkhai river) towards Jadugora, Jamshedpur, Jharkhand. The canal tunnel is passing under the road and railway track. The tunnel is 1.3 km long and highest crest height is 13 m. It is an irrigation canal constructed under Swarnarekha Multipurpose Project, Jamshedpur. It is designed using USBR monogram. The thickness of tunnel lining is 500 mm and the base slab is 600 mm in this project. Mandal and Singh [6] have done the analysis and designed the wall thickness 450 mm and base slab thickness 500 mm.

The model in the current study has also been validated with the model created by Mandal and Singh [6].

2.3 *Properties of Material*

Properties of material are given in Tables 1, 2 and 3.

3 Results and Discussion

In the current work, analysis of forces and moments acting upon 'D' shaped tunnel lining without geogrid and with geogrid has been analysed. The most critical loading

Table 1 Properties of soil

Properties	Soil
Model	Mohr- Coulomb
Elastic modulus	50000 kN/m^2
Cohesion	0.2 kN/m^2
Friction angle	40 degrees
Unit weight	22 kN/m^3
Poisson ratio	0.3

Table 2 Properties of concrete used for lining

Properties	Concrete
Model	Elastic
Elastic modulus	20000000 kN/m^2
Unit Weight	25 kN/m^3
Poisson Ratio	0.2

Table 3 Properties of reinforcement

Properties	Geogrid
Model	Elastic
Elastic modulus	150000 kN/m^2
Unit weight	0.5 kN/m^3
Poisson ratio	0.2

condition due to the empty condition of the canal tunnel with saturated backfill is considered. If the flow condition in the canal is considered, it will counterbalance the horizontal thrust acting on the lining. In the rainy season, the backfill gets fully saturated hence, increase the earth's pressure. Finite element analysis has been done for both unreinforced and reinforced cases and then compared with each other.

The results of this study are validated with the live project of Sundernagar canal tunnel (designed using USBR monogram).

3.1 Comparison of FEM Analysis of Tunnel Lining with Unreinforced and Reinforced Backfill

The breadth of the lining is 5.0 m, the height of the crest is 5.0 m, the radius of the arch is 2.5 m, the height of the overburden crest is 4.0 m, the width of the trench at crest level is 7.0 m, and saturated unit weight of backfill is 22 kN/m^3, depth of overburden = 8 m.

The comparison of Axial force distribution is given in Figs. 2 and 3.

As it is clear by comparing Figs. 2 and 3 that in the bottom slab (element no. 19, 25, 27, 30, 32, 38), the axial thrust decreases in the reinforced case compared to the unreinforced case. In the walls and crown of the tunnel, the axial force increases in reinforced backfill compared to unreinforced backfill.

The comparison of Shear force distribution is given in Figs. 4 and 5.

As it is clear from Figs. 4 and 5, the decrease in shear force in the bottom slab is slight in case of reinforced backfill compared to unreinforced backfill. This is because there is no geogrid in the bottom slab. In the walls and crown of the tunnel, the shear force decreases considerably in case of reinforced backfill compared to unreinforced backfill.

The comparison of Bending moment distribution is given in Figs. 6 and. 7.

As it is clear by Figs. 6 and 7, in every element (floor, walls, and crown), moment is decreasing in the case of reinforced backfill as compared to the case of unreinforced backfill. That will further lead to a reduction in the thickness of tunnel lining.

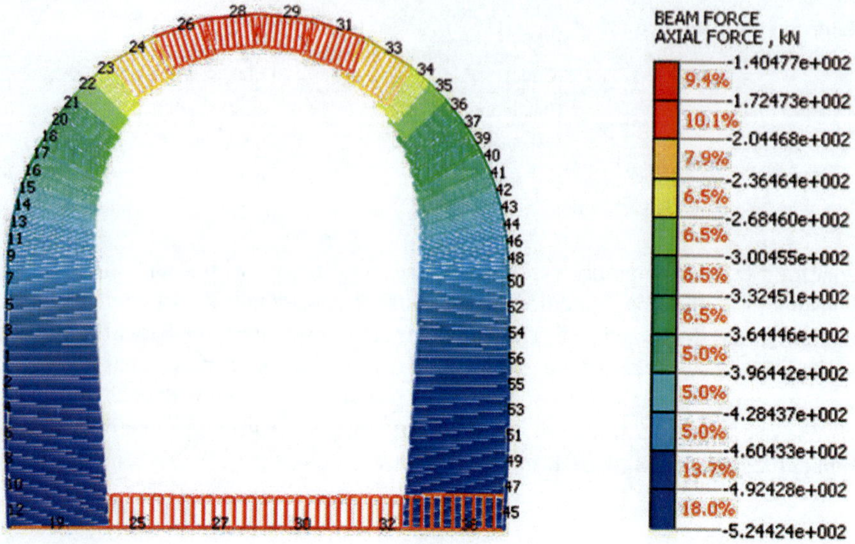

Fig. 2 Axial thrust distribution for unreinforced backfill

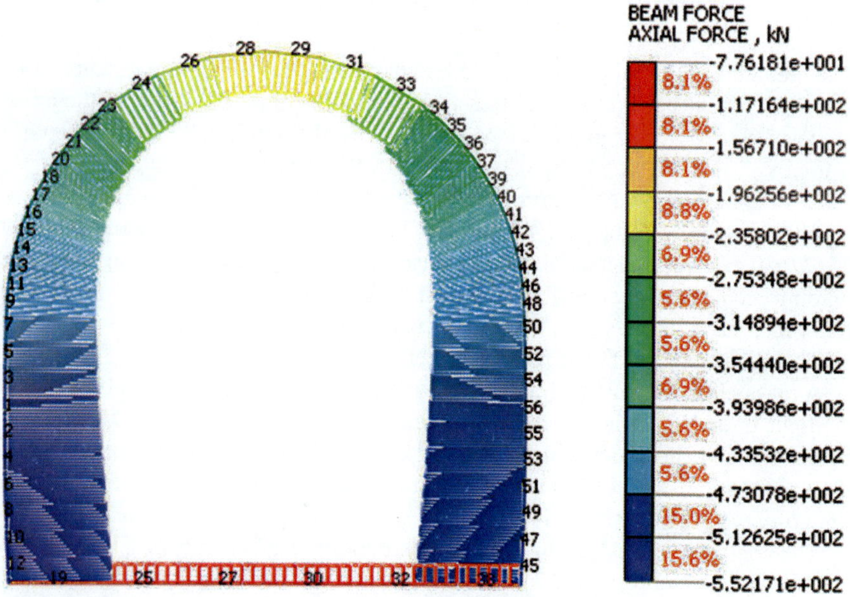

Fig. 3 Axial thrust distribution for reinforced backfill

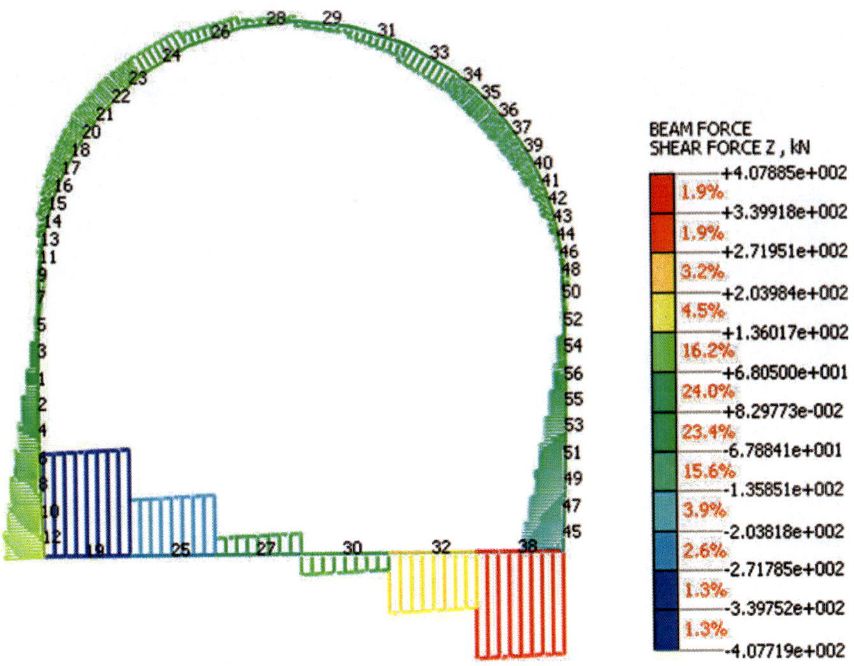

Fig. 4 Shear force distribution for unreinforced backfill

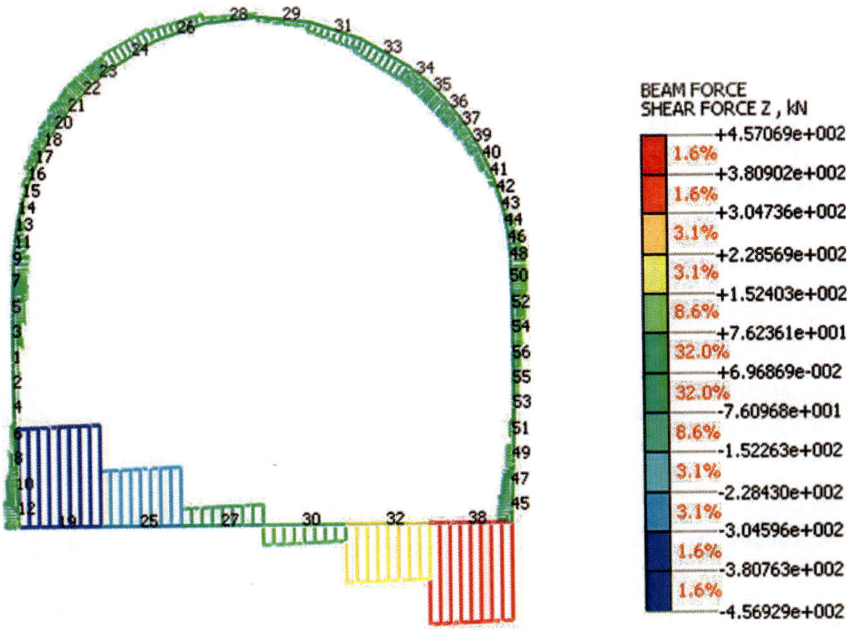

Fig. 5 Shear force distribution for reinforced backfill

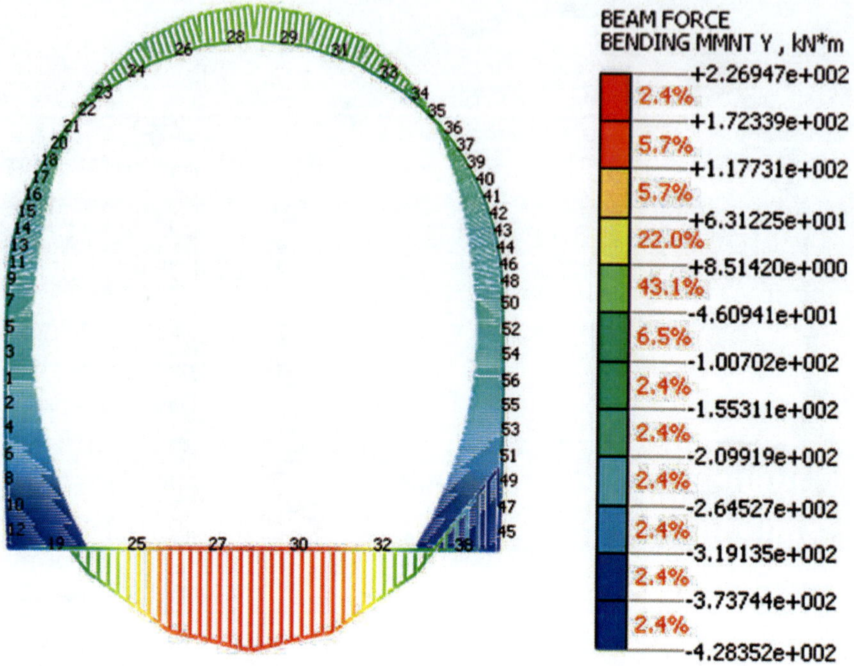

Fig. 6 Bending moment distribution for unreinforced backfill

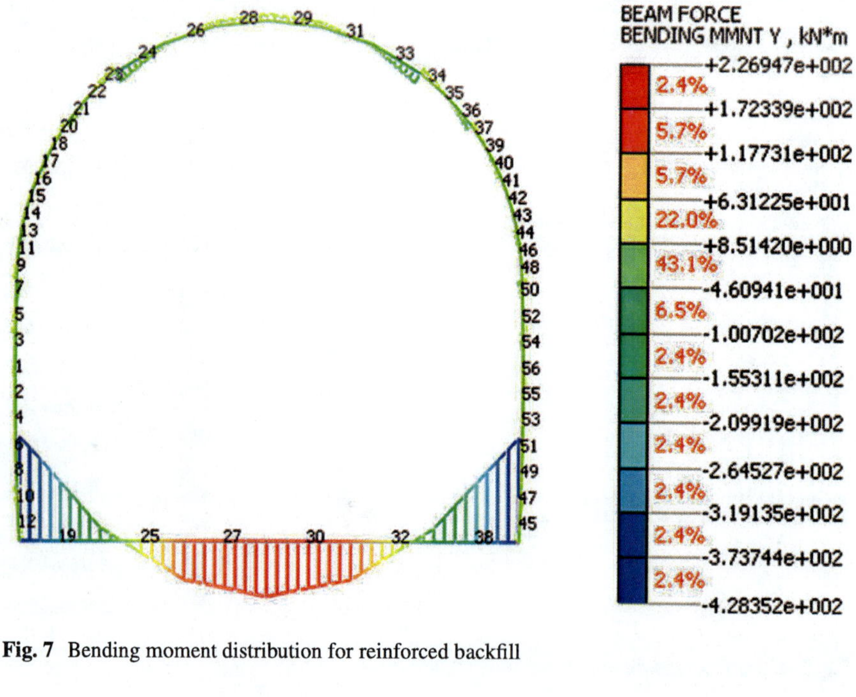

Fig. 7 Bending moment distribution for reinforced backfill

3.2 Graphical Representation of Results

The graphical representation of Axial thrust for both the cases (unreinforced and reinforced) is shown in Fig. 8.

Axial thrust is increasing as we move from upward to downward. The nature of the graph is zig-zag because elements have been numbered randomly as shown in Fig. 1a. Elements 2, 4, 6, 8, 10, 12 are in a downward position whereas, elements 3, 5, 7, 9. etc., are in the upward direction. Also, elements 19, 25 and 27 are in the base slab.

As it is clear from Fig. 8, the axial thrust in tunnel walls has been increased marginally in the case of reinforced backfill case compared to unreinforced backfill. Refer Fig. 1 for the location of elements.

The graphical representation of Shear Force for both the cases (unreinforced and reinforced) is shown in Fig. 9.

The nature of shear force can be seen in Fig. 4 and 5 (shear force diagram). As it is clear from Fig. 9, that shear force in vertical walls of tunnel decreases in reinforced backfill case compared to shear force in unreinforced backfill case. The zig-zag nature of the graph is zig-zag because elements are numbered randomly as shown in Fig. 1a. Refer Fig. 1 for the location of elements.

The graphical representation of moment for both the cases (unreinforced and reinforced) is shown in Fig. 10.

The nature of the Bending Moment is shown in Figs. 6 and 7 (bending moment diagram). The negative moment at the ends of the base slab and the positive moment in the centre of the base slab. A negative moment is developed in the walls of tunnel lining and a positive moment is developed at the crown.

As it is clear from the Fig. 10, the bending moment in the case of reinforced backfill decreases considerably compared to the bending moment in the unreinforced case. This will lead to a decrease in wall thickness of tunnel lining.

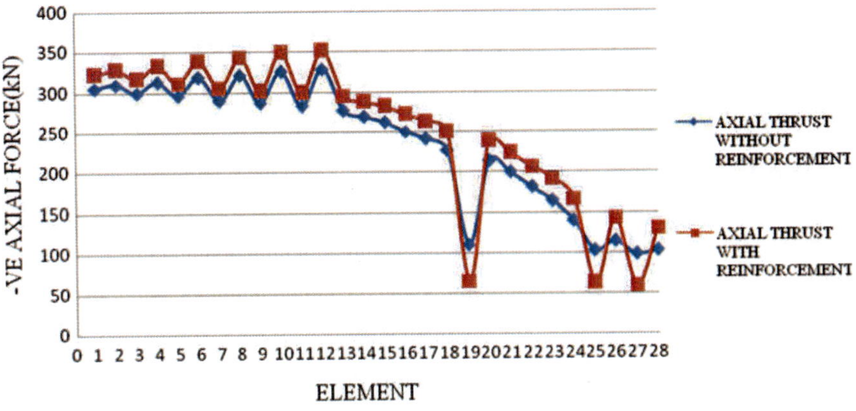

Fig. 8 Axial thrust for unreinforced and reinforced backfill

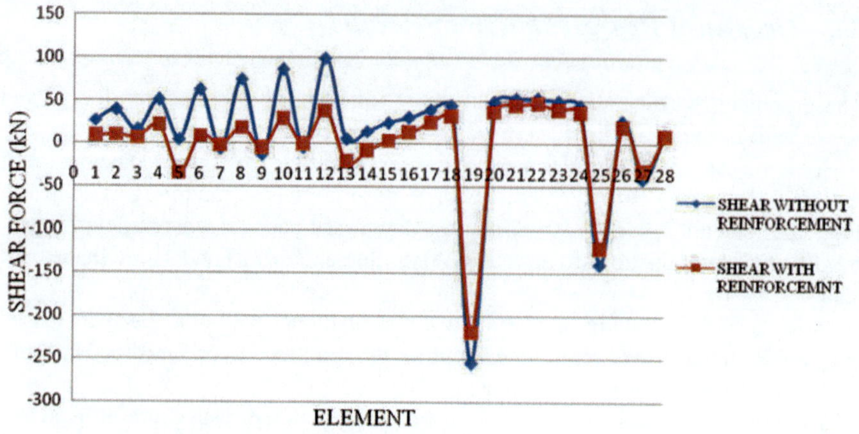

Fig. 9 Shear force for unreinforced and reinforced backfill

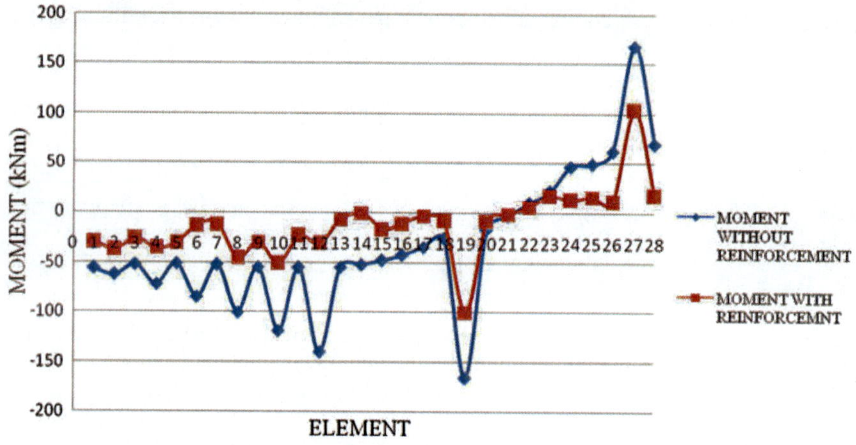

Fig. 10 Moment for unreinforced and reinforced backfill

3.3 Thickness of Lining Section

See Tables 4 and 5.

4 Conclusions

1. The values of thrust, shear, and moment obtained by Mandal and Singh [6] are compared with the present study's values. It has been seen that values obtained in both cases are matching to a great extent. The variation is due to the difference in modelling problems in both cases.

Table 4 Thickness and reinforcement for different overburden (without and with reinforcement)

Overburden (m)	Thickness of the section required (mm) (without reinforcement)	Thickness of the section required (with reinforcement)	Reinforcement details
4	350	300	12 mm ϕ @125 mm c/c
8	450	350	16 mm ϕ @125 mm c/c
11	550	450	20 mm ϕ @ 100 mm c/c

Table 5 Thickness and reinforcement for different soil parameters (without and with reinforcement)

Soil parameters γ (kN/m^3) ϕ (degrees)	Thickness of the section required (mm) (without reinforcement)	Thickness of the section required (with reinforcement)	Reinforcement details
$\gamma = 18, \phi = 30$	375	300	12 mm ϕ @125 mm c/c
$\gamma = 20, \phi = 35$	425	350	16 mm ϕ @125 mm c/c
$\gamma = 22, \phi = 40$	450	375	20 mm ϕ @ 100 mm c/c

2. The thickness of tunnel lining, which was found to be 450 mm, according to Mandal and Singh [6], is reduced to 350 mm in the present study.
3. The most critical loading condition is due to overburden, backfill, self-weight of structure and empty condition of the canal. Buoyant force, uplift pressure and seepage force do not play an important role in most critical loading conditions.
4. Shear force and moment on the lining have decreased in reinforced backfill compared to without reinforced backfill.
5. Thrust on the lining wall has increased marginally in case of reinforced backfill compared to the thrust in unreinforced backfill.
6. Lining thickness has been reduced by 25% (approx.) in the backfill, that is, reinforced compared to unreinforced backfill with different overburden pressure and soil parameters.

References

1. Karpurapu, R., Bathurst, R.J.: Behavior of geosynthetic reinforced soil retaining walls using the finite element method. Comput. Geotech. **17**, 219–299 (1995)
2. Tsukamoto, Y., Ishihara, K., Higuchi, T., Aoki, H.: Influence of geogrid reinforcement on lateral earth pressure against model retaining walls. Geosyn. Int. **6**, 195–218 (1999)
3. Zeigler, M.: Investigation of the confining effect of geogrid reinforced soil with plane starin model wall tests. Geotechnik Im Bauwesen
4. Ahmed, M.R, Tran, V.D.H., Meguid, M.A.: On the role of geogrid reinforcement in reducing earth pressure on buried pipes: experimental and numerical investigation. **55**(3), 588–599 (2015)

5. Banerjee, L., Chawla, S., Bhandari, G.: Experimental and 3-D finite element analysis on geocell reinforced embankments. J. Test. Eval. **47**(3), 1876–1899 (2018)
6. Mandal, G., Singh, A.K.: Finite element analysis of canal tunnel. IGC (2018)

Influence of Footing Size on Reinforcement Geometrical Parameters

B. Venkatesh and T. Thyagaraj

1 Introduction

Due to the increase in the demand for land for construction, it is necessary to improve the soil load-carrying capacity by applying low-cost ground improvement techniques. The soil reinforcement technique using geosynthetics plays a major role in improving the soil load-bearing capacity economically compared to other ground improvement techniques. In literature, several researchers performed a number of experimental and numerical studies using different types of geosynthetics such as geotextiles, geogrids and geocells. Based on the experimental findings, it was concluded that the improvement of soil load-carrying capacity depends on the optimum geometrical parameters of the reinforcements, such as placement depth of the topmost reinforcement layer from the footing base (u), the vertical spacing between the reinforcement layers (h), number of reinforcement layers (N), and size of the reinforcement layer (B_r), for example, Binquet and Lee [2, 3], Guido et al. [7], Yetimoglu et al. [15], Sitharam and Sireesh [12], Dash et al. [5], Ghosh et al. [6], Basudhar et al. [1], Latha and Somwanshi [9], Naderi and Hataf [10], Tavangar and Shooshpasha [14], and Buragadda and Thyagaraj [4].

Literature shows that the studies on the effect of footing size on optimum geometrical parameters of the reinforcement are limited [11, 14]. Therefore, the present study was conducted to understand the effect of footing size on optimum geometrical parameters of jute geotextile reinforcements in sand bed, that is, the placement depth of the first reinforcement layer (u) and spacing between the reinforcement

B. Venkatesh (✉) · T. Thyagaraj
Department of Civil Engineering, IIT Madras, Chennai 600036, India

T. Thyagaraj
e-mail: ttraj@iitm.ac.in

C. N. V. Satyanarayana Reddy et al. (eds.), *Ground Improvement and Reinforced Soil Structures*, Lecture Notes in Civil Engineering 152,
https://doi.org/10.1007/978-981-16-1831-4_60

Table 1 Properties of sand (after [4])

Property	Values
Specific gravity, G	2.68
Sand (%)	99
Fines (%)	1
D_{10} (mm)	0.25
D_{30} (mm)	0.42
D_{60} (mm)	0.95
C_u	3.8
C_c	0.743
e_{max}	0.666
e_{min}	0.466
Soil classification	SP

layers (h). A circular footing was used to perform the model tests on both unreinforced and reinforced sand beds. Each test was repeated two times to ensure the repeatability of the test results.

2 Experimental Program

2.1 *Materials*

2.1.1 Sand

All the laboratory model tests were performed using clean and dry sand. The sand was collected from Chennai surrounding localities. The index and physical properties of sand are presented in Table 1.

2.1.2 Geotextiles

The present laboratory model tests were performed on reinforced sand beds using natural woven jute geotextile (JGT). The natural woven jute geotextile was procured from the National Jute Board (NJB) approved company—Ballyfabs International Ltd., Chennai, India. Table 2 presents the physical and mechanical properties of the jute geotextile.

Table 2 Properties of jute geotextile (after [4])

Properties	Value
Physical properties	
Thickness (mm)	1
Mass per unit area (g/m^2)	315
Mechanical property	
Ultimate tensile strength (kN/m)	
Machine direction (MD)	13.8
Cross-machine direction (CMD)	12.5
Failure strain (%)	
Machine direction (MD)	4.9
Cross-machine direction (CMD)	5.9
Axial stiffness, EA (kN/m)	246

2.2 *Laboratory Test Set-Up*

A steel tank with inner dimensions of 450 mm × 450 mm × 350 mm was used to perform the model tests. The depth of the sand bed was maintained as 300 mm. The sand beds were prepared by adopting an air-pluviation technique to a relative density (R_d) of 70%. A 50 mm diameter circular footing made with 25 mm thick rigid mild steel was used for the present model tests. In order to increase the roughness of the footing base, a thin layer of sand was affixed using an araldite. A hydraulic jack which was welded to the reaction frame was used for push the footings into the sand beds.

The laboratory model tests were performed on unreinforced sand and jute geotextile sand beds by placing the jute geotextile at various depths. To find out the effect of footing size on placement depth of first reinforcement layer from the footing base (u) and spacing between the reinforcement layers (h), the reinforcement was placed at different depths, that is, 0.2D, 0.3D and 0.4D. A 10kN precalibrated proving ring was used for the load application on the footing followed in accordance to [8]-1982. Figure 1 shows the schematic diagram of the reinforced sand bed. The footing settlements were determined by using dial gauges D_1 and D_2, as shown in Fig. 1.

3 Results and Discussion

Bearing capacity ratio (BCR) is used to evaluate the improvement of soil load-carrying capacity with reinforcement layers. The BCR is defined as per Binquet and Lee [2, 3] as:

$$\text{BCR} = \frac{\text{Reinforced soil bearing capacity } (q_{rs})}{\text{Unreinforced soil bearing capacity } (q_{us})} \tag{1}$$

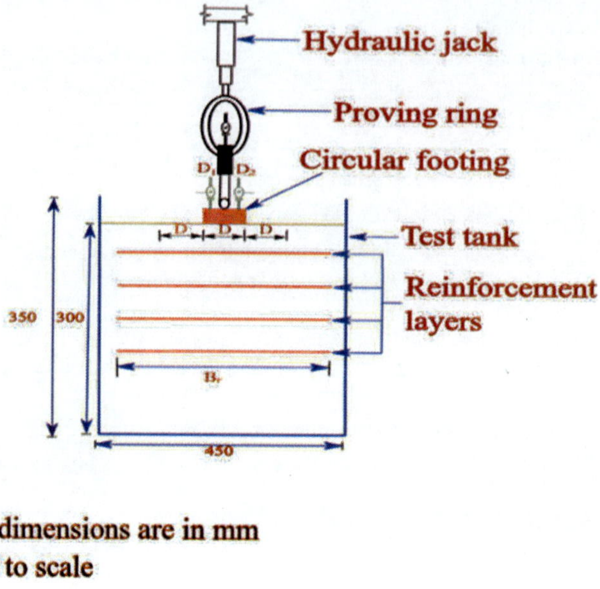

Fig. 1 Geometry of the multi-layered reinforced sand bed

where the bearing capacity of reinforced sand (q_{rs}) and unreinforced sand (q_{us}) are at the same settlement(s), respectively, however, at higher settlements, the ultimate bearing capacity of the unreinforced soil (q_{ult}) is used in place of q_{us}.

Figures 2 and 3 show the variation of pressure-settlement behavior of both unreinforced and reinforced sand beds for different depths of topmost reinforcement layer from the footing base (u) and different vertical spacing between the reinforcement layers (h) using single and four numbers of reinforcements, respectively. It can be observed from Fig. 2 that the load-settlement behavior of reinforced sand bed shows a pronounced peak for u = 0.3D and 0.4D. Similarly, Fig. 3 also shows the pronounced peak for vertical spacing of the reinforcement layers, h = 0.3D. This is due to the rupture of the jute geotextile, as shown in Fig. 4.

Figures 5 and 6 show the variation of bearing capacity ratio (BCR) with different ratios of u/D and h/D, respectively. For comparison, the results of BCR for 150 mm diameter (D) footing derived from Buragadda and Thyagaraj [4] are also presented in Figs. 5 and 6. Figures 5 and 6 show the variation of BCR for only footing settlement (s/D) of 12%. According to Tafreshi and Dawson [13], the footings should not be allowed to settle more than 15% of footing diameter (D). Hence, the variation of BCR is shown only for footing settlement (s/D) of 12%, irrespective of the footing size (D).

Table 3 presents the optimum placement depth of the first reinforcement layer (u/D) and optimum vertical spacing between the reinforcement layers (h/D) for footing sizes (D) of 50 mm and 150 mm. Table 3 shows the values of optimum u/D for two different sizes of the footings as the same, that is, 0.312D (D = 50 mm)

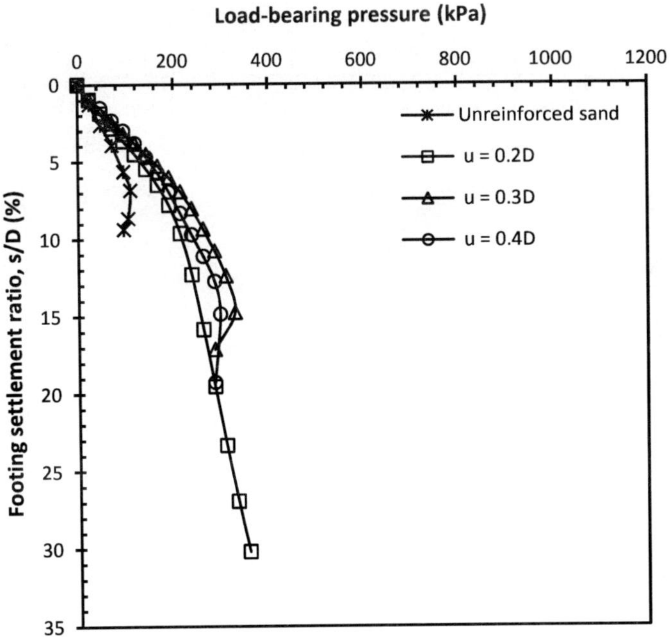

Fig. 2 Comparison of load-settlement curves of unreinforced and geotextile reinforced sand beds at different placement depths of the reinforcement layer (u/D)

and 0.31D (D = 150 mm). However, the values of optimum h/D for two different size footings vary, even though the variation is not much significant, that is, 0.29D (for D = 50 mm) and 0.3D (for D = 150 mm). The variation factor for two different diameter footings (50 mm and 150 mm) is almost 3. Also, the type of reinforcement used was the same in all the model tests. Finally, the variation is relatively higher for optimum vertical spacing of the reinforcement (h) than the optimum placement depth of the first reinforcement layer from the footing base (u).

4 Conclusions

The present experimental studies were conducted to understand the effect of footing size on the optimum geometrical parameters of the reinforcements, that is, the placement depth of the first reinforcement layer (u) and spacing between the reinforcement layers (h). The test results show that the effect of footing size (D) on optimum placement depth of the first reinforcement layer (u/D) and optimum vertical spacing between the reinforcement layers (h/D) is not significant even though the variation of footing size ratio is almost 3.

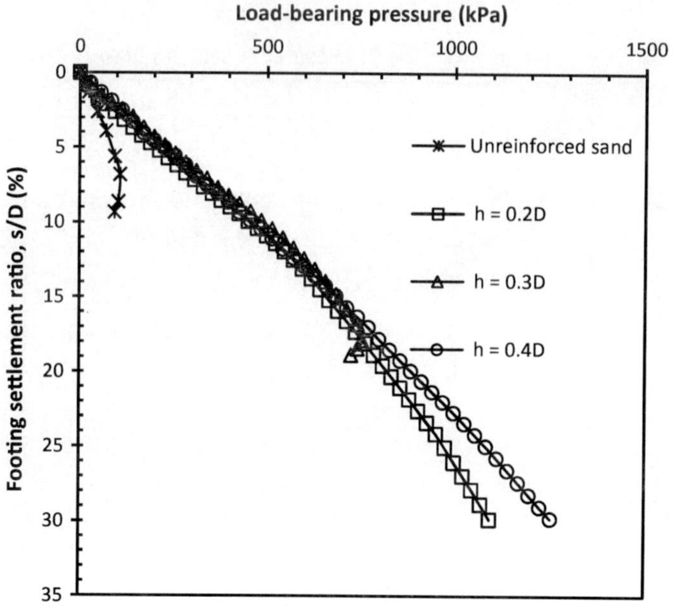

Fig. 3 Comparison of load-settlement curves of unreinforced and geotextile reinforced sand beds at a different vertical spacing of reinforcement layer (h/D)

Fig. 4 Image of ruptured jute geotextile after the plate load test

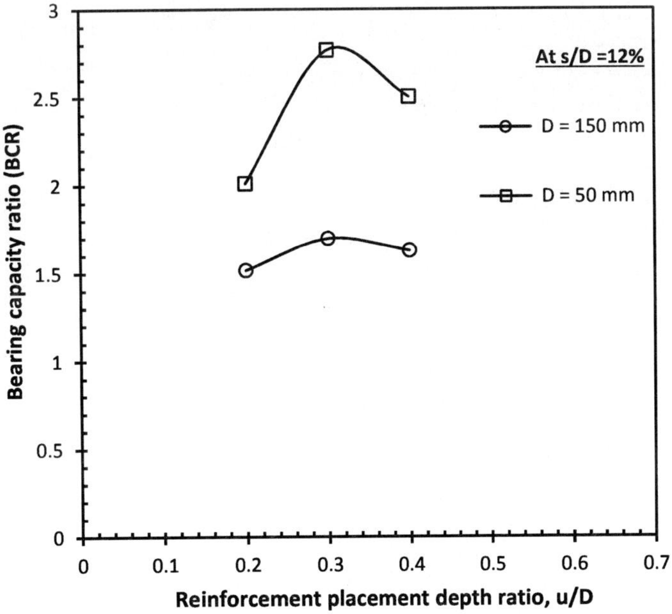

Fig. 5 Variation of the BCR with different ratios of u/D using footings of different size (D)

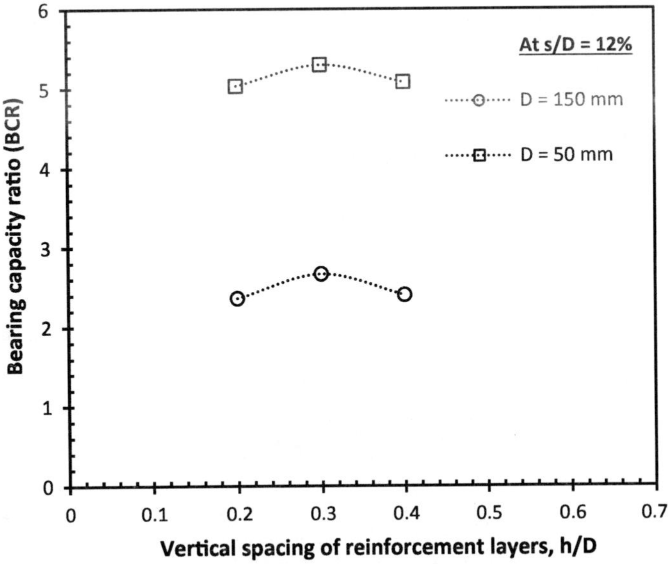

Fig. 6 Variation of the BCR with different ratios of h/D using footings of different size (D)

Table 3 Summary of test results in terms of optimum reinforcement geometrical parameters

Footing size (D)	Optimum geometrical parameter	
	u/D	h/D
50 mm	0.312	0.29
150 mm	0.31	0.3

References

1. Basudhar, P.K., Saha, S., Deb, K.: Circular footings resting on geotextile-reinforced sand bed. Geotex. Geomembr. **25**(6), 377–384 (2007)
2. Binquet, J., Lee, K.L.: Bearing capacity tests on reinforced earth slabs. J. Geotech. Eng. Div. ASCE **101**(12), 1241–1255 (1975)
3. Binquet, J., Lee, K.L.: Bearing capacity analysis of reinforced earth slabs. J. Geotech. Eng. Div. ASCE **101**(12), 1257–1276 (1975)
4. Buragadda, V., Thyagaraj, T.: Bearing capacity of jute geotextile-reinforced sand bed. Int. J. Geosyn. Ground Eng. **5**, 27 (2019)
5. Dash, S.K., Rajagopal, K., Krishnaswamy, N.R.: Performance of different geosynthetic reinforcement materials in sand foundations. Geosyn. Int. **11**(1), 35–42 (2004)
6. Ghosh, A., Ghosh, A., Bera, A.K.: Bearing capacity of square footing on pond ash reinforced with jute-geotextile. Geotext. Geomembr. **23**(2), 144–173 (2005)
7. Guido, V.A., Chang, D.K., Sweeney, M.A.: Comparison of geogrid and geotextile reinforced slabs. Can. Geotech. J. **23**, 435–440 (1986)
8. IS 1888: Method of load test on soils—Determination of bearing capacity of soils and its settlement. Bureau of Indian Standards, New Delhi (1982)
9. Latha, M., Somwanshi, A.: Effect of reinforcement form on the bearing capacity of square footings on sand. Geotext. Geomembr. **27**(6), 409–422 (2009)
10. Naderi, E., Hataf, N.: Model testing and numerical investigation of interference effect of closely spaced ring and circular footings on reinforced sand. Geotext. Geomembr. **42**(3), 191–200 (2014)
11. Omar, M.T., Das, B.M., Yen, S.C., Puri, V.K., Cook, E.E.: Ultimate bearing capacity of rectangular foundations on geogrid-reinforced sand. Geotech. Test. J. **16**(2), 246–252 (1993)
12. Sitharam, T.G., Sireesh, S.: Model studies of embedded circular footing on geogrid-reinforced sand beds. Proc. Inst. Civil Eng.-Ground Improv. **8**(2), 69–75 (2004)
13. Tafreshi, S.M., Dawson, A.R.: Comparison of bearing capacity of a strip footing on sand with geocell and with planar forms of geotextile reinforcement. Geotext. Geomembr. **28**(1), 72–84 (2010)
14. Tavangar, Y., Shooshpasha, I.: Experimental and numerical study of bearing capacity and effect of specimen size on uniform sand with medium density, reinforced with nonwoven geotextile. Arab. J. Sci. Eng. **41**(10), 4127–4137 (2016)
15. Yetimoglu, T., Wu, J.T.H., Saglamer, A.: Bearing capacity of rectangular footings on geogrid reinforced sand. J. Geotech. Eng. Div. ASCE **120**(12), 2083–2099 (1994)

Effect of Geosynthetic Reinforcement on CBR Strength of Soft Soil-Aggregate System

Gautam, M. Hussain, and D. Bhowmik

1 Introduction

A proper and well-established road network plays a vital role in socio-economic growth of a nation. Rural roads connect the villages and agricultural fields to the urban and developed city centres. The load coming on the pavement surface has to be transmitted finally to the subgrade. In unpaved roads during the rainy season, water percolates into the subgrade layer and reduces its bearing strength considerably. A weak subgrade cannot provide proper support to different pavement layers and will not be able to withstand heavy traffic loads. Therefore, a stable subgrade with optimum strength is the prime concern for pavement design. Various chemical and mechanical methods are adopted to stabilise the soil subgrade [1, 3]. Geosynthetics can be used to separate different pavement layers from intermixing and act as filter media and restrict the escape of soil particles with rainwater. They can facilitate the efficient drainage of rainwater and can improve the bearing strength of roads through reinforcement. With the use of geosynthetics, there is a reduction in plastic deformation and a decrease in the depth of base course [2, 13, 16, 21, 27]. The service life is increased, and pavements' construction and maintenance time is reduced with the inclusion of geosynthetics [9, 10, 15, 24].

The improvement in geosynthetic reinforced pavements' performance depends on the factors such as type and material of geosynthetic, bearing strength of subgrade, grading of aggregates, location of placement of geosynthetic, etc. [17, 8, 18]. In their study, Fannin and Sigurdsson [12] found geosynthetics to be most effective for sections with a thin base layer. Under dynamic loading, Geotextile, when used in pavement, does not allow pore pressure to develop and restricts the movement of soil into the aggregate layer in the form of the slurry [25]. Many researchers have done various tests to observe the effect of geosynthetics in pavement design. Plate load

Gautam (✉) · M. Hussain · D. Bhowmik
Department of Civil Engineering, National Institute of Technology Silchar, Silchar, India

C. N. V. Satyanarayana Reddy et al. (eds.), *Ground Improvement and Reinforced Soil Structures*, Lecture Notes in Civil Engineering 152,
https://doi.org/10.1007/978-981-16-1831-4_61

tests give more useful results as they better replicate the field conditions, but they are expensive to conduct. Extensive field tests [6, 20, 23], have been conducted to study the behaviour of unpaved road. CBR tests are easier and quick laboratory tests whose results can be used to correlate with the engineering properties of materials [22, 28].

Hence, the main objective of this work is to observe the effect on geosynthetic reinforcement on the soft soil-aggregate system. Limited studies have been done on the use of a combination of geotextile and geogrid for pavement reinforcement. Therefore, in this work combination of geogrid and geotextile was used as geocomposite to study the cumulative impact of different geosynthetics when used together in the pavement.

2 Materials

2.1 Soil and Aggregates

The clayey soil used in the study was collected from the NIT Silchar campus. The soil was oven-dried and all the lumps were broken using a wooden mallet. ASTM D 4318 [5] and ASTM D 2487 [4] were followed to find out the soil index properties as presented in Table 1. Aggregates of size 10 mm were used for the CBR test. The obtained values of different aggregate properties are shown in Table 2. and agreed with IS:2386 making it suitable for use in road construction.

Table 1 Properties of soil

Particulars	Soil
Specific gravity	2.72
Soil classification (USCS)	CH
Liquid limit (%)	55
Plastic limit (%)	29
Maximum dry unit weight (kNm^{-3})	14.6
Optimum moisture content (%)	23

Table 2 Properties of aggregates

Particulars	Aggregates
Specific gravity	2.64
Water absorption (%)	2.8
Aggregate Impact Value (%)	28.41
Aggregate crushing strength (%)	27.34

2.2 *Geosynthetic*

The nonwoven geotextile used for the test was manufactured from high-quality UV stabilized polyester. The fibres were mechanically bonded through the needle-punching process to form a strong, flexible and dimensionally stable fabric structure. Woven geotextile used was made of staple polypropylene fibre durable in adverse chemical and biological soil conditions. Biaxial geogrid, made of high-density polyethene, was used. Table 3 present the specification of geosynthetics.

3 Experimental Studies

CBR test is the most widely accepted test to know the quality of pavement about the resistance offered by the material against the applied load. The current experimental work consists of several CBR tests on the unreinforced and reinforced soil-aggregate system. Every part of North-East India receives rainfall of more than 1000 mm annually [11]. Due to adverse topographical conditions and tough terrain locations, the road infrastructure in the North-East is not adequately developed. Unsurfaced or unpaved roads have no asphaltic cover; the rainwater percolates into the subgrade and makes it very soft, leading to a reduction in load-bearing strength of the subgrade. According to IRC-SP 20 (2002), the CBR test for the subgrade of rural roads should be done for the critical water content. A series of soil samples were collected daily at the NIT Silchar campus during the monsoon season, that is, from June to August. The average moisture content found out in the soil for this period was around 35%. Therefore, 35% of water by weight of soil was considered for preparing the soil sample for the CBR test. The CBR value obtained for soil with 35% of moisture content is 2.08%, indicating that soil became soft and possessed low strength at high moisture content. In the soil-aggregate system, subgrade soil was compacted in the CBR mould up to the height of 125 mm in three equal parts and aggregates were

Table 3 Properties of geosynthetics

Particulars	Nonwoven Geotextile	Woven Geotextile	Geogrid
Mass per unit area (gsm)	250	240	190
Thickness (mm)	1.5	1	1.5
Tensile strength (kN/mm)	15	45	27.2
Apparent opening size (micron)	80	75	–
Mesh size (mm × mm)	–	–	10 × 10

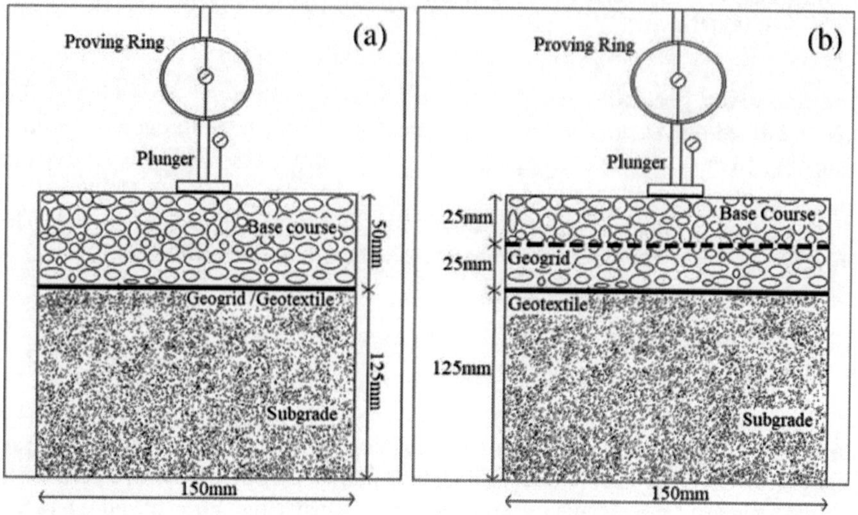

Fig. 1 Schematic set up of soil-aggregate system reinforced with **a** geogrid/geotextile **b** geogrid + geotextile

filled for 50 mm height in two parts. Geosynthetic when used alone for reinforcing, was placed at the junction of soil and aggregate layer.

A combination of geogrid with nonwoven and woven geotextile was also used for reinforcing. A gap of 25 mm was maintained between geotextile and geogrid when used together because when geogrid is placed in direct contact with geotextile then the optimum interlocking of aggregates is not achieved and geogrid may slide over the geotextile [16]. Therefore, when used as geocomposite, the geotextile was placed at the junction of soil-aggregate layer and geogrid was placed in the middle of the aggregate layer. The schematic setup of the soil-aggregate arrangement is presented in Fig. 1. The geosynthetic was cut into a circular shape and the diameter of the geosynthetic was kept 5 mm less than that of the CBR mould to prevent any frictional resistance between the geosynthetic and wall of the mould. The number of blows given per layer of soil and aggregate by hammer is such that the compaction energy equals the standard proctor energy. Therefore, the number of blows given per layer of soil and aggregate was 55 and 33 blows, respectively. The prepared arrangement was tested in a loading frame at a strain rate of 1.25 mm/min. Before that, a surcharge of 4 kg was placed over the system. The load-penetration data of the prepared systems was recorded for a displacement of 20 mm.

4 Results and Discussions

The CBR tests on different systems were performed and their load-penetration responses were studied. The various test combinations performed are mentioned in Table 4. The behaviour of the soil-alone, at a water content of 23% (OMC) and 35%

Table 4 Details of the tests carried

Designation	Test systems
S'	Soil alone (water content = 23%)
S	Soil alone (water content = 35%)
SA	Soil-aggregate system
SANWG	Soil-aggregate system reinforced with nonwoven geotextile
SAWG	Soil-aggregate system reinforced with woven geotextile
SAG	Soil-aggregate system with geogrid
SAGNWG	Soil-aggregate system reinforced with geogrid + nonwoven geotextile
SAGWG	Soil-aggregate system reinforced with geogrid + woven geotextile

under the applied load is presented in Fig. 2. The bearing capacity of the subgrade soil decreased with an increase in the moisture content. CBR value of soil reduced from 5.9% to 2.1%, with the increase in moisture content from 23% to 35%. Therefore, unpaved roads become very critical in the rainy season as rainwater percolating into the subgrade layer makes the soil soft and reduces the strength of the roads. The remaining unreinforced and reinforced arrangements were prepared and tested, with the soil having a moisture content of 35%. The addition of aggregate in the soil-aggregate system improved the load resistance capacity as aggregates uniformly staggered the coming traffic load to a larger area of subgrade. The soil-aggregate system yielded a CBR value of 3.25%. For the soil-aggregate system, there were undulations observed, unlike the smooth load-penetration response of the soil alone. These undulations were observed due to the crushing and rearrangement of the aggregates under the applied load. Figure 3 shows that there was intermixing of soil and aggregate during the test, which is not desirable for the roads.

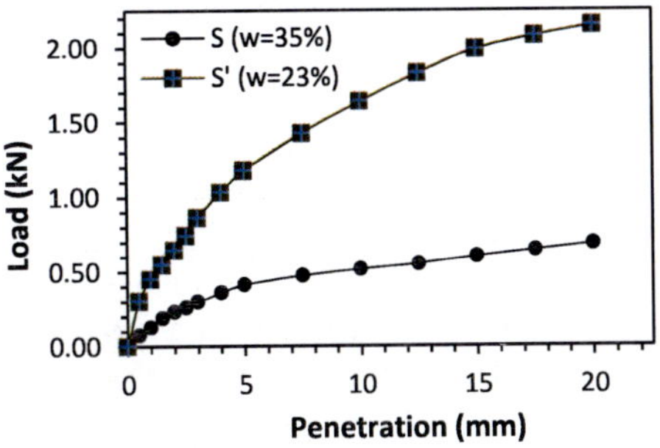

Fig. 2 Load-penetration response of the soil-alone system at water content 35% and 23%

Fig. 3 Intermixing of soil and aggregate layer in the unreinforced system

Fig. 4 Exhausted geogrid after the test

4.1 Effect of Geotextile Reinforcement

Both nonwoven and woven geotextile were tested as reinforcing material placed between soil and aggregate layers. From Fig. 5, it is clear that the nonwoven geotextile did not have any profound impact on the bearing strength due to its low tensile strength. The nonwoven geotextile failed to withstand high loads and was ruptured. Both nonwoven and woven geotextiles served the purpose as separator and prevented the movement of aggregates into the soil layer and vice versa. Geotextile can be preferred for its function of separation, stabilisation, filtration [13]. Since the permeability of nonwoven geotextile is more, it can very efficiently facilitate the drainage of rainwater from the unpaved roads while retaining the soil particles. Woven geotextile proved to be more effective in the reinforcement because of its high tensile strength. There CBR value obtained with the inclusion of nonwoven geotextile was

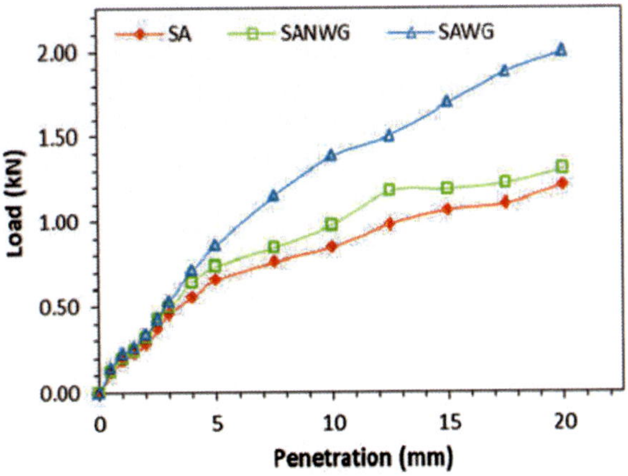

Fig. 5 Load-penetration graph of unreinforced and geotextile reinforced soil-aggregate system

Table 5 CBR values, Improvement factor (IF) and Secant modulus of different test systems

Test Systems	S	SA	SANWG	SAWG	SAG	SAGNWG	SAGWG
CBR (%)	2.1	3.25	3.67	4.27	4.64	5.12	6.02
IF	–	–	1.13	1.31	1.43	1.57	1.85
Secant modulus (kPa/mm)	–	56.85	75.41	87.77	95.00	105.07	123.62

3.67% which had a nominal increment from the unreinforced soil-aggregate system, whereas the CBR value for the woven geotextile reinforced system was 4.27%. The woven geotextile can be used in pavements for its better performance as a separator and reinforcement as well. Improvement factor (*IF*) was defined as the ratio of CBR value of the reinforced system to that of the unreinforced soil-aggregate system to evaluate the benefits of different reinforcing materials. Table 5 summaries the CBR values, improvement factor and secant modulus of different systems.

4.2 *Effect of Geogrid Reinforcement*

The mesh of geogrid helped in confining and interlocking aggregates. The confinement provided was dependent upon the grading of aggregate and aperture size of geogrid. The CBR value achieved with geogrid reinforcement was 4.64%, with an improvement factor of 1.43. The lateral restraint of aggregates provided better distribution of load over soil subgrade [24]. With better lateral restrains, the stiffness of the base layer improves, which causes a reduction in shear stresses. In the unreinforced system, the mode of failure was punching, whereas geogrid was a general failure [7]. A certain level of penetration is required to introduce the membrane effect in geosynthetic after which it applies an upward component of tensile force [14]. Figure 4 shows the exhausted geogrid after the test. The combination of geogrid with nonwoven and woven geotextile was also tried to study the effect of geocomposite where different geosynthetic materials perform different functions to improve the pavement characteristics collectively. The combination of woven geotextile and geogrid reinforcement yielded the best results with a CBR value of 6.02% and an improvement factor of 1.85. With the use of geocomposite, the functions of separation, confinement, lateral restrain, filtration, and reinforcement can be achieved. The ribs of geogrid facilitated the interlocking and friction at the interface, which enhanced the lateral restraint in the aggregate layer, whereas for lateral restrain geotextile depends on interface friction only [26].

The bearing strength achieved with the combination of woven geotextile and geogrid was more than that of the geocomposite of nonwoven geotextile and geogrid as shown in Fig. 6. The addition of nonwoven geotextile to the geogrid had benefits as nonwoven geotextile separated the soil and aggregate layer and stabilised the system; therefore, the penetration resistance achieved was more than that of the geogrid alone reinforced system, which can also be observed from the secant subgrade modulus as shown in Table 5. Further to quantify the impact of geosynthetic reinforcement on the soil-aggregate system *reinforcement ratio* was analysed. Koerner [19] defined reinforcement ratio at a particular penetration as, the ratio of load resistance offered by the geosynthetic reinforced system to that of load resisted by the unreinforced system. The membrane effect of geosynthetic became evident after a certain penetration level was achieved. For small penetration values, the role of geosynthetic reinforcement was not profound. Therefore, the reinforcement ratio was calculated from the penetration of 5 mm and henceforth. The reinforcement ratio versus penetration curves for the different reinforcing combinations is presented in Fig. 7. It was observed from the curve that with the addition of geosynthetic, the soil-aggregate system could sustain higher loads as the reinforcement ratio obtained was more than one.

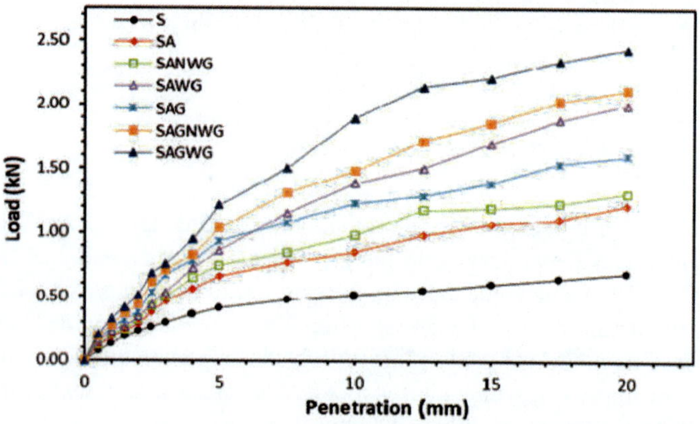

Fig. 6 Load-penetration graph for soil, unreinforced and geosynthetic reinforced soil-system

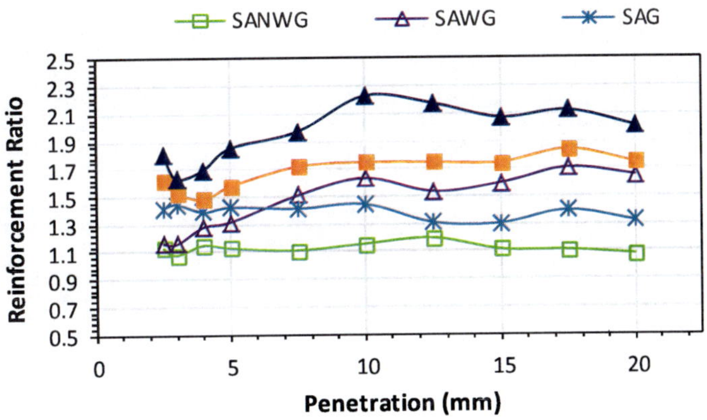

Fig. 7 Reinforcement ratio of reinforced soil-aggregate systems

4.3 *Impact of Reinforcement on the Base Layer Thickness*

The role of the base course is to disperse the coming load to a bigger area of subgrade and facilitate the structural support to the unpaved layer. It is prepared with hard and durable aggregates. The thickness of the base course is selected so as it can reduce the traffic load intensity through the depth to a level where the underlying subgrade can handle it. Generally, aggregates are acquired from the quarry near the construction site. In cases where there is no local quarry available or there are terrain restrictions in procuring aggregates, it becomes a difficult and costly task to arrange the amount of aggregates required to construct the base course. Geosynthetics help in restricting the depth of base course leading to a reduction in the use of construction materials. The use of geosynthetic improves the pavement parameters and reduces the cost and time of construction. IRC: SP 20 (2002) relates the CBR value with the thickness of pavement. From Fig. 8, it can be established that by improving the CBR value the thickness of pavement required for a particular traffic load reduces.

4.4 *Design Example*

Considering the design for an unpaved road with a total number of commercial vehicles passing daily to be around 400. The difference between the CBR value of the unreinforced and geogrid reinforced soil-aggregate system was 1.39%. From Fig. 8, the reduction obtained in the thickness of pavement by placing geogrid between subgrade and base course layer is 100 mm. Similarly, if the combination of geogrid and woven geotextile is used to reinforce the pavement, the pavement thickness can be reduced by around 200 mm. The depth of the base course can be associated with the construction cost of the road. Lesser the thickness of pavement layers, lesser

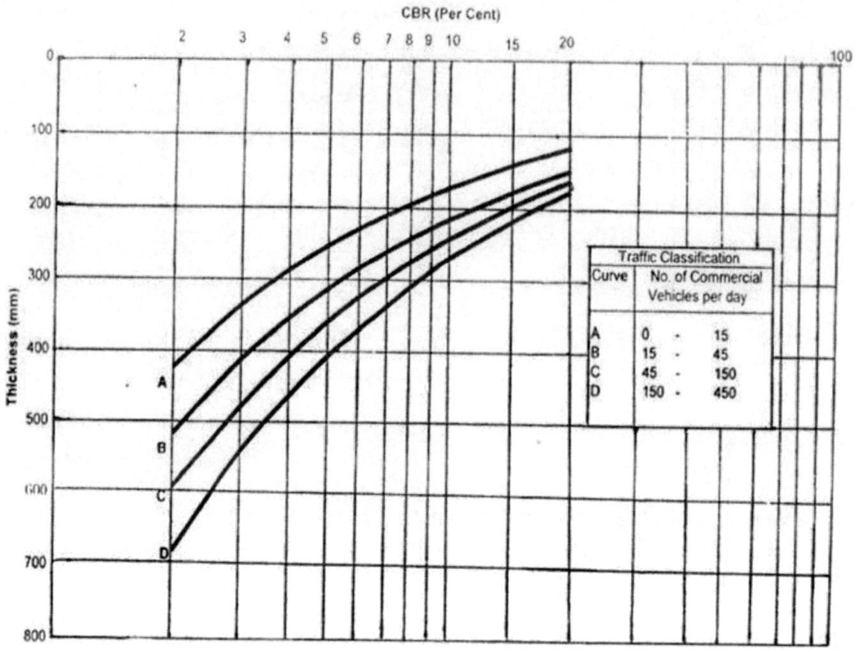

Fig. 8 CBR curves for flexible pavement design (IRC: SP:20–2002)

would be the construction cost. Reduction in the amount of aggregates required for the construction of the base course will reduce the construction cost and time.

5 Conclusions

A set of CBR tests were performed to understand the impact of geosynthetic reinforcement on unreinforced soil-aggregate systems. Different geosynthetics viz. nonwoven and woven geotextile and geogrid were employed for the reinforcement. Based on these studies the conclusions that are derived are as follows:

1. The inclusion of geosynthetic reinforcement increases the bearing resistance of the soil-aggregate system with an increment of 85% in CBR values for geocomposite consisting of geogrid and woven geotextile.
2. The load-bearing capacity became profound at higher penetration values due to the introduction of the membrane effect in geosynthetic.
3. The nonwoven geotextile prevented the movement of aggregate from the base layer into the subgrade layer and vice versa. Still, because of very low tensile strength, its impact on bearing strength was nominal. The woven geotextile was very effective as separator and significantly improved the CBR value because

of its high tensile strength. At higher penetration values, the woven geotextile reinforced system performed better than geogrid.

4. The geogrid restrains the aggregates with the proper interlocking of aggregates, but this is mainly dependent upon the gradation of aggregates and the mesh size of geogrid.
5. The geosynthetic can significantly decrease the pavement thickness. There is a reduction in the requirement of aggregates hence making the construction of pavement quicker, more economical and ecofriendly.

References

1. Afrin, H.: A review on different types soil stabilization techniques. Int. J. Transp. Eng. Technol. **3**(2), 19–24 (2017)
2. Al-Qadi, I., Brandon, T., Valentine, R., Lacina, B., Smith, T.: Laboratory evaluation of geosynthetic-reinforced pavement sections. Transp. Res. Record 1439 **21**(2), 25–31 (1994)
3. Amhadi, T.S., Assaf, G.J.: Overview of soil stabilization methods in road construction. In: El-Badawy, S., Valentin, J. (eds) Sustainable Solutions for Railways and Transportation Engineering. GeoMEast 2018. Sustainable Civil Infrastructures. Springer, Cham (2019)
4. ASTM D 2487: Standard practice for classification of soils for engineering purposes (Unified Soil Classification System). ASTM International, West Conshohocken, PA (2006)
5. ASTM D 4318: Standard test methods for liquid limit, plastic limit, and plasticity index of soils. ASTM International, West Conshohocken, PA (2005)
6. Bergado, D.T., Youwai, S., Hai, C.N., Voottipruex, P.: Interaction of nonwoven needle-punched geotextile under axisymmetric loading conditions. Geotext. Geomembr. **19**(5), 299–328 (2001)
7. Binquet, J., Lee, K.L.: Bearing capacity analysis of reinforced earth slabs. J. Geotechn. Eng. Division **101**(12), 1257–1276 (1975)
8. Broms, B.B.: Polyester fabric as reinforcement in soil, C. R. Coll. Int. Soils Text, Paris 1, pp. 129–135 (1977)
9. Cancelli, A., Montanelli, F.: In-ground test for geosynthetic reinforced flexible paved roads. Proceedings of the Conference Geosynthetics '99, pp. 863–878. Roseville, Minnesota, USA: IFAI (1999)
10. Chen, Q., Abu-Farsakh, M., Tao, M.: Laboratory evaluation of geogrid base reinforcement and corresponding instrumentation program. Geotech. Test. J. **32**(6), 516–525 (2009)
11. Dikshit K.R., Dikshit J.K.: Weather and climate of North-East India. In: North-East India: Land, People and Economy. Advances in Asian Human-Environmental Research, pp. 149–173. Springer, Dordrecht (2014)
12. Fannin, R.J., Sigurdsson, O.: Field observations in stabilisation of unpaved roads with geosynthetics. J. Geotechn. Eng. ASCE **122**(7), 544–553 (1996)
13. Giroud, J.P., Noiray, L.: Geotextile reinforced unpaved road design. J. Geotechn. Eng. ASCE **107**(9), 1223–1254 (1981)
14. Göbel, C.H., Weisemann, U.C., Kirschner, R.A.: Effectiveness of a reinforcing geogrid in a railway subbase under dynamic loads. Geotext. Geomembr. **13**(2), 91–99 (1994)
15. Goldfingle,G.: Float on Ground Engineering, British Geotechnical Association, UK, October issue, pp. 32–34 (2009)
16. Hufenus, R., Rueegger, R., Banjac, R., Mayor, P., Springman, S.M., Brönnimann, R.: Full-scale field tests on geosynthetic reinforced unpaved roads on soft subgrade. Geotext. Geomembr. **24**(1), 21–37 (2006)
17. Ismail, I., Raymond, G.P.: Geosynthetic reinforcement of granular layered soils. Geosynthetics '95, Nashville, TN, USA, IFAI, St. Paul, MN, USA 1, pp. 317–330 (1995)

18. Kinney, T.C., Abbott, J., Schuler, J.: Benefits of using geogrids for base reinforcement with regard to rutting. Transp. Res. Record: J. Transp. Res. Board **1611**(1), 86–96 (1998)
19. Koerner, R.M.: Designing with Geosynthetics, 5th edn., pp. 184–186. Prentice Hall, New Jersey (2005)
20. Miura, N., Sakai, A., Taesiri. Y.: Polymer grid reinforced pavement on soft clay grounds. Geotextoles and Geomembranes **9**, 99–123 (1990)
21. Moayed, R.Z., Nazari, M.: Effect of utilization of geosynthetic on reducing the required thickness of subbase layer of a two layered soil. World Academy of Science, Engineering and Technology 49, Article 175, pp. 963–967 (2011)
22. Nair, A.M., Latha, G.M.: Bearing resistance of reinforced soil aggregate systems. Proc. ICE Ground Improvement **164**(2), 83–95 (2010)
23. Palmeira, E.M., Antunes, L.G.S.: Large scale tests on geosynthetic reinforced unpaved roads subjected to surface maintenance. Geotext. Geomembr. **28**, 547–558 (2010)
24. Perkins, S.: Mechanical response of geosynthetic-reinforced flexible pavements. Geosynthetics Int. **6**(5), 347–382 (1999)
25. Qurishee M.A.: Application of goesynthetics in pavement design. Int. Res. J. Eng. Technol. **4**(07) (2017)
26. Steward, J., Williamson, R., Mohney, J.: Guidelines for use of fabrics in construction and maintainence of low-volume roads. Report No. FHWA-TS-78-205 (1977)
27. Webster, S.L., Watkins, J.E.: Investigation of Construction techniques for tactical bridge approach roads across soft ground: final report. (No. Tech Rpt. S-77-1). United States Army Corps of Engineers.; Soils and Pavement Laboratory (US) (1977)
28. Yeole, M.M., Patil, J.R.: Geotextile can be worth their cost in pavement. IOSR J. Eng. **3**, 45–48 (2013)

Influence of Rainfall on the Interface Shear Strength of Unsaturated Lateritic Soil with Geosynthetics

K. A. Dhanya, A. Musaib, and P. V. Divya

1 Introduction

The primary reason for failures in most of the geotechnical structures, such as embankments, reinforced soil walls, slopes, etc., is the development of high pore water pressure due to the lack of drainage. Failures can happen if water accumulates within the geotechnical structures faster than the drainage rate. Granular soils with fine content less than 15% are ideally used as fill material in various geotechnical applications. Of late, locally available marginal soils with poor drainage characteristics are being increasingly used in bulk geotechnical applications such as structural fill in embankments, reinforced soil slopes, etc [3, 6, 7, 9, 13]. This helps in significant cost savings when compared with conventional granular backfill materials [14, 19, 20].

The major part of Kerala is covered with lateritic soils, which are highly sensitive to the wet and dry climate. Lateritic soils are formed under sub-tropical and tropical monsoon climates with well-defined wet and dry climatic conditions. Chemical composition and mineralogical characteristics depend on the parent rock and the degree of weathering. In general, the soil stratification consists of hard laterite at the surface of about 3 m and a lithomargic lateritic layer which is generally classified as sandy silt/silty sand above the parent rock. Locally available lateritic soils are increasingly used in the construction of highway embankments.

K. A. Dhanya · A. Musaib · P. V. Divya (✉)
Indian Institute of Technology Palakkad, Palakkad, Kerala, India
e-mail: divyapv@iitpkd.ac.in

K. A. Dhanya
e-mail: 101804101@smail.iitpkd.ac.in

A. Musaib
e-mail: 102004008@smail.iitpkd.ac.in

C. N. V. Satyanarayana Reddy et al. (eds.), *Ground Improvement and Reinforced Soil Structures*, Lecture Notes in Civil Engineering 152,
https://doi.org/10.1007/978-981-16-1831-4_62

Several failures are reported in lateritic soil slopes in Kerala, associated with the rainfall. Rainfall is one of the most important triggering factors for failures in geotechnical structures. It can change the hydrological and geotechnical behaviour of the soil resulting in massive destruction. When rainfall infiltration occurs, negative pore water pressure or the matric suction decreases and shear strength of the soil reduces.

Many places of Kerala have been witnessing internal erosion along with landslides, in recent years, associated with intense rainfall. Dhanya and Divya [5] reported a recent evidence of soil piping due to internal erosion in a lateritic soil slope associated with heavy rainfall, in the northern part of Kerala. Drainage helps in the reduction of pore water pressure, which otherwise can affect the shear strength of the material. Slopes and embankments need to be provided with proper drainage facilities and erosion control measures.

It is well known that geosynthetics can be used to repair failed slopes and the construction of new embankments. Permeable geosynthetic inclusions help to dissipate the excess pore water pressure within the poorly draining fill material. Several authors [17, 10, 11, 16] reported the application of non-woven geotextile layers installed within the fill material to contribute to the stability of the soil structure by dissipating excess pore water pressures during the construction. They also minimize the build-up of pore pressure in the future, which can be caused by rainfall infiltration. Some investigators also report the application of composite geosynthetics which combine the function of reinforcement and drainage in poorly draining soils [4, 11, 15, 19, 21]. These composite geosynthetics generally have a non-woven geotextile attached with a strengthening component such as geogrids or polyester yarns.

In many applications, soil structural fill is compacted at its maximum dry density and optimum moisture content and remains practically unsaturated. The role of suction in increased shear strength of the soil is well recognized [8]. Seasonal variations in the soil moisture content can result in significant changes in soil suction. Rainfall infiltration causes loss of matric suction in the unsaturated lateritic soil. The present study investigates the influence of rainfall-induced wetting on shear strength of lateritic soil as well as lateritic soil-geosynthetics interface. The interface between geotextiles and soil fill has long been identified as an important factor affecting the stability of geosynthetic reinforced soil structures. This is particularly important in the design of geosynthetic reinforced soil structures with marginal fills with poor drainage properties.

2 Materials Used

Lateritic soil was collected from a site in Kerala, which was subjected to rainfall-induced slope failure. Figure 1 shows the soil used in the present study. Geotechnical properties of the soil were determined by conducting various laboratory tests as per standard procedures. Particle size analysis has been carried out to obtain the grain size distribution of the soil (ASTM D422). Figure 2 shows the grain size distribution

Fig. 1 Lateritic soil used for the study

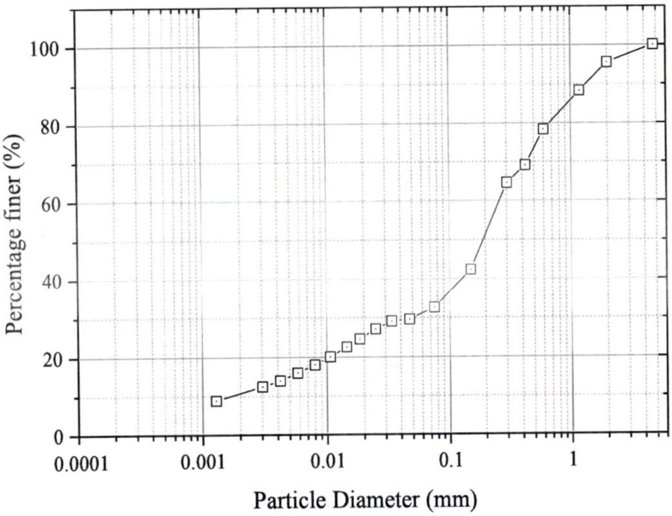

Fig. 2 Particle size distribution curve of soil

of the lateritic soil. Percentage fines in the soil were observed to be 33.4%. According to USCS classification, the soil was categorized as silty sand (SM).

The soil's maximum dry unit weight and optimum moisture content were 17.2kN/m^3 and 15%, respectively, as obtained from standard Proctor tests (ASTM D698). Compaction characteristics of the lateritic soil are as shown in Fig. 3.

Various geotechnical properties of the soil are summarized in Table 1.

Geosynthetics used in the present study were non-woven polypropylene geotextile. It has a tensile strength of 30.5 kN/m in the machine direction and 28.5 kN/m

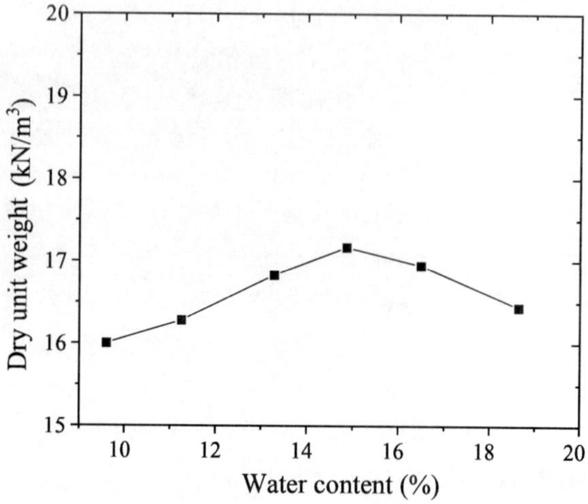

Fig. 3 Compaction curve of the soil

Table 1 Properties of the soil collected

Properties	Values
Specific gravity	2.58
Sand sized (%)	66.6
Percentage fines (%)	33.4
Silt sized (%)	21.63
Clay sized (%)	11.77
Liquid limit (%)	38
Plastic limit (%)	Non-plastic
Soil classification as per USCS	SM
Maximum dry unit weight	17.2 kN/m^3
Optimum moisture content, OMC (%)	15

in the cross-machine direction. Transmissivity of the geotextile at a normal stress of 20 kN/m^2 corresponds to 8.5 × 10^{-6} m^2/sec.

3 Testing Procedure

3.1 Influence of Rainfall-Induced Wetting on the Soil–Soil Shear Strength

A series of direct shear tests were carried out to evaluate the influence of rainfall-induced wetting on the shear strength properties of the lateritic soil. Soil samples for each test were mixed with a water content corresponding to optimum moisture

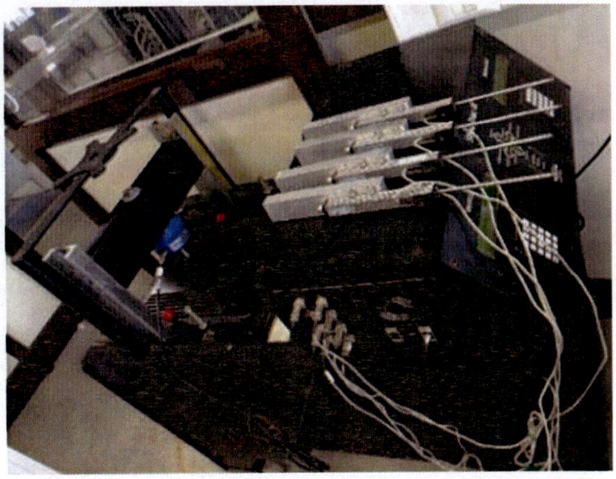

Fig. 4 Large direct shear testing equipment

content (15%) and stored in an airtight polyethylene bag wrapped with a wet cloth for at least 24 h to attain moisture equilibrium. This was verified by determining soil water content at the end of moisture equilibrium by the oven drying method.

The soil was compacted in three layers in a shear test box to attain 95% maximum dry density. Saturated direct shear tests were conducted on soil samples compacted at an optimum moisture content at three different normal stresses, that is, 50 kN/m^2, 100 kN/m^2, and 200 kN/m^2, to determine the shear strength parameters. As per ASTM D3080, soil samples were consolidated for 24 h under each normal stress prior to shearing. The shearing rate was computed according to ASTM D3080, from the time corresponding to 90% consolidation. This was done to ensure the dissipation of excess pore water pressure at failure. Figure 4 shows the equipment used for carrying out tests. Shear loading was applied with the help of a computer-controlled unit that utilizes a micro stepper motor. Displacement transducers of 100 mm travel were used for measuring the shear displacements as well as vertical displacements. Both horizontal and vertical displacement was measured using displacement transducers. The effective angle of internal friction of the soil was found to be 37°.

In order to study the influence of rainfall-induced wetting on the shear strength properties of the lateritic soil, constant water unsaturated direct shear tests were conducted on soil samples adopting the procedure reported by Jotisankasa and Mairaing [12]. Initial compacted conditions of all the samples were maintained uniform. The moist compacted samples were prepared similar to that of saturated testing. The samples were compressed one-dimensionally at a normal stress of 100 kN/m^2 for one day.

Water was sprinkled prior to shearing in order to induce wetting due to rainwater infiltration to simulate the field condition. Thus, the moisture content in the soil was altered prior to shearing. Samples were kept undisturbed and wrapped for 3 days such that the sprinkled water distributes uniformly within the samples and the suction equilibration is achieved. Samples were sheared at a rate of 0.5 mm/min. After completion of the test, gravimetric water content was determined by taking three

Fig. 5 Filter paper method for the determination of SWCC of the soil

representative samples from the top, middle, and bottom of the soil sheared specimen to estimate suction from the soil water characteristics curve (SWCC) of the soil. Filter paper method as per ASTM D5298-10 [2] was used to determine the soil water characteristics curve of the soil, as shown in Fig. 5.

The shear strength of unsaturated soil can be represented by Eq. 1 [18].

$$\tau = c' + (\sigma_n - u_a)\tan\phi' + c^s \quad (1)$$

where c' is the cohesion intercept of saturated soil, σ_n is the normal stress at failure plane, u_a is the pore-air pressure, ϕ' is the effective friction angle of soil and c^s is the apparent cohesion due to soil suction. It is a function of degree of saturation and soil suction.

3.2 *Influence of Rainfall-Induced Wetting on the Soil-Geosynthetics Interface Strength*

The interface testing was conducted as per ASTM D5321 [1]. Non-woven geotextile was laid over the substrate in the lower test box, as shown in Fig. 6.

Geotextile was clamped to the lower box. The soil was compacted in three layers in a shear test box to attain 95% maximum dry density. The sample preparation procedure and initial compacted condition of the soil were maintained the same as explained before.

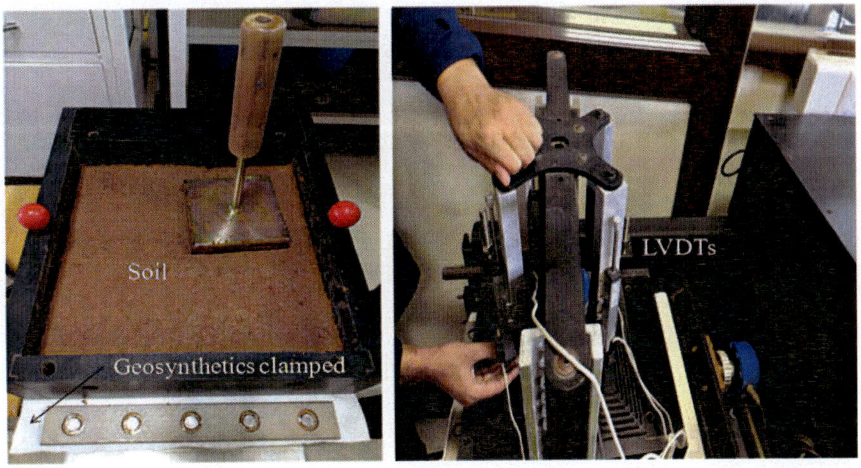

Fig. 6 Interface shear testing using large direct shear equipment

4 Results and Discussions

The results of constant water unsaturated direct shear tests in order to study the influence of rainfall-induced wetting on the shear strength properties of the lateritic soil are shown in Fig. 7. The wetting-induced altered water content is represented herein as *w*. Results show that apparent cohesion is highly influenced by water content. A strain-softening behavior was observed for the samples with lower moisture content due to breakage of bonding provided by water menisci [12].

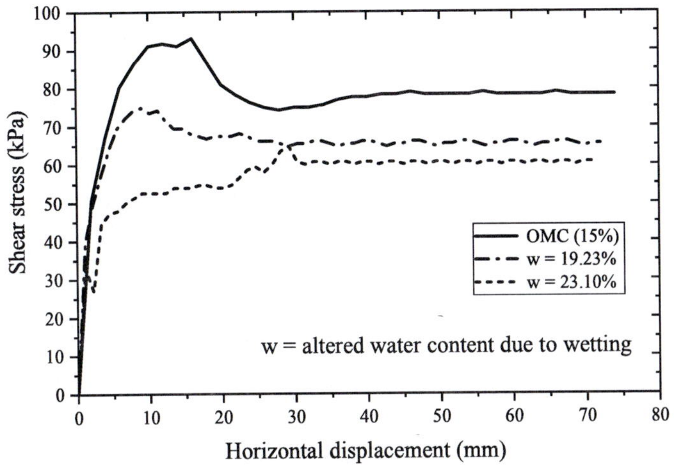

Fig. 7 Shear stress versus horizontal displacement graph for each water content

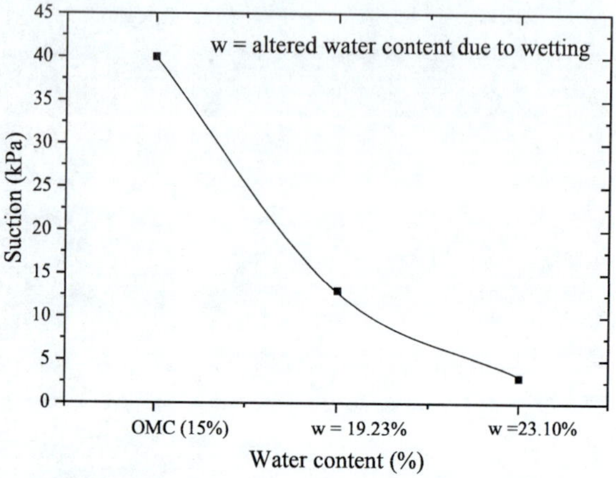

Fig. 8 Variation of soil suction due to wetting

As the wetting is induced, a reduction in shear stress was observed. This is because, for unsaturated soil conditions, initially, an apparent cohesion exists within the soil mass due to soil suction. This contributes to a higher shear strength for the soil. But as the degree of saturation increases, initial suction (at OMC) within the sample decreases, reducing its strength. Soil suction was determined from the soil water characteristics curve of the lateritic soil from the measured gravimetric water content. Figure 8 shows the variation of suction within the soil samples due to wetting. A reduction of 30% in strength was observed as the water content increased from 15% (OMC) to 23.10% (OMC + 8). Also, dilation of the samples was observed during shearing at higher suction (low water content). But as the suction decreases (high water content), contraction behavior was observed. Apparent cohesion, c^s was computed and the variation of c^s due to soil suction with water content is shown in Fig. 9.

It was observed that during rainfall infiltration, the degree of saturation in the soil is increased. Thus, the matric suction in the soil mass reduces and thereby, the apparent cohesion (contribution of matric suction to shear strength) decreases. Similarly, the results of interface shear testing were analyzed. The results are summarized in Table 2. Here, the adhesion factor corresponds to the ratio of shear strength of the soil-geosynthetics system to the shear strength of soil, and its value was observed to be varied between 0.7 and 1.2. As can be observed, the adhesion factor is increasing as the water content is increased.

Figure 10 shows the percentage reduction of shear strength for both soil and soil–geosynthetic system. A significant reduction in soil shear strength was observed as the water content is increased due to wetting.

When the moisture content was increased by 4% due to wetting, the shear stress of the lateritic soil was reduced by 20%. In contrast, the corresponding reduction in strength at the interface for the soil–geosynthetic system was only 3%. Similarly, an increase in moisture content by 8% due to wetting resulted in a reduction in strength by 30% and 4% for soil–soil and soil–geosynthetic systems, respectively.

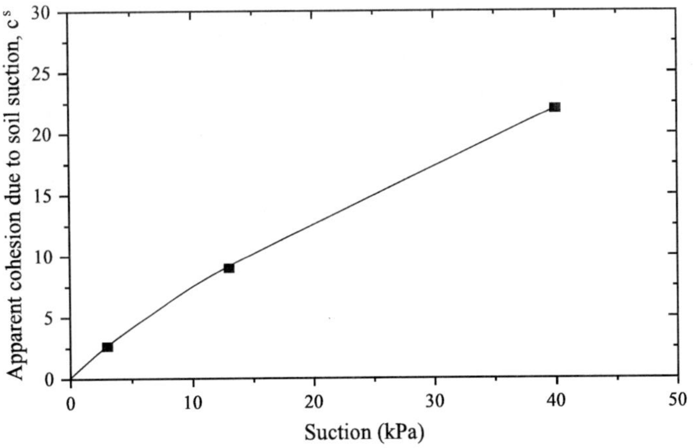

Fig. 9 Variation of apparent cohesion with suction

Table 2 Summary of interface shear strength properties

Parameter	OMC (15%)	w = 19.23%	w = 23.10%
Soil-Geotextile interface shear stress (kPa)	72	70	69
Adhesion factor	0.75	0.9	1.2

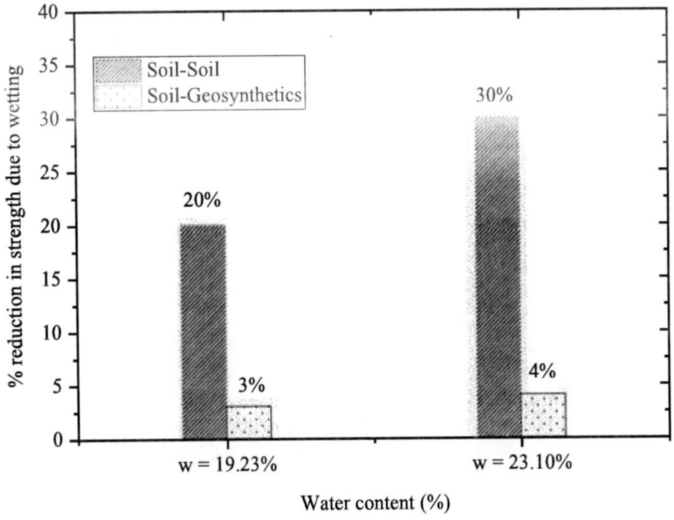

Fig. 10 Percentage reduction in strength due to wetting

The reduction in shear strength at the interface for a soil–geosynthetic system due to wetting-induced increase in water content was significantly less. This can be due to the transmissivity of the geotextile at the interface resulting in drainage of water during shearing. Thus, the bond strength can be increased. Thus, high transmissivity non-woven geotextiles can effectively drain and reinforce the poorly draining soil, even in undrained conditions. Thus, rainfall-induced wetting will have less influence in the interface properties of lateritic soil-non-woven geotextile interface owing to the high transmissivity.

5 Conclusions

The present study investigates the influence of rainfall-induced wetting on shear strength of lateritic soil and lateritic soil-geosynthetics interface. Apparently, cohesion due to the suction in lateritic soil decreased by 70% and 93% when the moisture content was increased by 4% and 8%, respectively. Reduction in suction can lead to loss of shear strength of unsaturated lateritic soil. It was observed that, when the moisture content was increased by 4% due to wetting, the shear strength of the lateritic soil was reduced by 20%. In contrast, the corresponding reduction in strength at the interface for the soil–geosynthetic system was only 3%. Similarly, an increase in moisture content by 8% due to wetting resulted in a reduction in strength by 30% and 4% for soil–soil and soil–geosynthetic systems, respectively. Thus, rainfall-induced wetting has less influence in the interface properties of lateritic soil-non-woven geotextile interface due to the high transmissivity.

References

1. ASTM D5321-17: Standard Test Method for Determining the Shear Strength of Soil-Geosynthetic and Geosynthetic-Geosynthetic Interfaces by Direct Shear. American Society for Testing and Materials International (2014)
2. ASTM D5298-10: Standard Test Method for Measurement of Soil Potential using Filter Paper. American Society for Testing and Materials International (2010)
3. Albino, U.R., Portelinha, F.H.M., Zornberg, J.G., Futai, M.M.: Numerical simulation of infiltration into the fill of a wall reinforced with nonwoven geotextiles. Comput. Geotechn. **108**, 27–39 (2019)
4. Bhattacherjee, D., Viswanadham, B.V.S.: Numerical studies on the performance of hybrid-geosynthetic-reinforced soil slopes subjected to rainfall. Geosynth. Int. **22**(6), 411–427 (2015)
5. Dhanya, K.A., Divya, P.V.: Geosynthetic filters to prevent soil piping and internal erosion. ISSMGE TC 107 Symposium on Laterites and lateritic soil, pp. 95–104 (2019)
6. Elias, V., Christopher, B.R., Berg, R.R.: Mechanically Stabilized Earth Walls and Reinforced Soil Slopes Design & Construction Guidelines. Federal Highway Administration Washington, DC, USA (2001)
7. Esmaili, D., Hatami, K., Miller, G.A.: Influence of matric suction on geotextile reinforcement-marginal soil interface strength. Geotextiles Geomembranes **42**, 139–153 (2014)

8. Fredlund, D.G., Morgenstern, N.R., Widger, R.A.: The shear strength of unsaturated soils. Can. Geotech. J. **15**, 313–321 (1978)
9. Hatami, K., Esmaili, D., Chan, E.C., Miller, G.A.: Moisture reduction factors for shear strength of unsaturated reinforced embankments. Int. J. Geomechan. **16**, 1–16 (2016)
10. Iryo, T., Rowe, R.K.: On the hydraulic behaviour of unsaturated nonwoven geotextiles. Geotextile Geomembranes **21**, 381–404 (2003)
11. Iryo, T., Rowe, R.K.: Infiltration into an embankment reinforced by nonwoven geotextiles. Canadian Geotechn. J. **42**(4), 1145–1159 (2005)
12. Jotisankasa, A., Mairaing, W.: Suction-monitored direct shear testing of residual soils from landslide-prone areas. J. Geotechn. Geo-environ. Eng. **136**(3), 533–537 (2010)
13. Jotisankasa, A., Rurgchaisri, N.: Shear strength of interfaces between unsaturated soils and composite geotextile with polyester yarn reinforcement. Geotextiles Geomembranes **46**, 338–353 (2018)
14. Keller, G.R.: experiences with mechanically stabilized structures and native soil Backfill. Transp. Res. Rec. **1474**, 30–38 (1995)
15. Mamaghanian, J., Viswanadham, B.V.S., Razeghi, H.R.: Centrifuge model studies on geocomposite reinforced soil walls subjected to seepage. Geosynthetics Int. **26**(4), 371–387 (2019)
16. Miki, H.: Reinforcing mechanism and analysis method of geotextile reinforced embankment. Report of public works research institute, Ministry of Construction, Tsukuba, Japan, 197 (1997)
17. Mitchell, J.K., Zornberg, J.G.: Reinforced soil structures with poorly draining backfills. Part II: Case histories and applications. Geosynthetics Int. **2**(1), 265–307 (1995)
18. Vanapalli, S.K., Fredlund, D.G., Pufahl, D.E., Clifton, A.W.: Model for the prediction of shear strength with respect to soil suction. Canadian Geotechn. J. **33**(3), 379–392 (1996)
19. Vibha, S., Divya, P.V.: Geosynthetic reinforced soil walls with construction and demolition waste backfill. Symposium on Recent Advances in Sustainable Geotechnics, Indian Institute of Technology, Kanpur (2019)
20. Vibha, S., Divya, P.V.: Geosynthetic-reinforced soil walls with sustainable backfills. Indian Geotechn. J., 1–10 (2020)
21. Vibha, S., Divya, P. V. : Performance of geosynthetic reinforced MSE walls with marginal backfills at the onset of rainfall infiltration. Int. J. Geosynthetics Ground Eng. **7**(1), 1–16 (2021)

The Behavior of Multi-tiered Mechanically Stabilized Earth (MSE) Retaining Wall

Ananya Srivastava, Sagar Jaiswal, and Vinay Bhushan Chauhan

1 Introduction

A mechanically stabilized earth (MSE) retaining wall is a complex structure comprising alternate layers of soil reinforcement elements and the compacted backfill, attached to a wall facing. The steadiness of such a wall system is a derivative achieved from the interaction between the backfill and soil reinforcements which further provides internal and external stability to the wall system. MSE retaining wall consists of reinforcements embedded into the soil mass to work as soil wall and able to sustain the lateral earth pressure and any other lateral load-induced due to natural or man-made activity in the vicinity of the wall structure.

Often, the favorable geographical and topographical conditions required for the construction of a tall MSE wall are hard to achieve as the empty land space might not available [1]. Thus in this situation, the MSE walls can be built in a tiered fashion, with an offset distance, *D* which is the distance between the modular block facings of the successive tiers of the tiered MSE wall, which in turn also reduces the maximum tensile stresses in the reinforcement layers lying in the lower section of the wall [2]. This develops the fundamental performance and finances of construction as well as permits the erection of walls with complex geometries. Furthermore, a single-tiered MSE wall possesses very high tensile stresses in the reinforcing layers positioned at the deeper depths due to a high overburden pressure exerted by the backfill column situated over it. Consequently, the preferred factor of safety (FOS) is accomplished by either ensuing a reinforcement that has a high load-bearing capacity or by the positioning of two reinforcing layers very closely i.e. decreasing the vertical spacing between the layers of reinforcement.

A. Srivastava · S. Jaiswal · V. B. Chauhan (✉)
Department of Civil Engineering, Madan Mohan Malaviya University of Technology, Gorakhpur 273010, India

C. N. V. Satyanarayana Reddy et al. (eds.), *Ground Improvement and Reinforced Soil Structures*, Lecture Notes in Civil Engineering 152,
https://doi.org/10.1007/978-981-16-1831-4_63

As suggested by FHWA [3], for tall MSE walls, the preference must be given to tiered walls from the perspective of constructability as a monolithic wall creates several difficulties related to the foundation soil and requires some additional safety measure to avoid the failure of the MSE wall. The configuration of a tiered wall system allows a new start with a new leveling pad. It also lessens the vertical stresses on the facing elements and authorizes better control of the vertical alignment of the wall face.

From the available literature on MSE retaining walls, [4–6] it is evident that multi-tiered walls render a better outcome in comparison to single-tiered tall walls, especially when it is necessary to erect an MSE retaining wall which is competent in terms of stability, fiscal concerns, and visual appeal.

The design of the MSE retaining wall as per AASHTO [7] recommendations and NCMA [8] design manual define the guidelines for up to two-tiered walls only. The criteria for the design of a two-tiered wall can be extended to walls with more than two tiers as recommended by FHWA [3]. For such structures, the compound and global stability evaluation become even more specific and critical. "Regarding the internal stability analysis, the conditions for additional vertical stress can be used for walls with more than two-tiers, provided that only the proximately superimposing tier is considered to subsidize the increase in vertical stress on the lower tier" [2]. There also subsists an elastic solution based on a hypothesis of "rigid" walls for appraising additional vertical stresses in a specified tier of a multi-tier wall due to the influence of all superimposing tiers. "Regardless of the approach used for estimating the increase in vertical stresses for evaluation of internal stability, the analysis of tiered walls should proceed from the top wall to the bottom wall so that the stresses are properly accumulated and accounted for in the design of the bottom-most wall" [9].

Therefore given the above, this study aims to evaluate the behavior and performance of tiered wall system and compare its outcomes with the single-tiered wall of similar height, while evaluating their stability in terms of FOS and presents an understanding of the possible failure modes of these complex infrastructures. For the aforementioned goal of this research work, a plane strain model is used for the numerical replication of the MSE walls, also evaluation of the same is performed by implying the strength reduction technique and incorporating the finite element method (FEM) for the assessment of the FOS of MSE retaining walls under several assumed variations of design parameters.

2 Numerical Modeling

Given the objectives set for the present study, a detailed and extensive numerical analysis has been performed with the help of a numerical computational tool OPTUM G2 [9], a finite element program dedicated to geotechnical deformation and stability analysis under plane strain conditions.

The MSE wall considered in the current study is a three-tiered wall system reinforced using a geogrid element and having a total height of the wall H, 12 m where the individual height of the wall at each tier is given as $H_1 = H_2 = 4.2$ m, $H_3 =$ 3.6 m, where, H_1 is the height of lowermost tier, H_2 is the height of middle-tier and H_3 is the height of uppermost tier having a rigid facing (concrete blocks). The configuration of the MSE wall numerical model is considered according to the FHWA [3] design recommendations. Another single-tiered MSE wall of similar height (12 m) is also constructed under the same guidelines to provide comparative results in this study. Figures 1 and 2 show the three-tiered MSE wall and single-tiered MSE wall considered in this study respectively.

For the system, a stiff leveling pad fabricated with concrete of the proportions $2 \times 0.2\ m^2$ (in elevation) is rested on the soil (as shown in Fig. 2). On top of this,

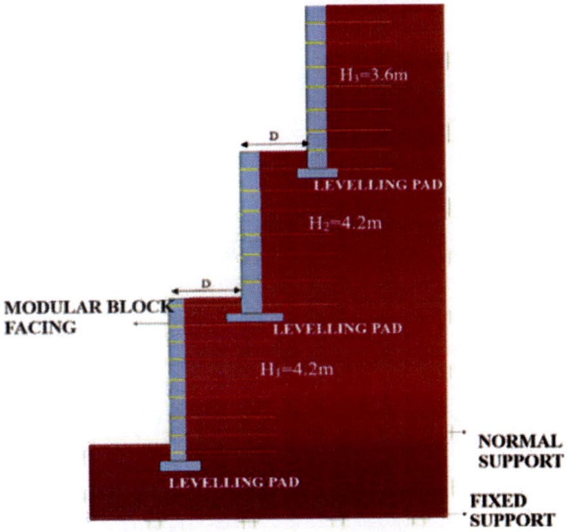

Fig. 1 A three-tiered MSE wall model (12 m) considered in the present study

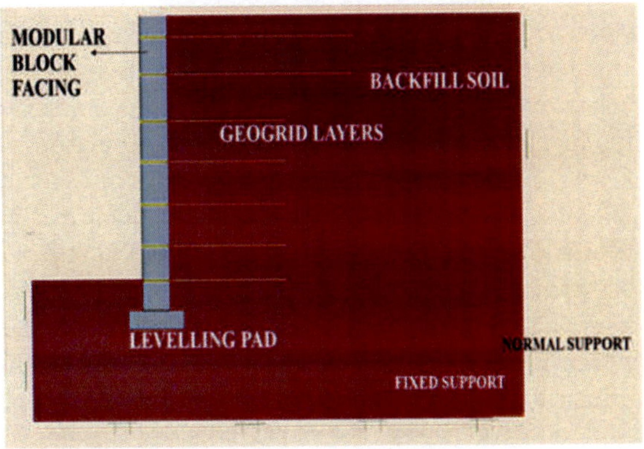

Fig. 2 A single-tiered MSE wall model (12 m) considered in the present study

a concrete facing having dimensions "1.5 m × 0.6 m" (in the vertical direction) is positioned, above which a backfill layer having a boost of 0.6 m is placed. To simulate the field-building procedure to model the MSE wall, before placing the next facing in the form of a concrete block, a reinforcing layer of the preferred length, L is positioned on backfill, such that a section of the reinforcement length is sealed amid the consecutive rigid facings. Afterward, these backfill material is again positioned over the geogrid layer and this process is continued until the required height of the wall is attained [10].

To simulated the numerical model, the lateral boundaries were secured by normal support to prevent the movement of the soil in the horizontal directions. Fixed supports were applied to the bottom of the soil foundations to avoid the soil movement in both the horizontal and vertical directions.

Furthermore, due to the absence of any similar study in the available literature using experimental evaluation (either small scale or full-scale model) or numerical/analytical modeling, the present numerical model could not be validated. However, every possible precaution has been taken while modeling the retaining wall to predict the behavior of the multi-tiered walls precisely.

The reinforcement length (L) is varied from 0.6 H to 0.8 H for the parametric study. The selection of reinforcement length of 0.6 H, which assists with the guidelines as recommended by NCMA [7], 0.7 H, is following the guidelines suggested by FHWA [2] and 0.8 H is considered for the evaluation of the fact that will any further change in the reinforcement length might affect the factor of safety or not and it also verifies the recommended length of the reinforcement by FHWA [3] and NCMA [8]. These variations in the length aim to represent the modification in the performance and behavior of the tiered wall in static loading circumstances.

The offset distance, D amid the successive consecutive tiers is kept constant in this study for the tiered MSE wall system. The offset distance, D can be determined using Eq. (1) as mentioned follows [2].

$$D \leq H_u \tan\left(45 - \frac{\phi_r}{2}\right) \tag{1}$$

where H_u is the height of the upper tiers in the tiered MSE wall and ϕ_r is the friction angle of the soil mass. The upper-tier exerts a surcharge over the lower tiers and the behavior of both the tiers is dependent on each other [3]. The concrete facing blocks, footing, leveling pad, are modeled as linearly elastic materials with a unit weight of 25 kN/m^3. Backfill soil is modeled as an elastoplastic material with a Mohr-Coulomb failure criterion. "The force transfer mechanism among two different materials is simulated as an elastoplastic interface and interface coefficient values like 0.7 and 0.8 for geogrid-soil and geogrid-block interfaces respectively, are assigned at all the interfaces in this MSE wall and footing system" [11].

Furthermore, MSE walls are constructed to resist the lateral earth pressures primarily from the retained backfill. Also, the calculation of the distribution and the magnitude of lateral forces on retaining wall systems under all the loading circumstances has always been a significant area of investigation for geotechnical engineers, as the total push on the wall is the crucial aspect to decide the sectional proportions of the retaining structures. By decreasing the lateral earth pressure on the wall, sectional dimensions of the wall can be significantly reduced, which would lead to the overall economy in the construction of retaining wall [12].

For the present numerical analysis, the six-node triangular elements with a three-point Gaussian integration rule is adopted for higher accuracy of the findings from the model. To estimate the efficiency of the numerical model to investigate the behavior of the MSE walls, a sensitivity analysis is carried out by ranging the number of elements in mesh from 2000 to 6000 and found that 4000 number of elements are sufficient for the considered mesh for simulation [9] and noted that the selected number of elements produces a very fine mesh and refined geometry is significant for achieving reasonable performance.

3 Results and Discussion

Based on the outcomes of the present analysis of the single-tiered and three-tiered MSE wall of the same height are analyzed, compared, and discussed in the present section. The effect of the reinforcement length on the maximum displacement of the wall facing, (Δ) the effect of the reinforcement length on the stability of the wall, effect of the variation of the stiffness of the geogrid on the FOS, the economic benefit of the tiered wall system, effect of tiers on the stability of the wall system and the potential failure planes are discussed below.

The reinforcement length has been varied from 0.6 to 0.8 H in the present study for both the single-tiered and three-tiered wall systems. Figure 3 illustrates the variation of the maximum displacement of the wall facing and the variation of FOS of the single-tiered and three-tiered wall systems w.r.t normalized reinforcement length. As evident from the observed variation of the lateral displacement of the wall facing (Δ/H) in the three-tiered wall system is maximum with a value of 1.71% with the least reinforcement length of 0.6 H and is minimum at $L = 0.8\ H$ having $\Delta/H =$ 1.32%. Also, it can be observed that the tiered wall system faces lesser Δ/H w.r.t single-tiered wall system.

The single-tiered wall system shows up to a 75% increment in the Δ/H values than the three-tiered wall system, thus highlighting the superiority of the tiered system of walls, in terms of lesser wall movement. This high value of lateral displacement is undesirable for the safe design of the MSE wall system and any structure existing in the vicinity of the retaining structure may prove fatal for MSE walls of greater heights.

The reinforcement length majorly contributes to the overall constancy of the MSE wall. Also, in both the single and three-tiered wall arrangement, the corresponding factor of safety w.r.t the varied reinforcing lengths (in ascending order) increases considerably, which is shown in Fig. 3. Despite the increment in reinforcement

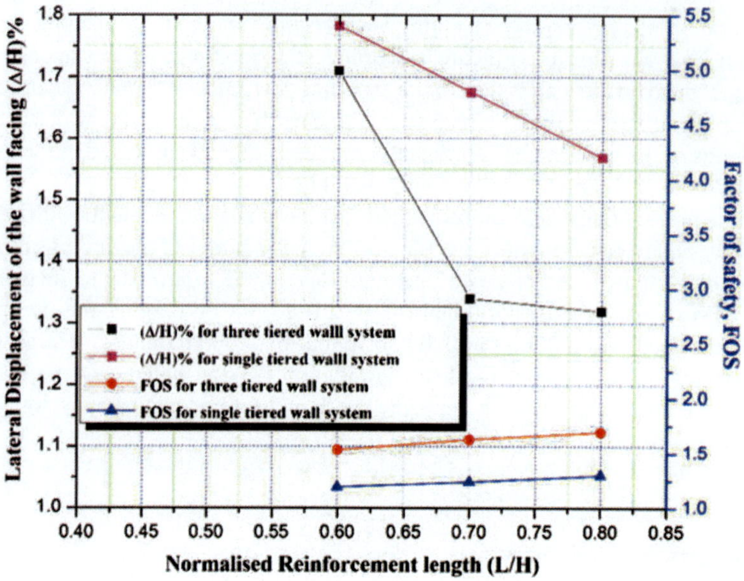

Fig. 3 Variation of normalized maximum displacement of the wall facing Δ/H and FOS of the single-tiered and three-tiered wall systems w.r.t normalized reinforcement length, L/H

lengths from 0.6 H to 0.8 H, the single-tiered wall system cannot even achieve the minimum criterion of a safer design (i.e. FOS = 1.5) [2]. Contrary to the behavior of the single-tiered wall system, the three-tiered wall system performs undeniably better in terms of FOS values. Even at the minimum length of the reinforcing layer, which is 0.6 H, the FOS achieved by the three-tiered wall system is 1.53, satisfying the required criterion of the safe design. Moreover, as the reinforcement length varies from 0.6 H to 0.8 H, there is a significant rise in the FOS, which is 29% greater than the conventional single-tiered MSE wall at these specific variations of the reinforcement lengths. The variation of reinforcement length causes the FOS to rise by 9.4% in the three-tiered wall system while varying the L/H from 0.6 to 0.8 at a given spacing of 0.6 m.

To investigate the reinforcement configuration for the single-tier wall to reach the minimum FOS, the vertical spacing between the successive layer was reduced to 0.4 m and it was noticed that the FOS improved and achieves the required minimum FOS of 1.5 for the safer design of the retaining structure [3].

For single-tier MSE walls, the use of reinforcing length $L/H = 0.6$, the FOS was noted as 1.44 and found to be 1.54 and 1.6 corresponding to $L/H = 0.7$ and 0.8, respectively. Although the FOS of the single-tiered wall system increases with L/H ratio, still it is lesser than the maximum FOS achieved by the three-tiered wall system. Moreover, the reduction in vertical spacing caused an additional introduction of 10 more layers of geogrid, which shall increase the cost of construction of the wall.

Thus from an economic point of view, the construction of the three-tiered MSE wall was found to be more suitable.

In the case of single-tiered walls the minimum design FOS is not achieved, therefore to attain the required FOS either the vertical spacing in between the reinforcing layers is reduced, the stiffness of the geogrid is increased or the length of the reinforcing layer is increased. Therefore given the above, some more numerical simulations have been conducted by increasing the length of the reinforcing layer, increasing the stiffness of the geogrid, and decreasing the vertical spacing between the geogrids. The results so obtained are discussed below.

The stiffness of the geogrid (*EA*) was increased from 500 kN/m to 2000 kN/m at intervals of 500 kN/m each while keeping the vertical spacing between the successive reinforcement layers as 0.6 m for both the wall systems. With the increase in stiffness of the geogrid, the FOS of the wall systems increased substantially. The three-tiered wall and single-tiered wall systems experienced an increment in FOS by 16% and 26%, respectively. Figure 4 represents the graph depicting the variation in FOS values w.r.t the change in axial stiffness of the geogrid. As evident with the increased stiffness, the single-tiered wall system exhibited a higher growth in the FOS values. But the cost of geogrid increases with the increment in its stiffness, thus suggesting that the viable option is to construct the wall in a three-tiered system, to attain the same stability.

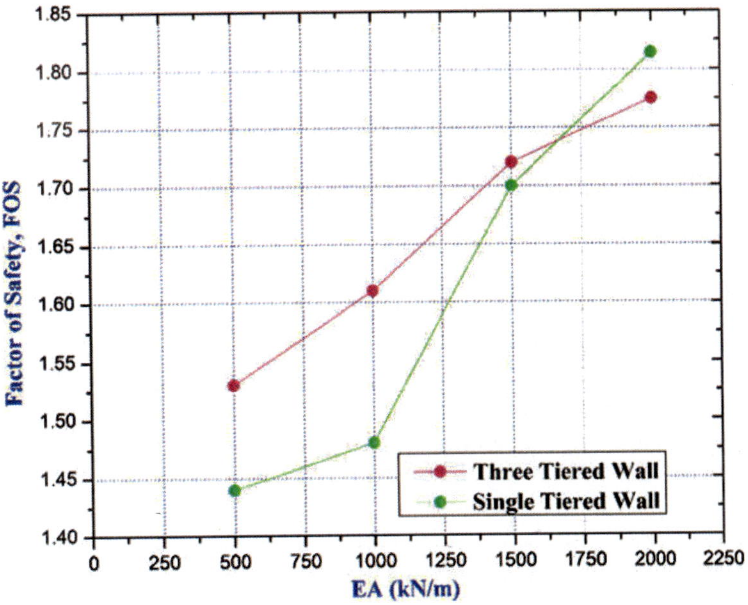

Fig. 4 Variation of the factor of safety w.r.t stiffness of the geogrid in single-tiered and three-tiered wall systems

For the single-tiered wall system, the reinforcement length was increased to achieve the minimum required FOS for the safe design. It was noted that at *L*/*H* equals 1.0, the FOS was found to be 1.56. During the simulations on the three-tiered wall systems, similar stability conditions were obtained at $L/H = 0.6$ whereas it took an increment of 40% in the *L*/*H* ratio for the single-tiered wall to achieve the same benchmark of stability. This percentage shall further increase and play a vital role in the tiered MSE walls of greater heights. Thus, it can be safely suggested that in terms of FOS, the construction of the three-tiered wall is a better substitute.

The analysis of the 12 m high, three-tiered wall system, and single-tiered wall system have shown that the number of tiers has a substantial impact on the stability of the wall system. The tiered wall system exhibits the highest FOS value of 1.7, which is quite appreciable w.r.t the serviceability of the structure. Also, the three-tiered wall satisfies the minimum desirable criterion of the factor of safety at all the varied reinforcement lengths in this study. Whereas in the single-tiered, the stability of the structure suffers a setback, and the maximum FOS achieved is 1.30 in this case at $L/H = 0.8$. The above evaluation suggests that the construction of the tiered wall structure is markedly more beneficial and feasible than the single-tiered conventional wall systems.

The precise assessment of the potential failure plane behind the retaining wall provides a desired outcome to understand the behavior of the wall and design the retaining structures efficiently [13, 14]. The analysis of the critical failure plane, for the single-tiered and three-tiered wall systems, has been carried out to understand its possible modes of failure of such walls and is discussed below by the shear dissipation diagram. The dissipation is often a good indicator of the intensity of plastic deformation.

From the noted failure planes, it is observed that for the single-tiered and three-tiered wall system (as shown in Figs. 5 and 6), the soil mass underneath the increased offset could not overlap with the potential failure plane. In the single-tiered wall system, there occurs disruption in the soil mass beneath the leveling pad instead of the reinforced zone sandwiched between the geogrid, due to which before any failure in the reinforced zone, the leveling pad descends downwards (as shown in Fig. 5). The single-tiered wall fails in overturning as one single entity with the concrete block facings being laterally dislocated. The failure plane occurs in a wedge outline, commencing from the leveling pad and spreading towards the end of the topmost reinforcement layer. This observed specific actions might be accredited to the statistic that the inclusive width of reinforced soil mass acts as the sole unit which causes the movement of supported mass laterally away from the backfill in this single-tiered wall system. In the three-tiered wall, under static loading, a combined potential failure plane passes below the walls and meets at the ends of the lowermost reinforcement layer in each wall, and further extending towards the backfill surface.

The potential failure plane doesn't interact with the uppermost tier at all, as the upper-tier wall acts as a stationary load to the lower walls due to its proximity, and subsequently upsurges the stresses in the bottom portions of the lower tiers of the MSE wall. The critical failure plane originates from the toe and transmits from the end of reinforcements provided at the second-tier and further extends towards the

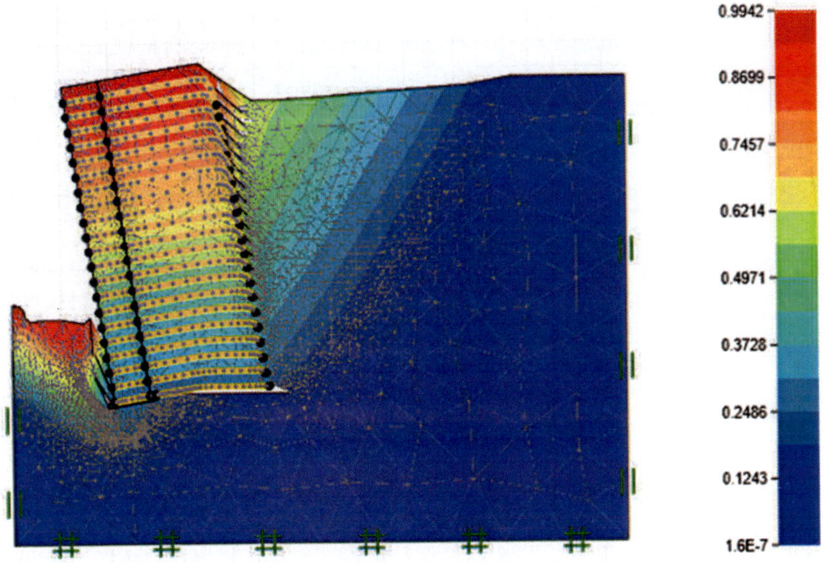

Fig. 5 Deformations and distribution of shear dissipation of the single-tiered MSE wall model (12 m) considered in the present study

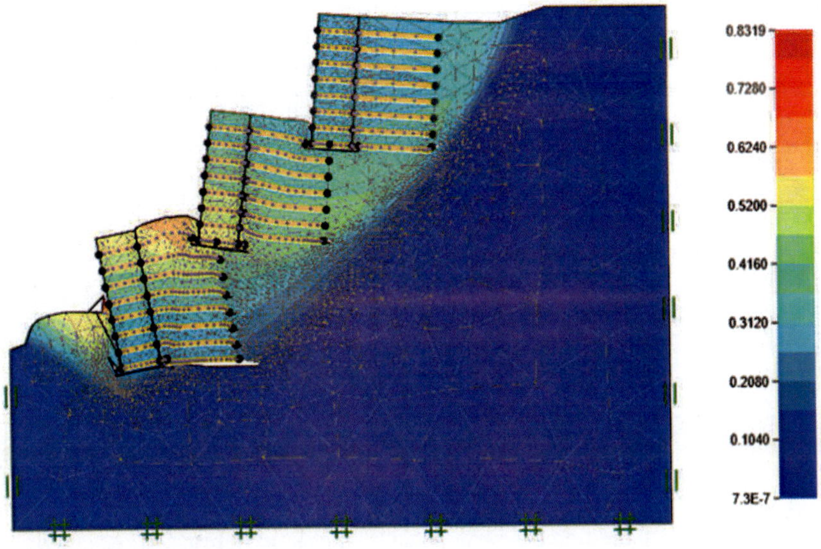

Fig. 6 Deformations and distribution of shear dissipation three-tiered MSE wall model (12 m) considered in the present study

backfill, as depicted in Fig. 6. In the case of a multi-tiered wall, each tier is disordered contrarily in its manner, not acting as a single entity as in the case of a conventional single-tiered wall system.

4 Conclusions

The present study assesses the stability of the MSE walls reinforced with geogrid, the effect of the reinforcement length, the effect of the number of tiers, and the modes of failure of a 12 m high (*H*) three-tiered and single-tiered wall system. The offset distance, *D*, is taken as 3 m so that the wall system acts one whole unit, and the upper tier of the three-tiered wall system, exerts a surcharge over the lower tiers. The reinforcement length *L* is varied from 0.6 to 0.8 *H* satisfying the design requirements laid by NCMA [8] and FHWA [3]. From the results obtained from the present study, it is apparent that the above-mentioned parameters play a significant role in improving the overall stability of MSE walls. The factor of safety (FOS) value for the three-tiered wall system is found to higher than the FOS for single-tiered tall MSE retaining wall systems for a given configuration of the reinforcement. Therefore, this study highlights the fact that whenever the need arises to construct a high MSE wall, it is safe and convenient to construct multi-tiered MSE retaining walls as a substitute.

Moreover, some distinctive conclusions drawn from the study are mentioned as follows:

1. The reinforcement length plays a major role in providing stability to the MSE structure. The stability of the three-tiered wall system in this study is up to 75% higher than the stability achieved in the single-tiered wall system. The individual growth of 9.4% and 8.4% in FOS values is observed in three-tiered and single-tiered wall systems, respectively, when the reinforcement length is varied from 0.6 to 0.8 *H*. This growth in the FOS values shall increase substantially when MSE walls of greater heights are erected. This proves that the effect of the tier is prominent and the tiered wall exhibits better performance than the conventional single-tiered wall of similar height.
2. With the increase in stiffness of the geogrid, the FOS of the wall systems increased substantially. The three-tiered wall and single-tiered wall systems experienced an increment of FOS by 16% and 26%, respectively. As evident with the increased stiffness, the single-tiered wall system exhibited a higher growth in the FOS values. However, the cost of geogrid increases with the increment in its stiffness, thus suggesting that the viable option is to construct the wall in a three-tiered system, to attain the same stability.
3. The maximum lateral displacement of the wall facing, Δ/H is 5.4% and 1.71% in the single-tiered and three-tiered wall system, respectively. Again, the tiered wall systems emerge as a viable alternative for situations in place of a single-tiered wall. The maximum lateral displacement of the wall facing is found to be inversely proportional to the length of reinforcement. As the reinforcement

length increases from 0.6 to 0.8 H, maximum lateral displacement decreases from 1.71% to 1.32% and from 5.4% to 4.2% in three-tiered and single-tiered walls, respectively.

4. A three-sided chunk shear failure configuration is observed in a three-tiered MSE wall system where the critical failure plane of the upper tiers combines with the failure plane of the lower tiers into a single failure plane propagating towards the horizontal backfill surface having a significant inclination with the vertical. It is also witnessed that the soil mass underneath the offset distance does not intersect the failure path. The single-tiered wall fails in overturning as one single entity with the concrete block facings being laterally dislocated. The failure plane occurs in a wedge outline, commencing from the leveling pad and spreading towards the end of the topmost reinforcement layer. This behavior may be due to the fact that the reinforced zone acts as a sole unit which causes the whole massive reinforced zone to move laterally away from the backfill in this single-tiered wall system.

References

1. Dasaka, S.M., Dave, T.N., Gade, V.K., Chauhan, V.B.: Seismic earth pressure on gravity retaining walls using EPS geofoam. ICPMG 1025–1030 (2014). ISBN 978-1-138-00152-7. https://doi.org/10.1201/b16200-145
2. Leshchinsky, D., Han, J.: Geosynthetic reinforced multitiered walls. J. Geotech. Geoenviron. Eng. **130**(12), 1225–1235 (2004). https://doi.org/10.1061/(ASCE)1090-0241(2004)130:12(1225)
3. FHWA: Mechanically Stabilized Earth Walls and Reinforced Soil Slopes Design and Construction Guidelines, vols. 1, 2. Publication No. FHWA-NHI-10-024. US Department of Federal Highway Administration (FHWA) (2010)
4. Hatami, K., Bathurst, R.J.: Numerical model for reinforced soil segmental walls under surcharge loading. J. Geotech. Geoenviron. Eng. **132**(6), 673–684 (2006). https://doi.org/10.1061/(ASCE)1090-0241(2006)132:6(673)
5. Tatsuoka, F., Tataeyama, M., Mohri, Y., Matsushima, K.: Remedial treatment of soil structures using geosynthetic-reinforcing technology. Geotext. Geomembr. **25**(4–5), 204–220 (2007). https://doi.org/10.1016/j.geotexmem.2007.02.002
6. Yang, G., Zhang, B., Lv, P., Zhou, Q.: Behaviour of geogrid reinforced soil retaining wall with concrete-rigid facing. Geotext. Geomembr. **27**, 350–356 (2009). https://doi.org/10.1016/j.geotexmem.2009.03.001
7. AASHTO: LRFD Bridge Design Specifications. Customary U.S. Unit. American Association of State Highway and Transportation Officials, Washington, DC, USA (2012)
8. NCMA: Design manual for segmental retaining walls. In: Bernardi, M. (ed.) 3rd edn. National Concrete Masonry Association, Herndon (2012)
9. Optum G2: Finite element program for geotechnical analysis, optum computational engineering (2020) http://www.optumce.com
10. Ojha, R., Chauhan, V.B.: Performance of geosynthetic reinforced segmental retaining walls. In: Shehata, H., Brandl, H., Bouassida, M., Sorour T. (eds.) Sustainable Thoughts in Ground Improvement and Soil Stability, GeoMEast, pp. 196–206 (2019). https://doi.org/10.1007/978-3-030-34184-8_13
11. Srivastava, A., Chauhan, V.B.: Numerical studies on two-tiered MSE walls under seismic loading. S. N. Appl. Sci. **2**(10), 1–7 (2020). https://doi.org/10.1007/s42452-020-03414-6

12. Chauhan V.B., Khan R., Dasaka S.M.: Reduction of surcharge induced earth pressure on rigid non-yielding retaining wall using relief shelves. In: Anirudhan, I.V., Maji, V. (eds.) Geotechnical Applications. Lecture Notes in Civil Engineering, no. 13. https://doi.org/10.1007/978-981-13-0368-5_23
13. Dasgupta, U., Chauhan, V.B., Dasaka, S.M.: Influence of spatially random soil on lateral thrust and failure surface in earth retaining Walls. In: Georisk: Assessment and Management of Risk for Engineered Systems and Geohazards, vol. 11, no. 3, pp. 247–256. https://doi.org/10.1080/17499518.2016.1266665
14. Khan, R., Chauhan, V.B., Dasaka, S.M.: Reduction of lateral earth Pressure on retaining wall using relief shelf: a numerical study. In: Proceedings of International Conference Soil Environment, vol. 117, pp. 1–8, Bangalore, India (2016)

Performance of Strip Footing on Sand Bed Reinforced with Multilayer Geotextile with Wraparound Ends

Sagar Jaiswal, Ananya Srivastava, and Vinay Bhushan Chauhan

1 Introduction

The concept of bearing capacity of the soil mass was put forward by Terzaghi [1] for the very first time, and thereafter numerous researchers have proposed various analytical models to predict the bearing capacity of a footing resting over an isotropic and homogenous soil mass. The usages of reinforced soil mass in the contemporary soil structures were proposed by Casagrande and its use as a composite material was presented by Henri Vidal for the first time in the mid nineteenth century [2]. Usages of geogrid for enhancing the bearing capacity of the ground i.e. for improved ultimate load-bearing capacity and reduced settlement of the foundation, as a reinforcing material for the soils having poor strength, have become a widely adopted solution for the utilization of poor strength ground as a foundation bed [3]. Moreover, apart from reinforcing action, geogrids may also serve the purpose of drainage control, erosion control, separation, etc. The tensile reinforcement in the form of geogrid can be used for the construction of load-bearing structures and also improves its overall stability when subjected to static as well as dynamic loadings [4]. The reinforced soil foundation has a considerable potential to support the shallow foundation as a cost-effective substitute for traditional approaches of construction [5].

In the past, various researchers have accompanied studies to assess the response of shallow foundation emphasized with metal strips and geogrid subjected to concentrated vertical loading [6]. The ultimate bearing capacity of circular and strip footings under vertical and inclined load was determined by using the limit equilibrium method [7]. In order to assess the performance of reinforced foundation, a series of large-scale testing was conceded to evaluate the influence of reinforcement on the

S. Jaiswal · A. Srivastava · V. B. Chauhan (✉)
Department of Civil Engineering, Madan Mohan Malaviya University of Technology, Gorakhpur 273010, India

C. N. V. Satyanarayana Reddy et al. (eds.), *Ground Improvement and Reinforced Soil Structures*, Lecture Notes in Civil Engineering 152,
https://doi.org/10.1007/978-981-16-1831-4_64

bearing capacities and settlement criterion by using biaxial geosynthetics as horizontal reinforcement [5]. Furthermore, various field tests and small-scale laboratory tests had recommended that reinforcing the soil mass with single or multiple layers of geogrids under vertical concentrated load may result in higher ultimate-load bearing capacity of the reinforced soil mass ($q_{u, R}$) and reduced settlement of the foundation (s) [8, 9]. The geometrical parameters of the reinforcing layers (length and width), their depth beneath the footing, the elastic stiffness of geogrid are the key factors that sincerely affect the $q_{u, R}$. The available literature suggested that the width of the reinforcement layer should be 4 to 6 times of the footing width to maximize the benefits from the provision of the reinforcement in the foundation soil [9, 10]. Nevertheless, sometimes the availability of land on both sides of footing restricts the use of sufficient width of reinforcement than required. The above circumstances have been addressed by Kazi et al. [11, 12] and Shukla [13], in which a rearrangement in the placement of geogrid is introduced with the wrapping ends of the reinforcement on both the ends along its width. The practice of wraparound ends has improved the bearing capacity of soil mass along with the saving in land space. In the recent past, the concept of using a reinforcing layer with full wraparound ends has been introduced, which does not only improve the bearing capacity of the soil mass additionally compared to a reinforced bed with reinforcement without wraparound ends but also confined the soil mass and restricted its lateral movement on the application of load over the footing [14].

From the widespread literature survey, it has been noted that the width of the reinforcing layer (b), the vertical length of the wrapping ends (d), lap length of the overlapping portion (L), placement of the first layer of reinforcement (u), the vertical spacing between consecutive layers (h), and the number of reinforcing layers (N) are the governing factor in controlling the bearing capacity and settlement criterion. From the available literature, it is perceived that the following parameters have been recommended in order to have maximum reinforcement benefits for a strip footing of width (B): $b/B = 6$ [9]; $L/B = 0.6$ [12]; $h/B = 0.2$–0.4 [8]; $d/B = 0.2$ [11]; $u/B = 0.25$–0.5 [9]; $N = 4$ [8, 15].

From the available research related to the numerical approach of determining the ultimate load-bearing capacity of the strip footing, no discussion had been done on the interaction parameters between the soil and geogrid. The force transfer mechanism between soil and reinforcing material is measured through the behavior of interface, and the assessment of the interface behavior is essentially required to understand the interaction between two dissimilar materials i.e. soil and geogrid. Numerical models for defining the performance based on the laboratory and/or field tests are called constitutive models, which are suitable for the realistic characterization of the mechanical behavior of the solid materials and interfaces, which is essential for the appropriate solution of the practical problems. "In the traditional plasticity models such as Drucker–Prager and Mohr–Coulomb, it is assumed that the behavior is elastic until the material reaches a certain yield point, often defined by the yield stress. Subsequently, the material reaches into a plastic region governed by conditions such as a yield criterion and flow rule that defines the plastic flow, like a liquid." Considering the above, the present research aims to analyze the effect of the interface parameter between soil and geogrid on the $q_{u,R}$ of the footing. Also, it is noticed that the discussion on the effect of reinforcement parameter i.e. elastic stiffness (*EA*, where *E* is young's modulus and *A* is the cross-sectional area) is absent in the available literature.

Most of the studies were relevant to the behavior of footing which was resting on the surface, which is not the actual situation in the practice, however, this is close to Terzaghi's [1] original derivation for the bearing capacity determination, where he has assumed that overburden due to the surrounding soil is zero. Concerning the above, align with the actual site condition all the foundations are embedded one, therefore this particular study is highlighting the behavior of embedded footing resting over the reinforced soil mass.

In view of the above, a detailed parametric study has been performed using a finite element method based program Optum G2 [16], for a strip footing of width *B*, by varying the width of reinforcement (*b*) and the results thus obtained from the present analysis are compared with the available data. Moreover, the elastic stiffness (*EA*) of the geogrid reinforcement has been varied to analyze its effect on the $q_{u,R}$ and the responsiveness of *EA* towards the settlement and the soil deformation has been discussed comprehensively in the subsequent section. The objective of conducting the present research work is to focus on the significance of interface parameters that guide the interaction between soil and geogrid, which has been discussed by varying the strength reduction factor (R_{int}) in between the soil and the geogrid reinforcement.

The previous findings stated that for simulation of the actual behavior of soil, the Mohr–Coulomb (M-C) yield criterion has been used frequently for the evaluation of the ultimate load-bearing capacity of the footing. Few researchers had reported that the Mohr–Coulomb criterion produces singularities due to an irregular pyramid in the bay of principal stresses while computing numerical equations intended for operating the plastic flow at the corner of the yield surface [17]. To overcome the above situation, the Drucker–Prager (D-P) yield criterion is taken into consideration in the present study. A comparison of the Drucker–Prager yield criterion and Mohr–Coulomb yield criterion has also been discussed in the consequent section of this paper. The Drucker–Prager yield criterion uses slightly different expressions for characterizing plastic and yield potential function. The following Eqs. 1 and 2 can be used to obtain the parameters for the Drucker–Prager model with the help of shear strength parameters.

$$M = \frac{3\sin\phi}{\sqrt{3+\sin^2\phi}} \tag{1}$$

$$k = \frac{3c\cos\phi}{\sqrt{3+\sin^2\phi}} \tag{2}$$

where c and ϕ are the cohesion and the internal friction angle of the soil that is used in the M-C model to define soil strength. Moreover, k and M are the Drucker–Prager parameters that signify the cohesion and the friction coefficient, respectively.

2 Numerical Modeling

A rigid strip footing has been analyzed by utilizing the Optum G2 program resting on a medium dense cohesionless soil [16]. The effect of reinforcement width, elastic stiffness of the geogrid material, and the interface parameters between the soil and geogrid have been studied for a footing having width $B = 2$ m. The vertical boundary of the model is restrained in the normal direction and the horizontal bottom is fully fixed (as shown in Fig. 1).

It has been reported in many studies that if sufficient length and width are not provided to the mesh, the failure plane is being intercepted by the boundaries and inappropriate bearing capacity determination is shown in numerical analysis. To avoid this situation, the boundary limits in the present study are considered at 5*B* in the vertical direction and 10*B* in the horizontal respectively. The soil mass has been modeled using six node plane-strain triangular elements for achieving higher accuracy in the results. An elastoplastic constitutive model considering the D-P yield criterion with zero dilatancy, following associated flow rule has been used to model the soil. The reinforcement material is modeled using the structural geogrid elements having axial elastic stiffness, *EA*.

It has been recommended by various researchers that the accuracy of any numerical model depends upon the appropriate number of elements present in the mesh as a lesser number of elements may lead to an inconsistent result and a high number of elements in a given mesh may pose problems to computational efficiency. Keeping this in mind, a sensitivity analysis has been performed for unreinforced soil mass, by varying the number of elements from 1000 to 8000 for the present mesh at an interval of 1000 each, and the results indicated that 5000 elements are sufficient for

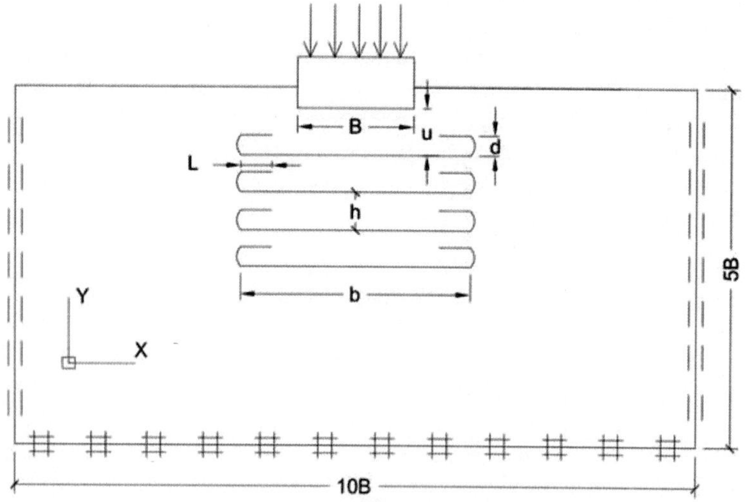

Fig. 1 Boundary conditions and geogrid reinforced soil using the wraparound ends technique

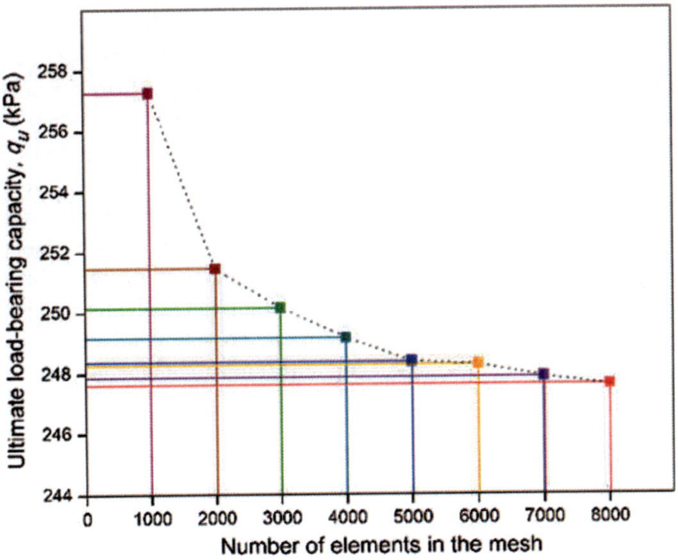

Fig. 2 Sensitivity analysis for the mesh considered in this study

the current mesh, as a negligible change is observed in the ultimate load-bearing capacity of the unreinforced footing (q_u) after increasing the number of elements from 5000 (as shown in Fig. 2) [18].

The vertical length of geogrid (d) (used for wrapping geogrids at both ends) has been chosen in such a manner that it should always be lesser than the value of spacing between consecutive layers of geogrid (h). Therefore, a reinforced mass lying between two consecutive geogrids is such that two parallel reinforcing layers exist along with two wrapped around ends and vertical length of the reinforcement of length, L and d, respectively, are present over there. A unit cell of the reinforcement comprises two horizontal reinforcement, two small vertical reinforcement, and two small horizontal reinforcement i.e. lap length.

3 Material Properties

The assigned soil properties are $k = 0$, and $M = 0.736$, corresponding to the cohesionless soil having a dry unit weight of $\gamma_{dry} = 15$ kN/m^3 with shear strength parameters $\phi = 26°$ and $c = 0$ as per Mohr–Coulomb failure criterion. To examine the geometrical parameters of reinforcement, a parametric study has been conducted by varying the reinforcement width (b) from 3 m to 6 m at an interval of 1 m. Following geometrical parameters of the geogrid are kept constant while performing the analysis: lap length (L) = 0.3B, vertical length of the wraparound ends (d) = 0.2B, depth of placement of first layer (u) = 0.3B, vertical spacing between consecutive geogrid layers (h) = 0.3B.

Also, the axial elastic stiffness (*EA*) of the geogrid has been varied from 500 kN/m to 2000 kN/m at an interval of 500 kN/m for analyzing the settlement behavior and improvement in the bearing capacity of the footing resting over corresponding reinforced soil mass. With the above-mentioned evaluation, parameters are determined and while keeping the optimum reinforcement width and elastic stiffness fixed, the strength reduction factor (R_{int}) has been varied at 0.7, 0.8, and 0.9 to study the effect of soil–geogrid interaction on the behavior of footing resting over reinforced soil mass. With the use of multiplier elastoplastic analysis, the load-settlement behavior of the footing is evaluated for all the cases studied in the present study. The multiplier elastoplastic analysis involves the application of an incremental vertical surcharge till the state of incipient failure of the footing reaches and the corresponding $q_{u,\,R}$ of the footing is determined from the load-settlement curve by applying the double tangent method [19].

4 Results and Discussion

As mentioned earlier in this study, numerical simulations have been carried out for the investigation of optimum widths of reinforcement, the effect of elastic stiffness of geogrid, and also the impact of strength reduction factor (i.e. interface parameter) on the ultimate load-bearing capacity ($q_{u,\,R}$) and settlement of the footing. The results have been presented in the form of plots of load-bearing pressure versus normalized settlement of the footing and discussed in the following section.

To ascertain the efficiency of the Drucker–Prager model, the current model has been compared with the Mohr–Coulomb model by using an established computational technique i.e. limit analysis. A strip footing of width 2 m has been considered resting over an unreinforced soil mass in both the model i.e. M-C and D-P yield criteria and a multiplier distributed load has been applied on the footing in the downward direction followed by limit analysis. The results obtained from the above analysis are shown in the form of potential failure patterns and principal stresses vectors σ_1 and σ_3 in Fig. 3, where σ_1 and σ_3 are major and minor principal stresses, respectively.

The soil parameters i.e. cohesion and angle of internal friction angle for both the models are kept similar by utilizing Eqs. (1) and (2). The results indicated that both the models produce a similar potential failure pattern as proposed by Terzaghi [1], which is in good agreement with each other. Also, the ultimate load-bearing capacity of the footing obtained for M-C and D-P yield criteria is 247.8 kPa and 247.6 kPa, respectively, in which the difference in ultimate values of bearing capacities is only about 0.08%, and the noted observation justifies the selection of the D-P model for analysis.

Furthermore, the ultimate load-bearing capacity ($q_{u,\,R}$) has been investigated for the parameters discussed in the trailing section for various width of reinforcement. The ultimate load-bearing capacity of the unreinforced case (q_u) is found to be 248.6 kPa, however, with the inclusion of geogrid in the soil bed, a severe increase in

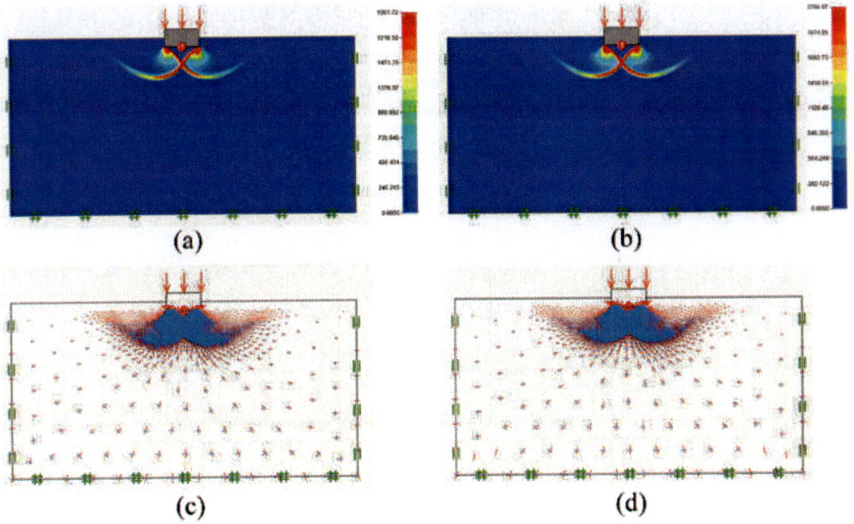

Fig. 3 Potential failure pattern (**a**) M-C yield criterion; (**b**) D-P yield criterion; and σ_1 and σ_3 vectors (**c**) M-C yield criterion; and (**d**) D-P yield criterion

the magnitude of ultimate bearing capacity has been noticed due to the presence of the tensile member beneath the footing. Moreover, to ensure the minimum width of reinforcement required for the sufficient increase in the $q_{u,R}$, b has been varied from 3 m to 6 m (as shown in Fig. 4). The results indicated that the maximum improvement in $q_{u,R}$ is observed at $b = 3$ m i.e. 403.3 kPa and no further significant improvement in the load-bearing pressure is observed beyond the 3 m wider reinforcing layer. The above-noted behavior may be attributed to the virtue of the confinement effect, which is being introduced by the wraparound ends of the geogrid, however, the maximum improvement in the bearing capacity with 3 m wide geogrid was due to the proximity of the wraparound ends of the reinforcing layer with the footing, which restrains the possible lateral movement of the stressed soil mass near the footing. With a further increase in the width of the reinforcing layer, the distance of the wrapping ends of geogrid from the footing increases, and the confinement effect decreases, and due to this a marginal reduction in the $q_{u,R}$ is observed in all the other cases. Based on the above observation, it can be easily concluded that for a strip footing of width $B = 2$ m, the width of reinforcing layers must be kept 3 m for the maximum benefit of reinforcement, which is 1.5 B only. Previously, the researchers had reported the optimum width of reinforcement based on the regression model for footing resting on pond ash to be 5B to 7B [20]. In a similar study, utilizing the full wraparound ends of geogrids, the optimum width of reinforcement is reported to be 2B [14]. The optimum width of reinforcement using the wraparound ends of geogrid, the present study suggests a reinforcement width of 1.5 B, where B is the width of footing.

For an economic design of any reinforced structure, the consideration of the overall cost of the geogrid material is an important consideration that depends upon the

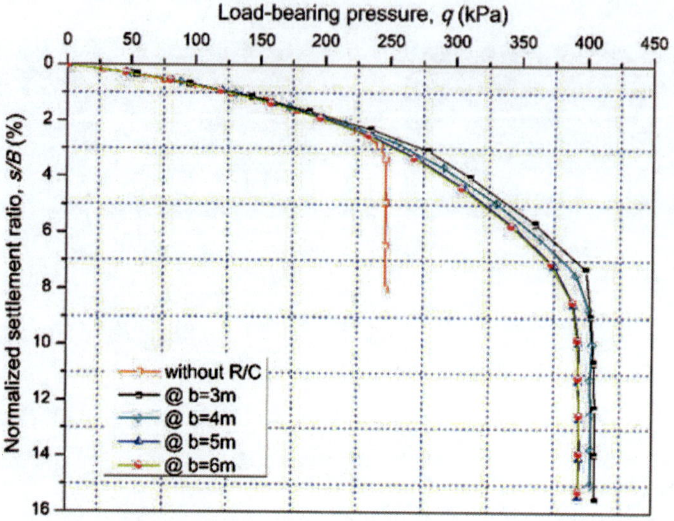

Fig. 4 Variation of load-bearing pressure versus normalized settlement ratio of the footing for varying width of the reinforcement

volume of reinforcing material, kind of product, durability, stiffness, cross-sectional area, and many other factors also. In this study, only the width and stiffness of the geogrid are considered. Also, higher stiffness leads to a higher cross-sectional area or a stiffer material being used for the production of the geogrid. So, it is always preferable to study the effect of stiffness of the geogrid on the overall bearing capacity to ensure the use of a minimum quantity of material with increased stiffness for a safer and cost-efficient design.

In the present study, to evaluate the influence of *EA*, it has been varied from 500 kN/m to 2000 kN/m at an interval of 500 kN/m each, keeping all the other parameters and properties constant and its effect on the ultimate load-bearing capacity ($q_{u,\,R}$) as well as the footing settlement have been studied (as shown in Fig. 5). It is evident from Fig. 5 that the inclusion of geogrid enhances the $q_{u,\,R}$, also it increases with an increase in *EA*. However, this improvement is noteworthy only when the stiffness of the geogrid changes from 500 kN/m to 1000 kN/m and the curve shown in Fig. 5 remains virtually constant for higher values of *EA*, a similar result was reported by Benmebarek et al. [14].

Based on the laboratory experiment, Martin et al. [21] had reported that the texture of the reinforcement material is the key element for the mobilization of the frictional forces in between the reinforcement material and surrounding soil. Based on the above, the R_{int} has been varied from 0.7 to 0.9 in an interval of 0.1 keeping all the other parameters invariable. This interprets the fact that the mobilization of shear stresses depends on the behavior of the geogrid–soil interface. The multiplier distributed load applied on the footing is being taken by the tensile forces generated in the geogrid and the fundamental cause for the generation of tensile stresses in the geogrid is due

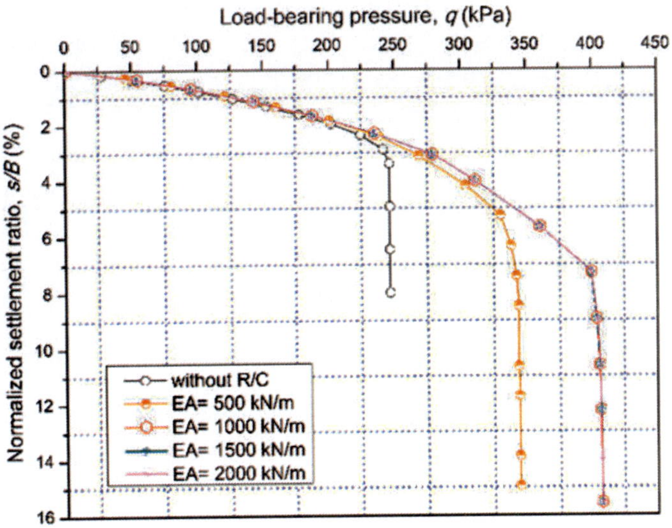

Fig. 5 Variation of load-bearing pressure versus normalized settlement ratio to investigate the effect of elastic stiffness, *EA*

to the mobilization of the shear stress (friction forces) at the interface of the soil and reinforcing material. This is the reason that R_{int} is an important characteristic to understand whether the alteration in the texture of a geogrid material having identical elastic stiffness can lead to substantial improvement in the ultimate bearing capacity and minimize the mobilization of soil along with the reinforcing material. Therefore, it is essential to understand the effect of R_{int} on the overall bearing capacity of the soil mass reinforced with geogrid.

It has been noted that at a given footing settlement ratio, the ultimate load-bearing capacity of the embedded footing resting on the reinforced soil mass ($q_{u,R}$) has been improved by 9% when the R_{int} has been changed from 0.7 to 0.8 and it is improved by 16.2% when R_{int} is varied from 0.7 to 0.9 (as shown in Fig. 6). This is a significant improvement in $q_{u,R}$, where the same geogrid material having only difference in the surface texture can lead to the maximum utilization of the reinforcing action. However, most of the studies recommend the value of R_{int}, which is $\delta = 2/3$ of ϕ, where δ is the interface friction angle of two dissimilar materials (in the present case, soil and geogrid) [11, 12]. Therefore, it can be concluded that the alteration in the surface texture of the geogrid can enhance the R_{int} and subsequently increase the efficiency of the reinforced soil bed to bear the load even for the same capacity of reinforcement for a given soil bed can be utilized by the reinforcing material. However, taking $R_{int} = 2/3$ is leading towards a very conservative design, where one cannot expect to utilize the full capacity of the reinforcement and that is why it gives rise to a very costly design. Though, before proceeding with designing any structure, it is recommended that the material intended to use for the reinforcement should

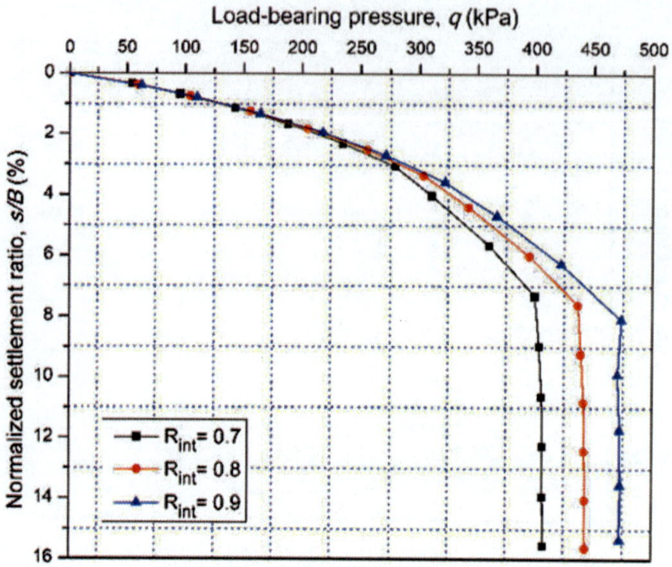

Fig. 6 Variation of load-bearing pressure versus normalized settlement ratio for varying R_{int}

always be tested in laboratory for the determination of R_{int} value in the presence of actual backfill material/foundation material.

The normal forces generated in the geogrid are shown in Fig. 7 for the various value of strength reduction factor (i.e. R_{int}). It is evident from the above observation that as the value of R_{int} increases, the ultimate load-bearing capacity of the reinforced bed also increases, and subsequently the normal forces generated in the geogrid increase. It can be noticed from Fig. 7; in all the cases, the maximum normal forces are generated in the bottom-most layer (farthest from the footing), which is due to the overburden pressure lying above the reinforcing layer contributed by the weight of overlying soil mass and the applied surcharge.

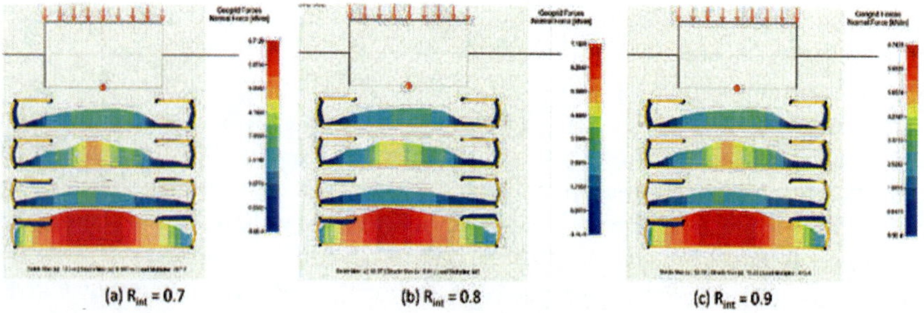

Fig. 7 Normal forces in reinforcing layer for interface strength reduction factor, R_{int} (**a**) 0.7; (**b**) 0.8; and (**c**) 0.9

5 Conclusions

This paper presents the numerical study of a strip footing to investigate the effects of wraparound ends of the geogrid on the ultimate load-bearing capacity ($q_{u,\,R}$) and footing settlement (*s/B*), in which the soil is modeled by using the Drucker–Prager yield criterion. A parametric study has been carried out to evaluate the influence of the width of reinforcement (b) and elastic stiffness (*EA*) of the geogrid. Furthermore, this study also highlights the effect of the interface between geogrid and soil mass, on the load-settlement response of a footing resting over a reinforced earth bed. Based on the findings and discussions of the present study, the following conclusions can be made:

1. The Drucker–Prager (D-P) model is an appropriate model for determining the load-settlement behavior and the ultimate load-bearing capacity of a foundation as the occurrence of singularity is avoided, which is seen in the case of the Mohr–Coulomb model.
2. The width of the geogrid layer (b) governs the overall load-bearing capacity of the reinforced soil mass system. The finding of the present study suggests an optimum width of the geogrid layers, b equals to 1.5 B for maximizing the effective utilization of the wraparound end technique.
3. The elastic stiffness, *EA* of a geogrid is an influential parameter, which affects the ultimate load-bearing capacity of the reinforced soil mass ($q_{u,\,R}$) and also the controls of the settlement of the footing. The maximum improvement in the $q_{u,\,R}$ is observed at $EA = 1000$ kN/m, and the curves between the load-bearing pressure and the normalized settlement remain virtually constant for higher values.
4. Strength reduction factor for the interface between geogrid and earth bed (R_{int}) creates a better resemblance of the numerical simulation with the actual site condition imparts an important role in deciding the $q_{u,\,R}$. This study recommends for the precise assessment of the $R_{int,}$ it is necessary to conduct laboratory testing of the material which is proposed to use below the footing for reinforcement purpose.

References

1. Terzaghi, K.: Theoretical Soil Mechanics. Wiley, New York, N.Y. USA (1944)
2. Yazdandoust, M.: Investigation on the seismic performance of steel-strip reinforced soil retaining walls using shaking table test. Soil. Dyn. Earthq. Eng. **97**, 216–232 (2017)
3. Maheshwari, P., Chauhan, V.B.: Beams on extensible geosynthetics and stone-column-improved soil. Proc. Inst. Civil Eng. Ground Improv. **166**(4), 233–247 (2013). https://doi.org/10.1680/grim.12.00005
4. Srivastava, A., Chauhan, V.B.: Numerical studies on two-tiered MSE walls under seismic loading. SN Appl. Sci. **2**(10), 1–7 (2020). https://doi.org/10.1007/s42452-020-03414-6

5. Adams, M.T., Collin, J.G.: Large model spread footing load tests on geosynthetic reinforced soil foundations. J. Geotech. Geoenviron. Eng. **123**(1), 66–72 (1997). https://doi.org/10.1061/(ASCE)1090-0241(1997)123:1(66)
6. Fragaszy, J., Lawton, E.: Bearing capacity of reinforced sand subgrades. J. Geotech. Eng. **110**(10), 1500–1507 (1985). https://doi.org/10.1061/(ASCE)0733-9410(1984)110:10(1500)
7. Meyerhof, G.G., Hanna, A.M.: Ultimate bearing capacity of foundations on layered soils under inclined load. Can. Geotech. J. **15**(4), 565–572 (1978). https://doi.org/10.1139/t78-060
8. Yetimoglu, T., Wu, J.T., Saglamer, A.: Bearing capacity of rectangular footings on geogrid-reinforced sand. J. Geotech. Eng. **120**(12), 2083–2099 (1994). https://doi.org/10.1061/(ASCE)0733-9410(1994)120:12(2083)
9. Khing, K.H., Das, B.M., Puri, V.K., Cook, E.E., Yen, S.C.: The bearing-capacity of a strip foundation on geogrid-reinforced sand. Geotext. Geomembr. **12**(4), 351–361 (1993). https://doi.org/10.1016/0266-1144(93)90009-D
10. Aria, S., Shukla, S.K., Mohyeddin, A.: Optimum burial depth of geosynthetic reinforcement within sand bed based on numerical investigation. Int. J. Geotech. Eng. (2017). https://doi.org/10.1080/19386362.2017.1404202
11. Kazi, M., Shukla, S.K., Habibi, D.: Behaviour of an embedded footing on geotextile-reinforced sand. Proc. Inst. Civil Eng. Ground Improv. **169**(2), 120–133 (2016). https://doi.org/10.1680/grim.14.00022
12. Kazi, M., Shukla, S.K., Habibi, D.: Behavior of embedded strip footing on sand bed reinforced with multilayer geotextile with wraparound ends. Int. J. Geotech. Eng. **9**(5), 437–452 (2015). https://doi.org/10.1179/1939787914Y.0000000085
13. Shukla, S.K.: An Introduction to Geosynthetic Engineering. CRC Press (2017)
14. Benmebarek, S., Djeridi, S., Benmebarek, N., Belounar, L.: Improvement of bearing capacity of strip footing on reinforced sand. Int. J. Geotech. Eng. **12**(6), 537–545 (2018). https://doi.org/10.1080/19386362.2017.1309136
15. Kolay, P.K., Kumar, S., Tiwari, D.: Improvement of bearing capacity of shallow foundation on geogrid reinforced silty clay and sand. J. Constr. Eng. 1–10 (2013). https://doi.org/10.1155/2013/293809
16. Optum G2: Finite Element Program for Geotechnical Analysis, Optum Computational Engineering (2020). www.optumce.com
17. Sloan, S.W., Booker, J.R.: Removal of singularities in Tresca and Mohr-Coulomb yield functions. Commun. Appl. Numer. Methods **2**(2), 173–179 (1986). https://doi.org/10.1002/cnm.1630020208
18. Pandey, A., Chauhan, V.B.: Evaluation of pull-out capacity of helical anchors in clay using finite element analysis. Geo-Congress 2020 GSP **317**(60), 60–68 (2020). https://doi.org/10.1061/9780784482803.007
19. Lovisa, J., Shukla, S.K., Sivakugan, N.: Behaviour of prestressed geotextile reinforced sand bed supporting a loaded circular footing. Geotext. Geomembr. **28**(1), 23–32 (2010). https://doi.org/10.1016/j.geotexmem.2009.09.002
20. Bera, A.K., Ghosh, A., Ghosh, A.: Regression model for bearing capacity of a square footing on reinforced pond ash. Geotext. Geomembr. **23**(3), 261–285 (2005). https://doi.org/10.1016/j.geotexmem.2004.09.002
21. Martin, J.P., Koerner, R.M., Whitty, J.E.: Experimental friction evaluation of slippage between geomembranes, geotextiles and soils. In: Proceedings of the International Conference on Geomembranes, vol. 1, pp. 191–196. IFAI, Denver, CO, USA (1984)

Probabilistic Analysis of Reinforced Soil Retaining Structures Using FORM and Surrogate-Based Monte Carlo Simulation

Ekansh Agarwal, Anindya Pain, and Shantanu Sarkar

1 Introduction

Reinforced soil retaining structures (RSRS) are widely used nowadays. RSRS have all the attributes to eliminate the drawbacks of their conventional counterparts and are economical, easy to construct, and offer a wide range of advantages to the designers and the architects. RSRS have become a crucial component of any slope stabilization project. These structures are durable and hence enhance the stability of the structure in which they are used. RSRS are analyzed using various analytical as well as numerical approaches for both, internal and external stabilities. The initial works on the design and analysis of RSRS were made by Leshchinsky et al. [1], Ling et al. [2], and Ling and Leshchinsky [3] using log-spiral failure surface in limit equilibrium (LE) framework. Another approach of designing RSRS is the horizontal slice method (HSM), which includes splitting the reinforced portion into a number of horizontal slices and then employing suitable equations of equilibrium to calculate the required reinforcement force. Different formulations of HSM are available depending upon many factors, the equations of equilibrium to be used also being the one [4–9]. Horizontal slice method has an advantage that it can consider the variation of acceleration along the wall height owing to the horizontal nature of slices rather than vertical alike the conventional methods.

Many researchers have followed the experimental approach to analyze the RSRS using 1-g shake table tests and centrifuge [10–18]. A detailed review of different experimental investigations can be obtained from Sabermahani et al. [19] and Srilatha et al. [20].

E. Agarwal (✉)
Academy of Scientific and Innovative Research (AcSIR), Ghaziabad 201002, India

E. Agarwal · A. Pain · S. Sarkar
Geotechnical Engineering. Division, CSIR-Central Building Research Institute, Roorkee 247667, India

C. N. V. Satyanarayana Reddy et al. (eds.), *Ground Improvement and Reinforced Soil Structures*, Lecture Notes in Civil Engineering 152,
https://doi.org/10.1007/978-981-16-1831-4_65

The conventional approach to analyze the RSRS is known as allowable stress design (ASD) or more precisely the deterministic approach. This approach increases the factor of safety of the structure under consideration to such a high extent that the structure becomes uneconomical. Hence, it has become obsolete and replaced by a probabilistic approach, which takes into account the uncertainties associated with the parameters involved, and a minimum factor of safety is guaranteed just by the satisfaction of limit state function.

The most common approach to perform the probabilistic analysis of RSRS is to calculate the reliability index associated with the slope, which gives an idea about its stability. Many researchers have worked on the methodology to find out the surface of the minimum reliability index. Li and Lumb [21] and Bhattacharya et al. [22] had found the critical deterministic surface and subsequently that surface was used as the initial surface for investigating the surface of minimum reliability index i.e. β_{min}.

Basha and Babu [23] had used the target reliability-based approach (TRA) to analyze the RSRS probabilistically. Three different modes of failure of the reinforcement were chosen for the same and the analysis was done using the first-order reliability method (FORM). However, no further validation of the above methodology using another technique such as meta-modeling/sampling-based methods was made. Metya and Bhattacharya [24] had computed the reliability of earth slopes using a computational procedure.

The literature mentioned above reveals the fact that there is a dire need to analyze the RSRS probabilistically to account for the uncertainty of geotechnical parameters along with the dynamic nature of the earthquake. The main aim of the present study revolves around the reliability analysis of RSRS using FORM and validating the same using a more efficient and accurate multivariate adaptive regression splines (MARS) [25] method.

2 Deterministic Analysis of Internal Stability of Geosynthetic RSRS

To analyze the RSRS deterministically, HSM (Fig. 1) is used in the present study in a pseudo-static framework. Soil is assumed to be dry, homogenous, and linearly elastic in nature. The backfill is cohesionless and the effect of pore water pressure is neglected. The failure surface is assumed as log-spiral. As discussed in the above section, Nouri et al. [6] had presented a comparison among different formulations of HSM. Hence, simple *(2N+1)* formulation is chosen in the present study, which satisfies the horizontal equilibrium for the whole wedge and vertical equilibrium for individual slices together. The number of equations and unknowns is reduced to *2N+1* in this case. The equations of equilibrium used in Shahgholi et al. [4] are used in the same manner to calculate the required tensile strength of the reinforcement (T_j) from Eq. (1).

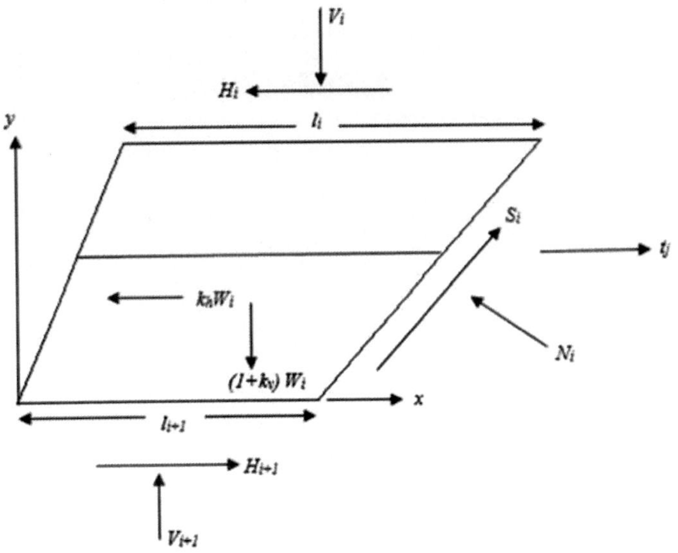

Fig. 1 Forces acting on a single horizontal slice with reinforcement

$$\sum_{j=1}^{m} t_j = T_j = \sum_{i=1}^{N} N_i \sin\alpha + \sum_{i=1}^{N} W_i k_h - \sum_{i=1}^{N} S_i \cos\alpha \tag{1}$$

where S_i = shear force upon base of slice, N_i = normal force upon base of slice, and W_i = Weight of slice

T_j may be expressed in a non-dimensional term as mentioned below:

$$K = T_j / 0.5\gamma H^2 \tag{2}$$

The value of K is optimized to find out the critical failure surface using sequential quadratic programming (SQP) in MATLAB.

The results from the present study have been validated by comparing them with Ling et al. [2] (Fig. 2) for $H = 5$ m, $\beta = 90°$, $\gamma = 18$ kN/m^3, $\phi = 30°$, $k_h = 0$–0.3 and $k_v = 0$ and they are in good adherence with the said literature.

3 Probabilistic Analysis

The methodology to calculate the probability of failure (P_f) associated with the critical failure surface is explained in this section. P_f is calculated for two different modes of failure (tension and pullout) followed by the calculation of system probability of failure using FORM and surrogate-based Monte Carlo Simulation (MCS).

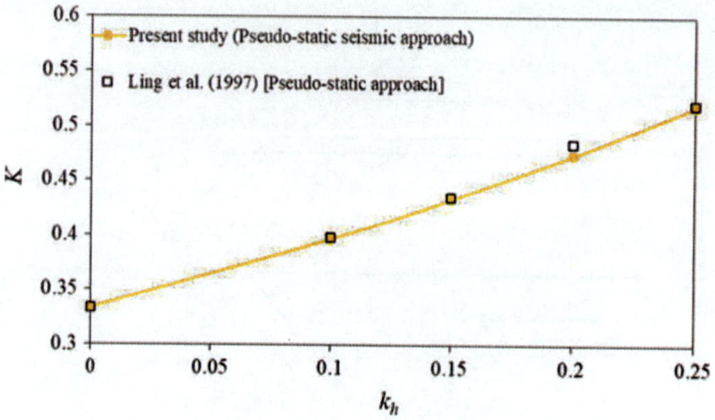

Fig. 2 Comparison of the non-dimensional total required tensile strength from the deterministic analysis

3.1 Reliability Analysis Using FORM

In the present study, the model presented by Metya and Bhattacharya [24] is used to find out the critical probabilistic failure surface and associated P_f. The critical probabilistic failure surface is found out in the same manner as the critical deterministic failure surface i.e.

$$\beta_{min} = min_x \beta(G, X) \tag{3}$$

where $\beta(G, X)$ is the reliability index for a given set of geotechnical parameters (G).

The location of the failure surface w.r.t. the origin of the reduced variates determines the reliability or more precisely, the safety of the system. Hence, the required minimum distance may be calculated as follows:

$$D = (X_1'^2 + \cdots + X_n'^2)^{1/2}. \tag{4}$$

where $X' = (X_1', X_2', \ldots, X_n')$ is a point on the failure surface with a minimum distance to the origin.

System reliability index (β_{sys})

In the present study, the series system reliability index is calculated by considering the failure modes to be connected in series. Failing of even one component disables the whole system when the modes are connected in series. P_f for a system connected in series is given by the following equation:

$$P_{fsys} = 1 - [\{P_{ft}\}\{P_{fpo}\}] \tag{5}$$

Table 1 Attributes of the random variables used in the present study

Random variable	Properties		
	Mean (μ)	Coefficient of variation (CoV) %	Distribution
γ (soil unit weight)	18 kN/m^3	5	Normal
ϕ (internal friction angle of the soil)	34°	5	Log-normal
T_u (ultimate tensile strength)	20 kN/m	5	Normal
L_{ei} (pullout length of the RF)	0.2 m	0.5	Normal

where, P_{ft} = probability of failure for tension failure mode and P_{fpo} = probability of failure for pullout failure mode.

Comparison of FORM results

The input variables used in the present study are tabulated in Table 1. All the random variables follow Gaussian distribution except the internal friction angle of soil, which follows the log-normal distribution.

To validate the formulation, the results from the present study have been compared with Basha and Babu [23] for $H = 5.5$ m, $\phi = 37°$, $\beta = 78.7°$, $\gamma = 18$ kN/m^3, $k_h = 0.216$, $N = 10$, $\delta/\phi = 0.8$, $T_u = 20$ kN/m and $k_v = 0$. The results are in accordance with the said literature with minor differences noted due to differences in methodologies used. The value of reliability index (β_t) for tension mode of failure from the present study is 2.806 in comparison to 2.230 from Basha and Babu [23].

3.2 *Reliability Analysis Using MARS-Based MCS*

MARS was primarily developed as a statistical tool to define the relationship between input and output variables [25–27]. MARS approximates the response function based on a forward and backward iterative approach, the main advantage being the automatic identification of basis functions and the parameters associated with them. A MARS model may be formulated as:

$$M = \sum_{q=1}^{n'} \psi_k B_q^f(z_i) \tag{6}$$

where ψ is the coefficient of expansion and $B_q^f(z_i)$ is the basis function.

The value of generalized cross-validation (GCV) defines the goodness of the fit of a MARS model, the best being the one having the lowest value of GCV.

To analyze the RSRS using MARS, training data sets were generated using Latin hypercube sampling (LHS) for the three random variables (T_u, ϕ, γ) used in the study (tension mode). The value of performance function is obtained using the

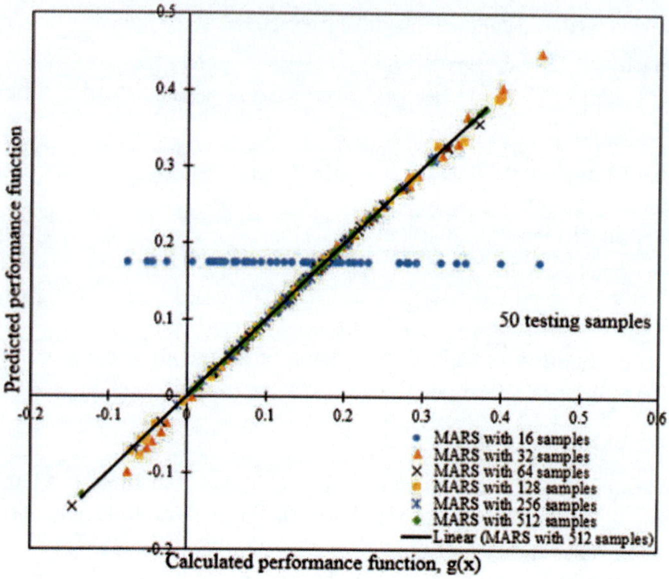

Fig. 3 Comparison between the values of performance function predicted by MARS models and the ones calculated by LEM

deterministic code. Further, a group of MARS models were constructed using these data sets where each model consisted of 2^n training samples. The performance of these constructed MARS models is tested using 50 testing samples chosen randomly. The value of coefficient of determination (R^2) is then calculated for each of these models and the optimum number of training samples is noted for the model having a maximum value of R^2. The optimum number of simulations is calculated by varying the sample size from 1000 to 80000 and noting the value of simulation after which the fluctuation in P_f is minimal. Moreover, a comparison is drawn between the values of performance function predicted by different MARS models and the ones calculated by the deterministic model as shown in Fig. 3. The results depict very good fitting and also validate the ability of MARS method to predict good results with the optimum number of samples taken in the present study. The optimum number of training samples for both the modes of failure is chosen as 512, having the maximum value of R^2 equal to 0.99. The optimum number of simulations is chosen as 50000.

4 Results and Discussions

In the contemporary study, RSRS is analyzed using the two methods namely first-order reliability method (FORM) and MARS-based Monte Carlo simulations (MCS). The probability of failure (P_f) is calculated for both, tension and pullout modes of

Table 2 P_f for different modes of failure using FORM

Mode of failure	P_f (%)	β (Reliability index)
Tension mode	8.08	1.40
Pullout mode	1.16	2.27
System probability of failure	9.14	1.33

failure using the aforementioned methods. Further, the system probability of failure is calculated and a comparison is made between the two methods.

4.1 *Results Using FORM*

Probabilistic analysis of RSRS is done using FORM for $H = 5$ m, $\phi = 34°$, $\beta = 90°$, $\gamma = 18$ kN/m^3, $k_h = 0.2$, $N = 10$, $\delta/\phi = 1$, $T_u = 20$ kN/m and $k_v = 0$. The results have been tabulated below (Table 2).

The results show that the most critical mode of failure is the tension mode having the least value of reliability index as 1.40. However, the probability of failure of system is higher than the tension mode, emphasizing the need to perform system probabilistic analysis.

4.2 *Results Using MARS-Based MCS*

Probabilistic analysis of RSRS is done using MARS-based MCS for the same parameters mentioned in Sect. 4.1. The results have been tabulated below.

The results from Table 3 for MARS-based MCS are in good accordance with the results obtained using FORM (Table 2). Moreover, the surrogate method follows the same trend as FORM, where tension mode is the critical mode among the two mentioned modes of failure (Fig. 4). The above-mentioned results show the ability of MARS-based MCS to predict good results. The main advantage of surrogate-based modeling is its accurate and efficient performance when the probability of failure is very small, unlike the FORM that suffers from the assumption of linearization of failure surface at the design point.

Table 3 P_f for different modes of failure using MARS-based MCS

Mode of failure	P_f (%)	β (Reliability index)
Tension mode	7.85	1.41
Pullout mode	1.39	2.20
System probability of failure	9.13	1.33

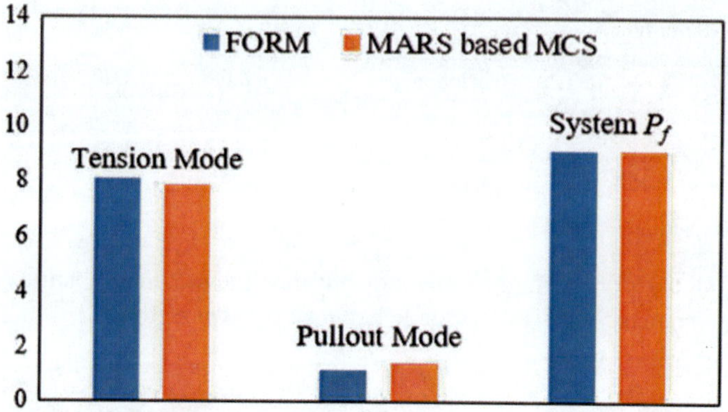

Fig. 4 Comparison between FORM and MARS-based MCS

5 Conclusions

In the present study, probabilistic analysis of reinforced soil retaining structure (RSRS) is done in a pseudo-static framework. For this, *(2N+1)* formulation of horizontal slice method (HSM) is used. Two different modes of failure are considered for the calculation of reliability indices using FORM. Non-linear programming (SQP) is used to calculate the required tensile strength of the reinforcement. System reliability of failure is calculated by considering the modes of failure to be connected in series. The tension mode of failure is found to be the most critical one, however, the P_f of system is found to be the highest among all. A surrogate-based MCS is also performed to validate the results obtained from FORM. Latin hypercube sampling (LHS) is employed to construct the MARS-based model. It is found that MARS-based MCS is an efficient, accurate, and robust method for the analysis of RSRS and it outperforms FORM in terms of assumptions of the failure surface and computational efficiency. The present study reveals that FORM and MARS-based MCS, both are capable of handling the uncertainties associated with the statistical parameters of soil and reinforcement, the latter being more efficient. The probabilistic analysis provides an economical and efficient way of designing the RSRS.

It is worthy to note that the present study is performed for reinforced soil retaining wall (slope inclination angle >70°). For this case, the simple *(2N+1)* formulation provides the most efficient and accurate results [6]. However, as the slope of inclination of RSRS decreases, the results from simple *(2N+1)* incline towards a conservative and uneconomical side due to the non-satisfaction of the moment equilibrium equation. Therefore, a more rigorous *(5N-1)* method, which satisfies all the equilibrium equations, is recommended for lower inclination angles of RSRS.

References

1. Leshchinsky, D., Ling, H., Hanks, G.: Unified design approach to geosynthetic reinforced slopes and segmental walls. Geosynth. Int. **2**(5), 845–881(1995)
2. Ling, H.I., Leshchinsky, D., Perry, E.B.: Seismic design and performance of geosynthetic-reinforced soil structures. Geotechnique **47**(5), 933–952 (1997)
3. Ling, H.I., Leshchinsky, D.: Effects of vertical acceleration on seismic design of geosynthetic-reinforced soil structures. Geotechnique **48**(3), 347–373 (1998)
4. Shahgholi, M., Fakher, A., Jones, C.J.F.P.: Horizontal slice method of analysis. Geotechnique **51**(10), 881–885 (2001)
5. Fakher, A., Nouri, H., Shahgholi, M.: Limit equilibrium in reinforced soil walls subjected to seismic loads. In: Proceedings of the 3rd Iranian International Conference on Geotechnical Engineering and Soil Mechanics, pp. 281–286. Tehran (2002)
6. Nouri, H., Fakher, A., Jones, C.J.F.P.: Development of horizontal slice method for seismic stability analysis of reinforced slopes and walls. Geotext. Geomembr. **24**(3), 175–187 (2006)
7. Nouri, H., Fakher, A., Jones, C.J.F.P.: Evaluating the effects of the magnitude and amplification of pseudo-static acceleration on reinforced soil slopes and walls using the limit equilibrium horizontal slices method. Geotext. Geomembr. **26**(3), 263–278 (2008)
8. Khosravizadeh, M., Dehestani, M., Kalantary, F.: On the seismic stability and critical slip surface of reinforced slopes. Soil Dyn. Earthq. Engi. **85**, 179–190 (2016)
9. Mehdipour, I., Ghazavi, M., Ziaie Moayed, R.: Stability analysis of geocell-reinforced slopes using the limit equilibrium horizontal slice method. Int. J. Geomech. **17**(9), 06017007 (2017)
10. Murali Krishna, A., Madhavi Latha, G.: Seismic response of wrap-faced reinforced soil-retaining wall models using shaking table tests. Geosynth. Int. **14**(6), 355–364 (2007)
11. Latha, G.M., Krishna, A.M.: Seismic response of reinforced soil retaining wall models: influence of backfill relative density. Geotext. Geomembr. **26**(4), 335–349 (2008)
12. El-Emam, M.M., Bathurst, R.J.: Facing contribution to seismic response of reduced-scale reinforced soil walls. Geosynth. Int. **12**(5), 215–238 (2005)
13. El-Emam, M.M., Bathurst, R.J.: Influence of reinforcement parameters on the seismic response of reduced-scale reinforced soil retaining walls. Geotext. Geomembr. **25**(1), 33–49 (2007)
14. Ling, H.I., Mohri, Y., Leshchinsky, D., Burke, C., Matsushima, K., Liu, H.: Large-scale shaking table tests on modular-block reinforced soil retaining walls. J. Geotech. Geoenvironmental Eng. **131**(4), 465–476 (2005)
15. Ling, H.I., Liu, H., Mohri, Y.: Parametric studies on the behavior of reinforced soil retaining walls under earthquake loading. J. Eng. Mech. **131**(10), 1056–1065 (2005)
16. Ling, H.I., Leshchinsky, D., Wang, J.P., Mohri, Y., Rosen, A.: Seismic response of geocell retaining walls: experimental studies. J. Geotech. Geoenvironmental Eng. **135**(4), 515–524 (2009)
17. Cowland, J.W., Wong, S.C.K.: Performance of a road embankment on soft clay supported on a geocell mattress foundation. Geotext. Geomembr. **12**(8), 687–705 (1993)
18. Sitharam, T.G., Hegde, A.: Design and construction of geocell foundation to support the embankment on settled red mud. Geotext. Geomembr. **41**, 55–63 (2013)
19. Sabermahani, M., Ghalandarzadeh, A., Fakher, A.: Experimental study on seismic deformation modes of reinforced-soil walls. Geotext. Geomembr. **27**(2), 121–136 (2009)
20. Srilatha, N., Latha, G.M., Puttappa, C.G.: Effect of frequency on seismic response of reinforced soil slopes in shaking table tests. Geotext. Geomembr. **36**, 27–32 (2013)
21. Li, K.S., Lumb, P.: Probabilistic design of slopes. Can. Geotech. J. **24**(4), 520–535 (1987)
22. Bhattacharya, G., Jana, D., Ojha, S., Chakraborty, S.: Direct search for minimum reliability index of earth slopes. Comput. Geotech. **30**(6), 455–462 (2003)
23. Basha, B.M., Babu, G.S.: Target reliability-based optimisation for internal seismic stability of reinforced soil structures. Geotechnique **62**(1), 55–68 (2012)
24. Metya, S., Bhattacharya, G.: Probabilistic critical slip surface for earth slopes based on the first order reliability method. Indian Geotech. J. **44**(3), 329–340 (2014)

25. Friedman, J.H.: Multivariate adaptive regression splines. The Annals of Statistics, pp. 1–67 (1991)
26. Hastie, T., Tibshirani, R., Friedman, J.: The elements of statistical learning: data mining, inference, and prediction, 2nd edn. Springer Science & Business Media, New York (2009)
27. Dey, S., Mukhopadhyay, T., Adhikari, S.: Metamodel based high-fidelity stochastic analysis of composite laminates: a concise review with critical comparative assessment. Compos. Struct. **171**, 227–250 (2017)

Numerical Study of Multi-layered Geocell Confined Pavement Subgrade

Arghadeep Biswas and Haradhan Sarkar

1 Introduction

The competent subgrade is the primary need and has been a critical issue for ever-increasing road networks. Thus, it has enforced designers/practitioners to consider different ground improvement techniques/alternatives based on availability, feasibility, economics, and importantly different environmental issues for its durability. Amongst the few, geocell has complied with most of the issues and thus, preferred the most over other methods/alternatives.

Considerable researches are performed on the application of geocell in pavement structures. A brief is presented by Sarkar and Biswas [1], which may be a good read in this regard. Critically, it can be concluded that most of the studies have assumed geocell in the base or sub-base layer, whereas, the less reckoned weak subgrade may be the cause of failure. In this regard, except very few, studies/applications have used single layer geocell system regardless of thickness required in the laboratory and/or in practice. Recently, researchers have identified few critical issues of such adaptation, namely, buckling of geocell walls at the top-array/top part just under the loading, difficulties in achieving the desired degree of compaction at the bottom of geocell–soil matrix and non-utilization of a large portion of geocell mattress beyond the load dispersion [2–4] etc. In such condition, provision of multiple layers of geocell reinforcement would be a practicable way to overcome the drawbacks with greater performance [5–8]. In this regard, Tafreshi et al. [7] have confirmed the improved performance of multi-layered-geocell using cyclic plate load tests; whereas, Li et al. [8] have reported an increasing bearing capacity with the increase in a number of

A. Biswas (✉)
Jalpaiguri Government Engineering College, Jalpaiguri 735102, West Bengal, India

H. Sarkar
Ghani Khan Choudhury Institute of Engineering & Technology, Malda 732141, West Bengal, India

C. N. V. Satyanarayana Reddy et al. (eds.), *Ground Improvement and Reinforced Soil Structures*, Lecture Notes in Civil Engineering 152,
https://doi.org/10.1007/978-981-16-1831-4_66

Table 1 Material properties for numerical analysis

Material/layer	Material properties				
	E (kPa)	c_u (kPa)	φ (□)	υ	γ (kN/m^3)
Subgrade clay	18,000	30	0	0.45	17.3
[a]Base	414,000	0	38	0.30	22.0
[a]Asphalt layer	4,134,693	–	–	0.30	23.0
Geocell-sand matrix	92,719	93	38	0.30	16.4
Sand	13,000	0	40	0.30	16.4

Source [a]Saad et al. [12]

reinforced layers. Such research outcomes have encouraged authors to envisaged the objectives of present study.

In the present work, geocell layers are applied over the weak subgrade to check the benefits of multi-layer geocell in pavement applications as compared with conventional single layer of geocell and unreinforced pavement sections. The analysis is done in Plaxis-3D with dynamic analysis. It is found that multi-layer geocell system performs better in terms of greater stress distribution to the subgrades and reduced settlement of the overall geo-structure.

2 Materials and Methodology

In this study, clay, sand (for infilling, and, as cushion between geocell layers), and geocell properties are adopted from the laboratory test performed by Biswas [9, 10]. The stiffness and modulus parameters of geocell–soil matrix are calculated according to Latha and Rajagopal [11]. The summarized material properties, including the base and asphaltic layer [12], used in the numerical simulation are presented in Table 1.

3 Numerical Simulation

The evaluation of performance of reinforced flexible pavement on clay subgrade, with special attention to multiple layer geocell configuration, is the objective of this study. A three-layered (subgrade, base, and asphalt layer) pavement section (Fig. 1) is considered for the purpose and studied for different configurations, such as, unreinforced (Fig. 1), with single-layer geocell (Fig. 2a) and multiple-layer geocells over the subgrade (Fig. 2b).

The present study has referred to the reported dataset of Saad et al. [12] for the base and asphaltic layer properties. Saad et al. [12] have reported the results of a series of FE analyses, through finite element program ADINA having an implicit solution scheme for flexible pavement. In the present study, the properties, such as

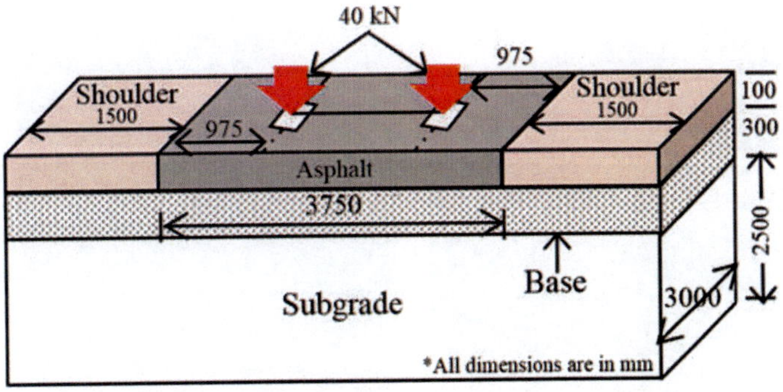

Fig. 1 Schematic diagram of the three-layer flexible pavement system considered

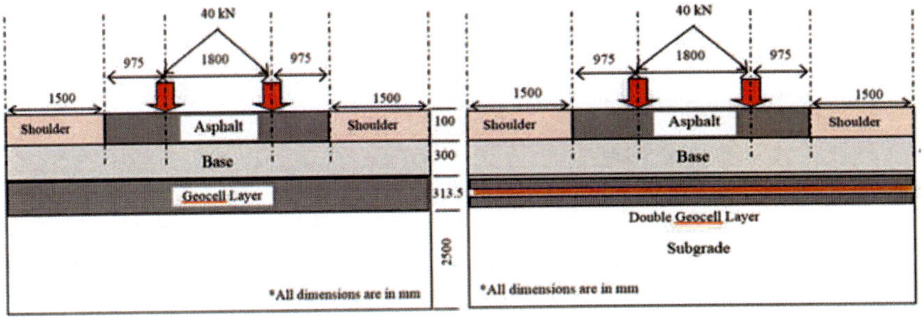

Fig. 2 Pavement structure with **a** single- and **b** double-layer geocell reinforcement

tire impact dimension (406×178 mm^2) and load–time relationship (a triangular 0.1 s load cycle with maximum 40 kN wheel load intensity at half of the cycle time) were validated in Plaxis. The result, as compared to Saad et al. [12], has satisfied the authors (Fig. 3) to adopt the parametric values for the present study.

It is referred that the Mohr–Coulomb and Drucker–Prager model are better in representing the soil nonlinearity. Hence, in the analysis, the subgrade and base are modeled using Mohr–Coulomb criteria; whereas, the asphalt layer is modeled as linear-elastic material. The Rayleigh coefficient, for considering the damping effect of soil and asphalt layers, is adopted as per Ju and Ni [13] Al-Qadi [14], and Hasheminejad et al. [15].

4 Result and Analysis

The study has compared the behavior of three different pavement sections as unreinforced, single layer geocell-sand and double-layer geocell-sand matrix overlying the weak clay subgrade. In the reinforced section, the subgrade was overlain by a 315 mm thick (single-layer configuration) and two numbers of 150 mm thick geocell

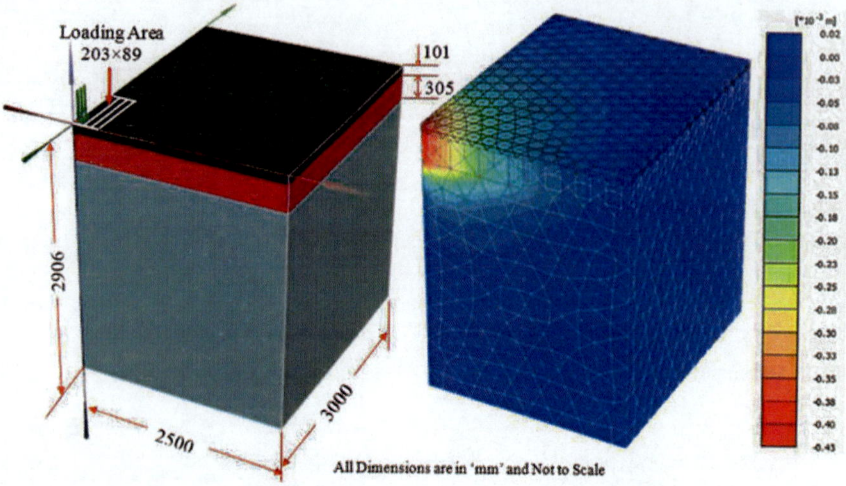

Fig. 3 Validation of Saad et al. [12] using Plaxis3D

mattresses (double-layer configuration with a 15-mm thick sand cushion in between layers) in-filled with sand [5, 10]. In all the cases, an axle load of 80 kN, with each wheel load of 40 kN, is applied. The subgrade clay strength (c_u) was kept 30 kPa for all the cases.

The unreinforced pavement section showed considerable surface deflection, about 2 mm, under the load. Such deformation has significantly reduced with single and double layer of geocells (about 1.2 and 0.7 mm respectively). As per the reinforced-section performance, about two-fold improvement in performance is achieved with double-layer configuration of similar material consumption (Table 2).

Typical pavement behavior with unreinforced and reinforced sections is presented in Fig. 4a–c. Responses demonstrate reduction in deformation with a greater load dispersion for reinforced sections. Besides, a reduced stress overlapping is also observed in case of higher degree of reinforcements. In addition, the reinforced pavements depicted considerable improvement in terms of dynamic time-displacement response and showed reduction in deflection under the load (Fig. 5).

The analysis indicated that response of multi-layer geocell reinforced pavement systems is better compared to the other two configurations (Fig. 5). A deflection profile, along the centerline of loads, presented in Fig. 6, depicting the effect of

Table 2 Improvement in surface deformation of different pavement sections

Sl. no.	Subgrade condition	Surface deflection (mm)	Improvement (%)
1	Unreinforced	1.97	–
2	Single layer geocell reinforcement	1.19	40
3	Double layer geocell reinforcement	0.73	63

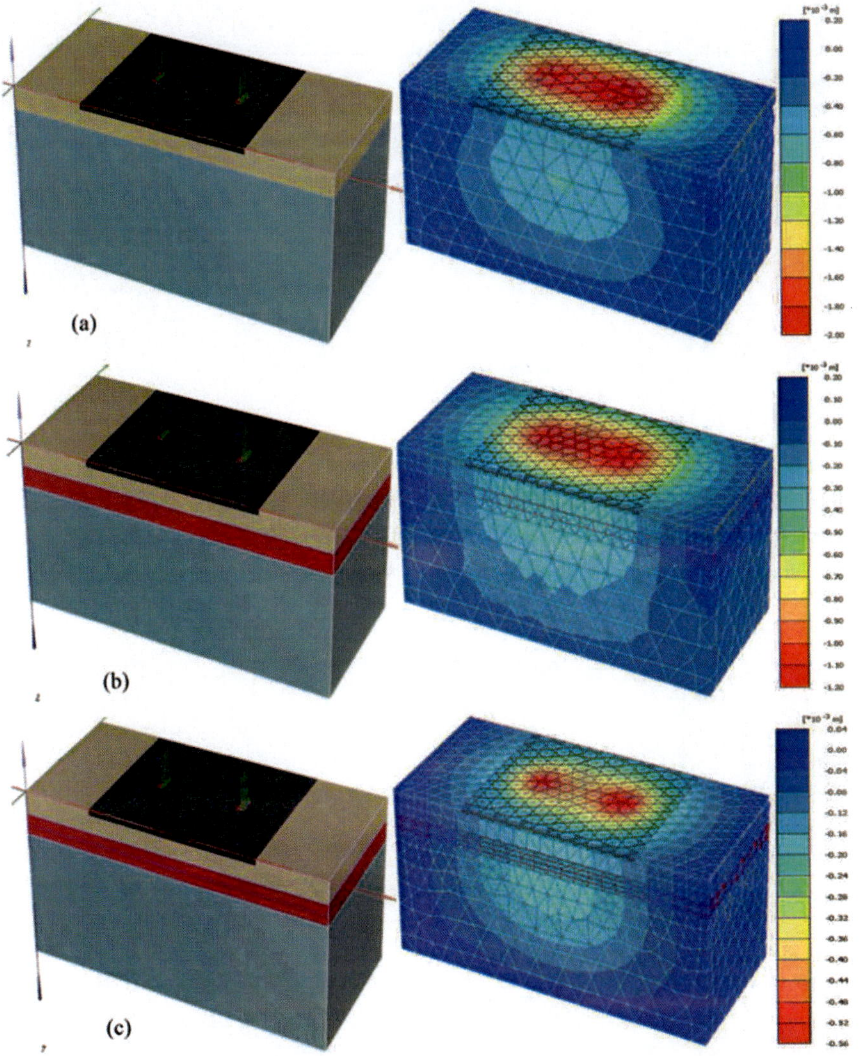

Fig. 4 Typical responses **a** unreinforced, **b** single-layer, and **c** double-layer geocell reinforced pavement sections

wheel-loads in the transverse direction of pavement sections. It reflects the influence of wheel load diminishes with the distance with a maximum deformation at and in between the wheels. Furthermore, it is also noticeable that, with geocells, the pavement sections between the wheels have behaved more elastically and showed deep beam actions compared to the unreinforced section.

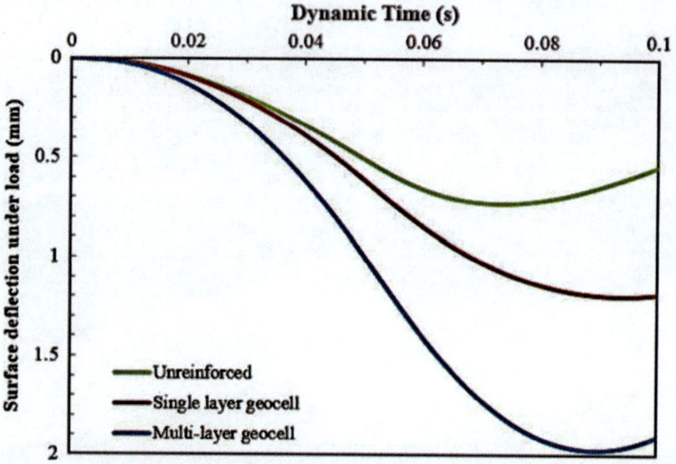

Fig. 5 Dynamic time vs. surface deflection under the wheel load

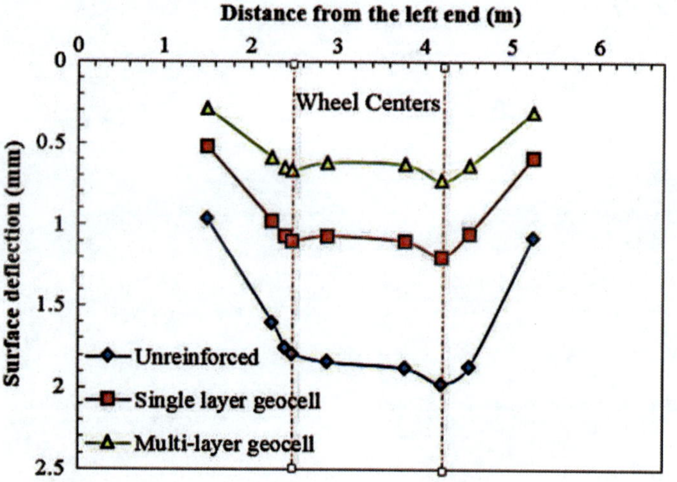

Fig. 6 Deflection at different locations of the pavement sections

5 Conclusion

The present study shows that flexible pavement confined with geocell layers gives considerably higher performance with geocell and enhances with the multi-layer of geocell system. It behaved much more elastically under the dynamic wheel load in terms of reduction and regain of deflections. This study also confirms the improved benefits in terms of reduction in material consumption when used in layers (geocell-mattress). The analytical outcome has encouraged and explored the scope of physical tests. The authors admit the limitation of the article with respect to validation, which was planned to be done through physical modeling in the laboratory; however, has not been possible yet due to unforeseen situation arose and restrictions imposed, therefore.

References

1. Sarkar, H., Biswas, A.: Application of geocell reinforcement in pavements-a brief review. In: CD Proceedings of Indian Conference on Geotechnical and Geo-Environmental Engineering (ICGGE-2019), pp. 1–6. MNNIT Allahabad, Prayagraj, India (2019)
2. Biswas, A., Krishna, A.M., Dash, S.K.: Influence of subgrade strength on the performance of geocell-reinforced foundation systems. Geosynth. Int. **20**(6), 376–388 (2013)
3. Biswas, A, Krishna, A.M., Dash, S.K.: Behavior of geosynthetic reinforced soil foundation systems supported on stiff clay subgrade. Int. J. Geomech. ASCE (2016). https://doi.org/10.1061/(ASCE)GM.1943-5622.0000559
4. Biswas, A., Krishna, A.M.: Geocell-reinforced foundation systems: a critical review. Int. J. Geosynth. Ground Eng. **3**(17), 1–18 (2017)
5. Sarkar, H., Biswas, A.: Behavior of multi-layer-geocell reinforced soil embankment. In: 7th Indian Young Geotechnical Engineers Conference (7IYGEC 2019), pp. 48–51. NIT Silchar, Assam, India (2019)
6. Sarkar, H., Biswas, A.: Response of multi-layered-stepped geocell reinforcement in soil structures. In: Proceedings of Indian Geotechnical Conference 2019, pp. 1761–1770. SVNIT Surat, India (2019)
7. Tafreshi, S.N.M., Khalaj, O., Dowson, A.R.: Pilot scale load tests of a combined multilayered geocell and rubber-reinforced foundation. Geosynth. Int. **20**(3), 143–161 (2013)
8. Li, L., Cui, F., Xiao, H.: Experimental study on the properties of geocell-reinforced embankments. In: China-Europe Conference on Geotechnical Engineering, pp. 1160–1163 (2018)
9. Biswas, A.: Influence of subsoil strength on performance of geosynthetic-reinforced foundations. Ph.D. Thesis submitted to IIT Guwahati, Assam, India (2016)
10. Biswas, A.: Comparative performance of different geosythetics on sandy soil overlying clay subgrade of varying strengths. Innov. Infrastruct. Solut. **4**(18), 1–16 (2019)
11. Latha, G.M., Rajagopal, K.: Parametric finite element analyses of geocell supported embankments. Can. Geotech. J. **44**, 917–927 (2007)
12. Saad, B., Mitri, H., Poorooshasb, H.: 3D FE analysis of flexible pavement with geosythetic reinforcement. J. Transp. Eng. **132**, 402–415 (2006)
13. Ju, S.H., Ni, S.H.: Determining Rayleigh damping parameters of soils for finite element analysis. Int. J. Numer. Anal. Meth. Geomech. **31**, 1239–1255 (2007)
14. Al-Qadi, I.L., Wang, H., Yoo, P.J., Dessouky, S.H.: Dynamic analysis and in-situ validation of perpetual pavement response to vehicular loading. Transp. Res. Rec. **2067**, 29–39 (2008)
15. Hasheminejad, N., Vuye, C., Bergh, W.V., Dircks, J., Leysen, J., Sels, S., Vanlanduit, S.: Identification of pavement material properties using vibration measurements. Proc. ISMA **2016**, 2217–2231 (2016)

Use of Geosynthetics as a Soft Structural Measure to Mitigate Flood Hazard and Bank Erosion Problem

Suresh Maurya, Manish Gupta, and R. Chitra

1 Introduction

Conventional structural measures like boulders and concrete revetments are often used on slope of river banks to prevent soil erosion. These armored systems, when placed directly on the soil, have reduced the magnitude of flood hazards to some extent. River water fluctuates between low water level and high flood level. This causes seepage in and out of the bank slopes. If the seepage is blocked by placing impermeable protection materials, it leads to the development of pore pressure, which further destabilizes the bank slope. Further seepage water also removes the fine soil particles, especially with receding flows. This leads to undermining the armored protection, leading to failure of bank slope. Similarly, armored protection is also required on the river bed close to the bank toe portion. The armored layer for protection of river bed also called as apron may consist of crated geotextiles bags or boulders in case of availability. During high flows, the discharge intensity increases, causing erosion of river bed even when it is protected. Fine particles may get removed between the voids of the armored layer. Hence, a properly designed armored layer and filter media are to be provided on the bank revetment, toe-portion and apron area as per IS 14262 [1].

Satisfying the filter and drainage criterion for conventional graded design is extremely expensive, often difficult to obtain, time-consuming to install and involves the problem of segregates during placement, thus compromising its filter ability. Many of the shortcomings of the traditional filter can be overcome using geosynthetics to perform both drainage and filtration. The purpose of geosynthetics filter is to allow water to flow through it while preventing fine soil particles from being washed away. The design of filter is based upon the particle size distribution of the

S. Maurya (✉) · M. Gupta · R. Chitra
CSMRS, Ministry of Jal Shakti, Department of WR, RD & GR, New Delhi 110016, India
e-mail: maurya_suresh@yahoo.co.in

C. N. V. Satyanarayana Reddy et al. (eds.), *Ground Improvement and Reinforced Soil Structures*, Lecture Notes in Civil Engineering 152,
https://doi.org/10.1007/978-981-16-1831-4_67

soil on the bank. Similarly, the armored layer/apron provided on river beds also needs filter below it to prevent the removal of soil fines of the bed materials.

The Central Soil and Materials Research Station, New Delhi has provided quality control support in evaluating geosynthetics material properties to numerous projects. Geosynthetics materials in the form of geobags, geotextile filter media, geomattress, geotubes, geobags filled PP rope gabions etc. have been implemented in various projects to mitigate flood hazard and reduce bank erosion, are discussed below.

2 Application of Geosynthetics in Managing Flood and Bank Erosion Problem Along Various Rivers

2.1 Restoration and Bank Protection Works Along the Rohmoria Reach of River Brahmaputra

Rohmoria area is about 20 km northeast of Dibrugarh town on the south bank of Brahmaputra in Dibrugarh district of Assam. The area has witnessed erosion for the last 60 years and more than 25 villages have been wiped out by erosion. During the period from 2009 to 2013, due to heavy floods and erosive forces of the flowing river, the bank line along the Rohmoria reach has shifted by as large as 400 m. A reach of approximately 9 km was identified as highly affected zone and bank erosion and flood protection measures were planned for a stretch of 2.6 km using non-woven geotextile filter materials and geobags (see Fig. 1).

The embankment construction was a part of bank protection while the launching apron and key were part of bed protection. This was followed by well-dressed bank slope at an inclination of IV: 3H. The height of slope was approximately 5.5 m. The bank and bed protection were carried with multiple layer of non-woven geotextile bags placed over non-woven geotextile filter layer along the bank length of

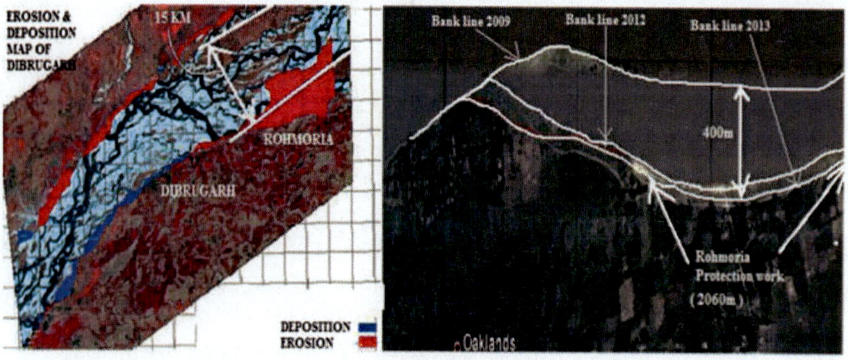

Fig. 1 Bank Erosion and Rohmoria protection work

Fig. 2 Placement of the filter media, Geobags and Polypropylene Rope Gabions

2600 m. Strips of steel gabions and polypropylene rope gabions filled with non-woven geotextile bags were placed at regular intervals to impart further stability to the scour protection measure. Figure 2 shows the placement of the filter media, Geobags and Polypropylene Rope Gabions [2].

2.2 *Engineered Anti-Erosion Works Along the Banks of River Brahmaputra in Sonitpur District, Assam*

The project is located in the Sonitpur District of Assam from Brahmaputra CH: 282 km (Silamari) to 373 km (Borgaon). The erosion in above said reaches so severe that the river bank line migrates 300–400 m each year. At present, the difference between the dyke and the river at Bisawnath–Panpur reach is only 35 m at some locations. The river has already touched the only road used for plying vehicles to N.H-52. In Borgaon reach, River has migrated more than 5 km on its R/B for a reach of about 5 km during the last 10–11 years.

The quick and effective preventive measure adopted by the Tezpur Water Resources Division was to implement Geosynthetics materials and Gabions under Flood Management Programme. The scheme protected Biswanath–Panpur reach including upstream Silamari and far downstream Bhumuraguri to Borgaon against erosion of the River Brahmaputra. Figure 3 shows the cross-section of launching apron; bank pitching and embankment work at CH: 10 m of Biswanath Reach [3].

Bank Revetment with Launching Apron. River bank of length 15,600 m is dressed to the inclination of 1 V:2H and over this, a layer of non-woven geotextile of 400 gsm is laid as filter media, anchored at upper end and lower ends junction of bank

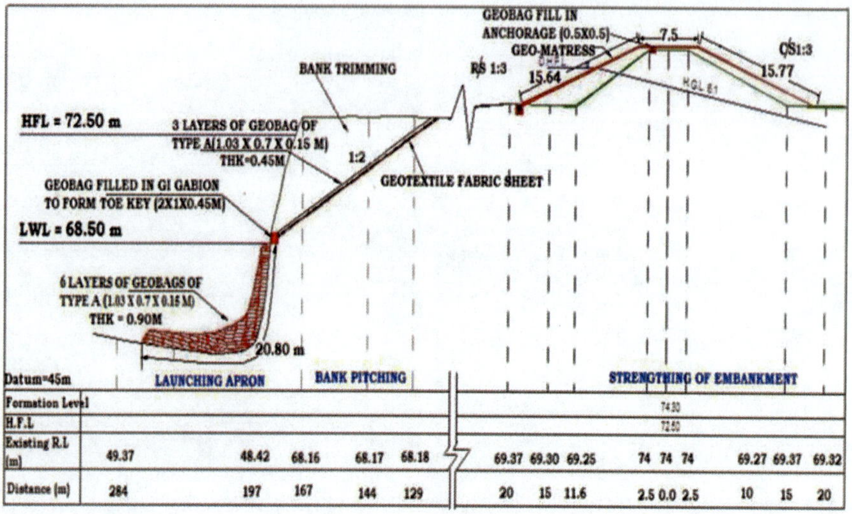

Fig. 3 Cross-section of apron, bank pitching and embankment work at Biswanath Reach

and apron to form a toe-key of gabion box filled with sand geotextile bags. After placement of geotextile filter, sand-filled geotextile bags of Type-A (size 1.03 × 0.70 m) made of non-woven geotextile are placed all along the length of bank. Width of the apron is kept 20.80 m and top height of the revetment is maintained with respect to HFL of 72.5 m. Thickness of slope pitching on bank is 0.45 m and launching apron is 0.90 m carried out with multiple layers of geotextile bags. Strips of zinc-coated wire mesh Gabion box (size 2 × 1 × 0.45 m) filled with geotextile bags are placed along the length in toe-key of the bank revetment and also at specified intervals across the length of launching apron to impart further stability to the scour protection measure. Total quantity of filter material applied is 1,67,650 m^2, geotextile bags for the protection works is approximately 38,08,147 Nos. Gabion box used in apron and toe-key is 49428 Nos.

Raising and Strengthening of the Embankment. It is carried out for the total length of 8758 m. Crest width is kept 7.50 m and top height is maintained at RL 74.30 m with respect to HFL of 72.50 m with freeboard of 1.80 m. Filling of earthwork is done in uniform layers not exceeding 22.50 cm thick with profiling to achieve a slope of 1 V:3H. Country side slope is protected by turfing with grass sods and river side slope is protected by a sand-filled geo-mattress. It is a double-layered composite geotextile fabricated to form a three-dimensional mattress after filling sand through pump at design slope of affected reach, the upper layer of the mattress is made from polypropylene woven geotextile needle-punched with a mixture of UV stabilized green fibers and cut tape yarns and the lower layer of the mattress is also a UV stabilized polypropylene woven fabric. Geo-mattress is anchored at upper and lower ends of embankment slope by bending the mat into the trench filled with sand-filled geotextile bags. Total quantity of geo-mattresses is approximately

Fig. 4 Implementation of Geotextile bags, Geo-mattress and PSC Porcupine works

161,737 m^2. Figure 4 shows the implementation of Geotextile bags, Geo-mattress and PSC Porcupine works.

2.3 *River Training Works Along the Banks of River Swan, Himachal Pradesh*

Swan River originates from Joh-Marwari Village near Daulatpur Chowk in Tehsil Amb of District Una and flows through the inter mountainous valley of Una District. Swan River is known as "Sorrow of District Una" as it creates havoc during monsoon almost every year due to floods. Approximately, 10,000 ha of agricultural land is affected by floods in Swan River and nearly 2000 ha of fertile land is not being cultivated due to fear of floods.

In Phase 1, the main Swan River has been provided with embankments on both banks from Jhalera Bridge (RD 19,160) to Santokhgarh Bridge (RD 2500) in a length of 16.67 km. The work of this phase was started in 2000 and was completed in 2009. In this phase, 2260 hectares of land has been reclaimed.

In Phase 2, the main Swan River has been provided with embankments on its both banks from Gagret Bridge (RD 47,500) to Jhalera Bridge (RD 19,160) in a length of 28.34 km. The work of this phase was started in 2008 and was completed in 2012. In this phase, 5000 hectares land has been reclaimed (Fig. 2). Channelization work is extended from downstream of Santokhgarh bridge up to Himachal boundary in Phase 3. Keeping in view encouraging result of Swan Channelization Phase 1 and Phase 2 the State Govt. decided to extend the Swan River Flood Management Programme in Phase 4 to cover balance reach of main Swan River from Daulatpur bridge (RD

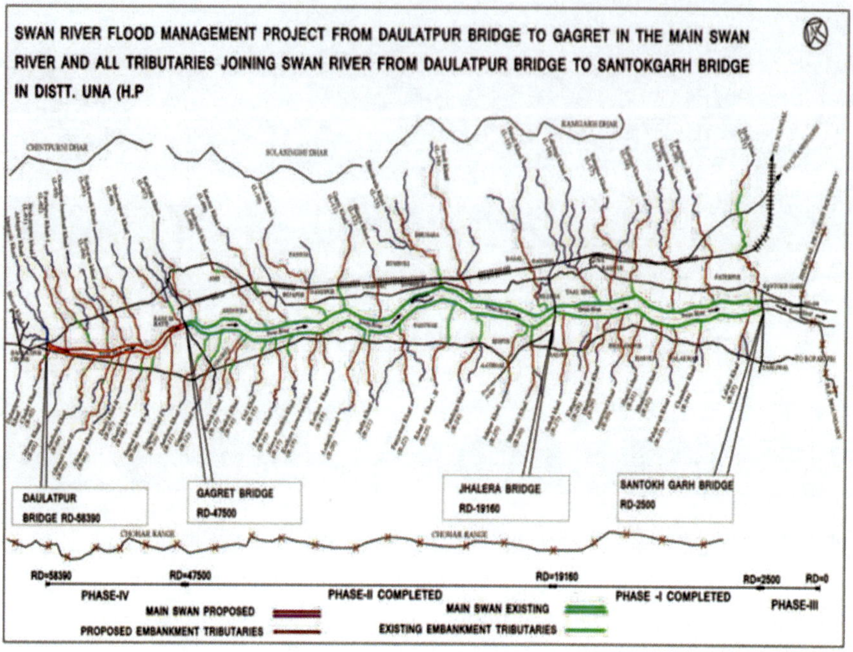

Fig. 5 Executed and proposed Swan River Flood Management Projects

58,390) to Gagret bridge (RD 47,500), total 10.89 km in main Swan River along with 55 number major tributaries, which are joining the main Swan River from Daulatpur bridge to Santokhgarh bridge so as to get the full benefits of the flood protection scheme. Figure 5 shows the executed and proposed Swan River Flood Management Projects [4].

Embankment and Apron. Low height heterogeneous type of embankment with Central clay core for seepage control is dressed to an inclination of 1 V:1H and local sand/river bed material in the outer core of embankment is dressed to a slope of 1 V:2H. The height of embankment is on the basis of HFL with freeboard of 1.5 m in Main Swan River and 0.75 m in tributaries. Figure 6 shows the completed embankment in Phase II.

Geotextile fabric filter forms an integral part of protection works. It is provided below the wire crated apron filled with stone pitching and below toe wall to prevent soil erosion. This filter material now replaces traditional granular material used in the past.

Fig. 6 Completed embankment in Phase II

2.4 Anti-Erosion and River Training Works Along the Banks of River Ranganadi

The River Ranganadi is an aggrading and meandering type of river by nature. As a result, it creates sand chores in one bank and causing erosion at the other bank. In addition to gradual rise of river bed, it is a matter of great concern that during every flood season, due to an increase in the inflow of water from the upper catchment areas, the surplus/excess discharge has to be allowed to spill over the Ranganadi Dam of the NEEPCO hydel power project, which ultimately increases the discharge at the downstream reaches. Due to combined effect of increase in the river discharge along with gradual rise in the river bed, flood lift occurs in the River Ranganadi within a very short period and threatens both the banks' embankment system. Every year, the rise in flood lift decreases the free board and causes overtopping. Now, for safety of the embankments against the anticipated flood lift, the embankment system at both the banks of River Ranganadi is raised by North Lakhimpur Water Resources Division under the Flood management scheme. It consists of raising and strengthening of embankment, bank revetment and launching apron. Figure 7 shows the installation of geotextile mattress at various stages [5].

Bank Revetment with Launching Apron. Construction is carried out for left and right bank of Ranganadi River at various vulnerable reaches for a total length of 7200 m where the river bank is dressed to the inclination of 1 V: 2H and over this a layer of geotextile tubular mattress of 0.3 m fill height is laid and anchored at the top and toe of bank slope by bending the mat into the key trench of size 1.0 m × 0.75 m. Geotextile mattress is a double-layered composite geotextile fabricated to form a three-dimensional mattress after filling sand through pump at design slope of affected reach. Total quantity of geo-mattresses applied is 1061939 m^2.

Launching of apron of size 9 m width and thickness 0.9 m all along the left and right bank is carried out with six layers of sand filled non-woven geotextile bags, which include two sets of three layers of sand-filled geotextile bags of Type A (size 1.03 m × 0.70 m) in one layer of gabion box (size 2 m × 1 m × 0.45 m). At the junction of the bank and apron, toe-key is formed from two layers of strips of zinc-coated

Fig. 7 Installation of Geotextile mattress at various stages

wire mesh gabion box (size 2 m × 1 m × 0.45 m) filled with a set of three layers of sand filled non-woven geotextile bags. Total quantity of non-woven geotextile bags used for the protection works is approximately 20,45,583 numbers and total quantity of gabion box applied is 82700 numbers. Figure 8 shows the cross-section of bank revetment and apron.

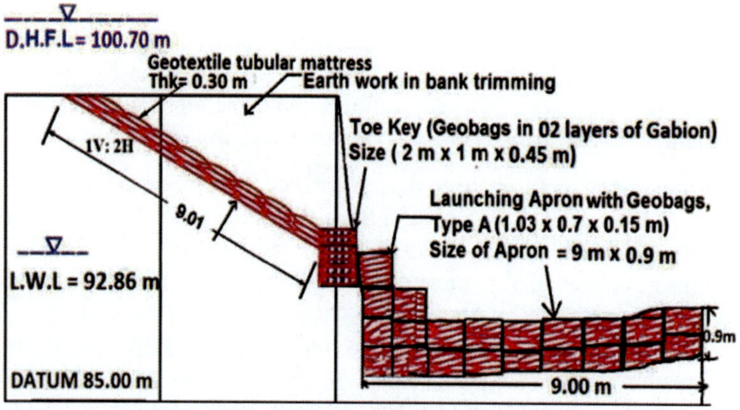

Fig. 8 Cross-section of bank revetment and apron

2.5 Use of Geobags for Restricting the Flow of River Dibang

The combined flow of Dibang and Lohit Rivers was originally falling into Dihang River at about 25 km downstream of Dibru-Saikhowa National Park at Kobo. Avulsion of left bank of River Lohit occurred at confluence of Lohit with Dibang near Dhola-Hatighuli, area of Tinsukia District in year 1989. As a result, River flow of Lohit diverted towards south. The breach was plugged in 1990 but erupted again. Initially, the breach developed in a length of 1 km in 1992 and then increased to 1.3 km in 1993. This problem was further aggravated due to pushing of flow of Lohit towards south by River Dibang flowing from the opposite direction (see Fig. 9).

Increased scarcity of available good quality stone aggregates and boulders led the Brahmaputra Board to implement flood management and erosion control works by using non-woven geotextile bags. The Tie-bund constructed, restricted the flow of River Dibang in its original course and avoided pushing of Lohit River toward bank and habitat area. Pitching of Tie-bund and launching apron with multiple layers of geotextile bags was a part of preventive measures. This was followed by well-dressed tie-bund at a stable inclined slope of 1 V:1H. The Tie-bund construction was a part of flood protection work while apron was part of bed protection work (see Fig. 10) [6].

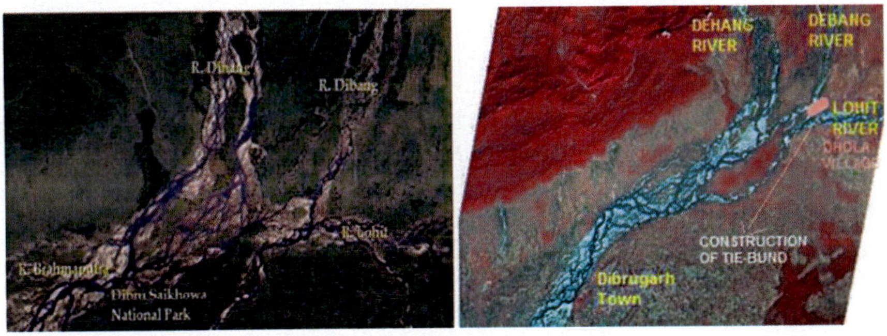

Fig. 9 Confluence of rivers and location of Tie-bund

Fig. 10 Pitching of Tie-bund and launching apron

Fig. 11 Anti-erosion works at Farakka Barrage Project, West Bengal

2.6 *Anti-Erosion Works at Farakka Barrage Along the Banks of River Ganga*

The River Ganga with its large mean discharge and a flat slope of about 1 in 20,000 near Farakka has been meandering through ages causing continuous bank erosion at different places. The Farakka Barrage across River Ganga is designed to serve the need of preservation and maintenance of Calcutta Port by improving the regime and navigability of river system. Jurisdiction of anti-erosion works has been extended 40 km u/s and 40 km d/s of barrage axis since 2004. The anti-erosion measures mainly involve laying of woven geotextile fabric filter under Tarza mat in water as well as on the land depending upon earth profile and over that boulder-filled G.I. wire crates of 1.6 m height (two layers of 1.0 and 0.6 m each) along 34 m apron is laid to arrest scour due to high discharge intensities. These anti-erosion works were very effective in inducing siltation in 2007 proving the efficacy of the design (see Fig. 11). Lot of anti-erosion works with this design have since then been executed at vulnerable reaches during past years. Typical c/s of anti-erosion works is shown in Fig. 12 [7].

2.7 *Flood Protection Works Along Left and Right Bank of River Sarada in District Lakhimpur Kheri*

Flood protection work has been taken up by Flood Division Sarda Nagar, UP to protect sensitive cluster of villages along left and right bank of river Sarda in district Lakhimpur Kheri at an estimated cost of Rs. 2504 lakhs in 2010–2011 [8]. In order to provide protection to the vulnerable reaches, high strength woven geotextile bags (Tensile strength 55kN/m) have been used in construction of Bank revetment and launching apron for a total length of 2410 m and PSC porcupines at 6o m c/c in three rows as an additional protection to the banks in order to reduce the velocity of flow. Such application is rapidly deployed to achieve maximum benefit to the community,

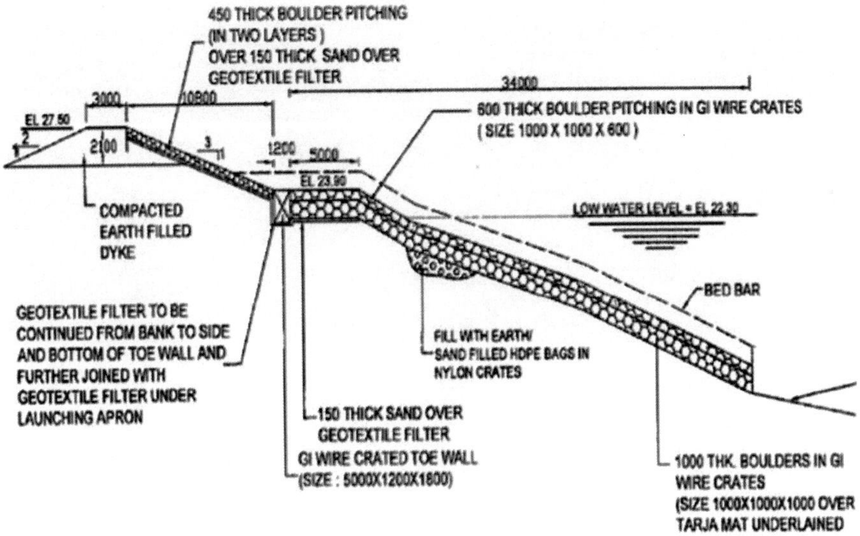

Fig. 12 Typical cross-section of anti-erosion works at Farakka Barrage Project

typically through the use of on-site river bed materials, innovative woven geotextile materials and construction techniques. This scheme has benefitted population and protected 140 ha of cultivated and homestead land including thickly populated villages, other public and private properties.

Bank Revetment with Launching Apron. River bank is dressed to the inclination of 1 V:2H and over this, a systematic pitching of sand-filled woven geotextile bags of 200 gsm is carried out all along the length of 2410 m. Thickness of slope pitching on bank is 1.0 m. Construction of launching apron is carried out with the same geotextile bags filled with local river bed material. Width of the apron is kept 8.0 m wide and 4.0 m in depth and top height of revetment is maintained with respect to HFL of 129.34 m. Thickness of slope pitching on bank and launching apron is carried out with multiple layers of geotextile bags in staggered manner to ensure stability of protection works. Figure 13 shows the implementation of geotextile bags in revetment and launching apron.

3 Conclusions

The paper presents the problems and the remedial works carried out along the vulnerable reaches using soft structural measures. Such application replaces all other conventional methods (e.g. Boulders, RCC etc.) for immediate protection where flood is a regular phenomenon and construction is to be completed in a limited time period. The use of geosynthetics materials permitted to carry out the protection works at a

Fig. 13 Implementation of geotextile bags in revetment and launching apron

faster rate. The use of the mechanically zinc-coated wire mesh Gabion box ensured the stability of the geotextile bags by providing peripheral confinement to the bank structures. Such an arrangement can be rapidly deployed to achieve maximum benefit to the community, typically through the use of on-site river bed materials, innovative geosynthetics materials and construction techniques.

Considering the advantages of geosynthetics materials, their use may rapidly increase in the future and the importance of material evaluation should therefore be emphasized to ensure that the geosynthetics materials and gabions meet the qualifying criteria.

Protection work increases the resistance of river banks to erosion and deflecting the current away. These generally shift the problem in the u/s or the d/s and necessitate further works to safeguard the land against erosion. Against this changing landscape, it is no longer acceptable for flood management practices to simply focus on reducing flooding and reducing the susceptibility to flood damage. Wider opportunities need to be considered to eliminate root causes.

Progressive damage to rivers and floodplains due to floods has increased substantially during the twentieth century. Climatic models also suggest that this trend will continue into the future [8]. Increased flood damage is associated with not only just increased precipitation but also with increasing population size and density and increased infrastructure residing within floodplains.

References

1. Ministry of Water Resources, River Development and Ganga Rejuvenation, Government of India: Guidelines for Use of Geotextiles/Geotextile bag/Geotextiles tubes in Construction of Flood Management Works (2016)
2. Maurya, S., Gupta, M., Chitra, R., Kumar, N.: Restoration and Bank Protection works along the Rohmoria reach of River Brahmaputra using Geosynthetics. In: International Symposium Geosynthetics India, pp. 159–168. Indian Chapter of IGS, New Delhi (2013)

3. Maurya, S., Gupta, M., Chitra, R.: Engineered anti erosion works along the Banks of Brahmaputra River in Sonitpur District, Assam. In: 6th Asian Regional Conference 2016, Publication no. 333, pp. 124. Indian Chapter of IGS and CBIP, New Delhi (2016)
4. Maurya, S., Gupta, M., Chitra, R.: Flood protection works along the Banks of River Swan, Himachal Pradesh. In: 50th Indian Geotechnical Conference 2015, pp. 332. IGS Pune Chapter, Pune (2015)
5. Maurya, S., Gupta, M., Chitra, R.: Anti-erosion and river training works along the Banks of Ranganadi River, Assam. Int. J. Eng. Appl. Sci. Technol. **2**(10)(03), 11–16 (2018)
6. Maurya, S., Kumar, N., Gupta, M.: Use of Geobags for restricting the flow of river Dibang, a tributary of River Brahmaputra. In: Indian Geotechnical Conference 2013, sr. no. 284, pp. 229. IGS Roorkee Chapter, Roorkee (2013)
7. Gupta, M., Chitra, R., Ratnam, M.: Geosynthetics in river bank erosion control. In: Advances in Geosynthetics, sr. no. 25, pp. 383–394. SAGES, Hyderabad (2012)
8. Milly, P.C.D., Wetherald, R.T., Dunne, K.A., Delworth, T.L.: 2002: Increasing risk of great floods in a changing climate. Nature **415**, 514–517 (2002)

Sustainability Study on Geosynthetic Reinforced Retaining Wall Construction

Sateesh Pisini, Swetha Thammadi, and Sanjay Shukla

1 Introduction

Geotechnical engineering being the most resource-intensive due to its involvement in the early stages of a project potentially influences the sustainability of all civil engineering projects. It is generally unnoticed because the energy used is indirect in nature, that is, in the form of materials and natural resources (e.g., land use, steel and concrete) as reported by Basu and Misra [1]. Resources utilized in the process are accounted for by the thermodynamics-based energy accounting methods of exergy, emergy and embodied energy. Exergy accounting methods are mostly used in procedures involving chemical reactions in industrial manufacturing. Emergy is used as an environmental engineering accounting instrument. Embodied energy is basically heat energy. Materials that are low in embodied energy should be used for a sustainable process. Environmental sustainability in geotechnical construction is often equated to resource efficiency parameterized by the embodied energy or embodied CO_2 of the materials used in a project by Chau et al. [2].

In this paper, the study is conducted on two retaining walls—conventional reinforced cement concrete cantilever retaining wall and geotextile-reinforced retaining wall. Conventional retaining walls are abundantly seen in the field. Construction of geosynthetic-reinforced retaining walls has become a common practice globally in the past few decades. They have been proven to be a cost-effective solution to traditional retaining wall construction as mentioned by Bathrust [3]. Woven

S. Pisini (✉) · S. Thammadi · S. Shukla
Department of Civil Engineering, Fiji National University, Suva, Fiji
e-mail: sateesh.pisini@fnu.ac.fj

S. Thammadi
e-mail: swetha.thammadi@fnu.ac.fj

S. Shukla
e-mail: s.shukla@fnu.ac.fj

C. N. V. Satyanarayana Reddy et al. (eds.), *Ground Improvement and Reinforced Soil Structures*, Lecture Notes in Civil Engineering 152,
https://doi.org/10.1007/978-981-16-1831-4_68

geotextiles and geogrids are commonly used as components of soil reinforcement in geosynthetic-reinforced retaining walls, also known as mechanically stabilized earth (MSE) walls. The geosynthetic layers are included in the backfill at designed vertical spacing. The backfill is compacted at the desired density. Compared to conventional concrete retaining walls, the construction of geotextile-reinforced retaining walls is simple and cost-effective. The performance of a geosynthetic-reinforced wall depends heavily on how the facing components are designed and built. The facing components can be mounted, as the wall is being built or after it is built. Shukla [4] specified that the various kinds of face panels include geosynthetic wraps, pre-cast segmental reinforced concrete panels, segments/modular concrete blocks (MCBs), full-height panels of pre-cast concrete, wire mesh panels and gabion baskets. A wraparound faced retaining wall has been chosen for the purpose of this study. The resource indicator is used for establishing the sustainable option of choice of best-retaining wall in terms exergy, emergy and embodied energy.

2 Case Study

A retaining wall with silty sand base material and dense gravel backfill material has been considered for this study. The retaining wall design with all details for in situ base soil, backfill, masonry, concrete, reinforcement and loading details are shown in Tables 1 and 2 as well as Figs. 1 and 2. Alternatively, a geotextile wraparound-faced retaining wall design with the reinforcement length, lap length and spacing of geotextile has been proposed. Exergy, emergy and embodied energy techniques of energy accounting are used to model the energy flow process (Tables 3 and 4).

Study on several types of retaining walls shows that the supported structures use less embodied energy throughout the models than the cantilever structures and that the material energy occupies the biggest proportion of energy consumption within the design developed by Chau et al. [2]. Although the amount of steel used was much lower than that of concrete, steel was the dominant material energy contributor. They have proposed that the use of recycled steel would reduce the consumption of material energy. It has been reported that the embodied energy consumption can be used as an indicator of environmental impact, although other parameters such as

Table 1 Retained soil properties

Soil type	Dense gravel
Soil condition	In situ
Moist density	17.5 kN/m^3
Saturated density	20.8 kN/m^3
Effective internal friction angle	36°
External wall friction angle	18°

Table 2 Base soil properties

Soil type	Stiff clay
Soil conditions	In situ
Soil density	19 kN/m^3
Effective cohesion	30 kN/m^2
Effective internal friction angle	18°
External wall friction angle	9°
External base friction angle	12°
Ultimate design bearing capacity	150 kN/m^2
Loading details	
Live surcharge load	5 kN/m^2

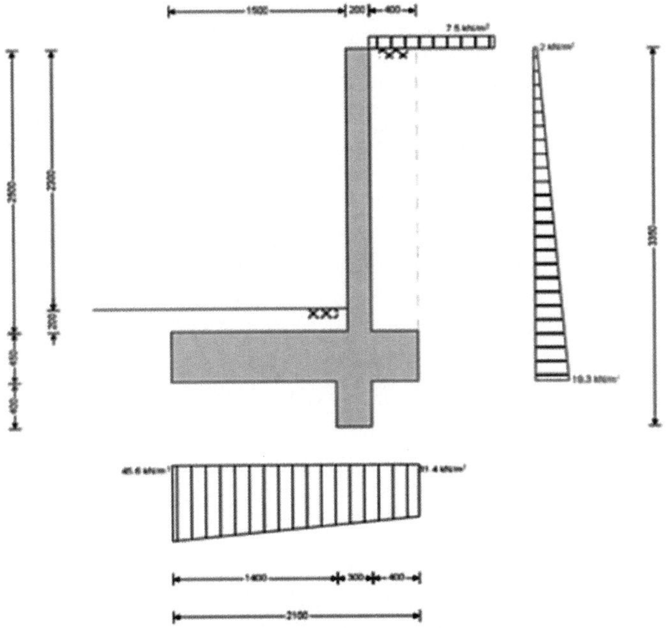

Fig. 1 Conventional cantilever retaining wall

carbon dioxide emissions should also be studied for an extensive assessment. In this research, an attempt has been made to compare the conventional retaining wall with geotextile-reinforced retaining wall as stated in Table 5.

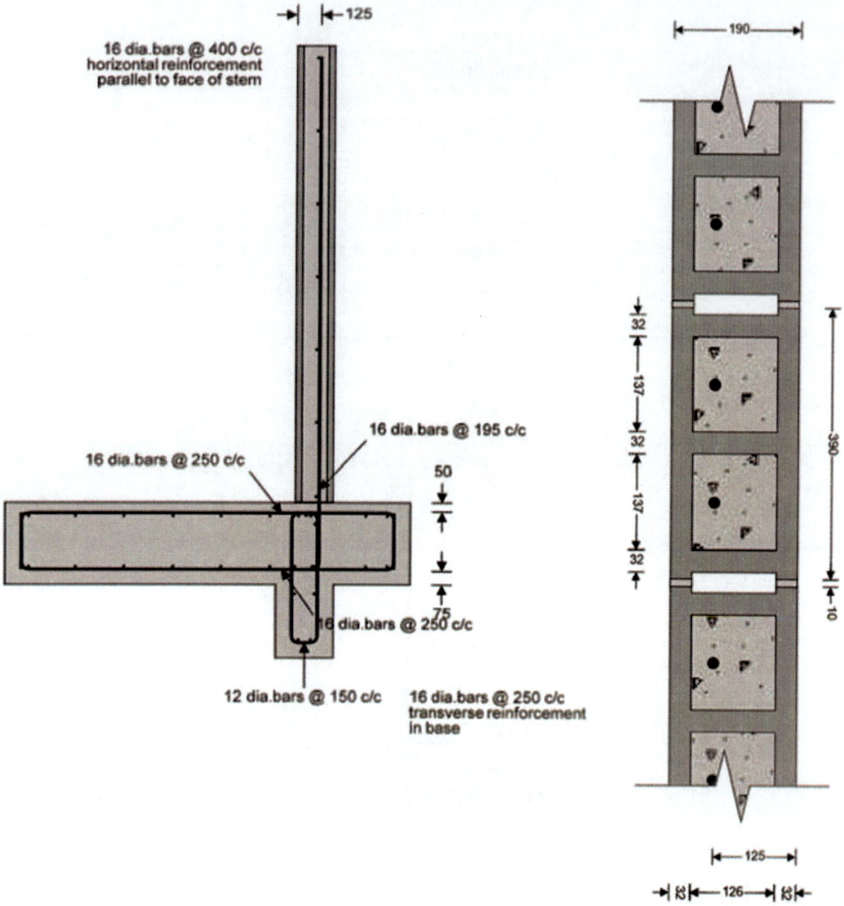

Fig. 2 Cross-sectional details of reinforcement and masonry

2.1 Design Details of Cantilever Retaining Wall

The present case study is a comparison of cantilever retaining wall (CRW) with that of geotextile-reinforced retaining wall with wrap-around facing. The necessary calculations for design of CRW have been made as per the AS 4678-2002 [5]. The geometric details of conventional retaining wall are mentioned in Fig. 1.

The retained and base soil properties, loading details of CRW are well detailed in Tables 1 and 2. The properties of concrete and reinforcement are mentioned in Fig. 2 based on specifications as per AS 3600-2009 [6] and AS 3700 [7] published by Standards Australia.

Table 3 Properties of granular backfill and design parameters of GRRW

Total unit weight (γ_b)	17.5 kN/m^2
Angle of internal friction (ϕ'_b)	36°
Effective angle of soil–reinforcement interface (ϕ'_i)	$\frac{2}{3}\phi'_b$
Allowable tensile strength (σ_{all})	20 kN/m
Factor of safety against geotextile rupture (F_R)	1.5
Factor of safety against geotextile pullout (F_P)	1.5
Coefficient of active earth pressure (K_a)	0.259
Spacing of geotextile layers (S_v)	$\frac{\sigma_{\text{all}}}{(K_a\gamma_b z)F_R}$
Effective length of geotextile layer (L_e)	$\frac{S_v K_a F_p}{2\tan(\phi'_i)}$
Length of reinforcement layer (L_r)	$\frac{H-z}{\tan(45+\phi'_b/2)}$
Lap length (L_l)	$\frac{S_v K_a F_p}{4\tan(\phi'_i)}$

Table 4 Properties of geotextile layers for this case study

Layer no.	Depth, z (m)	σ_v (kN/m^2)	σ_h ($= K_a \cdot \sigma_v$) (kN/m^2)	S_v (m)	L_e (m)	L_r (m)	L ($= L_e + L_r$) (m)	L_l (m)
1	0.5	8.75	2.275	5.86	2.57	1.02	3.59	1.28
2	1	17.5	4.55	2.93	1.28	0.77	2.05	0.64
3	1.5	26.25	6.825	1.95	0.86	0.51	1.37	0.43
4	2	35	9.1	1.47	0.64	0.26	0.90	0.32
5	2.5	43.75	11.375	1.17	0.51	0.00	0.51	0.26

2.2 *Design of Geotextile-Reinforced Retaining Wall*

The method for design of wraparound-faced geotextile-reinforced retaining wall (GRRW) as explained by Shukla [4] has been adapted in the current study with the details on properties of granular material and design parameters of GRRW as in Table 3 along with formulae used for design of GRRW. Figure 3 represents the schematic sketch of geotextile-reinforced retaining wall with wrap-around facing. On the other hand, Table 4 shows the calculations for properties of geotextile layers for this case study. The design of GRRW has been based on analysis for internal stability, external stability and analysis for the facing system. The actual height of the conventional retaining wall is 2.5 m from the case study.

Further, keeping the field aspects and construction simplicity in view, it is recommended to use the spacing of geotextile layers (S_v) = 0.5 m, length of the geotextile layers (L) = 1 m and lap length (L_l) = 1 m for $z \leq 4$ m. In general, a minimum

Table 5 Resource use indicator calculations for CRW and GRRW

Resource consumption calculation for conventional retaining wall

Sl No.	Materials	Volume (m^3)	Density (Kg/m^3)	Mass (Kg)	Emergy		Embodied energy		Cumulative exergy	
					Emergy intensity (× 1011) (sej/Kg)	Total emergy (× 1011) (sej)	Embodied energy intensity (MJ/Kg)	Total embodied energy (MJ)	Unit exergy (MJ/Kg)	Total exergy (MJ)
		1	2	3 = 2 * 1	4	5 = 3 * 4	6	7 = 6 * 3	8	9 = 8 * 3
1	Soil	1.785	1784.5	3185.3325	28	89189.31	0.45	1433.3996	0.02	63.70665
2	Cement	355	2400	852,000	19.7	16,784,400	4.6	3,919,200	5.35	4,558,200
3	Steel	0.00733	7850	57.5405	41.3	2376.42265	36.4	2094.4742	41	2359.1605
Total						16,875,966		3,922,728		4,560,623

Resource consumption calculation for geotextile-reinforced retaining wall with_wrap around face

Sl no.	Materials	Surface area (m^2)	Surface density (g/m^2)	Mass (kg)	Emergy		Embodied energy		Cumulative exergy	
					Emergy intensity (× 1011) (sej/kg)	Total emergy (× 1011) (sej)	Embodied energy intensity (MJ/kg)	Total embodied energy (MJ)	Unit exergy (MJ/kg)	Total exergy (MJ)
1	Soil	1.785	1784.5	3185.3325	28	89189.31	0.45	1433.3996	0.02	63.70665
4	Plastic (PVC)	0.0025	400	0.001	58.5	0.0585	77.2	0.0772	67	0.067
Total						89,190		1433		64

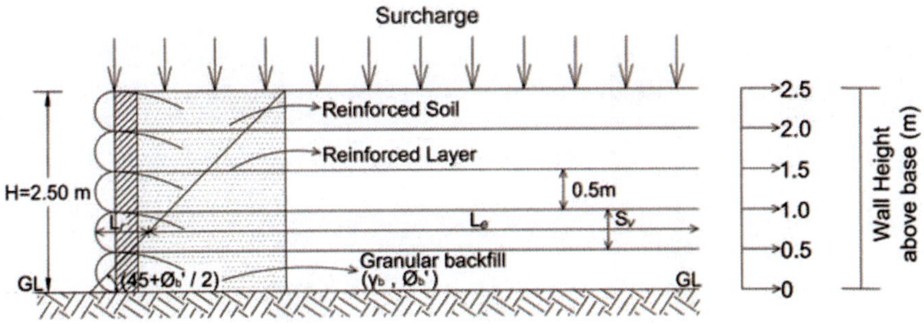

Fig. 3 Schematic sketch of GRRW with wrap-around facing

reinforcement length of 2.4 m, regardless of the wall height, has been recommended, predominantly due to limitations in size during conventional spreading as well as for compaction equipment. Therefore, in this case, the reinforcement length of 2.5 m has been considered for resource consumption calculation for GRRW.

3 Methodology

The calculations for land use are based on the volume of soil excavated for the construction of retaining wall using parameters mentioned in Fig. 1. Volume of cement and steel has been calculated based on the reinforcement details in Fig. 2.

The Exergy values of unit emergy for cement and steel have been adopted from Brown and Buranakaran [8] and Pulselli et al. [9] while the values of unit emergy for land are used from the emergy folios of Odum [10]. The embodied energy values per unit mass are adopted from ICE Database version 1.6a [11]. The exergy values of cement and steel used in the calculations are the same as those used by Berthiume and Bouchard [12] and are originally based on the values calculated by Szargut et al. [13]. Table 5 shows the resource indicator calculation for conventional retaining wall and geotextile-reinforced retaining wall with wraparound face. Also, Fig. 4 illustrates the percentage projection of emergy/embodied energy/Exergy to total resource consumption (emergy + embodied energy + exergy) of CRW and GRRW, respectively.

Although the calculations for exergy and emergy have been done, conventionally embodied energy consumption is used to represent energy use of the materials chosen. The reason being that LCA of buildings and building materials is traditionally done using embodied energy as stated by Chau et al. [2] and Storesund et al. [14].

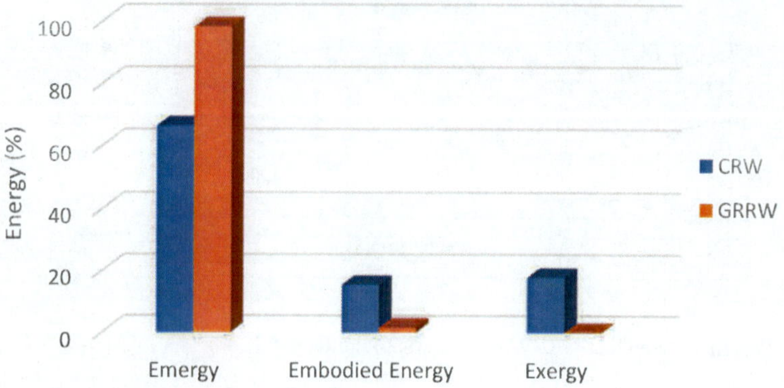

Fig. 4 Resource consumption of emergy, embodied energy and exergy to total energy consumption for CRW and GRRW

4 Conclusions

Considering the design of the proposed alternatives and results of this research, the conclusions of this research are summarized below:

1. The percentage contribution of emergy is greater in the case of GRRW compared to CRW. However, it is to be noted that emergy, embodied energy and exergy values for GRRW are negligible when compared to that of CRW.
2. The results of the resource consumption calculation show that GRRW is more resource-use efficient compared to CRW.
3. Height of retaining wall considered for this study is 2.5 m whereas, if the height of the wall is more than 4 m, then the compilation of results would allow GRRW to be substantially resource-efficient and highly economical as compared to CRW.
4. Calculation of the resource consumption would be effective when comparing similar materials for the best alternative among available options of design and construction of retaining walls.

References

1. Basu, D., Misra, A.: A quantitative sustainability indicator system for pile foundations. In: Geo Congress, American Society of Civil Engineers (2012)
2. Chau, C., Soga, K., Nicholson, D., O' Riordan, N., Inui, T.: Embodied energy as environmental impact indicator for basement walls. In: Geo Congress, Geotechnical Special Publication No. 178, pp. 867–874 (2008)
3. Bathurst, J.R.: Reinforced soil retaining wall testing, modeling and design. Proc. of Indian Geotechnical Conference, Chennai, India, December (2006)
4. Shukla, S.K.: An Introduction to Geosynthetic Engineering. Taylor and Francis, London (2016)

5. AS4678: Australian Standard of Earth-Retaining Structures. Standards Australia, Sydney (2002)
6. AS3600: Australian Standard for Design and Construction of Concrete Building Structures. Standards Australia, Sydney (2009)
7. AS3700: Australian Standard for Masonry Structures. Standards Australia, Sydney (2011)
8. Brown, M.T., Buranakaran, V.: Emergy indices and ratios for sustainable material cycle and recycle options. Resour. Conserv. Recycl. **38**, 1–22 (2003)
9. Pulselli, R.M., Simoncini, E., Pulselli, F.M., Bastianoni, S.: Emergy analysis of building manufacturing, maintenance and use: em-building indices to evaluate housing sustainability. Energy Build. **39**, 620–628 (2007)
10. Odum H.T., Brown M.T., William-Brandt S.: Handbook of emergy evaluation, Folio #1. Introduction and global Budget (2000)
11. Hammond, G., Jones C.: Inventory of Carbon and Energy (ICE) Version 1.6b. Carbon Vision Building Program, University of Bath, UK (2009)
12. Berthiaume, R., Bouchard, C.: Exergy analysis of the environmental impact of paving material manufacture. Trans. Can. Soc. Mech. Eng. **23**(1B), 187–196 (1999). https://doi.org/10.1139/tcsme-1999-0013
13. Szargut, J., Morris, D.R., Stewart, F.R.: Exergy Analysis of Thermal, Chemical and Metallurgical Processes. Hemishere Publishing Company (1988)
14. Storesund, R., Messe, J., Kim, Y.: Life cycle impacts for concrete retaining walls vs. bioengineered slopes. In Proceedings of the Geo Congress 2008, Geotechnical Special Publication No. 178, pp. 875–882 (2008)

Geocomposite Drain for Capillary Cut-Off and Horizontal Subsurface Drainage in High Altitude Roads in Uttarakhand—A Case Study

D. Roychowdhury

1 General

Geosynthetics has found its way into the design and construction of road infrastructure in a big way in recent years. These materials come in different forms, such as geotextiles, geogrids, geonets, geomembranes and geocomposites, each serving a specific function. Geotextiles, also known as filter fabrics, find wide and distinct uses in geotechnical engineering applications in

(i) Separation of dissimilar materials placed contiguously to avoid intermixing
(ii) Reinforcement of weak soils (or other load-bearing materials like fly ash, slag, etc.)
(iii) Filtration, where the water flows across the textile and works to retain the fines
(iv) Drainage, where the water flows within the geotextile (in-plane) or in cross-plane direction

In most cases, the geotextile performs more than one function. Only the textile is to be designed for each application separately considering the engineering requirements.

2 Introduction

Geocomposites are manufactured as a combination of any two or more synthetic materials like geotextiles, geogrids, geonets, geomembranes, etc. in laminated or composite form, to achieve desired engineering properties. One of the more popular and extensively used geocomposite is a drainage geocomposite, which is formed

D. Roychowdhury (✉)
Central Public Works Department, Nizam Palace, Kolkata 700020, India

C. N. V. Satyanarayana Reddy et al. (eds.), *Ground Improvement and Reinforced Soil Structures*, Lecture Notes in Civil Engineering 152,
https://doi.org/10.1007/978-981-16-1831-4_69

by combining a geotextile with a drainage core in the form of extruded geonet, or cuspated sheets and yield much higher flow properties than even very thick geotextiles.

In the present context, we are essentially looking at the capillary break and drainage function of the geocomposite material. The drainage function can be achieved by using plain geotextiles for low volumes of flow water. For higher water flow conditions, different types of drainage geocomposites are better suited. The drainage core provides a large in-plane flow capacity whereas the geotextile provides separation, filtration, some drainage and capillary break actions. The capillary break is provided by draining out the capillary water through the drainage core. The selection of geocomposites depends on site-specific soil and moisture conditions, availability and cost considerations.

Flow performance of geocomposites is found to deteriorate over time due to:

(i) Elastic deformation of the adjacent geotextile intruding into the drainage core space
(ii) Creep deformation of the drainage core itself and/or creep deformation of the adjacent geotextile intruding into the drainage core space
(iii) Chemical clogging of the geotextile and/or drainage core
(iv) Biological clogging of the geotextile and/or drainage core

Usually, manufacturers provide reduction factors for different range of products and different application areas so as to provide allowable flow rate to be used in the design. The short-term flow rate is determined from short-term tests as per relevant codes.

3 Subsurface Drainage Requirement in Flexible Pavements

It is well known that the ingress of water into the pavement crust is one of the major causes of distress in flexible pavements. Water in the asphalt surface leads to loss of tensile strength, stripping of bitumen from aggregate and reduction of stiffness modulus of the order of 30%. Moisture in the unbounded base and sub-base layers leads to loss of stiffness of around 50%, thereby causing large deformation under load in flexible pavements. It causes erosion of fine aggregates/soil, erosion of shoulders, and eventually ruts and pot-holes [9].

In hill roads in cutting, the source of water can be from rainfall, snow melt, seepage of water from hill side, waterfalls, capillary rise of water from the soil subgrade, etc. Water enters the crust through cracked/rutted surface, joints and edges, pervious shoulders, pervious/damaged side drains, etc. Surface drainage as well as subsurface drainage is required for keeping the structural elements of the pavement in well-drained conditions—the latter having been mostly neglected in practice. Subsurface drainage requirement is both vertical (behind retaining and breast walls) as well as horizontal (open-graded granular sub-base). However, with the advent of geotextiles and geocomposites, good quality vertical as well as horizontal subsurface drainage options are available.

In case of hill roads in the Himalayan region, many situations are seen where the phreatic line comes out of the cut slope surface, leading to seeping surfaces. This situation is more prolific in the summer seasons when the snow melts in the upper reaches and otherwise dry hillsides start seeping water. Perforated pipe drains in the hillsides [7] may not be feasible due to the presence of boulders, or fractured rocks with irregular drainage paths.

4 Conventional Subsurface Drainage Systems

A typical flexible pavement consists of an asphaltic layer, a base layer of Wet Mix Macadam or Crusher Run Macadam, a granular sub-base (GSB) layer for drainage and the available or compacted subgrade. In case of hill roads in cutting, the subgrade available is part of a hill slope and has to be checked for stability. In case of high altitude roads, a capillary barrier is also provided in the form of a sand layer, to ameliorate the effects of freeze-thaw action in frost susceptible soils [10]. The GSB layer is provided in two parts—one upper sub-base (drainage layer) and a lower sub-base (separation layer). The drainage layer (Grading 2 of IRC 37) has a typical permeability of 15–45 m/day [9].

One of the major problems faced by field engineers in use of conventional GSB drainage layers, especially in hill roads are poor quality control of GSB material and its placement, blockage of longitudinal drains and cross drainage facilities by debris leading to pooling of water over the pavement, large variations in discharge from streams leading to design failure, etc. Furthermore, high altitude hill roads also suffer from additional factors like—

- Steep slopes in catchment area leading to sudden floods,
- Snowfall accumulating on pavement surface and its clearing operations leading to severe damage to top bituminous layers,
- Diurnal freeze-thaw on road surface creates cracks on road surface,
- Seasonal waterfalls damaging the pavement structure as well as spraying the shoulders and road surface continuously during monsoons.

5 Geocomposite Sheet Drains for Subsurface Drainage

A geocomposite sheet drain having sufficient in plane flow capacity is an excellent drainage alternative, as being a manufactured item, variations in quality and engineering properties are minimal. Sheet drains do not suffer from any day lighting problems (less UV stability may result in loss of material at the exposed edge over time, which is inconsequential). It works as a separation and filtration material as

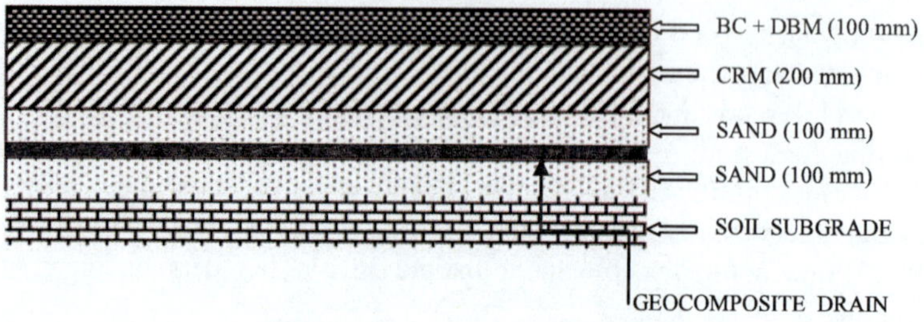

Fig. 1 A typical section showing placement of geocomposite drain

well as a capillary break. The placement of GCD is usually over the soil subgrade [4] (Fig. 1).

The requirements of a good horizontal geocomposite drain (GCD) are

(i) Sufficient stiffness to support traffic without significant deformation under dynamic loading, (ii) Inflow capacity greater than infiltration from adjacent layers, (iii) Sufficient transmissivity to rapidly drain the pavement section and prevent saturation of the base, (iv) Sufficient air voids within the geocomposite to provide a capillary break, (v) Prevent fines from the subgrade or sub-base to enter the drainage core.

For the GCD to act as a capillary barrier, it is important that it has a capacity for drainage that leaves an air gap at the top of the drainage core even with traffic load [2]. In hill road sections where peak flow is much higher than the average flow, it is not always possible to provide such high-capacity subsurface drains. In such cases, the sand layer provided below the GCD also helps as a capillary break.

5.1 Design Methodology

The design of a GCD is done to evaluate the flow requirement of the drainage system and deciding on a suitable GCD whose allowable rate is such that a suitable FOS is achieved with reference to the required flow rate to provide effective drainage. The Geosynthetic Research Institute (GRI) Standard GC8 [5] methodology has been followed to arrive at the factor of safety [4]. The design process consists of:-

Determination of the required drainage capacity of the system (q_{reqd})—The longitudinal and cross drainage provided in hill roads take away most of the stormwater flow. Only a fraction of the total discharge is required to be carried by the subsurface drainage system, and that too, at specific vulnerable locations [9, 10]. In a hill road, the water entering the subgrade may come from adjacent hill faces, rainfall in the catchment area, percolation of water through the wearing surface, capillary rise from the water table, or from snow melt. It is only possible to identify locations and estimate the flow from site observations.

Short-term or basic flow capacity (100 h testing condition is required by GRI GC 8) of the GCD (q_{basic})—It depends on the type of GCD core chosen (e.g. bi-planar/triplanar geonet, geomat, cuspated core, etc.), hydraulic gradient, normal stress on the GCD, testing boundary conditions (soft/soft contact, rigid/soft contact), test seating time, etc. The short term or basic flow rates of the GCD can be determined in accordance with ISO 12958—2010 (or ASTM D4716). However, most of the time design is done as per technical data sheets provided by the reputed manufacturers. The GCD used in this pilot project is realized by thermo-bonding an extruded monofilament drainage core, with two similar non-woven geotextiles on both sides, to act as filtering/separation layers. The 3-D draining core has longitudinal parallel channels. The technical details and calculations are given in subsequent sections.

Estimation of Normal Stress on the GCD

The normal stress on the GCD is essentially due to tyre pressure of traffic. IRC 37:2012 recommends that for stress analysis, a surface pressure of 0.56 MPa with point load of 40kN be adopted for pavement design [7].

$$P = 40,000\,N\,q = 0.56\ \text{N}/\text{mm}^2$$

But $q = P/\pi a^2$ Hence $a = \sqrt{}(P/\pi q)$ We get $a = 151$ mm say 150 mm.

The stress levels at the top of subgrade level are worked out based on Burmister's two layer theory [6]. Here each layer is assumed to be homogeneous, isotropic and linearly elastic with an Elastic Modulus (E) and Poisson Ratio (ν). The top layer has a finite thickness and the bottom layer has infinite depth. The stresses in a two layer system depend on the Elastic Modulus Ratio (E_1/E_2) and the thickness to radius ratio (h_l/a); where (Fig. 2)

h_1 = thickness of top layer a = radius of surface pressure,

q = surface pressure and σ_c = interface stress at center line below loaded area of radius 'a'

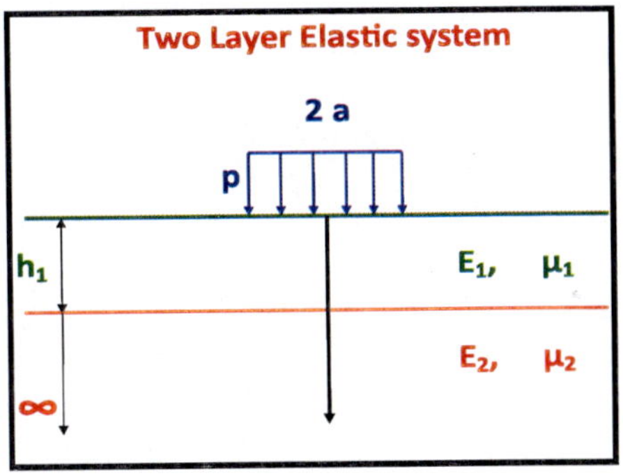

Fig. 2 The two-layer pavement system

It can be seen that higher E_1/E_2 ratio for a given a/h_1 leads to a lower σ_c/q meaning less load transfer to the subgrade and hence on to the GCD.

It is revealed in various literature that the typical E values for asphalt, aggregate base courses and subgrade of silty sand are 3500 MPa, 450 MPA and 20 MPa respectively [1]. Although the pavement crust is made of three layers, i.e. GSB, CRM and DBM/BC layers, considering a composite top layer hardly makes a difference. In fact by neglecting the impact of higher E value of the bituminous layer, we err on the side of caution. We can take the E value for the aggregate base courses as the governing E value and find the vertical stress at the top of the subgrade level below the loaded area. At best, an equivalent E value considering all layers above the subgrade may be calculated using the relation [3]:

$$E_{eq} = \sum H_i E_i \Big/ \sum H_i.$$

Now E_1/E_2 is calculated by taking a suitable value of E for the soil subgrade. This method, however, neglects the impact of lower E-value of the subgrade on the E-value of the pavement crust.

Alternatively, we can use the design method given in IRC 37—2012 to arrive at the ratio E_1/E_2 [8]. When both sub-base and the base layers are made up of unbound granular layers, the composite resilient modulus of the granular sub-base and the base is given as:

$$M_{R\_granular\ overlay} = 0.2 * h_1^{0.45} * M_{R\_subgrade}$$

where h_1 = total thickness of granular sub-base and base (in mm)

M_R = modulus of resilience of respective layer (in MPa)

Poisson's ratio for granular bases and sub-bases is recommended as 0.35.

The modulus of resilience is the elasticity modulus of a material under repeated loads. The pavement layers are normally not elastic and will show plastic deformations in every load cycle. But if the traffic load is less than the strength of the material, after a certain number of load repetitions (approximately 100–200 cycles), the strain is almost completely recoverable and can be considered elastic [15]. In such case, the Elastic Modulus can be taken as the moment of resilience. In a triaxial test, moment of resilience is defined as the ratio of deviator stress to elastic strain of the soil.

As such, if we assume that $M_R = E$, then

$$E_1 \big/ E_2 = M_{R\_granular\ overlay} \big/ M_{R\_subgrade} = 0.2 * h_1^{0.45}.$$

It can be seen that the modulus ratio, in this case, does not depend on the individual M_R values but is empirically related to the thickness of the pavement layer. For typical values of E, h_1, etc. the IRC method yields higher load transfer at the subgrade level and leads to a conservative design for the GCD.

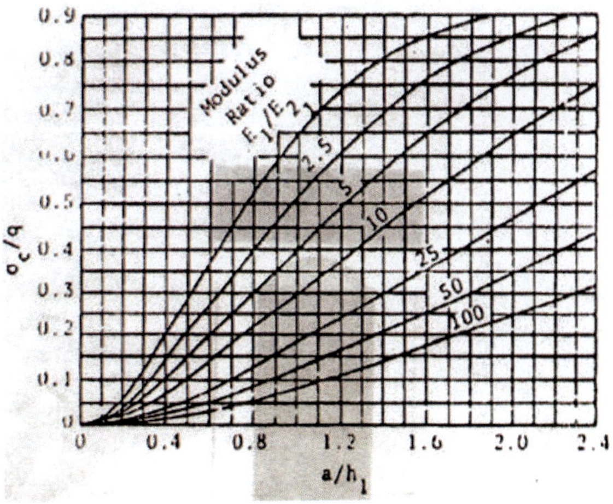

Fig. 3 Variation of interface stress with E_1/E_2. The curves assume $\nu = 0.5$ (after Huang [6])

Having arrived at the E_1/E_2, and a/h_1 values, σ_c/q can be read off the graph in Fig. 3.

It is also seen that even an increase in surface pressure to 0.80 MPa has little effect on the absolute normal stress levels on the GCD.

Assessment of Allowable Flow Rate (q_{allow})

The allowable flow rate is achieved by reducing the basic flow rate for various factors detailed below:

(a) Reduction factor for creep to account for long term behavior (RF_{CR})
(b) Reduction factor for chemical clogging (RF_{CC})
(c) Reduction factor for biological clogging (RF_{BC})

It is pertinent to mention here that ASTM D7931 2018 also envisages another reduction factor for geotextile intrusion into the core for continued stress exposure (RF_{GI}). This is beyond what is expected during the 100 h testing of the product and will also depend, in case of subsurface drainage application in roads, on the type and intensity of traffic, structure and material properties of the GCD core, material and properties of the geotextile, wetting conditions of the geotextile, etc.

$$q_{allow} = q_{basic}\left[\frac{1}{RF_{CR} \times RF_{CC} \times RF_{BC} \times RF_{GI}}\right]$$

and

$$FOS = q_{allow}\big/q_{reqd}.$$

The FOS depends on reliability of the allowable flow rate (product dependent) as well as the required flow rate (site dependent). The Compilation of Codes, Rules and

Regulations of the State of New York Department of Environmental Conservation Part 363 Subpart 6 (6 CRR-NY 363-6.12 Geosynthetic Drainage Layers) specifies that for hydraulic flow capacity calculations in landfill leachate collection or drainage systems, the designer must use a factor of safety of at least three, in addition to reduction in flow rate due to creep, biological and chemical clogging [14]. Any higher FOS would result in the requirement of very large flow capacity GCDs and consequently cost as well as availability may be restrictive.

5.2 *Selection of Geocomposite, Placement and Causes of Failure*

The performance of geocomposites depends on the performance of the component geotextile filters and the performance of the drainage core [12, 17]. As seen from the typical section adopted, the sand layer is kept adjacent to the geocomposite for preventing installation damage to the GCD as well as to ensure that it is compatible for filtration/separation function with the veneer geotextile provided. For the latter objective, the formulation of Koerner [12, 13] may be used. Alternately, Carrols simplified formulation $AOS_{geotextile} < 2.5 * D_{85}$ may be used, where D_{85} is the size of soil particles in mm, of which 85% of provided sand is finer. AOS is the Aperture Opening Size (O_{95}) of the veneer geotextile. Sandwiching the geocomposite with at least 100 mm of sand on both sides gives us the freedom to tweak the sand properties as well as the geotextile properties, considering availability, cost, lead time, etc. Another important aspect of the textile is that it spans over the openings of the drainage core. As such it should be strong and stiff enough not to intrude into the core under load and thereby increase the flow reduction factor for core intrusion (RF_{GI}).

In so far as the selection of drainage core is concerned, it should have adequate flow capacity. It should not yield under traffic loads, including impact. In case of hill road applications requiring large flows, cuspated core material with lower pitch may be desirable provided they are available satisfying other properties. In case of large volumes, it is preferable to have samples tested in GAILAP accredited laboratories (IRC:113:2018). In case, the GCD is expected to function as a capillary barrier, the drainage core should have air gap at the top, i.e. the top geotextile should not saturate from capillary rise of water.

5.3 *Care During Construction*

The following steps during work execution are necessary to maintain integrity and effectiveness of the GCD:

1. The GCD has exposed geotextile and should always be stored away from sunlight.
2. While laying the GCD, the main drainage direction should be in the flow direction and should always be placed with an outward gradient for easy drainage.
3. One edge of the GCD roll is generally provided with 150 mm extra geotextile. The contiguous GCD on that side is placed inside the overlap, butting the core of its neighbor. This ensures lateral continuity of the drainage core while ensuring proper separation. While placing the GCD in cold windy conditions, it is preferable to use wide packing tape to hold down the GCD panels to each other.
4. It is imperative that while laying the sand cover on the GCD, care should be taken that the construction equipment does not move over the GCD directly. No tracked vehicle like dozer should be allowed on the material either.

In a study in the USA, where geocomposites used for drainage have been exhumed after 1–15 years of use [17]. It was found that though age did not seem to affect the results, failure was seen in filter function of geotextile, as a result of which soil finer than AOS of the textile had clogged the drainage core entirely. Installation-related damage and excessive core deformation were also seen in some cases. A control layer of sand would, therefore, go a long way in keeping good contact with the GCD and also work better for retention.

6 Case Study—A Pilot Project in High Altitude Region of Himalayas (Uttarakhand)

The instant case is a road built by CPWD in the Uttarakhand region, which is around 9000 ft. Two small patches of approximately 20 m road length with a properly designed horizontal geocomposite drain (GCD) were at locations where subsurface water was seen rising to the top of base course. Road is through cutting in hard rock. The crust consists of 200 mm Crusher Run Macadam, 60 mm Dense Bituminous Macadam and 40 mm Bituminous Concrete. The geocomposite drain (GCD) has been placed over a sand bed of 100 mm overlaid on a rocky granular subgrade.

The base course was removed and the subgrade exposed for this stretch. The GCD has been laid in a 3% cross gradient. It was decided to provide GCD sandwiched between 100 mm sand layers for this stretch before laying the CRM and compacting. The sand layers are provided to avoid construction damage. The work was executed in the month of October 2017 under very cold and windy conditions. After considering the available options in the market and very small working window available, a geocomposite consisting extruded monofilament UV stabilized PP core laid in longitudinal channels in the drainage direction, with two needles punched thermally bonded non-woven PP geotextiles on both sides, was chosen for the particular site. It is the first of its kind application in hill roads in India.

6.1 Design of GCD

Determination of the required drainage capacity of the system **(q_{reqd})**—In the instant case, two locations were identified in cut-sections where water was flowing in from the cut-face and also by capillary rise (as evident from wet side slopes of the hill on valley side). The water was seen to flow in small channels at two or three locations in the approximately 20 m stretches. Only the water flowing in the main surface channel was measured. Measurements were done for 5 min periods three times during the day, at around 3 h interval. The maximum discharge collected was 8 L during a 5 min period. The water was found to affect a road length of 2–4 m. at critical locations or less. Therefore, the flow per meter width per sec works out to 0.009 L. This discharge can be increased to account for seasonal variations, site-specific subsurface flow, extent of run-off management provided in the stretch, level of maintenance anticipated, etc. In this case, a factor of 3 is applied to account for the above effects. Thus design/required discharge capacity (q_{reqd}) works out to **0.03 L/(m · s).**

Short Term or Basic Flow Capacity (100 h Testing Condition) of the GCD (q_{basic})

Normal stress on the GCD: The stress ratio (stress on GCD/stress on road surface) on the top of lower sand layer using the formulation above works out to 0.14. Hence the stress on GCD resting on lower sand layer (100 mm above subgrade) is equal to 0.14 * 560 kPa = 78 kPa (say 100 kPa for design).

$$\text{Hydraulic gradient} = 0.03$$

Basic flow rate from manufacturer's technical data sheet: 0.20 L/(m · s).

Assessment of allowable flow rate (q_{allow})—The allowable flow rate is achieved by reducing the basic flow rate for various factors detailed below:

(a) Reduction factor for creep deformation of drainage core (RF_{CR}) = 1.3†
(b) Reduction factor for chemical clogging (RF_{CC}) = 1.1@
(c) Reduction factor for biological clogging (RF_{BC}) = 1.1@
(d) Reduction factor for geotextile intrusion into the core (RF_{GI}) = 1.4†

†This factor has been adopted from Koerner [12].

@These reduction factors have been taken from GRI-GC8 [5] and ISO 12228-4 as reported by Blond [2]

$$\text{Total Reduction Factor} = 1.3 * 1.1 * 1.1 * 1.4 = 2.2$$

$$(q_{allow}) = 0.20/2.2 = 0.091 \text{ L}/(\text{m} \cdot \text{s})$$

Hence FOS = (q_{allow})/(q_{reqd}) = 0.09/0.03 = 3 (Table 1 and Figs. 4, 5, 6 and 7).

Table 1 Important properties of the geocomposite drainage material [7, 11, 16]

S. no.	Material	Standard	Unit	Value	Tolerance	IRC:34-2011 (Para 4.6.2)
1.	Tensile strength (MD)	EN ISO 10319	kN/m	18	–	≥16
2 (a)	In plane flow capacity (MD)	EN ISO 12958	L/(m · s)	For $i = 0.03$ @ 100 kPa = 0.20 l/(m · s) @ 200 kPa = 0.10 l/(m · s)	±30%	For $i = 0.03$ Not provided
2 (b)	In plane flow capacity (MD)	EN ISO 12958	L/(m · s)	For $i = 1.0$ @ 100 kPa = 1.40 l/(m · s) @ 200 kPa = 0.80 l/(m · s)	±30%	For $i = 1.0$ @ 100 kPa ≥ 0.55 l/(m · s) @ 200 kPa ≥ 0.45 l/(m · s)
B. GEOTEXTILE (UV stabilized polypropylene)						
1.	Static puncture resistance	EN ISO 12236	N	1400	±20%	≥3000
2.	Permittivity	EN ISO 11058	l/(m^2 · s)	100	(−) 30%	≥100
3.	Apparent opening size (AOS)	EN ISO 12956	Micron	110	±50	≤150

Note In case of provision of conventional drainage system, the GCD and the 100 mm sand layers on both sides would be replaced by 200 mm ($t = 0.2$ m) granular sub-base (GSB). The permeability (k) of granular sub-base of Grading 2 of IRC:37 is around 30 m/day or 0.35×10^{-3} m/s. Discharge per meter width for hydraulic gradient of 1.0 ($i = 1.0$) works out as under

$Q_{\text{per m width}}$ (m^3/s) $= k * i * (W * t) = 0.35 \times 10^{-3} * 1.0 * (1.0 * 0.2) = 0.07 \times 10^{-3}$ m^3/s = 0.07 L/s

This compares well with the GCD capacity of 0.8 L/s even at 200 kPa normal stress on GCD at hydraulic gradient of 1.0 as per manufacturer's data sheet

Fig. 4 Geocomposite MacDrain™ W-1071

Fig. 5 Seepage water visible on CRM surface before installation

Fig. 6 Geocomposite drain being rolled out for cutting

Fig. 7 GCD laid in slope and being covered by sand layer

The installation stretches are being monitored during the working season when the roads are in operation, to gauge the effectiveness of the horizontal subsurface drainage system. In the snow melting season of 2018 and 2019, no distress or water accumulation on pavement surface has been reported.

6.2 Drawbacks of the Study

This pilot project was done when surface water was observed on the pavement surface before black topping. A more detailed survey of other vulnerable locations, at various points of time during the working season, including more measurements, would have given more insight into the quantity and variation of water seepage as well as capillary rise.

Similarly, a better market survey of available GCD options could have been done, which was obviated by the impending close of the working season.

It is difficult to assess the subsurface flow during the most critical monsoon period, as precipitation saturates all catchment area, obliterating the difference between surface flows and subsurface flows.

The simplistic methodology adopted for estimating E_1/E_2 of the composite layer needs validation by instrumented studies to assess the stresses on the GCD due to traffic.

Also after few years of use, especially with monsoon wetting, the pavement crust deteriorates leading to reduction of moment of resilience of the top layers and therefore increased vehicular loads on the GCD. This may cause crushing of the drainage core, intrusion of textile into the drainage core, loss of integrity of the textile, etc.—leading to functional failure of the GCD.

7 Conclusions

In this case, we could provide a FOS of 3, which is quite sufficient to provide drainage of subsurface water as well as act as a capillary barrier.

A large variety of GCDs are available in the market for which selection guidelines and design codes are required for different applications. More testing of GCDs, especially independent third-party testing by accredited laboratories, is required for higher volumes of application. For this testing, facilities are required to be developed.

The various reduction factors being used in design calculations are generic values taken from literature. These may require to be revisited for specific applications for different climatic zones and environmental conditions, especially in a hot and humid country like India.

Acknowledgments The untiring efforts of Shri Vinay Sheel Saxena, Assistant Engineer (Civil) and Shri Net Ram, Junior Engineer (Civil) of Border Road Sub-Division, Harshil, Uttarakhand (CPWD) for getting the work executed in extremely cold and windy conditions at an altitude of about 14,000 feet are commendable and acknowledged. Shri Ankit Kachhal, Senior Engineer from M/s Maccaferri Environmental Solutions Pvt. Ltd. was present at site during installation and gave valuable inputs regarding handling, storage and installation of the product.

References

1. Bahador, M., Evans, T.M., Gabr, M.A.: Modelling effect of geocomposite drainage layers on moisture distribution and plastic deformation of road sections. ASCE J. Geotech. Geoenviron. Eng. 1407–1418 (2013)
2. Blond, E.: Webinar on Testing and Design of Drainage Geocomposites, Geosyntheti News Alerts (GNA), vol. 12, April 2020. https://www.geosyntheticnews.com.au/news/april-2020-vol-12/. Accessed 8 Aug 2020
3. Brahma, P., Mukherjee, S.P.: A realistic way to obtain equivalent young's modulus of layered soil. In: Proceedings of Indian Geotechnical Conference (GEO*trendz*), pp. 305–308 (2010)
4. Christopher, B.R., Hayden, S.A., Zhao A.: Roadway base and subgrade geocomposite drainage layer. In: Goddard, J.B., Suits, L.D., Baldwin, J.S. (eds.) Testing and Performance of Geosynthetics in Subsurface Drainage, ASTM STP 1390, West Conshohocken, PA, USA (2000)
5. GRI Standard GC8: Standard Guide for Determination of the Allowable Flow Rate of a Drainage Geocomposite, Geosynthetic Institute, PA (2013)
6. Huang, Y.H.: Pavement Analysis and Design, 2nd edn. Pearson Education Inc., USA (2004)
7. IRC:SP:34-2011: Recommendations for Road Construction in Areas Affected by Waterlogging, Flooding and/or Salts Infestation, 1st Revision. Indian Roads Congress, New Delhi (2011)
8. IRC: 37-2018: Guidelines for the Design of Flexible Pavements, 4th Revision. Indian Road Congress, New Delhi (2018)
9. IRC:SP:42-2014: Guidelines of Road Drainage, 1st Revision. Indian Roads Congress, New Delhi (2014)
10. IRC:SP:48-1998: Hill Road Manual, Indian Roads Congress, New Delhi (1998)
11. IRC:SP:50-2013: Guidelines on Urban Drainage, 1st Revision. Indian Roads Congress, New Delhi (2013)
12. Koerner, R.M.: Designing with Geosynthetics, vols. I & II, 6th edn. Published by Xlibris Corporation (2012)
13. Koerner, R.M. (ed.): Geotextiles from Design to Applications, Woodhead Publishing Series in Textiles: Number 175 (2016)
14. Official Compilation of Codes, Rules and Regulations of the State of New York, 6 CRR-NY 363-6.12 © Thomson Reuters, USA (2020). from https://govt.westlaw.com/nycrr/Document/Id4d71958dfe911e7aa6b9b71698a280b?viewType=FullText&originationContext=documenttoc&transitionType=CategoryPageItem&contextData=(sc.Default). Accessed 8 Aug 2020
15. Saglik, A., Gungor, A.G.: Resilient modulus of unbound and bituminous bound road materials. In: Proceedings of 5th Eurasphalt & Eurobitume Congress, Istanbul (2012)
16. Technical Data Sheet for MACDRAIN™ W 1071 (Rev.04). Maccaferri Environmental Solutions Pvt. Ltd., India (2015)
17. Vlotman, W.F., Willardson, L.S., Dierickx, W.: Envelope Design for Subsurface Drains. Publication by International Institute for Land Reclamation and Improvement, Wageningen, The Netherlands (2000)